T0190126

Fatigue of Materials

Written by a leading researcher in the field, this revised and updated second edition of a highly successful book provides an authoritative, comprehensive and unified treatment of the mechanics and micromechanisms of fatigue in metals, nonmetals and composites.

The author discusses the principles of cyclic deformation, crack initiation and crack growth by fatigue, covering both microscopic and continuum aspects. The book begins with discussions of cyclic deformation and fatigue crack initiation in monocrystalline and polycrystalline ductile alloys as well as in brittle, semi-crystalline and noncrystalline solids. Total-life and damage-tolerant approaches are then introduced in metals, nonmetals and composites along with such advanced topics as multiaxial fatigue, contact fatigue, variable amplitude fatigue, creep-fatigue, and environmentally assisted fatigue. Emphasis is placed upon scientific concepts and mechanisms and the basic concepts are extended to many practical cases wherever possible. The book includes an extensive bibliography and a problem set for each chapter, together with worked-out example problems and case studies.

The book will be an important reference for students, practicing engineers and researchers studying fracture and fatigue in materials science and engineering, mechanical, civil, nuclear and aerospace engineering, and biomechanics.

Fatigue of Materials

SECOND EDITION

S. SURESH

Massachusetts Institute of Technology

CAMBRIDGE
UNIVERSITY PRESS

CAMBRIDGE UNIVERSITY PRESS
Cambridge, New York, Melbourne, Madrid, Cape Town, Singapore, São Paulo

Cambridge University Press
The Edinburgh Building, Cambridge CB2 2RU, UK

Published in the United States of America by Cambridge University Press, New York

www.cambridge.org
Information on this title: www.cambridge.org/9780521570466

First published 1998
Reprinted 2001 (with corrections), 2003, 2004

A catalogue record for this publication is available from the British Library

Library of Congress Cataloguing in Publication data
Suresh, S. (Subra)
    Fatigue of materials / S. Suresh. – 2nd ed.
       p.   cm.
    Includes bibliographical references and index.
    ISBN 0 521 57046 8 (hardcover). – ISBN 0 521 57847 7 (pbk.)
    1. Materials – Fatigue.   2. Fracture mechanics.   I. Title.
    TA418.38.S87   1998
    620.1′126–dc21   98-15181 CIP

ISBN-13 978-0-521-57046-6 hardback
ISBN-10 0-521-57046-8 hardback

ISBN-13 978-0-521-57847-9 paperback
ISBN-10 0-521-57847-7 paperback

Transferred to digital printing 2006

To
My wife Mary and our children,
Nina and Meera

Contents

Preface to the second edition

The first edition of this book was written primarily as a research monograph for the Solid State Science Series of Cambridge University Press. Since its first publication, however, the book has found wide readership among students and practicing engineers as well as researchers. In view of this audience base which evolved to be much broader than what the book was originally intended for, it was felt that now would be an appropriate time for the preparation of an updated and revised second edition which includes newer material, example problems, case studies and exercises. In order to have the greatest flexibility in the incorporation of these new items in the book, it was also decided to publish the second edition as a 'stand-alone' book of Cambridge University Press, rather than as a research monograph of the Solid State Science Series.

In writing the second edition, I have adhered to the objectives which are stated in the preface to the first edition. In order to structure the expanded scope coherently, the book is organized in the following manner. The introduction to the subject of fatigue, the overall scope of the book and background information on some of the necessary fundamentals are provided in the first chapter. The book is then divided into four parts. Cyclic deformation and fatigue crack nucleation in ductile, brittle and semi-crystalline or noncrystalline solids are given extensive coverage in Part One. This is followed by discussions of stress-based and strain-based approaches to fatigue in Part Two. Principles of fracture mechanics and the characteristics of fatigue crack growth in ductile, brittle and semi-crystalline or noncrystalline solids are then taken up in separate chapters in Part Three. Part Four comprises advanced topics where, in addition to separate chapters on crack growth retardation under constant and variable amplitude fatigue, small cracks, and environmental interactions, a new chapter on contact fatigue is included. In each chapter, updated material and references, as well as case studies and worked-out and practice problems are included.

Since the publication of the first edition, numerous students, research colleagues and engineers from many countries have provided valuable feedback, constructive criticisms and suggestions. The first edition was translated into the Chinese language in 1993 under the sponsorship of the Chinese Academy of Sciences. Z. G. Wang and his colleagues of the State Key Laboratory for Fracture and Fatigue in Shenyang deserve special thanks for diligently going over the entire manuscript of the first edition and offering many helpful suggestions. Every attempt has been made to respond to these comments in the second edition, while trying to adhere to the main objectives of writing this book. The first edition as well as drafts of the second edition were also used by the author as text material for subjects taught for graduate students at the Massachusetts Institute of Technology, Harvard University and Brown University.

I am thankful to many individuals who supplied me with information and original micrographs for the second edition. They include H. Azimi, U. Essmann, W. Milligan, T. Nakamura, A. Pineau, R.O. Ritchie and D.F. Socie. A number of colleagues, especially J. Dominguez, T.C. Lindley, F.A. McClintock, T. Nicholas and L. Pruitt, kindly read drafts of some chapters and offered helpful suggestions. During the preparation of this second edition, I have had the good fortune to interact extensively with four colleagues, A.E. Giannakopoulos, L.P. Kubin, H. Mughrabi and C.F. Shih, who read many sections of the book, and provided helpful criticisms, suggestions on key references, solutions to some practice problems and advice on improvement of presentation. I am most grateful to them for their scholarly feedback and strong interest in this book.

I wish to thank S. Capelin of the Cambridge University Press for his strong and continued support of this book project and for giving me considerable flexibility in the preparation of the manuscript, and to M. Patterson for her efficient copy-editing of the manuscript. K. Greene, D. LaBonte, G. LaBonte, C.-T. Lin and L. Ward deserve special mention for their cheerful help with the preparation of figures and references.

My research work on fatigue has been supported over the years by the U.S. Department of Energy, Office of Naval Research, Air Force Office of Scientific Research, and National Science Foundation. This support is gratefully acknowledged. I thank the Department of Materials Science and Engineering at the Massachusetts Institute of Technology for giving me the flexibility and the time for the preparation of this second edition through the award of the R.P. Simmons Endowed Professorship. In addition, I thank the Swedish Research Council for Engineering Sciences for awarding me the Swedish National Chair in Mechanical Engineering for the period 1996–98 at the Royal Institute of Technology (Kungl Tekniska Högskolan, KTH), Stockholm, where a significant portion of the new material for the second edition was written. I thank the colleagues in the Departments of Solid Mechanics and Materials Science and Engineering at KTH for their hospitality during my stay in Stockholm.

Finally, I express my deepest gratitude to my wife Mary and our daughters Nina and Meera for all their patience and support during the time I spent writing this book. Without their care, affection and tolerance, this project could not have been completed.

S. Suresh

Preface to the first edition

Fatigue of materials refers to the changes in properties resulting from the application of cyclic loads. Research into the deformation and fracture of materials by fatigue dates back to the nineteenth century. This branch of study has long been concerned with engineering approaches to design against fatigue cracks initiation and failure. However, along with the development of 'science of materials' and 'fracture mechanics' in recent decades, *fatigue of materials* has also emerged as a major area of scientific and applied research which encompasses such diverse disciplines as materials science (including the science of metals, ceramics, polymers, and composites), mechanical, civil and aerospace engineering, biomechanics, applied physics and applied mathematics. With the increasing emphasis on advanced materials, the scope of fatigue research continues to broaden at a rapid pace.

This book is written with the purpose of presenting the principles of cyclic deformation and fatigue fracture in materials. The main approach adopted here focuses attention on scientific concepts and mechanisms. Since fatigue of materials is a topic of utmost concern in many engineering applications, this book also includes discussions on the extension of basic concepts to practical situations, wherever appropriate. In writing this book, I have attempted to achieve the following objectives:

(i)   To present an integrated treatment, in as quantitative terms as possible, of the mechanics, physics and micromechanisms of cyclic deformation, crack initiation and crack growth by fatigue.

(ii) To provide a unified scientific basis for understanding the fatigue behavior of metals, nonmetals and composites.

(iii) To develop a balanced perspective of the various approaches to fatigue, with a critical analysis of the significance and limitations of each approach.

The topics addressed in this book are developed to the extent that the presentation is sufficiently self-explanatory. The scope of the book is spelled out in Chapter 1. Some background information on the relevant topics is also provided in the first chapter to set the scene for later developments. This book could serve as a state-of-the-art reference guide to researchers interested in the fatigue behavior of materials and as a text for a graduate course on fatigue. Sections of the book could also be used for an introductory course on fatigue for practicing engineers. Senior undergraduate and graduate students taking courses on mechanical behavior of materials or fracture mechanics may find this monograph useful as a supplement to their textbooks.

Since there exists a vast amount of published information on fatigue of materials which spans a time period of well over a hundred years, it is not feasible to cover all the topics and results on this subject in a single monograph. This situation is further compounded by the fact that *fatigue* is a research area in which one encounters considerable empiricism and conflicting viewpoints. Therefore, I had to rely on my own judgement in the selection of topics and references in order to balance the aforementioned objectives of writing this book. My interpretations of this research area have been shaped over the past twelve years through collaborations, discussions and correspondence with numerous colleagues in the United States and abroad. Their contributions are cited throughout the text and in the bibliography.

In organizing various sections of this book, I have resisted the temptation to categorize materials systems in terms of their individual composition or microstructure, as commonly done in many books on materials. Instead, I have chosen to present a unified treatment of fatigue in different broad classes of materials while, at the same time, pinpointing the significant role of microstructure in influencing cyclic deformation and fracture. The mechanistic and mechanics aspects of fatigue are developed in the early chapters primarily in the context of ductile metals and alloys. These concepts are then extended to brittle solids (such as ceramics and ceramic composites) and to semi-crystalline and noncrystalline solids (such as polymers and organic composites) in later chapters of the book.

I am thankful to many colleagues who supplied me with information and original photographs. These individuals include R.W. Hertzberg, H. Kobayashi, T.C. Lindley, T. Nicholas, T. Ogawa, C.M. Rimnac, R.O. Ritchie and M.T. Takemori. I am indebted to J.R. Brockenbrough, C. Laird, M. Miller, H. Mughrabi, P. Neumann, R.M.N. Pelloux and A.K. Vasudevan for reviewing drafts of various chapters and for offering helpful comments. A special note of gratitude is extended to A.S. Argon who provided insightful suggestions on the presentation of fundamental concepts and to C.F. Shih who kindly read drafts of all the sections dealing with fracture mechanics.

My research work on fatigue has been sponsored over the years by U.S. Department of Energy, National Science Foundation and Office of Naval Research. This support is very much appreciated. My colleagues and students in the Solid Mechanics and Materials Science Groups in the Division of Engineering at Brown University have greatly contributed to my research by providing an intellectually stimulating and friendly environment.

This project was initiated with the encouragement and support of R.W. Cahn. His enthusiasm and interest in this work provided a strong motivation throughout the preparation of the manuscript. A special mention should also be made of R. Bentley, S. Capelin and I. Pizzie of Cambridge University Press for their efficient editorial work on the manuscript and for promptly responding to my queries and concerns. T. Judd skillfully helped with the preparation of figures.

I express my profound gratitude to my wife Mary for her support, devotion and patience throughout this project. The company of our little daughters Nina and Meera provided joyful breaks from long hours of concentrated work.

Finally, I wish to dedicate this book to my mother Lakshmi on the occasion of her sixtieth birthday. Her countless sacrifices for the sake of my education are gratefully acknowledged.

S. Suresh

CHAPTER 1

Introduction and overview

The word *fatigue* originated from the Latin expression *fatīgāre* which means 'to tire'. Although commonly associated with physical and mental weariness in people, the word *fatigue* has also become a widely accepted terminology in engineering vocabulary for the damage and failure of materials under cyclic loads. A descriptive definition of fatigue is found in the report entitled *General Principles for Fatigue Testing of Metals* which was published in 1964 by the International Organization for Standardization in Geneva. In this report, fatigue is defined as a term which 'applies to changes in properties which can occur in a metallic material due to the repeated application of stresses or strains, although usually this term applies specially to those changes which lead to cracking or failure'. This description is also generally valid for the fatigue of nonmetallic materials.

Fatigue failures occur in many different forms. Mere fluctuations in externally applied stresses or strains result in *mechanical fatigue*. Cyclic loads acting in association with high temperatures cause *creep-fatigue*; when the temperature of the cyclically loaded component also fluctuates, *thermomechanical fatigue* (i.e. a combination of thermal and mechanical fatigue) is induced. Recurring loads imposed in the presence of a chemically aggressive or embrittling environment give rise to *corrosion fatigue*. The repeated application of loads in conjunction with sliding and rolling contact between materials produces *sliding contact fatigue* and *rolling contact fatigue*, respectively, while *fretting fatigue* occurs as a result of pulsating stresses along with oscillatory relative motion and frictional sliding between surfaces. The majority of failures in machinery and structural components can be attributed to one of the above fatigue processes. Such failures generally take place under the influence of cyclic loads whose peak values are considerably smaller than the 'safe' loads estimated on the basis of static fracture analyses.†

1.1 Historical background and overview

Fatigue is a branch of study which encompasses many scientific disciplines and which offers a rich variety of phenomena for fundamental and industrial research. From published reports, research on the fatigue of materials can be traced back to the

† Failing to recognize this fact was the primary cause of catastrophic accidents involving the first commercial jet aircraft, the Comet. A case study of the Comet aircraft failures is provided in Section 1.1.1.

1

first half of the nineteenth century. Since that time, scores of scientists and engineers have made pioneering contributions to the understanding of fatigue in a wide variety of metallic and nonmetallic, brittle and ductile, monolithic and composite, and natural and synthetic materials. It is not feasible to present, in a few pages, a comprehensive survey of the historical development of these contributions to fatigue. Nevertheless, in an attempt to highlight the salient topics of fatigue and to set the scene for the scope of this book, an overview of major advances and of key areas in fatigue is given in this section. The chapters to follow provide detailed discussions of various fatigue phenomena along with the relevant historical backgrounds wherever appropriate.

The expression *fatigue* has been in use for a very long time. In the days of long distance travel aboard sailing vessels, the straining of masts due to the frequent hoisting of sails was referred to as fatigue. The first study of metal fatigue is believed to have been conducted around 1829 by the German mining engineer W.A.J. Albert (see Albert, 1838). He performed repeated load proof tests on mine-hoist chains made of iron. One end of the chain was loaded while the chain was supported on a 360-cm (12-ft) disc. The chain links were repeatedly subjected to bending, at a rate of 10 bends per minute up to 100 000 bends, by a crank coupling which oscillated the disc through an arc.

Interest in the study of fatigue began to expand with the increasing use of ferrous structures, particularly bridges in railway systems. The first detailed research effort into metal fatigue was initiated in 1842 following the railway accident near Versailles in France which resulted in the loss of human lives (*The Times* of London, May 11, 1842; for a comprehensive description of this accident, see Smith, 1990). The cause of this accident was traced to fatigue failure originating in the locomotive front axle. As early as 1843, W.J.M. Rankine, a British railway engineer who later became famous for his contributions to mechanical engineering, recognized the distinctive characteristics of fatigue fractures and noted the dangers of stress concentrations in machine components. The Institution of Mechanical Engineers in Britain also began to explore the so-called 'crystallization theory' of fatigue. It was postulated that the weakening of materials leading to eventual failure by fatigue was caused by the crystallization of the underlying microstructure. In 1849, the British Government commissioned E.A. Hodgkinson to study the fatigue of wrought and cast iron used in railway bridges. The report of this commission (Hodgkinson, 1849) described alternating bending experiments on beams whose midpoints were repeatedly deflected by a rotating cam. In this time period, research on fatigue fracture was also documented in the work of Braithwaite (1854) who employed the term *fatigue* exclusively to denote the cracking of metals under repeated loading. (Braithwaite (1854), however, credits one Mr. Field for coining this term. Poncelet (1839) is also generally given credit for introducing the term *fatigue* in connection with metal failure, although it had been used earlier in the context of other phenomena.)

A. Wöhler conducted systematic investigations of fatigue failure during the period 1852–1869 in Berlin, where he established an experiment station. He observed that the strength of steel railway axles subjected to cyclic loads was appreciably lower than their static strength. Wöhler's studies involving bending, torsion and axial loading included fatigue tests on full-scale railway axles for the Prussian Railway Service and on a variety of structural components used in small machines. His work (e.g., Wöhler, 1860) also led to the characterization of fatigue behavior in terms of stress amplitude–life (S–N) curves and to the concept of fatigue 'endurance limit'. The rotating bending machine widely used today for cyclically stressing metals is conceptually the same as the one designed by Wöhler. Although his rotating bending apparatus had a maximum speed of only 72 revolutions per minute, one of his fatigue test specimens was subjected to 132 250 000 stress cycles without producing fracture.

Another well known fatigue researcher of this era was W. Fairbairn who performed tests on riveted wrought iron girders for the British Board of Trade; in some cases, as many as 3 100 000 load cycles were applied. On the basis of his experiments, Fairbairn (1864) concluded that the wrought iron girders subjected to cyclic stresses with a maximum of only one-third of the ultimate strength would fail. In 1874, the German engineer H. Gerber began developing methods for fatigue design; his contribution included the development of methods for fatigue life calculations for different mean levels of cyclic stresses. Similar problems were also addressed by Goodman (1899).

The notion that the elastic limit of metals in reversed loading can be different from that observed in monotonic deformation was popularized by Bauschinger (1886). His work essentially identified the occurrence of cyclic softening and cyclic strain hardening. Bauschinger also confirmed many of the results reported earlier by Wöhler. By the end of the nineteenth century, some eighty papers on fatigue had been published in such diverse application areas as railway rolling stock axles, crankshafts, chains, wire ropes, and marine propeller shafts (see, for example, the survey by Mann, 1958).

Interpretations of fatigue mechanisms based on the old crystallization theory were laid to rest by the pioneering work of Ewing & Rosenhain (1900) and Ewing & Humfrey (1903). These researchers investigated the fatigue of Swedish iron and published optical micrographs of cyclic damage on the specimen surface. It was convincingly shown that slip bands developed in many grains of the polycrystalline material. These slip bands broadened with the progression of fatigue deformation and led to the formation of cracks; catastrophic failure of the specimen was instigated by the growth of a single dominant flaw. They showed that the slip bands intersecting the polished surface caused slip steps in the form of elevations and depressions which we now commonly refer to as 'extrusions' and 'intrusions', respectively. (The micromechanisms of fatigue damage and crack nucleation in metals form the topics of discussion in Chapters 2–4 of this book.)

In 1910, O.H. Basquin proposed empirical laws to characterize the S–N curves of metals. He showed that a log–log plot of the stress versus the number of fatigue cycles resulted in a linear relationship over a large range of stress. Significant contributions to the early understanding of cyclic hardening and softening in metals were also made by Bairstow (1910). Using multiple-step cyclic tests and hysteresis loop measurements, Bairstow presented results on the hysteresis of deformation and on its relation to fatigue failure. In France, Boudouard (1911) conducted fatigue experiments on steel bars which were subjected to vibrations by means of an electro-magnetic apparatus similar to the one designed earlier by Guillet (1910). The effect of heat treatments on the fatigue resistance of steels was the subject of Boudouard's study. Other notable contributions of this time period included those of Smith (1910), Bach (1913), Haigh (1915), Moore & Seeley (1915), Smith & Wedgwood (1915), Ludwik (1919), Gough & Hanson (1923), Jenkin (1923), Masing (1926) and Soderberg (1939). In 1926, a book entitled *The Fatigue of Metals* was published by H.J. Gough in the United Kingdom. A year later, a book bearing the same title was published by H.F. Moore and J.B. Kommers in the United States. By the 1920s and 1930s, fatigue had evolved as a major field for scientific research. Investigations in this time period also focused on corrosion fatigue of metals (Haigh, 1917; McAdam, 1926; Gough, 1933), damage accumulation models for fatigue failure (Palmgren, 1924; Miner, 1945), notch effects on monotonic and cyclic deformation (e.g., Neuber, 1946), variable amplitude fatigue (Langer, 1937), and statistical theories of the strength of materials (Weibull, 1939). A prolific researcher of this period was Thum (e.g., Thum, 1939) who, along with many German colleagues, reported experimental results on such topics as fatigue limits, stress concentration effects, surface hardening, corrosion fatigue and residual stresses in numerous publications. Gassner (1941) was another prominent German researcher whose studies of variable amplitude fatigue found applications in the German aircraft industry. (A comprehensive survey of the contributions of German engineers and scientists to the field of fatigue, particularly during the period 1920–1945, can be found in Schutz, 1996.) Chapters 7 and 8 discuss the key features of these developments in the context of total-life approaches.

The occurrence of *fretting* was first documented by Eden, Rose and Cunningham (1911) who reported the formation of oxide debris between the steel grips and the fatigue specimen that was contacted by the grips. Tomlinson (1927) performed the first systematic experiments on fretting fatigue by inducing repeated small amplitude rotational movement between two contacting surfaces, and introduced the term 'fretting corrosion' to denote the oxidation due to this repeated contact. The deleterious effects of fretting damage on the fatigue properties of metals, as reflected in the formation of pits on the fretted surface and in the marked reduction in fatigue strength, were reported by Warlow-Davies (1941) and McDowell (1953). Discussions of fatigue failures arising from fretting as well as sliding and rolling are considered in Chapter 13.

The notion that plastic strains are responsible for cyclic damage was established by Coffin (1954) and Manson (1954). Working independently on problems associated with fatigue due to thermal and high stress amplitude loading, Coffin and Manson proposed an empirical relationship between the number of load reversals to fatigue failure and the plastic strain amplitude. This so-called Coffin–Manson relationship (described in Chapter 8) has remained the most widely used approach for the strain-based characterization of fatigue.

Although the fatigue of metals by the development of slip bands and the slow growth of microscopic flaws was documented in the work of Ewing and Humfrey in the early 1900s, the mathematical framework for the quantitative modeling of fatigue failure was not available. The stress analyses of Inglis (1913) and the energy concepts of Griffith (1921) provided the mathematical tools for quantitative treatments of fracture in brittle solids. However, these ideas could not be directly employed to characterize the fatigue failure of metallic materials. Progress in this direction came with the pioneering studies of Irwin (1957) who showed that the amplitude of the stress singularity ahead of a crack could be expressed in terms of the scalar quantity known as the stress intensity factor, K. With the advent of this so-called *linear elastic fracture mechanics* approach, attempts were made to characterize the growth of fatigue cracks also in terms of the stress intensity factor. Paris, Gomez & Anderson (1961) were the first to suggest that the increment of fatigue crack advance per stress cycle, da/dN, could be related to the range of the stress intensity factor, ΔK, during constant amplitude cyclic loading. Although their original paper on this topic was not accepted for publication by the leading journals in this field, their approach has since been widely adapted for characterizing the growth of fatigue cracks under conditions of small-scale plastic deformation at the crack tip. The major appeal of the linear elastic fracture mechanics approach is that the stress intensity factor range, determined from remote loading conditions and from the geometrical dimensions of the cracked component, uniquely characterizes the propagation of fatigue cracks; this method does not require a detailed knowledge of the mechanisms of fatigue fracture (Paris, Gomez & Anderson, 1961; Paris & Erdogan, 1963). (A detailed review of the fracture mechanics concepts and of their applications to fatigue is provided in Chapter 9.)

The effects of various mechanical, microstructural and environmental factors on cyclic deformation as well as on crack initiation and growth in a vast spectrum of engineering materials have been the topics of considerable research in the past four decades. In this period, substantial progress has been made in the understanding of cyclic deformation and crack initiation mechanisms in fatigue, thanks to advances in optical and electron microscopy. Notable among these developments are studies which identified key microscopic features of fatigue deformation and fracture. Thompson, Wadsworth & Louat (1956) demonstrated that slip bands along which deformation was concentrated in fatigued metals persistently reappeared at the same locations during continued cycling even after some material was removed near the surface; they termed these surface markings 'persistent slip bands'. Zappfe & Worden (1951) documented observations of characteristic ripple markings on fatigue fracture surfaces which are now known as *fatigue striations*. Correlations of the spacing between adjacent striations with the rate of fatigue crack growth, first published by Forsyth & Ryder (1960), became central to the development of various theories for fatigue crack growth (Chapters 10 and 12) and to the analysis of fatigue failures in engineering structures. Studies in this area by many researchers have provided valuable information on substructural and microstructural changes responsible for the cyclic hardening and softening characteristics of materials (Chapters 2 and 3) and on the role of such mechanisms in influencing the nucleation (Chapters 4–6) and growth (Chapters 10–12) of fatigue cracks.

With the application of fracture mechanics concepts to fatigue failure, increasingly more attention was paid to the mechanisms of subcritical crack growth. Conceptual and quantitative models were developed to rationalize the experimentally observed fatigue crack growth resistance of engineering materials (e.g., Laird & Smith, 1962; McClintock, 1963; Weertman, 1966; Laird, 1967; Rice, 1967; Neumann, 1969; Pelloux, 1969). Concomitant with this research, there was expanding interest in understanding the processes by which the stress intensity factor range could be altered by the very history of crack advance. An important contribution in this direction came from the experimental results of Elber (1970, 1971) who showed that fatigue cracks could remain closed even when subjected to cyclic tensile loads. This result also implied that the rate of fatigue crack growth might no longer be determined by the nominal value of stress intensity factor range, ΔK, but rather by an effective value of ΔK which accounted for the details of fracture surface contact in the wake of the advancing fatigue crack tip. Elber's work focused on the influence of prior plastic deformation on crack closure during fatigue fracture. Although Elber's conclusion about the role of crack closure in influencing fatigue crack growth has since remained controversial, it also became evident from the studies of a number of researchers in the 1970s and early 1980s that Elber's arguments of premature contact between the crack faces (based on the effects of prior crack tip plasticity) represented just one mechanism associated with the phenomenon of fatigue crack closure. From a survey of published information and on the basis of new results obtained from their own investigations, Ritchie, Suresh & Moss (1980), Suresh, Zamiski & Ritchie (1981), and Suresh & Ritchie (1984a) categorized the basic features and implications of various types of crack closure and coined the expression 'plasticity-induced crack closure' for Elber's mechanism of crack face contact due to prior plastic deformation. Further contributions to crack closure may arise from fracture surface oxidation, viscous environments trapped within the crack walls and stress-induced phase transformations. In addition, periodic deflections in the path of a fatigue crack can cause reductions in the effective driving force for fatigue fracture (Suresh, 1983a, 1985a) by partially 'shielding' the crack tip from applied stresses. A discussion of the historical development of these concepts is presented in Chapter 14.

A significant outcome of the investigations of different types of crack shielding processes is the realization that the rate of fatigue crack growth is not only affected by the instantaneous value of imposed ΔK, but also by prior loading history and crack size. As the mechanics of fatigue fracture become dependent on the geometrical conditions, the similitude concept implicit in the nominal use of fracture mechanics, namely the notion that cracked components of different dimensions exhibit the same amount of crack growth when subjected to the same value of ΔK, is no longer applicable. This breakdown of the similitude concept is further accentuated by the experimental observations that small fatigue flaws (typically smaller than several millimeters in length and amenable to be characterized in terms of linear elastic fracture mechanics) often exhibit growth rates which are significantly faster than those of longer flaws (typically tens of millimeters in length), when subjected to identical values of far-field ΔK. Furthermore, fatigue flaws of dimensions comparable to or smaller than the characteristic microstructural size scale often exhibit rates of crack growth which diminish with an increase in crack length. Such crack growth cannot be satisfactorily analyzed in terms of available theories of fracture mechanics. This so-called 'short crack problem', apparently first identified by Pearson (1975), most severely affects the development of design methodology for large structural components on the basis of experimental data gathered from smaller-sized laboratory test specimens. It is, therefore, not surprising to note that a significant fraction of research effort since the late 1970s has been devoted to the study of crack closure phenomena and crack size effects on the progression of fatigue fracture. Associated with this research effort are attempts to develop characterization methodology for the propagation of fatigue flaws in the presence of large-scale plastic deformation and in the vicinity of stress concentrations (see Chapters 4, 7, 8 and 15).

Although fatigue failure under fixed amplitudes of cyclic stresses generally forms the basis for fundamental studies, service conditions in engineering applications invariably involve the exposure of structural components to variable amplitude spectrum loads, corrosive environments, low or elevated temperatures and multiaxial stress states. The development of reliable life prediction models which are capable of handling such complex service conditions is one of the toughest challenges in fatigue research. Although major advances have been made in these areas, the application of fatigue concepts to practical situations often involves semi-empirical approaches. Available models for fatigue involving conditions of multiaxial stress conditions, complex load spectra and detrimental environments are discussed in Chapters 3, 7, 8, 10 and 16.

The majority of fatigue research reported in the open literature pertains to metallic materials. There has, however, been a surge in interest aimed at nonmetallic materials and composites which offer the potential for mechanical, thermal and environmental performance hitherto unobtainable in conventional metals. This growing interest has also generated a corresponding increase in research into the fatigue behavior of advanced ceramics (e.g., Suresh, 1990a,b; Roebben *et al.*, 1996), polymers (e.g., Hertzberg & Manson, 1980; Hertzberg, 1995), and their composites. While the existence of cyclic slip has traditionally been considered a necessary condition for the occurrence of fatigue failure in ductile solids, it is now recognized that mechanical fatigue effects in nonmetallic materials can arise in some cases from the kinematic irreversibility of microscopic deformation under cyclic loads even in the absence of cyclic dislocation motion. Mechanisms which impart kinematic irreversibility of deformation during fatigue can be as diverse as microcracking, stress-induced phase transformations, dislocation plasticity, creep, interfacial sliding or 'craze' formation. An extension of existing knowledge on the fatigue of metallic systems to these advanced materials and the identification of new mechanistic phenomena associated with the fatigue of nonmetals and metal–nonmetal composites form the basis for many ongoing research efforts. Chapters 3, 5, 6, 11, and 12 in this book provide a detailed description of the fatigue characteristics of a wide variety of brittle and noncrystalline materials, including many nonmetallic composites and layered solids.

1.1.1 Case Study: Fatigue and the Comet airplane

The dramatic effect of subcritical crack growth by fatigue on the mechanical integrity of aircraft structures was clearly brought to light by the series of crashes involving the first commercial jet aircraft, the Comet. This airplane was manufactured by the de Havilland Aircraft Company, England, and was designed to capture the rapidly growing long-distance air travel business, spurred by the economic recovery of Great Britain and Continental Europe after World War II. The fatigue failures of the Comet cabin structures, which led to several accidents in the 1950s, also obstructed the prominent role played by the British in the commercial jet aircraft industry. It is widely believed that the fatigue problem of the Comet may have served as a catalyst in the eventual emergence of the rival Boeing Aircraft Company in the United States as a world leader in commercial aviation.

The use of the jet engine, pioneered by Sir Frank Whittle in Great Britain, for propelling the commercial jet aircraft was still an untested proposition at the time of design of the Comet. The fuel consumption rate of a jet engine was more than twice that of a piston engine. In order to limit the fuel consumption rate of a jet aircraft to a level no higher than that of one propelled by a piston engine, the plane had to travel twice as fast.† This meant flying in the rarefied atmosphere of high altitude, typically around 12 000 m (or 40 000 ft), which was more than double the altitude at which the World War II vintage aircraft flew. The speed of sound is approximately 1200 km/h (or 760 miles per hour) at sea level; at an altitude of 12 000 m, it drops to about 1060 km/h. At this high altitude, termed the lower stratosphere, the coldest air with a temperature of approximately −56 °C exists, and clouds

† There are several basic requirements for the optimum performance of a passenger jet aircraft. Firstly, it must travel as fast as possible for optimizing such factors as fuel efficiency, number of flights per unit time period, and return on capital expenses. Secondly, it must fly below the speed of sound in order to avoid a precipitous rise in the specific energy consumption, which is related to the ratio of thrust to weight or to the ratio of drag to lift. Thirdly, the colder the air through which the aircraft flies, the greater the efficiency of the jet engine. Fourthly, an aircraft should not fly at an altitude higher than what is necessary because flying through rarefied atmosphere requires oversized wings. For passenger jet aircraft, the first two requirements suggest a maximum cruising speed typically in the neighborhood of 90% of the speed of sound or Mach 0.9, and the last two requirements suggest an optimum altitude of 10 000–12 000 m (Tennekes, 1996).

and thunderstorms are rare so that meteorological conditions do not impede flight schedules during cruising. The colder outside air at such altitudes also enhances the efficiency of the jet engines as the difference between the intake temperature and the combustion temperature is raised.

A particularly important issue for high altitude flights was the design of the cabin wherein the temperature and pressure had to be at near-ground levels for the comfort of the passengers and the crew. The aircraft fuselage would have to be repeatedly stressed from no pressure differential between the inside and the outside whilst on the ground to a large pressure difference between the inside passenger cabin and the rarefied atmosphere outside during cruising. The fuselage, therefore, had to be capable of withstanding high stresses arising from cabin pressurization during such high altitude flights in thin air. It would turn out that the fatigue stress cycles induced on the metal skin of the fuselage by the repeated pressurization and depressurization of the cabin during each flight contributed to the catastrophic fracture in several Comet airplanes (e.g., Dempster, 1959; Petroski, 1996).

On the first anniversary of commercial jet aircraft operation, May 2, 1953, a de Havilland Comet airplane disintegrated in mid-air soon after take-off from the airport in Calcutta, India. The crash occurred during a heavy tropical thunderstorm. The official organization investigating the crash concluded that the accident was the result of some form of structural fracture, possibly arising from higher forces imposed on the airframe by the stormy weather, or from the overcompensation by the cockpit crew in trying to control the plane in response to such forces. Consequently, the design of the aircraft structure was not viewed as a cause for concern.

On January 10, 1954, another Comet aircraft exploded at an altitude of 8230 m (27 000 ft) in the vicinity of Elba Island in the Mediterranean Sea, after taking off from Rome in good weather. Once again, no flaws in design were identified, and the aircraft was placed back in service only weeks after this second crash.

The third accident took place soon afterwards on April 8, 1954, when a Comet exploded in mid-air upon departure from Rome, after a brief stopover during a flight between London and Cairo. The wreckage from the crash fell in deep sea water and could not be recovered. This led investigators from the Royal Aircraft Establishment (RAE), Farnborough, England, to renew efforts to recover pieces from the second crash over Elba. Evidence began to emerge indicating that the tail section was intact from the Elba crash, and that the pressurized cabin section had torn apart before fire broke out.

In order to probe into the origin of cabin explosion, RAE engineers retired a Comet airplane from service and subjected its cabin to alternate pressurization and depressurization, to about 57 kPa (8.25 psi) over atmospheric pressure, by repeatedly pumping water into it and then removing it. During such simulated cabin pressurization, the wings of the aircraft were also stressed by hydraulic jacks to mimic wing loading during typical flight conditions. After about 3000 pressurization cycles, a fatigue crack originating in a corner of a cabin window advanced until the metal skin was pierced through. Figure 1.1 schematically shows the location of cracks in a failed Comet airplane.

The Comet, being the first commercial jetliner, was designed and built at a time when the role of fatigue in deteriorating the mechanical integrity of airframe components was not appreciated, and when subcritical fatigue crack growth had not evolved into a topic

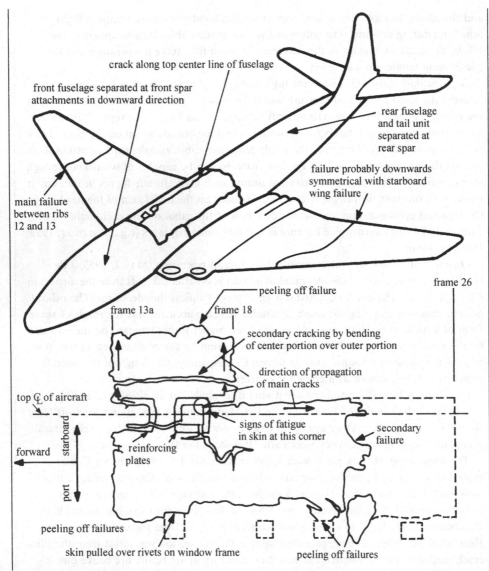

crack along top center line of fuselage

front fuselage separated at front spar
attachments in downward direction

rear fuselage
and tail unit
separated at
rear spar

failure probably downwards
symmetrical with starboard
wing failure

main failure
between ribs
12 and 13

frame 26

peeling off failure

frame 13a   frame 18

secondary cracking by bending
of center portion over outer portion

direction of propagation
of main cracks

top $\mathcal{C}_L$ of aircraft

signs of fatigue
in skin at this corner

secondary
failure

forward

reinforcing
plates

starboard

port

peeling off failures

skin pulled over rivets on window frame   peeling off failures

Fig. 1.1. Schematic diagram illustrating the location of fatigue cracks in a failed Comet
airplane. (After Petroski, 1996.)

of extensive research. It was assumed that the possibility of one fatigue cycle per flight,
due to cabin pressurization upon take-off and depressurization during landing, would not
be significant enough to advance any flaws in the fuselage to catastrophic proportions.
The cabin walls were designed to contain a pressure of 138 kPa (20 psi), two and a half
times the service requirements. As an added demonstration of safety, the passenger cabin
of each Comet was pressurized once to 114 kPa (16.5 psi) in a proof test, before the plane
was placed in service. The investigative report of the Court of Inquiry into the Comet
failures noted that the de Havilland designers believed '... that a cabin (which) would

survive undamaged a test to double its working pressure ... would not fail in service under the action of fatigue...'. This notion was proven erroneous, at a significant cost to de Havilland and to the British commerical aircraft industry.

The RAE tests revealed that the cabin failures in the first three Comet accidents were due to fatigue cracking which was aided by stress elevation at the rivet holes located near the window openings of the passenger cabin. In subsequent designs of the new Comet 4 models, which facilitated trans-Atlantic commercial jet travel for the first time, the window sections were replaced with a new reinforced panel which had much greater resistance to fatigue failure.

> No aircraft has contributed more to safety in the jet age than the Comet. The lessons it taught the world of aeronautics live in every jet airliner flying today.
>
> D.D. Dempster, 1959, in *The Tale of the Comet*

1.2 Different approaches to fatigue

There are different stages of fatigue damage in an engineering component where defects may nucleate in an initially undamaged section and propagate in a stable manner until catastrophic fracture ensues. For this most general situation, the progression of fatigue damage can be broadly classified into the following stages:

(1) Substructural and microstructural changes which cause nucleation of permanent damage.

(2) The creation of microscopic cracks.

(3) The growth and coalescence of microscopic flaws to form 'dominant' cracks, which may eventually lead to catastrophic failure. (From a practical standpoint, this stage of fatigue generally constitutes the demarkation between crack initiation and propagation.)

(4) Stable propagation of the dominant macrocrack.

(5) Structural instability or complete fracture.

The conditions for the nucleation of microdefects and the rate of advance of the dominant fatigue crack are strongly influenced by a wide range of mechanical, microstructural and environmental factors. The principal differences among different design philosophies often rest on how the crack initiation and the crack propagation stages of fatigue are quantitatively treated.

It is important to note here that a major obstacle to the development of life prediction models for fatigue lies in the choice of a definition for crack initiation. Materials scientists concerned with the microscopic mechanisms of fatigue are likely to regard the nucleation of micrometer-size flaws along slip bands and grain boundaries, and the roughening of fatigued surfaces as the crack inception stage of fatigue failure. A practicing engineer, on the other hand, tends to relate the limit of resolution of the (nondestructive) crack detection equipment (typically a fraction of a millimeter) with the nucleation of a fatigue crack and with the initial crack size used for design. Scattered within the limits of this broad range of choices, there lies a variety of definitions for crack nucleation which are specific to certain classes of fatigue-critical engineering applications. The total fatigue life is defined as the sum of the number of cycles to initiate a fatigue crack and the number of cycles to propagate it subcritically to some final crack size. In light of the foregoing discussion on what constitutes crack initiation, making a clear demarkation between crack initiation and crack propagation can become a critical task.

1.2.1 Total-life approaches

Classical approaches to fatigue design involve the characterization of *total fatigue life* to failure in terms of the cyclic stress range (the S–N curve approach) or the (plastic or total) strain range. In these methods, the number of stress or strain cycles necessary to induce fatigue failure in initially uncracked (and nominally smooth-surfaced) laboratory specimens is estimated under controlled amplitudes of cyclic stresses or strains. The resulting fatigue life incorporates the number of fatigue cycles to initiate a dominant crack (which can be as high as some 90% of the total fatigue life) and to propagate this dominant flaw until catastrophic failure occurs. Various techniques are available to account for the effects of mean stress, stress concentrations, environments, multiaxial stresses and variable amplitude stress fluctuations in the prediction of total fatigue life using the classical approaches (see Chapters 7, 8 and 14). Since the crack initiation life constitutes a major component of the total fatigue life in smooth specimens, the classical stress-based and strain-based methods represent, in many cases, design against fatigue crack initiation. Under high-cycle, low stress fatigue situations, the material deforms primarily elastically; the failure time or the number of cycles to failure under such *high-cycle fatigue* has traditionally been characterized in terms of the stress range. However, the stresses associated with *low-cycle fatigue* are generally high enough to cause appreciable plastic deformation prior to failure. Under these circumstances, the fatigue life is characterized in terms of the strain range. An example of a situation, where the classical (short-life) strain-based approach (also referred to as the *low-cycle fatigue* approach) has found much appeal, involves the prediction of fatigue life for the initiation and early growth of a crack within the strain field associated with the fully plastic region ahead of a stress concentration (see Fig. 1.2). The low-cycle approach to fatigue design has found particularly widespread use in ground-vehicle industries.

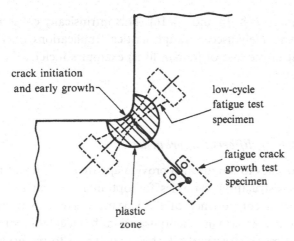

Fig. 1.2. Schematic diagram illustrating the various stages of fatigue in an engineering component and the approaches used to estimate the fatigue life. (After Coffin, 1979.)

1.2.2 *Defect-tolerant approach*

The fracture mechanics approach to fatigue design, on the other hand, invokes a 'defect-tolerant' philosophy. The basic premise here is that all engineering components are inherently flawed. The size of a pre-existing flaw is generally determined from nondestructive flaw detection techniques (such as visual, dye-penetrant or X-ray techniques or the ultrasonic, magnetic or acoustic emission methods). If no flaw is found in the component, proof tests are conducted whereby a structure, such as a pressure vessel, is subjected to a simulation test *a priori* at a stress level slightly higher than the service stress. If no cracks are detected by the nondestructive test method and if catastrophic failure does not occur during the proof test, the largest (undetected) initial crack size is estimated from the resolution of the flaw detection technique. The useful fatigue life is then defined as the number of fatigue cycles or time to *propagate* the dominant crack from this initial size to some critical dimension. The choice of the critical size for the fatigue crack may be based on the fracture toughness of the material, the limit load for the particular structural part, the allowable strain or the permissible change in the compliance of the component. The prediction of crack propagation life using the defect-tolerant approach involves empirical crack growth laws based on fracture mechanics. In terms of the requirements of linear elastic fracture mechanics, the defect-tolerant method is applicable under conditions of small-scale yielding (i.e. away from the plastic strain field of any stress concentrators), where the crack tip plastic zone is small compared to the characteristic dimensions of the cracked component (including the crack size) and where predominantly elastic loading conditions prevail. Various methods are available to incorporate the effects of mean stresses, stress concentrations, environments, variable amplitude loading spectra and multiaxial stresses in the estimation of useful crack growth life (see Chapters 7, 8, 10 and 14–16). This intrinsically conservative approach to fatigue has been widely used in fatigue-critical applications where catastrophic failures will result in the loss of human lives; examples include the aerospace and nuclear industries.

1.2.3 A comparison of different approaches

The different approaches to fatigue also provide apparently different guidelines for the design of microstructural variables for optimum fatigue resistance. These differences are merely a consequence of the varying degrees to which the role of crack initiation and crack growth are incorporated in the calculation of useful fatigue life. For example, in many structural alloys the resistance to the growth of long fatigue cracks generally increases with an increase in grain size (or a decrease in yield strength) at low ΔK values where a significant portion of subcritical crack growth life is expended. On the other hand, the total fatigue life estimated on the basis of stress-life plots generally exhibits the opposite trend; higher strength materials and finer grained microstructures usually lead to a longer fatigue life. The apparent contradiction between the two approaches can be reconciled by noting that the former approach to fatigue deals primarily with the resistance to fatigue crack growth, while the latter approach based on nominally defect-free laboratory specimens focuses mainly on the resistance to fatigue crack initiation. The choice of a particular microstructural condition for improved fatigue life is then predicated upon the design philosophy for a specific application. Optimization of microstructural characteristics for improved resistance to both crack initiation and crack growth would require a trade-off between the recommendations of the two approaches.

1.2.4 'Safe-life' and 'fail-safe' concepts

The safe-life and fail-safe design approaches were developed by aerospace engineers. In the safe-life approach to fatigue design, the typical cyclic load spectra, which are imposed on a structural component in service, are first determined. On the basis of this information, the components are analyzed or tested in the laboratory under load conditions which are typical of service spectra, and a useful fatigue life is estimated for the component. The estimated fatigue life, suitably modified with a factor of safety (or an ignorance factor), then provides a prediction of 'safe life' for the component. At the end of the expected safe operation life, the component is automatically retired from service, even if no failure has occurred during service (and the component has considerable residual fatigue life). Although an estimate of life may be obtained from practical tests on the actual component, the safe-life method is intrinsically theoretical in nature. This procedure invariably has to account for several unknowns, such as unexpected changes in load conditions, errors in the estimates of typical service load spectra, scatter in the test results, variation in properties among different batches of the same material, existence of initial defects in the production process, corrosion of the parts used in the component, and human errors in the operation of the component. By selecting a large margin of safety, a safe operating life can be guaranteed, although such a conservative approach may not be desirable from the viewpoints of economy and performance. On the other hand, if fatigue cracks are nucleated in the component during service, the component may well fail catastrophically. As noted by Gurney (1968), the safe-life approach depends on achieving a specified life without the development of a fatigue crack so that the emphasis is on the prevention of crack initiation. The fail-safe concept, by contrast, is based on the argument that, even if an individual member of a large structure fails, there should be sufficient structural integrity in the remaining parts to enable the structure to operate safely until the crack is detected. Components which have multiple load paths are generally fail-safe because of structural redundancy. In addition, the structure may contain crack arresters to prevent undesirable levels of crack growth. (The case study in Section 9.12 provides an example of this concept.) The fail-safe approach mandates periodic inspection along with the requirement that the crack detection techniques be capable of identifying flaws to enable prompt repairs or replacements. Whatever philosophy is employed in design, it is often preferable (and even required in some safety-critical situations, e.g., aircraft and nuclear industries) that the critical components of a structure be inspected periodically. This step eliminates dangerous consequences arising from false estimates and errors in the design stage, especially with the safe-life approach.

1.2.5 Case study: Retirement for cause

The design and maintenance of gas turbine engine components have traditionally involved estimates of fatigue life on the basis of low-cycle fatigue concepts. In this approach, entire classes of components are retired from service when some pre-determined design life is reached. For typical rotor components, such as discs which have been serviced and maintained by the United States Air Force in the past several decades, this approach meant that 1000 discs could be retired from service when, statistically, only one among them had developed a small fatigue crack (typically 0.75 mm or less in depth). In other words, 99.9% of the discs, which had considerable residual fatigue life, were retired prematurely.

The development of a new fatigue control methodology for gas turbine engine components provides an example of the practical use of the fail-safe concept. In 1985, the United States Air Force began implementing the so-called 'retirement for cause' (RFC) component life management methodology for its existing F100 aircraft engines, principally for cost savings, in lieu of the traditional approaches based on pre-determined life (Harris, 1987). In this program, the retirement of a gas turbine engine component from service occurs when the unique fatigue life of that particular component (as opposed to the predicted life for the entire population of the same component) is considered to be utilized. The individual component is retired from service when there is a specific reason or cause for removal from service, such as the existence of a flaw of a certain (maximum) allowable or detectable size. This system replaced classical low-cycle fatigue approaches where an entire population of components of a certain type were retired, regardless of the condition, when a pre-determined time or number of cycles was expended.

Analyses of crack growth and damage tolerance were carried out to identify which, if any, components were candidates for RFC. Subsequently, a list of components and combinations of inspection sizes and intervals were identified.† The F100 engine, for which the RFC program has initially been implemented, is currently in service in the twin-engine F-15 military aircraft built by the McDonnell Douglas Corporation and in the single-engine F-16 fighter aircraft built by the General Dynamics Corporation. There are over 3200 such engines in the operational inventory of the US Air Force. A total of twenty-three components used in such parts as fan, compressor and low pressure turbine rotors are being managed under this philosophy. The F100 engine overhaul manuals have been revised such that the RFC procedure replaces the classical time to retirement guidelines.

There were several developments for damage tolerance analysis and maintenance which evolved from the RFC program. One such development is the refinement of a nondestructive inspection method, based on eddy current monitoring, which is now commonly used under the RFC program management. Another outcome of this program is a cryogenic spin pit test for some titanium alloy discs. In this procedure, the component is spun at a low temperature where the fracture resistance drops. If the disc does not burst during the spin and safe operation is guaranteed above a certain 'inspection' size, the disc is placed back in service until the next inspection. It was also demonstrated that if cracks below the inspection size were to be present, the cryogenic spin pit test would not further extend these cracks or cause the damage zone dimensions at the crack tip to be altered. Whereas RFC was implemented on an F100 engine which was already designed and built, improvements to damage-tolerant design were implemented by another effort termed the ENgine Structural Integrity Program (ENSIP). Under ENSIP (Nicholas, Laflen, & VanStone, 1986; Cowles, 1988), all critical structural components in an engine had to be designed such that they could be inspected and, based on the inspection flaw size, could be flown safely until the next inspection. Thus, all components became, by definition, candidates for RFC. Since the inception of such damage tolerance procedures in the design stage, failure incidents due to low-cycle fatigue have been essentially eliminated from the US Air Force jet engine inventory.

The economic implications of such changes in failure control philosophy are also substantial. Initial estimates by the US Air Force reveal that, over the time period 1986–2005, the (life cycle) cost savings realized from the implementation of the RFC methodology to the F100 engines alone will amount to nearly $1 billion. Additional savings, amounting to as much as $655 million, are projected over this time period as a consequence of reductions in labor and fuel costs arising from the extension of maintenance intervals for the upgraded F100 core engines.

† All of the components which were analyzed did not meet the criteria for RFC; the ones which failed the criteria are still being retired after their design life, predicated upon low-cycle fatigue estimates, is expended.

Another example of the fail-safe approach is the so-called leak-before-break criterion, which was first proposed by Irwin (1964). This methodology is widely used in the structural design of pressure vessels and pipes. The leak-before-break criterion is developed as a means of ensuring that a pressure vessel or a pipe has the necessary structural integrity and service use even if a surface crack propagates through the thickness of the pressure vessel or pipe wall. Thus, the vessel or pipe would first 'leak' before any catastrophic fracture occurs so that the fatigue flaw could be easily detected and repaired. The implementation of this criterion would, therefore, require that the critical crack size at the design stress level of the material be greater than the wall thickness of the vessel. Examples of fracture mechanics analyses used in conjunction with the leak-before-break criterion can be found in any book on the subject (e.g., Barsom & Rolfe, 1987).

1.3 The need for a mechanistic basis

Fatigue of materials is a branch of study which provides a broad variety of complex mechanistic processes for scientific investigation. The size scales of observation that are of interest in this research area range from submicrostructural (even atomistic) levels to dimensions of structural components spanning tens or hundreds of meters. Implications of fatigue failure encompass many aspects of our lives. The practical significance of fatigue has even generated fictional stories and motion pictures whose themes have centered around fatigue failures (e.g., the novel *No Highway* by Nevil Shute, 1948, and its motion picture adaptation entitled *No Highway to Heaven*). The consequences of fatigue failure become most apparent when stories of disasters, such as aircraft accidents involving the loss of human lives, are publicized.†

Although considerations of fatigue failure are intimately tied to the practical aspects of structural integrity in engineering components, the mechanistic and scientific basis for the study of fatigue cannot be ignored because of the following reasons:

(1) The size scale over which permanent damage occurs at the tip of a fatigue crack is generally comparable to the characteristic microstructural dimension of the material, *even if the component dimensions and the crack size are orders of magnitude larger than the scale of the microstructure.*

(2) The total life and fracture mechanics approaches provide methods for characterizing the resistance of the material to crack initiation and growth under cyclic loads. However, these concepts alone cannot offer a quantitative description of the intrinsic resistance of the material to fatigue. This information can be obtained only if there exists a thorough understanding of the

† The 1985 crash of a Japan Airlines Boeing 747 due to a catastrophic fatigue failure of the rear pressure bulkhead and the resulting loss of 520 human lives is a case in point. A failure analysis of this accident is presented in Chapter 10.

micromechanisms of failure. There is ample evidence from published work on a wide variety of materials that subtle changes in the microstructure (and the environment) can lead to drastic alterations in the extent of cyclic damage and failure life. Thus, optimizing the microstructural characteristics of a material for improving the fatigue resistance inevitably requires a scientific knowledge of failure mechanisms.

(3) A significant portion of the fatigue crack growth life is spent at low ΔK levels where the maximum crack tip opening displacement during a loading cycle is typically smaller than a micrometer for most structural components. Since this dimension is smaller than the characteristic microstructural size scale in most materials, microstructural effects can markedly affect the resistance to fracture even if the crack size is significantly larger than the microstructural dimension.

(4) Even when a structural component is designed conservatively, fatigue failure may occur because of unexpected changes in service conditions. 'Post-mortem' analyses of fatigue failures often involve tracing the origin of fatigue failure *via* microscopic features present on the fracture surfaces, such as 'clam shell' markings and striations. These features can provide valuable information about the location where fracture initiated as well as about the magnitude of loads imposed upon the failed component. (See the various case studies in this book for an illustration of this point.) A fundamental knowledge of the link between the characteristic features observed on the fatigue fracture surfaces, the microscopic mechanisms of failure and the macroscopic rates of crack advance is vital to the success of such post-mortem analyses.

1.4 Continuum mechanics

Continuum descriptions of cyclic deformation and fatigue failure require an understanding of elasticity and plasticity theories. In this section, we present a brief review of the theories of linear elasticity and plasticity which provide the foundation for the derivations and discussions presented in various chapters of this book. The contents of this section are intentionally brief and confined only to those topics that are pertinent to the scope of this book. More elaborate details and derivations on these topics can be found in standard textbooks (e.g., Hill, 1950; Malvern, 1969).

Consider the infinitesimal element of volume in a stressed solid, shown in Fig. 1.3. The stress at a point is the local area intensity of the force which is transmitted between adjacent parts of the solid through an imaginary surface that divides the solid, such as the faces of the cubic element in Fig. 1.3. The stress can be represented by its scalar components, σ_{ij}, normal and tangential to the surface. For the cartesian

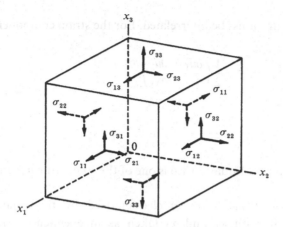

Fig. 1.3. Definition of stress components on an infinitesimal volume element.

coordinate system shown in Fig. 1.3, the stress components at a point are contained in the matrix

$$[\sigma_{ij}] \equiv \begin{pmatrix} \sigma_{11} & \sigma_{12} & \sigma_{13} \\ \sigma_{21} & \sigma_{22} & \sigma_{23} \\ \sigma_{31} & \sigma_{32} & \sigma_{33} \end{pmatrix} \equiv \begin{pmatrix} \sigma_{xx} & \sigma_{xy} & \sigma_{xz} \\ \sigma_{yx} & \sigma_{yy} & \sigma_{yz} \\ \sigma_{zx} & \sigma_{zy} & \sigma_{zz} \end{pmatrix}, \tag{1.1}$$

where $i, j = 1, 2, 3$, and σ_{ij} is the ith component of the force per unit area on a plane whose outward normal points toward the positive x_j direction; $x_1 = x$, $x_2 = y$, and $x_3 = z$. Each infinitesimal volume element in the body must be in mechanical equilibrium. Since there can be no net torque on the element, $\sigma_{ij} = \sigma_{ji}$. Similarly, no net force can act on the element, so that

$$\frac{\partial \sigma_{1i}}{\partial x_1} + \frac{\partial \sigma_{2i}}{\partial x_2} + \frac{\partial \sigma_{3i}}{\partial x_3} + b_i = 0, \qquad i = 1, 2, 3, \tag{1.2}$$

where b_i is the ith component of the body force per unit volume. In the absence of body forces, Eq. 1.2 can be expressed as

$$\frac{\partial \sigma_{ij}}{\partial x_j} = 0, \tag{1.3}$$

where summation over j is implied. Equations 1.2 and 1.3 are known as the *equilibrium equations*.

Under the influence of applied forces, let u_i be the components of displacement at a point in the body. The components of infinitesimal strain are defined as

$$\epsilon_{ij} = \frac{1}{2} \left(\frac{\partial u_i}{\partial x_j} + \frac{\partial u_j}{\partial x_i} \right). \tag{1.4}$$

When $i \neq j$ in this equation, the shear strains are obtained. However, it is important to note that, for $i \neq j$, Eq. 1.4 provides only one-half of the shear strains that are commonly defined in engineering where $\gamma_{ij} = 2\epsilon_{ij}$. In Eq. 1.4, six components of small strain, ϵ_{ij}, have been expressed in terms of three components of the displacement. This implies that the strains must be interrelated. For the strain components in the x_1–x_2 plane,

$$\epsilon_{11} = \frac{\partial u_1}{\partial x_1}, \quad \epsilon_{22} = \frac{\partial u_2}{\partial x_2}, \quad \epsilon_{12} = \frac{1}{2}\left(\frac{\partial u_1}{\partial x_2} + \frac{\partial u_2}{\partial x_1}\right). \tag{1.5}$$

From Eqs. 1.5, one finds that

$$\frac{\partial^2 \epsilon_{11}}{\partial x_2^2} - 2\frac{\partial^2 \epsilon_{12}}{\partial x_1 \partial x_2} + \frac{\partial^2 \epsilon_{22}}{\partial x_1^2} = 0, \tag{1.6}$$

for deformation on the x_1–x_2 plane. Equation 1.6 is one of the so-called *compatibility equations*.

A state of *plane stress* or *plane strain* is characterized by the conditions $\partial(\ )/\partial x_3 = 0$ and $\sigma_{13} = \sigma_{23} = 0$, with x_1 and x_2 taken as independent variables. For plane stress, the stress–strain relationships (to be discussed in the next subsection) are used along with the additional condition that $\sigma_{33} = 0$. Similarly, for plane strain, the stress–strain relationships are applied with $\epsilon_{33} = 0$. An *anti-plane* state is characterized by the conditions that $\sigma_{11} = \sigma_{22} = \sigma_{33} = \sigma_{12} = 0$.

Standard procedures of coordinate transformation (see, for example, Malvern, 1969) are used to derive the equilibrium and compatibility equations and strain–displacement relationships for the cylindrical coordinate system from the results discussed above for the cartesian reference system, Eqs. 1.3 and 1.6. These results for the cylindrical coordinate reference are presented in Section 9.3.2.

The equilibrium conditions are fulfilled automatically if the stresses are expressed in terms of the so-called Airy stress function χ, which is defined by the relationships

$$\sigma_{11} = \frac{\partial^2 \chi}{\partial x_2^2}, \quad \sigma_{22} = \frac{\partial^2 \chi}{\partial x_1^2}, \quad \sigma_{12} = -\frac{\partial^2 \chi}{\partial x_1 \partial x_2}, \tag{1.7}$$

in the cartesian coordinate system. Similarly, the compatibility condition, Eq. 1.6, when expressed in terms of the Airy stress function, becomes

$$\frac{\partial^4 \chi}{\partial x_1^4} + 2\frac{\partial^4 \chi}{\partial x_1^2 \partial x_2^2} + \frac{\partial^4 \chi}{\partial x_2^4} = 0. \tag{1.8}$$

Similar expressions for the equilibrium and compatibility conditions in terms of polar coordinates are given in Chapter 8. A complete solution is obtained by satisfying Eq. 1.8 and the prescribed boundary conditions.

1.4.1 Elements of linear elasticity

When the material undergoes only elastic deformation, the stresses and strains are related by *Hooke's law*,

$$\sigma_{ij} = C_{ijkl}\epsilon_{kl}, \tag{1.9}$$

where C_{ijkl} are the elastic constants which, for isotropic material response, are

$$C_{ijkl} = \lambda \delta_{ij}\delta_{kl} + G(\delta_{ik}\delta_{jl} + \delta_{il}\delta_{jk}). \tag{1.10}$$

λ is known as the Lamé constant and G is the shear modulus. δ_{ij} is the Kronecker delta with the property that $\delta_{ij} = 0$ for $i \neq j$ and that $\delta_{ij} = 1$ for $i = j$. The isotropic elastic constants defined by Young's modulus E and Poisson's ratio ν are related to G and λ by the expressions

$$E = \frac{G(3\lambda + 2G)}{G + \lambda} = 2G(1 + \nu), \qquad \nu = \frac{\lambda}{2(G + \lambda)}. \tag{1.11}$$

Young's modulus E is the ratio of the axial stress to the axial strain, whereas Poisson's ratio ν represents the ratio of transverse contraction to (axial) elongation in simple tension. The strains can be related to the stresses in terms of the elastic constants,

$$\epsilon_{11} = \frac{1}{E}[\sigma_{11} - \nu(\sigma_{22} + \sigma_{33})], \qquad \epsilon_{12} = \frac{1}{2G}\sigma_{12},$$

$$\epsilon_{22} = \frac{1}{E}[\sigma_{22} - \nu(\sigma_{11} + \sigma_{33})], \qquad \epsilon_{23} = \frac{1}{2G}\sigma_{23},$$

$$\epsilon_{33} = \frac{1}{E}[\sigma_{33} - \nu(\sigma_{11} + \sigma_{22})], \qquad \epsilon_{31} = \frac{1}{2G}\sigma_{31}. \tag{1.12}$$

When the volume element shown in Fig. 1.3 deforms reversibly by an infinitesimal strain increment $d\epsilon_{ij}$, the stresses do work on the element by the amount

$$dw = \sigma_{ij}d\epsilon_{ij} = C_{ijkl}\epsilon_{kl}d\epsilon_{ij}. \tag{1.13}$$

Under conditions of reversible and isothermal elastic deformation, the differential work per unit volume dw in Eq. 1.13 is also equal to the change in the Helmholtz free energy dF, which is an exact differential. For the conditions stated, the work of deformation is a single-valued function in strain space, such that

$$w = \frac{1}{2}C_{ijkl}\epsilon_{ij}\epsilon_{kl}, \qquad \text{(or)} \qquad \sigma_{ij} = \frac{\partial w}{\partial \epsilon_{ij}}, \tag{1.14}$$

which is useful for later discussion.

1.4.2 Stress invariants

For any general three-dimensional stress state given by Eq. 1.1, one can find *via* coordinate tranformation three normal stresses, known as *principal stresses*, which act on orthogonal planes that are free of shear stresses. The principal stresses, σ_1, σ_2 and σ_3, are the roots of the cubic equation,

$$\lambda^3 - I_1\lambda^2 - I_2\lambda - I_3 = 0, \tag{1.15}$$

where

$$I_1 = \sigma_{ii} = \sigma_1 + \sigma_2 + \sigma_3,$$

$$I_2 = \tfrac{1}{2}(\sigma_{ij}\sigma_{ij} - I_1^2) = -(\sigma_1\sigma_2 + \sigma_2\sigma_3 + \sigma_3\sigma_1),$$

$$I_3 = \det[\sigma_{ij}] = \sigma_1\sigma_2\sigma_3. \tag{1.16}$$

The coefficients I_1, I_2 and I_3 are independent of the orientation of the coordinate system chosen to describe the stress components. These coefficients are termed *stress invariants* because the principal stresses are physical quantities at the point in the solid under consideration. Any combination of these stress invariants also results in an entity which is an invariant.

The normal mean stress or the hydrostatic stress, σ_H, is defined as

$$\sigma_H = \tfrac{1}{3}(\sigma_{11} + \sigma_{22} + \sigma_{33}) = \tfrac{1}{3}\sigma_{kk} = \frac{I_1}{3}, \tag{1.17}$$

where the subscript $k = 1, 2, 3$. The hydrostatic stress σ_H causes a change only in volume (and not in shape) in an isotropic continuum. From Eqs. 1.11, 1.12 and 1.17, the compressibility ς, the ratio of the negative of the dilatation $e = \epsilon_{11} + \epsilon_{22} + \epsilon_{33}$ to the pressure $p = -\sigma_H$, can be written as

$$\varsigma = -\frac{e}{p} = \frac{3}{3\lambda + 2G} = \frac{1}{B}, \tag{1.18}$$

where B is the bulk modulus.

The deviatoric components of stress are defined as

$$s_{ij} = \sigma_{ij} - \frac{\delta_{ij}I_1}{3}. \tag{1.19}$$

These deviatoric stress components, unlike the hydrostatic stress, bring about a change of shape in the body and influence the plastic deformation. Analogous to the stress invariants I_1, I_2 and I_3, a new set of scalar invariants, J_1, J_2 and J_3, based on the principal components of the deviatoric stress tensor, s_1, s_2 and s_3, can be defined:

$$
\begin{aligned}
J_1 &= \sigma_{ii} - I_1 = 0, \\
J_2 &= \frac{1}{2}s_{ij}s_{ij} = \frac{1}{2}\{s_1^2 + s_2^2 + s_3^2\}, \\
J_3 &= \frac{1}{3}\left(s_{ij}s_{jk}s_{ki}\right) = s_1 s_2 s_3.
\end{aligned}
\tag{1.20}
$$

1.4.3 Elements of plasticity

When a ductile metallic material is loaded beyond the elastic limit, it undergoes permanent plastic deformation. Plastic flow in metals can be assumed incompressible. The total strain imposed on an elastic–plastic solid ϵ_{ij} can be written as the sum of the elastic and plastic strains ϵ_{ij}^e and ϵ_{ij}^p, respectively. The elastic strains are related to the stresses by Hooke's law, Eq. 1.9. The assumption of incompressibility of the material during plastic deformation leads to the condition that $\epsilon_{ii}^p = 0$.

The plastic behavior under complex multiaxial loading conditions is described by invoking constitutive laws which relate either the total strain to current stress state (so-called *deformation theories*) or the increment of plastic deformation to the stress and strain increment for a given state of the material (so-called *flow* or *incremental theories*). Any general theory of plasticity has the following major components:

(1) A *yield condition*, which specifies the onset of plastic deformation for different combinations of applied stresses.

(2)   A *hardening rule*, which prescribes the work hardening of the material and the change in yield condition with the progression of plastic deformation.

(3)   A *flow rule*, which relates the increments of plastic deformation or the components of plastic rate-of-deformation to the stress components.

Plastic deformation is determined by a yield condition which is a function $f(\sigma_{ij})$ of the current stress state. In most cases, the associative flow rule is used which assumes that the plastic strain increments are proportional to a function $f(\sigma_{ij})$ which depends on the current plastic state of the material. All rate-independent plasticity theories postulate that the material response is elastic for $f(\sigma_{ij}) < 0$. Elastic unloading occurs from a plastic state when $f(\sigma_{ij}) = 0$ and $(\partial f/\partial \sigma_{ij})d\sigma_{ij} < 0$, where $d\sigma_{ij}$ is the stress increment. Plastic deformation occurs when $f(\sigma_{ij}) = 0$ and $(\partial f/\partial \sigma_{ij})d\sigma_{ij} > 0$. If the material deformation is isotropic, the yield function is an isotropic function of stress such that $f(\sigma_{ij}) = f(I_1, I_2, I_3) = f(\sigma_1, \sigma_2, \sigma_3)$. Since the deformation of metallic materials is insensitive to moderate levels of hydrostatic stress σ_H, the yield function depends only on the deviatoric stress s_{ij}. If the yield response of the material is the same in tension and compression, i.e. if the material does not exhibit the Bauschinger effect (Chapter 3), $f(s_{ij}) = f(-s_{ij})$. Then f is an even function of J_3.

The von Mises and Tresca yield conditions are the most widely used flow criteria for metals. The von Mises condition states that

$$f \equiv J_2 - k^2 \equiv \tfrac{1}{2}s_{ij}s_{ij} - k^2 = 0, \tag{1.21}$$

or, in terms of principal stresses,

$$f = \tfrac{1}{6}\left[(\sigma_1 - \sigma_2)^2 + (\sigma_2 - \sigma_3)^2 + (\sigma_3 - \sigma_1)^2\right] - k^2 = 0. \tag{1.22}$$

k is a constant for an elastic–perfectly plastic solid and it is influenced by the prior strain history in a work-hardening material. Since J_2 is a measure of the *distortional energy* for an isotropic material, implied in the von Mises yield condition is the assumption that plastic flow occurs when the distortional energy reaches a critical value. For uniaxial tension, $\sigma_1 = \sigma_y$, $\sigma_2 = \sigma_3 = 0$, and $k = \sigma_y/\sqrt{3}$, where σ_y is the tensile yield stress. For pure shear, $\sigma_1 = -\sigma_3 = \tau_y$, $\sigma_2 = 0$, and $k = \tau_y$, where τ_y is the shear yield stress.

The Tresca yield condition states that the material will yield when the maximum shear stress reaches a critical value:

$$\tfrac{1}{2}|\sigma_1 - \sigma_3| = k, \tag{1.23}$$

where the principal stresses are arranged in the order $\sigma_1 \geq \sigma_2 \geq \sigma_3$. For uniaxial tension, the Tresca condition predicts that $k = \sigma_y/2$.

The yield condition $f(\sigma_{ij}) = 0$ for an isotropic material is represented in three-dimensional stress space with the principal stresses σ_1, σ_2 and σ_3 as the coordinate axes. In this principal stress space, Fig. 1.4(*a*), the shape of the *yield surface* representative of the von Mises yield condition, Eq. 1.22, is a right circular cylinder, while the Tresca yield condition, Eq. 1.23, is represented by the surface of a regular hexagonal prism. The surfaces are parallel to the hydrostatic stress line *OG*, which

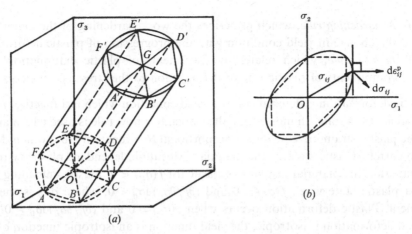

Fig. 1.4. (*a*) The yield surface drawn in three-dimensional principal stress space. The von Mises and Tresca conditions for yield are represented by the right circular cylinder and the inscribed hexagonal prism, respectively. (*b*) von Mises ellipse and Tresca hexagon for a state of biaxial stress.

denotes the condition $\sigma_1 = \sigma_2 = \sigma_3$.[†] The surfaces of the von Mises cylinder or the Tresca hexagonal prism are perpendicular to the deviatoric plane or Π plane (such as the plane $ABCDEF$ in Fig. 1.4(*a*)), which describes the condition $\sigma_1 + \sigma_2 + \sigma_3 = 0$. In a biaxial stress state represented by the principal stress coordinates σ_1 and σ_2, the von Mises yield condition is represented by an ellipse and the Tresca condition is shown by the inscribed hexagon, Fig. 1.4(*b*). Since net work has to be expended on the body during plastic deformation, the rate of energy dissipation is nonnegative, such that

$$\sigma_{ij}d\epsilon_{ij}^{\mathrm{p}} \geq 0. \tag{1.24}$$

Geometrically, this condition implies that the yield surface must be convex. Thus the yield condition requires the stress point to be on the yield surface and to be directed outward from the surface. At a smooth point on the yield surface, Eq. 1.24 implies that the incremental plastic strain vector $d\epsilon_{ij}^{\mathrm{p}}$ must be normal to the yield surface, and that the vector denoting the incremental change in stress $d\sigma_{ij}$ must have an acute angle with the strain vector. At a corner on the Tresca yield surface, this criterion must be applied separately to the two surfaces intersecting the corner.

This normality of the plastic strain increment to the yield surface is reflected by the *flow rule*,

$$d\epsilon_{ij}^{\mathrm{p}} = d\lambda \frac{\partial f}{\partial \sigma_{ij}}, \tag{1.25}$$

where $d\lambda$ is a positive scalar that can be taken to denote an *effective stretching*. By analogy with Eq. 1.14, f can be regarded as a plastic potential.

[†] This is so because if the stress state characterized by $(\sigma_1, \sigma_2, \sigma_3)$ lies on the yield surface, so does $(\sigma_1 + \sigma_{\mathrm{H}}, \sigma_2 + \sigma_{\mathrm{H}}, \sigma_3 + \sigma_{\mathrm{H}})$, where σ_{H} is any value of the hydrostatic stress defined in Eq. 1.17.

Given the representation of the flow criterion in terms of the yield surface, it is important to know how the yield surface changes during plastic deformation. The theory of perfectly (or ideally) plastic solids assumes that the yield function is unaffected by plastic deformation. If the material exhibits increasing resistance to plastic deformation with plastic straining, the simplest approach to handle such *strain hardening* is to invoke the so-called *isotropic hardening* model. During isotropic hardening, the yield surface expands uniformly, but it has a fixed shape and its center remains fixed in stress space. The dependence of the size of the yield surface on deformation can be determined by developing a universal stress–strain relationship

$$\sigma_e = h\left(\int d\bar{\epsilon}^p\right), \tag{1.26}$$

which relates two scalar quantities through the function h: the effective stress σ_e (which measures the size of the yield surface) and the effective plastic strain increment $d\bar{\epsilon}^p$. If one uses the von Mises criterion for yield,

$$\sigma_e = \sqrt{3J_2} = \sqrt{\tfrac{3}{2}s_{ij}s_{ij}}. \tag{1.27}$$

The numerical constant on the right hand side of this equation is chosen such that $\sigma_e = |\sigma_{11}|$ in uniaxial loading. The effective plastic strain increment is defined as

$$d\bar{\epsilon}^p = \sqrt{\tfrac{2}{3}d\epsilon_{ij}^p d\epsilon_{ij}^p}, \tag{1.28}$$

where the numerical factor is chosen such that in a state of uniaxial stress σ_{11}, $d\bar{\epsilon}^p = d\epsilon_{11}^p = -2d\epsilon_{22}^p = -2d\epsilon_{33}^p$.

An alternative statement of the isotropic hardening rule is obtained from the argument that the size of the yield surface is a function F only of the total plastic work, such that

$$\sigma_e = F(w_p), \qquad w_p = \int \sigma_{ij}d\epsilon_{ij}^p, \tag{1.29}$$

where the integration is carried out over the actual strain path. This condition provides results equivalent to Eq. 1.26.

The use of isotropic hardening for tension–compression cyclic deformation in many metallic materials does not rationalize the differences in elastic limit commonly found between forward and reverse loading. In an attempt to account for this so-called Bauschinger effect, an alternative hardening rule, known as kinematic hardening, has been proposed. In the classical models of kinematic hardening, the yield surface does not change its shape and size, but simply translates in stress space in the direction of its normal. The application of isotropic hardening and various kinematic-type hardening rules to cyclic deformation is described in Chapter 3.

The discussions up to this point have focused on the *incremental* or *flow* theories of plasticity. There is also a different approach, known as the *deformation* or *total strain* theory, which is adopted in plasticity problems mainly in view of its mathematical simplicity. The deformation theory, where the total strain ϵ_{ij} is taken to be a function of the current stress, is merely a nonlinear elasticity theory. It is assumed here that

$$\epsilon_{ij}^{p} = \upsilon s_{ij}, \tag{1.30}$$

where υ is a positive scalar function during loading and a negative scalar function during unloading. From Eq. 1.30, it is readily seen that

$$\upsilon = \tfrac{3}{2} \frac{\bar{\epsilon}^{p}}{\sigma_e}. \tag{1.31}$$

σ_e takes the value given in Eq. 1.27 and $\bar{\epsilon}^{p} = \sqrt{(2/3)\epsilon_{ij}^{p}\epsilon_{ij}^{p}}$. This can be used in conjunction with a universal stress–strain relationship of the form given in Eq. 1.26. Under proportional loading (i.e. for loading in which the components of the stress tensor vary in constant proportion), the incremental and deformation theories of plasticity exhibit agreement, while substantial differences are encountered between the two approaches for severely nonproportional loading.

1.4.4 Elements of linear viscoelasticity

Deformation which exhibits such features as stress–strain hysteresis, stress relaxation, creep, or dynamic response to stresses which fluctuate sinusoidally with time represents a material behavior which is both elastic and viscous. Such a deformation is known as *viscoelastic*. Examples of viscoelastic behavior are commonly found in high polymers.

Continuum models characterizing the viscoelastic constitutive behavior commonly involve series and parallel arrangements of springs and dashpots. Figures 1.5(a)–(c) show three simple examples of such models where the viscoelastic solid is subjected to a force F. The Maxwell element, Fig. 1.5(a), comprises a series arrangement of a linear spring, which produces an instantaneous displacement u in response to F, and a dashpot, whose velocity $\dot{u}$ is instantaneously proportional to F. The Kelvin–Voigt element is a linear spring and a dashpot in parallel, Fig. 1.5(b). The spring constant and the coefficient of viscosity of the dashpot are denoted as $\mathcal{K}$ and η, respectively. These idealized models qualitatively capture the relaxation, decay and creep phenomena, although their quantitative predictions can deviate considerably from the behavior of real viscoelastic materials. Further refinements to these models can be made by incorporating additional elements. Figure 1.5(c) shows an example of a so-called standard linear solid where an extra spring is placed in parallel with a Maxwell element.

The applied force F and the displacement u at the point of application of the force are related in the following way.

$$\dot{u} = \frac{\dot{F}}{\mathcal{K}} + \frac{F}{\eta}, \qquad F(0) = \mathcal{K}u(0), \qquad \text{for the Maxwell model,} \tag{1.32}$$

$$F = \mathcal{K}u + \eta\dot{u}, \qquad u(0) = 0, \qquad \text{for the Kelvin–Voigt model,} \tag{1.33}$$

Here, (t) after a variable denotes the value of the variable at time t, and $(\dot{\ })$ denotes the rate of change with t.

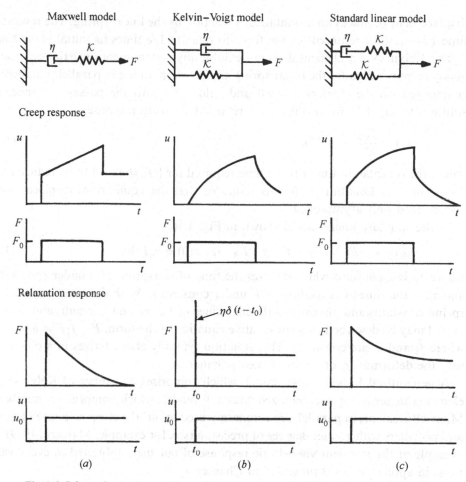

Fig. 1.5. Schematic arrangement (top row), creep response (middle row) and relaxation response (bottom row) of (*a*) the Maxwell element, (*b*) the Kelvin–Voigt element, and (*c*) a standard linear model. The creep response includes loading and unloading.

The Maxwell element thus characterizes *steady creep* under a constant load F_0 after an initial glassy response with the displacement $F_0/\mathcal{K}$ in the following manner:

$$u(t) = \frac{F}{\mathcal{K}} + \frac{F_0 t}{\eta}. \tag{1.34}$$

If a displacement $u(t)$ is suddenly applied such that $u(t) = 0$ when $t < 0$ and $u(t) = u_0$ when $t \geq 0$, *stress relaxation* characterized by the Maxwell solid takes the form:

$$F(t) = \mathcal{K}e^{-(\mathcal{K}/\eta)t}u_0. \tag{1.35}$$

The ratio $\tau = \eta/\mathcal{K}$ denotes the *relaxation time* which is the rate of decay of the force according to the exponential law in Eq. 1.35 after the sudden imposition of the displacement produces an instantaneous reaction by the linear spring. The relaxation time τ is the time required for the force to relax to $1/e$ times its initial value Ku_0.

The Kelvin–Voigt element does not exhibit any instantaneous elastic response or glassy response because the linear spring and the dashpot are in parallel. If a force F_0 is imposed on the element at $t = 0$ and held fixed with the passage of time, the solution to Eq. 1.33 for $u(t)$ gives the retarded elasticity response:

$$u(t) = \frac{F_0}{\mathcal{K}}(1 - e^{-t/\tau}),\tag{1.36}$$

where the retardation time τ is the time required for $\{(F_0/\mathcal{K}) - u\}$ to be reduced by a factor of $1/e$. Equation 1.36 thus indicates that the equilibrium displacement is approached only asymptotically.

In the standard linear model shown in Fig. 1.5(c),

$$E_R(u + \tau_F \dot{u}) = F + \tau_u \dot{F}, \qquad E_R \cdot \tau_F \cdot u(0) = \tau_u F(0),\tag{1.37}$$

where τ_F is a constant which denotes the time of relaxation of u under constant F, and τ_u is the time of relaxation of F under constant u. With different values of the spring constants and viscosity, different values of E_R, τ_F and τ_u result, and thus the model may be described by a constitutive equation of the form: $\dot{F} + \tilde{f}_1 F = \tilde{u}_1 \dot{u} + \tilde{u}_2 u$, where $\tilde{f}_1$ and $\tilde{u}_i$ are constants. This equation typically characterizes in a qualitative way the deformation of a cross-linked polymer.

A generalized Kelvin–Voigt model, which comprises a number of Kelvin–Voigt elements in series, or a generalized Maxwell model, which comprises a number of Maxwell elements in parallel, are commonly used to 'fit' the creep response of metals and polymers with greater degrees of precision (see, for example, Malvern, 1969). An example of the transient viscoelastic response of polymers subjected to cyclic variations in applied stress is presented in Chapter 6.

1.4.5 Viscoplasticity and viscous creep

A fluid exhibiting Newtonian viscous behavior develops a straining rate which is proportional to the applied shear stress, from the very onset of the application of the stress. On the other hand, a viscoplastic (or Bingham-plastic) material resists straining until a critical value of the shear stress is reached, beyond which the rate of plastic straining is in proportion to the stress. Sour dough, clay, and paste are materials which provide examples of such viscoplastic response.

Consider the case of simple shear with $\sigma_{xy} = \sigma_{yx} = \tau$ and $\dot{e}_{xy} = \dot{\epsilon}^c_{yx} = \dot{\epsilon}^c$; all other components of stress σ_{ij} and creep strain rate $\dot{\epsilon}^c_{ij}$ vanish. In this case of the so-called Bingham material (Bingham, 1922), the onset of straining occurs when the absolute value of τ exceeds a critical value k, such that

$$2\eta\dot{\epsilon}^c = 0, \text{ if } f_y < 0; \quad 2\eta\dot{\epsilon}^c = f_y\tau, \text{ if } f_y \geq 0; \quad f_y = \left(1 - \frac{k}{|\tau|}\right).\tag{1.38}$$

Here, η is the coefficient of viscosity and f_y is a yield function. Assuming incompressibility, Hohenemser and Prager (1932) generalized Bingham's model to the general state of stress where

$$2\eta\dot{\epsilon}_{ij}^c = 0, \text{ if } f_y < 0; \quad 2\eta\dot{\epsilon}_{ij}^c = f_y s_{ij}, \text{ if } f_y \geq 0; \quad f_y = \left(1 - \frac{k}{\sqrt{J_2}}\right). \tag{1.39}$$

Recall that s_{ij} is the component of the deviatoric stress defined in Eq. 1.19, and that J_2 is the second invariant of the deviatoric stress, Eq. 1.20. Further extensions of this approach, including the assumption of compressibility, have been introduced in the literature.

Most commonly, tensile tests are carried out at different strain rates and temperatures to characterize the viscoplastic creep response of materials. The results of such tests are commonly analyzed using empirical approximations of the form:

$$\sigma = \sigma_0(\dot{\epsilon}^c)^m(\epsilon^c)^n, \tag{1.40}$$

where σ is the uniaxial tensile stress, ϵ^c is the viscous part of the true strain, $\dot{\epsilon}^c$ is the viscous strain rate, m is the strain-rate sensitivity parameter, n is the strain hardening exponent, and σ_0 is a material parameter. In steady-state creep, where the strain rate remains roughly constant, the constitutive response is given by the so-called Norton–Odqvist law:

$$\dot{\epsilon}_{ij}^c = \frac{3}{2}\dot{\epsilon}_y\left(\frac{\sigma_e}{\sigma_y}\right)^{(n_c-1)}\frac{s_{ij}}{\sigma_y}, \tag{1.41}$$

where σ_e is the von Mises effective stress, Eq. 1.27, σ_y is the yield strength, $\dot{\epsilon}_y$ is the yield strain rate and n_c is the power-law creep exponent.

Since creep deformation is a thermally activated process, the mechanistic origins of the evolution of strain rate are tied to an Arrhenius-type equation,

$$\dot{\epsilon}^c = Ae^{-(Q/RT)}, \tag{1.42}$$

where Q is the activation energy, T is the absolute temperature and R is the universal gas constant. Note that Eq. 1.42 essentially denotes the temperature-dependence of viscosity η which is implicit in Eq. 1.38.

1.5 Deformation of ductile single crystals

The most comprehensive mechanistic details of cyclic deformation and fatigue fracture have been obtained on ductile single crystals. In an attempt to develop some background information for the discussions of fatigue in metallic crystals (to be presented in the next two chapters), this section focuses on the terminology and methods used to describe the deformation of ductile single crystals. Further details on this subject can be found in any introductory text on physical or mechanical metallurgy.

1.5.1 Resolved shear stress and shear strain

The anisotropic deformation of single crystals is characterized in terms of the resolved shear stress, τ_R, and resolved shear strain, γ_R, acting on a specific slip plane along a specific slip direction. Figure 1.6 shows a cylindrical specimen of a single crystal of cross-sectional area A, which is subjected to a uniaxial tensile or compressive load P. Let the slip plane normal and the slip direction be oriented at an angle ϕ_0 and λ_0, respectively, with respect to the loading axis. **n** and **b** are the unit vectors normal to the slip plane and along the slip direction, respectively. χ_0 ($= 90° - \phi_0$) is the complementary of ϕ_0.

The load P can be resolved along **b** to give a shear force, $P\cos\lambda_0$, acting in the slip direction. The corresponding initial shear stress, τ_{R_0}, resolved on the slip plane is the shear force, $P\cos\lambda_0$, divided by the area of the slip plane, $A/\cos\phi_0$, so that

$$\tau_{R_0} = \frac{P}{A}\cos\phi_0 \cos\lambda_0 = \sigma\cos\phi_0 \cos\lambda_0. \tag{1.43}$$

The condition for the onset of plastic deformation is given by the so-called *Schmid law* which states that *a crystalline solid flows plastically when the resolved shear stress, τ_{R_0}, acting along the slip direction in the slip plane reaches a critical value, τ_c*:

$$\sigma\cos\phi_0\cos\lambda_0 = \sigma\sin\chi_0\cos\lambda_0 = M\sigma = \tau_c. \tag{1.44}$$

M is known as the Schmid factor, which has a maximum value of 0.5 corresponding to the orientation $\phi_0 = \lambda_0 = 45°$.

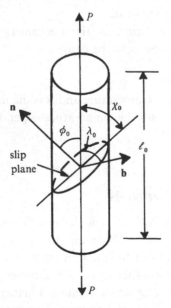

Fig. 1.6. A schematic diagram of a single crystal showing the orientations of the slip plane and the slip direction.

Consider the geometrical changes in the slip system as the initially cylindrical crystal deforms plastically (Fig. 1.7). A reference gage length vector $\boldsymbol{\ell}_0$ along the axis of the cylinder prior to deformation changes in both magnitude and direction with the progression of slip. Let $\boldsymbol{\ell}$ be the instantaneous gage length vector at any point during plastic deformation. It can be shown by simple geometrical arguments that

$$\boldsymbol{\ell} = \boldsymbol{\ell}_0 + \gamma_R(\boldsymbol{\ell}_0 \cdot \mathbf{n})\,\mathbf{b}, \tag{1.45}$$

where $\cdot$ denotes a dot product. In order to express $\boldsymbol{\ell}$ in terms of $\boldsymbol{\ell}_0$, consider

$$\boldsymbol{\ell} \cdot \boldsymbol{\ell} = \boldsymbol{\ell}_0 \cdot \boldsymbol{\ell}_0 + \gamma_R^2(\boldsymbol{\ell}_0 \cdot \mathbf{n})^2 (\mathbf{b} \cdot \mathbf{b}) + 2\gamma_R(\boldsymbol{\ell}_0 \cdot \mathbf{n})\,\boldsymbol{\ell}_0 \cdot \mathbf{b}). \tag{1.46}$$

Equivalently,

$$\ell^2 = \ell_0{}^2 (1 + 2\gamma_R \cos\phi_0 \cos\lambda_0 + \gamma_R^2 \cos^2\phi_0). \tag{1.47}$$

If γ_R is expressed in terms of the instantaneous and original gage lengths, ℓ and ℓ_0, respectively, and the initial orientation of the slip plane and slip direction ϕ_0 and λ_0, we find that

$$\gamma_R = \frac{1}{\cos\phi_0} \left\{ \sqrt{\left(\frac{\ell}{\ell_0}\right)^2 - \sin^2\lambda_0} - \cos\lambda_0 \right\}. \tag{1.48}$$

In an actual tensile test on a single crystal, the ends of the crystal are gripped to the testing machine and the gage length vectors $\boldsymbol{\ell}_0$ and $\boldsymbol{\ell}$ are confined to remain parallel to the initial longitudinal axis of the cylinder (i.e. along the loading axis). Consequently, the single slip displacement schematically illustrated in Fig. 1.7 is tantamount to the rotation of both the slip plane and the slip direction toward the tensile axis as deformation proceeds in a real experiment. The initial orientation angles ϕ_0 and λ_0 will decrease with increasing plastic deformation.

The Schmid factor also relates the increment of shear strain $d\gamma_R$ on a slip system to the increment in longitudinal strain $d\epsilon$. For instantaneous orientation angles ϕ and λ and small strain increments,

$$\frac{d\gamma_R}{d\epsilon} = \frac{\sigma}{\tau_R} = \frac{1}{M} = \frac{1}{\cos\phi\cos\lambda}. \tag{1.49}$$

When $\phi = \lambda = 45°$, $M = 0.5$, $\tau_R = 0.5\sigma$.

The instantaneous value of resolved shear stress, τ_R, may be obtained by noting that the glide plane, with the cross-sectional area $A/\cos\phi_0$, remains undistorted by slip. Since the longitudinal axis at any point during slip is inclined at an angle λ to the slip direction,

$$\tau_R = \frac{P}{A}\cos\phi_0 \cos\lambda = \sigma\cos\phi_0 \cos\lambda. \tag{1.50}$$

From Eq. 1.45, we note that

$$\cos\lambda = \boldsymbol{\ell} \cdot \mathbf{b}/\ell = \sqrt{1 - (\ell_0 \sin\lambda_0/\ell)^2}. \tag{1.51}$$

Combining Eqs. 1.50 and 1.51, the instantaneous value of resolved shear stress can be written as

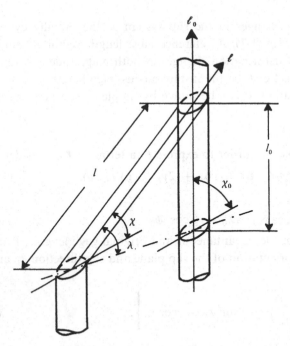

Fig. 1.7. A schematic diagram of change in the gage length vector from ℓ_0 to ℓ due to slip.

$$\tau_R = \frac{P}{A} \cos\phi_0 \sqrt{1 - (\ell_0 \sin\lambda_0/\ell)^2}. \qquad (1.52)$$

The measurement of the applied force P and the instantaneous length of the crystal ℓ during a uniaxial test producing single slip in a single crystal can thus be converted into a plot of the resolved shear stress τ_R and resolved shear strain γ_R using Eqs. 1.52 and 1.48, respectively. It is seen from Eq. 1.52 that τ_R increases with the elongation of the crystal (for a fixed value of the applied force P). This phenomenon is known as 'geometrical softening'. It is important to note that, while the slip system rotates toward the tensile axis in a monotonically loaded crystal, fully reversed fatigue loads do not cause any orientation change.

Figure 1.8 shows a typical stress–strain curve at room temperature for an FCC single crystal, oriented initially for single slip and subjected to uniaxial tension. Here the variation of the resolved shear stress τ_R with the resolved shear strain γ_R exhibits three distinct stages. Stage I commences at the critical stress τ_0 after an initial elastic deformation. This region of 'easy glide' is characterized by primary slip and by straight and uniformly spaced slip lines. As the crystal deforms and the slip systems rotate with respect to the loading axis, secondary slip is activated. This onset of secondary slip and the attendant decrease in mean slip distance generally marks the beginning of stage II where the crystal exhibits a significant increase in the rate of work hardening. With the progression of plastic deformation in stage II, the increase in dislocation density and the propensity for cross slip promote the formation of

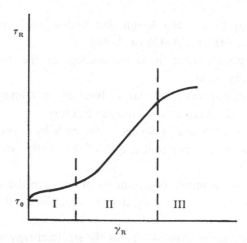

Fig. 1.8. A typical stress–strain curve for an FCC single crystal exhibiting three distinct stages of deformation.

dislocation cell structures. With an increase in applied stress (and/or temperature), the dislocations become increasingly more capable of circumventing (by cross slip) barriers which were generated during the high-hardening stage II deformation. The onset of cross slip in stage III is accompanied by a reduction in work hardening. Materials, such as aluminum, with a high stacking fault energy (i.e. with a high propensity for cross slip), exhibit pronounced stage III deformation even at low applied stresses. A detailed discussion of the mechanisms of monotonic deformation in ductile single crystals, for different conditions of crystallographic orientation, test temperature and strain rate, can be found in any textbook on mechanical metallurgy. Several example problems illustrating the methods to identify active slip systems in cubic crystals are presented in the next chapter.

Exercises

1.1   The design of the twin-engine Boeing 777 aircraft was launched in the late 1980s with the objective of capturing the long-distance air travel market. The aircraft was designed to fill a gap between the four-engine 747 with a passenger capacity of around 400 and the twin-engine 767 with a passenger capacity of approximately 200. During the development of the 777, Boeing had to face competition for the 300–350 passenger 777 aircraft from the newer generation of planes introduced by rival manufacturers, i.e. the 323-seat MD-11 from McDonnell-Douglas which replaced the DC-10 with more efficient engines, and the four-engine A-340 and the twin-engine A-330 from Airbus. The airlines which worked with Boeing in the development of the

777 forged an agreement early on in the design that its fuselage would be much wider than that of the MD-11, A-330 or A-340.

(a) Speculate about the implications of the larger fuselage on the strength, weight and power requirements.

(b) Speculate about the implications of the larger fuselage on damage tolerance considerations and design against fatigue fracture.

1.2   A circular shaft of length l and radius a is twisted at the ends by a torque T, which results in an angle of twist α per unit length of the shaft. The shear modulus of the material is G.

(a) Find the magnitude of the nonzero components of stress in the shaft.

(b) Derive expressions for the strain energy density and the total strain energy.

1.3   For an isotropic elastic solid, derive expressions for the strain energy density in terms of the components of stress or strain tensors for (*a*) plane strain and (*b*) plane stress.

1.4   For a homogeneous and isotropic elastic medium, show that the principal axes of the stress and strain tensors coincide.

1.5   The *octahedral plane* is defined as the plane which makes equal angles with the principal stress directions.

(a) Show that the *octahedral shear stress*, which is the shear stress on the octahedral plane, is

$$\sigma_{\text{oct}} = \tfrac{1}{3} \sqrt{(\sigma_1 - \sigma_2)^2 + (\sigma_2 - \sigma_3)^2 + (\sigma_3 - \sigma_1)^2},$$

where σ_1, σ_2 and σ_3 are the principal stresses.

(b) Show that $(3/2)\sigma_{\text{oct}}^2 = J_2$, where J_2 is the second invariant of the deviatoric stress tensor defined in Eq. 1.20.

(c) Describe the von Mises yield condition, Eq. 1.22, in terms of the octahedral shear stress.

1.6   Equations which relate the plastic strain increments to the stress deviator components are known as the *Prandtl–Reuss* equations (i.e. $d\epsilon_{ij}^{\text{p}} = s_{ij} d\lambda$ where s_{ij} and $d\lambda$ are defined in Eqs. 1.19 and 1.25, respectively).

(a) Show that when $f(\sigma_{ij}) = J_2$, the plastic potential equations 1.25 become the Prandtl–Reuss equations. f is the plastic potential function defined in Eq. 1.25 and J_2 is the second invariant of the deviatoric stress tensor defined in Eq. 1.20.

(b) Show that the Prandtl–Reuss equations correspond to incompressible plastic deformation.

1.7   Identify the slip systems with the same Schmid factor in a face-centered cubic crystal where the stress axis for push–pull fatigue loading is along the [110] direction.

1.8   A cubic crystal contains a screw dislocation. The Burgers vector of the
       dislocation is parallel to [001]. A crack propagates rapidly in the crystal
       along the (001) plane in the [110] direction. Sketch the geometry and orien-
       tation of the steps formed on the fracture surface.

Part one

CYCLIC DEFORMATION AND FATIGUE CRACK INITIATION

CHAPTER 2

Cyclic deformation in ductile single crystals

Studies which link the origin of fatigue damage to microscopic deformation processes date back to the work of Ewing & Rosenhain (1900) and Ewing & Humfrey (1903) who reported cracking along traces of active slip planes in fatigued iron. With the invention of electron microscopes, considerable progress has been made in developing a detailed understanding of substructural and microstructural changes induced by cyclic straining. Research work in the past several decades has clearly established the existence of a rich variety of fundamental mechanisms that are specific to cyclic loading conditions. A thorough knowledge of these phenomena is essential for microstructural design for fatigue resistance in engineering materials.

The most conclusive results of deformation mechanisms in fatigue have been obtained on high purity materials, in particular single crystals of face-centered cubic (FCC) metals.[†] In commercial materials, on the other hand, microstructural complexities often preclude clear identification and quantitative treatment of fatigue mechanisms. Cyclic deformation mechanisms in commercial materials are also strongly influenced by processing methods and impurity content.

In this chapter, attention is focused on the mechanisms and micromechanics of deformation in single-crystalline metals and alloys with FCC crystal structures. Also included are brief descriptions of known cyclic-deformation characteristics of body-centered cubic (BCC) and hexagonal close-packed (HCP) crystals. Cyclic deformation in some rock salt crystals is addressed in Chapter 5. This discussion is followed by a treatment of continuum formulations for uniaxial and multiaxial fatigue deformation in polycrystalline ductile solids in Chapter 3. Fatigue crack nucleation in monocrystalline and polycrystalline metals and alloys is considered in Chapter 4. Cyclic deformation and crack initiation in brittle solids, such as ceramics and ceramic composites, and in semi-crystalline and noncrystalline solids, such as polymers and organic composites, are considered in Chapters 5 and 6, respectively. Cyclic damage has also been extensively investigated using stress–life or strain–life approaches where the deformation, crack initiation and crack growth events are all incorporated within the context of phenomenological continuum descriptions; these methods are examined in Chapters 7 and 8.

[†] Engineering alloys are used occasionally in monocrystalline form in some fatigue-prone structures. An example is a gas turbine engine for aircraft in which the turbine blades are made of nickel-base super-alloy single crystals. See Section 2.10 for the cyclic deformation mechanisms in this alloy system.

2.1 Cyclic strain hardening in single crystals

When well-annealed FCC single crystals, suitably oriented for single slip, are subjected to cyclic strains under fully reversed loading, rapid hardening (that is, a rapid increase in flow stress with increasing number of cycles) is noticed even in the initial few cycles. With continued cyclic straining, the rate of hardening progressively diminishes and a quasi-steady state of deformation, known as 'saturation', is reached. Once saturation occurs, the variation of the resolved shear stress τ_R with the resolved shear strain γ_R is not altered by further cycling and the stress–strain hysteresis loop develops a stable configuration. (See Section 1.5.1 for definitions of the resolved shear stress and the resolved shear strain, and Section 2.3.1 for an example problem on the identification of active slip systems.)

It is well established that plastic strains are necessary for the inducement of fatigue fracture in ductile single crystals. Under typical cyclic loading conditions involving low values of imposed strains, the resolved plastic shear strain amplitude γ_{pl} is only a fraction of the total strain amplitude γ_t because of *reversible* slip, and it decreases with cyclic work hardening. Consequently, fatigue tests conducted with a fixed total strain amplitude do not provide a clear description of the progression of cyclic deformation. Instead, the most common method of performing strain-controlled tests involves fixed amplitudes of plastic strains. Figure 2.1 shows typical hysteresis loops associated with initial hardening and the attainment of a stable hysteresis loop upon saturation after N_s cycles. The maximum value of the resolved shear stress at saturation is denoted as τ_s at the total applied shear strain of γ_{ts}.

A *nominal* measure of damage accumulation under cyclic loads is the *cumulative plastic strain* which is defined as

$$\Gamma = 4 \sum_{i=1}^{N} \gamma_{pl,i}, \tag{2.1}$$

where $\gamma_{pl,i}$ is the resolved plastic shear strain amplitude in the ith cycle and N is the total number of cycles. For fully-reversed straining under a fixed plastic strain amplitude γ_{pl}, $\Gamma = 4\gamma_{pl}N$. It is important to note that the parameter Γ is only an approximate measure of global fatigue damage and that it does not accurately capture the extent of permanent damage at low plastic strain amplitudes where a significant fraction of dislocation motion can be reversible.

2.2 Cyclic saturation in single crystals

Experiments of fully reversed fatigue under fixed amplitudes of resolved plastic shear strain, γ_{pl}, point to the existence of a saturation stress after initial (rapid) cyclic work hardening. A plot of the peak resolved shear stress at saturation, τ_s, as a function of γ_{pl} provides the 'cyclic stress–strain curve' for a single crystal, which was first experimentally measured by Mughrabi (1978). Figure 2.2 schemati-

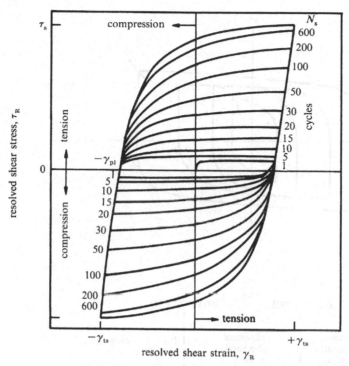

Fig. 2.1. Schematic illustration of typical variations in resolved shear stress, τ_R, as a function of the resolved shear strain, γ_R, for an FCC single crystal oriented for single slip and fatigued at a fixed value of resolved plastic shear strain amplitude $= \gamma_{pl}$. τ_R increases rapidly during the initial fatigue cycles. After N_s cycles, the hysteresis loops saturate with a peak tensile stress of τ_s at a total applied shear strain of γ_{ts}.

cally shows such a curve for FCC single crystals oriented for single slip. Actual stress–strain data obtained for a number of FCC metals fatigued at different temperatures are listed in Table 2.1.

There are three regions, marked *A*, *B* and *C*, which exhibit distinctly different strain hardening characteristics in Fig. 2.2. At low values of plastic strain amplitude ($\gamma_{pl} < \gamma_{pl,AB}$), denoted region *A*, work hardening occurs during cyclic loading. (This hardening behavior is measured in terms of the *saturation stress* and is different from the rapid cyclic hardening *prior to saturation, at a fixed value of* γ_{pl}, discussed in Section 2.1.) Region *A* is followed by a 'plateau' in the stress–strain curve (region *B*). This latter regime, where the saturation stress, τ_s^*, is independent of the plastic strain, extends until $\gamma_{pl,BC}$. A further increase in γ_{pl} results in an increase in τ_s^*, stage *C*.

In region *A*, work hardening displayed by the cyclic stress–strain curve is almost entirely due to the accumulation of primary dislocations. Microscopically, saturation of the hysteresis loops corresponds to a state where a dynamic equilibrium is achieved between bundles of edge dislocations and the surrounding matrix plied

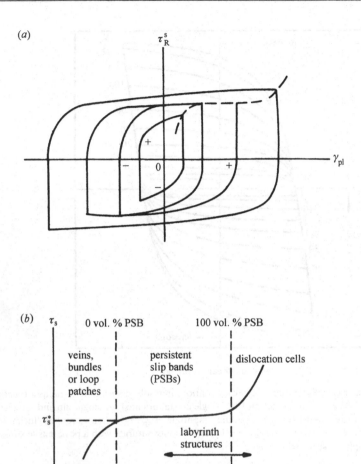

Fig. 2.2. (*a*) Hysteresis loops with the resolved shear stress at saturation, τ_R^s, plotted against the resolved plastic shear strain, γ_{pl}. The stress–strain curve is drawn through the tips of stable hysteresis loops. (*b*) A schematic diagram showing different regimes of the saturation stress–strain curve.

by screw dislocations. Under these conditions, fine slip markings are observed on the free surfaces; the specimen could withstand an infinite number of fatigue cycles because the cyclic plastic strain does not cause progressively degenerating damage.

One of the most visible features of cyclic saturation is the localization of slip along bands. This process is nucleated at strain amplitudes corresponding to the beginning of region *B* in the cyclic stress–strain curve (Fig. 2.2) and is intensified as the applied plastic strain is increased. Early observations, e.g., Ewing & Humfrey (1903) and Gough (1933), showed that fatigue failure initiated as a fine crack along those bands where slip was particularly intense. These slip lines were termed 'persistent slip

Table 2.1. *Cyclic stress–strain characteristics of FCC single crystals.*

Material	$\gamma_{pl,AB}$	$\gamma_{pl,BC}$	τ_s^* (MPa)	Reference
Cu (523 K)	1.0×10^{-4}	1.0×10^{-3}	14.0	Lisiecki & Weertman (1990)
Cu (295 K)	6.0×10^{-5}	7.5×10^{-3}	27.5	Mughrabi (1978)
Cu (77.4 K)	—	8.0×10^{-3}	48.0	Basinski, Korbel & Basinski (1980)
Cu (4.2 K)	—	—	73.0	Basinski, Korbel & Basinski (1980)
Cu–2.0 Al (at.%) (295 K)	1.0×10^{-4}	3.0×10^{-3}	33.0	Wilhelm & Everwin (1980)
Cu–5.0 Al (at.%) (295 K)	—	—	32.0	Woods (1973)
Cu–16.0 Al (at.%) (295 K)	—	—	20.0–25.0	Li & Laird (1994)
Cu–2.0 Co (at.%) (295 K)	3.0×10^{-4}	5.0×10^{-3}	27.5	Wilhelm & Everwin (1980)
Ni (295 K)	1.0×10^{-4}	7.5×10^{-3}	52.0	Mughrabi, Ackermann, & Herz (1979)
Ni (293 K)	1.0×10^{-4}	8.0×10^{-3}	50.0	Bretschneider, Holste & Tippelt (1997)
Ni (600 K)	7.5×10^{-5}	5.0×10^{-3}	20.5	Bretschneider, Holste & Tippelt (1997)
Ni (750 K)	—	—	12.0–16.0	Bretschneider, Holste & tippelt (1997)
Ag (295 K)	6.0×10^{-5}	7.5×10^{-3}	17.5	Mughrabi, Ackermann & Herz (1979)
Al–1.6 Cu (at.%) (295 K)	1.5×10^{-5}	1.5×10^{-3}	95.0	Lee & Laird (1983)
Fe–11Ni–16Cr–2Mo (wt%) (295 K)	—	—	59.0	Yan *et al.* (1986)
Fe–19Ni–11Cr–2Mo (wt%) (295 K)	—	—	59.0	Kaneko, Morita & Hashimoto (1997)

bands' (PSBs) by Thompson, Wadsworth & Louat (1956) who found that in Cu and Ni, the bands persistently reappeared at the same sites during continued cycling even after a thin layer of the surface containing these bands was removed by electropolishing. Numerous studies, including Laufer & Roberts (1966), Lukáš, Klesnil & Krejčí (1968), Watt, Embury & Ham (1968), and Woods (1973), have conclusively demonstrated that the PSBs form through the bulk of the single crystals, and that the bands of coarse slip merely mark their egress at the specimen surfaces.

Static deformation experiments after fatigue loading (Broom & Ham, 1959) and microhardness measurements on fatigue-induced slip bands (Helgeland, 1965) reveal that the PSBs are much softer than the matrix. These results imply that, during saturation in the plateau region of the cyclic stress–strain curve, essentially the entire deformation is carried by the PSBs. In fact, the very formation of the PSBs appears to be closely related to the occurrence of the plateau. These slip bands first appear at $\gamma_{pl} \sim \gamma_{pl,AB}$, and their volume fraction, f, increases *linearly* to nearly 100% as γ_{pl} is raised to a value of $\gamma_{pl,BC}$. For Cu, Ni and Ag, fatigued at room temperature, the ratio of the threshold saturation stress for PSB formation ($\tau_{PSB} \approx \tau_s^*$) to the shear modulus ($G$) is approximately the same ($\approx 6.5 \times 10^{-4}$), Mughrabi, Ackermann & Herz (1979). Figure 2.3 shows the markings exhibited by PSBs on the surface of a Cu single crystal subjected to 15 000 cycles of fatigue at γ_{pl} values of 1.5×10^{-3} and

(a)   (b)

Fig. 2.3. PSB markings on the surface of a Cu single crystal subjected to 15 000 cycles of fully reversed fatigue loads at two different values of resolved plastic shear strain amplitude, γ_{pl}. (a) $\gamma_{pl} = 1.5 \times 10^{-3}$. (b) $\gamma_{pl} = 4.5 \times 10^{-3}$. Width of the specimen = 4.7 mm. (From Winter, 1974. Copyright Taylor & Francis, Ltd. Reprinted with permission.)

4.5×10^{-3}. Note the increase in the volume of the crystal covered by the PSBs due to the increase in γ_{pl}.

2.2.1 Monotonic versus cyclic plastic strains

An approximate, but direct, comparison of strain hardening under monotonic and cyclic loading can be made if the cumulative plastic shear strain amplitude in a fatigue test is taken to be analogous to the resolved shear strain in a monotonic tensile test. Figure 2.4 shows such a comparison for Cu. It becomes evident here that strain hardening under monotonic tension occurs much faster than that under cyclic loading. Whereas cumulative plastic strains $\Gamma \sim 10$ are necessary to achieve the saturation stress, $\tau_s^* = 28$ MPa (at $\gamma_{pl} = 4.5 \times 10^{-3}$), a shear strain of only 0.3 produces the same stress level in a monotonic load test. This difference becomes even more substantial at lower values of γ_{pl}.

2.3 Instabilities in cyclic hardening

Cyclic strain hardening under a constant value of plastic strain amplitude results in a monotonically rising stress–strain curve (Figs. 2.1 and 2.4). However, if the single crystal is subjected to an increasing stress amplitude (ramp loading), the resulting strain amplitude exhibits a sequence of maxima and minima (Fig. 2.5), which were termed *strain bursts* by Neumann (1968). Although the absolute values of stress amplitudes at which strain bursts take place are not well defined, their

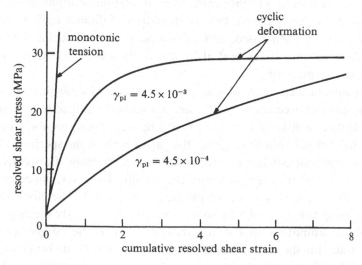

Fig. 2.4. Resolved shear stress plotted against cumulative resolved shear strain for copper single crystals subjected to tension and fatigue. (After Wilkens, Herz & Mughrabi, 1980.)

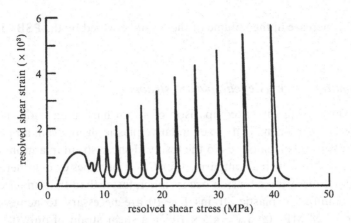

Fig. 2.5. Strain bursts observed in a copper single crystal subjected to increasing stress amplitude at 7.1 kPa/cycle at 90 K. (After Neumann, 1968.)

relative occurrence seems to be perfectly periodic in that they appear whenever the stress amplitude is raised by at least 11.5% within about fifty fully reversed fatigue cycles. These strain bursts have been observed in single crystals of Cu, Ag, Mg and Zn (Neumann, 1968) and of Cu–Al alloys (Desvaux, 1970).

The effects of loading mode on the possible occurrence of strain bursts can be rationalized from the arguments presented by Neumann (1974, 1983). If the crystal is subjected to sufficient fatigue cycles (at a given stress amplitude), hardening due to the mutual trapping of dislocations progressively reduces γ_{pl}. This effect continues until the mean free path for dislocation motion is smaller than their mean spacing. Consequently, the probability of close encounters among dislocations decreases. If the stress amplitude is slowly raised, the disintegration of dipoles leads to an avalanche of free dislocations. Macroscopically, this process is manifested in the form of a strain burst. At the higher stress level, the released dislocations are trapped again during subsequent deformation.

This discontinuous hardening trend is also known to occur locally during the very early hardening stage of a constant plastic strain amplitude fatigue test. Neumann suggests that whether such *local* strain bursts can be seen macroscopically depends strongly upon their interactions throughout the gage length of the specimen. Under increasing stress amplitude conditions, experiments appear to show the occurrence of strain bursts (coherently) through the entire gage length. However, coherent strain bursts are not compatible with constant plastic strain control. Therefore, the discontinuous hardening behavior adds up to asynchronously or slowly varying macroscopic plastic strain amplitude in a strain-controlled test. If the fatigue specimens which exhibit strain bursts are subjected to monotonic tensile deformation, they undergo easy glide of zero strain hardening where coarse slip steps are formed (Broom & Ham, 1959; Neumann, 1968).

2.3.1 Example problem: Identification of active slip systems

The discussions presented in the preceding sections and those to follow in this and subsequent chapters in the context of cyclic deformation in ductile single crystals require a quantitative understanding of the nature of microscopic slip. Here, the procedures used to identify active slip systems are illustrated with the aid of a worked example.

Problem:

Consider a face-centered cubic crystal that is subjected to tension loading. Initially, the crystal with a circular cylindrical geometry and cross-sectional area A, has its tensile axis oriented along $[\bar{1}23]$.

(i) Identify the slip system which is activated initially.
(ii) As plastic deformation occurs, show how the relative orientation between the stress axis and the glide direction changes.
(iii) Calculate the applied tensile strain at which duplex slip involving the primary and the conjugate slip systems is activated.

Solution:

(i) According to the Schmid law, the slip system which operates first is the one for which the resolved shear stress, defined in Eq. 1.44, is a maximum. The product $\cos\phi_0 \cos\lambda_0$ for each of the twelve slip systems in an FCC crystal is estimated from the 'cosine formula'. This formula states that the cosine of the angle, ϕ, between the normals, $\langle h_1 k_1 l_1 \rangle$ and $\langle h_2 k_2 l_2 \rangle$, to two planes in a cubic crystal is:†

$$\cos\phi = \frac{h_1 h_2 + k_1 k_2 + l_1 l_2}{\sqrt{(h_1^2 + k_1^2 + l_1^2)} \cdot \sqrt{(h_2^2 + k_2^2 + l_2^2)}}. \tag{2.2}$$

For example, cosine of the angle ϕ between the normal to the plane (111) and the direction $[\bar{1}23]$ is: $(-1 + 2 + 3)/(\sqrt{14}\sqrt{3}) = 4/(\sqrt{14}\sqrt{3})$. Similarly, cosine of the angle λ between the slip direction $[1\bar{1}0]$ and the loading direction $[\bar{1}23]$ is: $(-1 + 2 + 0)/(\sqrt{14}\sqrt{2}) = -3/(\sqrt{14}\sqrt{2})$.

For the FCC crystals which glide along the $\{111\}\langle 110 \rangle$ slip system, the three [UVW] ($\langle 110 \rangle$) slip directions located on each of the four (hkl) ($\{111\}$) planes can be identified by recourse to the Weiss zone law which states:

$$hU + kV + lW = 0. \tag{2.3}$$

With the above information, the cosine of the angle between the slip plane normal and the tensile loading axis, $\cos\phi$, and the cosine of the angle

---

† More generally, if $\mathbf{n}_1$ and $\mathbf{n}_2$ are the unit vectors normal to the two planes, then $\mathbf{n}_1 \cdot \mathbf{n}_2 = |\mathbf{n}_1| \cdot |\mathbf{n}_2| \cdot \cos\phi$. See Section 2.8.1 for further discussion.

Table 2.2. *Crystallographic geometric relationships for the FCC crystal with its tensile axis initially along* [$\bar{1}$23].

Slip system	(Plane) [Direction]	Notation	$\cos\phi$	$\cos\lambda$	$\cos\phi\cos\lambda$
Primary	(111)[1$\bar{1}$0]	A2	$\dfrac{4}{\sqrt{14}\sqrt{3}}$	$\dfrac{-3}{\sqrt{14}\sqrt{2}}$	$\dfrac{-12}{14\sqrt{6}}$
	(111)[$\bar{1}$01]	A3	$\dfrac{4}{\sqrt{14}\sqrt{3}}$	$\dfrac{4}{\sqrt{14}\sqrt{2}}$	$\dfrac{16}{14\sqrt{6}}$
	(111)[0$\bar{1}$1]	A6	$\dfrac{4}{\sqrt{14}\sqrt{3}}$	$\dfrac{1}{\sqrt{14}\sqrt{2}}$	$\dfrac{4}{14\sqrt{6}}$
Conjugate	($\bar{1}$$\bar{1}$1)[1$\bar{1}$0]	B2	$\dfrac{2}{\sqrt{14}\sqrt{3}}$	$\dfrac{-3}{\sqrt{14}\sqrt{2}}$	$\dfrac{-6}{14\sqrt{6}}$
	($\bar{1}$$\bar{1}$1)[101]	B4	$\dfrac{2}{\sqrt{14}\sqrt{3}}$	$\dfrac{2}{\sqrt{14}\sqrt{2}}$	$\dfrac{4}{14\sqrt{6}}$
	($\bar{1}$$\bar{1}$1)[011]	B5	$\dfrac{2}{\sqrt{14}\sqrt{3}}$	$\dfrac{5}{\sqrt{14}\sqrt{2}}$	$\dfrac{10}{14\sqrt{6}}$
Cross-slip	(1$\bar{1}$1)[110]	C1	0	$\dfrac{1}{\sqrt{14}\sqrt{2}}$	0
	(1$\bar{1}$1)[$\bar{1}$01]	C3	0	$\dfrac{4}{\sqrt{14}\sqrt{2}}$	0
	(1$\bar{1}$1)[011]	C5	0	$\dfrac{5}{\sqrt{14}\sqrt{2}}$	0
Critical	($\bar{1}$11)[110]	D1	$\dfrac{6}{\sqrt{14}\sqrt{3}}$	$\dfrac{1}{\sqrt{14}\sqrt{2}}$	$\dfrac{6}{14\sqrt{6}}$
	($\bar{1}$11)[101]	D4	$\dfrac{6}{\sqrt{14}\sqrt{3}}$	$\dfrac{2}{\sqrt{14}\sqrt{2}}$	$\dfrac{12}{14\sqrt{6}}$
	($\bar{1}$11)[0$\bar{1}$1]	D6	$\dfrac{6}{\sqrt{14}\sqrt{3}}$	$\dfrac{1}{\sqrt{14}\sqrt{2}}$	$\dfrac{6}{14\sqrt{6}}$

between the glide direction on the slip plane and the tensile loading axis, $\cos\lambda$, are determined for each of the twelve possible slip systems, as shown in Table 2.2. For a given tensile load P and crystal cross-sectional area, A, the last column of this table gives the initial Schmid factors, M, for the different slip systems, and the maximum value of $|\cos\phi \cdot \cos\lambda|$ identifies the slip system with the highest resolved shear stress, as shown in Section 1.5.1. The last column of Table 2.2 reveals that for the given orientation of the crystal, single slip occurs initially along the A3 slip system.

(ii) As slip deformation occurs along the A3 slip system, the axis of the cylindrical crystal begins to rotate. Consequently, the stress axis moves towards the [$\bar{1}$01] glide direction. This process is visualized most

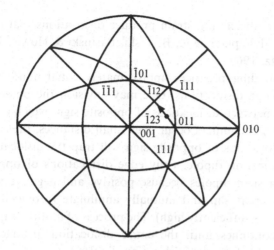

Fig. 2.6. A stereographic projection showing the rotation of the loading axis, initially at $[\bar{1}23]$, towards the primary glide direction $[\bar{1}01]$ until $[\bar{1}12]$. This rotation is marked by the arrow.

conveniently by examining the [001] stereographic projection, which is shown in Fig. 2.6. (Basic descriptions of stereographic projection can be found in an introductory text on mechanical metallurgy.) After this rotation, when $[\bar{1}12]$ is reached, the stress axis lies between the zones [001] and $[\bar{1}11]$, and both the primary and conjugate slip systems are equally stressed.

(iii)   The applied tensile strain, e_{app}, at which duplex slip involving the primary and conjugate systems is activated is readily calculated with reference to Fig. 1.7:

$$e_{app} = \frac{\ell - \ell_0}{\ell_0} = \frac{\ell}{\ell_0} - 1 = \frac{\sin \lambda_0}{\sin \lambda} - 1. \tag{2.4}$$

In the present case,

$$\cos \lambda_0 = \cos \angle\{(\bar{1}23)[\bar{1}01]\} = \frac{4}{\sqrt{14}\sqrt{2}} \Longrightarrow \lambda_0 = 40.89°. \tag{2.5}$$

Likewise,

$$\cos \lambda = \cos \angle\{(\bar{1}12)[\bar{1}01]\} = \frac{3}{\sqrt{6}\sqrt{2}} \Longrightarrow \lambda = 30°. \tag{2.6}$$

Combining Eqs. 2.4–2.6, one finds that $e_{app} = 0.31$.

2.3.2 *Formation of dislocation veins*

Transmission electron microscopy (TEM) of cyclic strain hardening in Cu single crystals, oriented for single slip and subjected to equal tension–compression cyclic straining, show the following general trends.

- The initial few cycles of alternating strain produce dislocations that accumulate on the primary glide plane (e.g., Basinski, Basinski & Howie, 1969; Hancock & Grosskreutz, 1969).

- Fully reversed cyclic loading produces approximately equal numbers of positive and negative edge dislocations. One may envision the possibility of frequent encounters between dislocations of opposite sign, which attract one another. When such encounters occur over small distances, the strong attraction between dislocations of opposite sign will trap the dislocations, thereby creating a dislocation dipole. Only edge dislocations of opposite sign are likely to form such dipoles because positive and negative screw dislocations can easily cross slip and mutually annihilate (provided that the stacking fault energy is sufficiently high). The process of mutual trapping of edge dislocations continues until the entire dislocation arrangement within the veins is composed of edge dislocation dipoles.

- The generation of primary dislocations is a consequence of the geometrical condition that, during fully reversed fatigue straining, there is no rotation of the slip system with respect to the loading axis. Therefore, the primary slip system remains the most highly stressed. Microscopy results by many researchers, which document the absence of any lattice rotation between adjacent channels, suggest that the average Burgers vector within the veins is close to zero because of equal numbers of positive and negative edge dislocations (e.g., Mughrabi, 1980). Thus, the veins do not produce long-range internal stresses. One of the most notable distinctions between monotonic and cyclic hardening of FCC single crystals at *low amplitudes of imposed strains* is this absence of long-range internal stresses under fatigue deformation. (The stresses due to the edge dislocation dipoles in the veins are significantly more short-ranged ($\propto 1/r^2$) than those arising from the pile-up of dislocations ($\propto 1/\sqrt{r}$), where r is the radial distance from the core of dislocation; see, for example, Neumann, 1983.) The absence of long-range internal stresses has also been identified by X-ray measurements (Hartman & Macherauch, 1963; Wilkens, Herz & Mughrabi, 1980).

- With continued cycling, accumulation of dislocations occurs predominantly in the form of mutually trapped primary edge dislocation dipoles.

- These networks of edge dislocation dipoles are commonly referred to as *veins*, *bundles*, or *loop patches*. The veins have an elongated shape; their long axis is parallel to the primary dislocation lines and their cross section normal to the long axis is equi-axed. The veins are separated by *channels*, which are relatively dislocation-free and are of a size comparable to that of the veins. The width of the veins is about 1.5 μm in copper fatigued at 20 °C. A reduction in temperature promotes a finer dislocation structure.

- The dislocation density within the vein is of the order of 10^{15} m^{-2} which corresponds to a mean dislocation spacing of 30 nm. The dislocation density within the channels is three orders of magnitude smaller, which implies that the average dislocation spacing in the channels is comparable to the width of the channels.

- The veins contribute to rapid hardening in the early stage of fatigue by partially impeding dislocation motion on the primary slip system. Increasing the number of cycles leads to an increase in both the dislocation density within the veins and the number of veins per unit volume. This results in an enlargement of the network of interconnected veins packed tightly with primary edge dislocations which occupy up to 50% of the volume of the material. In the dislocation-poor regions between the veins, screw dislocations ply back and forth during cyclic straining.

Figure 2.7 is an example of the vein structure in a copper single crystal, oriented for single slip under γ_{pl} control. This micrograph shows the view of the primary slip plane which, in the notation used by Basinski, Korbel & Basinski (1980), is $(11\bar{1})$; the primary slip vector is along [101] (which is normal to the vein structure). Primary screw dislocations can be seen in the channels dividing the veins.

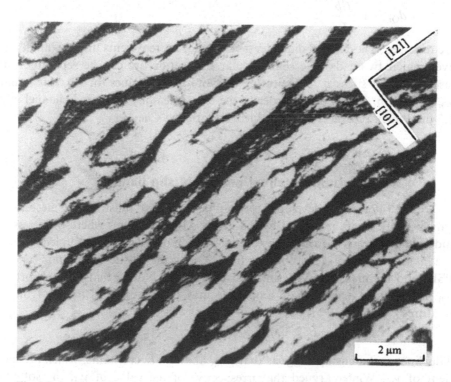

Fig. 2.7. A transmission electron micrograph of matrix vein structure in a section parallel to the primary glide plane of a single crystal of Cu fatigued to saturation at 77.4 K. (From Basinski, Korbel & Basinski, 1980. Copyright Pergamon Press plc. Reprinted with permission.)

2.3.3 Fundamental length scales for the vein structure

The characteristic dimensions associated with the dislocation structure of the veins can be rationalized in terms of the physical mechanisms, Neumann (1983). The most fundamental characteristic length associated with the vein structure, i.e. the mean spacing of the edge dislocations in the veins, can be related to the trapping distance, d_{trap}, of the edge dislocations at the saturation stress, τ_s:

$$d_{trap} \sim \frac{Gb}{8\pi(1-v)\tau_s},\qquad(2.7)$$

where G is the shear modulus, b is the magnitude of the Burgers vector and v is Poisson's ratio. Taking $Gb = 11\ \mathrm{J\,m^{-2}}$, $v = 0.33$ and $\tau_s = 28\ \mathrm{MPa}$ (the maximum value of the saturated peak stress for the vein structure) for Cu at room temperature, one obtains $d_{trap} \leq 24\ \mathrm{nm}$. This compares reasonably well with the mean dislocation spacing of 30 nm estimated from the measurement of dislocation densities.

For cyclic hardening to take place, there is a need for significant dislocation multiplication. In a high purity single crystal, this is only possible by the bowing out of dislocations by a Frank–Read mechanism. In this process, the channel areas adjacent to the veins with a diameter equal to the Orowan–Frank–Read length

$$d_{Or} \sim 1.5\frac{Gb}{\tau_s}\qquad(2.8)$$

are required. The channel diameter given in Eq. 2.8 represents the maximum distance over which two ends of a semi-circular dislocation arc can be separated under a resolved shear stress τ_s. Substituting the appropriate values for Cu, it is found that $d_{Or} \approx 0.6\ \mathrm{\mu m}$. This is of the order of the experimentally measured channel width of 1.5 μm which separates adjacent veins in fatigued Cu.

2.4 Deformation along persistent slip bands

Winter (1974) developed a two-phase model to illustrate the mechanisms by which plastic strain is carried by the PSBs. At equilibrium, where the stress amplitude, the hysteresis loop and the volume fraction of the crystal occupied by the PSBs are invariant during saturation (for fixed γ_{pl}), let the plastic strain amplitudes in the PSBs and the matrix be γ_{PSB} and γ_m, respectively. Noting that f is proportional to γ_{pl}, the average plastic strain amplitude may be expressed in terms of the law of mixtures,

$$\gamma_{pl} = f\gamma_{PSB} + (1-f)\gamma_m.\qquad(2.9)$$

This linear relationship suggests that γ_{PSB} and γ_m are constants, which are independent of γ_{pl}. Winter argued that irrespective of the value of γ_{pl}, the softer phase comprising the PSBs is fatigued at a strain amplitude of γ_{PSB} and the matrix (the harder phase) at γ_m. The crystal deforms at the appropriate γ_{pl} by adjusting the relative amounts of the two phases (i.e. by changing f) within the plateau (region B) of the saturation cyclic stress–strain curve. The parameters, γ_m and γ_{PSB}, in Eq. 2.9 correspond to $\gamma_{pl,AB}$ and $\gamma_{pl,BC}$, respectively, in Fig. 2.2 and Table 2.1. This line of reasoning, albeit simplistic, provides an appealing rationale for the similarities in the conditions that govern the formation of PSBs in single crystals of a variety of metals and alloys.

Winter's two-phase model also has limitations (e.g., Brown, 1977). Firstly, the model implies reversibility, contrary to the reality of a fatigue experiment where a decrease in γ_{pl} does not lead to the disappearance of the PSBs. Secondly, the model is not applicable for materials where fatigue loads alter the microstructure; for example, precipitation-hardened alloys (in which cyclic straining can shear the strengthening precipitates) or work-hardened materials (in which fatigue softening occurs). Thirdly, the two-phase model is not applicable to cyclic deformation involving dislocation climb, i.e. above about half the melting point.

As the PSBs penetrate through the bulk of the crystal, the strain carried by them is macroscopically reversible in that the local strain at the maximum value of the applied tensile stress is identical to that at the compressive maximum (Finney & Laird, 1975). On a polished surface, slip steps visible to the naked eye disappear and reappear every quarter cycle. Although the surface slip steps form in proportion to γ_{pl}, the displacements *within the band* are not fully reversed. This leads to the formation of slip offsets and a rough topography within the bands which appears to be the precursor to crack nucleation (see Chapter 4). There is considerable variation in the thickness and distribution of slip, with the coarse PSBs consisting of narrower slip bands known as micro-PSBs.

It is generally recognized that the PSBs are a pre-requirement for the formation of fatigue cracks in pure crystals. Therefore, the observation of threshold stress (or strain) for the formation of PSBs automatically implies the existence of a fatigue (stress or strain) limit below which no cyclic failure occurs (Laird, 1976; Mughrabi, 1978).

2.5 Dislocation structure of PSBs

A PSB is composed of a large number of slip planes (~ 5000 in Cu at 20 °C) which form a flat lamellar structure and span the entire cross section of the single crystal. A periodic array of dislocation ladders or walls further divides the PSB lamellae into channels, Fig. 2.8. Note that this figure is an oversimplification insofar as the edge dislocations are not infinitely long, but are elongated dislocation dipole loops (debris). The PSB walls, predominantly made up of edge dislocations, are normal to the primary Burgers vector, **b**. The plastic deformation in the channels in-between the walls occurs mainly by the glide of screw dislocations.

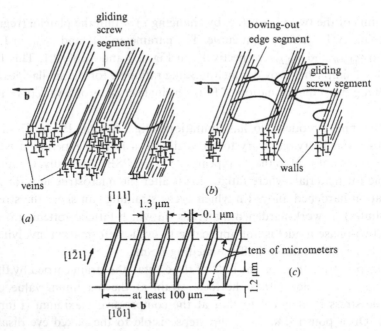

Fig. 2.8. Dislocation arrangements in FCC metals. (*a*) Vein structure in the matrix. (*b*) An enlarged view of the dipolar walls and dislocation debris within a PSB. (*c*) A three-dimensional sketch of PSB geometry formed in Cu at 20 °C. (After Mughrabi, 1980, and Murakami, Mura & Kobayashi, 1988.)

The dislocation structure found within the PSBs is considerably different from that of the matrix. The matrix contains, about 50% by volume, vein-like structures consisting of dense arrays of edge dislocations. On the other hand, the PSB structure is generated due to the mutual blocking of glide dislocations and the formation of parallel wall (ladder) structures which occupy about 10%, by volume, of the PSBs (Laufer & Roberts, 1966; Woods, 1973; Mughrabi, 1980). The walls are 0.03–0.25 µm in thickness with a spacing of about 1.3 µm and they consist of dipoles. The dislocation densities in the matrix veins and PSB walls are 10^{11}–10^{12} cm^{-2} and are two to three orders of magnitude greater than those in the in-between channels containing screw dislocations. Weak-beam TEM studies of copper crystals reveal the presence of about two-thirds vacancy dipoles and one-third interstitial dipoles of primary edge dislocations (Antonopoulos, Brown & Winter, 1976; Antonopoulos & Winter, 1976). A schematic of the dislocation arrangements in the veins and in the PSBs is given in Fig. 2.8.

Figure 2.9 is an arrangement of several electron micrographs showing the three-dimensional geometry of the matrix veins and PSB structures in Cu. The ($1\bar{2}1$) plane clearly shows the ladder-like arrangement of the walls in the PSB, which are oriented perpendicularly to the primary Burgers vector **b**. Figure 2.10 is a TEM micrograph of a ($1\bar{2}1$) section of a Cu crystal which shows the ladder structure in the PSB and the matrix vein structure consisting of dislocation bundles.

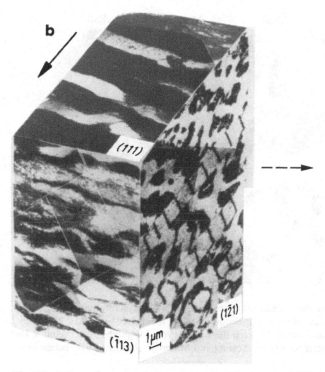

Fig. 2.9. A three-dimensional view, constructed from several TEM images, of the matrix vein and PSB dislocations in Cu cycled to saturation at 20 °C at $\gamma_{pl} = 1.5 \times 10^{-3}$. The specimen loading axis, indicated by the dashed line, is almost on the $(1\bar{2}1)$ plane and it makes an angle of 47° with the primary Burgers vector **b**. (From Mughrabi, Ackermann & Herz, 1979. Copyright American Society for Testing and Materials. Reprinted with permission.)

The matrix veins accommodate plastic strains only of the order of 10^{-4}, and hence they undergo only microyielding. On the other hand, PSBs support high plastic shear strains of the order of 0.01 and undergo macroyielding. This can involve dislocation multiplication by the bowing-out of the edge dislocations (from the walls) and by their transport along the channels; the screw dislocations in the channels may also draw the edge dislocations out of the walls (see the schematic in Fig. 2.8(*b*)). Figure 2.11 is an electron micrograph of a section parallel to the primary glide plane, (111), of monocrystalline Cu at 20 °C. The primary edge dislocations bowing out of the walls are also evident. The same sense of curvature is exhibited by dislocations of the same sign.

Two different mechanisms have been proposed to rationalize the quasi-steady state deformation of the saturation stress–strain curve. Deformation *within the matrix veins* is believed to be accomplished by the back and forth ('flip-flop') motion of dislocation loops that are produced by jogs during cross slip of screw dislocations under cyclic straining (Feltner, 1965; Finney & Laird, 1975). It has been suggested by Grosskreutz & Mughrabi (1975) that this flip-flop mechanism can accommodate plastic strains of the order of 10^{-4}, and hence accounts for the saturation behavior

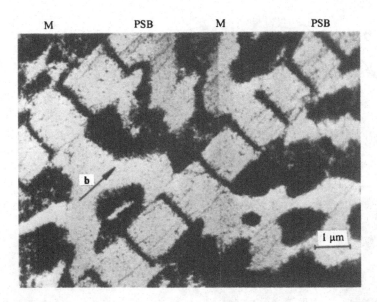

Fig. 2.10. TEM image of dislocation structures in a Cu crystal fatigued at $\gamma_{pl} = 10^{-3}$ at room temperature. A view of the (1$\bar{2}$1) section revealing the matrix veins (M), PSB walls and screw dislocations in the channels between the walls. (From Mughrabi, Ackermann & Herz, 1979. Copyright American Society for Testing and Materials. Reprinted with permission.)

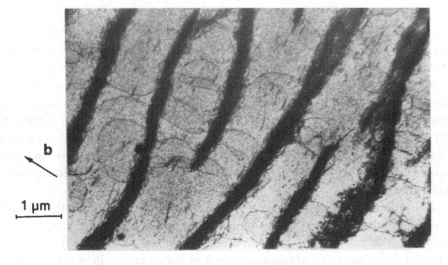

Fig. 2.11. TEM of a section parallel to the primary slip plane of single crystal Cu fatigued to saturation at $\gamma_{pl} = 5 \times 10^{-3}$. Fast neutron irradiation was employed to pin the dislocations at the peak tensile stress of the fatigue cycle. (From Mughrabi, Ackermann and Herz, 1979. Copyright The American Society for Testing and Materials. Reprinted with permission.)

in the veins either during rapid hardening or during saturation in region A. On the other hand, a dynamic equilibrium between dislocation multiplication and annihilation is considered responsible for saturation *within the PSB* in the plateau regime of the cyclic stress–strain curve. Consequently, the local densities of edge and screw dislocations are kept constant (Essmann & Mughrabi, 1979). Dislocation multiplication occurs by the bowing out of edge dislocations between the walls, whilst annihilation occurs by climb of edge dislocations of opposite sign on glide planes 1.6 nm apart in the wall structure of PSBs. Furthermore, the annihilation of screw dislocations (on glide planes 50 nm apart) at room temperature is also believed to occur in the dislocation-poor channels of the PSBs.

2.5.1 Composite model

Mughrabi (1981, 1983) proposed a composite model for fatigued crystals containing dislocation walls and cell structures. His analysis examines the development of long-range internal stresses as a natural consequence of deformation compatibility among the substructures. As large differences exist between the local dislocation densities in the walls and channels of the PSBs, let τ_w and τ_c denote the local flow stresses in the walls and channels, respectively. Under an imposed far-field stress, the deformation of the composite, consisting of walls and channels, occurs in the following sequence: (i) elastic straining of both the walls and channels, (ii) elastic straining of the walls, plastic yielding of the in-between channels, and (iii) plastic yielding of the channels and walls. Processes (ii) and (iii) denote plastic micro-yielding and macro-yielding, respectively. The composite yields plastically under the action of an applied stress τ:

$$\tau = f_w \tau_w + f_c \tau_c, \tag{2.10}$$

where f_w and f_c are the area fractions of the walls and channels, respectively, in the glide plane. If it is assumed, for strain continuity at the interface between walls and channels, that the total strain γ_T is the same in the walls and channels,

$$\gamma_T = \gamma_{pl,w} + \gamma_{el,w} = \gamma_{pl,c} + \gamma_{el,c}, \tag{2.11}$$

where $\gamma_{pl,w}$ and $\gamma_{el,w}$ are the plastic and elastic shear strains, respectively, in the walls, and $\gamma_{pl,c}$ and $\gamma_{el,c}$ are the corresponding strains in the channels. From Eqs. 2.10 and 2.11 and the condition that $f_c + f_w = 1$,

$$\tau_w = \tau + f_c(\tau_w - \tau_c) = \tau + Gf_c(\gamma_{pl,c} - \gamma_{pl,w}), \tag{2.12}$$

where G is the shear modulus, and $\gamma_{pl,c} - \gamma_{pl,w} = \gamma_{el,w} - \gamma_{el,c}$. Similarly,

$$\tau_c = \tau - f_w(\tau_w - \tau_c) = \tau - Gf_w(\gamma_{pl,c} - \gamma_{pl,w}). \tag{2.13}$$

An outcome of Eqs. 2.12 and 2.13 is that at the macroscopic yield stress τ, the local stress in the walls is larger than τ by

$$\Delta\tau_w = Gf_c(\gamma_{pl,c} - \gamma_{pl,w}), \tag{2.14}$$

and the local stress in the channels is smaller than τ by

$$\Delta\tau_c = -Gf_w(\gamma_{pl,c} - \gamma_{pl,w}). \tag{2.15}$$

Equations 2.14 and 2.15 together must satisfy the condition that

$$f_c \Delta\tau_c + f_w \Delta\tau_w = 0. \tag{2.16}$$

The local stresses, Eqs. 2.14 and 2.15, are long-range internal stresses that develop due to different amounts of deformation in the walls and channels which have different dislocation distributions. Thus, a crystal with a heterogeneous dislocation distribution exhibits local deformation similar to that of a composite. This effect has important implications for a wide variety of fatigue phenomena, such as cell formation (Sections 2.8 and 2.9), the Bauschinger effect (Chapter 3), cyclic slip irreversibility and fatigue crack nucleation (Chapter 4).

The composite model discussed above rationalizes, in a simple manner, the accommodation of plastic strains in the plateau regime of the cyclic stress–strain curve. However, the development of quantitative criteria for the inception of a fatigue crack requires detailed constitutive formulations of the inelastic deformation in the PSBs.

2.5.2 Example problem: Dislocation dipoles and cyclic deformation

Problem:

Consider an elongated edge dislocation dipole loop, which is shown schematically in Fig. 2.12(a). The length of the dipole loop is l_d and the separation distance of the dipole dislocations from glide plane to glide plane is s_d.

(i)   Examine the condition for the stability of the dipole loop under the influence of an applied shear stress, by identifying the ratio, l_d/s_d, for which the dislocation bowing stress, τ_{bow}, and the dipole passing stress, τ_{dipole}, are just equal.

(ii) Now consider a fatigued Cu crystal whose dislocation arrangement consists essentially of uniformly distributed edge dislocation dipoles. The length of the dipoles, $l_d = 0.5\,\mu m$, and the glide plane separation, $s_d = 20\,nm$. Assume that the dipole density $\rho_{dipole} = 10^{20}\,m^{-3}$, and that the cyclic plastic strains are accommodated exclusively by the dipole

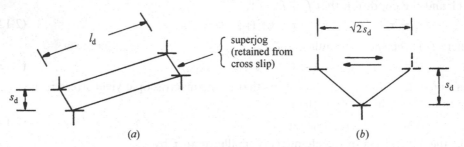

Fig. 2.12. (a) Schematic illustration of the dislocation dipole. (b) Glide path for dipole flip-flop.

flip-flop mechanism. Find the cyclic stress amplitude which would be required for the operation of the flip-flop mechanism.
(iii)   What would be the maximum shear strain amplitude?

Solution:

(i)   The dipole passing stress† can be estimated from Eq. 2.7, from which it is seen that

$$\tau_{dipole} = \frac{Gb}{8\pi(1-v)s_d}. \tag{2.17}$$

The dislocation bowing stress is

$$\tau_{bow} \approx \frac{Gb}{l_d}. \tag{2.18}$$

Equating these two stresses, we find the condition for the stability of the dipole loop to be

$$l_d \approx 8\pi(1-v)s_d, \qquad \text{or} \qquad \frac{l_d}{s_d} \approx 15. \tag{2.19}$$

(ii)   Substituting $G = 42$ GPa, $b = 0.25$ nm, $v = 0.33$, and $s_d = 20$ nm (which are typical values for Cu) in Eq. 2.17, we find that the amplitude of the stress necessary for the flip-flop mechanism is

$$\tau_{dipole} = \frac{Gb}{8\pi(1-v)s_d} \approx 31 \text{ MPa}. \tag{2.20}$$

This is of the order of the stresses seen for Cu at room temperature.

(iii)   The total dislocation density is

$$\rho = \rho_{dipole} \times l_d = 10^{20} \times 0.5 \times 10^{-6} \text{ m}^{-2} = 0.5 \times 10^{14} \text{ m}^{-2}. \tag{2.21}$$

The average glide path, schematically sketched in Fig. 2.12(b), is $(\sqrt{2} \cdot s_d)$. Since there are two dislocations in the dipole, the average glide path per dislocation is,

$$L_{gl} = \frac{1}{2} \cdot \sqrt{2} \cdot s_d = 0.7 \times 20 = 14 \text{ nm}. \tag{2.22}$$

The shear strain amplitude is:

$$\gamma_{pl} = \pm \rho \cdot b \cdot L_{gl}$$
$$= \pm 0.5 \times 10^{14} \text{ (m}^{-2}\text{)} \times 0.25 \times 10^{-9} \text{ (m)} \times 14 \times 10^{-9} \text{ (m)}$$
$$= \pm 1.75 \times 10^{-4}. \tag{2.23}$$

---

† Strictly speaking, the dipole passing and dislocation bowing processes should be treated as functions of dislocation displacement.

2.6 A constitutive model for the inelastic behavior of PSBs

In this Section, we examine a constitutive model for PSBs in FCC single crystals which is predicated upon the work of Repetto & Ortiz (1997), where a complete list of references to prior studies on relevant crystal elasticity and plasticity theories can be found. The formulation discussed here is amenable for implementation in finite-element simulations of fatigue crack initiation; some examples of such simulations will be presented in Chapter 4.

2.6.1 General features

As a starting point to the analysis, assume that the plastic strains are carried entirely within the PSBs, primarily by single glide, with the surrounding matrix taken to be elastic. At the free surfaces where the PSBs exit the crystal, however, considerations of multiple slip are necessary. Since transient cyclic deformation prior to cyclic saturation constitutes only a small portion of the fatigue life, attention is focused on deformation occurring after saturation. Consequently, the dislocation densities in all the slip systems are assumed to remain constant.

The total deformation in the FCC crystal is envisioned as being composed of three mechanisms which allow a multiplicative decomposition of the total deformation gradient $\mathbf{F}$ (Lee, 1969):

$$\mathbf{F} = \mathbf{F}^e\mathbf{F}^p\mathbf{F}^v. \tag{2.24}$$

Here, $\mathbf{F}^e$ is the elastic part accounting for lattice distortion including rotation,[†] $\mathbf{F}^p$ is the plastic part accounting for the cumulative effect of slip in the PSBs (which leaves the lattice undistorted as well as unrotated), and $\mathbf{F}^v$ is a factor representing the effects of vacancy generation.

Let τ_R^α be the resolved shear stress acting in the slip system α along the direction $\bar{\mathbf{s}}^\alpha$ (i.e. a $\langle 110 \rangle$ direction) on a slip plane (i.e. a $\{111\}$ plane) whose normal is $\bar{\mathbf{m}}^\alpha$. If $\boldsymbol{\tau}$ is the Kirchhoff stress tensor,

$$\tau_R^\alpha = \mathbf{s}^{\alpha T}\boldsymbol{\tau}\mathbf{m}^\alpha; \qquad \mathbf{s}^\alpha = \mathbf{F}^e\bar{\mathbf{s}}^\alpha \quad \text{and} \quad \mathbf{m}^\alpha = \mathbf{F}^{e-T}\bar{\mathbf{m}}^\alpha, \tag{2.25}$$

where the superscript T refers to the transpose of the tensor. The flow rule derived from the kinematics of slip is written as (Rice, 1971):

$$\dot{\mathbf{F}}^p\mathbf{F}^{p-1} = \sum_\alpha \dot{\gamma}^\alpha\, \bar{\mathbf{s}}^\alpha \otimes \bar{\mathbf{m}}^\alpha, \tag{2.26}$$

where $\dot{\gamma}^\alpha$ is the shear strain rate on the slip system α and $\otimes$ denotes a vector dyadic product. This constitutive formulation is completed by writing an equation for the evolution of γ^α on the basis of the type of loading (e.g., fully reversed and γ_{pl} controlled) and the material behavior (e.g., rate-independent plastic response).

[†] Details for extracting $\mathbf{F}^e$ for FCC crystals can be found in Teodosiu (1982).

2.6.2 Hardening in the PSBs

As the resolved shear stress τ_R^α causes dislocations to move within the PSB along the primary slip system α, it is conceivable that this motion is restricted by the mutual trapping of edge dislocations of opposite sign. The plastic hardening due to this trapping process leads to a shear strain evolution which depends on τ_R^α as follows:†

$$\gamma^\alpha(\tau_R^\alpha) = b\rho^\alpha w \left\{ \frac{1}{1 - \tilde{P}(\tau_R^\alpha)} - 1 \right\}, \tag{2.27}$$

where b is the magnitude of the Burgers vector, ρ^α is the dislocation density in the primary slip system, and w is the distance of separation between the walls in the PSB. The term $[1 - \tilde{P}(\tau_R^\alpha)]$ denotes the subfraction of destabilized dislocations that are trapped by the nearest wall, and $1/[1 - \tilde{P}(\tau_R^\alpha)]$ signifies the average number of jumps taken by a dislocation, from one dislocation wall to a neighboring wall in the PSB, before being trapped in a stable position. $\tilde{P}(\tau^\alpha)$ can be determined directly from experimentally measured stable hysteresis loops of τ_R^s versus γ_{pl}, similar to the ones shown in Fig. 2.2(a).

2.6.3 Hardening at sites of PSB intersection with the free surface

While considerations of single slip are appropriate for describing the deformation within the ladder structure of a PSB, modeling of multiple slip is required at the site of egress of a PSB at the free surface of the crystal. A possible mechanism for hardening here is postulated on the basis of obstacles to the motion of primary dislocations by forest dislocations. When the obstacles are randomly distributed in the slip plane, the underlying kinetic equation can be analytically stated (Cuitiño & Ortiz, 1992) in the form of a hardening rule:

$$\dot{\gamma}^\alpha = \frac{\dot{\tau}_R^\alpha}{h^{\alpha\alpha}}, \qquad h^{\alpha\alpha} = h_c^\alpha \left(\frac{\tau_R^\alpha}{\tau_c^\alpha} \right)^3 \left\{ \cosh\left[\left(\frac{\tau_c^\alpha}{\tau_R^\alpha} \right)^2 \right] - 1 \right\}. \tag{2.28}$$

Here $h^{\alpha\alpha}$ is the self-hardening modulus, $h_c^\alpha = \tau_c^\alpha/\gamma_c^\alpha$ is the characteristic hardening modulus, $\tau_c^\alpha \equiv CGb\sqrt{\pi n^\alpha}$ is the characteristic flow stress (with G being the shear modulus and C, a nondimensional constant), and $\gamma_c^\alpha \equiv (b\rho^\alpha)/(2\sqrt{n^\alpha})$ is a characteristic glide strain. The density of obstacles created by forest dislocations, n^α, takes the form

$$n^\alpha = \sum_\beta a^{\alpha\beta} \rho^\beta, \tag{2.29}$$

where β denotes all relevant slip systems other than the primary system α. The interaction coefficients $a^{\alpha\beta}$, which depend on the nature of the dislocation junctions

---

† Equation 2.27 is an oversimplification in that it does not account for the link between the hardening process and the motion of screw dislocations in the channels or the increasing densities of loops and dislocation debris.

or jogs formed by the intersection of primary dislocations with forest dislocations, have been exprimentally determined for the 12 slip systems of FCC crystals (Franciosi & Zaoui, 1982). In the multi-slip regions of the PSB, the trapping mechanism in the primary system and the forest hardening mechanism should both be considered.

2.6.4 Unloading and reloading

The discussion up to this point deals with monotonic loading. Upon repeated reversals of the load, the kinetic equation (Ortiz & Popov, 1982) for unloading and reloading can be formulated as:

$$\tilde{P}(\tau_0) = \frac{\tilde{P}(\tau_{max})e^{-\gamma_{pl}/\gamma_{pl,c}}}{1 + \tilde{P}(\tau_{max})\left(e^{-\gamma_{pl}/\gamma_{pl,c}} - 1\right)}, \tag{2.30}$$

where τ_{max} is the maximum resolved shear stress attained during loading and τ_0 is the critical resolved shear stress for the onset of reversed slip. The function $\tilde{P}$ carries the same meaning as in Eq. 2.27. Implicit in this derivation is the notion that $\tau_0 \to 0$ when the resolved shear strain γ_{pl} significantly exceeds a characteristic value, $\gamma_{pl,c}$. Thereafter, the reversed slip gradually facilitates the dissolution of the dislocation/ obstacle arrangement. Any subsequent reloading returns the deformation to conform to the virgin stress–strain response. During reloading, elastic deformation prevails until $\tau_R \to \tau_0$. Thus, the model ostensibly accounts for the Bauschinger effect (see Chapter 3), although its predictions have not been rigorously checked with experimental observations.

2.6.5 Vacancy generation

Once cyclic saturation is attained in the plateau regime of the cyclic stress–strain curve, the density of dislocations within the PSBs is maintained at an approximately constant value ($\sim 6 \times 10^{15}$ m^{-2}) by the establishment of a dynamic equilibrium between dislocation multiplication and annihilation. It has been pointed out (Mughrabi, Ackermann and Herz, 1979) that the bowing of edge dislocation segments out of the walls, at radii greater than the critical Frank–Read radius, is the most probable mechanism for dislocation multiplication (see Fig. 2.11). Dislocation annihilation is facilitated when two edge dislocation segments of opposite sign get closer than the critical separation distance, $y_e \approx 1.6$ nm (Essmann & Mughrabi, 1979).[†] Figure 2.13 is a schematic of an elongated dislocation loop in a PSB in which the pure edge segments (which are parallel to the walls) are at the critical separation y_e. The length of the loop L may be identified with the width of the specimen. The reduction of the spacing y and the eventual annihilation of the

---

[†] Note that the above argument does not take into account the process of annihilation of screw dislocations in the channels. In addition, care should be exercised in analyzing the annihilation process since it may involve dislocation climb or merely a mechanical collapse due to very high local interaction stresses.

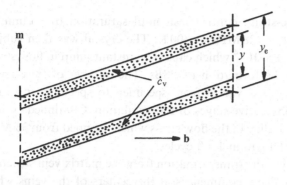

Fig. 2.13. A schematic showing the annihilation of dislocation loops and the attendant production of vacancies.

edge segments necessarily involves dislocation climb until the area of the loop Ly vanishes by generating point defects (which, as seen earlier, are predominantly vacancies). Vacancy generation also has the concomitant effect of promoting a steady elongation of the PSB along the nominal slip plane, which gives rise to surface protrusions. The rate of deformation induced by dislocation climb due to the annihilation of edge dislocations is found (Repetto & Ortiz, 1997) to be

$$\dot{\mathbf{F}}^{v}\mathbf{F}^{v-1} = \dot{c}_{v}\ \bar{\mathbf{s}}^{\alpha} \otimes \bar{\mathbf{s}}^{\alpha} = \left(\tfrac{1}{2}\rho^{\alpha}y_{e}^{2}\dot{\gamma}^{\alpha}\right)\ \bar{\mathbf{s}}^{\alpha} \otimes \bar{\mathbf{s}}^{\alpha}. \tag{2.31}$$

The rate of elongation of the PSB is thus established via the vacancy generation rate ($\dot{c}_{v}$) and the known slip rate. The implications of this PSB elongation process, along with that of vacancy diffusion, to fatigue crack initiation will be considered in Chapter 4.

2.7 Formation of PSBs

At the beginning of stage B in the saturation cyclic stress–strain curve, structural changes must take place within the matrix to accommodate high values of plastic strains because the dislocation veins in the matrix cannot accommodate strains in excess of approximately 10^{-4}. The ensuing formation of PSBs has been the subject of a number of electron microscopy studies (e.g., Mughrabi, Ackermann & Herz, 1979; Mecke, Blochwitz & Kremling, 1982; Jin, 1989; Holzwarth & Essmann, 1993).

2.7.1 Electron microscopy observations

Holzwarth & Essmann (1993) presented a study of the mechanism by which the matrix vein structure is transformed into the wall structure of a PSB. They started with a saturated matrix vein structure in a Cu single crystal at $\gamma_{pl} = 10^{-4}$, using a fully-reversed plastic-strain control test, until saturation to a cumulative plastic strain of $\Gamma = 15$ at $\tau_s = 28$ MPa at 300 K. The crystal was then subjected to a sudden increase in γ_{pl} to 4×10^{-4} which caused an instant jump in flow stress to 33 MPa and which initiated the formation of PSBs. The number of fully-reversed strain cycles at the higher γ_{pl} was varied from specimen to specimen in order to gather information on the consecutive stages of PSB evolution. Continued cycling at the higher γ_{pl} caused a rapid decline in the flow stress which dropped from 33 MPa to a steady-state value of 28 MPa within 1000 cycles.

These experiments reveal that the transformation from the matrix vein structure to the PSB wall structure most likely commences at the centers of the veins wherein exist small areas that are dislocation-poor. These soft areas are surrounded by a harder shell of higher dislocation density, wherein develop the first dislocation walls. In the plateau regime, the walls shift at a rate of 1–2 nm/cycle, and this shift plays an important role in establishing the typical ladder pattern in the PSBs.

Figure 2.14 shows the dislocation arrangements in a Cu crystal subjected uniformly to fully reversed plastic straining at $\gamma_{pl} = 4 \times 10^{-4}$ (without a change in γ_{pl} from a prior lower value). Here, an inhomogeneous matrix vein structure is

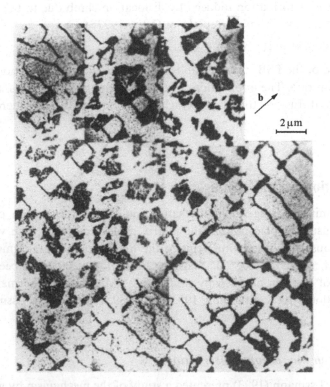

Fig. 2.14. The evolution of a PSB wall structure in the dislocation-poor region of the matrix veins (marked by arrows). $\mathbf{g} = (111)$. $\gamma_{pl} = 4 \times 10^{-4}$. (From Holzwarth & Essmann, 1993. Copyright Springer–Verlag. Reprinted with permission.)

evident, with dislocation-poor interior regions in the veins. Consider the PSB in this figure which cuts through a row of veins. From the geometry of the nascent wall structure and the surrounding vein structure, it is noted that the walls originate from the vein shells and that they have to move very little to establish their spacing during PSB evolution. The destruction of veins is seen to begin preferentially in the dislocation-poor regions where each vein forms two walls. The scenario emerging from Fig. 2.14 is essentially the same as that put forth for Ni by Mecke, Blochwitz & Kremling (1982), and for Cu by Jin (1989).

2.7.2 Static or energetic models

A quantitative description of the formation of dislocation structures within the veins and PSBs can potentially be obtained from calculations of the equilibrium positions of finite populations of dislocations. Although theoretical models of the evolution of PSBs are in their infancy, the few results available to date appear to provide some fundamental justification for the experimentally observed geometrical arrangement of dislocations in fatigued crystals. Detailed numerical simulations are necessary to establish a link between the conditions for the stability of dislocation structures and the geometries of dislocation arrangements that develop during fatigue. The basis for the earlier models (such as the low-energy dislocation structures (LEDS) model of Kuhlmann-Wilsdorf, 1979) was the Taylor–Nabarro lattice (Taylor, 1934; Nabarro, 1952). This lattice consists of regular arrangements (of parallel straight edge dislocations) extending to infinity. The infinite Taylor–Nabarro lattices cannot explain the existence of dislocation-free channels between the veins and the sharp transition in dislocation density from the veins to the in-between channels.

It has been shown by Neumann (1983, 1986) that finite sections of Taylor–Nabarro lattices of certain shapes are stable. Neumann has considered a regular planar array of positive and negative edge dislocations (say, on the x–y plane) with the dislocation lines extending along the z direction. The Burgers vector of a dislocation is then given by $\mathbf{b} = (\zeta_s b, 0, 0)$, $\zeta_s = \pm 1$, $b > 0$. The position of the dislocation, with respect to the origin of the coordinate system, is described by the position vector $\mathbf{r}_j = \mathbf{r}(x_j, y_j)$, $j = 1, ..., n$, such that the slip plane of the jth dislocation is $y = y_j$. When a stress τ_a is applied, the force per unit length on the ith dislocation is given by

$$F_i = b_i \left\{ \tau_a + B \sum_{j \neq i} b_j g(\mathbf{r}_i, \mathbf{r}_j) \right\}; \qquad B = \frac{G}{2\pi(1 - \nu)}, \qquad (2.32)$$

$$g(\mathbf{r}_i, \mathbf{r}_j) = \frac{(x_i - x_j)\{(x_i - x_j)^2 - (y_i - y_j)^2\}}{\{(x_i - x_j)^2 + (y_i - y_j)^2\}^2}. \qquad (2.33)$$

For mechanical equilibrium, all dislocation structures must be relaxed, i.e. the net stress on a dislocation line (due to the applied stress and due to all the other dislocations) must be zero. This can be accomplished by shifting the dislocations within their slip planes (i.e. with y_j fixed) into appropriate positions x_j. Mathematically, this process reduces to equating the right hand side of Eq. 2.32 to zero, which results in n nonlinear equations for the n unknowns x_j. In this way, the stability of veins can be modeled by using various shapes of the finite Taylor–Nabarro lattices as the starting configurations.

Figure 2.15(a) shows such a relaxed configuration of a diamond-shaped section of Taylor–Nabarro lattice in which all the dislocations are at their equilibrium positions at zero applied stress. It is found that, for this section, the application of a stress results in the emergence of dipolar walls of dislocations from the polarization of the initial configuration, Fig. 2.15(b). Figure 2.15(c) shows the equilibrium configuration of a wall structure of dislocations. Here the ratio of wall spacing to wall height is of the order of unity, in concurrence with the experimental observations of ladder

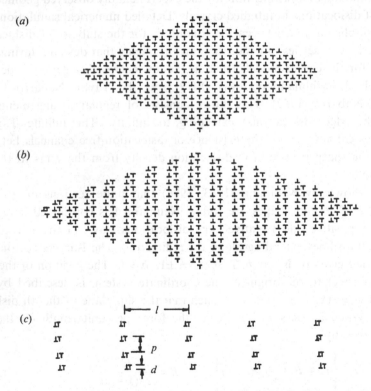

Fig. 2.15. (a) and (b) show equilibrium states of finite Taylor-Nabarro lattices at $\tau_a = 0$ and $\tau_a = 0.425\tau_{max}^d$, respectively. (c) is the computed equilibrium structure of dipole walls at $D_y = bd/lp$. Figure (a) is an example of a 9×9 quadrupole structure. Figure (c) is an example of walls (parallel columns) each of which consists of four dipoles. (From Neumann, 1986. Copyright Elsevier Sequoia S.A. Reprinted with permission.)

structures shown in Fig. 2.10. For the arrangements seen in Fig. 2.15(c), the dipole strength is bd/lp, where d is defined in Fig. 2.15(c) and the inset in Fig. 2.16. (It should be noted that for an isolated dipole, the dipole strength is defined by the relationship $D_y = b(y^+ - y^-)$, where y^+ is the y coordinate of the positive dislocation and y^- is the y coordinate of the negative dislocation in the dipole, and b is the magnitude of the Burgers vector.)

When the applied stress τ_a reaches a certain critical value, τ_{max}, the initial dislocation configuration (made up of multipoles) begins to decompose into dipolar walls. Figure 2.16 shows the variation of the decomposition stress, τ_{max}, calculated from the foregoing numerical simulation, as a function of the total number of dislocations used in the model. In this figure, τ_{max} is normalized by the decomposition stress, τ_{max}^d, of the most narrow elementary dipole of the configuration, with the separation distance between the dislocations in the dipole being d in a direction normal to the slip plane. τ_{max}^d ($\propto b/4d$) is the passing stress of two dislocations (say, the ith and kth dislocations which have opposite signs, with $y_i - y_k = d$). The veins in this

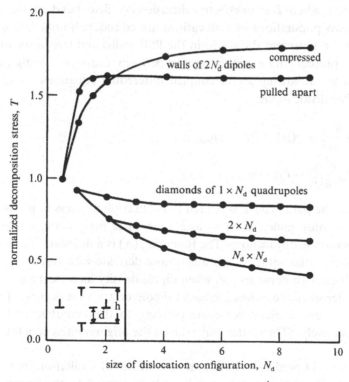

Fig. 2.16. Normalized decomposition stress, $T = \tau_{max}/\tau_{max}^d$, vs the size of dislocation configuration for diamond-shaped Taylor–Nabarro lattice and for the wall structure. The inset shows an elementary quadrupole for which $h = 2d$. The matrix vein structure is represented here by assemblies of elementary quadrupoles, as in Fig. 2.15(a). The PSB wall structure is modeled by columns of dipole walls, as in Fig. 2.15(c). (After Neumann, 1986.)

figure are composed of elementary quadrupoles (see inset in Fig. 2.16) which make up a diamond-shaped Taylor lattice.

Figure 2.16 indicates that the wall structure is more stable than the vein structure (at comparable values of τ_{max}^d). This result provides a rationale for the experimental observation that, at saturation, the decomposing veins cannot rearrange into a different vein structure with smaller d values, but transform into a ladder structure. This numerical model, albeit limited and very simplistic in its consideration of the arrangements of dislocations in fatigued crystals, provides encouraging possibilities for developing a physical basis for fatigue deformation.

2.7.3 Dynamic models of self-organized dislocation structures

Attempts have also been made to develop analytical models of continuous concentration fields of dislocations in fatigued crystals (Walgraef & Aifantis, 1985; Aifantis, 1987). In this approach, the dislocations are represented only by a density, $\rho(x, t)$, which is a function of space and time. The to-and-fro motion of dislocations under a cyclic stress is modeled in one dimension as a diffusion phenomenon with a flux term $D(\partial^2 \rho / \partial x^2)$, where D is an effective diffusion coefficient and ρ is the density of dislocations. Two populations of dislocations are considered: immobile dislocations of density ρ_i (to represent those within the PSB walls) and free dislocations of density ρ_m (to represent those within the in-between channels). Walgraef and Aifantis arrive at the following set of coupled differential equations for densities of trapped and free dislocations:

$$\frac{\partial \rho_i}{\partial t} = D_i \frac{\partial^2 \rho_i}{\partial x^2} + g(\rho_i) - \overline{P} \rho_i + c \rho_m \rho_i^2,$$

$$\frac{\partial \rho_m}{\partial t} = D_m \frac{\partial^2 \rho_m}{\partial x^2} + \overline{P} \rho_i - c \rho_m \rho_i^2. \tag{2.34}$$

In these equations, the parameter $\overline{P}$ is related to the maximum stress or plastic strain rate, x is the dislocation glide direction, and c measures the pinning rate of freed dislocations by immobile dislocations. The function $g(\rho_i)$ is a dislocation generation/ loss function whose initial value is zero and whose derivate with respect to time is positive at low dislocation densities (i.e. when dipole density increases) and becomes negative at high densities (i.e. when increased dipole destruction occurs). The subscripts 'i' and 'm' denote quantities corresponding to the immobile and mobile dislocations, respectively. (The reader may consult the original references for further details.)

Solution of Eqs. 2.34 predicts instabilities in the form of oscillations in time and spatial patterning. These two instabilities have been related to the occurrence of strain bursts and the formation of PSBs, respectively. However, this analysis relies on rather global assumptions on the diffusion of dislocations and ignores the specific dislocation geometries within dipolar PSB walls and channels.

Differt & Essmann (1993) have proposed a dynamic model for edge dislocation walls within a reaction–transport modeling framework. In this work, the reaction terms are consistent with the experimentally observed properties of fatigue-induced dislocation structures (which overcome some of the limitations of the Walgraef & Aifantis model). Two important length scales are introduced in the reaction terms: (i) the critical annihilation distance of a dipole under the influence of an applied stress and (ii) the critical distance for the spontaneous annihilation of closely spaced dipoles. The analysis shows how the walls move. Edge dislocations are deposited on the walls by the moving screws. If the fluxes on both sides of the wall are not balanced, the wall is destructed on one side and reconstructed on the other. The wall moves until the fluxes get balanced. This mechanism rationalizes why a freshly-formed PSB is less periodic and imperfect than a mature one.

In summary, the dislocation arrangements in fatigue can be broadly classified into two basic groups (e.g., Glazov, Llanes & Laird, 1995) : (i) structures in which equilibrium is maintained, and (ii) nonequilibrium self-organized dislocation structures. The former group includes the low-energy dislocation structure (LEDS) discussed earlier in this section and includes the Taylor-lattice calculations. The latter includes models of the type postulated by Walgraef & Aifantis. The nonequilibrium structures have been shown to provide a rationale for the instigation of fatigue instabilities such as the formation of ladder structures in PSBs and strain bursts.

We close this section by noting that several approaches have emerged to rationalize the patterning seen in fatigue by recourse to computer simulations. Hesselbarth & Steck (1992) have confirmed the Neumann model for equilibrium structures through two-dimensional cellular automata simulations of stress-induced patterning from an initial random configuration of edge dislocations. Devincre & Pontikis (1993) have studied the periodicity of dipolar walls as a function of applied stress and edge dislocation density, while full three-dimensional simulations of dislocation structures under cyclic stressing have been undertaken by Devincre & Kubin (1997).

2.8 Formation of labyrinth and cell structures

Electron microscopy of fatigued Cu crystals (oriented for single slip) provide further insights into the dislocation structures formed in the range of γ_{pl} $\approx 10^{-4}$–10^{-2} (Ackermann, Kubin, Lepinoux & Mughrabi, 1984). For $\gamma_{pl} < 10^{-3}$, the slip band structure is characterized by the hard matrix comprising veins and the softer PSBs with the wall structure. An increased contribution of secondary slip (which becomes more pronounced at higher γ_{pl}) and a gradual evolution of 'laby-

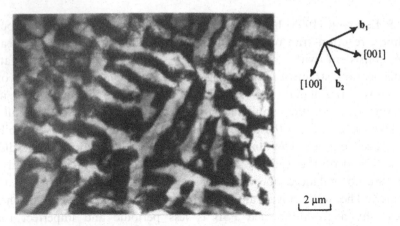

Fig. 2.17. A view of the (010) section of a Cu crystal that was cycled to saturation at $\gamma_{pl} = 5 \times 10^{-3}$, revealing the labyrinth structure. (From Ackermann *et al.*, 1984. Copyright Pergamon Press plc. Reprinted with permission.)

rinth' and 'cell' structures are noticed for $\gamma_{pl} > 2 \times 10^{-3}$. Figure 2.17 is a TEM image of a (010) section of the Cu crystal that was subjected to fatigue at $\gamma_{pl} = 5 \times 10^{-3}$. Note the formation of a labyrinth structure with its walls oriented parallel to the $\langle 100 \rangle$ direction with a mean spacing of about 0.75 μm. The labyrinth consists of two sets of orthogonal Burgers vectors; b_1 and b_2 denote the primary and conjugate Burgers vectors, respectively. Labyrinth walls are also known to be produced during cyclic deformation of Cu–Ni alloys (Charsley & Kuhlmann-Wilsdorf, 1981), of ionic crystals (Majumdar & Burns, 1982) and of BCC metals (e.g., Mori, Tokuwame & Miyazaki, 1979). The occurrence of walls of labyrinth structures has been rationalized on the basis of a geometrical argument by Dickson, Boutin & L'Espérance (1986), who considered a three-dimensional stacking of twin dislocation loops of rectangular shape. The directions in which the stacking of the loops was geometrically most favorable was shown to be consistent with the crystallographic orientations of labyrinth structures actually observed in a variety of metal crystals.

Ackermann *et al.* (1984) have suggested the following changes at the higher γ_{pl} end of the saturation stress–strain curve (Fig. 2.2) for FCC crystals: matrix phase with labyrinth structure → PSBs and labyrinth structure → cell structure. Secondary slip (prevalent in region *C*) originates at the PSB–matrix interface and spreads in the form of an expanding cell structure which fills the PSBs. The transformation of all the PSBs into a cell structure appears to occur after 10^6 cycles. This marks the beginning of secondary hardening in region *C*. The above structural changes reported for Cu have also been found in Ni (Mecke, Blochwitz & Kremling, 1982). Figure 2.18 shows an example of a cell structure formed in Cu in regime *C*.

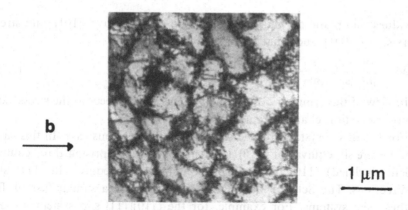

Fig. 2.18. A view of the (1$\bar{2}$1) section of a Cu crystal that was cycled to saturation at γ_{pl} = 1.45 × 10^{-2}, revealing cell structures formed by multiple slip in regime C. The primary Burgers vector is along **b**. (Photo courtesy of H. Mughrabi. Reprinted with permission.)

2.8.1 Example problem: Multiple slip

Problem:

(i) A single crystal of an FCC metal, with a ⟨001⟩ orientation along the stress axis, is deformed until the onset of plasticity. How many equivalent slip systems are activated? What is the Schmid factor for these slip systems? Is the deformation accompanied by a change of axial orientation?

(ii) Answer the above questions for a BCC single crystal with a ⟨111⟩ orientation which deforms on slip systems of the type {$\bar{1}$01}⟨111⟩.

Solution:

(i) A repeat of the calculations, similar to those listed in Table 2.2, readily reveals that there are eight slip systems for this case with $M = |\cos\phi \cdot \cos\lambda| = 1/(\sqrt{3}\sqrt{2}) = 0.408$. In other words, the slip systems A2, A3, B2, B4, C1, C3, D1 and D4 have the same Schmid factor, M. These eight slip systems comprise four {111} slip planes, each containing two ⟨110⟩ directions inclined equivalently to the stress axis [001]. The resolved shear stress on the other four slip systems (A6, B5, C5 and D6) is zero for this initial orientation because the four directions in the {111} plane are perpendicular to the stress axis. Therefore, these slip systems are not activated.†

The Schmid factor, M, can also be computed in a more concise manner, for the slip system (111)[$\bar{1}$01] for example, by ascribing the following

---

† In a real experiment, however, errors in alignment and orientation may lead to the activation of slip systems which differ from this predicted behavior. See Section 3.1 for a further discussion of this point.

values: slip plane normal, $\mathbf{n} = [111]$, slip direction $\mathbf{b} = [\bar{1}01]$, and stress axis, $\mathbf{s} = [001]$, such that

$$M = \frac{\mathbf{n} \cdot \mathbf{s}}{|\mathbf{n}| \cdot |\mathbf{s}|} \times \frac{\mathbf{n} \cdot \mathbf{b}}{|\mathbf{n}| \cdot |\mathbf{b}|} = 0.408. \qquad (2.35)$$

In view of the symmetry of the slip systems with respect to the stress axis, no orientation change occurs.

(ii)   For the BCC crystal, proceed similarly to the previous case. In this case, there are six equivalent {110}⟨111⟩ slip systems comprising three equivalent (inclined) {110} planes, each containing two equivalent ⟨111⟩ slip directions. The Schmid factor is also computed in a similar fashion for these slip systems. For example, for the (110)[1$\bar{1}$1] slip system, assign: $\mathbf{n} = [110]$, $\mathbf{b} = [1\bar{1}1]$, and $\mathbf{s} = [111]$. Substituting these values in Eq. 2.35 yields that $M = 0.27$. As in the previous case, in view of the symmetry of the slip systems with respect to the stress axis, no orientation change occurs.

2.9 Effects of crystal orientation and multiple slip

As shown in Section 2.2, the cyclic deformation of FCC single crystals oriented for single slip exhibits two prominent features: (i) the existence of a plateau region, and (ii) the formation of persistent slip bands with their characteristic wall structures. An extension of the fatigue mechanisms seen in monocrystalline FCC metals to polycrystalline metals and alloys, however, requires a knowledge of cyclic deformation in different orientations, including those involving multiple slip.

Table 2.3 summarizes the results from a number of experimental studies of the effects of duplex and multiple slip on cyclic deformation. It is evident that the possibility of occurrence of a plateau in the cyclic stress–strain curve, as well as the extent of the plateau region are strongly influenced by the crystallographic direction of the FCC crystal along which the fatigue loading is imposed. (See the worked example in Section 2.3.1 for the identification of single and multiple slip systems for a given crystal orientation.)

Gong, Wang & Wang (1997) have studied the effect of [001] multiple slip on the cyclic stress–strain curve in Cu crystals. They found an absence of a plateau regime, and in the strain range $\gamma_{pl} = 1.0 \times 10^{-4}$–$3.0 \times 10^{-3}$, no PSBs were found to occur. Fatigued [001] crystals comprise principally labyrinth structures of primary and critical dislocations, which can more easily accommodate multiple slip and cross slip than the ladder structures. The labyrinths consist of two types of mutually orthogonal wall arrangements of dislocations: one is always parallel to (001) (i.e. normal to the stress axis), while the other is along (100) or (210) or (120). Figure 2.19 shows an example of

Table 2.3. *Effects of crystallographic orientation and multiple slip on the cyclic stress–strain response of FCC single crystals at 20 °C.*

Metal	Loading axis	Observation	Reference
Cu	[001] [Ī11]	Higher hardening rate than in single slip	Kemsley & Paterson (1960)
	[Ī11]	No plateau	Lepisto & Kettunen (1986)
	[Ī12] [012] [Ī22] [001]	Higher cyclic hardening rate in multi-slip than in duplex slip	Jin & Winter (1984) and Jin (1983)
	[034]	Pseudo plateau than in single slip	Gong et al. (1995)
	[Ī17]	No plateau	Gong et al. (1995)
	[001]	No plateau, no PSBs	Gong, Wang & Wang (1997)
	[011]	A clear plateau over $\gamma_{pl} = 1.1 \times 10^{-4} - 7.2 \times 10^{-3}$; small, irreversible rotation of slip system during symmetric tension-compression loading which causes deformation bands to form	Li et al. (1998)
Ni	[001] [Ī11]	Pseudo plateau in regime B	Mecke & Blockwitz (1982)

such a structure. The interactions among different slip systems in the labyrinth and the attendant formation of Lomer–Cottrell locks causes a much higher cyclic hardening rate in multi-slip orientations than in single glide. The labyrinths accommodate different imposed plastic strains by appropriately adjusting their channel widths: an increase in γ_{pl} is accommodated by a reduction in the channel width of the labyrinth.

Another noteworthy feature of multiple slip during cyclic deformation is the apparent improvement in the fatigue limit. In the [001] specimens, the fatigue limit, defined as the critical value of γ_{pl} below which (nonpersistent) slip bands do not form, is approximately 1.7×10^{-4}, compared to 6.0×10^{-5} for single slip fatigue of Cu. Further implications of such multiple slip for fatigue deformation and for the cyclic response in polycrystals will be discussed in the next chapter.

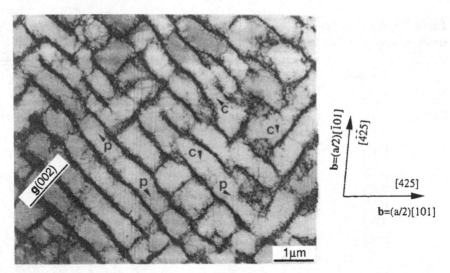

Fig. 2.19. A transmission electron micrograph of the labyrinth structure formed in a single crystal of Cu fatigued in equal tension–compression along the [001] stress axis at $\gamma_{pl} = 1.8 \times 10^{-3}$ at room temperature. The TEM foil is parallel to $(\bar{1}20)$. The symbols 'p' (for primary) and 'c' (for critical) denote the projections of $[\bar{1}01]$ and [101] screw dislocation segments on the (120) plane along the $[\bar{4}25]$ and [425] directions, respectively. (From Wang, Gong & Wang, 1997. Copyright Elsevier Ltd. Reprinted with permission.)

2.10 Case study: A commercial FCC alloy crystal

Face-centered cubic Ni-base alloys find important technological applications wherein high temperatures (up to about 1100 °C) are encountered. In some of these components where essentially uniaxial stresses exist, as in the turbine blades of aircraft jet engines, it is commonly known that the elimination of transverse grain boundaries promotes substantial improvements in creep and fatigue lives. As a result, directionally solidified and single crystalline superalloys are widely used as materials for turbine blades in gas turbine engines.†

Nickel-base superalloys typically contain a solid solution of 5–20 wt% Cr for enhancing oxidation resistance. In addition, Al and Ti (1–5 wt%) are added to increase strength. Precipitation hardening results in the formation of an ordered $Ni_3(Al,Ti)$ γ' phase which has the $L1_2$ structure. The volume fraction of the γ' phase can be varied over the range 20–70% by controlling the Al and Ti concentration, while its size and shape can be closely controlled by thermomechanical treatments. The excellent strength of Ni-base superalloys at elevated temperatures is attributable to the presence of γ' precipitates. In addition, elements such as Fe, Co, Mo, Nb, W and Ta, are added to monocrystalline Ni-base alloys to solid-solution-strengthen the γ and γ' phases.

In this section, we consider aspects of the microscopic cyclic deformation of a Ni-base superalloy ($\langle 001 \rangle$ orientation for loading). The concentration, in wt%, of this alloy, commercially known as PW (Pratt & Whitney) 1480, is Al (4.8%), Ti (1.3%), Ta (11.9%),

† Single crystal Ni-base superalloys are also used as blades in the turbopump of the Space Shuttle main engine.

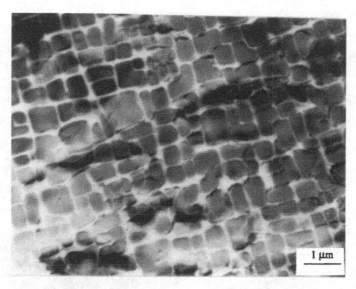

Fig. 2.20. TEM micrograph of the initial microstructure showing the γ/γ' precipitate structure. **g** = {200}. (From Milligan & Antolovich, 1987. Copyright *Metallurgical Transactions*. Reprinted with permission.)

Cr (10.4%), Co (5.3%) and W (4.1%), with the balance being Ni; the C content is 42 ppm. Figure 2.20 is the initial TEM micrograph showing the γ/γ' precipitate structure of this alloy prior to mechanical testing. This example also illustrates the evolution of dislocation networks and stacking faults as a result of uniaxial cyclic loading and of the interactions between dislocations and precipitates at different temperatures. The results summarized here are from the work of Milligan & Antolovich (1987, 1991).

Distinct changes occur in the monotonic and fully-reversed cyclic deformation of the PW 1480 alloy crystal in different temperature regimes. Orientation dependence of the critical resolved shear stress in the family of {111}⟨110⟩ slip systems, tension–compression asymmetry as well as anisotropic strain hardening result from a highly anisotropic octahedral slip, at temperatures typically below 760 °C. This trend has been attributed to the ease of cube cross-slip and to a multitude of active slip systems. Above about 800 °C, however, deformation is mostly isotropic, with the mechanism being largely governed by the by-pass of γ' particles by the dislocations which causes pronounced wavy slip. This behavior is strain-rate sensitive, with increasing strain rate extending slip planarity and anisotropic deformation of higher temperatures.

Figure 2.21(*a*) shows a typical dislocation structure of PW1480 subjected to 6 cycles of fully reversed strain-controlled fatigue. Dislocations and stacking faults with the γ' precipitates on the primary {111} slip planes are seen. The dislocation pairs (marked by arrows) were determined to be of primary $a/2\langle 110 \rangle$ Burgers vector. Figure 2.21(*b*) is from the same area, with the primary slip plane viewed edge on. It is seen here that (i) the dislocation pairs in the matrix are ostensibly in the same slip bands that contain the faulted dislocation loops within the precipitates, and that (ii) the slip deformation is highly planar. These micrographs also imply that the deformation mechanism involves the shearing of the γ' precipitates by dislocations with the primary Burgers vector.

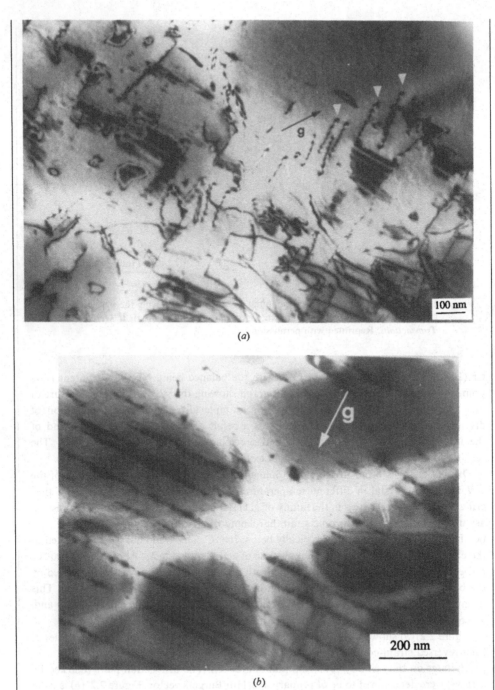

(a)

(b)

Fig. 2.21. Dislocation substructure in the superalloy crystal after six fully reversed cycles at 20 °C with $\gamma_{pl} = 1.1 \times 10^{13}$, and $\dot{\gamma}_{pl} = 8.7 \times 10^3 \text{ s}^{-1}$. (a) A bright-field TEM micrograph showing stacking faults and faulted loops in the precipitate, and matrix pairs (arrows) between the precipitates. $\mathbf{b} = \{200\}$. (b) A bright-field micrograph of the same area looking edge on. The planar nature of the structure is evident. $\mathbf{g} = \{111\}$. (From Milligan & Antolovich, 1991. Copyright *Metallurgical Transactions*. Reprinted with permission.)

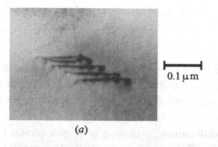

(a)

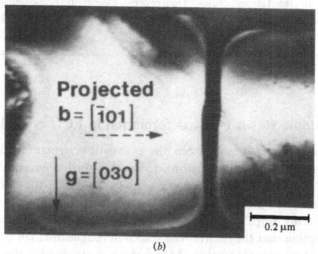

Projected
$\mathbf{b} = [\bar{1}01]$

$\mathbf{g} = [030]$

0.2 μm

(b)

Fig. 2.22. γ' precipitates in the PW1480 Ni-base superalloy sheared by $\{111\}a/2\langle110\rangle$ dislocations after a plastic strain of 0.08% at 705 °C. (a) A bright-field TEM micrograph. $\mathbf{g} = \{\bar{1}11\}$. (b) A dark-field micrograph using the (030) superlattice spot. $\mathbf{b}$ denotes the Burgers vector of the dislocation which is oriented at 45°. Dislocations are invisible here because the $\mathbf{g} \cdot \mathbf{b} = 0$ criterion is satisfied. (From Milligan & Antolovich, 1987. Copyright *Metallurgical Transactions*. Reprinted with permission.)

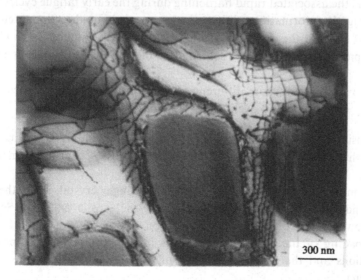

300 nm

Fig. 2.23. Dislocations by-passing the γ' precipitates after 765 fatigue cycles at $\gamma_{pl} = 2.3 \times 10^{-3}$ and $\dot{\gamma}_{pl} = 8.7 \times 10^{-3}$ s^{-1} at 1093 °C. $\mathbf{g} = \{200\}$. (Photograph courtesy of W.W. Milligan. Reproduced with permission.)

Faulted dislocation loops are left behind by this shearing process as deformation debris. A more direct evidence for such shearing is available in Figs. 2.22(*a*) and (*b*) which show the shearing of a γ' precipitate by $\{111\}\,a/2\,\langle 110 \rangle$ dislocations after a plastic strain of 0.08% at 705 °C.

Figure 2.23 shows an example of the dislocation by-passing of γ' precipitates in the same alloy subjected to 765 fatigue cycles. This process essentially homogenizes the macroscopic fatigue deformation response which becomes isotropic in tension and compression.

2.11 Monotonic versus cyclic deformation in FCC crystals

There are some similarities between the dislocation structures discussed above for cyclic deformation and those known to develop during uniaxial, monotonic tension in FCC single crystals. At very low γ_{pl} ($\leq 5 \times 10^{-4}$), the dislocation configurations first generated during rapid cyclic hardening correlate well with the substructures found during Stage I deformation of FCC monocrystals in monotonic tension, with the exception that the matrix 'veins' seen in fatigue are akin to the cell structures in Stage II of monotonic tension. At higher strain amplitudes, the dislocation structures in the cyclic work hardening stage are similar to the unidirectional Stage I configurations only during the first few cycles. The similarity then disappears with the progressive formation of persistent slip bands. At high plastic strain amplitudes, corresponding to regime *C* of the cyclic stress–strain curve, the formation of cell structures and the associated rapid hardening during the early fatigue cycles finds an analogy in Stage II deformation of FCC single crystals in unidirectional tension. The formation of PSBs at the onset of saturation is somewhat analogous to coarse slip band development during Stage III deformation of FCC single crystals in monotonic tension.

There are also some significant differences between the dislocation structures developed during monotonic and cyclic deformation:

- The density of dislocations produced during cyclic loading is significantly higher than that generated, at comparable stresses, during monotonic tension.
- During monotonic tensile deformation of a single crystal, both the slip plane and the slip direction rotate toward the tensile axis (see Section 1.5). However, there is no such orientation change during fully reversed cyclic loading of the crystal. This results in a preponderance of primary dislocations and the absence of long-range internal stresses during cyclic hardening.
- The evolution of persistent slip bands with their wall structure of edge dislocations is specific to cyclic deformation.

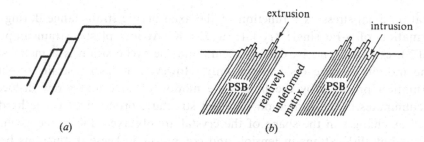

Fig. 2.24. (*a*) A series of steps resembling a staircase pattern produced by monotonic plastic strain. (*b*) A rough surface consisting of hills and valleys produced by cyclic plastic strain.

- A striking feature of fatigue deformation is the establishment of a saturated state where the peak resolved shear stress is independent of the plastic shear strain amplitude.
- Because of the short-range interactions among dislocations during to-and-fro motion in fatigue, there develops a high density of point defect clusters ($\sim 10^{15}$ cm^{-3}). The resultant friction stress can be as high as 25% of the peak shear stress at 20 °C.
- The flow stress of FCC crystals exhibits a stronger dependence on temperature and strain rate in fatigue than in tension.
- One of the most visible distinctions between monotonic and cyclic deformation is in the development of surface roughness. Whereas monotonic loading leads to the formation of surface slip steps which resemble a staircase geometry, cyclic deformation produces sharp peaks and valleys (known as 'extrusions' and 'intrusions', respectively) at sites where the persistent slip bands emerge at the specimen surface (Fig. 2.24). (Recall that the cumulative plastic strains in fatigue are significantly greater than those in monotonic tension, Fig. 2.4.) Details of the formation of extrusions and intrusions are discussed in Chapter 4.

2.12 Cyclic deformation in BCC single crystals

Pure BCC crystals such as α-Fe, Mo and Nb, when oriented for single slip are known to behave significantly differently from the FCC crystals. Generally, the core of the screw dislocation in BCC metals does not dissociate, and the particular nature of the screw dislocation core structure in BCC induces very high lattice friction (so-called Peierls stress). Effects such as strain-rate sensitivity, strong temperature dependence of cyclic deformation, relative mobility of edge and screw dislocations, as well as asymmetry of slip between tension and compression are a consequence of the special role of screw dislocations in the BCC metals.

Mughrabi, Herz & Stark (1976), Mughrabi & Wüthrich (1976) and Mughrabi, Ackermann & Herz (1979) report different regimes in the variation of mean saturation axial stress as a function of the axial plastic strain range during cyclic deformation of α-Fe single crystals at 295 K. At low plastic strain amplitudes ($\leq 10^{-3}$), essentially no hardening occurs and the cyclic strain is a manifestation of the motion of edge dislocations only. However, at higher strain amplitudes, deformation proceeds by the large-scale motion of edge and screw dislocations and culminates in the formation of a cell structure; pronounced cyclic hardening as well as changes in the shape of the crystal are observed due to the asymmetric slip of screw dislocations in tension and compression. These distinctions between low and high strain fatigue are peculiar to BCC crystals. Although no PSBs have been identified in either regime of plastic strain amplitudes, ill-defined bands of slip, which could lead to crack nucleation, have been noticed. In agreement with these findings, TEM investigations of dislocation structures ahead of fatigue cracks (described in Chapter 4) have identified the existence of PSBs in polycrystalline Cu but not in pure α-Fe.

The following differences between FCC and BCC crystals point to some causes for the distinctions in their fatigue response: (i) At 295 K and at low plastic strain amplitudes, thermally-activated glide of screw dislocations as well as dislocation multiplication are strongly suppressed in BCC α-Fe. (ii) Whereas FCC metals are only weakly strain rate-sensitive, the flow stress of BCC metals is strongly dependent upon the strain rate. For this reason, the cyclic stress–strain curves for BCC crystals should be obtained at constant values of imposed strain rates. It is generally seen that, as a consequence of dynamic strain-ageing,† high temperatures, very low strain rates and the addition of impurity atoms (such as carbon, nitrogen and oxygen) to the BCC metal promote cyclic damage that is more similar to that found in FCC metals. Mughrabi, Ackermann & Herz (1979) report that the addition of 30 weight ppm carbon to α-Fe single crystals leads to cyclic stress–strain curves closer to those measured for FCC metals and slip bands analogous to the PSBs. PSBs are also known to form in both the surface and interior grains of polycrystalline low carbon steels (Pohl, Mayr & Macherauch, 1980). These results show that any comparison of dislocation structures and cyclic slip characteristics reported by different investigators must be made with caution because even a small impurity content can lead to marked variations in fatigue micromechanisms.

2.12.1 Shape changes in fatigued BCC crystals

If slip occurs on different planes during tension and compression portions of fatigue, a crystal must undergo shape changes due to this slip asymmetry. This effect, observed in BCC crystals (Nine, 1973; Neumann, 1975; Mughrabi & Wüthrich, 1976;

---

† Dynamic strain-ageing refers to the phenomenon whereby certain materials (e.g., low carbon steels) generally exhibit an increase in yield and fatigue strengths over certain temperature ranges as a result of the interaction between dislocations and solute atoms (e.g., carbon and nitrogen). Details of the mechanisms of dynamic strain-ageing can be found in any textbook on mechanical metallurgy.

Guiu & Anglada, 1980), effectively transforms an initially circular cross section of a cylindrical crystal into an ellipse. Neumann showed that the shape change produced by cyclic straining can be correlated with slip irreversibility, which is an important factor for crack nucleation. Here we follow Guiu and Anglada for the derivation of slip irreversibility due to shape changes in BCC crystals.

Consider a BCC crystal in which it is assumed that slip occurs on several different planes in tension and compression in one direction defined by the unit vector **b**. Since all these planes are parallel to **b**, they can be represented in terms of two basic reference planes *of the same zone* with unit normal vectors $\mathbf{n}_1$ and $\mathbf{n}_2$, which can be arbitrarily chosen. The glide strain in tension can be represented by the total glide strain α_{1t} in plane $\mathbf{n}_1$ and by the total glide strain α_{2t} in plane $\mathbf{n}_2$. Similarly, the total glide strains in compression can be represented by α_{1c} and α_{2c} on planes $\mathbf{n}_1$ and $\mathbf{n}_2$, respectively.

In BCC crystals, any macroscopic slip plane can be visualized as being composed of microscopic slip steps on planes of the {110} and {112} types. Assume that the crystal is subjected to the same magnitude of plastic strain ϵ_p in both compression and tension; the length of the crystal can then be taken to be unchanged. If the slip direction for both cases is **b**, it can be shown that the *net* displacement after N fatigue cycles of a point in the crystal is

$$\Delta \mathbf{u}(\mathbf{r}) \;=\; 2N\gamma_1 \left| \frac{\mathbf{l} \times (\mathbf{n}_1 \times \mathbf{n}_2)}{\mathbf{l} \cdot \mathbf{n}_2} \right| (\mathbf{r} \cdot \mathbf{n}) \mathbf{b}, \qquad \mathbf{n} = \frac{\gamma_1 \mathbf{n}_1 + \gamma_2 \mathbf{n}_2}{\gamma}. \tag{2.36}$$

Here, **r** is the vector that locates the point under consideration with respect to the origin of the coordinate system (Fig. 2.25), $\gamma_1 (\equiv \alpha_{1t} - \alpha_{1c})$ is the net (irreversible) shear strain on plane $\mathbf{n}_1$, **l** is the unit vector along the cylindrical crystal axis (i.e. deformation axis), **n** is a vector which is perpendicular to both **b** and **l**, $\gamma \equiv |\gamma_1 \mathbf{n}_1 + \gamma_2 \mathbf{n}_2|$ and $\gamma_2 \equiv \alpha_{2t} - \alpha_{2c}$. The above result is most directly applicable to BCC crystals which undergo shape changes as a consequence of glide along $\langle 111 \rangle$. For such cases, a more convenient representation of Eqs. 2.36 is in terms of the angles ξ, ζ, ϕ_1 and ϕ_2; ξ is the angle between the tensile axis and the slip direction, and ϕ_1, ϕ_2 and ζ are the angles made by the reference planes $\mathbf{n}_1, \mathbf{n}_2$ and the maximum resolved shear stress plane, respectively, with the plane of the same zone whose axis is the slip direction. One finds that

$$\Delta \mathbf{u}(\mathbf{r}) = \frac{Ff_g \epsilon_{tot}}{\sin \xi} (\mathbf{r} \cdot \mathbf{n}) \mathbf{b}; \qquad F = \frac{\tan (\zeta - \phi_1) - \tan (\zeta - \phi_2)}{2 \cos \xi}, \tag{2.37}$$

and f_g is the fraction of glide strain in plane $\mathbf{n}_1$ in tension which is not reversed in compression, and $\epsilon_{tot} = 4N\epsilon_p$. If the initial diameter of the single crystal is d_0, it will undergo a shape change into an ellipse which has major and minor diameters of d_1 and d_2, respectively, after N fatigue cycles. The following relationship is also satisfied:

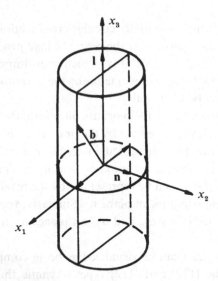

Fig. 2.25. Nomenclature for the determination of shape changes induced by tension–compression slip asymmetry during the cyclic straining of BCC single crystals. See text for details. (After Guiu & Anglada, 1980.)

$$Ff_g\epsilon_{tot} = \frac{(d_1/d_2) - 1}{\sqrt{(d_1/d_2)}}.$$ (2.38)

The slip irreversibility, Eqs. 2.36 and 2.37, is known to play an important role in the nucleation of fatigue cracks in BCC crystals. Furthermore, microcrack nucleation at the boundaries of surface grains in polycrystalline α-Fe has also been linked to the shape changes produced by the asymmetry of slip. Further discussions of the effects of slip irreversibility on crack initiation will be provided in Chapter 4.

2.13 Cyclic deformation in HCP single crystals

Hexagonal metals, such as Ti, Mg, Co, Zr and Be, are used in many engineering applications where deformation and cracking under cyclic loads are topics of interest. Compared to the volume of research done on the fatigue of FCC and BCC metals, less information is available on the cyclic deformation response of hexagonal closed-packed (HCP) metal crystals (see, for example, the earlier studies of Partridge (1969), Stevenson & Vander Sande (1974) and Kwadjo & Brown (1978) on magnesium crystals).

Among HCP crystals, the most comprehensive experimental studies of cyclic deformation have been carried out on Ti crystals. In this section, we review some results for Ti in order to compare and contrast its cyclic behavior with that of ductile FCC metals and to identify micromechanisms which are specific to hexagonal metals.

2.13.1 Basic characteristics of Ti single crystals

Titanium single crystals exhibit the following prominent features:

- Upon cooling, the BCC(β) → HCP(α) transformation occurs in Ti at approximately 880 °C.
- As with BCC crystals, the deformation characteristics of Ti crystals are strongly influenced by impurity and interstitial content.
- The stacking fault energy (SFE) of Ti is about 0.15 J m^{-2} (as compared to 0.04 J m^{-2} for Cu). Although the possibility of formation of stacking faults in Ti has been questioned in view of its high SFE (which is expected to promote cell formation), microscopy work on both monotonically and cyclically loaded Ti crystals has shown the presence of stacking faults.
- In transition HCP metals such as Ti, the choice of the operative slip system is strongly influenced by the electronic properties. The core of a screw dislocation in Ti has a three-dimensional structure that preferentially moves in the prismatic plane with a high Peierls force. Such an HCP crystal then behaves like a BCC metal under monotonic loading, while the divalent HCP metals, such as Be, Mg, Zn and Cd which have basal slip, behave more like FCC metals.
- In HCP metals, deformation twinning is a dominant mode of plastic deformation. Ti crystals deform by $\{11\bar{2}4\}$, $\{11\bar{2}2\}$, and $\{10\bar{1}1\}$ twins during compression along the c-axis, and by $\{10\bar{1}2\}$, $\{11\bar{2}3\}$, and $\{11\bar{2}1\}$ twins during tensile loading along the c-axis.

We now examine the effects of crystal orientation and equal tension–compression cyclic straining on substructure evolution in a high purity Ti single crystal (with oxygen and carbon concentrations of 116 and 22 ppm, respectively, and total other impurity content of 45 ppm). This discussion is derived from the work of Gu *et al.* (1994).

2.13.2 Cyclic deformation of Ti single crystals

The inset on the right side of Fig. 2.26 shows the unit triangle for the (0001) stereographic projection for Ti and the orientations of three α-Ti single crystals, marked A, B and C in the triangle, which were used for the fatigue tests. Specimen A was oriented on the (0001)–$(10\bar{1}0)$ border which is subject to duplex slip. Specimen

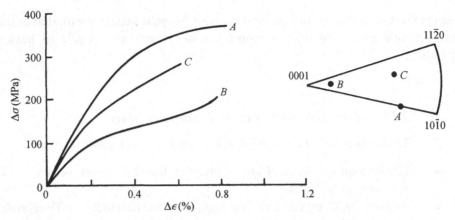

Fig. 2.26. Cyclic stress–strain curves for the three specimens of α–Ti single crystals. The inset shows the orientations of the specimens *A*, *B* and *C* in the unit triangle. (From Gu *et al.*, 1994.)

B was oriented near the (0001) corner, while specimen *C* was located near the middle of the triangle. The saturated cyclic stress–strain curves (CSSC) for the three specimens are plotted in Fig. 2.26. Orientations *A* and *B* represent the upper and lower bounds, respectively, among the three cases considered here, while specimen *C* exhibits an in-between response. Specimen B has a 'plateau-like' regime, similar to that of FCC crystals. The dislocation substructures in each of the three fatigue specimens were examined by TEM after the completion of the fatigue tests.

Gu *et al.* (1994) offer the following line of reasoning in an attempt to rationalize the effect of crystal orientation on cyclic deformation in the α-Ti single crystals. (1) Typically, the cyclic deformation is strongly influenced by the propensity for twin formation. An increase in the occurrence of cyclic twins (in the sequence $B \rightarrow C \rightarrow A$, among the three specimens considered here) causes a marked increase in the cyclic hardening rate. (Under cyclic loading, $\{10\bar{1}2\}$, $\{11\bar{2}1\}$, $\{11\bar{2}2\}$ and $\{112\bar{3}\}$ twins have been observed in Ti single and polycrystals.) (2) At fixed applied strain amplitudes, orientations (such as Specimen *C*) which promote single slip and cross slip give rise to planar dislocation dipole arrays and dislocation loops (similar to fatigued FCC crystals), whereas cell structures are found in the specimens oriented for duplex and multiple slip.

Exercises

2.1   Why does cyclic slip remain confined to the primary slip plane during the formation of vein structure or the PSB structure?

2.2   Why are the vein structures and PSB structures in FCC crystals composed mainly of edge dislocations?

2.3   What are the characteristic dimensions associated with the geometry of the vein structure and what is the physical basis for such dislocation configurations? What are the effects of changes in test temperature (either increase or decrease) on the geometry of dislocation arrangements within the vein structure and within the channels separating the veins?

2.4   Discuss the effects of stacking fault energy on the deformation of a ductile solid in monotonic tension and in tension–compression fatigue.

2.5   A long crystal is bent to a semicircular shape with a radius of 20 cm. The crystal has a square cross section (2 cm × 2 cm).

    (a) If it is assumed that all bending is accommodated by the generation of edge dislocations, calculate the total number of dislocations.

    (b) If the magnitude of the Burgers vector of the edge dislocations is 0.32 nm, calculate the dislocation density.

2.6   Discuss the mechanisms responsible for dynamic strain-ageing.

2.7   There is a similarity between the hexagonal close-packed (HCP) and body-centered cubic (BCC) crystal structures.

    (a) Show which plane in the BCC structure is similar to the basal plane in the HCP structure.

    (b) Draw the arrangement of atoms in this plane and determine the stacking arrangement normal to this plane in the BCC structure. Is the stacking the same as in the HCP structure?

CHAPTER 3

Cyclic deformation in polycrystalline ductile solids

Following the discussion of cyclic deformation in ductile single crystals, we now direct attention to polycrystalline metals and alloys. Firstly, experimental observations are discussed to show how the cyclic deformation of single crystals oriented for multiple slip can be correlated with that of polycrystalline aggregates. Continuum characterization of cyclic hardening and softening of polycrystals is addressed, along with the effects of slip characteristics, alloying and precipitation. The Bauschinger effect is then introduced, wherein both microscopic and continuum viewpoints are presented to rationalize the dissimilar yield responses seen in tension and compression during load reversals. Continuum models for uniaxial and multiaxial cyclic deformation, which are of interest for engineering structures, are the focus of subsequent sections. Lower bound and upper bound theorems for elastic shakedown under cyclic loading are presented, and the origins of cyclic creep and ratchetting are then examined. The chapter concludes with the derivation of critical temperatures which signify distinct transitions in the cyclic deformation characteristics of particle-reinforced metal-matrix composites and layered materials. While this chapter confines attention to ductile crystalline solids, the cyclic deformation characteristics of brittle solids as well as semi-crystalline and noncrystalline solids are taken up in Chapters 5 and 6.

3.1 Effects of grain boundaries and multiple slip

The mechanisms of cyclic damage observed in ductile single crystals are also known to be generally applicable to the deformation of *near-surface* grains in polycrystalline metals of high purity. However, the presence of precipitates, impurities, inclusions and grain boundaries in commercial materials gives rise to fatigue deformation characteristics which may significantly deviate from those discussed for monocrystalline solids. It is pointless to include here the vast amount of empirical information gathered to date on the cyclic stress–strain behavior of a wide spectrum of engineering alloys. However, in an attempt to develop an overall perspective on fatigue damage in many broad classes of materials, we present representative experimental data and mechanistic interpretations on the effects of grain boundaries, alloying and precipitation. Also included in these descriptions are the phenomenological constitutive laws for cyclic hardening and softening of metals and alloys.

Studies of polycrystalline metals have established that PSBs can form within the bulk of the material. Winter, Pedersen & Rasmussen (1981) investigated fatigue deformation within the bulk of polycrystalline Cu, of 100–300 μm grain size, by examining TEM foils taken at various depths below the free surface. At strain amplitudes of about 10^{-4}, PSBs confined to a single slip system and of a wall structure similar to that of single crystals were identified in the interior of the polycrystalline fatigue specimen. Pohl, Mayr & Macherauch (1980) have also observed PSBs in the interior sections of fatigued polycrystalline low carbon steel. Although PSBs can traverse through low angle boundaries, large angle grain boundaries are impervious to them. When the strain amplitude is raised to values beyond 10^{-3}, labyrinth and cell structures are observed in polycrystalline copper, as in the case of single crystals (Chapter 2).

Experimental studies of polycrystalline metals have also identified the existence of regimes in the cyclic stress–strain curve where PSB formation has a significant effect (Lukáš & Kunz, 1985). Specifically, the saturation stress–strain curves for coarsegrained Cu exhibit a region of low cyclic strain hardening (somewhat analogous to the plateau for the single crystal); the occurrence of this region corresponds to a continuous increase in the volume of the material occupied by the PSBs, which carry a higher plastic strain than the matrix.

3.1.1 Monocrystalline versus polycrystalline FCC metals

It was shown in Chapter 2 that single crystals of FCC metals, oriented for single slip, exhibit a plateau in the cyclic stress–strain curve which spans a wide range of resolved plastic shear strain amplitude, γ_{pl}, Fig. 2.2. In this plateau regime, with a fixed value of the saturated peak resolved shear stress $\tau_s = \tau_s^*$, deformation localizes along persistent slip bands whose volume fraction increases with increasing γ_{pl} in order to accommodate cyclic deformation at a fixed value of τ_s^*. In the FCC crystals oriented for multiple slip, however, well-defined PSBs are not commonly found; even crystals oriented for duplex slip show only a very weak plateau (Table 2.1). Instead, dislocation arrangements which promote multiple slip and cross slip, such as the labyrinth structure in the [001] Cu crystals, are observed.

Studies of monocrystalline Cu oriented for multiple slip provide insights into the cyclic response of polycrystals. As shown later in Section 3.3, the cyclic stress–strain curve for many fine-grained FCC metals can be approximated, in a limited regime of plastic strain amplitudes, by a power-law equation of the type:

$$\frac{\Delta\sigma}{2} = k\left(\frac{\Delta\epsilon_{pl}}{2}\right)^{n_f}, \tag{3.1}$$

where $\Delta\sigma/2$ and $\Delta\epsilon_{pl}/2$ are the applied uniaxial stress and plastic-strain amplitudes, respectively, k is an experimentally determined material constant, and n_f is the cyclic strain hardening exponent. The equation can be recast into a form amenable for use with single crystals by employing the Taylor factor, M_T (= 3.06 for randomly-textured polycrystalline FCC metals):

$$\tau_s = \frac{\Delta\sigma}{2M_T} = k'(\gamma_{pl})^{n_f}, \qquad \epsilon_{pl} = \frac{\gamma_{pl}}{M_T} \tag{3.2}$$

For fine-grained polycrystalline Cu, $k' = 146\,\text{MPa}$ and $n_f = 0.205$ (Lukáš & Kunz, 1985).

Figure 3.1 shows a comparison of the cyclic stress–strain curves for monocrystalline Cu, oriented for single slip and multiple slip, and polycrystalline Cu. The solid line is the CSS curve for Cu single crystals oriented for single slip which shows a plateau at a saturation stress of 28 MPa in regime B. The symbols are from the experiments of Gong, Wang & Wang (1997) for [001] Cu single crystals which have eight symmetric slip systems.† Note the absence of the plateau in the Cu multi-slip crystal. The dashed line in Fig. 3.1 is the power-law function, Eq. 3.2, for Cu polycrystals, corrected for the Taylor factor. It can be inferred from Fig. 3.1 that the absence of a plateau in the polycrystalline FCC metals is similar to that in the multiple-slip Cu single crystal.‡

As a general trend, two factors clearly distinguish polycrystalline FCC metals from single crystals oriented for single slip:

(1) Grains in a polycrystalline ensemble have a variety of slip orientations.
(2) The incompatibility of elastic and plastic deformation between adjacent grains in a polycrystalline ensemble promotes local multiaxial loading and predominantly multiple slip.

For fine-grained polycrystalline FCC metals, these two factors lead to multiple slip deformation which resembles the cyclic response of single crystals oriented for multiple slip. For coarse-grained FCC metals, however, these two factors may not dominate cyclic response since, statistically, the majority of the grains may not have their orientations along the edges of the standard triangle of the stereographic projection. Many of the crystal orientations here may fall within the triangle. Consequently, coarse-grained polycrystalline FCC metals may exhibit cyclic deformation which nearly resembles that of single-slip oriented monocrystals, with a low strain hardening or even a mild plateau (e.g., Polák, Klesnil & Lukáš, 1974).

---

† See the worked example in Section 2.8 for the identification of the symmetric slip system for the [001] orientation. Although all eight systems would, in principle, be expected to have the same Schmid factor, small errors during the alignment, machining or mounting of the fatigue specimen can lead to a preferentially higher Schmid factor in one of the eight systems. On the basis of their slip trace observations, Gong et al. (1997) identify the $B4$ system to operate first as the primary system for which the resolved shear stress and plastic shear strain are computed.

‡ It is also of interest to note here that the work of Mughrabi & Wang (1981) showed an excellent agreement between the cyclic deformation results at low plastic strain amplitudes (regime A) for Cu single crystals (single slip orientation, where the effects of incompatibility stresses were ignored) and polycrystals by employing the Sachs orientation factor. Above regime A, they found that the effective orientation factor was larger than the Sachs factor, but smaller than the Taylor factor.

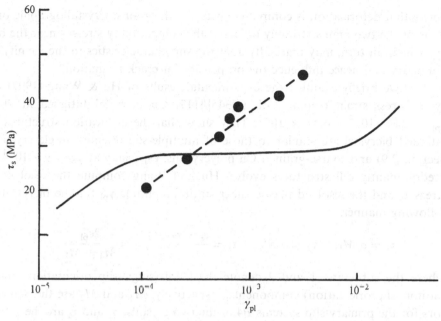

Fig. 3.1. A comparison of the cyclic stress–strain (CSS) curves of Cu single crystals (—) and polycrystals (- - -) at 20 °C. The dashed line is a plot of the stress–strain response of Cu polycrystals (from Lukáš & Kunz, 1985), corrected by the Taylor factor, M_T. The symbols (•) denote the CSS response of Cu single crystals oriented for multiple slip with the loading axis along [001]. (After Gong, Wang & Wang, 1997.)

3.1.2 Effects of texture

Differences in slip characteristics (i.e. single slip versus multiple slip deformation) within the individual grains of a polycrystalline FCC metal are also known to promote strong texture effects during cyclic loading (e.g., Mughrabi & Wang, 1981). The experiments of Llanes *et al.* (1993) and Peralta *et al.* (1995) on coarse-grained polycrystalline Cu (grain size $\sim$ 700 μm) reveal that a cyclically stronger ⟨111⟩–⟨100⟩ texture coaxial with the loading axis promotes a high cyclic hardening, whereas a 'random' texture of nearly the same grain size results in a much softer CSS curve as a result of dominant single slip. A comparison of the texture effects for the coarse-grained and finer-grained Cu ($\sim$ 700 μm and 100 μm grain size, respectively) showed that the effects of crystallographic texture were much stronger than the effects of grain size on the CSS curves.

3.2 Cyclic deformation of FCC bicrystals

When a bicrystal, whose component crystals are elastically and plastically anisotropic, is subjected to cyclic loads, stresses evolve at the grain boundary in such

a way that deformation is compatible between the grains. Crystallographic orientations of the two grains strongly influence the compatibility stresses near the boundary which, in turn, may markedly alter the slip characteristics in the vicinity of the
boundary and hence influence the propensity for crack formation.

Here we briefly examine the experimental results of Hu & Wang (1997) on the
cyclic stress–strain response of a $[\bar{3}45]/[\bar{1}17]$ Cu bicrystal fatigued at $20\,^{\circ}\mathrm{C}$ at
$\gamma_{\mathrm{pl}} = 2.8 \times 10^{-4} - 6.45 \times 10^{-3}$. They show that the dislocation structures of the
fatigued bicrystal are similar to those of multiple-slip-oriented single crystals (see
Section 2.9) and coarse-grained Cu polycrystals, especially at $\gamma_{\mathrm{pl}} > 9 \times 10^{-4}$ where
predominantly cell structures evolve. Hu and Wang compute the resolved shear
stress τ_{s} and the resolved plastic shear strain amplitude γ_{pl} for the bicrystal in the
following manner:

$$\tau_1 = \sigma_s M_1, \quad \tau_2 = \sigma_s M_2, \quad \tau_s = \frac{\tau_1 + \tau_2}{2}, \quad \gamma_{\mathrm{pl}} = \frac{2\epsilon_{\mathrm{pl}}}{M_1 + M_2}, \tag{3.3}$$

where the subscripts 1 and 2 denote the $[\bar{3}45]$ (single-slip orientation) and $[\bar{1}17]$
(double-slip orientation) components, respectively, M_1 and M_2 are the Schmid factors for the primary slip systems $B4$ of the two crystals, σ_s and τ_s are the saturation
values of the axial stress applied to the crystal and the resolved shear stress, respec-
tively, and ϵ_{pl} is the applied axial plastic strain amplitude. Figure 3.2 is a plot of τ_s vs
γ_{pl} for the $[\bar{3}45]/[\bar{1}17]$ Cu bicrystal for symmetric tension–compression loads with
plastic strain control. Note the absence of a plateau in the CSS curve. Also plotted
here are the stress–strain data for a coarse-grained Cu polycrystal with a grain size of

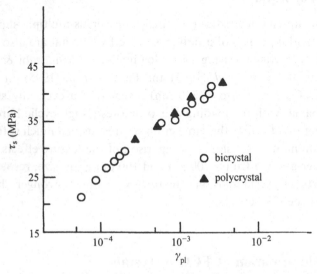

Fig. 3.2. Cyclic stress–strain curves of $[\bar{3}45]/[\bar{1}17]$ Cu bicrystal (after Hu & Wang, 1997) and
coarse-grained Cu polycrystal (after Lukáš & Kunz, 1985) appropriately modified by the Taylor
factor.

1.2 mm. For the Cu polycrystal, σ_s and ϵ_{pl} were converted into τ_s and γ_{pl} using the Taylor factor, $M_T = 3.06$ in the following manner: $\tau_s = \sigma_s/M_T$ and $\gamma_{pl} = M_T\epsilon_{pl}$.

The CSS curve of the bicrystal overlaps with that of the coarse-grained polycrystal, when modified appropriately by the Taylor factor, Fig. 3.2. (Recall that a similar overlap was demonstrated in Fig. 3.1 between a Cu single crystal oriented for multislip and a Cu polycrystal.) The foregoing results illustrate that under some conditions, it is possible to correlate the cyclic deformation response of polycrystals with the corresponding behavior seen in single crystals and bicrystals.

3.2.1 Example problem: Number of independent slip systems

Problem:

von Mises recognized that, in a deformed polycrystal, arbitrary shape changes of a crystallite are possible only if at least five independent slip systems operate. Provide a simple physical reasoning to justify this requirement.

Solution:

An arbitrary shape change is brought about by translations and rotations. There are six possible degrees of freedom, in general: three for translation and three for rotation. During plastic deformation of crystalline materials, however, the volume remains constant because deformation occurs by shear in preferential slip systems. This leaves five independent components for the strain tensor. Since shape changes must be accommodated by shear on the slip systems, arbitrary shape changes are possible only if at least five independent slip systems operate.

3.3 Cyclic hardening and softening in polycrystals

The uniaxial deformation of engineering alloys subjected to cyclic loads is usually characterized by the cyclic stress–strain (CSS) curve, similar to that discussed for single crystals. The transient phenomena typically associated with cyclic deformation are schematically illustrated in Fig. 3.3. In the case of constant amplitude, fully-reversed stress control, Fig. 3.3(a), cyclic hardening or softening of the material is reflected by a reduction or an increase, respectively, in the axial strain amplitude. Similarly, under constant amplitude, strain-controlled fatigue loading, cyclic hardening or softening of the material causes an increase or decrease, respectively, in the axial stress amplitude, Fig. 3.3(b).

In both stress-controlled and strain-controlled fatigue, the respective strain amplitude and stress amplitude reach a stable saturation value after an initial 'shakedown' period. This saturation state gives rise to stable hysteresis loops. During 'shakedown', there is a continual change in dislocation substructure until a stable config-

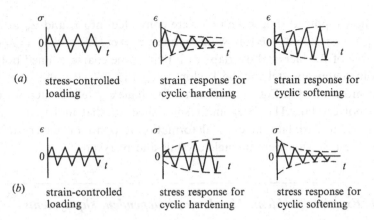

Fig. 3.3. Phenomena associated with transient effects in fatigue. σ, ϵ and t denote stress, strain and time, respectively.

uration representative of the saturated state is reached. Beyond this point, the hysteresis loop remains essentially the same cycle after cycle over the life of the test specimen. The parameters used to describe the salient features of cyclic hysteresis loops are defined in Fig. 3.4(a). The locus of the tips of stable hysteresis loops provides the cyclic stress–strain curve, Fig. 3.4(b).

Stress- and strain-controlled fatigue represent extremes of fully *unconstrained* and fully *constrained* loading conditions. In real engineering components, there is usually some structural constraint of the material at fatigue-critical sites. It thus seems appropriate to characterize fatigue response of engineering materials on the basis of data obtained under strain-controlled fatigue rather than cyclic stress-controlled conditions.

Strain-controlled tests have gained increasing use in the determination of CSS curves for engineering alloys. Three commonly used strain-controlled test methods are indicated in Fig. 3.4(c). In the constant amplitude test, the specimen is cycled within a constant plastic strain limit (until failure) to obtain a single stable hysteresis loop. Multiple test specimens are needed to determine the entire CSS curve using this method. In the multiple step method, a specimen is cycled between constant plastic strain limits until a saturation loop results. Then the plastic strain limits are incremented until another stable hysteresis loop is obtained. This process is continued until the entire CSS curve is measured from a single test specimen. In the incremental step method, the specimen is subjected repeatedly to a strain pattern comprising linearly increasing and decreasing amplitudes, from zero to a certain maximum total strain. The resulting stable hysteresis loop provides the CSS plot. In some alloys exhibiting planar slip deformation, the incremental step method provides a CSS response which is different from the other direct methods because of the variations in the development of dislocation structures.

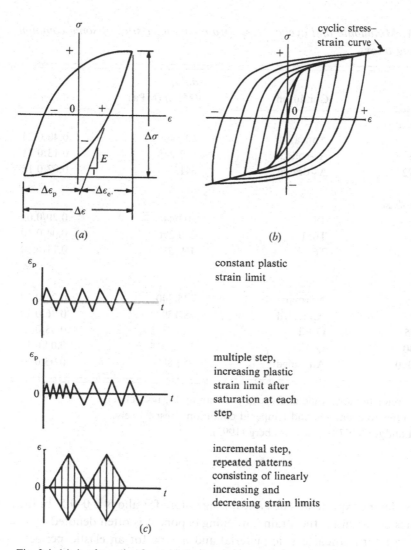

Fig. 3.4. (*a*) A schematic of a stable hysteresis loop and the nomenclature. $\Delta\epsilon_e$, $\Delta\epsilon_p$ and $\Delta\epsilon$ denote the elastic, plastic and total strain range, respectively. (*b*) Cyclic stress–strain curve drawn through the tips of stable hysteresis loops. (*c*) Procedures for obtaining cyclic stress–strain curves.

The monotonic stress–strain behavior of ductile solids under uniaxial tension is generally represented by a constitutive law, such as the Ramberg–Osgood relationship:

$$\epsilon = \frac{\sigma}{E} + \left(\frac{\sigma}{A}\right)^{1/n_m},$$

(3.4)

where E is Young's modulus, A is a constant commonly referred to as the monotonic strength coefficient, ϵ is the uniaxial strain, σ is the uniaxial stress, and n_m is the

Table 3.1. *Monotonic and cyclic stress–strain characteristics of some common engineering alloys.*

Alloy	Condition	σ_y/σ_y' (MPa)/(MPa)	n_m/n_f
Copper-base			
OFHC	Annealed	20/140	0.40/0.24
Brass 365	As-rolled	172/248	0.13/0.21
Cu–Be 172	As-drawn	641/490	0.02/0.15
Aluminum-base			
2024	T4	303/448	0.20/0.09
6061	T651	290/296	0.04/0.10
7075	T6	469/517	0.11/0.10
Iron-base			
SAE 1015	Normalized	225/240	0.26/0.22
Ferrovac E	Annealed	48/159	0.36/0.19
SAE 1045	Q+T	1365/825	0.08/0.15
AISI 4340	Q+T	1172/814	0.07/0.15
Mar M-300	Annealed	952/800	0.03/0.08

σ_y and σ_y' refer to monotonic and cyclic yield strengths, respectively.
Q and T refer to quenched and tempered conditions, respectively.
Source: Landgraf (1978) and Hertzberg (1995).

strain hardening exponent. The typical range of n_m for alloys is 0–0.5. In the applied mechanics community, the strain hardening exponent is often denoted by $n = 1/n_m$. Here $n = 1$ for a linear elastic material and $n = \infty$ for an elastic–perfectly plastic solid.

In an analogous fashion, the cyclic stress–strain response is characterized by the relationship

$$\frac{\Delta\epsilon}{2} = \frac{\Delta\sigma}{2E} + \left(\frac{\Delta\sigma}{2A'}\right)^{1/n_f}, \tag{3.5}$$

where A' is the cyclic strength coefficient and n_f is the cyclic strain hardening exponent. For most metals, n_f varies between 0.1 and 0.2 despite vast differences in their cyclic hardening and softening characteristics. Table 3.1 provides a list of the strain hardening characteristics in some common engineering alloys. As a general rule-of-thumb, well-annealed, polycrystalline metals of high purity exhibit cyclic hardening due to dislocation multiplication, as evidenced by an increase in the stress amplitude with fatigue cycles (at a fixed strain amplitude); work-hardened materials undergo strain softening under cyclic loading. The rearrangement of prestrain-induced dislocation networks due to fatigue causes cyclic softening.

3.4 Effects of alloying, cross slip and stacking fault energy

A clear demonstration of the relationship among the slip characteristics of the material, the development of dislocation structures, and the overall cyclic stress–strain behavior is found in the work of Feltner & Laird (1967a,b). They conducted an investigation of the effects of prior loading history, slip planarity, plastic strain amplitude and temperature on cyclic hardening and softening of polycrystalline Cu, Al and Cu–7.5 wt % Al. They illustrated that pure FCC metals harden or soften under fatigue loads to a characteristic steady-state saturation flow stress after an initial transient period. For wavy slip materials, with a high stacking fault energy and a strong propensity for cross slip, the steady-state cyclic flow stress is independent of load history and of the degree of prior cold work at large plastic strain amplitudes (see Fig. 3.5a); here only the plastic strain amplitude and temperature influence the flow characteristics in fatigue.† Pure FCC metals fatigued at high strain amplitudes exhibit a cell structure at steady state, regardless of the dislocation structure which existed before cyclic straining. The cell size, which is independent of the history of the material, increases with decreasing plastic strain amplitude and increasing temperature.

In contrast to this behavior, when a planar slip mode is introduced by the addition of 7.5 wt % Al in Cu (which reduces the stacking fault energy), the cyclic stress–strain curve becomes sensitive to both prior deformation and thermal history, Fig. 3.5(b). Inhibition of cross slip in this case precludes the attainment of a common steady-state dislocation structure in fatigue from an initially hard (cold-worked) or soft (annealed) condition. The low stacking fault energy Cu–7.5 Al alloy does not develop a cell structure during cyclic saturation, but exhibits dissociated dislocations arranged in planar bands. Furthermore, the saturation dislocation structures in the cold-worked conditions of the alloy were significantly different, with the latter microstructure consisting of fewer bands and a greater density of tangled interband dislocations that are apparently inherited from prior cold work.

Experimental studies of substructure developments during fatigue in different wavy slip metals have documented a one-to-one correspondence between the saturation stress and the deformation-induced cell size. Based on their own investigations

---

† It is now generally recognized that the ease of cross slip plays a decisive role in determining the slip mode. Stacking fault energy, however, is not the only parameter influencing cross slip. For example, Gerold & Karnthaler (1989) demonstrate that short range order can also strongly influence cross slip.

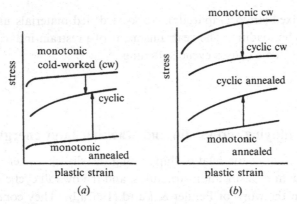

Fig. 3.5. (*a*) Schematic showing that the cyclic stress–strain curve is independent of prior deformation history in wavy slip materials. (*b*) History-dependent behavior for planar slip materials. (After Feltner & Laird, 1967a.)

and from a literature review, Plumtree & Pawlus (1988) developed the following empirical relationship for the dependence of saturation stress on cell size:

$$\frac{\sigma_s - \sigma_0}{E} = B\frac{b}{\lambda_s}. \tag{3.6}$$

In Eq. 3.6, σ_s is the saturation stress, σ_0 is the back stress, E is Young's modulus, b is the magnitude of the Burgers vector, λ_s is the linear intercept cell size and B is a material constant which increases in proportion to the stacking fault energy (SFE). The works of Pratt (1967), Feltner & Laird (1967b), Abdel-Raouf & Plumtree (1971), Saga, Hayashi & Nishio (1977) and Kayali & Plumtree (1982) collectively show that the constant B is 7.8 for Al (SFE $= 200 \times 10^{-3}$ J m^{-2}) and 3.6 for Cu (SFE $= 40 \times 10^{-3}$ J m^{-2}). Equation 3.6 does not appear to be a function of the type of deformation in that it describes the steady-state flow stress–cell size data for both monotonic and cyclic loading conditions.

For aluminum fatigued at a total strain range of 1% at room temperature, the cell size and the misorientation between neighboring cells remains constant from the onset of saturation until final failure despite noticeable changes in the cell morphology. One may then infer that the dislocation density remains relatively constant upon the attainment of saturation. A survey of relevant literature shows that the operative mechanism during cyclic saturation is the irreversible bowing of dislocation segments from the cell walls. After being emitted from the cell walls, the dislocation segments traverse the cell and enter the adjacent walls where they are annihilated by dynamic recovery. This equilibrium between work hardening (dislocation generation) and dynamic recovery (dislocation annihilation) is believed to result in a constant maximum stress and dislocation density during the saturation stage.

3.5 Effects of precipitation

Cyclic hardening and softening in precipitation-hardened alloys is promoted by a mechanism in which the precipitate geometry and distribution as well as dislocation–particle interactions can be altered by cyclic straining. Existing experimental evidence reveals that initial cyclic hardening occurs in these materials due to an increase in dislocation density and due to dislocation–precipitate interactions (e.g., Brett & Doherty, 1978). Subsequently, cyclic softening is highly favored if the precipitates in the age-hardened alloy are easily *sheared* by dislocations, i.e. if they are fine, closely spaced and coherent with the matrix. A particularly interesting feature of deformation in these alloys is that the formation of PSBs causes cyclic softening without reaching a saturation stage. Various mechanisms have been proposed to rationalize cyclic softening in precipitation-hardened alloys:

(*a*) *Reversion or re-solution* by which the metastable strengthening precipitates completely dissolve in the matrix after being cut by dislocations to a size smaller than the critical size for particle nucleation (Broom, Mazza & Whittaker, 1957).

(*b*) *Disordering of ordered precipitates* due to the motion of single dislocations through them which leads to a loss of strength (Calabrese & Laird, 1974).

(*c*) *Microstructural inhomogeneities* (due to pre-straining or quenching prior to ageing) in the form of slip bands, that are depleted of strenghening metastable precipitates and are replenished with stable particles (Laird & Thomas, 1967).

(*d*) *Over-ageing*, which leads to the substitution of metastable precipitates in the matrix by coarsely distributed, stable ones (Broom, Mazza & Whittaker, 1957).

(*e*) *Dissolution due to Ostwald ripening* of unsheared precipitates, adjacent to the sheared ones on the slip lines at the expense of the smaller precipitates in the slip band (Sargent & Purdy, 1974). (Ostwald ripening refers to the process by which the interfacial free energy between the precipitate and the matrix causes precipitate coarsening upon further annealing, even after precipitation from supersaturated solid solution is complete. The larger particles grow at the expense of the smaller ones, thereby releasing excess surface energy and causing the microstructure to coarsen; see, for example, Martin, 1980.) The difference between this and the reversion mechanism is that the critical radius for Ostwald ripening is much larger than the value for precipitate growth from the supersaturated solid solution.

3.6 The Bauschinger effect

The Bauschinger effect refers to the experimental result that, after a certain amount of forward plastic deformation in tension or compression, the material yields at a lower stress when the direction of loading is reversed than for continued forward deformation (Bauschinger, 1886). A knowledge of the mechanisms underlying the Bauschinger effect is essential for the development of the constitutive models for complex cyclic deformation, for a fundamental understanding of the work hardening phenomena, and for rationalizing such fatigue effects as mean stress relaxation and cyclic creep (Chapter 8). For example, many commercial aluminum alloys containing nonshearable strengthening precipitates (such as the peak-aged and over-aged 7075 alloys used in aircraft applications) are stretched prior to temper treatments to relieve thermal residual stresses. Since many of these alloys are known to exhibit Bauschinger effects, low flow stresses may result under service conditions if the material is loaded in a direction opposite to the stretching direction. It is also known that the Bauschinger effect in precipitation-hardened commercial alloys can persist even after the cyclic hysteresis loops are stabilized. On a more fundamental level, the Bauschinger effect can be used to identify the contributions to strain hardening from different kinds of dislocation mechanisms. The study of the Bauschinger effect is, therefore, commonly regarded as a 'litmus test' for the validity of strengthening theories in the sense that any complete hardening theory must be capable of quantitatively accounting for the Bauschinger effect.

3.6.1 Terminology

One of the common methods of quantifying the Bauschinger effect involves the definition of the reverse strain. This reverse strain is the magnitude of additional strain after load reversal which makes the reverse yield stress equal in magnitude to the maximum flow stress attained in the forward deformation. However, a realization of the differences between forward and reverse flow stresses at any strain value can be achieved by recourse to the construction schematically depicted in Fig. 3.6. In Fig. 3.6(*a*), *ABC* represents the forward deformation in uniaxial tension, with *C* being the point of unloading. *CD* is the tensile unloading segment and *DEF* is the reverse (compression) loading segment. The magnitudes of the stress and the accumulated strain are replotted in Fig. 3.6(*b*) irrespective of the direction of loading.

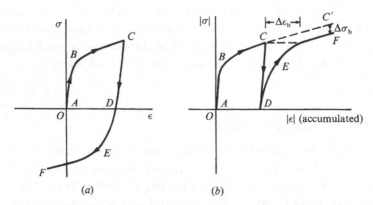

(a)   (b)

Fig. 3.6. (*a*) Schematic of the stress–strain curve for fully reversed loading. (*b*) Only the magnitudes of the stress and the accumulated strain are replotted to illustrate the Bauschinger effect.

Here, *ABCD* represents the tensile segment and *DEF* is the compression loading portion. If the forward hardening segment is extrapolated (until point C'), the instantaneous difference between the flow stress levels $\Delta\sigma_b$ in the (nearly parallel) segments CC' and *EF* provides an indication of the Bauschinger effect in terms of stress and characterizes the extent of 'permanent softening'. Similarly, the difference in strain values $\Delta\epsilon_b$ between the forward and reversed deformation at the maximum forward tensile stress amplitude (point *C*) characterizes the Bauschinger strain.

3.6.2 Mechanisms

The origins of the Bauschinger effect are related to the changes in dislocation substructure induced by reversed loading and in the changes in the internal stress systems. In polycrystalline metals where dislocation walls and subgrain boundaries form during forward straining, the dissolution of cell walls or sub-boundaries upon stress reversals is considered a contributing factor to the Bauschinger effect (Hasegawa, Yakou & Kocks, 1986). Furthermore, long-range internal stresses induced by strain incompatibility between the PSB walls and channels can lead to easier reverse flow in materials which form well-defined PSBs.

For particle-hardened alloys, the mechanistic basis for the Bauschinger effect is often provided in terms of the interaction of dislocations with the strengthening particles (Orowan, 1959). Age-hardened alloys can be broadly classified into two groups: (i) those containing precipitates which are coherent with the matrix and which can be sheared by dislocations, and (ii) those containing larger semi-coherent or incoherent particles which are not penetrable by dislocations. The two types of particles can give rise to very different hardening response in monotonic and cyclic loading. Furthermore, the Bauschinger effects seen in the two cases are also distinctly different.

When the matrix contains permeable particles, mobile dislocations cut through these barriers with essentially no dislocation pile-up. If the material is subjected to a compressive stress following a tensile stress, the impingement of the shearable barriers with dislocations once again leads to a reverse flow stress comparable in magnitude to that seen in forward deformation due to the paucity of dislocation pile-up. Consequently, there is little contribution to the flow stress from internal stresses (Wilson, 1965; Stoltz & Pelloux, 1976).

In the case of alloys with coarse, incoherent and impermeable particles, the monotonic stress–strain curve exhibits a large hardening rate. For these alloys, a simple model to rationalize the Bauschinger effect is often formulated in terms of the following three factors which influence the forward yield stress: (i) a contribution, σ_0, to strength from solid solution hardening and the stress to bow out the first set of dislocations past the obstacles, (ii) forest hardening, σ_d, due to the interactions of mobile dislocations with forest dislocations, and (iii) a mean internal or back stress,

σ_M^F, which is exerted on the matrix by the particles. The total flow stress during forward deformation is

$$\sigma_f = \sigma_0 + \sigma_d + \sigma_M^F. \tag{3.7}$$

At reverse strains of the order of the forward strain, the deformation must be capable of overcoming σ_0 and σ_d. However, σ_M^F now assists reverse deformation, rather than opposing it and hence the total reverse flow stress becomes

$$\sigma_r = \sigma_0 + \sigma_d - \sigma_M^F. \tag{3.8}$$

From Eqs. 3.7 and 3.8, $\Delta\sigma_b$ in Fig. 3.6(*b*) becomes

$$\Delta\sigma_b = \sigma_f - \sigma_r = 2\sigma_M^F. \tag{3.9}$$

The back stress which arises from the internal stress system in the material is very localized in the matrix, and it is believed to be a consequence of the inhomogeneity of plastic deformation on a microscopic scale. The term $\Delta\sigma_b$ is sometimes referred to as 'permanent softening'.

A large Bauschinger effect has been documented in Al–4Cu alloys containing nonshearable θ' (CuAl$_2$) precipitates, Al–Cu–Mg alloys containing S' (Al$_2$CuMg) precipitates, and Al–Cu–Zn–Mg alloys containing η (MgZn$_2$) precipitates (Wilson, 1965; Abel & Ham, 1966; Stoltz & Pelloux, 1976; Wilson & Bate, 1986). Using X-ray diffraction measurements of directional lattice strains in two-phase cubic alloys, Wilson (1965) demonstrated that reverse loading destroys the internal stresses, reducing them to zero at some strain value. At this point, the stress difference between the forward extrapolated and the knee portion of the reverse stress–strain curves, Fig. 3.6(*b*), was equal to $1.9\sigma_M^F$. Here the mean stress in the matrix, σ_M^F, was estimated using X-rays after unidirectional plastic deformation. Note that this experimental correlation is quantitatively consistent with the prediction of Eq. 3.8.

Some of the most comprehensive studies of the Bauschinger effect have been conducted on dispersion-strengthened metals. In their study of plastic deformation in Cu–SiO$_2$, Brown & Stobbs (1971a,b) and Atkinson, Brown & Stobbs (1974) related the internal stresses to the formation of dislocation shear (Orowan) loops around hard particles on many slip planes. In the unrelaxed state typical of low temperatures and low strains, there are two contributions to strengthening: (i) A homogeneous mean stress in the matrix is produced by the Orowan loops left around the particles. This stress state is amenable to a transformation strain analysis because the Orowan loops may be regarded equivalent to the transformation strains in Eshelby's classical inclusion problem (Eshelby, 1957). This mean internal stress, σ_M^F, in the matrix opposes continued forward deformation during tensile loading. (ii) The Orowan loops also give rise to an additional inhomogeneous stress locally at the particle because the presence of the loops around the particles repels successive dislocations. This increase in internal stress over and above the initial Orowan stress has been termed the 'source-shortening stress' by Atkinson, Brown & Stobbs (1974). The occurrence of a Bauschinger effect upon load reversal is then viewed as a consequence of plastic relaxation which can arise due to (a) the removal of Orowan loops by the formation of secondary dislocations and prismatic loops of primary Burgers vectors (Atkinson, Brown & Stobbs 1974), or (b) the shrinkage of Orowan loops by climb *via* pipe diffusion (Gould, Hirsch & Humphreys, 1974). These mechanisms have also been extended to rationalize the Bauschinger effect in metal-matrix composites reinforced with hard particles and fibers (e.g., Brown & Clarke, 1977; Lilholt, 1977).

3.7 Shakedown

Ductile metals are often subjected to cyclic loads in such a manner that the early fatigue cycles lead to the build-up of residual stresses. Repeated contact loading in ball bearings and railway rails, for example, commonly engenders plastic yielding. A consequence of this plastic flow is the generation of residual stresses, which can possibly be of such a magnitude that a steady-state is attained after some load reversals wherein a closed cycle of entirely elastic reversed deformation is promoted. For this situation, there is no net accumulation of plastic strain in subsequent cycles and the system is said to have undergone *shakedown*. In other words, the residual stresses generated during the early load reversals alleviate the applied loads by inhibiting further plastic deformation in subsequent cycles, and induce a state of reversible elastic response. Note that during shakedown, the maximum applied load alone is of a magnitude which violates the yield condition.

The limiting value of the applied load below which no continued accumulation of plastic strain is possible during cyclic loading is known as the *shakedown limit*. If the shakedown limit is exceeded, plastic strains continue to accumulate during each cycle and this phenomenon is commonly known as *ratchetting*, *cyclic creep* or *incremental collapse*.

The conditions governing the occurrence of shakedown can be formulated in terms of the so-called shakedown theorems, for elastic–perfectly plastic solids:

- **Statical, lower-bound theorem** (Melan, 1938): Shakedown occurs whenever a system of residual stresses, that satisfies equilibrium requirements, acts in conjunction with applied loads in such a manner that the yield criterion is not violated.

- **Kinematical, upper-bound theorem** (Koiter, 1956): If any kinematically admissible plastic strain cycle could be identified in which the internal plastic work is exceeded by the work done by the applied loads, then shakedown will not take place. Under these conditions, cyclic plasticity and incremental collapse are possible.

Elastic shakedown occurs when the development of residual stresses results in a steady state in which cyclic deformation is purely elastic. *Plastic shakedown*, on the other hand, refers to a state of deformation where a closed cycle of alternating plasticity occurs without any accumulation of plastic strains, ratchetting or incremental collapse. Examples of both types of shakedown are discussed in Chapter 13 in the context of rolling contact fatigue. The shakedown limits for the thermal fatigue of metal-matrix composites and for different types of contact fatigue are considered in Section 3.10.2 and Chapter 13.

3.8 Continuum models for uniaxial and multiaxial fatigue

The two strain hardening rules which are widely used in metal plasticity are those involving isotropic hardening and kinematic hardening. If the yield surface (which represents the yield condition in stress space) expands uniformly during plastic flow with no shape change or translation, isotropic hardening occurs. On the other hand, kinematic hardening takes place when the yield surface does not change its size or shape, but simply translates in stress space in the direction of the outward normal. These hardening models can be written in terms of the functions F and h:

$$F(\sigma_{ij} - \xi_{ij}) - h(\lambda_h) = 0, \tag{3.10}$$

where σ_{ij} are the components of the symmetric stress tensor, ξ_{ij} denote the translation of the initial yield surface, λ_h is a scalar function of the plastic strain and $h(\lambda_h)$ is a function quantifying the expansion of the yield surface. When $\xi_{ij} = 0$, isotropic hardening occurs and when $h(\lambda_h)$ is a constant, kinematic hardening is induced.

The hysteresis loop for an isotropically hardening solid is schematically sketched in Fig. 3.7(a). Here OAB represents the tensile loading segment with strain hardening. When the load is reversed, yielding in compression commences at C, with the stress at C having the same magnitude as at B. Continued compression loading causes further hardening. If the load is reversed at D, subsequent tensile yielding occurs at E when the stress reaches the same magnitude as at D. In descriptions of isotropic hardening, the relationship between the increase in flow strength and equivalent strain is independent of stress state. The Bauschinger effect is non-existent for an ideally isotropically hardening material.[†]

The solid lines in Fig. 3.7(b) schematically represent the uniaxial stress–strain curve for an elastic–plastic solid in which compressive yielding begins at point C as a consequence of the Bauschinger effect. The dashed lines in this figure denote the constitutive response derived from a kinematic hardening model (e.g., Prager, 1956; Ziegler, 1959). If a linear hardening assumption is used (i.e. if ξ_{ij} in Eq. 3.10 is determined from the law $d\xi_{ij} = c\,d\epsilon_{ij}^{p}$, where $d\epsilon_{ij}^{p}$ are the increments of the plastic strain and c is a constant), the plastic portion of tensile deformation in Fig. 3.7(b) is

---

[†] If the ductile solid is reinforced with a brittle phase, the overall Bauschinger effect for the composite can be rationalized *even when the plastic deformation of the matrix is described by an isotropic hardening rule*. The Bauschinger effect in this case arises from the constrained flow of the matrix material between the brittle particles (e.g., Llorca, Needleman & Suresh, 1990).

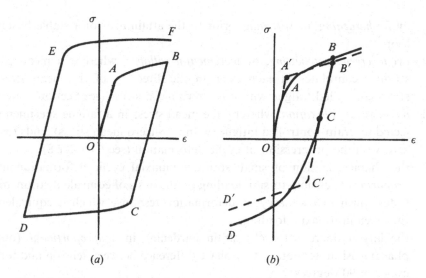

Fig. 3.7. (*a*) Shapes of hysteresis loops associated with (*a*) isotropic hardening and (*b*) kinematic hardening.

approximated by the straight line $A'B'$. Upon load reversal at B', the behavior follows elastic unloading line $B'C'$. Subsequent plastic compression is represented by the dashed line $C'D'$ which is parallel to $A'B'$. If c is assumed to be a function of the stress invariants, the slope of the stress–strain curve at C will depend on the value of the stress and hence on the location of C. Note that the kinematic hardening model does predict the existence of the Bauschinger effect. This theory has been implemented into many finite element codes.

Consider now the case of tension–compression loading over many cycles. The kinematic hardening rule predicts that a steady state of alternating plastic flow will set in after the first cycle of loading. On the other hand, the isotropic hardening model suggests that the specimen will 'shake down' to an elastic state. However, as seen in Figs. 3.3 and 3.4, a steady state of plastic strain amplitude is reached only after an initial transient behavior. Thus, it is noted that simple isotropic and kinematic hardening rules may not provide an adequate description of plasticity, when a material is subjected to periodic unloading and reloading along a different stress path (e.g., Mróz, 1967).

In this section, a brief summary of some prominent models for cyclic plasticity is presented, and their strengths and limitations are pinpointed. More comprehensive reviews of these formulations are available in the literature (Drucker & Palgen, 1981; Dafalias, 1984; Chaboche, 1986). Any complete model for cyclic plasticity must be capable of rationalizing the following fatigue characteristics:

(1) the *Bauschinger effect*,
(2) elastic *shakedown*,

(3) *cyclic hardening* or *softening* prior to the attainment of a stable hysteresis loop,

(4) *cyclic creep*, *ratchetting*, or *incremental collapse*, whereby a mean plastic strain accumulates in each cycle in the direction of the mean stress in stress-controlled fatigue with a nonzero mean stress (see Section 3.9),

(5) *mean stress relaxation*, whereby the mean stress in a fatigue specimen subjected to strain-controlled fatigue (with a nonzero mean stress) tends toward zero with the progression of cyclic deformation (see Chapter 8),

(6) the characterization of small strain, multiaxial cyclic deformation under *proportional* tension–torsion loading on the basis of equivalent strain amplitudes which provide a cyclic deformation response which is equivalent to that seen in uniaxial loading,

(7) the largest degree of cyclic strain hardening in *nonproportional* (out-of-phase) tension–torsion, with a phase difference between tension and torsion cycles of 90 degrees.

Many models for cyclic plasticity have been proposed over the years which rationalize the foregoing cyclic phenomena with varying degrees of success. In addition to the Prager–Ziegler kinematic hardening model described earlier in this section, the available approaches to modeling cyclic plasticity can be classified into the following groups: (i) the parallel sub-element model, (ii) field of work hardening moduli, (iii) two-surface models, and (iv) other developments involving combined isotropic–kinematic hardening models or internal variable concepts. The following sections describe the features of each of these approaches.

3.8.1 Parallel sub-element model

In the sub-element model of Masing (1926), Fig. 3.8(*a*), the solid is viewed as being composed of a set of n elastic–perfectly plastic elements of the same dimensions, but with different yield points. These elements are arranged in parallel and are symmetric in tension and compression. Consider the following sequence of events:

(1) When the material is loaded in tension, the weakest element yields first at a stress σ_1.

(2) If loading is continued further, the second weakest element yields at a tensile stress σ_2.

(3) The total load on the material is $\sigma_2 \cdot A$ (where A is the total cross-sectional area), whereas the total load on element 1 is $\sigma_1 A_1$, where A_1 is the area of element 1. Upon unloading, the residual stress on element 1 equals $-(\sigma_2 - \sigma_1)$ while the residual stress on each of the other $(n-1)$ elements (which are all elastic at this point) is $(\sigma_2 - \sigma_1)/(n-1)$.

(4) If forward loading were to be continued beyond σ_2 to a far-field tensile stress, σ_3, the residual stress on element 1 would become $-(\sigma_3 - \sigma_1)$ and

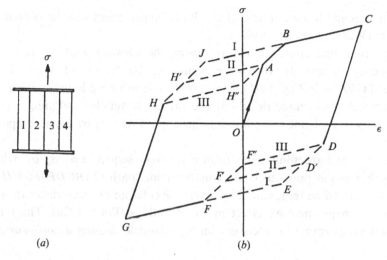

Fig. 3.8. (a) Masing's parallel sub-element model. (b) Three types of kinematic hardening behavior which can be extracted from the sub-element model. (After Asaro, 1975.)

on element 2, $-(\sigma_3 - \sigma_2)$ and so forth. For this case, reverse plastic straining would begin at $-\sigma_R$ where $|\sigma_R| = \sigma_1 + \sigma_2 - \sigma_3$.

(5)   If the forward tensile loading curve has the form $\sigma = f(\epsilon)$, then the reverse loading curve takes the form $\sigma = 2f(\frac{1}{2}\epsilon)$.

On the basis of the Masing sub-element approach, Asaro (1975) presented micro-mechanical arguments for three types of kinematic hardening. The premise here is that, at fixed values of strain rate $\dot{\epsilon}$ and temperature T, the applied stress required to achieve a certain mechanical state is a function of the internal structure such that

$$\sigma = \sigma(\alpha_1, \alpha_2, \alpha_3, \ldots). \tag{3.11}$$

It is assumed that the internal structures form in the indicated sequence $\alpha_1, \alpha_2, \alpha_3$ during forward loading and cause the yield surface to translate by the increment $d\xi_{ij}(\alpha_1, \alpha_2, \alpha_3, \ldots)$. The order in which the recovery of individual events α_i occurs during reverse loading determines the type of kinematic hardening.

If $\alpha_1, \alpha_2, \alpha_3$, etc. (which are the microstructural variables determining the flow strength) are viewed as being the plastic strains in the various sub-elements, then one has the situation where the recovery has the sequence $2\alpha_1, 2\alpha_2, 2\alpha_3$, etc. In other words, element 1 deforms plastically twice as much in compression before element 2 yields as it did originally in tension. It is also implied here that $(\partial\sigma/\partial\alpha_1)$ is not a function of α_2, α_3, etc. and likewise for $(\partial\sigma/\partial\alpha_2)$, $(\partial\sigma/\partial\alpha_3)$, etc. This sequence, when represented by a stress–strain plot, results in the type I behavior schematically shown as loop $OABCDEGHJB$ in Fig. 3.8(b). The type I kinematic hardening model ratio-nalizes the Bauschinger effect in a number of alloys with shearable precipitates. In these materials, dislocation cells are envisioned as the elastic–plastic elements which generate the internal stress systems. The type I sub-layer model has also been found to provide a reasonable description of the Bauschinger effect in spherodized plain carbon steels (Wilson & Bate, 1986).

If the structural operations occurring during the forward loading are reversed upon unloading in exactly the same sequence, hardening of type II, loop $OABCDD'F'FGHH'A$ in Fig. 3.8(b), is obtained. The resulting loops contain inflection points which are symmetric about the origin. Reversible twinning could be characterized by this scheme if the initial twins are not relaxed by the accompanying plastic flow.

The third type of hardening arises from a recovery sequence $\alpha_3, \alpha_2, \alpha_1$, which is opposite to that in the previous case. This behavior, path $OABCDF''FGHH''A$ in Fig. 3.8(b), appears to be represented by reversible collapse of linear dislocation pile-ups and by the shape memory effect in Cu–Zn alloys (Pops, 1970). The type III scheme manifests as reversible recovery on a microscale causing a *nonlinear elastic response*.

The Bauschinger effect in aluminum alloys containing nonshearable precipitates, dispersion-hardened copper crystals, and oxide-dispersed superalloys can be rationalized on the basis of type III kinematic hardening behavior. Typical characteristics of materials which exhibit type II or type III behavior involve: (i) a large fraction of elastic second phase, (ii) local dislocation reversibility, and (iii) stabilization of the dislocation array around the precipitates against plastic relaxation. Extensions of the Masing-type multicomponent models to thermo-mechanical fatigue have also been explored by Maier & Christ (1996).

3.8.2 Field of work hardening moduli

Mróz (1967, 1969) introduced the concept of a 'field of work hardening moduli' to generalize the known rules of kinematic hardening to cyclic loading. This field involves a configuration of surfaces of constant work hardening moduli in stress space. A similar concept was also proposed independently by Iwan (1967), who rationalized the Bauschinger effect and multiaxial cyclic deformation using a collection of yield surfaces. Although each one of the yield surfaces in such a model obeys a linear work hardening rule, the combined effect of the nested yield surfaces gives rise to a nonlinear work hardening law. In this section, we discuss in some detail the anisotropic hardening model due to Mróz.

As a starting point to this discussion, it is instructive to define the linear and nonlinear moduli in uniaxial deformation. The total strain increment $d\epsilon$ is composed of the elastic part $d\epsilon^e$ and the plastic part $d\epsilon^p$:

$$d\epsilon = d\epsilon^e + d\epsilon^p. \tag{3.12}$$

If $d\sigma$ is the stress increment, $d\epsilon = d\sigma/E_t$, $d\epsilon^e = d\sigma/E^e$, and $d\epsilon^p = d\sigma/E^p$, where E_t, E^p, and E^e are the tangent, plastic and elastic moduli, respectively. From Eq. 3.12, it is readily seen that

$$\frac{1}{E_t} = \frac{1}{E^e} + \frac{1}{E^p}.$$ (3.13)

During elastic deformation, $E_t = E^e$. As plastic deformation is initiated, there is a change in E^p from an initial value of infinity (in the vicinity of the elastic limit) to some finite value. The tangent modulus, E_t, can be used as a direct measure of the plastic modulus, E^p, during plastic flow because E^e is constant.

Consider a fatigue specimen of an initially isotropic material which is first subjected to tensile loading, as indicated by the segment $OABCDE$ in Fig. 3.9(a). Once plastic deformation begins at point A, the tangent modulus decreases. Let the stress–strain curve be approximated by n linear segments of constant tangent moduli $E_1, E_2, \ldots, E_n$, Fig. 3.9(a). In stress space, this approximation is represented by n hypersurfaces $S_0, S_1, \ldots, S_n$, where S_0 is the initial yield surface and $S_1, S_2, \ldots, S_n$ are the regions of constant work-hardening.

For the initially isotropic material, the yield surfaces $S_0, S_1, \ldots, S_n$ are similar and concentric, and centered on the origin O. In two-dimensional stress space, these surfaces form a family of circles, Fig. 3.9(b). Consider proportional loading, where different components of the stress tensor increase in constant proportion to one another. Assume first that all yield surfaces translate in stress space without changing their shape or orientation. When the applied stress reaches the elastic limit, point A in Fig. 3.9(a), the initial yield surface S_0 starts to move along the σ_2 axis until it touches the circle S_1 at B. During this period, all the other surfaces remain fixed. Between A and B, the plastic strain is defined by the tangent modulus E_1. Similarly, as the stress is increased from B to C, the surfaces S_0 and S_1 translate together until the point C is reached where S_0 and S_1 contact the surface S_2 which until now had been stationary. The tangent modulus between B and C has a constant value E_2. Beyond C, the surfaces S_0, S_1 and S_2 translate together as the stress is raised until point D is reached. Figure 3.9(b) shows the situation where the applied stress corresponds to the point E.

Now let the specimen be unloaded from the stress point E and reversed into compression. When point G is reached, reverse plastic flow occurs; the surface S_0 starts to move downward until it touches surface S_1. This corresponds to the stress point H. The difference between stresses at H and G is twice the difference between stresses B and A. Similarly, loading between the stress points HI and IJ corresponds to the previously described steps BC and CD. Note that the curve of inverse loading is obtained by symmetry with respect to the curve $OABCDE$. The curve of reverse loading $EGHIJK$ is uniquely defined by the primary loading curve $OABCDE$. If $\sigma = f(\epsilon)$ is the equation of the primary curve $OABCDE$ and if a new coordinate system $(\bar{\sigma}, \bar{\epsilon})$ is chosen with its origin at E and rotated 180°, the curve $EGHIJK$ is defined by

$$\bar{\sigma} = 2f(1/2\,\bar{\epsilon}).$$ (3.14)

This result is identical to that predicted by the Masing model. Equation 3.14 also indicates that $\bar{\sigma}_K = 2\sigma_E$; therefore, $\sigma_K = -\sigma_E$. From K, subsequent loading KNE is

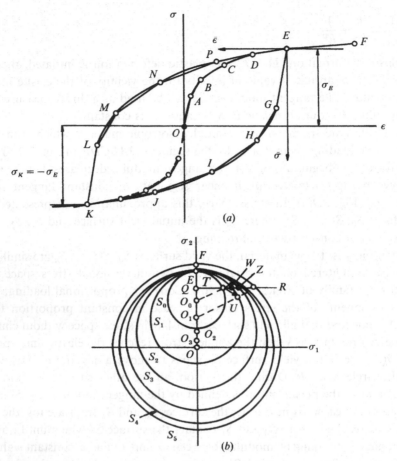

Fig. 3.9. (*a*) Representation of the stress–strain curve by regions of constant tangent moduli. (*b*) Representation in stress space. (After Mróz, 1967.)

identical to that of *EHK*. Upon reaching point *E*, deformation follows a steady cycle *EHKNE*.

Mróz also extended the above model to situations involving nonproportional multiaxial loading. As an example, consider the case where the specimen is unloaded partially to a stress point *Q* inside S_0 from the stress point *E*, Fig. 3.9(*b*), whereupon the stress point moves along *QR* contacting the surface S_0 at *T*. The surface S_0 must move such that the point *T* makes contact with the surface S_1 without intersecting it. (Therefore, translation will not occur along the normal at *T*, i.e., along the extrapolation of O_0T.) Thus *T* will move to a point *U* on S_1. As the stress point moves along *TR*, the surface S_0 will touch S_1 at *W* and subsequently they move together to touch S_2 at *Z* and so on. When the stress path meets the surface S_4, the work hardening modulus on the loading path becomes equal to that at *E*.

Mróz's geometrical model can be mathematically stated using a typical situation sketched in Fig. 3.10. S_l and S_{l+1} are two yield surfaces whose centers are at O_l and

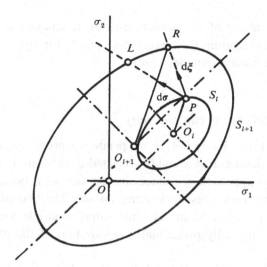

Fig. 3.10. Two yield surfaces S_l and S_{l+1}; the translation of surface S_l along PR is defined by its relative position with respect to S_{l+1}. (After Mróz, 1967.)

O_{l+1}, respectively. These centers are defined by the position vectors $\xi_{ij}^{(l)}$ and $\xi_{ij}^{(l+1)}$, respectively, from the origin O. S_l and S_{l+1}, respectively, are defined by the following equations:

$$f(\sigma_{ij} - \xi_{ij}^{(l)}) - \{\sigma_0^{(l)}\}^{n_1} = 0, \quad f(\sigma_{ij} - \xi_{ij}^{(l+1)}) - \{\sigma_0^{(l+1)}\}^{n_1} = 0, \tag{3.15}$$

where f is a homogeneous function of order n_1 of its arguments and σ_0^l and $\sigma_0^{(l+1)}$ are constants. If P lies on the surface S_l, the instantaneous translation of S_l will occur along PR, where R is a point on S_{l+1} corresponding to the same direction of outward normal. The position of R is determined by drawing from O_{l+1} a vector $O_{l+1}R$ parallel to $O_l P$. From the homogeneity of f in Eq. 3.15 and by denoting the stresses at P and R by $\sigma_{ij}^{(l)}$ and $\sigma_{ij}^{(l+1)}$, respectively, one obtains

$$\sigma_{ij}^{(l+1)} - \xi_{ij}^{(l+1)} = \frac{\sigma_0^{(l+1)}}{\sigma_0^{(l)}}\left(\sigma_{ij}^{(l)} - \xi_{ij}^{(l)}\right). \tag{3.16}$$

The translation of S_l is given by

$$d\xi_{ij}^{(l)} = \frac{dq}{\sigma_0^{(l)}}\left[\left(\sigma_0^{(l+1)} - \sigma_0^{(l)}\right)\sigma_{ij}^{(l)} - \left(\xi_{ij}^{(l)}\sigma_0^{(l+1)} - \xi_{ij}^{(l+1)}\sigma_0^{(l)}\right)\right]. \tag{3.17}$$

The parameter dq is obtained from the continuity condition that the stress point remain on the yield surface,

$$\left(d\xi_{ij}^{(l)} - d\sigma_{ij}\right)\frac{\partial f}{\partial \sigma_{ij}} = 0, \quad dq = \frac{(\partial f/\partial \sigma_{ij})d\sigma_{ij}}{(\partial f/\partial \sigma_{mn})(\sigma_{mn}^{(l+1)} - \sigma_{mn}^{(l)})}. \tag{3.18}$$

Mróz's model can be extended to situations where, in addition to translation, the yield surfaces expand or contract. An important difference between the Mróz model and the Prager–Ziegler model is in the direction in which the yield surface translates.

The Mróz formulation captures many of the salient features of uniaxial fatigue deformation described earlier in this section. However, it does poorly for asymmetric cyclic deformation (White, Bronkhorst & Anand, 1990).

3.8.3 *Two-surface models for cyclic plasticity*

Dafalias & Popov (1975) and Krieg (1975) independently proposed continuum models for nonlinearly hardening ductile solids subjected to random uniaxial and multiaxial cyclic loads. Their approach introduces the concept of a bounding surface in stress space which always encloses the loading surface. The two-surface model provides a practical simplification of the aforementioned multiple loading surface approach of Mróz by analytically prescribing the variation of the plastic modulus.

Consider a ductile solid which is subjected to a fully reversed stress cycle; the stress–strain curve for this loading history is shown in Fig. 3.11. The material deforms elastically along OA; path ABC denotes plastic deformation. At point C, the material is elastically unloaded along CD. This is followed by plastic loading in compression along DEF. FG denotes the path for elastic unloading at F and $GHBC$ represents subsequent plastic reloading. Beyond point B, the line HX' coincides with the prior loading line BC. Corresponding to each load reversal, three regions, which characterize the manner in which the plastic modulus E^p changes, can be identified. Consider, for example, the loading path starting at F. The first part FG represents elastic deformation. The second part GH represents the plastic portion in which the

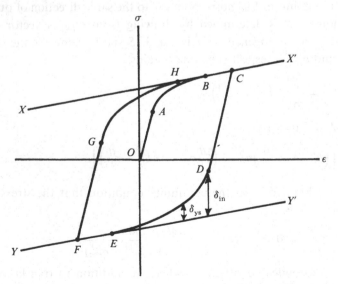

Fig. 3.11. Schematic illustration of the line bounds in stress–strain space. (After Dafalias & Popov, 1975.)

value of E^p changes from point G to point H. The third part HX' also represents plastic behavior; however, E^p has a constant value in the region. This third part lies on lines such as XX' or YY' which provide bounds.

Dafalias & Popov postulated that the instantaneous value of the plastic modulus E^p is determined by (i) the relative position of the current plastic loading point with respect to the bounding line (δ_{ys} in Fig. 3.11) and (ii) the amount of plastic work, $w^p = \int \sigma d\epsilon^p$, accumulated during plastic deformation prior to the elastic deformation preceding the current plastic state. The value of δ_{ys} at the initial yield point is δ_{in}. All loading states lying on a line drawn parallel to XX' in Fig. 3.11 have the same values of δ_{ys} and E^p.

The generalization of the above model to multiaxial loading conditions in stress space is schematically illustrated in Fig. 3.12. This figure shows two circles, where the inner circle with its center at k represents the loading surface and the outer circle with its center at r represents the bounding surface. Let point a on the loading surface represent the current plastic state of the material. Given point a, there is a corresponding point b on the bounding surface, and the distance between a and b, denoted as δ_{ys}, determines the value of the generalized plastic modulus at point a for the multiaxial case. Point b can be defined in a number of ways: (i) if the two surfaces are congruent, b can be taken as the point determined from the condition of congruency with respect to a; (ii) b can be obtained by the intersection of the normal to the loading surface at a with the bounding surface; or (iii) b can be taken as the point of intersection of the line drawn through ka with the bounding surface, as shown in Fig. 3.12. If σ_{ij} are the coordinates at a and σ_{ij}^* are the coordinates at b, the distance, δ_{ys}, can be defined as

$$\delta_{ys} = \sqrt{(\sigma_{ij}^* - \sigma_{ij})(\sigma_{ij}^* - \sigma_{ij})}. \tag{3.19}$$

Let K be the generalized plastic modulus for the multiaxial case. Assuming the associated flow rule, Eq. (1.25), and that the plastic strain increment $d\epsilon_{ij}^p$ is propor-

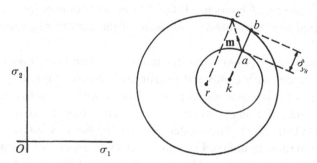

Fig. 3.12. Schematic illustration of the loading and bounding surfaces in stress space and of the motion of these surfaces during multiaxial loading. (After Dafalias & Popov, 1975.)

tional to the projection $\overline{d\sigma}$ of the stress increment $d\sigma_{ij}$ on the unit normal ω to the loading surface,

$$d\epsilon_{ij}^{p} = \frac{1}{K}\overline{d\sigma}\, w_{ij}. \tag{3.20}$$

In Fig. 3.12, c is the point on the bounding surface where the outward normal to it lies on the same direction as the normal to the loading surface. If **m** is the unit vector along the line that joins a and c, the translation of the bounding surface along **m** fulfills the condition that the point of contact is also the current stress state.

Dafalias & Popov formulated various conditions for the deformation and translation of the loading surface as well as of the bounding surface according to different hardening rules, particularly with reference to well known kinematic hardening rules. In general, the inner yield surface is taken to follow kinematic hardening rules while the outer yield surface follows isotropic hardening. The use of the two-surface model to rationalize cyclic creep is illustrated in Section 3.9.

Modifications of the two-surface models of cyclic plasticity have been examined by Moosbrugger & McDowell (1989) within the context of various isotropic and kinematic hardening formulations. These workers also show that the two-surface models for rate-independent plasticity offer superior correlation with experimental data on nonproportional cyclic deformation in stainless steels.

3.8.4 Other approaches

Bodner & Partom (1975) introduced an internal state variable approach to characterize the resistance of the material to plastic flow. The internal variable, referred to as the 'hardness', was taken to be a function of the plastic work. This approach was extended by Bodner, Partom & Partom (1979) to model uniaxial cyclic loading of elastic–viscoplastic materials. The evanescent hardening model is a phenomenological approach analogous to the Prager–Ziegler hardening rule, with the exception that the direction and amount of displacement of the yield surface are modified by the back stress (e.g., Chaboche, 1986). The endochronic or internal time approach uses an internal time variable as a measure of the history of deformation (e.g., Valanis, 1980).

White, Bronkhorst & Anand (1990) have formulated a phenomenological rate-independent plasticity model with combined isotropic–kinematic hardening. Here, the isotropic component of the resistance to plastic flow is shown to soften initially during a load reversal. This softening is taken to represent a dynamic recovery of the underlying dislocation substructure. The model captures the basic characteristics of small strain cyclic deformation in uniaxial loading, and some aspects of nonproportional strain cycling in tension–torsion. However, the predictions deviate significantly from experiments for large strain torsion.

3.9 Cyclic creep or ratchetting

As alluded to in the preceding section, when a fixed amplitude of cyclic stresses is imposed on a material, a phenomenon known as *cyclic creep* or *ratchetting* takes place if the plastic deformation during the loading portion is not opposed by an equal amount of yielding in the reverse loading direction. Figure 3.13(*a*) shows an example of this process for a fatigue softening material subjected to cyclic loads with a tensile mean stress, where cyclic creep occurs in the direction of increasing tensile strains. Here the accumulation of damage is accelerated by two processes: (i) an increase in cyclic plastic strain from cycle to cycle as a result of cyclic softening, and (ii) the displacement of the mean strain to higher tensile strain levels. Similarly, a compressive mean stress may enhance the conditions for buckling, as shown in Fig. 3.13(*b*). Cyclic creep can also be found, even under equal tension–compression loading, in materials in which pronounced yield anisotropy exists between tension and compression. Examples include cast iron, maraging steel and most composites.

From a continuum viewpoint, cyclic creep may be rationalized by recourse to the two surface models of cyclic plasticity (Section 3.8.3). Consider, for example, the case of cyclic creep under a tensile mean stress, Fig. 3.14. Assume that cyclic loading is imposed between the uniaxial stresses, $\sigma_2 > 0$ and $\sigma_1 < 0$, with $(\sigma_2 - |\sigma_1|)/2 > 0$. In Fig. 3.14, A_1, A_2, A_3, A_4, etc., denote points at which load reversal occurs. k_1 is the point in stress space where yielding initiates prior to approaching the bound σ_1 in compression. Similarly, k_2 is the corresponding point for the onset of yield prior to reaching the bound σ_2 in tension. The value of the plastic modulus E^p changes between k_1 and σ_1 and between k_2 and σ_2, according to the distance δ_{ys} of the instantaneous stress point from the bounding lines YY' or XX', respectively. As noted in Section 3.8.3, the smaller the value of δ_{ys} (i.e. the closer the stress point is to the bounding line), the smaller is the value of E^p.

Assume that at the onset of plastic deformation, the distance between the lines $\sigma = k_2$ and XX' is smaller than the distance between the lines $\sigma = k_1$ and YY'. In this case, the values of the distance δ_{ys} for tensile deformation are smaller than the corresponding values for compressive deformation. On the other hand, the plastic work done before each load reversal, $W^p = \int \sigma d\epsilon^p$, is approximately equal.

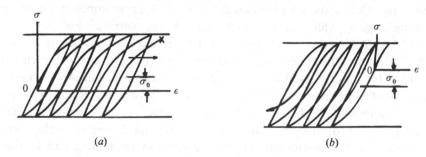

(*a*)   (*b*)

Fig. 3.13. Cyclic creep under (*a*) tensile and (*b*) compressive mean stress.

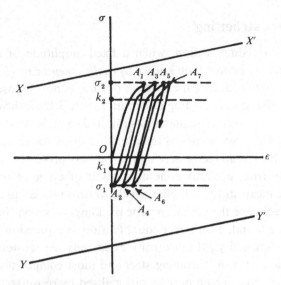

Fig. 3.14. A rationale for the occurrence of cyclic creep in terms of the bounding surface concept. (After Dafalias & Popov, 1975.)

Therefore, from Section 3.8.3, the plastic modulus is smaller for the stress states lying on the loading paths $k_2\sigma_2$ than for the ones on $k_1\sigma_1$. The cyclic loops, therefore, are not closed and they progressively shift to the right of the strain axis as shown in Fig. 3.14. This argument thus provides one rationale for the occurrence of cyclic creep. With continued shift to increasing mean strains, the distance between $\sigma = k_1$ and XX' and that between $\sigma = k_2$ and YY' tend to become equal. At this point, the plastic moduli also become equal and the loops become stabilized.

Cyclic creep or ratchetting has major implications for fatigue in a variety of engineering applications. For example, in rolling or sliding contact fatigue, combinations of high normal and shear tractions are imposed on a thin layer of material near the contacting surfaces. Within this layer, which is 10–50 μm thick in gears and bearings and about 2 mm thick in railway track, shear strains as high as 100% are accumulated with repeated applications of the load. Under conditions of free rolling or low traction (where the traction coefficient, i.e. the ratio of the shear to the normal traction, $\mu_t < 0.25$), an unsymmetrical cycle of shear is imposed on the highly deforming region which is confined beneath the surface. For high traction ($\mu_t > 0.25$), on the other hand, the critically stressed material lies at the surface. For this latter case, the material is subjected to a nonproportional load cycle which is made up of tension, followed by shear, followed by compression. An example of the application of cyclic creep models to contact fatigue problems is found in the work of Bower & Johnson (1989), who employed phenomenological nonlinear kinematic hardening models for analyzing the effects of strain hardening on cyclic creep contact fatigue. Further discussions of cyclic creep and ratchetting can be found in Section 13.4.4.

3.10 Metal-matrix composites subjected to thermal cycling

The deformation mechanisms of precipitation-hardened and particle-reinforced alloys under cyclic mechanical loads were discussed in the earlier sections of
this chapter. We now direct attention at continuum analyses of cyclic plastic deformation in metal-matrix composites. The problem considered here pertains to situations where the reinforced metal is subjected to cyclic variations in temperature. The
stresses and plastic strains in the composite arise as a result of thermal expansion/
contraction mismatch between the matrix and the reinforcement. The thermal
cycling conditions to be analyzed here are such that plastic flow in the matrix
dominates over any creep or thermal shock effects. We seek to derive closed-form
solutions for different characteristic temperatures which signify the onset and spread
of cyclic plasticity in the metallic matrix. As shown later in Chapter 8, these results
for cyclic plasticity are useful for estimating the fatigue lives of reinforced metals, by
invoking different damage and failure criteria. The results presented here are from
Olsson, Giannakopoulos & Suresh (1995).

3.10.1 Thermoelastic deformation

Consider a ductile alloy which is reinforced uniformly with spherical particles of a brittle material, e.g., a ceramic. Figure 3.15(a) shows a spherical volume
element of the composite, where the outer hollow sphere is the matrix alloy (phase 1)
and the concentric inner sphere is the ceramic reinforcement (phase 2). Each phase is
assumed to be homogeneous and isotropic, with the interface between them
mechanically well-bonded. For the volume element with spherical symmetry, the
field quantities are functions only of the radial position r, the material properties,
and the concentration of the brittle phase. The radial and tangential directions are
identified with subscripts r and θ, respectively. The inner region occupied by the
ceramic particle is $0 < r < r_i$, and the outer region occupied by the metal is

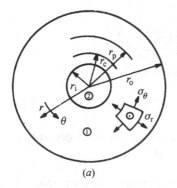

(a)

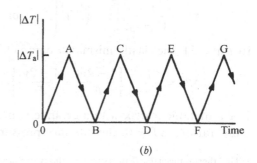

(b)

Fig. 3.15. (a) Geometry of the representative volume element for the thermal cycling analysis for
the metal-matrix composite. (b) Thermal loading history.

$r_i < r < r_o$. The volume fraction of the ceramic particle is, $f_p = (r_i/r_o)^3$, while that occupied by the metal is $(1 - f_p)$. The concentration of the particles, f_p, can be dilute or nondilute with only the proviso that the particles be well dispersed within the matrix without forming an interpenetrating composite. For the small strain analysis considered here, the radial and tangential strains are related to the radial displacement u by

$$\epsilon_r = \frac{du}{dr}, \qquad \epsilon_\theta = \frac{u}{r}, \qquad \epsilon_r = \frac{d}{dr}(r\epsilon_\theta). \tag{3.21}$$

The last of these three equations provides the condition for compatible strains. The equation for equilibrium in the radial direction is:

$$\frac{d\sigma_r}{dr} + \frac{2}{r}(\sigma_r - \sigma_\theta) = 0. \tag{3.22}$$

If both the matrix and the particle initially deform elastically,[†] the stress–strain relations are written as

$$\epsilon_r = \frac{\sigma_r - 2\nu\sigma_\theta}{E} + \alpha\Delta T, \qquad \epsilon_\theta = \frac{(1-\nu)\sigma_\theta - \nu\sigma_r}{E} + \alpha\Delta T, \tag{3.23}$$

where E is Young's modulus, ν is Poisson's ratio, and α is the thermal expansion coefficient, with the subscripts 1 and 2 denoting the matrix and the particle, respectively.

Consider a uniform change in temperature, ΔT, from an initial stress-free temperature. For an increase or a decrease in temperature, $\Delta T > 0$ or $\Delta T < 0$, respectively. For this pure thermal loading case, where the composite is free of any external kinematical constraint, the appropriate outer boundary condition for the volume element in Fig. 3.15(a) is that $\sigma_r = 0$. The stresses in the spherical particle are then readily determined from straightforward thermoelastic analysis to be

$$\sigma_{r2} = -p, \qquad \sigma_{\theta 2} = -p, \qquad p = \frac{2E_1}{3(1-\nu_1)} \frac{(\alpha_2 - \alpha_1)(1-f_p)\Delta T}{1 - \frac{2M_{el}}{3}(1-f_p)}, \tag{3.24}$$

and the stresses in the matrix are

$$\sigma_{r1} = \frac{pf_p}{(1-f_p)}\left\{1 - \left(\frac{r_o}{r}\right)^3\right\}, \qquad \sigma_{\theta 1} = \frac{pf_p}{(1-f_p)}\left\{1 + \frac{1}{2}\left(\frac{r_o}{r}\right)^3\right\}. \tag{3.25}$$

In Eq. 3.24, the elastic mismatch parameter M_{el} is defined as

$$M_{el} = \frac{E_1}{1-\nu_1}\left\{\frac{1-2\nu_1}{E_1} - \frac{1-2\nu_2}{E_2}\right\}. \tag{3.26}$$

For a metallic matrix with a ceramic reinforcement ($\alpha_1 > \alpha_2$) and for a decrease in temperature, $\Delta T < 0$, the interface pressure $\sigma_{r2} = -p < 0$.

[†] The thermomechanical properties of the two phases are assumed to be temperature-independent over the range of thermal excursions considered here. The temperature-dependence, however, can easily be incorporated within the context of this analysis by recourse to numerical solutions.

3.10.2 Characteristic temperatures for thermal fatigue

If the change in temperature ΔT from the initial stress-free temperature becomes sufficiently large, plastic yielding occurs in the ductile matrix. Since the effective stress in the matrix is

$$\sigma_e = |\sigma_{\theta 1} - \sigma_{r1}| = \frac{3}{2} \frac{f_p}{(1 - f_p)} |p| \left(\frac{r_0}{r}\right)^3, \tag{3.27}$$

the maximum effective stress develops at the interface which is the site for the onset of plastic flow. The temperature change necessary to initiate plastic flow at the interface, ΔT_1, is found by setting $(\sigma_e)_1 = \sigma_y$ at $r = r_i$, where σ_y is the yield strength of the matrix:

$$|\Delta T_1| = \frac{\sigma_y(1 - \nu_1)}{E_1 |\alpha_2 - \alpha_1|} \left\{ 1 - \frac{2M_{el}}{3}(1 - f_p) \right\}. \tag{3.28}$$

The plastic zone which commences at the interface spreads outwards.

The spread of the plastic zone is analyzed conveniently by assuming that the matrix is an elastic–perfectly plastic solid. Within the volume element shown in Fig. 3.15(a), three deformation regions can be identified: (i) the volume occupied by the reinforcement particle within which only elastic deformation occurs, (ii) an elastic–plastic region, $r_i < r < r_p$, within which $(\sigma_e)_1 = \sigma_y$, and (iii) an elastic outer region, $r_p < r < r_0$, with a stress-free boundary at $r = r_0$. Continuity of radial stress and displacement across the boundaries between these regions and the condition that $(\sigma_e)_1 = \sigma_y$, lead to

$$|(\alpha_2 - \alpha_1)\Delta T_1| =$$

$$\frac{\sigma_y(1 - \nu_1)}{E_1} \left[\left(\frac{r_p}{r_i}\right)^3 - M_{el} \left\{ 2\log\left(\frac{r_p}{r_i}\right) + \frac{2}{3}\left[1 - \left(\frac{r_p}{r_0}\right)^3\right] \right\} \right]. \tag{3.29}$$

This equation provides a direct connection between the size of the plastic zone r_p and the magnitude of thermal strain $(\alpha_2 - \alpha_1)\Delta T_1$, and reduces to Eq. 3.28 when $r_p = r_i$.

As the plastic zone continues to spread under a monotonic temperature change $|\Delta T|$, another critical value of $|\Delta T|$, which we shall term $|\Delta T_3|$, is reached at which the entire matrix becomes plastic. The magnitude of ΔT_3 is determined by setting $r_p = r_0$ in Eq. 3.29:

$$|\Delta T_3| = \frac{\sigma_y(1 - \nu_1)}{E_1 |\alpha_2 - \alpha_1|} \left\{ \frac{1}{f_p} + \frac{2M_{el}}{3} \log f_p \right\}. \tag{3.30}$$

Now consider temperature cycles, $0 \leq |\Delta T| \leq \Delta T_a$, commencing at the stress-free temperature,† Fig. 3.15(b). The first unloading in the thermal cycle corresponds to the excursion from point A to point B.

If the yield strength of the matrix is the same in tension and compression, it can be readily shown that there exists a characteristic temperature change, $\Delta T_2 = 2\Delta T_1$,

---

† Equations 3.28 and 3.30 pertain to the first thermal loading ramp from $|\Delta T| = 0$ to $|\Delta T| = \Delta T_a$ (point A in Fig. 3.15(b)). In general, ΔT and ΔT_a can be positive or negative. For simplicity, we consider positive values of the characteristic temperatures in subsequent discussion.

below which a zone of reverse plasticity does not occur. In other words, if $|\Delta T_a| < \Delta T_2$, there is no accumulation of plastic strains during thermal cycling. Therefore, ΔT_2 represents a *shakedown limit* for the thermal fatigue of a metal-matrix composite. For $|\Delta T_a| < \Delta T_2$, only elastic conditions prevail after the first unloading (i.e. after point B).

If, on the other hand, $|\Delta T_a| > \Delta T_2$, a zone of reversed plastic flow of radius r_c develops at the particle–matrix interface. Repeated thermal cycling causes r_c to expand outward. Referring to Fig. 3.15(*b*), if the temperature is first increased (decreased) from some initial stress-free temperature (point O) to a maximum (minimum) value at point A, and then decreased (increased) to the current temperature ΔT, the size of the reversed or cyclic plastic zone r_c is given by

$$|(\alpha_2 - \alpha_1)(\Delta T_a - \Delta T)| =$$

$$\frac{2\sigma_y(1 - \nu_1)}{E_1}\left[\left(\frac{r_c}{r_i}\right)^3 - M_{el}\left\{2\log\left(\frac{r_c}{r_i}\right) + \frac{2}{3}\left[1 - \left(\frac{r_c}{r_o}\right)^3\right]\right\}\right]. \tag{3.31}$$

The right hand sides of Eqs. 3.29 and 3.31 differ only by a factor of 2 and by a change from r_p to r_c. A remarkable feature of Eqs. 3.28–3.31 is that both the monotonic and cyclic plastic zones initiate at the particle–matix interface and spread monotonically outwards during cyclic variations in temperature.

As the cyclic plastic zone spreads outwards, there exists another critical temperature at which the entire matrix begins to undergo reversed plastic yielding. This characteristic temperature ΔT_4 is obtained by setting $r_c = r_0$ in Eq. 3.31, from which it is readily seen that $\Delta T_4 = 2\Delta T_3$. Figure 3.16 schematically shows the spread of r_p and r_c for different values of ΔT_a with respect to the various characteristic temperatures.

For the uniform spatial distribution of spherical particles, the conditions that $|\Delta T_1| < |\Delta T_2|$ and that $|\Delta T_1| < |\Delta T_3|$ always hold. However, the condition that $|\Delta T_2| < |\Delta T_3|$ need not always be true, i.e. the first monotonic temperature change from O to A in Fig. 3.15(*b*) may cause complete yielding of the matrix with no cyclic plastic zones induced upon subsequent unloading (from A to B). From Eqs. 3.28 and 3.29 and from the condition for shakedown ($\Delta T_2 = 2\Delta T_1$), it can be shown that there exists a critical concentration of the reinforcement, $f_p = f_p^*$, at which $\Delta T_2 = \Delta T_3$, which obeys the equation:

$$1 - 2f_p^* + \frac{2M_{el}}{3}f_p^*\{\log(f_p^*) + 2(1 - f_p^*)\} = 0. \tag{3.32}$$

The value of f_p^* is unique for a fixed combination of the matrix and reinforcement, and is bounded by $0.203 < f_p^* < 0.586$. If $f_p \geq f_p^*$, $\Delta T_2 \geq \Delta T_3$, and the composite matrix undergoes no plastic strain accumulation during thermal cycling, although the matrix is fully yielded during the first thermal excursion from the initial stress-free state.

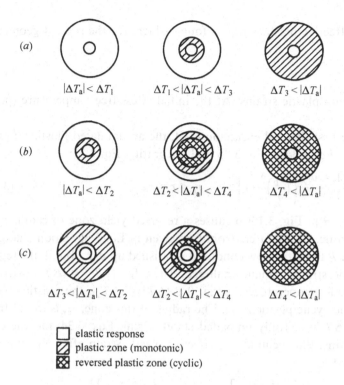

□ elastic response
▨ plastic zone (monotonic)
▩ reversed plastic zone (cyclic)

Fig. 3.16. The evolution of monotonic and cyclic plastic zones at different temperature amplitudes. (*a*) Monotonic thermal loading corresponding to point *A* in Fig. 3.15(*b*). (*b*) Temperature cycling for $\Delta T_2 < \Delta T_3$ corresponding to points *B*, *C*, *D*,... in Fig. 3.15(*b*). (*c*) Temperature cycling for $\Delta T_2 > \Delta T_3$ corresponding to points *B*, *C*, *D*,... in Fig. 3.15(*b*).

3.10.3 Plastic strain accumulation during thermal cycling

Consider the situation where an elastic–ideally plastic metal reinforced with ceramic particles is thermally cycled in such a way that $|\Delta T_2| < |\Delta T_a| < |\Delta T_3|$, which is a possible scenario for $f_p < f_p^*$. The general expressions for the tangential and radial components of the plastic strain in the matrix are

$$\epsilon_\theta^{pl} = -\overline{S}\frac{\sigma_y(1-\nu_1)}{E_1} + \frac{C_1}{r^3}, \qquad \epsilon_r^{pl} = 2\overline{S}\frac{\sigma_y(1-\nu_1)}{E_1} - 2\frac{C_1}{r^3}, \qquad (3.33)$$

where $\overline{S} = \text{sign}\{(\alpha_2 - \alpha_1)\Delta T\} = \{(\alpha_2 - \alpha_1)\Delta T\}/|(\alpha_2 - \alpha_1)\Delta T|$ is either $+1$ or -1, depending on whether the plastic flow is forward or reversed. The constant C_1 is obtained from the continuity conditions to be

$$C_1 = \overline{S}r_p^3\frac{\sigma_y(1-\nu_1)}{E_1} \qquad \text{for } |\Delta T_a| < |\Delta T_3|. \qquad (3.34)$$

Since the radial stress and the effective stress (and hence the tangential stress) are continuous across the plastic zone boundary, the plastic strains are also continuous.

From Eq. 3.33, the effective plastic strain is found which, for the present geometry, reduces to

$$d\epsilon_{\text{eff}}^{\text{pl}} = 2|d\epsilon_{\theta}^{\text{pl}}|, \tag{3.35}$$

because of proportional plastic strains. At the initial stress-free temperature (point O), $\epsilon_{\text{eff}}^{\text{pl}} = 0$.

Beginning with the first thermal excursion O–A, the accumulated plastic strain is found by substituting Eq. 3.34 in Eq. 3.35 and integrating Eq. 3.35:

$$\epsilon_{\text{eff}}^{\text{pl}}(r) = \frac{2\sigma_y(1 - \nu_1)}{E_1}\left\{\left(\frac{r_p}{r}\right)^3 - 1\right\}, \qquad r_i \leq r \leq r_p. \tag{3.36}$$

Unloading from A to B in Fig. 3.15(b) causes a reversed yield zone of radius r_c to develop. Inside this zone, $\epsilon_{\text{eff}}^{\text{pl}}(r)$ differs from that given in Eq. 3.36. Upon changing the temperature from B to C, the flow condition is satisfied up to r_p, but in the region $r_c \leq r \leq r_p$, the plastic strains remain unaltered. Thus, for $|\Delta T_2| \leq |\Delta T_a| \leq |\Delta T_3|$, and constant-amplitude temperature cycling, $0 \leq |\Delta T| \leq |\Delta T_a|$, the plastic strains change only inside the cyclic plastic zone. The radius of this zone, r_c, is found from Eq. 3.31 by setting $\Delta T = 0$ (fully unloaded state). Using Eqs. 3.34 and the continuity of plastic strains, the accumulated effective plastic strain after N_T temperature reversals is found as

$$\epsilon_{\text{eff}}^{\text{pl},N_T} = \frac{2\sigma_y(1 - \nu_1)}{E_1}\left\{\left[\left(\frac{r_p}{r}\right)^3 - 1\right] + 2(N_T - 1)\left[\left(\frac{r_c}{r}\right)^3 - 1\right]\right\}. \tag{3.37}$$

If $|\Delta T_3| < |\Delta T_a| < |\Delta T_4|$,

$$\epsilon_{\text{eff}}^{\text{pl},N_T} = \frac{2\sigma_y(1 - \nu_1)}{E_1}\left[\left(\frac{r_o}{r}\right)^3\left(1 + \frac{f_p E_1 |\alpha_2 - \alpha_1||\Delta T_a - \Delta T_3|}{\sigma_y(1 - \nu_1)}\right) - 1\right]$$
$$+ \frac{4\sigma_y(1 - \nu_1)}{E_1}(N_T - 1)\left[\left(\frac{r_c}{r}\right)^3 - 1\right]. \tag{3.38}$$

The reversed flow zone size r_c in Eq. 3.38 is given by Eq. 3.31. The first term on the right hand side of Eq. 3.38 gives $\epsilon_{\text{eff}}^{\text{pl},N_T}$ during the first temperature excursion from O to A, and the second term gives the accumulated value in the next $(N_T - 1)$ temperature reversals. It should be noted that the rate of plastic strain accumulation in Eqs. 3.37 and 3.38 is the same, i.e. the second terms on the right hand side of these two equations are identical. Thus, the difference in the magnitude of the accumulated plastic strain is only that due to the difference developed in the first temperature excursion.

3.10.4 Effects of matrix strain hardening

The preceding results were restricted to metallic matrices which do not show any strain hardening. If the matrix undergoes cyclic strain hardening, the following modifications to the results are anticipated.

(1) For cycling between fixed temperature limits, the four characteristic temperatures, ΔT_i, $i = 1, \ldots, 4$, are constants for an elastic–ideally plastic matrix. For a strain-hardening matrix, however, these temperatures change with the number of thermal cycles; they can still be defined based on the plastic strain distribution.

(2) The accumulation of plastic strain near the particle–matrix interface increases the flow strength of the hardening matrix. Further accumulation of plasticity is thereby suppressed.

(3) The result of the preceding section that plastic strain accumulation occurs only within the cyclic plastic zone, still applies to the hardening matrix. The rate of accumulation, however, is affected by the hardening rate.

Consider a matrix alloy with isotropic, linear strain hardening H, which captures the cyclic hardening characteristics of the matrix. The matrix yield strength σ_y is now modified to include hardening in the following way: $\sigma_y + H\epsilon_{\mathrm{eff}}^{\mathrm{pl}}(r)$. Let $\overline{H} = 2H(1 - \nu_1)/E_1$. It can be readily shown (Olsson, Giannakopoulos & Suresh, 1995) that the plastic strain increment during each load reversal is explicitly given by

$$\Delta\epsilon_{\mathrm{eff}}^{\mathrm{pl},N_T} = \epsilon_{\mathrm{eff}}^{\mathrm{pl},(N_T+1)} - \epsilon_{\mathrm{eff}}^{\mathrm{pl},N_T} = \frac{4\sigma_y(1-\nu_1)}{(1+\overline{H})E_1}\left[\left(\frac{r_c}{r}\right)^3 - 1\right]\left[\frac{1-\overline{H}}{1+\overline{H}}\right]^{N_T-1}. \tag{3.39}$$

For $\overline{H} = 0$ (no hardening), the results of the preceding section are recovered. An examination of Eq. 3.40 reveals that $\Delta\epsilon_{\mathrm{eff}}^{\mathrm{pl},N_T} \to 0$ as $N_T \to \infty$. Therefore, unlike the case of an ideally plastic matrix for which the plastic strain accumulation during thermal fatigue is unbounded, the accumulated plastic strain for a strain-hardening matrix asymptotically approaches a limiting value at large numbers of thermal cycles in the following manner:

$$\Delta\epsilon_{\mathrm{eff}}^{\mathrm{pl},\infty} = \Delta\epsilon_{\mathrm{eff}}^{\mathrm{pl},1}(r_c) \cdot \left(\frac{r_c}{r}\right)^3 + \frac{\sigma_y}{H}\left[\left(\frac{r_c}{r}\right)^3 - 1\right], \tag{3.40}$$

where $\Delta\epsilon_{\mathrm{eff}}^{\mathrm{pl},1}$ is the plastic strain accumulated after the first thermal excursion. The limiting strain in Eq. 3.40 has a maximum value at the particle–matrix interface. If the ductility of the matrix alloy exceeds this maximum value, an endurance limit for thermal fatigue is expected. This temperature endurance limit is higher than ΔT_2 because of strain hardening.

All of the foregoing analytical results pertain to situations where E_1, ν_1, α_1, α_2, σ_y and H are temperature-independent. The same approach can be adapted for analyzing thermal fatigue by including the temperature-dependence of these properties by recourse to numerical simulations (see, for example, Olsson *et al.*, 1995).

3.10.5 Example problem: Critical temperatures for thermal fatigue in a metal-matrix composite

Problem:

A 2024-T6 aluminum alloy, reinforced with 20 volume% of spherical SiC particles, is thermomechanically processed at 250 °C. At this temperature, the composite is almost free of internal stresses and its reinforcement is well-bonded to the matrix. The isotropic properties of the matrix are: $\alpha_1 = 22 \times 10^{-6} \, °C^{-1}$, $E_1 = 72$ GPa, $\nu_1 = 0.33$, $\sigma_y = 393$ MPa, and melting temperature, $T_{melt} = 660$ °C. (All the symbols are defined in Section 3.10.1.) The matrix exhibits essentially no strain hardening. The isotropic properties of the SiC particles are: $\alpha_2 = 4.7 \times 10^{-6} \, °C^{-1}$, $E_2 = 450$ GPa, $\nu_2 = 0.17$, and $T_{melt} = 660$ °C.

(i) If the reinforced alloy is uniformly cooled to room temperature (25 °C), what is the temperature at which plastic yielding begins during cooling from the processing temperature? (In order to obtain analytical results, assume that the mechanical properties of both phases of the composite are independent of temperature for the range of temperatures and heating/cooling rates considered.)

(ii) What is the volume fraction of the matrix which has undergone plastic yielding upon cooling?

(iii) If the composite is thermally cycled between room temperature and the processing temperature, compute the plastic strain accumulation per cycle.

(iv) Now suppose that the yield strength of the matrix decreases linearly with increasing temperature, and that the Poisson's ratios, Young's moduli and thermal expansion coefficients for both phases remain temperature-independent, for the range of temperatures considered. Explain briefly how the temperature dependence of yield strength of the matrix would affect (a) the critical temperatures ΔT_1 and ΔT_3 and (b) the monotonic plastic zone radius r_p.

Solution:

It is given that: $f_p = (r_i/r_o)^3 = 0.2$, and $T_{proc} = 250$ °C, $T_{room} = 25$ °C, and $|\Delta T| = T_{proc} - T_{room} = 225$ °C. Since $|\Delta T|$ is less than one half of the melting temperature, it may be assumed that creep effects are less dominant compared to plasticity effects (i.e. when long hold periods at temperatures are not involved). Denoting the properties of the matrix by the subscript '1' and those of the particle by the subscript '2' and substituting the appropriate values in Eq. 3.26, it is found that the elastic mismatch parameter $M_{el} = 0.35$.

(i)   Substituting the numerical values of the various terms in Eq. 3.28, we find that the temperature change, from the initial stress-free temperature (250 °C) at which plastic yielding begins is: $|\Delta T_1| = 172$ °C, i.e. at $250 - 172 = 78$ °C.

(ii)   To find the plastic zone radius r_p for a temperature change of $|\Delta T| = 225$ °C, replace ΔT_1 in Eq. 10.2 by $|\Delta T|$. Substituting the appropriate numerical values for $|\Delta T|$ and the various material parameters, it is seen that

$$1.0467 \left(\frac{r_p}{r_i} \right)^3 - 0.2333 \log \left(\frac{r_p}{r_i} \right)^3 = 1.0644, \quad \text{or} \quad \left(\frac{r_p}{r_i} \right)^3 \approx 1.025. \tag{3.41}$$

The volume fraction of the matrix which has undergone plastic yielding upon cooling from the processing temperature to room temperature is:

$$V_{pl} = \frac{r_p^3 - r_i^3}{r_o^3 - r_i^3} = \frac{f_p}{1 - f_p} \left\{ \left(\frac{r_p}{r_i} \right)^3 - 1 \right\}. \tag{3.42}$$

Substituting the value of r_p from Eq. (2) into this equation, it is seen that $V_{pl} \approx 0.63\%$.

(iii)   We note that the characteristic temperature change below which a zone of reversed plastic flow does not develop in the matrix is: $|\Delta T_2| = 2|\Delta T_1| = 344$ °C. Since the temperature range for thermal cycling $|\Delta T| = 225$ °C $< |\Delta T_2|$, it is apparent that repeated thermal cycling between room temperature and the processing temperature will not lead to any continued accumulation of plastic strains in the matrix.

(iv)   (a) As the yield strength decreases with increasing temperature, both ΔT_1 and ΔT_3 will be lowered compared to the case where the yield strength is independent of temperature. (b) The monotonic plastic zone size, r_p, increases because of the lower yield strength at higher temperatures. There is a corresponding increase in the volume of the matrix material which has undergone plastic yielding.

3.11 Layered composites subjected to thermal cycling

Layered materials represent a broad class of composites which find diverse applications in mechanical, thermal, electronic, magnetic, ferroelectric, optical and biomechanical components or devices. They also facilitate maintenance of clearance/tolerance or enhance appearance and biocompatibility. In many applications, fluctuations in externally imposed stresses or pulsations in internal stresses arising from such factors as the thermal expansion mismatch between the layers during repeated temperature excursions can lead to cyclic deformation and fracture. Examples of layered structures in which thermal cycling or thermo-mechanical fatigue is a topic of interest include: (i) laminated composites comprising organic or inorganic matrices and discontinuous or continuous reinforcements, (ii) thermal-barrier or wear-resistant coatings comprising a ceramic outer coating and an in-between metallic bond coat on a metallic substrate, and (iii) semiconductor devices with patterned metallic conduction lines, passivated by a glassy layer or a ceramic or a polymer, on the Si single-crystal substrates.

In this section, we consider the cyclic deformation response of layered structures subjected to fluctuations in temperature. This section begins with a basic treatment of the thermoelastic deformation of a bilayer by recourse to classical beam/plate theories. This is followed by the derivation of stresses in thin films on substrates which are subjected to thermal excursions. Attention is then directed at some critical temperatures for the occurrence of distinct transitions in the cyclic deformation response of a ductile layer in a metal–ceramic bilayer system which undergoes cyclic variations in temperature, similar to the analyses presented in the preceding section on metal-matrix composites.

3.11.1 Thermoelastic deformation of a bilayer

Consider a general bilayer, schematically shown in Fig. 3.17(a), which is subjected to uniform temperature changes at all times. The uniform thicknesses of layers 1 and 2 are h_1 and h_2, respectively. For a start, the interface between the two layers is assumed to be mechanically well bonded, and the two layers are taken to exhibit in-plane elastic and thermal isotropy with Young's moduli, E_1 and E_2, Poisson's ratios, v_1 and v_2, and coefficients of thermal expansion, α_1 and α_2. The in-plane shape of the bilayer is a rectangle with length L_x and width L_y.

Let the bilayer plate be flat and devoid of any internal stresses at some initial reference temperature T_0, which could be identified with the processing, diffusion-bonding or softening temperature. Let the temperature of the bilayer be now changed uniformly by ΔT to some instantaneous value T (with $T - T_0 = \Delta T > 0$ denoting heating and $\Delta T < 0$ denoting cooling) in such a way that the entire plate is always at the same temperature T. The thermal expansion or contraction mismatch between the layers causes gradients in stresses to develop through the thickness of each layer. As a result, the initially flat plate bends with a curvature κ_R (which is the inverse of the radius of bending). If $L_x \gg h_1 + h_2$ and $L_y \gg h_1 + h_2$, a state of equi-biaxial stresses exists in the plate such that $\sigma_{xx} = \sigma_{yy}$ and $\epsilon_{xx} = \epsilon_{yy}$ (i.e. the curved plate has the same state of stress as the skin of a balloon which is internally pressurized).†

---

† Near the free edges of the bilayer where the interface intersects the free surfaces, the stresses have to be modified locally so as to satisfy the stress-free boundary conditions at the free surfaces of the unconstrained bilayer plate. Consequently, there develops a three-dimensional stress state which comprises both in-plane and out-of-plane normal and shear stresses. The size of this 'edge zone', measured from each edge, is of the order of the total layer thickness. The analysis presented in this section is valid only away from this edge-zone.

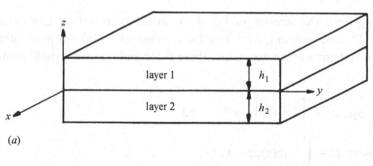

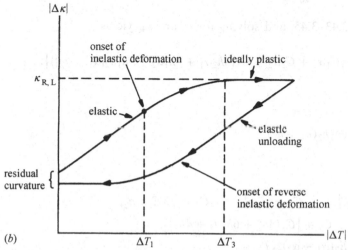

Fig. 3.17. (*a*) Geometry of the bilayer and the associated nomenclature. (*b*) Schematic representation of the variation of curvature with temperature for a metal–ceramic bilayer where the metallic layer is elastic–perfectly plastic. All material properties are assumed to be independent of temperature.

The thermoelastic strain in the bilayer, in regions away from the free edges, consists of two parts: (i) an in-plane normal strain, ϵ_0, which arises from a uniform stretch or contraction, and (ii) the strain due to bending, $\kappa_R z$, where z is the thickness coordinate:

$$\epsilon_{xx}(z) = \epsilon_{yy}(z) = \epsilon(z) = \epsilon_0 + \kappa_R z. \tag{3.43}$$

The only nonzero components of stress in the two layers are

$$\sigma_{xx,1}(z) = \sigma_{yy,1}(z) = \sigma_1(z) = \frac{E_1(z)}{1 - \nu_1(z)}\{\epsilon(z) - [\alpha_1(z)\Delta T]\},$$

$$\sigma_{xx,2}(z) = \sigma_{yy,2}(z) = \sigma_2(z) = \frac{E_2(z)}{1 - \nu_2(z)}\{\epsilon(z) - [\alpha_2(z)\Delta T]\}. \tag{3.44}$$

The subscripts '1' and '2' refer to quantities associated with layers 1 and 2, respectively, in Fig. 3.17(*a*). Static equilibrium dictates that the net force and the net moment arising from the stresses in Eq. 3.44 should be offset by any externally imposed force F^{ap} and moment, M^{ap}. For the unconstrained bilayer plate subjected only to a uniform temperature excursion, these force and moment equilibrium conditions give:

$$\int_{-h_2}^{0} \sigma_2(z)\mathrm{d}z + \int_{0}^{h_1} \sigma_1(z)\mathrm{d}z = 0, \quad \text{and}$$

$$\int_{-h_2}^{0} \sigma_2 z(z)\mathrm{d}z + \int_{0}^{h_1} \sigma_1 z(z)\mathrm{d}z = 0. \tag{3.45}$$

Combining Eqs. 3.43–3.45, and solving for ϵ_0 and κ_R yields

$$\epsilon_0 = \frac{\Delta T}{C_4}\{C_1\alpha_1 + C_2\alpha_2 + C_3(4h_2^2\alpha_1 + 4h_1^2\alpha_2 + 3h_1 h_2[\alpha_1 + \alpha_2])\}, \tag{3.46}$$

and

$$\kappa_R = \frac{D_1}{C_4} 6h_1 h_2(h_1 + h_2), \tag{3.47}$$

where

$$C_1 = \overline{E}_1^2 h_1^4, \quad C_2 = \overline{E}_2^2 h_2^4, \quad C_3 = \overline{E}_1\overline{E}_2 h_1 h_2,$$
$$C_4 = C_1 + C_2 + [C_3(4h_1^2 + 6h_1 h_2 + 4h_2^2)],$$
$$D_1 = \overline{E}_1\overline{E}_2(\alpha_1 - \alpha_2)\Delta T,$$
$$\overline{E}_1 = \frac{E_1}{1 - \nu_1}, \quad \text{and} \quad \overline{E}_2 = \frac{E_2}{1 - \nu_2}. \tag{3.48}$$

Substituting Eqs. 3.46–3.48 into Eq. 3.44, it is readily shown that the stresses at the outer surfaces of the two layers and at the interfaces are:

$$\sigma_1|_{z=h_1} = \frac{D_1}{C_4}\left\{\frac{C_3}{\overline{E}_2}(2h_1^2 + 3h_1 h_2) - \frac{C_2}{\overline{E}_2}\right\},$$

$$\sigma_1|_{z=0} = -\frac{D_1}{C_4}\left\{\frac{C_3}{\overline{E}_2}(4h_1^2 + 3h_1 h_2) + \frac{C_2}{\overline{E}_2}\right\},$$

$$\sigma_2|_{z=-h_2} = \frac{D_1}{C_4}\left\{-\frac{C_3}{\overline{E}_1}(2h_2^2 + 3h_1 h_2) + \frac{C_1}{\overline{E}_1}\right\},$$

$$\sigma_2|_{z=0} = \frac{D_1}{C_4}\left\{\frac{C_3}{\overline{E}_1}(4h_2^2 + 3h_1 h_2) - \frac{C_1}{\overline{E}_1}\right\}. \tag{3.49}$$

Note that the stresses vary linearly with z within each layer to a maximum value at the interface, and that there occurs a sharp jump in the magnitude of the stress at the interface between the two layers. While the foregoing analysis pertains to a bilayer, the same approach can be used to analyze the thermoelastic deformation of a multilayer comprising any number of layers. In such a case, the equilibrium equations, Eq. 3.45, can be solved numerically using a personal computer where additional effects involving temperature-dependence of material properties in the multilayer, plastic yielding and strain hardening in the ductile layers, in-plane anisotropy of elastic properties, continuous gradients in composition and properties through one or more layers, steady-state creep, as well as heat conduction arising from nonuniform temperature variations through the thicknesses of the layers can all be incorporated (e.g., Finot & Suresh, 1994).

3.11.2 Thin-film limit: the Stoney formula

In many practical applications, thin coatings are deposited on much thicker substrates. Examples include ceramic coatings on metal alloys for protection against elevated temperatures, environmental attack or contact fatigue damage. A very useful result, commonly known as the Stoney formula (after Stoney, 1909), for the estimation of the uniform stress in a thin film on a thick substrate can be derived from the above bilayer result.

Consider the thin-film limit of the bilayer shown in Fig. 3.17(a), where $h_1 \ll h_2$. With this geometric condition, Eq. 3.47 becomes

$$\kappa_R = 6 \frac{\overline{E}_1 h_1}{\overline{E}_2 h_2^2} (\alpha_1 - \alpha_2) \Delta T, \qquad h_1 \ll h_2. \tag{3.50}$$

In addition, Eqs. 3.48 and 3.49 give, for the thin-film limit,

$$\sigma_1|_{z=h_1} \approx -\overline{E}_1(\alpha_1 - \alpha_2)\Delta T, \quad \text{and} \quad \sigma_1|_{z=0} \approx -\overline{E}_1(\alpha_1 - \alpha_2)\Delta T. \tag{3.51}$$

In other words, the equi-biaxial stress in the thin film is essentially uniform. It is also readily seen from Eq. 3.49 that, for $h_1 \ll h_2$, the magnitude of the stresses in the thick substrate is very small.

Combining Eqs. 3.47 and 3.51, and denoting the various parameters associated with the film and the substrate with the subscripts 'film' and 'sub', respectively,

$$\sigma_{\text{film}} = -\left[\frac{1}{6} \cdot \left(\frac{E_{\text{sub}}}{1 - \nu_{\text{sub}}} \right) \cdot \frac{h_{\text{sub}}^2}{h_{\text{film}}} \cdot \left\{ \frac{1}{R_c} - \frac{1}{R_{c,0}} \right\} \right], \tag{3.52}$$

where $R_{c,0}$ is the initial radius of curvature of the thin-film/substrate system at some reference temperature and R_c is its radius of curvature after some temperature change ΔT. In other words, if the radius of bending before and after a temperature change can be measured experimentally, the stress in the film is computed easily using Eq. 3.52.† Note that once the curvature change is known for a thermal excursion, the film stress is determined with the knowledge only of the film and substrate thicknesses and of the elastic properties of the substrate. In other words, the elastic properties of the film, which are often very difficult to obtain, are not needed. The minus sign for the film stress in Eq. 3.52 simply implies that if the bilayer is convex

---

† In view of its simplicity, the Stoney equation, Eq. 3.52, is commonly used to determine the thin film stresses in the microelectronics and structural coatings industries where scanning laser interferometry methods are widely used to determine the radius of curvature of the thin-film/substrate system before and after thermal excursions.

shaped on the thin-film side (i.e. positive κ_R), then the stress in the film should be compressive. From Eq. 3.50, we see that such a situation arises when the coefficient of thermal expansion of the film is higher than that of the substrate and the bilayer is subjected to heating from a stress-free initial reference temperature.

3.11.3 Characteristic temperatures for thermal fatigue

Four different characteristic temperatures, which signify different distinct transitions in the evolution of cyclic deformation, were defined in Section 3.10.2 for the thermal cycling of a metal reinforced with brittle spherical particles. Four such characteristic temperatures for the thermal fatigue can also be analytically derived for the bilayer in Fig. 3.17(a) (for any arbitrary combination of layer thicknesses h_1 and h_2) where one or both layers is a metal (Suresh, Giannakopoulos & Olsson, 1994).

Let layer 1 in Fig. 3.17(a) be a metal and layer 2 be a brittle solid, such as a ceramic. With this choice, we see that in general, $(\alpha_1 - \alpha_2) > 0$. Let the plastic deformation of the metallic layer be characterized by an elastic–perfectly plastic deformation behavior, with a yield strength, σ_{y1} (which is the same in magnitude but opposite in sign for tension and compression). Layer 2 remains a linear elastic solid throughout any changes in temperature. Let the bilayer be subjected to uniform temperature changes which fluctuate with a constant amplitude ΔT_a.

Starting with the stress-free temperature T_0, consider the first uniform change in temperature $|\Delta T|$. Initially both layers deform elastically. When $|\Delta T|$ reaches a critical value $|\Delta T_1|$, the metallic layer begins to yield plastically at the interface. Setting the stress at the interface in layer 1 to equal σ_{y1} for this critical condition, it is found that

$$|\Delta T_1| = \frac{\sigma_{y1} \overline{E}_2 C_4}{D_1 \cdot [C_2 + C_3(4h_1^2 + 3h_1 h_2)]}, \tag{3.53}$$

where the various parameters are defined in Eq. 3.48.

If now the temperature continues to change monotonically beyond $|\Delta T_1|$, the plastic zone which initiates at the interface spreads outwards to the free surface of the metallic layer. After a further temperature change, there occurs a critical condition, $|\Delta T| = |\Delta T_3|$ that the entire metallic layer becomes fully plastic, where

$$|\Delta T_3| = \frac{\sigma_{y1}}{(\alpha_1 - \alpha_2)} \left\{ \frac{1}{\overline{E}_1} + \frac{1}{\overline{E}_2} \left[4\frac{h_1}{h_2} + 9\left(\frac{h_1}{h_2}\right)^2 + 6\left(\frac{h_1}{h_2}\right)^3 \right] \right\}. \tag{3.54}$$

If the metallic layer does not strain harden, this limit temperature for full yielding also signifies the condition for a limiting curvature, $\kappa_{R,L}$. That is, any further change in temperature beyond $|\Delta T_3|$ does not cause any change in the curvature of the bilayer. This limiting curvature is

$$\kappa_{R,L} = \frac{6\sigma_{y1}}{h_2 \bar{E}_2} \left\{ \left(\frac{h_1}{h_2}\right)^2 + \frac{h_1}{h_2} \right\}. \tag{3.55}$$

If the applied temperature amplitude for thermal cycling $|\Delta T_a|$ is always smaller than $|\Delta T_1|$, plastic deformation will not occur in the metallic layer. If $|\Delta T_1| \leq |\Delta T_a| < |\Delta T_3|$, the metallic layer will not undergo complete plastic yielding. Figure 3.17(b) schematically shows the evolution of curvature as a function of temperature for a metal–ceramic bilayer where the metallic layer exhibits elastic–perfectly plastic response.

For fluctuations in temperature, there are two additional limiting possibilities for the evolution of cyclic plastic zones in the metallic layer: (i) there occurs no reversed plastic yielding in layer 1, or (ii) the limiting curvature ($\kappa_R = \kappa_{R,L}$) which develops at $|\Delta T_a| = |\Delta T_3|$ is exactly reversed during each temperature reversal. Consider case (i) for which a characteristic temperature, $|\Delta T_2| = 2|\Delta T_1|$, exists. When $|\Delta T_a| < |\Delta T_2|$, no reversed plastic zone develops and thermal cycling does not lead to the accumulation of plastic strains in the metallic layer. It can be shown (e.g., Suresh, Giannakopoulos & Olsson, 1994) that the highest characteristic temperature corresponding to the complete reversal of the limiting curvature from $\kappa_R = \kappa_{R,L}$ to $\kappa_R = -\kappa_{R,L}$, case (ii), is: $|\Delta T_4| \leq 2|\Delta T_3|$. For a general bilayer with any general combination of h_1 and h_2, it is typically seen that $|\Delta T_4| \geq |\Delta T_3| \geq |\Delta T_2| \geq |\Delta T_1|$. For the thin-film limit,

$$\Delta T_1 = \Delta T_3 = \frac{\sigma_{y1}}{\bar{E}_1(\alpha_1 - \alpha_2)}, \qquad h_1 \ll h_2,$$

$$\Delta T_2 = \Delta T_4 = 2\Delta T_3, \qquad h_1 \ll h_2. \tag{3.56}$$

The effects of temperature-dependence of material properties, strain hardening in the metallic layer, and free edges (where singular fields and multiaxial stress states develop) on the evolution of monotonic and cyclic plastic deformation in the metal–ceramic bilayers are addressed in Suresh, Giannakopoulos & Olsson (1994).

Exercises

3.1 It was shown in Chapters 1 and 2 that the resolved shear stress in a single slip system is related to the applied stress by the Schmid factor, M. For a polycrystal, $\sigma = M_T \tau$, where M_T is commonly referred to as the Taylor factor. The value of M_T is computed such that the continuity of slip at the grain boundary satisfies the requirement for five independent slip systems to operate in each grain. The values of M_T are computed by invoking the principle of virtual work. Convince yourself, by consulting appropriate references and standard textbooks on mechanical metallurgy, that for random orientations of grains,

(a) For FCC polycrystals, $M_T = 3.06$.

(b) For BCC polycrystals, $M_T = 2.00$.

(c) Taylor-type calculations are not possible for HCP polycrystals where twinning and nonbasal slip influence deformation.

3.2 The Taylor factor M_T is based on the assumption of multiple slip. Sachs formulated an alternative orientation factor M_S ($= 2.24$ for FCC polycrystals), assuming that individual grains deform in single slip. (Consult standard textbooks on mechanical metallurgy where appropriate references can be found.)

(a) What are the limitations of this assumption in terms of plastic strain?

(b) It has been observed that the orientation factor in general is closer to the Sachs factor for moderately deformed crystals, but then assumes larger values approaching M_T at higher imposed strains. Why?

3.3 Silver solder joints can withstand a much higher normal stress than a tensile specimen made entirely of the solder material. Why?

3.4 Discuss possible effects of reinforcement volume fraction and shape (i.e. spheres, particles with sharp corners, and whiskers) on the development of the Bauschinger effect in metal-matrix composites.

3.5 Two engineering materials are subjected to fully-reversed, strain-controlled, cyclic deformation at room temperature. The first, a substitutional solid soution alloy, exhibits an isotropic hardening response. The second is a composite whose matrix is identical to the first alloy; spherical ceramic particles, 25% by volume, are uniformly dispersed in this matrix.

(a) Speculate on the dislocation–particle interactions in the second material during cyclic plastic deformation.

(b) Schematically sketch the shape of the cyclic stress–strain curve during repeated tension–compression loading.

(c) Discuss the type of hardening rule you would use to model the cyclic constitutive behavior of the composite.

3.6 Using the bounding surface concept and a rationale similar to that used in connection with Fig. 3.14, provide an explanation for the occurrence of cyclic creep in the direction of increasing compressive strain.

3.7 Discuss the cyclic deformation of a material in terms of the two-element Masing model, assuming elastic–perfectly plastic response.

(a) Construct the hysteresis loops for a small amplitude at which only the softer elements yield plastically and for a larger amplitude at which both elements yield plastically.

(b) Show that the so-called 'permanent softening' is expected only in one of the two cases and that the amount of permanent softening is equal to twice the internal back stress in the soft elements.

(c) Considering both cases, can permanent softening be considered an unambiguous measure of the back stress?

(d) Describe the occurrence of the Bauschinger effect in terms of deformation-induced internal stresses and the yielding of the softer elements upon load reversal.

3.8 Discuss the effects of cyclic strain hardening on the shakedown limit derived in Section 3.10.2 for the metal-matrix composite subjected to thermal cycling.

3.9 An aluminum film, 1 μm in thickness, is deposited onto a silicon substrate which is 500 μm thick and 100 mm in diameter. The isotropic properties of Al and Si are: $E_{Al} = 66$ GPa, $\nu_{Al} = 0.33$, $\alpha_{Al} = 23 \times 10^{-6} \,°C^{-1}$, $E_{Si} = 130$ GPa, $\nu_{Si} = 0.28$, $\alpha_{Si} = 3 \times 10^{-6} \,°C^{-1}$. At some reference stress-free temperature, this thin-film/substrate system is flat (i.e. zero curvature). This bilayer is now uniformly cooled by 50 °C.

(a) What is the average stress in the aluminum film?

(b) Is it tensile or compressive?

(c) Describe the direction in which the bilayer bends during the above thermal excursion.

(d) If the yield strength of the thin aluminum film is 140 MPa, what is the temperature change needed to cause plastic yielding in the aluminum film?

(e) If the thin-film/substrate system is thermally cycled between the stress-free initial temperature and some high temperature T_{cycle}, what is the minimum value of T_{cycle} needed to induce fully reversed plastic flow in the entire aluminum film?

3.10 A 50 mm × 50 mm bilayer plate is made by diffusion bonding a plate of pure Ni, 3 mm in thickness, to a plate of pure Al_2O_3, 2 mm in thickness, at 827 °C. The isotropic properties of Ni are: $E_{Ni} = 214$ GPa, $\nu_{Ni} = 0.31$, and $\alpha_{Ni} = 17.8 \times 10^{-6} \,°C^{-1}$ at 827 °C and $13.4 \times 10^{-6} \,°C^{-1}$ at 20 °C. The yield strength of Ni as a function of temperature, $\sigma_{y,Ni}(T)$, are: 148 MPa (20 °C), 140 MPa (227 °C), 115 MPa (427 °C), 69 MPa (627 °C) and 45 MPa (827 °C). The isotropic properties of the polycrystalline Al_2O_3 plate are: $E_{Al_2O_3} = 380$ GPa, $\nu_{Al_2O_3} = 0.25$, and $\alpha_{Al_2O_3} = 9.4 \times 10^{-6} \,°C^{-1}$ at 827 °C and $5.4 \times 10^{-6} \,°C^{-1}$ at 20 °C. You may ignore the strain hardening characteristics of Ni for the purpose of this problem.

(a) If the bilayer is uniformly cooled to room temperature (20 °C), what is the temperature at which plastic yielding begins during cooling from the bonding temperature?

(b) Is it possible to cause yielding of the entire Ni layer before reaching the room temperature?

CHAPTER 4

Fatigue crack initiation in ductile solids

The initiation of fatigue cracks is an event whose very definition is strongly linked to the size scale of observation. For example, materials scientists are likely to consider the nucleation of flaws along persistent slip bands as the initiation stage of fatigue failure, whilst a mechanical engineer may associate the resolution of crack detection with the threshold for crack nucleation. Between this wide range of viewpoints lies a variety of failure mechanisms that are affiliated with the inception of microscopic flaws at grain boundaries, twin boundaries, inclusions, microstructural and compositional inhomogeneities, as well as microscopic and macroscopic stress concentrations. The differences in the approaches to fatigue crack initiation constitute the fundamental distinction between the fatigue design philosophies currently practiced in industry. From a scientific standpoint, developing a quantitative understanding of crack initiation processes must be regarded as one of the most important tasks.

In this chapter, attention is first directed at the mechanisms of fatigue crack initiation in nominally defect-free (unnotched) pure metals and alloys, and commercial materials. Models for fatigue crack initiation are described and their significance and limitations are pinpointed. Also addressed are the mechanisms by which fatigue cracks initiate ahead of stress concentrations under fully compressive cyclic loads. Continuum aspects of crack initiation based on stress–life and strain–life approaches are addressed in Chapters 7 and 8, respectively, where the initiation of fatigue cracks at stress concentrations under tension and tension–compression fatigue are treated. Possible ways in which the seemingly conflicting viewpoints based on crack initiation and crack growth can be brought together are described in Chapter 15. Fatigue crack initiation in brittle solids and noncrystalline solids are considered in Chapters 5 and 6, respectively.

4.1 Surface roughness and fatigue crack initiation

The origin of fatigue cracks in metals and alloys of high purity is often rationalized by mechanisms of the type first proposed by Wood (1958). The basic premise of Wood's postulate is that repeated cyclic straining of the material leads to different amounts of *net* slip on different glide planes. The irreversibility of shear displacements along the slip bands then results in the 'roughening' of the surface of the material. This roughening is manifested as microscopic 'hills' and 'valleys' at sites where slip bands emerge at the free surface. The valleys so generated function as micronotches and the effect of stress concentration at the root of the valleys promotes additional slip and fatigue crack nucleation.

4.1.1 Earlier observations and viewpoints

The first documentation of slip-induced surface roughening during fatigue was by Forsyth (1953). He reported that, in a solution-treated Al–4 wt % Cu alloy, thin ribbons of the metal (0.1 μm thick and 10 μm long) 'extruded' at the specimen surface from the persistent slip bands. Similar extrusions were also seen in single crystals and polycrystals of silver chloride (Forsyth, 1957). The valleys and hills formed on the fatigued surface are commonly referred to as 'intrusions' and 'extrusions', respectively. The formation of an intrusion–extrusion pair during fatigue was identified by Forsyth & Stubbington (1955) in Al–4.5 wt % Cu and by Cottrell & Hull (1957) in Cu. Figures 4.1(*a*) and (*b*) show a pair of intrusions and extrusions, which were formed within 1% of the expected fatigue life, along slip bands in Cu cyclically strained at −183 °C. Both micrographs were taken from the surface of the same fatigue specimen.

Wood's hypothesis on the creation of surface roughness due to the to-and-fro motion of slip bands does not explain why the intrusions progressively deepen. In an attempt to rationalize the development of *net* slip offsets, Mott (1958) proposed a qualitative model. His suggestion was that screw dislocations, moving along different paths in the slip bands of a crystal during forward and reverse glide, repeat their paths by cross slip. The screw dislocations complete a circuit during a fatigue cycle; the volume encompassed by the circuit is then translated parallel to the dislocation by a distance equal to its Burgers vector. This displacement manifests itself in the

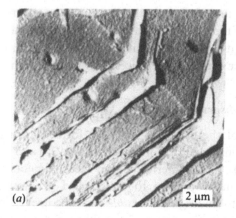
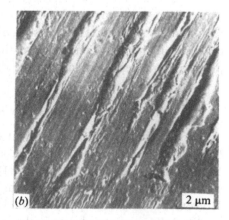

Fig. 4.1. (*a*) Intrusions and (*b*) extrusions along slip bands in polycrystalline Cu fatigued at −183 °C. (From Cottrell & Hull, 1957. Copyright The Royal Society, London. Reprinted with permission.)

form of an extrusion at the specimen surface. Although Mott's proposal formed the basis for a number of subsequent models for surface roughening, its basic feature that screw dislocations travel in a closed circuit has not been convincingly substantiated by experiments. Kennedy (1963) emphasized the need for a *gating mechanism* which would modify the forward–reverse oscillations of screw dislocations into irreversible displacements. The formation of obstacles to dislocation motion, such as creation of jogs as a consequence of edge–screw intersections and the intersection of two screw dislocations with a third dislocation at a node in a free surface, have all been suggested as possible gating mechanisms which would provide net irreversible slip during fatigue (see Lin & Lin, 1979 for a critical review).

The first quantitative statistical model for random slip leading to the formation of hills and valleys on fatigued surfaces was published by May (1960a,b) who adapted a variation of Mott's cross slip mechanism. The assumption here is that the reverse glide of dislocations is shifted from the forward path in a random manner. If the amount of this shift is comparable to the width of the slip band, this is tantamount to stating that random distribution of slip in each half-cycle is independent of the distribution in the previous cycles. Thus, hills and valleys are formed on the surface, with subsequent slip concentrating in the valleys in proportion to their depths. May assumed that if $f(z, N)$ is the fraction of the valleys (of width w) with a depth between z and $z + dz$ after N cycles of fatigue, $f(z, N)$ obeyed a diffusion equation

$$\frac{\partial f}{\partial N} = D\frac{\partial^2 f}{\partial z^2}, \qquad D = kbw\gamma\left(1 + \frac{\beta z}{w}\right), \tag{4.1}$$

where k and β are factors of the order of unity, b is the magnitude of the Burgers vector, and γ is the plastic strain in the slip band which is equivalent to γ_{PSB} in Eq. 2.9. Solving Eq. 4.1, May showed that

$$f = F\exp\left(-\left[\left\{\sqrt{(z + w)} - \sqrt{w}\right\}^2/b\gamma N\right]\right), \tag{4.2}$$

where F is a slowly varying function of z and N. For small values of N, the exponential factor is very small and the deepening of the valleys would not be significant. The main criticism of this statistical model is that it does not contain a sufficient number of physical variables which incorporate the random slip process in a realistic form.

4.1.2 Electron microscopy observations

A quantitative measurement of the height of slip steps formed during the fatigue of Cu single crystals has been conducted by Finney & Laird (1975) and Laird, Finney & Kuhlmann-Wilsdorf (1981). Using interferometric measurements, they showed that surface slip steps form in proportion to the applied plastic strain. However, individual steps were found to undergo *irreversible* slip. The irreversibility of slip steps was noticed even in the inert environment of dry nitrogen. Later studies by Cheng & Laird (1981) revealed that crack nucleation occurred preferentially at the site of the PSBs with the highest slip offset and the largest strain localization. For

Cu single crystals, γ_{pl} and the slip offset (denoted by $n_d b$, where n_d is the number of dislocations and b is the modulus of Burgers vector) were related by the expression,

$$\gamma_{pl} \propto (n_d b)^{m_p}. \tag{4.3}$$

For Cu single crystals subjected to a plastic strain amplitude in the range of 10^{-3} to 10^{-2}, $m_p \approx 0.78$ and $n_d b = 0.3$–$3\,\mu m$.

A technique, known as the taper-sectioning method, introduced by Wood (1958), has allowed high-resolution imaging of surface morphology in fatigued crystals. In this method, the specimen is sectioned along a plane which is oriented at a small angle α (i.e. a few degrees) to the specimen surface. The profile of the surface, as observed on the sectioned plane, is magnified by a factor of $(1/\sin\alpha)$, over and above the magnification obtained by other visual methods such as scanning electron microscopy, optical microscopy and optical interferometry. Hunsche & Neumann (1986) have refined this technique to obtain sections with sharp edges normal to the specimen surface so that the surface features can be recorded to a resolution of 20 nm in the scanning electron microscope. Basinski & Basinski (1984), Hunsche & Neumann (1986), and Ma & Laird (1989a,b) have utilized this method to examine the details of surface roughening in fatigued Cu. Their results collectively indicate the following general trends:

(1)   The surface of the fatigued crystal is covered with PSB extrusions, intrusions and protrusions. A *protrusion* is a surface uplift (a large extrusion), many micrometers in height, where a macro-PSB, tens of micrometers wide and containing tens of matrix and/or PSB lamellae, emerges at the free surface. A protrusion may contain several intrusions and extrusions. Figure 4.2 is a scanning electron micrograph obtained using the sectioning method which

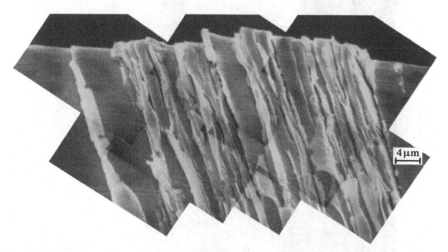

4μm

Fig. 4.2. Protrusions with extrusions and intrusions on the surface of a Cu crystal fatigued at room temperature for 120 000 cycles at $\gamma_{pl} = 0.002$. (From Ma & Laird, 1989a. Copyright Pergamon Press plc. Reprinted with permission.)

shows PSB protrusions with superimposed extrusions and intrusions in a Cu crystal fatigued for 120 000 cycles at a constant resolved plastic shear strain amplitude of 0.002.

(2) The extrusions, with a triangular cross section (base width $\approx 1 - 2\,\mu m$ and height $\approx 2 - 3\,\mu m$), grow at rates of $1 - 10\,nm\,(cycle)^{-1}$, whereas the growth rate of protrusions is an order of magnitude smaller.

(3) The lamallae within the protrusions undergo constrained slip in the direction of the primary Burgers vector. Concomitantly with the formation of protrusions, there develops a population of negative protrusions or encroachments.

(4) The protrusion height increases in proportion to the width of the macro-PSBs.

(5) Optical interferometry indicates that a greater than average local strain develops at the PSB–matrix interface.

The contour of a PSB profile, imaged directly by focusing on the specimen edge, is provided in Fig. 4.3. This figure shows a protrusion on the side surface of a Cu specimen fatigued at 77 K for 35 000 cycles at $\gamma_{pl} = 0.002$. The loading axis is along the vertical direction and **b** is the direction of the primary Burgers vector. A similar extrusion/protrusion is found on the opposite side of the crystal (not shown in Fig. 4.3).

Numerical simulations of the random slip processes have also been attempted to quantify the extent of surface protrusions produced by fatigue. Differt, Essmann &

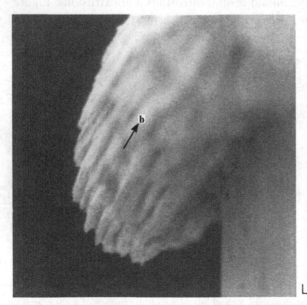

Fig. 4.3. The contour of a PSB profile created in a Cu crystal. (From Differt, Essmann & Mughrabi, 1986. Copyright Taylor & Francis, Ltd. Reprinted with permission.)

Mughrabi (1986) considered cyclic slip of a stack of planes by a random distribution of irreversible microscopic displacements. The mechanism for the development of slip irreversibility was taken to be the dynamic equilibrium between dislocation multiplication (at a Frank–Read source) and annihilation within the PSBs (see Section 4.2), which produces microscopic extrusions and depressions on the specimen surface. For an individual PSB, the mean width (peak to valley) of the roughness profile was predicted to be

$$\bar{w} = 2F \sqrt{Nb\gamma_{\text{PSB}}p_{\text{PSB}}h_{\text{p}}}, \tag{4.4}$$

where F is a nondimensional factor, b is the magnitude of the Burgers vector for the dislocations within the PSB, N is the number of fatigue cycles, p_{PSB} is the ratio of the irreversible strain to the total strain within the PSB, and h_{p} is the thickness of the PSB. For Cu single crystals, taking $b = 0.256$ nm, $h_{\text{p}} = 1000$ nm (for a PSB lamella composed of 5000 atomic planes with a mean spacing of 0.2 nm), $p_{\text{PSB}} \approx 0.3$, $\gamma_{\text{PSB}} = 0.0075$ (from Section 2.4) and $F \sim 1$, the mean height of the extrusion at $N = 10^6$ strain cycles is found to be $\sim 0.76\,\mu\text{m}$. This value is of the order of the slip step height observed in experiments. It should be pointed out that this model, where the quantitative estimates for the parameters for slip within a PSB are only rough estimates, cannot account for the interaction among closely spaced PSBs or for the formation of intrusions below the average level of the surface.

4.2 Vacancy-dipole models

The nucleation of PSBs at the beginning stages of cyclic saturation is also accompanied by the formation of extrusions. This surface roughening appears to be instigated when the average dislocation distance in the fatigued matrix approaches the annihilation distance for dislocations. Transmission electron microscopy studies reveal that there is a critical spacing between dislocations, below which their annihilation is favored. Consider two screw dislocations of opposite signs (left hand screw S_{LH} and right hand screw S_{RH}), gliding in the PSB channels (see Fig. 4.4(a)). When the distance between this pair of dislocations becomes less than a critical spacing, $y_{\text{s}} \approx 50$ nm at room temperature, they annihilate each other by cross slip (Mughrabi, Ackermann & Herz, 1979; Essmann, Gösele & Mughrabi, 1981). Similarly, a dipole consisting of edge dislocations of opposite sign will annihilate to form a vacancy if the spacing of the edge dislocations, y_{e}, becomes smaller than about 1.6 nm, as shown in Fig. 4.4(a). (Note that the edge dislocation pair in Fig. 4.4(a) is a vacancy dipole. An interstitial dipole is one where the signs of the edge dislocations on the two parallel planes, 1 and 2, are opposite to those shown in this figure.) It has been found in TEM that the majority of the dislocation dipoles that are observed after fatigue are of vacancy-type (Antonopoulos, Brown & Winter, 1976). This result also finds direct experimental support in the electrical conductivity measurements by Johnson & Johnson (1965) and Polák (1970), which show that

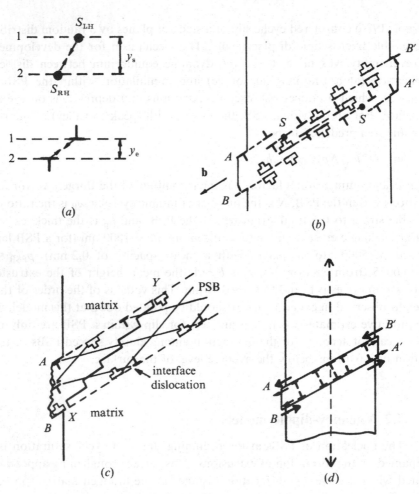

Fig. 4.4. (*a*) The critical annihilation distance for screw and edge dislocations. (*b*) Mechanism of extrusion formation by combined glide and dislocation annihilation. (*c*) Irreversible slip in the PSB creating effective interfacial dislocations which put the slip band in a state of compression. (*d*) The combined effects of applied stresses and internal stresses. Bigger arrows indicate repulsive forces on interfacial dislocations and smaller arrows denote forces caused by the applied load. (After Essmann, Gösele & Mughrabi, 1981.)

vacancies, of concentration $\sim 10^{-4}$, are generated during fatigue of metals. Such vacancy generation has also been clearly demonstrated by Argon & Godrick (1969) in LiF crystals fatigued at elevated temperature (see Chapter 5). It therefore appears reasonable to suppose that individual or clusters of vacancies produced by cyclic slip are responsible for the swelling of the material which produces protrusions and extrusions in fatigue, as in Fig. 4.3.

Essmann, Gösele & Mughrabi (1981) developed a model for surface roughening and crack nucleation on the basis of the hypothesis that the annihilation of the dislocations within the slip bands is the origin of slip irreversibility. Without annihilation, the to-and-fro motion of dislocations within the walls and channels of the PSB would be reversible and no permanent changes in surface topography would result. Essmann *et al.* propose the following sequence of events:

(*a*) Dislocations that are generated at dislocation sources (e.g., point *S* in Fig. 4.4(*b*)) are terminated by mutual annihilation before the reversal of strain.

(*b*) The annihilation of vacancy-type dipoles, shown in Fig. 4.4(*b*), is the dominant point defect generation process. Dislocations moving during the tensile portion of the fatigue cycle are denoted by solid symbols and those moving during the compression portion by open symbols. During tensile loading, slip is transmitted across the specimen by the sequence of microscopic processes extending from *A* to *A'*.

(*c*) At locations where the edge dislocations are annihilated (e.g., the PSB walls in Cu), the plane on which slip is dominant is changed because of the annihilation process. Therefore, the effective slip plane *A–A'* is not parallel to the primary Burgers vector **b**, but is slightly inclined to **b**. Slip steps are created at surface locations *A* and *A'* during the tensile portion of fatigue.

(*d*) On reversing the strain into compression, slip steps are formed at *B* and *B'* by a similar process. The steps *A–B* and *A'–B'* thus constitute an extrusion. If interstitial-type dipoles, rather than vacancy-type dipoles, are considered, then intrusions, rather than extrusions, form by a process analogous to that described in Fig. 4.4(*b*).

(*e*) The extrusion in Fig. 4.4(*b*) ceases to grow when the concentration of vacancies formed by edge dislocation annihilation attains a saturation value, $[\overline{C}_V]_{sat}$ ($\approx 3 \times 10^{-4}$ for Cu), within the slip band. At saturation, the effective slip plane no longer deviates from **b**.

(*f*) While Fig. 4.4(*b*) illustrates the formation of extrusions by two microscopic slip processes, Fig. 4.4(*c*) schematically shows this phenomenon for the situation involving the superposition of multiple slip processes. The path *X–Y* denotes the zig-zag glide of a combined slip process aided by annihilation, just as in Fig. 4.4(*b*). For clarity, the other paths are merely represented by straight lines. These lines join edge dislocations which have either survived the annihilation process and arrived at the free surface or have been deposited at the PSB–matrix interface. These latter interface dislocations have the same sign and give rise to internal stresses. Thus, the net result of the irreversible slip process after one cycle is a row of edge dislocations at the PSB–matrix interface. The extra half planes of atoms of these edge dislocations face into the PSB.

(*g*) If all the interface dislocations emerge at the free surface, the resulting surface roughness of the PSB lamellae in the direction of the active slip vector is given by

$$e = [\overline{C}_V]_{sat} \cdot d_0 / \cos\theta, \tag{4.5}$$

where d_0 is the diameter of the crystal and θ is the angle between **b** and the specimen surface. If all the dislocations remain at the PSB–matrix interface, the mean separation s_i of the interface edge dislocations is

$$s_i = \frac{b}{[\overline{C}_V]_{sat}} = \frac{b d_0}{e \cos\theta}. \tag{4.6}$$

Taking typical values for b and $[\overline{C}_V]_{sat}$ for Cu, one obtains $s_i \approx 1\,\mu m$.

(*h*) The arrangement of interface dislocations in Fig. 4.4(*c*) leads to an elastic compressive stress within the PSB acting along **b** and to a tensile stress in the matrix adjoining the PSB. Essmann *et al.* estimate that this compressive stress in the PSBs is of the order of 2 MPa for Cu.

(*i*) The combined effect of the applied stress and the internal stress produced by the interface dislocations is shown in Fig. 4.4(*d*). The bigger arrows denote the internal stress arising from the mutual repulsions of the interface dislocations. The smaller arrows refer to the stress resolved in the direction of the PSB due to the far-field axial load, which reverses sign during every half fatigue cycle. Thus, A and A' serve as stress concentration points in tensile loading where the internal stress and the applied stress combine to produce high local stresses. B and B' are stress concentrating sites during compression where the two stresses oppose each other.

Starting with a set of initial assumptions similar to those of Essmann *et al.*, Antonopoulos, Brown & Winter (1976) developed a model of dislocations at PSB–matrix interfaces by considering a continued increase in the density of vacancy dipoles during fatigue. Their approach is different from the foregoing model of Fig. 4.4 in that Antonopoulos *et al.* did not consider the annihilation of vacancy dipoles (which are continually replenished) to form vacancies. They concluded that the material in the PSB is *fiber-loaded in tension* parallel to **b** after the attainment of cyclic saturation. This prediction is contradictory to that of Essmann *et al.* A possible clue to this apparent contradiction can be obtained by noting that both models assume the sign of the internal stress to remain unchanged during fully reversed loading. Brown & Ogin (1985) have pointed out that if the effective slip plane does not switch (from $A–A'$ to $B–B'$ in Fig. 4.4) as the applied load is changed from tension to compression, then the internal stress also changes sign. Despite their differences, both models clearly pinpoint the significant role of vacancy dipoles and interfacial dislocations in promoting surface roughness. Both groups of authors also propose that cracks initiate at the surface steps created at the PSB–matrix interface. This prediction is consistent with a variety of experimental observations, to be described in Section 4.3.

Micromechanical models have also been proposed to describe the formation of intrusions and extrusions on the surfaces of metals due to glide on parallel planes (Lin & Ito, 1969; Lin & Lin, 1979; Tanaka & Mura, 1981). A feature common to these analyses is the assumption that the forward and reverse slip displacements during a fatigue cycle are accommodated within two closely-located, parallel layers, i.e. the most favorably oriented slip planes. Such an assumption apparently finds its basis from the experiments of Forsyth (1953) and Charsley & Thompson (1963), who found that the slip plane accommodating plastic deformation during the forward (tensile) loading and the one during reversed (compressive) loading are closely spaced, but distinct from each other. Both slip planes are still part of the same slip band.

If the dislocations piled up on a slip plane (layer I) under the maximum (positive) shear stress glide in the opposite direction along the same slip band upon load reversal, there is no net dislocation buildup. In order to incorporate slip irreversibility in the model, it is postulated that the irreversibility of dislocation motion in the two adjoining layers arises from the different levels of back stress during slip in the forward direction on layer I and in the reversed direction on layer II (Tanaka & Mura, 1981). The first tensile loading causes dislocation pile-up on layer I. The positive back stress (which opposes the stress causing dislocation motion) due to positive dislocations on layer I facilitates the pile-up of negative dislocations on layer II during reversed loading. The back stress due to dislocations on layer II helps further pile-up of dislocations on layer I during the following forward cycle. This process leads to dislocation pile-up with increasing number of fatigue cycles.

This model has been modified by Venkataraman *et al.* (1990) to represent the matrix–PSB interfaces as an array of vacancy dipoles. The analysis shows that there exists a critical number of fatigue cycles above which the accumulation of interface dislocations becomes energetically unfavorable as a result of an increase in stored elastic strain energy. Crack initiation occurs in the PSB at this point.

4.3 Crack initiation along PSBs

The interface between the PSB and matrix is a plane of discontinuity across which there are abrupt gradients in the density and distribution of dislocations. One may then expect these interfaces to serve as preferential sites for fatigue crack nucleation. As noted earlier, this was also the inference derived from the model of Essmann, Gösele & Mughrabi (1981). Direct experimental evidence of crack initiation at the interface has also been obtained by Hunsche & Neumann (1986) and Ma & Laird (1989a,b).

Figure 4.5 shows a fatigue crack nucleated along the leading edge of a macro-PSB protrusion in Cu fatigued at a constant plastic strain amplitude of 0.002. Concomitant interferometric observations of this test specimen revealed that the strains within the PSB are highly inhomogeneous and localized at the PSB–matrix interface. These results imply that fatigue crack initiation is strongly biased by the roughening of the surface. The population of these fatigue cracks increases linearly with the number of cycles and the applied strains. Furthermore, the statistical distribution of the crack sizes corresponds to the distribution of localized strains measured at the PSBs using white light interferometry.

Direct evidence of crack initiation and early crack growth along PSBs has been obtained in fatigued, polycrystalline Cu by Katagiri *et al.* (1977). With high voltage TEM observations of dislocation arrangements ahead of pre-cracks, Katagiri *et al.*

Fig. 4.5. Fatigue crack initiation (denoted by an arrow) at a PSB–matrix interface in a Cu crystal fatigued for 60 000 cycles at $\gamma_{pl} = 0.002$ at 20 °C. (From Ma & Laird, 1989b. Copyright Pergamon Press plc. Reprinted with permission.)

clearly established that, in materials that form PSBs, crack nucleation and early crack growth occur in the PSB. Figure 4.6 is a TEM image showing two PSBs (of length $\sim 100\,\mu m$, which is comparable to the grain size) in a Cu specimen where a surface layer, approximately $2\,\mu m$ thick, containing intrusions has been removed by electropolishing. A nascent crack is seen within one of the PSBs and the presence of the crack does not appear to modify their dislocation structure. An optical micrograph of the crack is also shown in the inset. A similar TEM image of the Cu specimen obtained by Katagiri *et al.* also revealed cracks nucleating along the PSB–matrix interface at the root of a surface intrusion.

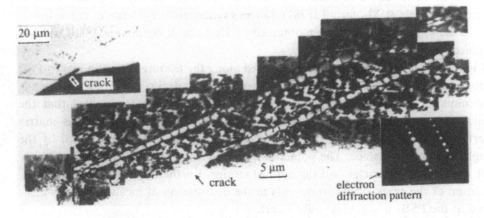

Fig. 4.6. A nascent fatigue crack along the ladder structure of a PSB in fatigued, polycrystalline copper. Inset at left is an optical micrograph showing the location of the crack with respect to the free surface. (From Katagiri *et al.*, 1977. Copyright *Metallurgical Transactions*. Reprinted with permission.)

4.4 Role of surfaces in crack initiation

As discussed in Chapter 2, it is now well accepted that a single PSB can extend throughout the cross section of a ductile monocrystal and that, in polycrystalline metals, slip bands form even within interior grains. It is, therefore, of interest to investigate whether a rough surface topography created by the presence of intrusions alone is responsible for crack nucleation or whether internal dislocation structures can play an equally significant role in determining the nucleation events and the overall fatigue life. In polycrystals (see Chapter 3), PSBs formed within interior grains produce slip that is confined to the individual grain. The large transfer of material causing the creation of rough topography is possible at surface grains, but not in the interior ones because of constraint from the surrounding matrix.

Experiments dating back to the work of Thompson, Wadsworth & Louat (1956) have suggested that removing the intrusions and extrusions by electropolishing the specimen surface increased fatigue life. More conclusive evidence for the argument that the surface geometry determines fatigue life is found in the study of Basinski, Pascual & Basinski (1983). These researchers demonstrated that, even in single crystals where coarse PSBs traverse through the bulk, elimination of surface roughness by electropolishing leads to a drastic enhancement in total fatigue life. This rejuvenation takes place at both low and high plastic strain amplitudes, i.e. at both the beginning and end of regime B in Fig. 2.2.

4.5 Computational models for crack initiation

Finite element simulations of fatigue crack initiation have been carried out by recourse to the constitutive model for the inelastic deformation of PSBs, which was discussed in Section 2.6. The roughening of the surface of the crystal due to vacancy diffusion and the ensuing crack initiation process, as revealed by computational simulations based on these models, are considered here.

4.5.1 Vacancy diffusion

The concentration of vacancies produced within the PSBs is markedly larger than that in the vicinity of the free surface. The annihilation of edge dislocation pairs, schematically sketched in Fig. 2.13, also causes the vacancy concentration in the PSB to surpass that in the matrix. The attendant steep gradient in vacancy concentration provides a driving force for the diffusion of vacancies within the PSBs in the direction of the free surface. This mechanism occurs by lattice diffusion in the crystal matrix as well as by pipe diffusion along the dislocation lines. Since high-cycle fatigue necessarily involves a sufficiently long period of time for the nucleation of a critical fatigue flaw in initially smooth-surfaced high-purity single crystals, one would expect sufficient time to be available for such diffusive processes, even at room temperature.

The evolution of vacancy concentration can then be estimated on the basis of the following considerations and approximations (Repetto & Ortiz, 1997). (1) The stress concentrations arising at the free surface by the roughening induced by the egress of the PSB produces strain energy density gradients, which in turn promote stress-assisted diffusion. (2) The vacancy concentration at the free surface remains at its equilibrium value given by

$$c_{eq} = \exp^{-\Delta G_v/kT}, \tag{4.7}$$

where ΔG_v is the incremental change in free energy per vacancy, k is Boltzman's constant, and T is the absolute temperature. (3) An effective diffusion coefficient for vacancy mobility by both pipe and lattice diffusion is determined from a rule-of-mixture type approximation based on the dislocation density. (4) Since the screw dislocations in the channels between the walls are parallel to the direction of vacancy flux to the free surface, they are postulated to contribute more strongly to vacancy diffusion than edge dislocations.

With the above line of reasoning, the flux of vacancies from the PSBs to the free surface is expected to follow the diffusion equation:

$$\frac{\partial c_v}{\partial t} = \nabla_0 \cdot \left\{ \left[D_{lat} \mathbf{I} + (b^2 \rho_s D_{pipe} \mathbf{s} \otimes \mathbf{s}) \right] \left(\nabla_0 c_v + \frac{c_v}{RT} \Delta W \right) \right\} + \frac{1}{2} \rho^\alpha y_e^2 \dot{\gamma}^\alpha, \tag{4.8}$$

where D_{lat} and D_{pipe} are the lattice and pipe diffusion coefficients, respectively, $\mathbf{I}$ is the identity tensor, b is the magnitude of the Burgers vector, ρ_s is the density of screw dislocations in the channels, $\mathbf{s}$ is the direction of the Burgers vector in the undeformed configuration of the crystal, R is the universal gas constant, and W is the elastic strain energy density. The last term on the right hand side of Eq. 4.8 is the vacancy generation rate defined in Eq. 2.31. The symbol ∇_0 is the material gradient operator defined such that $(\nabla_0 f)_I = \partial f/\partial X_I$, where (X_1, X_2, X_3) refer to the material reference frame defined with respect to the undeformed configuration of the crystal. The term within the square brackets in Eq. 4.8 equals $-\mathbf{J}_v$, where $\mathbf{J}_v$ is the vacancy flux through the crystal.

The net outward flow of vacancies causes the surface of the crystal to move inward, thereby forming an intrusion. The outward velocity of the vacancies, relative to the undeformed configuration of the crystal, is $\mathbf{V}_v = \mathbf{J}_v/c_v$. If one considers a small area on the undeformed surface of the crystal with a unit outward normal $\mathbf{N}$, the inward velocity of the surface region due to the egress is $V_N = \mathbf{J}_v \cdot \mathbf{N}.$†

---

† It is worth noting that the possibly significant process of annihilation of screw dislocations in the channels is not taken into account in this formulation.

4.5.2 Numerical simulations

The constitutive model discussed in Section 2.6 and the fully-coupled vacancy diffusion model discussed in Section 4.5.1 have been incorporated into a finite-element program by Repetto & Ortiz (1997) with the aim to simulate surface roughening in a Cu single crystal. Figure 4.7 shows the deformed mesh from the finite-element simulation where a large surface protrusion is clearly visible. A comparison of Fig. 4.7 with Fig. 4.3 shows that experimental trends of protrusion formation are captured by the numerical model. The shape of the protrusion is determined by a competition between two mechanisms: lengthening of the PSB which causes material to be pushed out of the matrix, and vacancy flux which causes the surface to recede. The former process dominates in the center of the PSB which facilitates the formation of a protrusion. It is also seen that PSB formation, multiple slip at corners and vacancy flux cause grooves to form at both the corners of the protrusion. These grooves serve as critical sites for the nucleation of fatigue cracks.

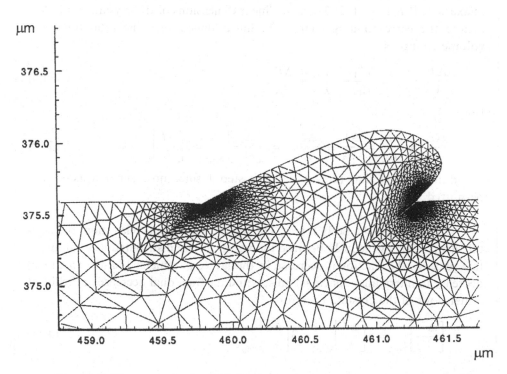

Fig. 4.7. Surface roughening in a [125] Cu single crystal oriented for single slip under symmetric tension–compression loading with $\gamma_{pl} = 6 \times 10^{-3}$. The figure shows the deformed mesh from the finite-element simulation after about 65 000 cycles. (After Repetto & Ortiz, 1997.)

4.5.3 Example problem: Effects of vacancies

Problem:

Repeated cyclic loading on an FCC crystal can produce a high density of vacancies. Show that if a cubic crystal is filled randomly with vacancies to a concentration x_v,

$$x_v = 3 \left(\frac{\Delta L}{L} - \frac{\Delta a}{a} \right),$$

where $\Delta L/L$ is the fractional change of the external edge dimension of the crystal due to the introduction of the defects, and $\Delta a/a$ is the fractional change of the lattice parameter (as measured by X-ray diffraction) due to the defects.[†]

Solution:

In order to arrive at the desired result, carry out the operation of introducing vacancies in the crystal in the following steps:

Step 1:

Create the vacancies inside the crystal by taking atoms from the interior and placing them on various free surfaces of the crystal. Apply, however, a set of body forces around the vacant sites so that no strains develop due to atomic relaxation. If L_i ($i = 1, 2, 3$) are the linear dimensions of the crystal, and if ΔL_i denote the corresponding changes in linear dimensions, the total fractional volume change is

$$\frac{\Delta V}{V} = x_v = \frac{\Delta L_1}{L_1} + \frac{\Delta L_2}{L_2} + \frac{\Delta L_3}{L_3}. \tag{4.9}$$

But,

$$\frac{\Delta L_1}{L_1} = \frac{\Delta L_2}{L_2} = \frac{\Delta L_3}{L_3} \equiv \frac{\Delta L}{L}, \quad \Rightarrow \quad \left[\frac{\Delta V}{V} \right]_{\text{step1}} = 3 \left[\frac{\Delta L}{L} \right]_{\text{step1}} = x_v. \tag{4.10}$$

In addition, note that $(\Delta a/a) = 0$ for step 1 since no atomic relaxation is permitted.

Step 2:

Now release the body forces. During the subsequent relaxation, the crystal shrinks homogeneously because the vacancies are distributed at random. In this process, the fractional decrease in the lattice parameter (i.e. the fractional decrease in interplanar spacings) must equal the fractional decrease in the linear dimensions of the crystal. Therefore,

$$\left[\frac{\Delta V}{V} \right]_{\text{step2}} = 3 \left[\frac{\Delta L}{L} \right]_{\text{step2}} = 3 \left[\frac{\Delta a}{a} \right]_{\text{step2}}. \tag{4.11}$$

[†] The author thanks Professor R.W. Balluffi of the Massachusetts Institute of Technology for bringing this problem to his attention.

Combining Eqs. 4.10 and 4.11, it is seen that

$$3\left[\frac{\Delta L}{L}\right]_{step1} + 3\left[\frac{\Delta L}{L}\right]_{step2} = 3\left[\frac{\Delta a}{a}\right]_{step2} + x_v. \tag{4.12}$$

However,

$$3\left[\frac{\Delta L}{L}\right]_{step1} + 3\left[\frac{\Delta L}{L}\right]_{step2} = 3\left[\frac{\Delta L}{L}\right]_{total} \equiv 3\left[\frac{\Delta L}{L}\right],$$

$$3\left[\frac{\Delta a}{a}\right]_{step1} + 3\left[\frac{\Delta a}{a}\right]_{step2} = 3\left[\frac{\Delta a}{a}\right]_{total} \equiv 3\left[\frac{\Delta a}{a}\right]. \tag{4.13}$$

Using Eqs. 4.10–4.13, it is seen that

$$x_v = 3\left(\frac{\Delta L}{L} - \frac{\Delta a}{a}\right).$$

4.6 Environmental effects on crack initiation

Surface roughening and fatigue crack initiation can occur in pure materials *in vacuo* and at temperatures down to 4.2 K (McCammon & Rosenberg, 1957). However, the fraction of fatigue life at which crack nucleation occurs can be significantly affected by the test environment. There is a wealth of experimental evidence indicating that the environment plays an important role in dictating the extent of slip irreversibility and fatigue life. For example, it was demonstrated a long time ago by Gough & Sopwith (1932) and by Thompson, Wadsworth & Louat (1956) that fatigue life is markedly improved *in vacuo* or in dry, oxygen-free media as compared to moist laboratory air.

Consider the case of fatigue in pure metals *in vacuo* or in an inert environment. *Single slip* during the tensile loading cycle produces slip steps at the surface. The extent of surface slip offset can be diminished by reverse slip during unloading or subsequent compression loading in fully reversed fatigue. In inert environments, surface roughening during fatigue occurs primarily by a random process (see Section 4.1.1). On the other hand, when slip steps form during the tensile portion of a fatigue cycle in laboratory air or in a chemically aggressive medium, the chemisorption of the embrittling species (such as oxygen or hydrogen) or the formation of an oxide layer on the freshly formed slip step makes reverse slip difficult on the same slip plane upon load reversal. In the embrittling medium, this process (schematically shown in Fig. 4.8) can provide a mechanism of enhanced surface roughening as well as easier transport of the embrittling species to the bulk of the material preferentially along the persistent slip bands, thereby facilitating crack nucleation.

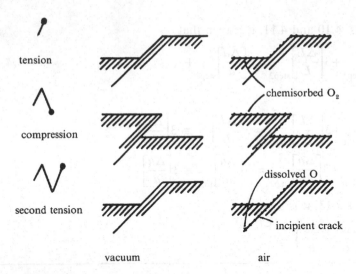

tension

compression

second tension

chemisorbed O_2

dissolved O

incipient crack

vacuum   air

Fig. 4.8. A model for fatigue crack nucleation near a free surface by the synergistic effect of single slip and environmental interactions. (After Thompson, Wadsworth & Louat, 1956, and Neumann, 1983.)

4.7 Kinematic irreversibility of cyclic slip

The preceding discussions focused on the irreversibility of displacements during the forward and reverse slip of dislocations. The various mechanisms of cyclic slip irreversibility can be summarized as follows:

(1) Cross slip of screw dislocations and different paths for their forward and reverse glide during a complete fatigue cycle.

(2) The extrusions, with a triangular cross section (base width $\approx 1 - 2\,\mu m$ and height $\approx 2 - 3\,\mu m$), grow at rates of $1 - 10\,nm\,(cycle)^{-1}$, whereas the growth rate of protrusions is an order of magnitude smaller.

(3) Random distribution of slip (independent of prior history) with the progressive, preferential deepening of valleys at surfaces.

(4) Dislocation–dislocation interactions leading to the formation of nodes, jogs or dislocation locks which impede motion during part of a fatigue cycle.

(5) Production of point defects during saturation, due to the dynamic equilibrium between dislocation generation and annihilation.

(6) Irreversibility due to shape changes (see Section 2.12.1) as well as differences in dislocation back stress due to slip on different glide planes during the tension and compression portions of fatigue.

(7) Reduction in slip displacement during unloading due to the adsorption of an embrittling species or due to the oxidation of slip steps, and the attendant creation of *net* slip irreversibility.

In brittle solids and semi- or noncrystalline materials, kinematic irreversibility of cyclic deformation can come about even in the complete absence of dislocation motion. The underlying mechanisms are examined in Chapters 5 and 6.

4.8 Crack initiation along grain and twin boundaries

The nucleation of fatigue cracks at grain boundaries occurs under the influence of embrittling environments (which preferentially attack grain boundaries and the particles, if any, on them) and elevated temperatures (at which grain boundary cavitation and sliding are promoted). Intergranular failure is also commonly observed in brittle solids due, at least in part, to the residual stresses induced by thermal contraction mismatch between adjacent grains or to the presence of grain boundary glass phases. The occurrence of grain boundary fatigue crack nucleation in a ductile solid, in the absence of grain boundary particles, creep deformation or environmental influences, is relatively less common. There have been some documented cases of purely mechanical fatigue failure along grain facets (e.g., Porter & Levy, 1960; Kim & Laird, 1978; Figueroa & Laird, 1983; Watanabe, 1985).

Using optical interferometric measurements of slip step heights at grain facets in fatigued copper, Kim & Laird (1978) noted that fatigue cracks may nucleate at grain boundaries if: (i) the grain boundaries separate highly misoriented grains, (ii) the active slip system of at least one of the grains is directed at the intersection of the boundary with the specimen surface, and (iii) the traces of the high angle grain boundaries in the free surface make a large angle (30–90°) with the tensile stress axis.

In general, grain boundary cracking may arise from one of two mechanisms during cyclic loading:

(1) At low to intermediate plastic strain amplitude, the impingement of PSBs at grain boundaries causes cracking (e.g., Figueroa & Laird, 1983; Mughrabi, et al., 1983). Figure 4.9(a) shows an example of this crack initiation mechanism after 7000 cycles at a plastic strain range of $\pm 5 \times 10^{-4}$ in copper at sites where primary slip bands intersect a grain boundary.

(2) At high plastic strain amplitudes, grain boundary cracking occurs as a consequence of surface steps formed at the boundary. Figure 4.9(b) is an example of this process. It shows optical interferograms of a slip-step of height 0.9 µm formed in polycrystalline Cu after 60 cycles of fully reversed fatigue loading at a plastic strain of $\pm 7.6 \times 10^{-3}$. Fringes shifted from left to right indicate depressions.

In BCC metals such as commercially pure iron, intergranular crack nucleation has also been observed under reversed bending and push–pull axial loading over the cyclic frequency range 0.01–1000 Hz (Guiu, Dubniak & Edward, 1982). It was mentioned earlier (Section 2.12) that the asymmetry of slip associated with the glide of screw dislocations in tension and compression can induce shape changes

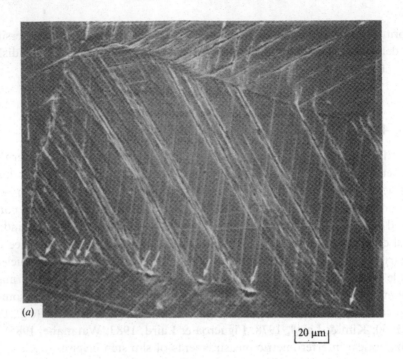

Fig. 4.9. (*a*) Nucleation of flaws (denoted by arrows) along a grain boundary. (From Figueroa & Laird, 1983. Copyright Elsevier Sequoia, S.A. Reprinted with permission.) (*b*) White light interferograms showing slip-step formation at grain boundary in fatigued Cu. (From Kim & Laird, 1978. Copyright Pergamon Press plc. Reprinted with permission.) The dark diagonal lines parallel to the arrow are fiducial markers whose separation is 100 μm.

in BCC single crystals. The surface roughness created by similar shape changes in the near-surface grains of polycrystalline BCC metals, such as α-iron, can cause intergranular crack nucleation (Mughrabi, Herz & Stark, 1981).

The process of slip involves a simple translation of atoms across a glide plane such that a rigid block of solid on one side of the slip plane moves with respect to the other in the direction of slip. Slip occurs by translations in whole multiples of the Burgers vector, so that the relative crystallographic orientation of different regions in a slipped material remains the same. On the other hand, a twin boundary is a surface where the atom positions in the twin on one side of the boundary are a mirror image of those in the untwinned matrix material on the other side of the boundary. Therefore, one observes a shape change in a twinned body.

Although twin boundaries are grain interfaces with the lowest energy, their role in crack nucleation has long been known (e.g., Thompson, Wadsworth & Louat, 1956; Boettner, McEvily & Liu, 1964; Neumann & Tönnessen, 1988). In FCC metals, twin boundaries are parallel to the slip planes so that the PSBs can fit into the region of high local stresses at the boundary. Therefore, the geometric relationship between the boundary and the slip plane may provide possible clues to the role of twins in fatigue crack nucleation. An intriguing aspect of fatigue crack nucleation at twin boundaries is that, in a stack of lamellar twins, there is a propensity for slip bands and cracks to form only at *every other* twin boundary (Boettner, McEvily & Liu, 1964). This trend has also been studied in greater detail in the context of annealing twins formed in polycrystalline Cu, Ni and austenitic stainless steel (Neumann & Tönnessen, 1988). Figure 4.10 shows an example of fatigue crack nucleation at every other twin boundary in Cu fatigued at room temperature.

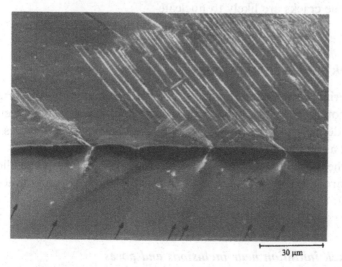

Fig. 4.10. Nucleation of fatigue cracks along every other twin boundary (indicated by arrows) in polycrystalline Cu fatigued at room temperature. (From Neumann & Tönnessen, 1988. Reprinted with permission from P. Neumann.)

Using the taper sectioning technique (discussed in Section 4.1.2) and electron channeling method in the SEM, Neumann & Tönnessen have detected the orientations of grains and the nucleation of microcracks at twin boundaries in FCC metals. They found that, at low imposed stress amplitudes, PSB formation within the grains was suppressed. However, PSBs were found exclusively parallel to and coincident with twin interfaces. Careful grain orientation measurements revealed that, even when slip activity ceases within the interior of grains, slip bands are activated at some twin interfaces as a consequence of local stress concentrations.

Neumann & Tönnessen have rationalized the observations of fatigue crack formation at every other twin boundary utilizing a mechanism which relies on the elastic anisotropy of the material containing the twins. For example, the shear modulus of Cu varies with direction by as much as a factor of 3.2. In order to ensure strain compatibility at twin boundaries in the elastically anisotropic material, internal stresses must be generated in the vicinity of the twin boundaries. Consider a stack of lamellar twins, where the crystallographic orientation of the lamellae changes back and forth from that of the matrix to the twin to the matrix as one traverses across the boundaries. With the change in orientation, there is also a change in the direction of internal stresses. The internal stresses act in concert with the resolved stresses from the applied loads at every other twin boundary. When the resultant stress is of sufficiently high magnitude, a PSB is formed near the twin–matrix interface and eventually develops into a fatigue crack. At the alternate boundaries, the internal stresses oppose the resolved stresses so that slip is obstructed. Neumann & Tönnessen used this approach to predict the internal stress field in the vicinity of twin boundaries using elasticity theory. Their simulation, in conjunction with electron channeling measurements of orientation of grains, correctly predict the twin boundaries at which fatigue cracks are likely to nucleate.

4.9 Crack initiation in commercial alloys

In engineering components made of commercial materials, the principal sites of heterogeneous fatigue crack nucleation include voids, slag or gas entrapments, inclusions, dents, scratches, forging laps and folds, macroscopic stress concentrations, as well as regions of microstructural and chemical nonuniformity. While surface grains are the most likely locations for crack initiation in metals and alloys of high purity, the formation of fatigue cracks is feasible at both near-surface and interior locations in commercial alloys.

4.9.1 Crack initiation near inclusions and pores

The fatigue lifetime and the maximum fatigue strength of commercial alloys are decreased by the presence of inclusions and pores (generally classified as defects).

The mechanisms of fatigue crack initiation at defects depend upon a number of mechanical, microstructural and environmental factors. These factors involve the slip characteristics of the matrix, the relative strength values of the matrix and the defect, the strength of the matrix–inclusion interface and the relative susceptibility of the matrix and the inclusion to corrosion in the fatigue environment (e.g., Cummings, Stulen & Schulte, 1958; Bowles & Schijve, 1973).

The effect of inclusions on fatigue crack initiation is often specific to the alloy system. Here we consider three examples.

(*a*) In high strength steels containing MnS particles, the initial stage of fatigue damage is the debonding of the inclusion from the matrix. This occurs by a quasi-static mode of failure in that the interfacial separation is induced during the very first tensile loading at far-field stress levels close to the threshold stress range for infinite fatigue life. Figure 4.11 shows an example of a partial debonding of an MnO–SiO_2–Al_2O_3 inclusion from the matrix of a 4340 steel and the nucleation of a fatigue crack normal to far-field tension.

(*b*) In aluminum alloys, constituent particles such as the S-phase (Al_2CuMg) and β-phase (Al_7Cu_2Fe), typically 1–10 µm in diameter, provide sites for fatigue crack nucleation. The type of cracking, however, is a function of microstructure and test conditions. (i) In 2024-T4 aluminum alloy, Grosskreutz & Shaw (1969) noticed the debonding of the particle–matrix interface *after* cyclic damage within the matrix over a large number of fatigue cycles. (ii) Another type of crack nucleation in the 2024-T4 alloys involves cracking along slip bands emanating from or terminating at the inclusions. Figure 4.12 presents the experimentally determined relative

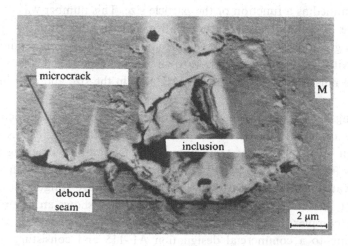

Fig. 4.11. Scanning electron micrograph showing the nucleation of a fatigue crack normal to the tensile axis (vertical direction) at the site of an MnO–SiO_2–Al_2O_3 inclusion which is partially debonded from the 4340 steel matrix denoted M. (From Lankford & Kusenberger, 1973. Copyright *Metallurgical Transactions*. Reprinted with permission.)

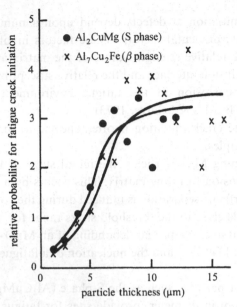

Fig. 4.12. Relative probability of crack initiation versus the constituent particle thickness normal to the stress axis for S and β inclusions in 2024-T4 aluminum alloy. (After Kung & Fine, 1979.)

probability for the initiation of fatigue cracks at two different types of constituents, S-phase and β-phase particles, in a commercial 2024-T4 aluminum alloy as a function of the particle thickness, measured in the direction normal to the stress axis. Here, the number of particles in the vicinity of which matrix cracks were initiated was determined as a function of the particle size. This number was divided by the particle size distribution to give the relative fatigue crack initiation probability curves (Kung & Fine, 1979). Note the precipitous increase in crack nucleation probability with an increase in the size of the inclusion and the scatter associated with the measurements due to the variability in the size and distribution of the inclusions.

(c) In high strength nickel-base superalloys, crack initiation has been identified with the existence of large defects, either pores or nonmetallic inclusions (e.g., Hyzak & Bernstein, 1982). At room temperature, crystallographic cracking at or near the surface is initiated at the sites of the defects at both low and high strain ranges. At an elevated temperature of 760 °C, low strain range fatigue results in crack nucleation at the interior of the specimen. Figure 4.13 shows an example of this process in a complex nickel-base superalloy (made by powder metallurgy methods conforming to a commercial designation AF-115 and consisting principally of Ni–Cr–Co–W–Ti–Al–Mo–Hf–Cb–C) where subsurface crack initiation occurs at an HfO_2 inclusion. At high strain range values, however, near-surface crack nucleation is dominant.

Fig. 4.13. Subsurface fatigue crack initiation at an HfO_2 inclusion in an AF-115 nickel-base superalloy at 760 °C. (From Hyzak & Bernstein, 1982. Copyright *Metallurgical Transactions.* Reprinted with permission.)

4.9.2 Micromechanical models

Several analytical models which estimate the number of fatigue cycles required for crack nucleation in the vicinity of defects have been developed over the years. These models, which have been compared with experimental results with varying degrees of success, can be broadly categorized into the following groups:

(*a*) Analyses wherein the voids or debonded inclusions are regarded as notches and the fatigue limit is estimated from the reduction in fatigue strength due to these defects (e.g., Tanaka & Mura, 1982). Notch effects on crack initiation are addressed in Chapters 7 and 8.

(*b*) Analyses wherein the pile-up of dislocations at the inclusion is considered responsible for cracking of the inclusion or interfacial debonding with the attendant inception of a fatigue flaw within the matrix.

(*c*) Formulations wherein the threshold stress intensity factor range is related to the square root of the (statistical average) projected area of the defects on a plane normal to the maximum tensile stress (e.g., Murakami & Endo, 1986).

In a set of models developed for fatigue crack nucleation from intermetallic particles, Chang, Morris & Buck (1979) and Morris & James (1980) considered a dislocation pile-up process which views crack initiation as being composed of two successive events: cracking inside the brittle particle and the onset of crack advance from the brittle particle into the ductile matrix. The first stage of particle cracking was assumed to occur when a critical elastic strain energy was reached inside the particle near the site of dislocation pile-up. The second stage, i.e. crack advance inside the matrix, was assumed to occur when the total energy of the system attained a minimum. The total energy comprises four terms: the elastic strain energy from the stress field of the piled-up dislocation array, the effective surface energy needed for crack advance, the work done by the applied stress in opening the crack, and the elastic strain energy of the crack under the applied stress field.

Tanaka & Mura (1982) presented an extension of the parallel layer model for surface roughening and slip band fracture to include crack initiation from intermetallic particles in high strength steels and aluminum alloys. The initiation of the crack was assumed to be determined by the energy criterion that the fatigue flaw initiates when the self strain energy of dislocation dipoles accumulated at the inclusion reaches a critical value. Tanaka & Mura considered three different processes of fatigue crack initiation: a slip band crack emanating from a debonded inclusion, a slip band crack initiating from an uncracked inclusion and inclusion cracking due to the impingement of slip bands. For the last mechanism, which is believed to be representative of inclusion cracking in aluminum alloys, the solution was obtained using Eshelby's equivalent inclusion method (Eshelby, 1957).

4.10 Environmental effects in commercial alloys

If a cyclically loaded engineering component is exposed to a chemically aggressive medium during service, preferential attack of the environment at select locations on the material surface may provide nucleating sites for fatigue cracks (Fig. 4.14). These sites are generally corrosion pits which form at surface locations where:

(1) Slip steps or intrusions are created at the surface.
(2) Grain boundaries, either embrittled by temper treatments (as in alloy steels) or surrounded by precipitate-free zones (as in age-hardened alloys) intersect the surface.
(3) The protective oxide layer on the surface is partially broken, exposing the underlying fresh metal to preferential chemical attack.
(4) Inclusions, such as MnS particles in steels, debond from the surrounding matrix at near-surface locations.
(5) One of the constituent phases in a multiphase alloy is preferentially corroded.

Corrosion pits are typically smaller than a millimeter in depth and serve as micronotches which locally elevate the stress level. Furthermore, the pH level of the corrosive medium inside the pit can be more acidic than that in the bulk, causing possible acceleration in the rate of fatigue crack growth. Experimental results obtained for Ni–Cr–Mo–V steels, austenitic stainless steels and aluminum alloys

Fig. 4.14. A fatigue crack initiated at corrosion pits in stress relief groove in a low pressure turbine rotor made of Ni–Cr–Mo–V steel. (From Lindley, 1982. Reprinted with permission from T.C. Lindley.)

have established that the formation of corrosion pits on the initially smooth surfaces of the fatigue specimen results in a significant reduction in the fatigue strength.

Figure 4.14 shows an example of fatigue cracks initiated at corrosion pits in stress relief groove in a low pressure turbine rotor made of a Ni–Cr–Mo–V steel. The rotor consists of stepped shafts with shrunk-on discs which carry the turbine blades. A combination of the steady mean stress due to the shrink fit and the pulsating stresses caused by the self-weight bending resulted in the initiation of fatigue cracks at corrosion pits. In the case of the turbine component shown in Fig. 4.14, the fatigue cracks led to the complete failure of the low pressure turbine shaft.

4.11 Crack initiation at stress concentrations

The initiation of fatigue cracks at stress concentrations is a topic of considerable interest in a wide variety of engineering applications. The issue of crack initiation at notches and other stress raisers subjected to cyclic tension or tension–compression fatigue has been investigated extensively in the context of stress–life approach, strain–life approach and fracture mechanics approach. These approaches will be addressed in Chapters 7–9.

Here we consider the processes by which fatigue cracks initiate ahead of stress concentration under fully compressive cyclic loads, primarily because the mechanistic origins of crack inception in this case can be treated without recourse to the total life approaches or defect-tolerant approach. Fatigue crack initiation in ceramics and polymers subjected to cyclic compression loading is examined in Chapters 5 and 6, respectively.

4.11.1 Crack initiation under far-field cyclic compression

The analysis of fatigue cracking is generally based on the premise that fully compressive cyclic loads or compressive stresses superimposed periodically or randomly upon tensile cyclic loading patterns do not significantly modify the inception and growth of fatigue cracks. This notion arises from the tacit assumption that fatigue cracks remain closed during compression loading or even below a certain tensile load known as the crack closure load (see Chapter 14). There is, however, a wealth of information available in the literature which has unequivocally established that far-field compressive stresses can have a marked effect on both the initiation and propagation of fatigue cracks in both brittle and ductile solids. A neglect of the influence of compression cycles in fatigue can lead to a nonconservative estimate of the useful fatigue life.

The initiation and growth of fatigue cracks under the influence of imposed cyclic compressive stresses are problems of significant practical importance in a wide variety of engineering applications. Examples of metallic structural components periodically subjected to compressive stress cycles include: landing gear and wing-root sections of aircraft,† coil springs, shellings and fillets joining the web and head of rails in railroad applications, near-surface regions of shot-peened materials, weldments containing residual compressive stresses, and components of deep-diving submersibles. The occurrence of this phenomenon can easily be tested in cast iron by subjecting it to a few cycles of zero–compression–zero loading. Small fatigue cracks, initiating around the defects and cavities in this material, are readily visible after such compression fatigue.‡

It has long been recognized (e.g., Gerber & Fuchs, 1968) that the application of uniaxial cyclic compressive loads to notched plates of metallic materials causes the nucleation and growth of fatigue cracks along the plane of the notch, in a direction normal to the far-field compressive stress. The cracks propagate at a progressively slower rate until complete crack arrest takes place at a fatigue crack length, a^* (Fig. 4.15). Figures 4.16(a) and (b) are examples of cracks initiated ahead of stress concentrations in an intermetallic and a metal-matrix composite, respectively, under fully compressive cyclic loads. Although this phenomenon of noncatastrophic crack growth from notches under cyclic compression is apparently similar to the

---

† The compression fatigue failure of upper spar cap on the wing of an F-15 military aircraft is a typical example (see Rich, Pinckert & Christian, 1986).
‡ A case study of fatigue cracking under cyclic compression in a total hip femoral component is presented in Chapter 10.

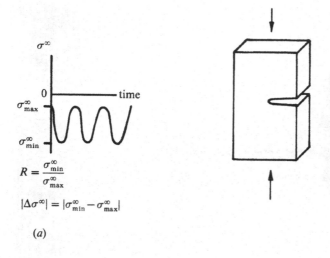

(a)

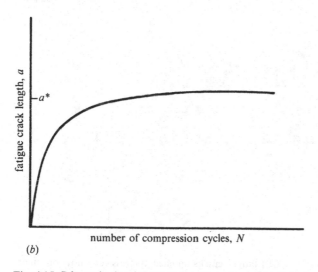

(b)

Fig. 4.15. Schematic showing (a) the loading of a notched specimen in cyclic compression and (b) typical variation of crack length, measured from the notch tip, as a function of the number of compression cycles.

behavior of 'nonpropagating' cracks nucleated under cyclic tension (Chapter 7), the mechanisms underlying the two processes are different.

Observations of controlled crack initiation in cyclic compression of notched metallic materials were first reported in the 1960s. Hubbard (1969) reported a fracture mechanics-based study of crack growth in a center-notched plate of 7075-T6 aluminum alloy. He found that fatigue cracks grew from the tip of the notch over distances of several millimeters. Similar observations have been made subsequently in a variety

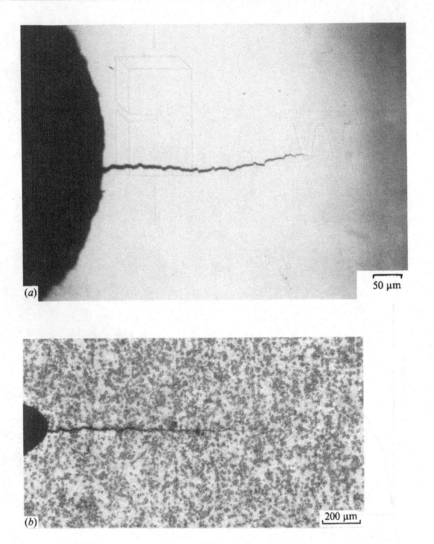

Fig. 4.16. Examples of mode I fatigue cracks initiated at stress concentrations under far-field cyclic compression:(*a*) Ti–48 Al intermetallic with a predominantly γ-phase microstructure. (*b*) Al–3.5 Cu alloy reinforced with 20 volume% of SiC particles. The cyclic compression loading axis is vertical in both cases. (Photographs courtesy of P.B. Aswath and Y. Sugimura, respectively.)

of ferrous and nonferrous alloys (Saal, 1971; Reid, Williams & Hermann, 1979; Suresh, 1985b; Pippan, 1987).

The mechanism by which a fatigue crack initiates and advances in a direction normal to the imposed compression axis is dictated by the development of a cyclic plastic zone ahead of the notch tip upon unloading from the far-field compressive stress. In zero–tension fatigue, there develops a region of reversed flow ahead of a tensile crack within which residual stresses comparable in magnitude to the flow stress in compression exist.† If one considers the case of a 'sharp' nonclosing notch which is subjected to a zero–compression–zero fatigue cycle, it is seen that the reverse flow induced within the monotonic plastic zone ahead of the notch tip upon unloading from the maximum compressive stress generates a zone of residual tensile stresses at the notch tip (Fig. 4.17). Full field finite element simulations of the generation of residual tensile stresses in notched plates of ferrous alloys have been reported by Holm, Blom & Suresh (1986). Quantitative and *in-situ* measurements of the evolution of residual tensile stresses ahead of stress concentrations subjected to cyclic compression loading have been reported by Pruitt & Suresh (1993); see Chapter 6.

Residual stresses are induced ahead of the notch tip during unloading from the far-field compressive stress because there is no contact (closure) in the wake of the notch tip. (If a long, sharp fatigue crack, rather than a notch, is subjected to cyclic compression, such a residual tensile field may not be generated because of the complete closure of the crack during far-field compression loading and unloading.) Once a fatigue crack emanates from the notch tip, the faces of the crack tend to remain partially or fully closed during some portion of the loading cycle. Experimental

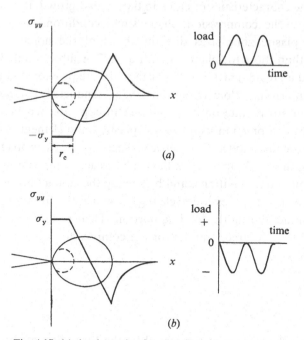

Fig. 4.17. (*a*) A schematic of a zone of residual compression ahead of a sharp notch (with a small included angle at the notch tip) subjected to cyclic tension in an elastic–perfectly plastic solid. r_c is the cyclic plastic zone defined in Eq. 9.74. (*b*) A zone of residual tension for the nonclosing notch subjected to cyclic compression.

---

† As shown in Section 9.6, the size of the cyclic plastic zone ahead of a stationary fatigue crack in an elastic–perfectly plastic solid subjected to zero–tension cyclic loading in plane stress is about one-quarter the size of the monotonic plastic zone.

measurements of crack closure (Suresh, Christman & Bull, 1986) reveal that as the length of the fatigue crack increases, the fraction of the loading cycle during which the crack remains open progressively diminishes. This increasing closure causes the crack to arrest completely after growth over a distance a^*, as shown in Fig. 4.15. The total distance of crack growth ahead of the notch is a complex function of such variables as the size of the residual tensile zone created during the first cycle, the stress state, load range, notch tip geometry and the microscopic roughness of the fatigue crack faces. The distance a^* is dependent upon the rate of exhaustion of the residual tensile zone and the rate of increase in closure stress with increasing fatigue crack length.

As crack advance under far-field cyclic compression is governed by local *tensile* stresses, the influence of microstructure on crack growth behavior is found to be similar to that seen under far-field cyclic tension. However, the mechanisms of fracture surface contact and abrasion are different for the two cases.

Since the size and shape of the monotonic and cyclic plastic zones and the extent of crack closure are strongly influenced by whether plane stress or plane strain conditions prevail (see Fig. 9.15 and Chapter 14), the stress state is expected to have a marked effect on the characteristics of crack initiation and growth in notched components subjected to cyclic compression. Plane stress conditions, where the monotonic and reversed plastic zone sizes directly ahead of the notch tip are about three times greater than those for plane strain, Eq. 9.77, a substantially faster rate of crack initiation and a greater crack growth distance a^* are promoted in plane stress than in plane strain. Holm, Blom & Suresh (1986) conducted a numerical simulation of the effects of stress state on crack growth in edge-notched plates of a bainitic steel under cyclic compression using elastic–plastic finite element calculations with an isotropic hardening model. Geometrical aspects of crack propagation were modeled by releasing the crack tip node at the peak of each compression cycle (σ_{max}), by changing the boundary conditions, and by solving the contact problem to determine the stress level of the compression cycle (σ_{cl}) at which the freshly formed crack faces would first contact during the loading portion. Their results reveal that the crack remains open during a larger fraction of the compression cycle in plane stress than in plane strain.

Exercises

4.1   The roughening of the surface of a material by the formation of slip steps plays an important role in the nucleation of fatigue cracks. Consider a cube which is made of an FCC single crystal. One of the corners of the cube is located at the origin of the Cartesian coordinate system (with axes x, y and z, and unit vectors along the axes **i**, **j** and **k**, respectively). A closed disloca-tion loop lies on the plane $x = a/2$, where a is the edge length of the cube.

The loop consists of dislocation segments of pure edge, pure screw and mixed edge–screw characters. If the loop expands in its own plane under the influence of an applied stress, slip steps will be formed on some faces of the crystal when all segments of the loop intersect the faces of the crystal. Indicate the faces of the cube on which the slip steps will form in the following cases:

(a) The Burgers vector of the dislocation loop is $\mathbf{b} = b\mathbf{j}$.

(b) The Burgers vector of the dislocation loop is $\mathbf{b} = b\mathbf{k}$.

4.2 An edge dislocation ($\perp_1$) is located on the plane $x = x_1$; x, y and z are the axes of the Cartesian coordinate system and $\mathbf{i}$, $\mathbf{j}$ and $\mathbf{k}$ are the corresponding unit vectors along these coordinate axes. The Burgers vector of $\perp_1$ is $\mathbf{b}_1 = b\mathbf{j}$ and the dislocation line is parallel to $\mathbf{k}$. A second edge dislocation line $\perp_2$, also oriented parallel to $\mathbf{k}$, is at the origin. Calculate the glide and climb forces $\perp_1$ would experience due to the presence of $\perp_2$, in terms of x_1 and θ (i.e. the angle between the x-axis and the line on the x–y plane connecting the two dislocations) for the case where the Burgers vector of $\perp_2$ is $\mathbf{b}_2 = b\mathbf{i}$.

4.3 A crystal contains a single edge dislocation (Burgers vector, $\mathbf{b} = b\mathbf{i}$ and dislocation line vector parallel to $\mathbf{k}$, where $\mathbf{i}$, $\mathbf{j}$ and $\mathbf{k}$ are the unit vectors along the Cartesian coordinate axes, x, y and z, respectively).

(a) The crystal is subjected to a tensile stress $\sigma_t = \sigma_{xx}$. Calculate the force on the dislocation. Is it a glide or a climb force?

(b) The stress is now reversed into compression with the compressive stress $\sigma_c = -2\sigma_{xx}$. Recalculate the magnitude and direction of the force on the dislocation.

(c) Find the magnitude and direction(s) of the maximum force on a screw dislocation (oriented parallel to $\mathbf{k}$ with a Burgers vector, $\mathbf{b} = b\,\mathbf{k}$) in a crystal subjected to a shear stress, σ_{xz}.

4.4 Consider two parallel screw dislocations of the same sign. Obtain an expression for the force on the dislocations as a function of their relative positions. Comment on the stability of this arrangement.

4.5 Describe the process of point defect production by the annihilation of two edge dislocations of opposite signs which are separated by two atomic planes. The extra planes of atoms in both the dislocations are on the outside of the slip planes for the dislocations. Use schematic diagrams to show the orientations, glide directions, and atomic arrangements before and after annihilation. Are the point defects vacancies or interstitials?

4.6 Consider two parallel edge dislocations of the same sign. Let θ be the angle between the line connecting the two dislocations and the direction of their Burgers vectors.

(a) Calculate the variation of the glide force and the climb force for the two dislocations as a function of the relative positions of the dislocations (i.e. as a function of θ).

(b) Which of the following configurations is more stable and why? *Configuration A*: The two dislocations align themselves one *below* the other along a direction normal to their slip planes. *Configuration B*: The two dislocations align themselves one *beside* the other along a direction normal to their slip planes.

(c) Give examples of at least two practical situations where the more stable configuration in part (b) is observed.

(d) Repeat part (a) for the situation where the two parallel edge dislocations are of opposite sign.

4.7 A dislocation dipole contains two coplanar edge dislocations of the same magnitude of Burgers vector but opposite signs. The dislocation lines are separated by a distance δ.

(a) Calculate the shear stress due to the dipole on a parallel plane a distance k away from the plane of the dipole, for $k \gg \delta$.

(b) Determine the positions of equilibrium of a positive edge dislocation on the $y = k$ plane.

CHAPTER 5

Cyclic deformation and crack initiation in brittle solids

In this chapter, we examine the mechanics and mechanisms of cyclic damage and crack nucleation in a wide range of brittle materials, including ceramics, glasses and ionic crystals. The fatigue behavior of brittle polymers is considered in the next chapter. Although the discussion of cyclic deformation and fatigue crack initiation for ductile materials was provided earlier in separate chapters, the corresponding descriptions for brittle materials warrant a single, unified presentation because of the nebulous demarkation between deformation mechanisms and flaw nucleation. For example, crack nucleation along grain boundaries can be regarded as the initial stage of the cyclic deformation process in some brittle materials. Fatigue crack growth in brittle ceramics and polymers are considered in Chapters 11 and 12, respectively.

It is also pertinent at this juncture to clarify the terminology used in the description of cyclic failure of brittle solids. In the metallurgy, polymer science and mechanical engineering communities, the word *fatigue* is a well accepted term for describing the deformation and failure of materials under cyclic loading conditions. However, in the ceramics literature, the expression *static fatigue* refers to stable cracking under sustained loads in the presence of an embrittling environment (which is commonly known as stress corrosion cracking in the metallurgy and engineering literature). The expression *cyclic fatigue* is used in the ceramics community to describe cyclic deformation and fracture. In keeping with the well-established universal conventions, and in an attempt to avoid confusion, we use the term *fatigue* in this book to denote deterioration and fracture of both metals and nonmetals *due only to cyclic loads*.

Earlier discussions of fatigue in metallic materials (Chapters 2–4) underscored the role of kinematically irreversible cyclic slip in promoting permanent damage and crack nucleation. Since cyclic stress effects in ductile metals are intimately related to the to-and-fro motion of dislocations, it has traditionally been assumed that dislocation motion (slip) is a necessary condition for fatigue. A broader examination of the requirements for cyclic fracture in both ductile and brittle solids clearly indicates that *kinematically irreversible* microscopic deformation can arise not only from cyclic slip, but also from microcracking, martensitic transformation, interfacial sliding or creep. The processes which impart kinematic irreversibility to microscopic deformation during the fatigue of brittle materials include:

(1) Frictional sliding of the mating faces of microcracks that are nucleated at grain boundaries (in single phase systems), at interphase regions (in multiphase systems), and along the interfaces between the matrix and the reinforcement (in brittle composites) under the influence of the applied loads.

(2) Progressive wear and breakage, under repeated cyclic loading, of bridging
     ligaments which connect the faces of microcracks and long flaws in brittle
     solids at low and elevated temperatures.
(3) Wedging of the mating surfaces of microscopic and macroscopic flaws by
     debris particles which are formed as a consequence of repeated contact
     between the crack faces, especially under fully compressive or tensile–com-
     pressive cyclic stresses.
(4) Microcracking due to the release of thermal residual stresses at grain bound-
     aries and interfaces, which gives rise to a permanent transformation strain.
(5) The inelastic strain arising from shear or dilatational transformations such
     as mechanical twins or martensitic lamellae.
(6) The viscous flow of glassy phases that are introduced during processing and/
     or formed as a result of environmental interactions at elevated temperature,
     and the associated interfacial cavitation in ceramics and ceramic composites
     during high temperature fatigue. The strain-rate dependence of viscous
     deformation causes the fatigue response of the brittle solid with glassy
     films to be both time-dependent and cycle-dependent.

5.1 Degrees of brittleness

Before embarking on a discussion of the mechanisms of cyclic damage in
brittle solids, it is essential to examine as to what constitutes brittle behavior. In
silicate glass, which is a good example of a highly brittle solid, the strength of the
atomic bonds primarily determines the resistance to fracture. Other examples of very
brittle materials include crystals of diamond structure, quartz and sapphire, and
many ceramics at room temperature (Table 5.1). As noted by Lawn & Wilshaw
(1975), flaws in highly brittle solids are characterized by three features: (i) the
flaws need not be large in size to affect the strength significantly, (ii) the flaws are
induced mainly on the surface of the material, as a result of such mechanisms as
contact damage, although flaws in the bulk may also be introduced during processing (see next section), and (iii) the flaws exhibit wide variations in size, location and
orientation, unless they are intentionally introduced by a controlled process. Design
of brittle materials on the basis of Weibull statistics (Weibull, 1939) separates flaw
populations into surface and volume flaws.

Nonbrittle or ductile solids, such as FCC metals, in which (dislocation) plasticity
plays the dominant role in controlling cyclic deformation and fracture, are at the
other end of the spectrum in this classification of brittleness. In-between these two
distinct groups of brittle and nonbrittle solids lies a category of materials called semi-
brittle solids (Table 5.1). In this class of materials, limited plastic flow occurs prior to
the growth of a brittle crack. Consequently, the component of the resolved shear
stress on the active slip plane becomes as important a factor in deformation and

Table 5.1. *Classification of the degree of brittleness according to crack initiation mechanisms and examples of material types which fit this clasification at room temperature.*

Classification	Main factors	Material
Highly brittle	Bond rupture	Diamond structure, zinc blende structure, silicates, alumina, mica, B, W, carbides, nitrides
Semi-brittle	Bond rupture, dislocation mobility	Sodium chloride structure, other ionic crystals, HCP metals, most BCC metals, glassy polymers
Nonbrittle or ductile	Dislocation mobility	FCC metals, nonglassy polymers, silver halides, some BCC metals

Source: Lawn & Wilshaw, 1975.

fracture as the component of the normal tensile stress on the ensuing crack plane. This inference is also supported by experimental observations which show that cracks in semi-brittle solids are nucleated as easily in compression as in tension. (This behavior is distinctly different from that of brittle solids, which are many times stronger in compression than in tension.) Semi-brittle solids generally do not contain a sufficient number of slip systems to accommodate plastic strains. Recall from Chapter 3 that five independent slip systems are needed to ensure strain compatibility in a polycrystalline solid. Therefore, the initiation of a brittle crack may be the principal factor responsible for relieving the strains accommodated by a limited amount of slip in these materials. Damage evolution by craze formation is another mode of semi-brittle deformation in polymers, which is discussed in the next chapter.

Increases in temperature tend to reduce the degree of brittleness in a material. While the classification in Table 5.1 is primarily intended for quasi-static loading conditions, it should be noted that strain rate and temperature strongly affect the degree of brittleness in many materials.

5.2 Modes of cyclic deformation in brittle solids

In ductile metals and alloys, the occurrence of cyclic deformation and fatigue fracture is clearly identified with to-and-fro dislocation motion either in the bulk or in the immediate vicinity of the crack tip. In brittle solids, however, it is generally not so straightforward to identify a clear cyclic effect. This difficulty arises because fluctuations in applied loads can lead to a large influence on the cyclic stress–strain curve and in the crack initiation/propagation life (compared to static loads of the same peak value) even when there are no discernable differences in the deformation and damage mechanisms between the static and cyclic loading cases. In this chapter, and in Chapters 6, 11 and 12, we demonstrate the following effects of cyclic loading on deformation and cracking in brittle solids.

(1) Cyclic loading resulting in an increased hysteresis or gradual shift in the stress–strain curve as a result of progressive damage evolution involving microcracking, crazing, cavitation or phase transformations at both low and high temperatures.

(2) Cyclic response of semi-brittle ionic crystals at elevated temperatures wherein the dislocation structures are apparently similar to those of fatigued ductile FCC crystals at room temperature.

(3) Elevated temperature cyclic response of ceramic composites in which the cyclic deformation characteristics at the tip of a crack are distinctly different from those immediately ahead of a statically loaded crack.

In addition, it is shown in this chapter and in Chapters 6, 11 and 12 that an identifiable fatigue effect, as seen by an enhanced or reduced time to failure or crack propagation rate compared to a static load of the same peak value or mean value, can occur in many brittle solids at room and elevated temperature, *even when the mechanisms of deformation and damage are the same under both static and cyclic loads*. The following processes which lead to such fatigue effect are considered in this connection.

(1) The formation of crazes during cyclic loading results in cyclic softening in brittle polymers (next chapter).

(2) The inducement of a permanent deformation (such as microcracking or crazing) within the cyclic damage zone ahead of the notch in a brittle ceramic or polymer at room temperature results in a mode of crack initiation and growth under cyclic loads which is distinctly different from that observed under monotonic compression loads (see Chapters 6, 11 and 12).

(3) The progressive breakdown of bridging ligaments in the wake of a fatigue crack results in a much higher crack velocity at room temperature under cyclic loads than under sustained loads of the same peak stress intensity (Chapter 11).

(4) The bridging of the wake of a fatigue crack by glassy ligaments (formed from the viscous flow of the glassy films left from the processing additives or formed in-situ as a result of environmental interactions) leads to a much lower crack velocity at high temperature under cyclic loads than under sustained loads of the same peak stress intensity (Chapter 11).

5.3 Highly brittle solids

In highly brittle solids with strong covalent or ionic bonding and very little mobility of point defects and dislocations, no distinct differences are known to exist between static and cyclic loading conditions as far as the mechanisms of microscopic deformation or microcrack nuleation are concerned. However, once microscopic flaws are induced in a 'static mode', frictional sliding of the faces of the cracks can impart an apparent plasticity to the deformation of the material, most notably under compression loading conditions. Consequently, one may envision the possibility of observing differences in the manner and rate of growth of the internal flaws under monotonic and cyclic loading conditions.

5.3.1 Mechanisms

There exist several possibilities for introducing crack nuclei during the fabrication and service of a brittle solid:

(*a*) Brittle solids contain a population of small microscopic flaws, commonly known as *Griffith flaws*. The presence of these flaws can cause marked reductions in fracture strength. On the free surface, the Griffith flaws can be initiated due to the impingement of hard dust particles, such as quartz, which are prevalent in the atmosphere. Within the bulk, defects such as pores, inclusions or gas bubble entrapments are likely to develop in a commercially-processed material. These internal defects serve as potential sites for the nucleation of a dominant crack.

(*b*) The free surface is almost invariably 'rough' on an atomic level. Surfaces typically consist of steps, grooves, ridges, pits, etc., as a result of crystal growth, dissolution, cleavage or ion bombardment, even when the surface preparation techniques and the specimen surfaces are 'clean'. These atomically rough surface features can serve as local stress raisers in brittle solids where atomic bond rupture is the principal mode of failure.

(*c*) In materials such as ceramics, rocks, cement mortar and concrete, distributed microcracking in the bulk is known to occur along grain facets and/or interfaces. In noncubic, single phase brittle solids and brittle composites, residual stresses generated at grain boundary facets and interfaces give rise to microcracking during cooling from the processing temperature as a result of thermal contraction mismatch between adjacent grains or phases. Furthermore, the residual stresses may aid in the nucleation of intergranular flaws under the influence of an external stress. In transformation-toughened ceramics (to be discussed in Section 5.5), microcracking occurs in conjunction with stress-induced martensitic transformations.

(*d*) Brittle materials, when exposed to certain embrittling environments, can suffer strength degradation and increased susceptibility to flaw nucleation. For example, the large sodium ion in the glass network is replaced by a smaller species, such as H^+ in an acid solution or Li^+ in a molten salt. Furthermore, local devitrification of glass can provide a preferential site for crack nucleation.

Although it is experimentally difficult to identify and quantify the extent of stable microcracking in smooth specimens of brittle solids at room temperature, especially under tensile loading conditions, observations of confined microcracking damage ahead of stress concentrations and cracks have been reported for a wide variety of brittle solids. Figure 5.1(*a*) shows a zone of grain boundary microcracks formed ahead of a notch in a single edge-notched specimen of Al_2O_3 (average grain size $\approx 18\,\mu m$) subjected to compression loading. Figure 5.1(*b*) documents a zone of microcracks, as observed in a TEM foil taken $0.5\,\mu m$ below the tensile fracture surface, in a transforming ceramic which contains an Al_2O_3 matrix with 23% metastable tetragonal ZrO_2. In this figure, microcracks were nucleated around the ZrO_2 particles following their transformation from a tetragonal to a monoclinic phase (see Section 5.5 for details of phase transformations).

5.3.2 *Constitutive models*

In a linear elastic solid containing a population of open microcracks, the flaws are generally characterized by a nondimensional density β, which depends on their geometry and spatial distribution:

$$\beta = \frac{2N_c}{\pi} \left\{ \frac{S^2}{P} \right\}, \tag{5.1}$$

where N_c is the number of microcracks per unit volume, S is the area of a microcrack and P is its perimeter. The symbol $\{\ \}$ denotes a volume average.

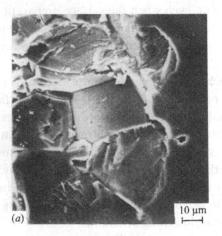
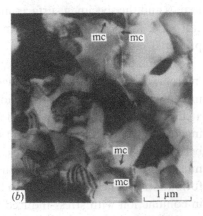

Fig. 5.1. (*a*) A zone of grain boundary microcracks formed ahead of a single edge-notch in an Al_2O_3 subjected to uniaxial compressive stresses in a direction normal to the plane of the notch. (From Suresh & Brockenbrough, 1988. Copyright Pergamon Press plc. Reprinted with permission.) (*b*) Microcracks ('mc'), denoted by arrows, as seen in a TEM foil taken 0.5 μm from the tensile fracture surface in ZrO_2-toughened Al_2O_3. (From Rühle, Clausen & Heuer, 1986. Copyright American Ceramic Society. Reprinted with permission.)

The change in elastic moduli as a function of the microcrack density, geometry and orientation has been investigated. Some representative results for a select number of crack systems are given in Fig. 5.2 for an initially isotropic material with an initial Poisson's ratio, $\nu_0 = 0.25$. The indicated moduli E/E_0 are measured in a direction normal to the crack planes and the loading is such that all microcracks are open. Figure 5.2 compares the results for various crack systems in terms of the crack density β, as defined in Eq. 5.1. For a random distribution of equal-sized penny-shaped cracks, $\beta = N_c a^3$, where a is the crack radius (Budiansky & O'Connell, 1976).

Generalized constitutive models for multiaxial loading conditions are well developed for elastic–plastic solids (see Chapter 3). A number of formulations, analogous to the deformation and incremental theories of plasticity (Section 1.4), have evolved from both experimental and theoretical attempts to characterize the constitutive response of brittle solids, such as concrete and rocks, in monotonic compression (e.g., McClintock & Walsh, 1962; Nemat-Nasser & Horii, 1982).†

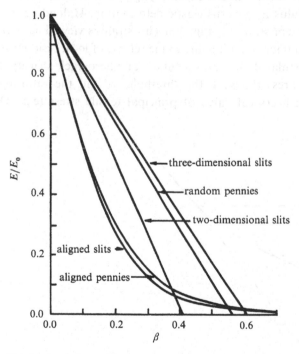

Fig. 5.2. Variation of the elastic modulus E for a microcracked solid, normalized by its modulus E_0 when no microcracks exist, is plotted as a function of the microcrack density β for the indicated crack systems. (After Laws & Brockenbrough, 1987.)

† The preponderance of research effort directed at monotonic compression deformation is a consequence of the fact that brittle solids are many times stronger in compression than in tension and that they are most commonly used in compression-dominated applications.

The development of constitutive models for a microcracking solid in monotonic and cyclic deformation requires the assumption of a microcrack nucleation criterion. The nucleation of microcracks is dictated by such complex factors as the composition of the matrix and grain boundary phases, the occurrence of any stress-induced phase changes in the microstructure, characteristic microstructural dimensions such as grain size, prior processing history (and the attendant thermal residual stresses generated upon cooling from the processing temperature), stress state and test temperature.

A characteristic stress–strain curve commonly used for the constitutive behavior of a microcracking ceramic in one cycle of zero–tension–zero loading is schematically shown in Fig. 5.3. Below a certain threshold stress σ_0, the material deforms as if it were an uncracked solid with Young's modulus E_0 and Poisson's ratio ν_0. Beyond σ_0, microcracking is assumed to increase continuously until a certain saturation stress σ_s is reached. A completely saturated state of microcracking, independent of the applied stress, is assumed based on the notion that the sites for microcrack nucleation become exhausted above some applied stress level, σ_s, when the local (thermal) residual stresses dictate microcracking (e.g., Hutchinson, 1987). Above σ_s, a reduced elastic modulus E_s governs elastic deformation. Unloading from tension gives rise to a permanent strain ϵ^T, Fig. 5.3. This strain is viewed as a transformation strain arising from microcracking due to the release of local residual stresses.

Criteria have been postulated for the evolution of microcrack density β as a function of the applied stress (between the threshold σ_0 and the saturation σ_s). Microcrack nucleation at a critical value of principal tensile stress (e.g., Ortiz &

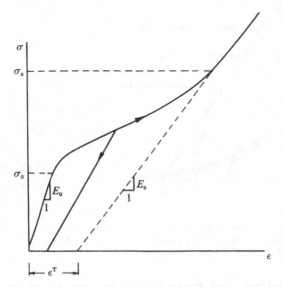

Fig. 5.3. Stress–strain curve for a microcracking material subjected to a zero–tension–zero load cycle. The strain axis is enlarged for clarity.

Giannakopoulos, 1990) and at a critical mean stress (e.g., Hutchinson, 1987) have been considered. Other constitutive models for isotropic microcracking have also been proposed where microcrack nucleation between the threshold and saturation stages is considered to be governed by an effective stress $\hat{\sigma}$. For example, Charalambides & McMeeking (1987) use the following small strain, nonlinear elastic constitutive relation to characterize the deformation of a microcracking solid:

$$E\epsilon_{ij} = [h(\hat{\sigma}) + \nu]\sigma_{ij} - \nu\sigma_{kk}\delta_{ij}, \tag{5.2}$$

where σ_{ij} and ϵ_{ij} are components of the stress and strain tensors, respectively, ν is Poisson's ratio, δ_{ij} is Kronecker's delta (defined in Section 1.4),

$$h(\hat{\sigma}) = \left[1 - \frac{16}{9}\beta(\hat{\sigma})\right]^{-1}, \tag{5.3}$$

for a random distribution of penny-shaped microcracks, and

$$\hat{\sigma} = \sqrt{\sigma_{ij}\sigma_{ij}}. \tag{5.4}$$

With respect to the different regimes of the loading curve in Fig. 5.3, Charalambides & McMeeking assumed that

$$\begin{aligned}
\beta &= 0 & \text{for} & & \hat{\sigma} &< \sigma_0, \\
\beta &= k(\hat{\sigma} - \sigma_0) & \text{for} & & \sigma_0 &\leq \hat{\sigma} \leq \sigma_s, \\
\beta &= \beta_s = k(\sigma_s - \hat{\sigma}) & \text{for} & & \hat{\sigma} &> \sigma_s, \tag{5.5}
\end{aligned}$$

where k is a factor which governs the rate of microcracking with stress.

In an independent study, Brockenbrough & Suresh (1987) used a similar assumption of penny-shaped flaws to develop a constitutive model for their numerical simulation of compression fatigue crack nucleation in highly brittle solids. Using *in situ* video photography and acoustic emission measurements, they measured the threshold stress σ_0 for microcrack nucleation in edge-notched specimens of monolithic alumina loaded in uniaxial cyclic compression (Fig. 5.1a shows notch tip microcracking from one such test). Microcrack nucleation was assumed to follow the relationship:

$$\begin{aligned}
\beta &= A\left(\frac{\hat{\sigma}}{\sigma_0} - 1\right)^{\bar{n}} & \text{for} & & |\hat{\sigma}| &> |\sigma_0|, \quad \beta < \beta_s, \\
\beta &= 0 & \text{for} & & |\hat{\sigma}| &\leq |\sigma_0|, \tag{5.6}
\end{aligned}$$

where A is a factor which, analogous to k in Eq. 5.5, controls the rate of microcracking with stress, $\bar{n}$ is an exponent of order unity and $\hat{\sigma}$ is as defined in Eq. 5.4. $\beta = \beta_s$ for $\hat{\sigma} > \sigma_s$. Unlike the earlier approaches, the assumption of a saturation in microcrack density is not essential for the numerical procedure employed for modeling crack initiation in cyclic compression. Figure 5.4 illustrates the stress–strain curve (path A) for compression loading.

Let the total density of open microcracks which exist at the maximum far-field compressive stress of the fatigue cycle be $\beta_{\max}$. Out of this total population, let a certain fraction of microcracks, of density β_u, close fully upon complete unloading

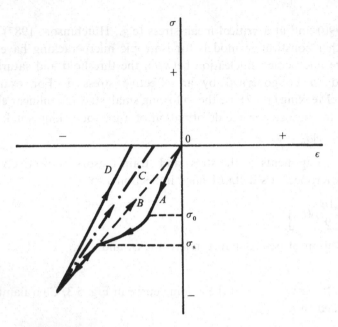

Fig. 5.4. Constitutive behavior of a microcracking brittle solid in cyclic compression. The strain axis is enlarged for clarity.

from the far-field compressive stress. The development of permanent strains after one compression cycle can then be quantified by an unloading parameter,

$$\lambda = 1 - \frac{\beta_u}{\beta_{max}}. \tag{5.7}$$

The unloading path B in Fig. 5.4 corresponds to the idealized situation where all the microcracks, of density β_{max} ($= \beta_s$ if saturation occurs) existing at the maximum far-field compressive stress, gradually close upon unloading. In this case, $\beta_u = \beta_{max}$, $\lambda = 0$ and no permanent strains exist in the fully unloaded state. Path D represents the other extreme case where all the microcracks nucleated during compression loading are blocked by the presence of debris within them or they are locked in friction. Here, $\lambda = 1$ and unloading occurs with the same slope as the initial loading portion of curve A. If the frictional sliding taking place between the faces of the microcracks during compression loading is partially reversed upon unloading, $0 < \beta_u < \beta_{max}$, $0 < \lambda < 1$, and unloading occurs along path C. Thus the linear unloading parameter λ conveniently characterizes permanent strains representing the entire range of linear unloading paths. The elastic secant moduli during unloading are given by

$$E_u = E_0\lambda + E_m(1 - \lambda), \qquad \nu_u = \nu_0\lambda + \nu_m(1 - \lambda). \tag{5.8}$$

E_m and ν_m are the values of Young's modulus and Poisson's ratio, respectively, at the maximum far-field compressive stress.

Using a finite element analysis, Brockenbrough & Suresh (1987) showed that when the unloading path leads to permanent strains, i.e. for $\lambda > 0$, the resulting residual stresses that arise in the matrix material in the vicinity of a stress concentration are distinctly different from those induced under monotonic loading conditions. This effect of permanent strains causes a mode of failure which is unique to cyclic loading conditions. More significantly, the results imply that kinematic irreversibility of microscopic deformation (in this case, the differences in the opening and closing of microcracks) occurring during fatigue in a brittle solid is qualitatively similar to the development of stress–strain hysteresis due to slip irreversibility in metal fatigue. Note the similarity of the above formulation for the compression fatigue of brittle solids to the process of compression fatigue in metals discussed in Section 4.11. A similar conceptual framework was used by Lawn *et al.* (1994) to rationalize the progressive evolution of microcracking during cyclic indentation of brittle ceramics.

A constitutive model similar to that shown in Fig. 5.4 has also been used to predict the size and shape of the cyclic damage zones developing ahead of tensile fatigue cracks in microcracking ceramic materials, Suresh & Brockenbrough (1990). These cyclic damage zones are taken up for discussion in Chapter 11.

5.3.3 On possible effects of cyclic loading

Cyclic load experiments conducted on smooth-surfaced specimens of brittle solids generally exhibit considerable scatter in number of cycles to failure or time to rupture. This difficulty arises as a consequence of the variability associated with initial flaw populations which, as discussed earlier, depend strongly on such factors as processing methods and specimen surface preparation techniques. In situations where a small artificial surface flaw is introduced to expedite failure, the results often are a strong function of the initial flaw size.

Available experimental results generally point to some effects of cyclic loading on time to rupture *vis-à-vis* that under static loads on account of the following possible mechanisms:†

(1) When environmental effects primarily dictate the rupture behavior, the life-time is controlled by time-dependent damage and failure mechanisms (see Chapter 16) and cyclic frequency or waveform has no effect the time to rupture (see Fig. 5.5).

(2) When purely mechanical cyclic loading leads to a progressive evolution of damage as, for example, in the exacerbation of intergranular microcracking during repeated cyclic loading (especially in the compression portion of tension–compression loading), a detrimental effect of cyclic loading can result. Under such circumstances, higher cyclic frequencies are generally more damaging.

† Usually, for comparison of times to rupture under static and cyclic loads, the maximum nominal stress corresponding to the fatigue cycle is taken to be the same as that of the static stress for smooth specimens.

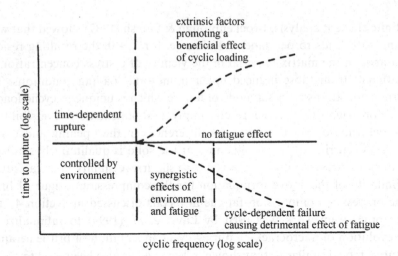

Fig. 5.5. A schematic illustration of possible beneficial or detrimental effects of cyclic loading on the time to rupture in brittle solids.

(3)   When extrinsic factors, such as deflection of microscopic flaws, wedging of the faces of microcracks by debris particles which are produced as a result of repeated contact or the closure of microcrack faces, prominently influence cyclic deformation behavior, an apparently beneficial effect of cyclic loading, vis-à-vis static loading, is feasible, as shown in Fig. 5.5.

5.3.4 Elevated temperature behavior

An elevation in temperature tends to lower the degree of brittleness in most materials. In some ceramics, several micromechanisms concurrently play an active role in dictating the overall creep deformation. These mechanisms include: atomic diffusional processes within the bulk of the grains and along grain boundaries and interfaces, grain boundary sliding, and even dislocation plasticity. In some other systems, such as many aluminum oxides subjected to certain combinations of temperatures and strain rates, apparent ductility at elevated temperatures may simply arise from the growth of microcracks along the facets of elastically deforming grains. The growth of intergranular flaws may be caused by *local* diffusion processes or by viscous flow of any amorphous films deposited along grain facets from the additives or impurities introduced during fabrication.

This latter mechanism of interfacial microcracking by viscous flow of an amorphous phase in an otherwise elastic medium can be illustrated with the example of a SiC whisker-reinforced Al_2O_3 matrix composite. It is known that SiC, when exposed to oxygen-containing media at temperatures typically in excess of 1250 °C, oxidizes to form an amorphous SiO_2 glass phase. The viscous flow of this phase at the elevated temperature leads to the nucleation of cavities and microcracks along the interfaces between the alumina matrix and the SiC particles. Figure 5.6(a) shows an example of interfacial cavities formed at every corner of a SiC particle in an Al_2O_3–33 volume% SiC whisker composite fatigued in 1400 °C air. The presence of the amorphous interfacial film and the kinematically irreversible cyclic displacements associated with the opening and closing of the interfacial microcracks promote tensile fatigue deformation mechanisms which can be different, in some cases, from those seen under sustained tensile loads. The reinforcement phase also breaks under cyclic loading conditions; Fig. 5.6(b) is a micrograph of a SiC whisker broken by cyclic tensile loads in 1400 °C air. In this figure, the meniscus of the silica glass phase can be seen within the broken whisker.

Environmental effects at high temperatures can influence deformation and fracture in brittle solids subjected to cyclic loads in a manner which is different from that observed under static stresses. If the material does not contain any macroscopic cracks or stress concentrations, the fatigue mechanisms described above are confined to the near-surface region, where oxygen is available. If through-thickness stress concentrations are present in the material, the transport of the environment to the tip of the defect causes a damage zone to develop under the influence of an applied stress. In the example shown in Fig. 5.6, the oxidation of SiC whiskers giving rise to interfacial microcracking is essentially the same for monotonic and cyclic loading conditions. However, mechanisms associated with microcrack opening and closure, bridging of the flaws by reinforcement particles and the breaking of debonded whiskers are affected in different ways depending upon whether the composite is subjected to monotonic or cyclic loads.

Direct tensile fatigue tests at 1000–1200 °C on hot-pressed Si_3N_4 unidirectionally reinforced (along the tensile axis) with 30 vol.% SiC (SCS-6) fibers also reveal a gradual reduction in the elastic modulus with the progression of fatigue damage (Holmes, 1991). No such changes in compliance were observed in specimens which are subjected to low stress cycles below the endurance limit (at 2×10^6 cycles). Fatigue-induced changes in elastic properties were observed at maximum stress levels that were above the monotonic proportional limit, with both mechanical fatigue and creep influencing progressive damage. This regime of fatigue is characterized by decreasing stiffness, increasing stress–strain hysteresis, and strain ratchetting (cyclic creep). However, when the maximum stress level in fatigue is below the monotonic proportional limit, only creep deformation occurs. This creep regime of loading is characterized by strain ratchetting, but no change in elastic properties. Holmes has also shown that the fatigue life deteriorates markedly with decreasing tensile load ratio R. Figure 5.7 contains experimental data on the changes in the cyclic stress–strain hysteresis loops with increasing number of fatigue cycles. Note the occurrence of cyclic creep (as in metals, Section 3.9) and of increase in specimen compliance due to repeated stress cycles.

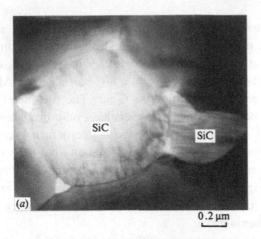

Fig. 5.6. Microscopic deformation mechanisms ahead of a crack tip in an Al_2O_3–33 volume% SiC whisker composite fatigued at a stress ratio, $R = \sigma_{min}/\sigma_{max} = 0.15$ and a load frequency of 0.15 Hz (sinusoidal waveform) in 1400 °C air. (*a*) TEM photograph showing nucleation of cavities at the interfaces between SiC whiskers and matrix Al_2O_3 grains. (*b*) A SiC whisker broken under cyclic loads. The meniscus of the amorphous glass phase can be seen within the broken whisker. (From Han & Suresh, 1989. Copyright American Ceramic Society. Reprinted with permission.)

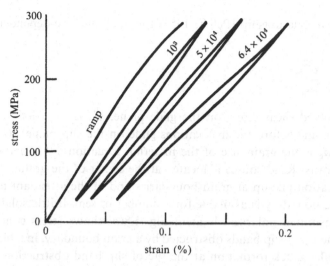

Fig. 5.7. Uniaxial tensile cyclic stress–strain behavior of a Si_3N_4–SiC fiber composite at maximum stress levels above the proportional limit (196 MPa). The number of stress cycles is indicated along with each hysteresis loop. Test temperature, $T = 1200$ °C, laboratory air environment, $R = 0.1$, and $v_c = 10$ Hz. Number of cycles to failure, $N_f = 6.5 \times 10^4$. (After Holmes, 1991.)

5.4 Semi-brittle solids

Among nonmetallic crystalline materials conforming to the definition of semi-brittle behavior, the fatigue deformation characteristics of ionic crystals have been the subject of much investigation. Experimental studies of monotonic deformation in rock-salt-type materials have shown that the pile-up of dislocations at obstacles or at other dislocations can lead to the nucleation of cracks. It is also known that, under some restricted conditions of temperature and strain rate, cyclic loading can produce slip bands and cellular structures in rock salt crystals similar to those observed in FCC metals.

5.4.1 Crack nucleation by dislocation pile-up

When the resolved shear stress exceeds a critical value on a favorably oriented low index plane of a semi-brittle crystal, such as MgO, dislocation sources of the Frank–Read type are activated. If the glide of the dislocation loops, moving outwardly from the source, is impeded by an obstacle such as a grain boundary or an inclusion, a pile-up of the dislocations occurs at the barrier. Since the piled-up coplanar dislocations are of the same sign, the mutually repulsive forces between them cause a concentration of stress at the obstacle. One may then postulate (Zener, 1948) that a *brittle* crack nucleates at the obstacle if the energy associated with the pile-up of dislocations is sufficient to compensate for the surface energy term γ_s associated with the creation of new crack surfaces. This scenario is the basis of the well known 'Petch' relationship (Petch, 1953) for the grain size-dependence of yield strength:

$$\tau_{xy} - \tau_{xy}^0 = \sqrt{\frac{\pi G \gamma_s}{d_g}},$$ (5.9)

where τ_{xy} is the resolved shear stress on the glide plane, τ_{xy}^0 is the friction stress (which is to be overcome before the dislocations glide on the slip plane), G is the shear modulus and d_g is the grain size of the material. Equation 5.9 is derived by assuming that the Frank–Read source is located at the center of the grain.

Evidence of dislocation pile-up at grain boundaries and of the attendant nucleation of a crack at the boundary is available for a number of semi-brittle solids. The birefringence of transmitted polarized light in Fig. 5.8(*a*) shows stress concentrations, generated at the tip of slip bands obstructed by a grain boundary, in a bicrystal of MgO. Transgranular crack formation at the site of slip band obstruction at the grain boundary is evident in the micrograph of the etched MgO bicrystal, Fig. 5.8(*b*).

Figure 5.9 illustrates some mechanisms by which the pile-up of dislocations can nucleate a brittle crack. The nucleation of a microcrack or a wedge-shaped cavity by the pile-up of dislocations at an obstacle such as a grain boundary (GB) is shown in Fig. 5.9(*a*). The stress gradients ahead of a dislocation pile-up are similar to those found in front of a shear (mode II) crack. If the interface in Fig. 5.9(*a*) is weak, preferential cracking occurs along the interface.

The crack nucleation mechanism shown in Fig. 5.9(*b*), which was first proposed by Cottrell (1958), is a process in which two intersecting slip planes provide the nucleus for a crack, even in the absence of a pre-existing obstacle to slip. This process has also been suggested as a mechanism for the initiation of subsurface fatigue cracks along {001} cleavage planes of β phase in Ti alloys composed of α–β duplex microstructures (Ruppen *et al.*, 1979).

5.4.2 Example problem: Cottrell mechanism for sessile dislocation formation

Problem:

Consider the dislocation reaction involving a [$\bar{1}\bar{1}1$] dislocation on the (101) slip plane and the [111] dislocation on the ($\bar{1}01$) plane in iron. This process is similar to that schematically sketched in Fig. 5.9(*b*).

(i)   Write down the dislocation reaction and find the product dislocation.
(ii) Show that this dislocation reaction is energetically feasible.
(iii) Show that the product of this dislocation reaction is a sessile dislocation.

Solution:

(i)   The Burgers vector of the product dislocation is the vectorial sum of the Burgers vectors of the dislocations on the two intersecting planes:

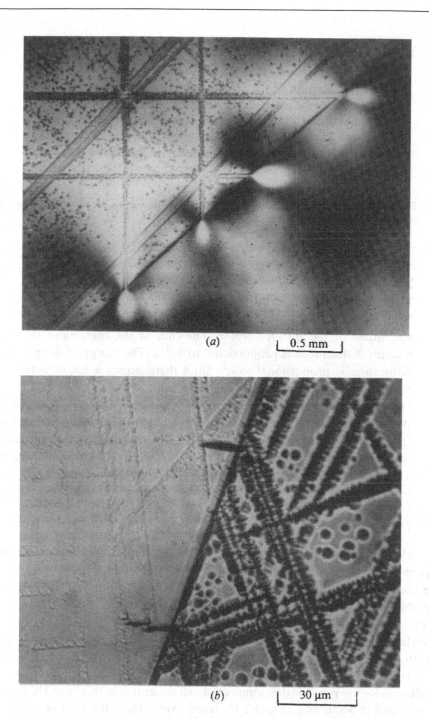

Fig. 5.8. (a) Stress concentrations revealed by the birefringence of transmitted polarized light in an MgO bicrystal at a location where a grain boundary obstructs slip. (From Ku & Johnston, 1964. Copyright Taylor & Francis, Ltd. Reprinted with permission.) (b) Transgranular cracks formed at sites where slip is impeded by the grain boundary in the MgO bicrystal. (From Johnston, Stokes & Li, 1962. Copyright Taylor & Francis, Ltd. Reprinted with permission.)

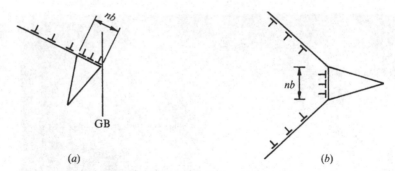

Fig. 5.9. Mechanisms for the nucleation of cracks by (*a*) dislocation pile-up at a grain boundary (GB) and (*b*) dislocation reactions.

$$\frac{a}{2}\left[\bar{1}11\right]_{(101)} + \frac{a}{2}\left[111\right]_{(\bar{1}01)} \rightarrow a[001]. \tag{5.10}$$

Thus, the Burgers vector of the product dislocation is $a[001]$.

(ii)   Note that the energy of a dislocation is proportional to the square of the magnitude of the Burgers vector. The sum of the energies of the two reacting dislocations is proportional to $3a^2/2$. The energy of the product dislocation is proportional to a^2. Since there occurs a net reduction in energy, this reaction is energetically favorable.

(iii) The product dislocation is of pure edge character, whose extra plane of atoms lies parallel to the (001) cleavage plane of BCC α-iron. Since this dislocation is not favorably oriented for slip (i.e. {110} plane and $\langle 111 \rangle$ direction), it cannot glide. It is a sessile dislocation and forms an obstacle to other dislocations gliding down the (101) and ($\bar{1}01$) planes.

5.4.3 Cyclic deformation

Under appropriate cyclic loading conditions, semi-brittle solids can exhibit characteristics of fatigue damage which are apparently similar to those found in ductile metal crystals. In his work on surface roughening, Forsyth (1957) demonstrated that intrusions and extrusions form on the surfaces of cyclically strained AgCl crystals in much the same way as the development of surface roughness in FCC metals and alloys (Chapter 4). Figure 5.10 is a micrograph, prepared under transmitted light, showing surface extrusions and crevices in AgCl. Forsyth found that the crevices eventually developed into fatigue cracks.

Conditions for the possible development of cell structures in BCC and HCP metals were examined in Sections 2.12 and 2.13, respectively. There has long been a search for similar ductile modes of fatigue deformation in semi-brittle crystal structures. Earlier investigations (e.g., McEvily & Machlin, 1961; Subramanium & Washburn, 1963; Argon & Godrick, 1969) of reversed fatigue in a rotating bend configuration in NaCl, LiF and MgO at room temperature did not reveal any substructural devel-

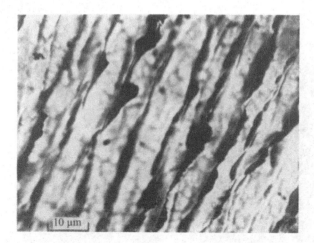

Fig. 5.10. Intrusions and extrusions formed on the surface of fatigued AgCl crystal. (From Forsyth, 1957. Copyright The Royal Society, London. Reprinted with permission.)

opments comparable to those found in cyclically strained FCC metals. It has, however, been shown by Argon & Godrick (1969) that the elevated temperature fatigue deformation of LiF is similar to the room temperature cyclic deformation characteristics of ductile FCC metals. They found that above 673 K (which is 59% of the absolute melting temperature), cross slip and dislocation climb were favored. Cyclic loading in this temperature regime produced pores throughout the highly strained volume of the crystal. Continued cyclic straining resulted in a gradual change in the density of the crystal as a consequence of cavitation. Argon and Godrick measured a fractional density change of 3×10^{-8} per cycle in LiF crystals fatigued at 783 K at a strain amplitude of 2.5×10^{-3}. Similar pore development was also observed in AgCl crystals which were fatigued above 423 K (58% of the absolute melting temperature).

Majumdar & Burns (1981) also employed direct push–pull fatigue loading in smooth specimens of LiF crystals at elevated temperatures and showed that dislocation banding occurred within subgrains. Microcracking also appeared to take place along the $\langle 110 \rangle$ directions at low strain amplitudes and at large numbers of fatigue cycles. Associated with the bands are alternate regions of high and low dislocation density which seem to be sites where dynamic recovery occurs. As the temperature is raised, the behavior of the LiF crystal appears to resemble more closely that of FCC metals. A 'PSB-like' slip pattern emerges in the fatigued crystal at a temperature of 573 K, and at low strain rates and strain amplitudes. Ladder-like dislocation structures appear to exist within the PSBs, Fig. 5.11. The arrangement and the spacing of dislocations within the ladders of PSBs, however, are different from those of ductile crystals. The fatigued crystal also exhibits a plateau where the saturation value of the shear stress is independent of the plastic strain amplitude as in FCC crystals (Fig. 2.2). Further increases in test temperature and strain amplitude cause a cellular structure to form, similar to the trend seen in regime C of Fig. 2.2.

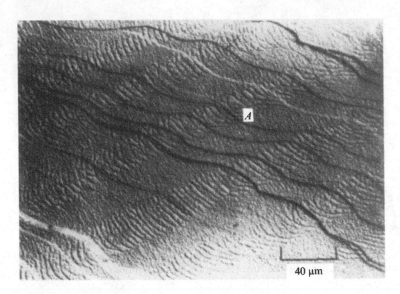

Fig. 5.11. PSB-like dislocation structure in LiF single crystal fatigued at 573 K. $\Delta\epsilon_p/2 = 0.5\%$, and $\dot\epsilon = 10^{-3}$ s^{-1}. Note that the rungs of the PSB ladder structure are bent, as, for example, at location A. The curved dark lines running across the micrographs are cleavage steps and their positions oscillate with the PSB structure, probably as a result of local stresses. (From Majumdar & Burns, 1982. Copyright Pergamon Press plc. Reprinted with permission.)

Majumdar & Burns (1987) also conducted fully reversed fatigue tests on MgO single crystals at 743 K. They found that dense bundles of dislocations developed as a consequence of reversed straining, similar to the vein structure evolving from the early stages of fatigue in FCC metals (Fig. 2.2). These bundles were aligned normal to the Burgers vector. Bowed out screw dislocations were observed between the edge dislocation bundles suggesting that the screws were largely mobile.

In summary, one of the principal contributing factors for the pronounced brittleness of ionic crystals at low temperatures is the limited possibility for cross slip. However, when certain combinations of temperature and strain rate favor cross slip, it is not surprising to observe the aforementioned similarities between their fatigue deformation characteristics and those of ductile FCC metals.

5.5 Transformation-toughened ceramics

Transformation-toughened ceramics constitute a special class of brittle solids in which 'plasticity' can be introduced by means of phase changes under the influence of an applied stress. The phase change imparts a propensity for nonlinear deformation and for stable fracture in monotonic and cyclic loading. Like the TRIP steels which exhibit TRansformation-Induced Plasticity, transforming ceramics offer the possibility of optimizing strength and ductility by the proper dispersion of a metastable phase in a stable matrix.

5.5.1 *Phenomenology*

Tetragonal (t) to monoclinic (m) phase changes occur as a martensitic transformation in ceramics which contain metastable tetragonal ZrO_2. This tetragonal phase may be present in a stable cubic matrix phase in the form of a precipitate (as in partially stabilized zirconia, PSZ), or a dispersoid (as in ZrO_2-toughened alumina, ZTA), or may be formed as the fine matrix phase in the (nearly) 100% t-ZrO_2 polycrystals, TZP. Since the seminal paper of Garvie, Hannick & Pascoe (1975) on this topic, it has been acknowledged that the dilatational and shear strains accompanying the t to m transformation can account for the remarkable toughening properties of ZrO_2-containing ceramics at low temperatures (typically below 700 °C).

Among the various transforming ceramic microstructures, a large body of research has centered around the monotonic and cyclic deformation and fracture characteristics of ZrO_2, partially stabilized with MgO (commonly referred to as Mg–PSZ). Figure 5.12(a) is a microstructure of a peak-aged (maximum strength) Mg–PSZ containing 9 mol.% MgO. This material is composed of cubic phase zirconia grains, of 50 μm average diameter, with the MgO in solid solution. Lens-shaped tetragonal precipitates, which are 300 nm long and metastable at room temperature, populate the interior of the grains. The tetragonal c-axis of the precipitates is parallel to their smallest dimension. The precipitates are oriented within the cubic phase grains in such a way that their c-axis is parallel to one of the three cubic axes.

Experimental studies by Chen & Reyes Morel (1986) and others reveal that the mechanism by which microscopic strains induced by martensitic transformation are converted to macroscopic plastic strains is via shear localization. Figure 5.12(a) shows an interior section of the material, deformed under a hydrostatic compressive stress of 200 MPa, showing shear bands within the grains. It is seen that the shear bands, which span the entire grain, have different orientations in different grains; the bands within individual grains appear to be parallel. In order for the transformation of individual tetragonal particles to cause a macroscopic shear strain via shear banding, it appears necessary that the particle transformation be correlated, as shown in Fig. 5.12(b).

Shear bands in 'transformation plasticity' can thus be deemed functionally equivalent to the slip bands in 'dislocation plasticity' of polycrystalline metals. Although the shear characteristics of the two phenomena are similar in this respect, there are also some major differences between them. While dislocation plasticity is volume-preserving, transformation plasticity induces both microscopic and macroscopic dilatational strains. For unconstrained $t \rightarrow m$ transformation in Mg–PSZ, the maximum amounts of volumetric and shear strains are 0.04 and 0.16, respectively. When embedded in an elastic matrix, correlated transformation of the particles leads to the

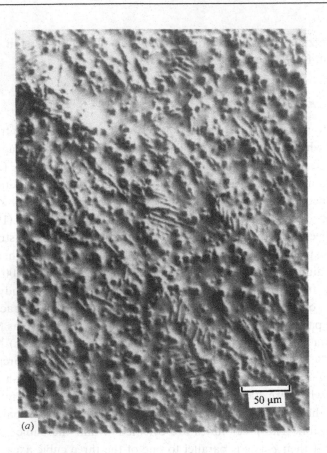

(a)

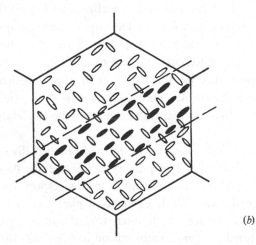

(b)

Fig. 5.12. (*a*) Transgranular shear band formation due to martensitic transformation. (From Chen & Reyes Morel, 1986. Copyright American Ceramic Society. Reprinted with permission.) (*b*) Correlated transformation of particles leading to the formation of a shear band.

nucleation of shear bands, whose orientation varies from grain to grain. Thus, the average shear strain over the particle for constrained transformation is less than 0.16, although the dilatant transformation strain is still 0.04 for the particle. It is also known that martensitic transformation can lead to microcracking as a consequence of the intersection of shear bands and grain boundaries, as well as by the decohesion of the transformed particle from the surrounding matrix. The opening of a population of microcracks also produces macroscopic dilatational strains.

Permanent phase changes cause stress–strain hysteresis in transforming ceramics upon loading and unloading, similar to the behavior found in metallic materials. Figure 5.13 is a stress–strain diagram of the peak-aged Mg–PSZ subjected to one cycle of uniaxial tensile loading. Beyond a tensile stress of about 275 MPa, fully irreversible phase transformation takes place; the resulting constitutive response becomes highly nonlinear. Elastic unloading is observed in the nonlinear regime; reloading occurs with an elastic modulus which is identical to that of the initial elastic regime. These results indicate how nonlinear effects associated with martensitic phase transformations can provide a mechanism for stable fatigue damage to occur in nominally brittle solids.

5.5.2 *Constitutive models*

Constitutive models for general multiaxial loading conditions have been proposed to quantify the monotonic and cyclic deformation response of transforming ceramics in uniaxial loading. Initial developments in this direction have come from attempts aimed at estimating the extent of toughening due to phase changes ahead of a crack tip (e.g., McMeeking & Evans, 1982; Budiansky, Hutchinson &

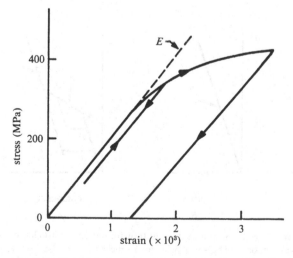

Fig. 5.13. Uniaxial tensile loading and unloading behavior of peak-strength Mg–PSZ. (After Marshall, 1986.)

Lambropoulos, 1983). These models, which are described in some detail below, neglect the shear component associated with the transformation and invoke the assumption that the dilatant transformation occurs at a critical mean stress σ_m.

In the analysis of Budiansky *et al.*, the constitutive behavior is formulated for a linear elastic matrix with embedded metastable particles which undergo irreversible inelastic volume expansion. Assume that the matrix material deforms linearly under both hydrostatic tension and compression with bulk modulus B, according to

$$\sigma_m \equiv \frac{1}{3}\sigma_{kk} = B\epsilon_{kk}, \tag{5.11}$$

where σ_{ij} and ϵ_{ij} are stress and strain tensors, respectively, and ϵ_{kk} is the total dilatation.

When the mean stress due to a monotonically increasing load is less than a critical value σ_m^c, the particles satisfy Eq. 5.11 with the same bulk modulus B, as shown in Fig. 5.14. However, once $\sigma_m > \sigma_m^c$, the incremental response of the particle is governed by B' where

$$\dot{\sigma}_m = B'\dot{\epsilon}_{kk}. \tag{5.12}$$

The inelastic or transformed portion of the dilatation θ_p is the difference between the total and elastic dilatation:

$$\theta_p = \epsilon_{kk} - \frac{\sigma_m}{B}. \tag{5.13}$$

Upon complete transformation, the incremental response is again governed by the bulk modulus B at large values of ϵ_{kk}, as per the relation

$$\dot{\sigma}_m = B\dot{\epsilon}_{kk}. \tag{5.14}$$

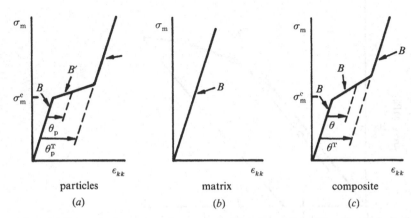

Fig. 5.14. Stress–strain response due to purely dilatant transformation of particles in the elastic matrix of the two-phase composite. The shear behavior is linear and the shear modulus is the same for both the particle and the matrix. (After Budiansky, Hutchinson & Lambropoulos, 1983.)

In this model, the particle and the matrix are assumed to exhibit the same shear behavior; the shear modulus of the composite is G at all strains. The incremental response of the composite for $\sigma_m > \sigma_m^c$ is

$$\dot{\sigma}_m = \overline{B}\dot{\epsilon}_{kk}, \qquad \frac{1}{\overline{B} + 4G/3} = \frac{f_p}{B' + 4G/3} + \frac{1 - f_p}{B + 4G/3}, \qquad (5.15)$$

where f_p is the volume fraction of the toughening particles. In the intermediate segment of Fig. 5.14, the dilatation of the transformed composite is

$$\theta = \left(1 - \frac{\overline{B}}{B}\right)\left(\epsilon_{kk} - \frac{\sigma_m^c}{B}\right), \qquad (5.16)$$

with $\theta = f_p\theta_p$. The maximum dilatation upon complete transformation is $\theta^T = f_p\theta_p^T$. If $B' \leq -4G/3$ in Eq. 5.15, a particle in an infinite elastic matrix with bulk and shear moduli B and G, respectively, transforms completely to θ_p^T as soon as the critical mean stress is imposed. Even if such a transformation occurs at a critical mean stress, a distribution of critical mean stresses may exist in a composite consisting of a distribution of particle sizes. In this case, the incremental bulk modulus may never drop below $4G/3$. If the particle distribution is sufficiently wide, supercritical transformation, i.e. $\overline{B} < -4G/3$, may never happen. Budiansky *et al.* term $\overline{B} = -4G/3$ as the condition denoting a critically transforming composite and $\overline{B} > -4G/3$ as the condition denoting a subcritical transformation. This latter case, with $G > 0$, is the condition for a real longitudinal wave speed.

Budiansky *et al.* use an incremental formulation, similar to that employed for an elastic–plastic solid (Section 1.4.3), where loading occurs on the transforming branch of the stress–strain curve if $\dot{\epsilon}_{kk} > 0$, and

$$\dot{\sigma}_m = \overline{B}\dot{\epsilon}_{kk} \qquad \text{and} \qquad \dot{\theta} = \left(1 - \frac{\overline{B}}{B}\right)\dot{\epsilon}_{kk}. \qquad (5.17)$$

Unloading occurs if $\dot{\epsilon}_{kk} < 0$ and

$$\dot{\sigma}_m = B\dot{\epsilon}_{kk} \qquad \text{and} \qquad \dot{\theta} = 0. \qquad (5.18)$$

Upon unloading to zero stress, the permanent dilatation is θ. If $\theta < \theta^T$, only partial (subcritical) transformation is seen. For this purely dilatant transformation, the stress–strain relations in three dimensions are

$$\epsilon_{ij} = \frac{1}{2G}s_{ij} + \frac{1}{3B}\sigma_m\delta_{ij} + \frac{1}{3}\theta\delta_{ij}, \qquad \sigma_{ij} = 2Ge_{ij} + B(\epsilon_{kk} - \theta)\delta_{ij}, \qquad (5.19)$$

where s_{ij} are the components of the stress deviator (Section 1.4.2) and e_{ij} ($= \epsilon_{ij} - [\delta_{ij}\epsilon_{kk}]/3$) are the components of the strain deviator, respectively.

In an attempt to address the issue of combined dilatational and shear effects on the multiaxial constitutive response, an experimentally based formulation has been suggested by Chen & Reyes Morel (1986). They carried out unconstrained and constrained compression experiments on cylinders of Mg–PSZ at room temperature. Their experimental results, showing the variation of axial plastic compression strain (denoted $-A$), radial strain (denoted R) and volumetric strain (denoted V) as a function of the differential axial compressive stress Σ for a superimposed pressure

$P = 200$ MPa, are plotted in Fig. 5.15. A, R and V scale with the applied differential stress in the ratio of $-2:3:4$.

The yield condition for transformation plasticity, which can be derived from the multiaxial compression experiments, is

$$f = \sigma_e + \left(\frac{\sigma_e^*}{\sigma_m^*}\right)\sigma_m - \sigma_e^* \geq 0, \qquad (5.20)$$

where σ_e is the effective stress, Eq. 1.27, σ_m is the mean pressure, and σ_e^* and σ_m^* are measures of hardness.

The experiments of Chen & Reyes Morel also suggest the following yield criterion consistent with Eq. 5.20:

$$Y^c = Y_0^c + \alpha^c P, \qquad (5.21)$$

where Y_0^c and Y^c are the compressive yield stress values at pressures of 0 and P, respectively, and α^c is a constant, which is about two for a wide range of strain values, except for very small and very large strains. Experiments on Mg–PSZ suggest that $\sigma_e^* = 3Y_0^c/5$ and $\sigma_m^* = 0.5Y_0^c$. The numerical value $\alpha^c = 2$, along with the assumption of normality flow, produces a ratio $A:R:V$ of $-2:3:4$, which is consistent with the results of Fig. 5.15. Furthermore, since this formulation is phenomenological, it also accounts for the effects of microcracking (induced by transformation) on deformation.

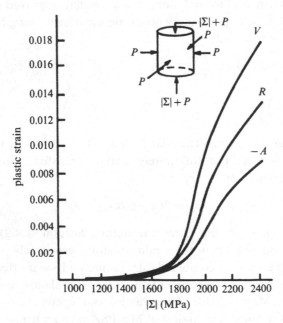

Fig. 5.15. Experimentally determined variation of axial (A), radial (R) and volumetric (V) strains plotted as a function of the differential axial compressive stress Σ for Mg–PSZ (maximum strength condition) under an imposed hydrostatic pressure, $P = 200$ MPa and a strain rate, $\dot{\epsilon} = 10^{-4}$ s^{-1}. (After Chen & Reyes Morel, 1986.)

This experimentally based model has been used to rationalize observations of tension and compression yield anisotropy in Mg–PSZ. Suresh & Brockenbrough (1988) have also implemented this constitutive formulation, in conjuction with the incremental (flow) theory of plasticity, to model the development of transformation zones and residual stresses ahead of stress concentrations in Mg–PSZ subjected to cyclic compression loads. For compression fatigue, the model (incorporating elastic unloading and reloading behavior in fatigue) provided accurate predictions of the direction of crack nucleation as well as the maximum distance of fracture.

5.6 Fatigue crack initiation under far-field cyclic compression

The possibility that fatigue crack initiation and stable crack growth, attributable solely to *cyclic variations* in applied loads, can occur at room temperature (even in the absence of an embrittling environment) in single phase ceramics, transformation-toughened ceramics, and ceramic composites was first demonstrated for cyclic compression loading of notched plates (Ewart & Suresh, 1986, 1987). This fatigue crack growth phenomenon is a true mechanical fatigue effect on account of the following experimental observations: (i) Fatigue crack initiation and growth occur *perpendicularly* to the compression axis in brittle solids subjected only to cyclic loads. Monotonic compressive stresses promote a splitting mode of failure *parallel* to the stress axis. (ii) There is a gradual increase in crack length with an increase in the number of compression cycles. (iii) Although an embrittling environment can modify the rate of crack growth, fatigue fracture in cyclic compression is not a consequence of environmental effects in that it can take place even *in vacuo*. (iv) The rate of crack growth is strongly influenced by such mechanical fatigue variables as mean stress, stress range, and stress state. (v) The overall crack growth characteristics of fatigue crack initiation and growth at a macroscopic level are qualitatively similar to those observed in ductile metals and alloys (see Section 4.11) and semi- or noncrystalline polymers and polymeric composites (see next chapter).

Brittle solids such as ceramics, rocks or concrete are known to fracture in *monotonic* compression in one of several mechanistically dissimilar modes (McClintock & Walsh, 1962; Nemat-Nasser & Horii, 1982):

(1) The frictional sliding of pre-existing flaws produces opening displacements at the crack tip which are sufficient to initiate a locally tensile mode of failure. Consequently, in unconstrained axial or radial compression, cracks extend across planes of local maximum principal tension and curve in a direction parallel to the maximum principal axis of compression. This process causes the material to 'split' parallel to the compression axis.

(2) The coalescence of pre-existing microcracks or weak interfaces results in a *shear failure* under small confining pressures.

(3)   Large confining pressures promote a relatively more homogeneous, pseudo-
      ductile deformation by limiting the coalescence of micro-defects.

Under *cyclic* compression loading conditions, however, brittle solids with stress
concentrations exhibit a completely different mode of crack initiation which is
macroscopically similar to that found in metallic materials (Section 4.11). When
notched plates of brittle solids are subjected to some combinations of (cyclic compression) stress amplitude and mean stress, confined microcracking occurs at the tip
of the notch (see Fig. 5.1a). If even a fraction of this microcracking deformation at
the notch tip is permanent (i.e. when microcracks do not open and close the same
way during compression loading and unloading), residual *tensile* stresses are created
within the microcrack zone upon unloading from the maximum far-field compression stress. Similarly, permanent strains associated with stress-induced phase
changes, frictional sliding, or creep can also cause residual tensile stresses to develop
ahead of stress concentrations under far-field cyclic compression. Examples of fatigue crack growth in cyclic compression ahead of stress concentrations in brittle
solids are shown in Figs. 5.16(a) and (b).

Brockenbrough & Suresh (1987) conducted a finite element simulation of near-
tip fields in notched plates of a polycrystalline aluminum oxide subjected to
uniaxial cyclic compression. They used a constitutive model for microcracking
which was described earlier in Fig. 5.4 and in Eqs. 5.4–5.8. Their predictions
of the variation of residual stresses normal to the plane of the notch, $\bar{\sigma}_{yy}$
(which is the stress σ_{yy} perpendicular to the plane of the notch, normalized by
the product of the maximum far-field compressive stress σ^∞ and the elastic stress
concentration factor for the notch tip K_t), are plotted in Fig. 5.17 as a function
of the distance directly ahead of the notch tip x, normalized by the notch tip
radius ρ. These computations were made using the following values for the vari-
ables in Eq. 5.6: $\sigma_0/(\sigma^\infty K_t) = 0.0076$, $A = 0.004$, $\bar{n} = 1$, and $\beta_s = 0.4$ for penny-
shaped microcracks. When the microcracking deformation ahead of the notch tip
leaves no permanent strains, i.e. when $\lambda = 0$, (see Eq. 5.7 and Fig. 5.4), essentially
no residual stresses are induced ahead of the notch tip after one compression
cycle. On the other hand, when $\lambda > 0$, a region of large residual tension is created
at the notch tip after the far-field compressive stress is removed. Note that
residual stresses are self-equilibrating. A zone of residual tension in the immediate
vicinity of the notch tip is accompanied by a zone of residual compression away
from the notch tip, Fig. 5.17.

The effect of this residual stress zone is that residual tensile stresses easily exceed
the tensile strength of the brittle solid over a distance of the order of the notch root
radius. This implies that a mode I crack will develop over this distance upon unloading from the far-field compression stress, an inference which is supported by experimental observations. Once a fatigue crack initiates from the notch tip, its rate of
growth during subsequent compression cycles is dictated by such factors as the
exhaustion of the residual stress zone created during the first cycle, formation of

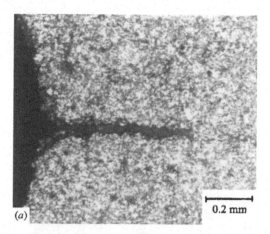

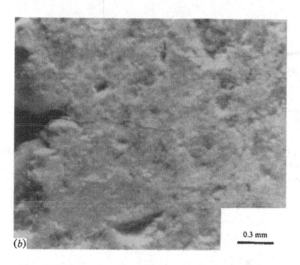

Fig. 5.16. Examples of mode I fatigue cracks initiated at stress concentrations under far-field cyclic compression: (*a*) Polycrystalline Al_2O_3. (b) Cement mortar. The compression axis is vertical in both cases. (From Ewart & Suresh (1986) and Suresh, Tschegg & Brockenbrough (1989), Reprinted with permission.)

debris particles of the brittle solid due to repeated contact between the crack faces, the generation of a residual stress field in subsequent cycles, and the development of closure due to an increase in crack length. Experiments by Ewart & Suresh (1987) on polycrystalline alumina show that the high frequency contact between the crack faces generates debris particles of the ceramic, typically of the order of 1 μm in size, within the crack. The presence of the debris particles promotes a wedging effect. Periodic removal of the debris particles by ultrasonic cleaning of the crack leads to a significant increase in the total distance of crack growth (Fig. 5.18). In ceramics with grain sizes in the range 1–30 μm, crack growth over a distance of the order of 1 mm

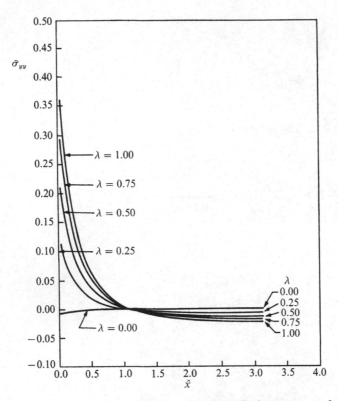

Fig. 5.17. Finite element predictions of the normalized stress $\bar{\sigma}_{yy}$ as a function of the normalized distance x/ρ directly ahead of the notch tip for different unloading paths, λ. (After Brockenbrough & Suresh, 1987.)

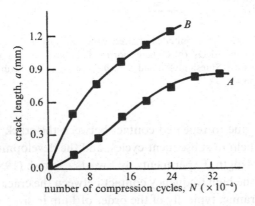

Fig. 5.18. Variation of fatigue crack length a, measured from the notch tip, as a function of the number of compression cycles in α-alumina of grain size $= 18$ µm (curve A). Curve B shows the increase in crack growth rates as a consequence of reducing crack closure by the removal of debris particles from the crack after every 5000 compression cycles using ultrasonic cleaning. (After Ewart & Suresh, 1986.)

has been observed under fully compressive far-field cyclic loads. The fatigue cracks arrest after growth over this distance because of the development of crack closure in cyclic compression and the exhaustion of the residual tensile field.

At the maximum far-field compressive stress, there exists a state of compression immediately ahead of the notch tip. Complete unloading, however, gives rise to a zone of residual tension at the notch tip. Numerical simulations by Brockenbrough & Suresh (1987) reveal that residual tensile stresses are generated only after unloading occurs below a certain critical value, σ_{cr}, of compressive stress, i.e. when $\sigma_{cr} < \sigma^\infty < \sigma_{max}$, Fig. 5.17. This result implies that the mean stress of the compression fatigue cycle will have an important effect on the magnitude and extent of the residual stress field. As the mean stress is pushed far below the zero level, the extent of residual tension will decrease and, consequently, crack initiation from the notch tip will become more difficult. Such predictions have been confirmed independently by experimental studies of the effects of cyclic compressive mean stress on crack initiation in β-alumina (James, Tait & Mech, 1991) and in brittle polymers (Pruitt & Suresh, 1993).

An important feature of the phenomenon of crack initiation from notches under cyclic compression is that both monolithic and composite (brittle or ductile) solids with vastly different microscopic deformation modes exhibit a macroscopically similar mode I fatigue crack growth behavior. The principal reason for this universal trend is that residual tensile stresses are induced within the notch tip damage zone during cyclic compression as long as permanent deformation occurs ahead of the notch tip; such permanent deformation can be a consequence of dislocation plasticity, phase transformation, microcracking, interfacial sliding or creep. Since the zone of residual tension is embedded within a compressive residual stress field, crack initiation and growth in cyclic compression is intrinsically stable even in brittle solids. Results of numerical analysis of the residual stress fields ahead of notches subjected to cyclic compression have been reported by Suresh & Brockenbrough (1988) and Suresh (1990a) for single phase ceramics, transforming ceramics, creeping solids, and cement mortar.

Experimental studies of crack initiation in cyclic compression in ceramic composites have pointed out the potentially deleterious effects that fatigue loads can have on the service life of a structural component (see, for example, Suresh, 1990a). Consider, for example, the case of hot-pressed Si_3N_4 which is reinforced with SiC whiskers. It is known that the addition of 20–30 volume% SiC to the Si_3N_4 matrix can lead to an increase in fracture toughness by more than a factor of two over that of the unreinforced matrix material. However, when the composite contains stress concentrations, the application of cyclic compressive loads causes fatigue cracks to initiate more easily in the composite than in monolithic Si_3N_4. This effect can be rationalized by noting that the stress–strain curve for the composite is more nonlinear than that for the matrix material. Unloading from a far-field compressive stress can promote a higher degree of permanent deformation which, in turn, may lead to larger residual tensile stresses ahead of the stress concentration (see Fig. 5.17).

Crack initiation ahead of stress concentrations subjected to cyclic compression also offers a capability to introduce controlled fatigue pre-cracks in brittle materials prior to the determination of fracture toughness, creep crack growth, or fatigue crack growth in tension. An advantage of this technique is that fatigue cracks can be introduced in brittle solids such as ceramics and ceramic composites and in ductile metals using similar cyclic compression test conditions and specimen geometries. Furthermore, the cyclic compression pre-cracking method is the only known technique for introducing fatigue pre-cracks in circumferentially notched cylindrical rods of brittle solids (which are used for quasi-static and dynamic mode I and mode III fracture tests).

5.6.1 Example problem: Crack initiation under far-field cyclic compression

Problem:

A polycrystalline ceramic plate containing a through-thickness edge notch is subjected to uniaxial cyclic compression loading at room temperature. The specimen is placed between two perfectly parallel surfaces and the cyclic compressive loads are applied in a direction normal to the plane of the (nonclosing) notch. The peak load of the fatigue cycle is σ_{max} and the lowest load is σ_{min}. (For fully compressive cyclic loading, where σ_{max} and σ_{min} are both negative, $\sigma_{min} < \sigma_{max}$, $|\sigma_{max}| < |\sigma_{min}|$ and $R = |\sigma_{min}|/|\sigma_{max}| > 0$.) Assuming that grain boundary microcracking due to the applied loads is the primary source of inelastic deformation, answer the following questions.

(i)   Four experiments are conducted where σ_{min} is held fixed and constant-amplitude compressive stress cycles with the following R ratios are used: (a) 2, (b) 10, (c) 30, and (d) ∞ (which corresponds to zero–compression cyclic loading with $\sigma_{max} = 0$). How would you expect the initial rate of crack growth and the maximum distance of crack growth under cyclic compression (before complete crack arrest) to vary as a function of R for these four cases?

(ii)   Three additional experiments are conducted where σ_{max} is held fixed and the following R ratios are used: (a) 2, (b) 10, and (c) 30. How would you expect the initial rate of crack growth and the maximum distance of crack growth under cyclic compression (before complete crack arrest) to vary as a function of R for these three cases?

Solution:

On the basis of the information provided, assume that the trends predicted by the simulations in Fig. 5.17 apply to the polycrystalline ceramic considered here.

(i)   It is evident from the preceding section and the given information that, for fixed σ_{min}, the extent of permanent damage developed ahead of the notch during compression loading is the same in all four cases. The extent of residual tensile stresses generated upon unloading then is dependent on the σ_{max}. That is, the closer is the value of σ_{max} to zero far-field stress (and hence the higher the R ratio), the larger would be the tensile stress generated at the notch tip. Therefore, the maximum tensile residual stress field (or a distance ahead of the notch tip over which a tensile stress of some level is developed) would be expected to increase with an increase in the applied R ratio. Consequently, the initial rate of crack growth and the maximum distance of crack growth under cyclic compression (before complete crack arrest) would both increase with an increase in R ratio.

(ii) For a fixed value of applied σ_{max}, σ_{min} increases with increasing R. Larger the value of R, greater would be the amount of permanent deformation (in this case, microcracking) at the notch-tip. As a simple approximation, one would then expect λ to be higher for higher values of R. From Fig. 5.17, it could then be argued again that the extent of tensile residual stress field, the initial rate of crack growth and the maximum distance of crack growth under cyclic compression (before complete crack arrest) would all increase with an increase in R ratio.

Exercises

5.1   Two dislocations in BCC α-iron with Burgers vectors $a_0/2\,[111]$ and $a_0/2\,[1\overline{1}\overline{1}]$ (where a_0 is the lattice dimension) glide along two intersecting {110}-type slip planes meet.
    (a) Write down the dislocation reaction and show that it is energetically favorable.
    (b) Discuss the orientation of a possible cleavage crack which could be triggered by this Cottrell mechanism.

5.2   A cubic crystal contains a mixed dislocation (i.e. a dislocation with combined edge and screw components) with a Burgers vector parallel to [001]. What would be the orientation and geometry of the steps on the cleavage fracture surface for a crack propagating along the cleavage plane (001)? Discuss various possibilities for the crack–dislocation intersection.

5.3   Consider a brittle elastic solid which is subjected to cyclic loading. The material is first loaded in uniaxial tension. At a threshold tensile stress σ_T, microcracks initiate in the material; a further increase in tensile stress to a peak value of σ_{Tmax} results in a progressive increase in both the number of microcracks per unit volume and the average size of the microcracks, and in the release of the residual thermal strain. The specimen is unloaded at σ_{Tmax}. During unloading, the microcracks remain open until a far-field tensile stress of σ_{Tcont}. Below σ_{Tcont}, lowering the far-field stress to zero and subsequent load reversal into compression results in a gradual recovery of stiffness until all the microcrack surfaces are closed (where the far-field stress is σ_{Cclose}). Further increases in compressive stress to a peak value of σ_{Cmax} are borne by the fully dense elastic solid. Unloading from the far-field compressive stress results in the elastic recovery of a fully dense solid until the microcracks reopen at a far-field compressive stress of σ_{Copen}, with $|\sigma_{Copen}| > |\sigma_{Cclose}|$.

(a) Plot the variation of the far-field stress against the average inelastic strain (i.e. total strain minus the elastic strain for the fully dense solid) for one full fatigue cycle involving zero–tension–zero–compression–zero loading.

(b) How is the stress–strain curve in (*a*) affected by the presence of a population of pre-existing microcracks prior to fatigue loading.

(c) Suppose that, in (*a*), frictional sliding occurs between the faces of the microcracks (as, for example, along grain boundaries or shear bands) upon unloading from the far-field compressive stress. Schematically sketch the stress–strain curve by incorporating the sliding mechanism.

5.4   Cyclic indentation of a brittle solid exhibits some features which are qualitatively similar to those discussed in connection with the cyclic compression loading of a notched brittle solid. Discuss the similarities and differences between the initiation of a fatigue crack in a notched brittle solid subjected to cyclic compression loading and the cracking that occurs beneath the tip of an indenter repeatedly pressing against the surface of a brittle solid.

5.5   The mechanism for the nucleation of a crack by the pile-up of dislocations, Fig. 5.9(*a*), at an obstacle (such as a grain boundary) was proposed by A.N. Stroh (*Proceedings of the Royal Society, London* **A223**, p. 404, 1954, and **A232**, p. 548, 1955). Since the dislocation pile-up leads to stress concentration in much the same way as a mode II crack, the near tip stress fields ahead of a mode II crack, Eqs. 9.47, can also be used to model the fields ahead of piled-up dislocations. Stroh postulated that cleavage fracture will nucleate at an obstacle when the maximum local tensile stress, $\sigma_{\theta\theta}$ reaches a critical value.

(a) Using Eq. 9.47, show that $\sigma_{\theta\theta}$ is a maximum at an angle of $-70.5°$ from the plane of dislocation pile-up.

(b) Following a procedure similar to that used in the previous problem on the Petch relationship, show that the cleavage fracture stress takes the form:

$$\sigma_F = \sigma_i + k_F d_g^{-1/2},$$

where σ_i is the friction stress associated with lattice resistance, k_F is a material constant, and d_g is the mean spacing of the obstacles.

5.6 List three reasons why the crack growth characteristics of brittle ceramics and ceramic composites from stress concentrations subjected to cyclic compression are qualitatively similar to those seen in notched metallic plates.

5.7 A silicon nitride ceramic matrix is reinforced with 10, 20 and 30 volume% of SiC short fibers. Uniaxial tensile tests conducted on the unreinforced as well as the reinforced ceramics show that the stress–strain curve exhibits greater nonlinearity with an increase in the concentration of SiC. An increase in SiC content also leads to an apparently higher overall fracture toughness. Discuss possible effects of SiC reinforcement and the effects of SiC concentration on the propensity of notched silicon nitride matrix ceramic to subcritical crack growth under cyclic compression.

CHAPTER 6

Cyclic deformation and crack initiation in noncrystalline solids

The cyclic deformation and fatigue crack initiation characteristics of semicrystalline and noncrystalline solids are the subjects of discussion in this chapter. Attention is devoted to a consideration of the response of polymers and organic composites. The principal objective of this chapter will be to document the prominent mechanisms of fatigue in amorphous materials and, in particular, to present a detailed description of the similarities and differences between the fatigue characteristics of crystalline and noncrystalline materials. As in the case of brittle ceramics and composites, the discussion of cyclic deformation and crack initiation is presented here in a single chapter because in many instances, microscopic cracking processes (such as crazing) constitute prominent mechanisms which influence fatigue response. Also addressed in this chapter is the topic of fatigue crack initiation at stress concentrations in polymers. Stress-based and strain-based approaches to fatigue life of polymers are considered in Chapters 7 and 8, respectively. Fatigue crack growth in semi-/noncrystalline solids is taken up in Chapter 12.

6.1 Deformation features of semi-/noncrystalline solids

6.1.1 Basic deformation characteristics

A noncrystalline material can be a metallic glass (e.g., Pd–20 at.% Si alloy), an inorganic glass (e.g., silicate glass), or an organic glass (i.e. a polymer). Amorphous materials exhibit a variety of permanent deformation modes depending on their basic structural unit. This fundamental unit is a single atom for a metallic glass whereas, for an inorganic glass, the building block is a SiO_4 tetrahedron. A long-chain molecule built upon a tetravalent carbon atom is the basic structural unit for an organic glass. The deformation of an amorphous solid can take place homogeneously through the bulk of the material or heterogeneously along shear bands or crazes. On a macroscopic scale, the deformation of an amorphous solid can be as varied as linear elastic, plastic or Newtonian viscous depending on the temperature, applied stress and strain rate. With the increasing technological importance of materials such as plastics in structural applications, the fatigue of amorphous solids has also become a topic of major practical interest.

The formation of long-chain molecules is the main feature of the structural constitution of polymers that differs from the basic building block of atomic solids (such as ceramics and metals) or space networks (such as oxide glasses). The obstacles to complete crystallization in polymeric solids arise from the very existence of the long-chain molecule and of the chain branches and side groups. Consequently, the structure of a polymer can be fully amorphous or of a semi-crystalline arrangement in which folded molecular chains are formed amid an amorphous phase.

With the exception of some modes of deformation such as craze formation or the rotation of molecular chains, the mechanisms of deformation and failure in polymeric solids exhibit many similarities to those discussed earlier in this book for metals and ceramics. Under cyclic loading, polymers display deformation modes (such as stress–strain hysteresis and cyclic softening) and subcritical crack growth analogous to atomic solids. It was shown earlier in the context of fatigue in metals and alloys (Chapters 2–4) and in ceramic materials (the preceding Chapter) that kinematic irreversibility of microscopic deformation is an important reason for the onset of true mechanical fatigue effects. In polymeric solids, such kinematically irreversible deformation can be manifested as crazing, shear band formation, rotation or other changes in the orientation of molecular chains, or a combination of these mechanisms.

6.1.2 *Crazing and shear banding*

Homogeneous deformation in glassy polymers is effected by the displacement of segmented molecular chains and by the gradual orientation of the long axis of the initially unaligned molecules with the tensile loading axis. *Crazing* and *shear flow* are the two most common modes of deformation during the fatigue of polymers. Crazing generally has the connotation of brittle failure, whereas deformation by shear banding represents a more ductile process. The relative dominance of these two processes during fatigue is dictated by many mutually competitive mechanisms which depend on the molecular structure, specimen geometry, processing methods, test temperature, loading rate, stress state, fiber reinforcements, and the degree of plasticization.

Crazes occur in glassy polymers, such as polystyrene (PS), polymethylmethacrylate (PMMA) and polysulfone (PSF) and in some semi-crystalline ones such as polyethylene (PE), polypropylene (PP), polyethyleneteraphthalate (PETP), and polyoxymethylene (POM) (see, e.g., Kramer (1983) and Kramer & Berger (1990)). When these polymeric solids are subjected to tensile stresses at low temperatures, fine crack-like features appear which apparently resemble surface flaws in ceramics. There is a continuity of material across a craze, whereas the faces of a Griffith crack in a brittle solid are fully separated. Crazes are always oriented perpendicularly to the maximum principal tensile stress, while the Griffith flaws in a brittle solid with an equi-axed grain structure essentially have a random distribution. However, crazes can be regarded as the polymer analogue of dilatant transformations arranged perpendicularly to the maximum principal tensile stress.

In this sense, crazes lead to inelastic strains in much the same way as the shear and/or dilatant transformations in metals and ceramics (such as mechanical twins or martensitic lamallae). Furthermore, purposely promoting the formation of crazes at lower stress levels in an attempt to develop appreciable dilatational plasticity is viewed as a possible means of toughening polymers (e.g., Argon, 1989). This approach is conceptually similar to the mechanisms of microcrack or transformation toughening in brittle solids (see Chapter 5). Cyclic deformation and the subcritical advance of fatigue fracture in many polymers are dictated by the nucleation, growth and breakdown of crazes.

A craze contains fibrils of highly oriented molecules (*craze matter*) separated by porous regions. The density of the craze matter is only 40 to 60% of the matrix density and the molecular fibrils are oriented along the direction of the maximum tensile stress. Figures 6.1(*a*) and (*b*) are electron micrographs of a newly formed craze

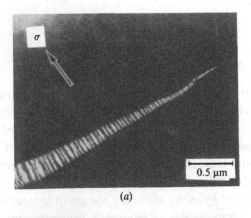

(*a*)

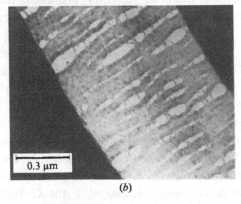

(*b*)

Fig. 6.1. (*a*) A newly formed craze in a thin slice of PS. Note the orientation of the craze normal to the tensile stress axis (indicated by the arrow) and of the fibrils parallel to the tensile stress axis. (*b*) An electron micrograph of the central section of another craze. (From Beahan, Bevis & Hull, 1971. Copyright Taylor & Francis, Ltd. Reprinted with permission.)

in a thin slice of PS which was loaded on a mini-tensile straining device. Note that the craze is oriented normal to the uniaxial tensile stress and that the fibrils within the craze are aligned with the tensile axis. (The craze geometry is discussed in further detail in Chapter 12.) It is generally known that the criterion for the nucleation of a craze in a three-dimensional stress state is of the form:

$$\sigma_{max} - \sigma_{min} \geq \Lambda(T) + \frac{\Omega_1(T)}{\sigma_H}, \tag{6.1}$$

where σ_{max} and σ_{min} are maximum and minimum principal stresses, respectively, $\Lambda(T)$ and $\Omega_1(T)$ are material constants which depend on the temperature, and σ_H is the hydrostatic stress (Section 1.4). Even if the overall hydrostatic stress is negative, a craze can advance if there exists one tensile stress component (with the craze extending normal to the tensile stress direction). The nucleation of crazes in crazable polymers has been investigated in detail, and the criteria for their formation have been developed in terms of the tensile hydrostatic stress component and the deviatoric stress component of the imposed stress state (e.g., Argon & Hannoosh, 1977; Argon, 1989).

The growth of crazes is also a topic of considerable interest because cracking in many polymers is preceded by the formation of a craze. In this sense, the craze zone ahead of a crack in a polymer is somewhat analogous to the plastic zone in front of a crack in a ductile metallic material. Under an imposed stress, the craze extends by drawing more polymer from its surface into the fibril while the fibrils themselves deform by creep, Kramer & Hart (1984); see Section 12.2 for further details.

Shear localization is also a prominent feature of deformation and failure in many polymers. At stress levels lower than the tensile strength of the glassy polymer, 'plastic' deformation can be initiated by the formation of shear bands. In polymers susceptible to shear banding, the onset of 'yielding' is associated with the inception of shear bands. Figure 6.2 shows shear bands in plastically stretched PETP. Shear bands are always oriented along the direction of the maximum shear stress.

6.1.3 Cyclic deformation: crystalline versus noncrystalline materials

There are many apparent differences in the cyclic deformation characteristics of metals and polymers at room temperature. These distinctions stem from the differences in their homologous temperature (in relation to the test temperature), thermal diffusivity and structural constitution.

(1) Whether a metallic material undergoes cyclic hardening or softening depends on such factors as the degree of prior cold work, composition, and heat treatment. Polymeric materials, on the other hand, exhibit only *cyclic softening* behavior. Drastic changes in composition, molecular structure, temperature, and strain rate alter only the degree to which cyclic softening occurs in polymers. Even cyclic stability is rarely observed; it may occur only under conditions of small strain amplitudes.

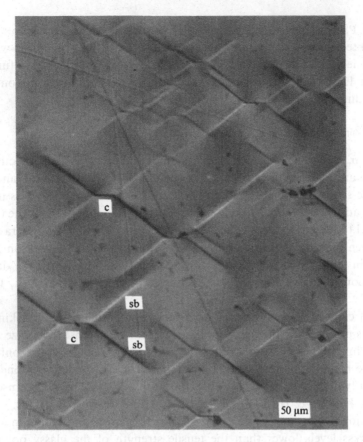

Fig. 6.2. Shear bands (sb) and crazes (c) in plastically stretched PETP. (From Argon, 1980. Copyright Academic Press. Reprinted with permission.)

(2) The second major distinction between fatigue of metals and polymers is the degree to which the loading rate influences the stress–strain characteristics and failure modes. For many polymeric materials, even room temperature is a significant fraction of the homologous temperature. Furthermore, hysteretic damping effects are substantial in thermoplastics as a result of their strong nonlinear viscous behavior, which is not the case for metals. These factors, coupled with the poor thermal diffusivity of polymers, can result in marked increases in temperature during cyclic deformation at sufficiently high strain rates. The attendant thermal softening becomes an important consideration and may even dominate over any intrinsic mechanical fatigue effects. These interactions between adiabatic thermal effects and mechanical fatigue are complex. For certain combinations of stress amplitudes and strain rates, thermal effects exacerbate the fatigue failure process while, in other cases, the combination of the thermal and mechanical effects is not deleterious to fatigue life. The imposition of low strain rates to circumvent

thermal softening effects may promote other time-dependent deformation processes such as creep.

(3) Repeated dislocation glide along crystallographic slip systems is the dominant mode of fatigue deformation in metallic materials. However, there exists a variety of microscopic deformation modes in polymeric materials. The underlying mechanisms may include homogeneous deformation involving normal and shear yielding, disentanglement, reorientation and slip of chain segments, crystallization, or heterogeneous deformation arising from shear banding and crazing.

6.2 Cyclic stress–strain response

Comprehensive studies of the cyclic stress–strain characteristics of polymeric materials have been reported by Rabinowitz & Beardmore (1974) and Beardmore & Rabinowitz (1975). These investigators have shown that cyclic softening is the general fatigue phenomenon in ductile polymers, *irrespective of the underlying molecular structure*, although cyclic softening becomes more pronounced with increasing ductility. Amorphous and semi-crystalline polymers as well as polymer–matrix composites exhibit cyclic softening. Changes in the extent of crystallinity mainly affect the degree and rate of cyclic softening. For example, when subjected to low strain amplitudes, amorphous polymers exhibit an incubation period prior to strain-softening; this incubation period diminishes with increasing values of imposed strain amplitudes. The cyclic stress–strain curves of homopolymers develop stable hysteresis loops. However, polymer–matrix composites undergo cyclic softening throughout the fatigue life without ever attaining a saturated state. This behavior may result from the gradual development of microscopic damage in the matrix, in the reinforcement or along the matrix–reinforcement interface.

6.2.1 Cyclic softening

Figure 6.3(*a*) shows the cyclic softening behavior of polycarbonate (PC) subjected to strain-controlled fatigue at 298 K. While the stress–strain hysteresis loops are generally symmetric about the origin in metallic materials, the initial hysteresis loop (in the first full cycle of loading) for polycarbonate exhibits a propeller-like shape at room temperature as a consequence of inelastic deformation. By repeating this strain-controlled fatigue experiment at several different strain amplitude levels, the cyclic stress–strain curve shown in Fig. 6.3(*b*) is obtained (which is the same as the strain-controlled test method described in Chapter 3 for metals). In the low strain regime, the monotonic and cyclic tension responses are identical. The amount of cyclic softening increases with increasing strain values. A compressive

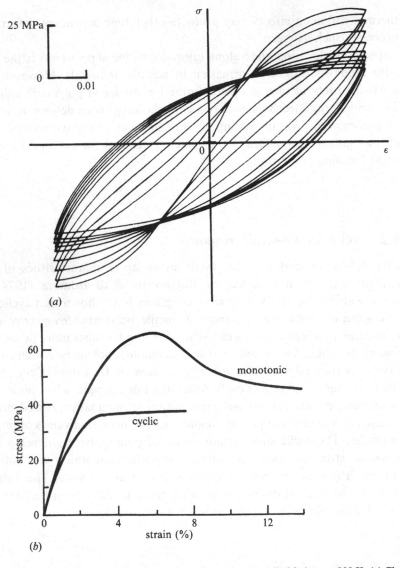

(a)

(b)

Fig. 6.3. Cyclic softening in PC subjected to strain-controlled fatigue at 298 K. (*a*) Change in the size and orientation of the hysteresis loops with the progression of fatigue deformation. (*b*) Stress–strain response in monotonic tension and fatigue. (From Rabinowitz & Beardmore, 1974. Copyright Chapman & Hall. Reprinted with permission.)

stress–strain curve in monotonic and cyclic loading can also be obtained in a similar fashion. The origin of the differences between monotonic and cyclic response is linked to the molecular rearrangements. During cyclic deformation, the strains are accommodated by molecular rearrangements at the microscopic level, whereas monotonic loading promotes more macroscopic permanent deformation associated with molecular rearrangements.

6.2.2 Thermal effects

As noted earlier, the local temperature rise associated with hysteretic heating in fatigue-loaded polymers can result in thermal softening even in room temperature fatigue for certain combinations of strain rates and cyclic stresses/strains. The typical differences between the thermal and mechanical fatigue effects are illustrated in the photographs of failed specimens of PMMA shown in Fig. 6.4.

6.2.3 Example problem: Hysteretic heating

Problem:

The temperature increase in the polymer is precipitated by the accumulation of hysteretic heat during each fatigue cycle. Consider a polymer subjected to a sinusoidal variation of cyclic stress,

$$\sigma = \sigma_0 \sin \omega t, \tag{6.2}$$

where σ is the stress at time t, σ_0 is the peak value of the stress cycle, and $\omega = 2\pi\nu_c$ is the angular frequency, ν_c being the number of stress cycles per unit time. If the viscoelastic behavior of the polymer is linear (see Section 1.4.4), the strain will also fluctuate sinusoidally, but it will be out of phase with the stress by a phase angle δ, which is a function of ω. The peak values of the stress and strain, σ_0 and ϵ_0, respectively, are related by the complex modulus E^*, where

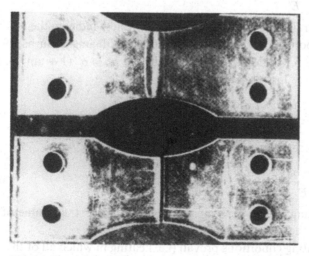

Fig. 6.4. Examples of failure due to cyclic thermal softening (top) and mechanical fatigue (bottom) in PMMA. The total length of the test specimen is 110 mm. (From Constable, Williams & Burns, 1970. Copyright Council of the Institution of Mechanical Engineers. Reprinted with permission.)

$$E^* = \tilde{E}' + i\tilde{E}'', \qquad |E^*| = \frac{\sigma_0}{\epsilon_0} = \sqrt{(\tilde{E}')^2 + (\tilde{E}'')^2}. \tag{6.3}$$

$\tilde{E}'$ is known as the *storage modulus* and it denotes the ratio of the stress in phase with the strain to the strain. The imaginary number $i = \sqrt{-1}$. $\tilde{E}''$ is the *loss modulus* and it is the ratio of the stress which is 90° out of phase with the strain to the strain. The linear viscous deformation due to sinusoidal fatigue loading can also be characterized in terms of the complex compliance,

$$D^* = \frac{1}{E^*} = D' - iD'', \tag{6.4}$$

where, analogous to the storage modulus and loss modulus, D' and D'' denote the storage compliance and loss compliance, respectively. The entity $\tan\delta = \tilde{E}''/\tilde{E}' = D''/D'$ is known as the *loss tangent*.

(i)   Find the rate of hysteretic energy dissipated per fatigue cycle.
(ii)   If H represents the heat transfer coefficient for loss of heat from the specimen surface to the surroundings, and T and T_0 are the instantaneous specimen temperature and the ambient temperature, respectively, derive an expression for the rate of temperature increase assuming adiabatic heating conditions.

Solution:

(i)   The rate of hysteretic energy dissipated as heat, $\dot{Q}$, per unit volume of the material during fatigue loading is given by

$$\dot{Q} = \sigma''\dot{\epsilon}'' = \frac{2\pi\nu_c\sigma'^2\tan\delta}{E'}, \tag{6.5}$$

where $\dot{\epsilon}$ is the strain rate and the superscripts $'$ and $''$ indicate the real and imaginary components, respectively (Constable, Williams & Burns, 1970). Combining Eqs. 6.2 and 6.5 with the result that $\sigma^2 = \sigma'^2(1 + \tan^2\delta)$, one obtains the energy loss per stress cycle,

$$Q = \int_0^{2\pi/\omega} \dot{Q}dt = \frac{\pi\sigma_0^2}{\tilde{E}'}\frac{\tan\delta}{1 + \tan^2\delta}. \tag{6.6}$$

The average energy dissipation rate per unit volume is

$$\dot{Q} = Q \cdot \frac{\omega}{2\pi} = \frac{\pi\nu_c\sigma_0^2}{\tilde{E}'}\frac{\tan\delta}{1 + \tan^2\delta} = \pi\nu_c D''\sigma_0^2. \tag{6.7}$$

For cyclic loading with a zero mean stress, the peak stress value of the fatigue cycle is the same as the stress amplitude $\Delta\sigma$.

(ii)   If adiabatic heating conditions prevail (i.e. heating in which all of the heat generated within the polymer is manifested as a temperature rise and none is lost to the surroundings), the time rate of change of temperature dT/dt is given by

$$\frac{dT}{dt} = \frac{\dot{Q}}{\rho c_p} = \frac{\pi v_c D'' \sigma_0^2}{\rho c_p},$$ (6.8)

where ρ is the mass density and c_p is the specific heat. In reality, however, some heat is lost to the surroundings. Equation 6.8 may be modified to account for heat loss:

$$\frac{dT}{dt} = \frac{\pi v_c D'' \sigma_0^2}{\rho c_p} - \frac{HA}{\rho c_p V}(T - T_0).$$ (6.9)

Here A and V are the surface area and volume of the fatigue test specimen, respectively. Note that the loss compliance D'' depends strongly on both tempeature and strain rate. As the specimen temperature increases and approaches a critical softening temperature, the specimen becomes too soft to support the load and suffers catastrophic fracture.

6.2.4 Experimental observations of temperature rise

The experimentally determined rise in temperature T for polytetrafluoroethylene (PTFE) subjected to fatigue at a frequency of 30 Hz in the room temperature environment is plotted in Fig. 6.5 as a function of the number of fatigue cycles for several different values of the imposed stress range $\Delta\sigma$. The endurance limit $\Delta\sigma_e$ of this polymer (i.e. the stress range below which fatigue failure does not occur for at least 10^6 stress cycles) is 6.5 MPa for the conditions of the experiment. Figure 6.5 reveals that, when $\Delta\sigma > \Delta\sigma_e$, a rapid increase in temperature occurs with increasing

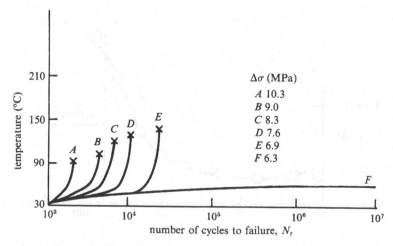

Fig. 6.5. Effect of the applied stress range $\Delta\sigma$ on temperature rise in PTFE subjected to stress-controlled fatigue. The symbol × denotes failure of the specimen. (After Riddell, Koo & O'Toole, 1966.)

number of cycles. The rate of initial elevation in temperature is higher for the larger values of $\Delta\sigma$. However, when $\Delta\sigma < \Delta\sigma_e$ (i.e. for curve F), the temperature increase is not sufficient to cause thermal failure. Consequently, the temperature stabilizes after prolonged fatigue loading, and the specimen is essentially capable of sustaining an infinite number of fatigue cycles.

Koo, Riddell & O'Toole (1967) performed fatigue experiments in an attempt to quantify the effect of thermal softening on fatigue life in three fluoropolymers: PTFE, polychlorotrifluoroethylene (PCTFE), and polyvinylidenefluoride (PVF$_2$). They derived estimates of the loss compliance D'' during fatigue on the basis of dynamic modulus measurements using a torsional pendulum. It was found that hysteretic heating caused a significant increase in loss compliance (Fig. 6.6). With an elevation in material temperature, the specimen became too soft to fail by purely mechanical fatigue, but instead suffered a loss in fatigue strength by thermal softening. This damage was at least partially recoverable by annealing the specimen in the early and intermediate stages of temperature rise (see Fig. 6.5). The most significant effect of temperature rise on fatigue damage was observed in the stress cycles just prior to final failure.

6.2.5 Effects of failure modes

In addition to the aforementioned effects of homogeneous deformation on cyclic response, the microscopic failure modes responsible for inhomogeneous deformation can also strongly influence the cyclic stress–strain behavior. Figure 6.7 illustrates the cyclic deformation characteristics of PC at 77 K. At this temperature, PC is susceptible to craze formation. Since crazes form only under local tensile stresses, cyclic softening is observed only in the tensile portion of fatigue, and the hysteresis

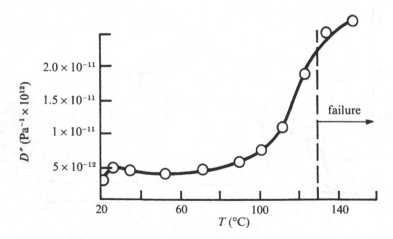

Fig. 6.6. Change in loss compliance D'' due to temperature rise in PTFE subjected to stress-controlled fatigue at room temperature at a cyclic frequency of 30 Hz. (After Koo, Riddell & O'Toole, 1967.)

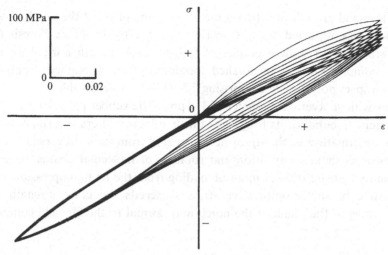

Fig. 6.7. Anomalous fatigue deformation of PC at 77 K where craze formation leads to cyclic softening only in the tensile portion of cyclic loading. (From Rabinowitz & Beardmore, 1974. Copyright Chapman & Hall. Reprinted with permission.)

loop remains stable in the compression portion. This type of anomalous softening is also seen in the stress–strain response of polymers when the deformation mechanism involves the stable growth of microscopic cracks under fatigue loading (Beardmore & Rabinowitz, 1975).

The overall fatigue response of a polymer is dictated by a combination of factors involving the molecular structure, deformation modes, and cyclic loading conditions (Hertzberg & Manson, 1980, 1986), which include:

(1) polymer composition, molecular weight and distribution, and thermody-namic state,

(2) structural and morphological changes induced by the mechanical loads and the environment, such as bond breakage, molecular alignment and disen-tanglement, or crystallization,

(3) the type of deformation, such as elastic, linear viscoelastic or nonlinear viscoelastic response,

(4) the mode of microscopic failure, such as crazing or shear banding,

(5) thermal softening, and

(6) the time-scale of the experiment *vis-à-vis* the kinetic rate of the processes causing structural changes.

6.3 Fatigue crack initiation at stress concentrations

It was shown in Sections 4.11 and 5.6 that the application of uniaxial cyclic compressive loads to notched plates of metals and ceramics, respectively, resulted in the nucleation and growth of fatigue cracks along the plane of the notch. A similar phenomenon was reported for polymeric materials by Pruitt & Suresh (1993). Figures 6.8(a) and (b) show examples of fatigue crack growth normal to the far-field cyclic compression axis in notched specimens of an untoughened polystyrene and a high-impact polystyrene comprising 7.5 wt% butadiene rubber in the form of gel particles with an average diameter of 1–2 µm. (The rubber particles are added to the polystyrene to enhance its toughness through increased craze formation.)

Inelastic deformation at the tip of notch, arising from such irreversible deformation processes as shear localization and rotation of molecular chains, generates a zone of residual tensile stresses upon unloading from the first compression cycle. If the magnitude of such compressive stresses exceeds the craze strength, crazes oriented parallel to the plane of the notch and normal to the far-field compression

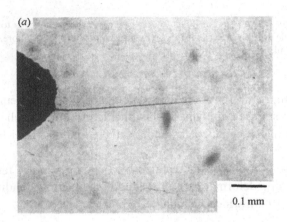

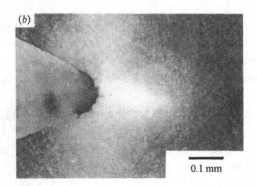

Fig. 6.8. Fatigue crack initiation and growth normal to the cyclic compression axis in (a) an untoughened PS (weight-average molecular weight, M_w = 300 000 and polydispersity = 2.4), and (b) a high-impact polystyrene comprising 7.5 wt% butadiene rubber in the form of gel particles with an average diameter of 1–2 µm (M_w = 240 000 and polydispersity = 2.8). The compression axis is vertical. (From Pruitt & Suresh, 1993. Copyright Taylor & Francis, Ltd. Reprinted with permission.)

axis (i.e. normal to the local maximum tensile stress) are induced ahead of the notch-tip. This provides a strong kinematically irreversible damage mechanism for the generation of residual tensile stresses in the subsequent cycles and for the advance of the crack.

The white region immediately ahead of the notch in Fig. 6.8(*b*) is approximately indicative of the region in which crazing occurs during cyclic compression. Figure 6.9(*a*) shows a transmission electron micrograph of a typical craze observed within this region during the cyclic compression loading of the toughened polystyrene. Figure 6.9(*b*) shows the craze penetrating through the rubber particle and the matrix. The features of the craze formed under imposed cyclic compression are the same as those produced during monotonic or cyclic tension (Pruitt & Suresh, 1993).

In-situ photoelastic and laser interferometric measurements have also been carried out in notched plates of a photoelastic resin to quantify the evolution of residual tensile stresses ahead of the stress concentration upon unloading from the far-field compression axis. Figure 6.10 shows the evolution of an increasing tensile residual stress field at the notch tip upon unloading from a maximum applied compressive stress of -16.5 MPa. Plotted in Fig. 6.10 are the contours of constant σ_{yy} (i.e. stress normal to the plane of the notch) at different stages of unloading. At an applied compressive stress of -2.76 MPa, tensile stresses with a magnitude in excess of $+4.96$ MPa span a distance of 0.07 mm ahead of the notch tip. Upon unloading to -0.92 MPa, the stresses exceed $+7.6$ MPa over a distance of 0.11 mm. Further unloading to -0.55 MPa causes the residual tensile stresses to exceed $+9.1$ MPa over a distance of nearly 0.1 mm. Given that the tensile strength of the brittle photoelastic resin is only 6.7 MPa, the stress measurements shown in Fig. 6.10 provide a justification for the nucleation of a fatigue crack ahead of the notch under cyclic compression loading. The overall initiation and growth characteristics of compression fatigue cracks in polymers are qualitatively similar to those of metals and ceramics (see Sections 4.11 and 5.6).

6.4 Case study: Compression fatigue in total knee replacements

Compression-dominated fatigue is a common occurrence in orthopedic polymer inserts used for total knee replacements. Figure 6.11 schematically shows the geometry of interest and the associated nomenclature for a total knee replacement. Ultrahigh molecular weight polyethylene (UHMWPE) has been successfully used as a material for total knee and hip orthoplasty. Quantitative analyses of the polymer damage in the total knee replacements have shown that the degree of degradation is linked to compression-dominated fatigue processes which stem from cyclic contact between the metal and the UHMWPE components of the artificial joints (Wright, Burstein & Bartel, 1985).

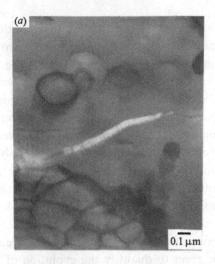

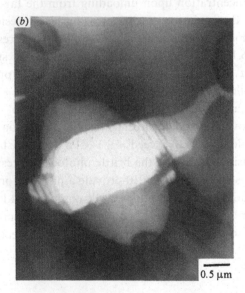

Fig. 6.9. (*a*) A transmission electron micrograph of a typical craze produced ahead of the notch during the cyclic compression loading of the toughened polystyrene. (*b*) shows the craze penetrating through the rubber particle and the matrix. The far-field compression axis is approximately perpendicular to the craze in both figures. (From Pruitt & Suresh, 1993. Copyright Taylor & Francis, Ltd. Reprinted with permission.)

As the articulating surfaces of the knee joint move during flexion, the polymer component is subjected to complex stress distributions within and at the surface of the UHMWPE insert. In a total knee replacement, the polymer insert is compressed by the metal component and results in compressive stresses perpendicular to the articulating surface (Bartel, Bicknell & Wright, 1986). The delamination and pitting of the tibial insert

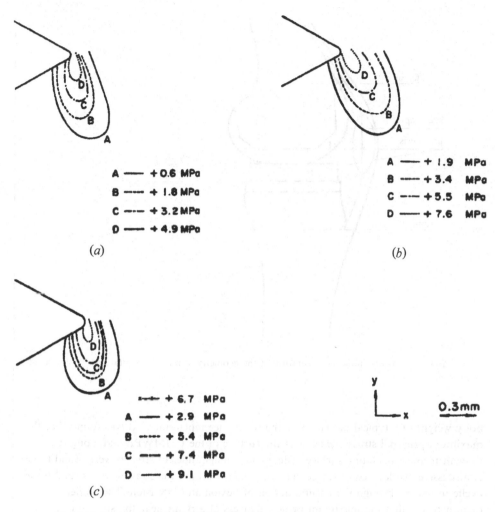

A — +0.6 MPa
B ---- + 1.8 MPa
C --- +3.2 MPa
D — +4.9 MPa

(a)

A — + 1.9 MPa
B ---- +3.4 MPa
C --- +5.5 MPa
D — +7.6 MPa

(b)

-·- + 6.7 MPa
A — +2.9 MPa
B ---- +5.4 MPa
C --- +7.4 MPa
D — +9.1 MPa

(c)

y

x   0.3mm

Fig. 6.10. *In-situ* photoelastic and laser interferometric measurements of the evolution of tensile residual stresses in a photoelastic resin during different stages of unloading from a maximum far-field compressive stress of –16.5 MPa. Far-field compressive stress magnitudes: (*a*) –2.75 MPa, (*b*) –0.92 MPa, and (*c*) –0.55 MPa. (After Pruitt & Suresh, 1993.)

also produces polymer debris which is known to cause osteolysis, infection and loosening of the implants (Mirra, Marder & Amstuz, 1982).

Bartel, Bicknell & Wright (1986) have used a three-dimensional finite-element model (FEM) to analyze the stresses in condylar-type knee replacements, Fig. 6.12. This model comprises a metal-backed UHMWPE with a uniform concave surface and the arrangement is loaded by a metal component with a convex surface. The contact surfaces are defined by two radii of curvature as the knee prosthesis has distinct radii for extension and flexion. For the analysis, the contact forces were chosen to simulate *in-vivo* conditions with femoral–tibial contact forces ranging from 4.3–4.9 times the

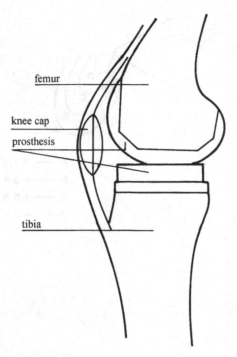

femur

knee cap

prosthesis

tibia

Fig. 6.11. A schematic representation of the geometry of interest in total knee replacement.

body weight of a typical patient. For the total knee replacement, it was found that the maximum principal stress occurred at the surface of the UHMWPE and along the tangent to the articulating surface. The greatest magnitudes of stresses were found to be compressive and located at the center of contact. Further, the stresses were found to be cyclic in nature through the natural action of flexion and extension. The model demonstrated that the maximum principal stress at a point near the surface of a total condylar type tibial knee component can range from 10 MPa of tension to more than 20 MPa of compression as the contact area sweeps across the surface in the action of knee flexion, Fig. 6.12. From this work, it was established that the primary damage mechanisms in the total knee replacement are driven by compressive or compression-dominated cyclic loading.

In an attempt to simulate the potentially deleterious effects of cyclic compressive loads in nucleating fatigue cracks at corners and stress concentrations in knee prosthesis, Pruitt & Suresh (1993) and Pruitt *et al.* (1995) carried out systematic experiments of compression fatigue cracking at notches in sterilized and unsterilized UHMWPE. On the basis of these laboratory experiments and from the information available from knee replacement components, it was concluded that the inception and growth of sharp fatigue cracks which have initiated under cyclic compressive loading can be further extended by subsequent action of cyclic tensile stresses to such a critical crack length that pitting, delamination and fatigue failure can occur at the UHMWPE surface.

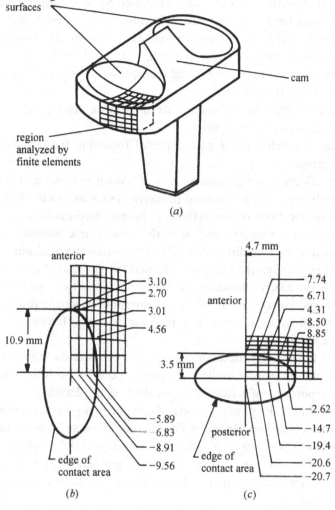

Fig. 6.12. (*a*) A schematic of the condylar-type tibial insert made of UHMWPE. The finite element mesh is also superimposed in this figure to illustrate the geometry analyzed. (*b*) Magnitudes of the maximum principal stress (in units of MPa) at the surface of the knee replacement during extension. (*c*) Magnitudes of the maximum principal stress (in units of MPa) at the surface of the knee replacement during flexion. (After Bartel, Bicknell & Wright, 1986.)

Exercises

6.1   A mechanical test is conducted on three different materials. In this experiment, a constant tensile stress is applied instantly to a specimen. The first material, a rubber, deforms rapidly during the application of the stress, with the extent of deformation becoming progressively less until it reaches an equilibrium gage length. The second, a thermosetting material, reacts to the applied stress by deforming almost instantaneously to an equilibrium extension. The third, a thermoplastic material, initially deforms to substantial extension in response to the stress and then deforms further with the extension varying linearly with time.

    (a) Describe the behavior of each material with an appropriate combination of springs and dashpots.

    (b) Describe the behavior of each material following the removal of the tensile stress.

6.2   Discuss the effects of (a) quenching rate of molten polymer and (b) plasticization on the dynamic mechanical properties (such as loss modulus). Also discuss the implications of such effects to fatigue deformation.

6.3   Two organic composites are made with the same matrix material, same size and concentration of unidirectional fiber reinforcements, and same interfacial properties between the fibers and the matrix. The spatial distribution of the fibers in the matrix, however, is different for the two cases. If crazing occurs in the polymeric matrix of the two composites, would you expect the monotonic and cyclic stress–strain response to be different for the two materials? Explain.

6.4   If crazing is the primary mode of damage in a polymer during cyclic deformation, would you expect cyclic softening to occur in both the tensile and compressive portions of a fully-reversed stress cycle? Explain.

6.5   If shear banding is the primary mode of damage in a polymer during cyclic deformation, would you expect cyclic softening to occur in both the tensile and compressive portions of a fully-reversed stress cycle? Explain.

6.6   It was shown in Section 6.3 that mode I fatigue crack growth occurs in notched specimens of polymers when subjected to fully compressive far-field cyclic loads.

    (a) What are the mechanisms responsible for this effect?

    (b) Discuss the possible effects of compressive stress amplitude and load ratio on the characteristics of crack initiation and growth from notches in polymers subjected to fully compressive cyclic loads.

    (c) Can a fatigue crack be induced along the plane of a notch in a unidirectionally reinforced graphite–epoxy composite subjected to uniaxial cyclic compression, with the axis of the graphite fiber parallel to the compression axis and with the plane of the notch perpendicular to the compression axis?

6.7   Repeat the example problem in Section 5.6.1 for the case of a notched plate of a polymer which is subjected to applied cyclic compression loading normal to the plane of the notch.

Part two

TOTAL-LIFE APPROACHES

CHAPTER 7

Stress–life approach

The preceding chapters were concerned with the evolution of permanent damage under cyclic deformation and with the attendant nucleation of a fatigue crack. While these discussions pertain to micromechanical processes, phenomenological continuum approaches are widely used to characterize the *total fatigue life* as a function of such variables as the applied stress range, strain range, mean stress and environment. These stress- or strain-based methodologies, to be examined in Part Two, embody the damage evolution, crack nucleation and crack growth stages of fatigue into a single, experimentally characterizable continuum formulation. In these approaches, the fatigue life of a component is defined as the *total* number of cycles or time to induce fatigue damage *and* to initiate a dominant fatigue flaw which is propagated to final failure. The philosophy underlying the cyclic stress-based and strain-based approaches is distinctly different from that of defect-tolerant methods to be considered in Part Three, where the fatigue life is taken to be only that during which a pre-existing fatigue flaw of some initial size is propagated to a critical size.

The stress–life approach to fatigue was first introduced in the 1860s by Wöhler. Out of this work evolved the concept of an 'endurance limit', which characterizes the applied stress amplitude below which a (nominally defect-free) material is expected to have an infinite fatigue life. This *empirical* method has found widespread use in fatigue analysis, mostly in applications where low-amplitude cyclic stresses induce primarily elastic deformation in a component which is designed for long life, i.e. in the so-called *high-cycle fatigue* (HCF) applications.† When considerable plastic deformation occurs during cyclic loading as, for example, a consequence of high stress amplitudes or stress concentrations, the fatigue life is markedly shortened. Here, fatigue design inevitably calls for the so-called *low-cycle fatigue* (LCF) approach. Realizing the important role of plastic strains in inducing permanent fatigue damage, Coffin (1954) and Manson (1954) independently proposed a plastic strain-based continuum characterization of LCF.

This chapter deals with the stress–life approach to fatigue where the effects of stress concentrations, mean stresses, surface modifications, variable amplitude cyclic loads, and multiaxial loads are also discussed. Total-life characterization of fatigue in nonmetallic materials is also addressed wherever appropriate. The stress-based and strain-based approaches have found widespread application, most notably in the design of many fatigue-critical components for automobiles and other surface vehicles. Strain-based approaches to total fatigue life are considered in the next chapter.

---

† A case study of the HCF fatigue problem in aircraft gas turbine engines is presented in Section 7.6.

7.1 The fatigue limit

Methods for characterizing the fatigue life in terms of nominal stress amplitudes using experimental data obtained from rotating bend tests on smooth specimens emerged from the work of Wöhler (1860) on fatigue of alloys used for railroad axles. In this approach, smooth (unnotched) test specimens are typically machined to provide a waisted (hour-glass) cylindrical gage length, which is fatigue-tested in plane bending, rotating bending, uniaxial compression–tension (push–pull) or tension–tension cyclic loading. Test methods for determining the stress–life response are spelled out in detail in ASTM Standards E466–E468 (American Society for Testing and Materials, Philadelphia).

From such an experiment, the stress amplitude σ_a for fully reversed loading (equal to one-half of the stress range from the maximum tension to maximum compression), is plotted against the number of fatigue cycles to failure, N_f, Fig. 7.1. The solid line illustrates the stress–life plot (also known as the S–N curve) observed for mild steels and other materials which harden by strain-ageing. Under constant amplitude loading conditions, these alloys exhibit a plateau in the stress–life plot typically beyond about 10^6 fatigue cycles. Below this plateau level, the specimen may be cycled indefinitely without causing failure. This stress amplitude is known as the *fatigue limit* or *endurance limit*, σ_e. The value of σ_e is 35% to 50% of the tensile strength σ_{TS} for most steels and copper alloys. The intercept of the stress–life curve with the ordinate is σ_{TS} at one-quarter of the first fatigue cycle. (For single crystals of FCC metals, the steady state value of the peak saturation stress τ_s^*, Fig. 2.2, may be regarded as an endurance limit.)

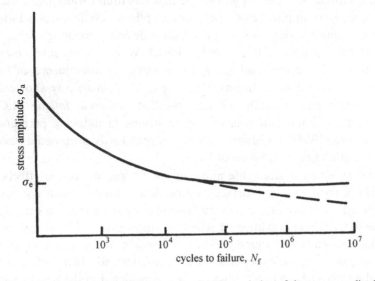

Fig. 7.1. Typical S–N diagram showing the variation of the stress amplitude for fully reversed fatigue loading of nominally smooth specimens as a function of the number of cycles to failure for ferrous and nonferrous alloys.

Table 7.1. *Cyclic endurance limit of some common engineering alloys.*

Material	Condition	σ_{TS} (MPa)	σ_y (MPa)	σ_e (MPa)
Al alloys[a]				
2024	T3	483	345	138
6061	T6	310	276	97
Steels[b]				
1015	Annealed	455	275	240
1015	60% CW	710	605	350
1040	Annealed	670	405	345
4340	Annealed	745	475	340
4340	Q&T[†] (204 °C)	1950	1640	480
4340	Q&T[†] (538 °C)	1260	1170	670
HY140	Q&T[†] (538 °C)	1030	980	480

[a] Endurance limit based on 5×10^8 cycles. Source: *Aluminum Standards and Data*, The Aluminum Association, New York, 1976.
[b] Endurance limit based on 10^7 cycles. Source: *Structural Alloys Handbook*, Mechanical Properties Data Center, Traverse City, Michigan, 1977.
[†] Refers to quenched and tempered condition; the data within parentheses refer to tempering temperature.

Many high strength steels, aluminum alloys and other materials do not generally exhibit a fatigue limit (see dashed line in Fig. 7.1). For these materials, σ_a (or $\Delta\sigma$) continues to decrease with increasing number of cycles. An *endurance limit* for such cases is defined as the stress amplitude which the specimen can support for at least 10^7 fatigue cycles. Table 7.1 lists the fatigue endurance limits for a variety of engineering alloys along with σ_{TS} and the monotonic yield strength, σ_y.

If Fig. 7.1 is redrawn on a log–log scale, with the (true) stress amplitude plotted as a function of the number of cycles or load reversals† to failure, a linear relationship is commonly observed. The resulting expression relating the stress amplitude, $\sigma_a = \Delta\sigma/2$, in a fully-reversed, constant-amplitude fatigue test to the number of load reversals to failure, $2N_f$, is (Basquin, 1910)

$$\frac{\Delta\sigma}{2} = \sigma_a = \sigma_f'(2N_f)^b, \tag{7.1}$$

where σ_f' is the fatigue strength coefficient (which, to a good approximation, equals the true fracture strength σ_f, corrected for necking, in a monotonic tension test for most metals) and b is known as the fatigue strength exponent or Basquin exponent

† A constant amplitude cycle is composed of two load reversals. As shown later in this chapter and the next one, the use of the number of load reversals, instead of the number of fatigue cycles, is helpful in analyzing variable amplitude fatigue.

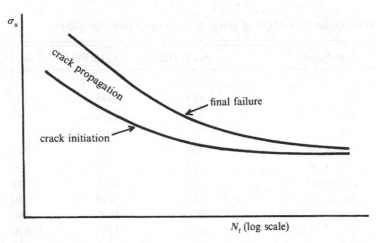

Fig. 7.2. Contributions of crack initiation and crack propagation processes to total fatigue life in a nominally smooth specimen.

which, for most metals, is in the range of −0.05 to −0.12. Typical values of σ_f' for many engineering alloys are tabulated in the next chapter.

The S–N curve schematically shown in Fig. 7.1 strictly pertains to the *total* fatigue life of a nominally smooth-surfaced, 'defect-free' material. Here total life implies the number of cycles to initiate fatigue cracks in the smooth specimen *plus* the number of cycles to propagate the dominant fatigue crack to final failure. This two-stage process involving initiation and propagation is represented in the S–N curve shown in Fig. 7.2. The fraction of the fatigue life which is expended in nucleating a dominant fatigue crack of engineering size (typically a fraction of a mm) may vary from essentially 0%, for specimens containing severe stress concentrations, rough surfaces or other surface defects, to as high as 80% in very carefully prepared, nominally defect-free, smooth specimens of high purity materials.

7.2 Mean stress effects on fatigue life

The aforementioned empirical descriptions of fatigue life pertain to fully reversed fatigue loads where the mean stress of the fatigue cycle σ_m is zero. However, fully reversed stress cycles with a zero mean stress are not always representative of many applications. The mean level of the imposed fatigue cycle is known to play an important role in influencing the fatigue behavior of engineering materials. Figure 7.3 schematically shows a fatigue cycle of sinusoidal waveform with a nonzero mean stress. In this case, the stress range, the stress amplitude and the mean stress, respectively, are defined as

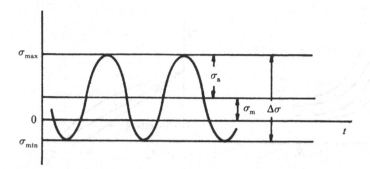

Fig. 7.3. Nomenclature for stress parameters which affect fatigue life. The variation of stress σ with time t is shown.

$$\Delta\sigma = \sigma_{max} - \sigma_{min}, \quad \sigma_a = \frac{\sigma_{max} - \sigma_{min}}{2}, \quad \sigma_m = \frac{\sigma_{max} + \sigma_{min}}{2}. \tag{7.2}$$

The mean stress is also characterized in terms of the load ratio, $R = \sigma_{min}/\sigma_{max}$. With this definition, $R = -1$ for fully reversed loading, $R = 0$ for zero-tension fatigue, and $R = 1$ for a static load.

When the stress amplitude from a uniaxial fatigue test is plotted as a function of the number of cycles to failure, the resultant S–N curve is generally a strong function of the applied mean stress level. Figure 7.4(*a*) shows the typical S–N plots for metallic materials as a function of four different mean stress levels, σ_{m1}, σ_{m2}, σ_{m3} and σ_{m4}. One observes a decreasing fatigue life with increasing mean stress value.

Mean stress effects in fatigue can also be represented in terms of constant-life diagrams, as shown in Fig. 7.4(*b*). Here, different combinations of the stress amplitude and mean stress providing a constant fatigue life are plotted. Most well known among these models are those due to Gerber (1874), Goodman (1899),[†] and Soderberg (1939). The life plots, represented in Fig. 7.4(*b*), are described by the following expressions:

$$\text{Soderberg relation}: \quad \sigma_a = \sigma_a|_{\sigma_m=0}\left\{1 - \frac{\sigma_m}{\sigma_y}\right\}, \tag{7.3}$$

$$\text{Modified Goodman relation}: \quad \sigma_a = \sigma_a|_{\sigma_m=0}\left\{1 - \frac{\sigma_m}{\sigma_{TS}}\right\}, \tag{7.4}$$

$$\text{Gerber relation}: \quad \sigma_a = \sigma_a|_{\sigma_m=0}\left\{1 - \left(\frac{\sigma_m}{\sigma_{TS}}\right)^2\right\}, \tag{7.5}$$

where σ_a is the stress amplitude denoting the fatigue strength for a nonzero mean stress, $\sigma_a|_{\sigma_m=0}$ is the stress amplitude (*for a fixed life*) for fully-reversed loading ($\sigma_m = 0$ and $R = -1$), and σ_y and σ_{TS} are the yield strength and tensile strength of the material, respectively.

[†] The modified Goodman equation, Eq. 7.4, is generally considered to be a modification of methods originally proposed by a number of different engineers.

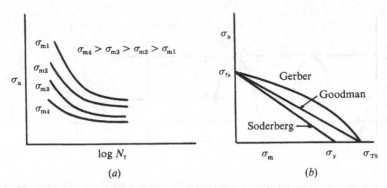

Fig. 7.4. (*a*) Typical stress amplitude–life plots for different mean stress values. (*b*) Constant life curves for fatigue loading with a nonzero mean stress.

As a general rule-of-thumb, the following observations can be made about the foregoing models for the effects of mean stress on fatigue life.

(1) Equation 7.3 provides a conservative estimate of fatigue life for most engineering alloys.

(2) Equation 7.4 matches experimental observations quite closely for brittle metals, but is conservative for ductile alloys. For compressive mean stresses, however, it is generally nonconservative. To circumvent this problem, one may assume that compressive mean stresses provide no beneficial effect on fatigue life.

(3) Equation 7.5 is generally good for ductile alloys for tensile mean stresses. It clearly does not distinguish, however, between the differences in fatigue life due to tensile and compressive mean stresses.

The constant life diagram for different mean stress levels, also commonly referred to as the Haigh diagram (Haigh, 1915, 1917), is schematically represented as shown in Fig. 7.5. In this figure, the maximum and minimum stresses of the fatigue cycle, both normalized by the tensile strength, are plotted. The dashed lines denote experimentally determined values of combinations of maximum and minimum stress levels (representing different mean stresses) which represent constant fatigue lives for the indicated number of cycles. This figure affords a convenient graphical representation of the effects of mean stress on S–N fatigue response, although considerable experimental effort is needed to determine empirically the information needed for this plot. Similar diagrams are also developed for notched members where the net-section stresses are used.

While the Basquin relation, given by Eq. 7.1, is valid only for zero mean stress, Morrow (1968) has presented a modification of the Basquin relation which accounts for mean stress effects (for any σ_m) in the following form:

$$\sigma_a = (\sigma_f' - \sigma_m)(2N_f)^b. \tag{7.6}$$

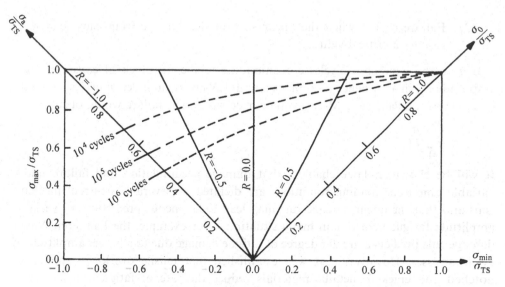

Fig. 7.5. A schematic representation of the Haigh diagram showing constant life curves for different mean stress levels in terms of the maximum and minimum stresses of the fatigue cycle.

The number of cycles to fatigue failure for any nonzero mean stress, N_f, can then be written as

$$N_f = \left\{1 - \frac{\sigma_m}{\sigma_f'}\right\}^{1/b} \cdot N_f|_{\sigma_m=0}, \tag{7.7}$$

where $N_f|_{\sigma_m=0}$ is the number of cycles to failure for zero mean stress.

7.3 Cumulative damage

The principles of stress-based characterization of total fatigue life are only relevant for constant amplitude fatigue loading. In reality, however, engineering components are invariably subjected to varying cyclic stress amplitudes, mean stresses and loading frequencies.

A simple criterion for predicting the extent of fatigue damage induced by a particular block of constant amplitude cyclic stresses, in a loading sequence consisting of various blocks of different stress amplitudes, is provided by the so-called Palmgren–Miner cumulative damage rule (Palmgren, 1924; Miner, 1945). Implicit in this *linear* damage rule are the assumptions that:

(1) The number of stress cycles imposed on a component, expressed as a percentage of the total number of stress cycles of the same amplitude necessary to cause failure, gives the fraction of damage.

(2) The order in which the stress blocks of different amplitudes are imposed does not affect the fatigue life.

(3) Failure occurs when the linear sum of the damage from each load level reaches a critical value.

If n_i is the number of cycles corresponding to the ith block of constant stress amplitude σ_{ai} in a sequence of m blocks, and if N_{fi} is the number of cycles to failure at σ_{ai}, then the Palmgren–Miner damage rule states that failure would occur when

$$\sum_{i=1}^{m} \frac{n_i}{N_{fi}} = 1. \tag{7.8}$$

It will be shown in later chapters that damage accumulation and failure under variable amplitude loading conditions are dictated by several concurrent mechanisms and that the linear damage rule may lead to erroneous predictions of variable amplitude fatigue behavior in many situations. For example, the Palmgren–Miner damage rule predicts a greater degree of fatigue damage due to a higher amplitude of cyclic stresses. However, it is well established that tensile overloads applied to notched and cracked metallic materials reduce the rate of fatigue crack growth and that the application of compressive overloads generally has the opposite trend (see Chapter 14).

Even for smooth specimens, the linear damage rule may provide incorrect results because of its omission of load sequence effects. Consider a smooth fatigue specimen which is subjected to n_1 and n_2 cycles of two different amplitudes of cyclic stresses, σ_{a1} and σ_{a2}, respectively, which have the same R ratio. As shown in Fig. 7.6, $\sigma_{a1} > \sigma_{a2}$. Let the fatigue life (number of cycles to failure) at σ_{a1} and σ_{a2} be N_{f1} and N_{f2}, respectively.

Case 1:

First consider the loading sequence in which σ_{a2} is applied after σ_{a1}. The extent of damage induced by this lower stress level may be in excess of the damage rule prediction of n_2/N_{f2}, if the preceding application of σ_{a1} for n_1 cycles had nucleated cracks or significantly contributed to the number of cycles necessary at the lower stress level to nucleate cracks. Consequently, one may observe for this case that $\sum n_i/N_{fi} < 1$. In this case, damage may occur even if σ_{a2} is below the endurance limit.

Case 2:

Now consider the reverse situation where the application of σ_{a2} precedes that of σ_{a1}. If the material strain-ages, the application of σ_{a2} prior to σ_{a1} may enhance the fatigue limit even if σ_{a2} is lower than the fatigue limit. This phenomenon is known as *coaxing*. For this case, $\sum n_i/N_{fi} > 1$.

7.4 Effects of surface treatments

The free surface of a component is a common site for the nucleation of a fatigue crack. Therefore, the manner in which the surface is prepared during man-

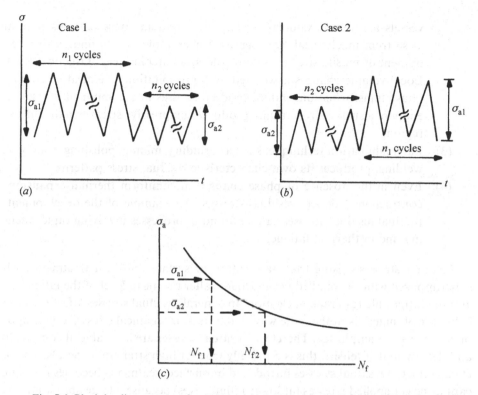

Fig. 7.6. Block loading sequence for (*a*) case 1 and (*b*) case 2. (*c*) Fractional life expended as estimated by the Palmgren–Miner rule.

ufacturing of the component has a decisive role in dictating the initiation life for fatigue cracks.

There exists a variety of surface treatments, such as carburizing, nitriding, flame hardening, induction hardening and shot-peening, which are designed to impart high strength, wear resistance or corrosion resistance locally in the near-surface regions of the material. Furthermore, common machining operations such as grinding, polishing and milling cause different degrees of surface roughness to develop. The valleys on the rough surface serve as stress concentrations, which, in turn, induce different levels of resistance to fatigue crack nucleation (see Chapters 4 and 15 for further discussions on this issue).

In addition to the roughness of the surfaces, the residual stresses that are induced by the surface treatments have an important effect on the fatigue life. Residual stresses are generated in a component as a consequence of thermal, chemical or mechanical treatments:

(1) Mechanical working, which causes nonuniform plastic deformation, may be used to provide a favorable residual stress distribution. Shot-peening of surfaces, surface rolling of fillets, and auto-frettage of gun tubes or pressure

vessels are some examples. Deleterious residual stress patterns may also arise from mechanical working, as, for example, in the undesirable development of tensile stresses due to cold straightening (e.g., Osgood, 1982).

(2) Local volume changes associated with precipitation, chemical reactions, or phase transformations induce residual stresses in a component. Case hardening of a surface by nitriding produces compressive stresses in the diffusion region.

(3) Each fabrication technique, such as grinding, milling, polishing, rolling, and welding, produces its own characteristic residual stress patterns.

(4) Even in the absence of phase changes, nonuniform thermal expansion or contraction produces residual stresses. An example of the development of thermal residual stresses can be found in processes involving rapid quenching and in thermal fatigue.

Residual stresses arising from fabrication or surface and heat treatments, when superimposed with the applied fatigue loads, alter the mean level of the fatigue cycle and the fatigue life for crack nucleation. In general, residual stresses affect the fatigue behavior of materials in the same way as the static mechanical stresses superimposed on a cyclic stress amplitude. Therefore, residual stresses are favorable, if compressive, and detrimental, if tensile; this is especially true for high strength materials. The beneficial effect of residual stresses introduced by surface treatments becomes less significant at larger applied stresses (at lower fatigue lives) because a large amplitude of the pulsating stress easily 'relaxes' the residual stress, particularly in softer materials.

Consider, for example, the case of shot-peening, which is widely used to improve the fatigue life of many engineering structural parts. Examples of shot-peened components include chassis, valve springs, gears and shafts for automobiles, and exhaust stack for aircraft engines. In the shot-peening process, a stream of small, hard spheres (typically 0.1 to 1 mm in diameter) is shot at a surface which is to be treated. Depending on the diameter of the shots, the velocity of their impingement on the surface and the duration of the process, the maximum (long range) compressive residual stress generated by the localized plastic deformation of the surface layer can reach about one-half the yield strength of the material. The compressive residual stress zone spans a depth of about one-quarter to one-half the diameter of the shots. As the shot-peened surface layer has a compressive mean stress, it acts to enhance significantly the total fatigue life by reducing the damaging effect of the tensile portion of fully reversed cyclic loads. Figure 7.7 shows an example of the beneficial effect of shot-peening on the endurance limit of steels with different levels of surface finish. A worked example quantifying the benefits of shot-peening in contact fatigue applications is presented in Chapter 13, where the combined effects of surface roughness, peening and lubrication on S–N fatigue behavior and the endurance limit are also examined.

Caution should be exercised in designing the parameters for shot-peening. If shot-peening is done for a longer time span than necessary, it may induce cracks along slip

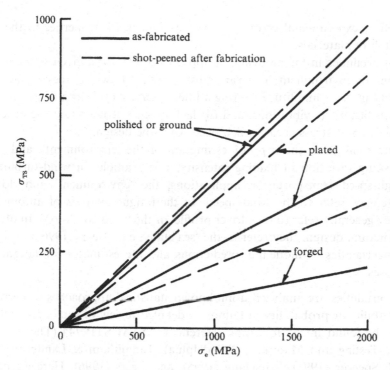

Fig. 7.7. Beneficial effect of shot-peening on the fatigue life of steels of different strength levels. The endurance limit σ_e (defined at 2×10^6 stress cycles) is plotted against the tensile strength, σ_{TS}. (After Mann, 1967.)

bands in such materials as Ni-base superalloys. In addition, the surface roughness induced by shot-peening can also, in some cases, make it easier for fatigue cracks to initiate, thereby countering the beneficial effects of compressive residual stresses.

Laser shock peening is another surface treatment method where a high intensity, short pulse duration laser beam is impacted on a surface to induce reversed plastic flow and compressive residual stresses. Here, residual compressive stresses extend deeper into the material (typically up to 2.5 mm) than in shot-peening. As an example, consider the turbine engine for aircraft in which the leading edge of the fan blade is laser shock peened to improve the resistance to fatigue cracking in the presence of foreign object damage (see Section 7.6).

7.5 Statistical considerations

There are a number of sources of uncertainty in the analysis of fatigue results, in general, and in the use of the stress–life approach. These arise from:

(1) Uncertainties and/or errors in the estimation of material properties which include microstructural variability from one specimen or batch to another as

well as experimental errors in the measurement of properties in the same batch of materials.

(2) Uncertainties in the modeling of applied stresses, for a given service condition and environment. This variability stems from two sources: (i) the variability in stress amplitudes during a known service cycle as a consequence of such factors as vibrations, and (ii) lack of knowledge about the exact distribution of stress cycles which occur over the design.

(3) Uncertainties in the *a priori* estimation of the 'environment', and in the ensuing variation in loading intensity. For example, in wind-turbine and high-speed transportation applications, the 'environment' and 'loading intensity' refer to the wind speed. In the fatigue analysis of automobiles, they generally refer to the 'driver profile' or the 'road condition'. In offshore structural design, they refer to the 'sea state', e.g., Veers, 1996.

(4) Uncertainties in modeling, predictions and life estimates for fatigue processes.

Such uncertainties are analyzed using known statistical approaches to derive the level of reliability or probability of failure. A detailed review of such approaches to fatigue can be found in the following references: ASTM STP 744 (The American Society for Testing and Materials, Philadelphia), Tangjitham & Landgraf (1993), Sobczyk & Spencer (1992), Wirsching (1995), and Veers (1996). Here we present some basic concepts commonly adopted for the statistical characterization of fatigue data.

A parameter which is widely used to describe the uncertainty is the coefficient of variation (C_{Var}) which is defined as:

$$C_{Var} = \frac{\sigma_{x,rv}}{m_{x,rv}}; \quad m_{x,rv} = \frac{1}{N_s} \sum_{i=1}^{N_s} x_{rvi}; \quad \sigma_{x,rv}^2 = \frac{1}{N_s} \sum_{i=1}^{N_s} (x_{rvi} - m_{x,rv})^2. \tag{7.9}$$

Here, N_s denotes the number of samples of a random variable x_{rv} (such as the stress amplitude in an S–N fatigue test), and $m_{x,rv}$ and $\sigma_{x,rv}$ denote the mean and the standard deviation, respectively.

A distribution in the value of the random variable is usually characterized in terms of normal distribution, log-normal distribution, or Weibull distribution. For example, the probability density function for normal distribution is defined as

$$f_n(x_{rv}) = \frac{1}{\sqrt{2\pi}(\sigma_{x,rv})} \exp\left\{ -\frac{1}{2} \left(\frac{x_{rv} - m_{x,rv}}{\sigma_{x,n}} \right)^2 \right\}, \tag{7.10}$$

where the mean $m_{x,rv}$ and the standard deviation $\sigma_{x,rv}$ are as defined previously.

Usually, the fatigue strength or the endurance limit values listed from experiments represent the arithmetic mean derived from multiple experiments. In brittle solids, such as ceramics and polymers, and in metallic alloys with considerable microstructural variability arising from processing, the extent of scatter in fatigue data may be large as a result of a large scatter in microscopic flaw size distribution. Consequently, different sets of experiments conducted on the same material may not give the same arithmetic mean of the critical strength parameter. To address this issue, Weibull (1939) proposed the concept of a probability of failure, P, at a given failure strength, σ_f normalized by an average value of a critical stress $\sigma_{cr,ave}$ (which may be identified with the average value of tensile fracture strength σ_u for a brittle solid or with the fatigue endurance limit, σ_e for a metallic alloy). At low values of σ_f, $P \to 0$, and at very high values of σ_f, $P \to 1$. Weibull defined the failure probability as follows:

$$P = 1 - P_s = 1 - \exp\left\{-\left(\frac{\sigma_f}{\sigma_{f,0}}\right)^{m_W}\right\}, \tag{7.11}$$

where m_W is known as the Weibull modulus, and $\sigma_{f,0}$ is a reference strength. Note that P in Eq. 7.11 represents the fraction of the total number N_s of identical test specimens in a batch for which the failure strength falls below σ_f. Similarly, P_s, the probability of survival, represents the fraction of the total number N_s of identical test specimens in a batch for which the failure strength exceeds σ_f. When $\sigma_f/\sigma_{f,0} = 1$, $P = 1 - \exp(-1) = 0.633$. In other words, the reference stress $\sigma_{f,0}$ represents the stress level for which the cumulative probability of failure of all specimens in the same batch at a stress level σ_f or lower, is 63.3%. Equation 7.11 can be rearranged to give

$$\frac{1}{1-P} = \exp\left\{\left(\frac{\sigma_f}{\sigma_{f,0}}\right)^{m_W}\right\} \quad \text{(or)} \quad \ln\frac{1}{1-P} = \left(\frac{\sigma_f}{\sigma_{f,0}}\right)^{m_W}. \tag{7.12}$$

Taking the logarithm of both sides once again,

$$\ln\left[\ln\frac{1}{1-P}\right] = m_W \ln\left(\frac{\sigma_f}{\sigma_{f,0}}\right). \tag{7.13}$$

It is readily apparent that a plot of the double-logarithm of $1/(1 - P)$ (ordinate) against the logarithm of σ_f (abscissa), based on experiments conducted on a number of identical specimens, yields a straight line whose slope is m_W and whose intercept with the ordinate is $-m_W \ln\sigma_{f,0}$. Such a plot is referred to as *the Weibull diagram*. Figure 7.8(a) schematically shows the probability of failure, Eq. 7.11, and Fig. 7.8(b) illustrates the Weibull diagram, Eq. 7.13. In practice, the Weibull modulus is commonly determined using the following steps. (1) A data sample of N_s identical specimens from the same batch of materials is tested to failure under static and/or cyclic loading conditions, and the appropriate strength value σ_f for each specimen is measured. (2) The values of σ_f are ordered from the smallest to the largest. The smallest strength value is ascribed an index value i of 1, the next higher value an index value of 2, and so forth. The highest strength is then assigned an index of N_s. (3) Calculate $P = i/(N_s + 1)$, for $1 \le i \le N_s$. (4) Calculate $\ln\{\ln[1/(1 - P)]\}$ and plot it against $\ln\sigma_f$. (5) The slope of this plot gives the Weibull modulus m_W.

The overall objective of incorporating probabilistic analyses into fatigue design is to ensure that a low probability exists for a combination of higher than average cyclic stress amplitude and a lower than average fatigue endurance limit (or stress

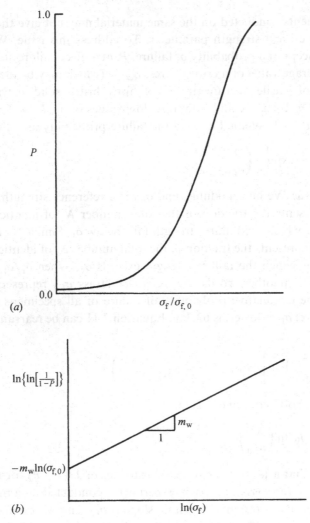

Fig. 7.8. (*a*) A Weibull plot of failure probability against the normalized failure strength. (*b*) The Weibull diagram, based on Eq. 7.13, whose slope is the Weibull modulus, m_W.

amplitude at a fixed life) to cause failure. In practical design involving the stress-based approach to total fatigue life, however, an endurance limit is first established on the basis of experiments conducted on carefully prepared smooth test specimens. This limit is then lowered by applying modifying factors (commonly known as the 'knock-down factors') to account for such effects as surface finish, size effects and constraints, temperature, stress corrosion, fretting, and numerous unknown effects. The damaging effects of repeated contact and corrosion at surfaces (arising, for example, from fretting fatigue), and the mitigation, at least in part, of such deleterious effects by recourse to shot-peening and lubrication are addressed in Chapter 13.

7.6 Practical applications

7.6.1 Example problem: Effects of surface treatments

Problem:

A high strength steel is to be used as a material for leaf springs in a ground transportation vehicle. During service, cyclic loading is expected to result in a load ratio, $R = 0$ (i.e. zero–tension cyclic loading). The 'as-heat-treated' condition of this alloy has an endurance limit, σ_e, of 400 MPa which was determined from laboratory tests conducted under fully reversed cyclic loading conditions. The tensile strength of the steel is 1500 MPa. The surface finish of the final product is expected to have an average roughness which, from prior experience in fatigue design, is known to 'knock down' the endurance limit by 40% from that measured in 'smooth' laboratory specimens. Shot-peening of the product prior to the assembly of the vehicle is known to introduce a maximum surface compressive stress of −550 MPa. Determine the maximum stress amplitude which the surface of the leaf spring, which has to be designed for an infinite fatigue life, can sustain.

Solution:

The endurance limit of the product with a rough surface finish is 40% lower than the value measured in the laboratory. The design endurance limit then is: $400 \times 0.6 = 240$ MPa. This value pertains to fully reversed cyclic stressing, i.e. for $R = -1$ or $\sigma_m = 0$.

The applied loading involves $R = 0$ and a stress amplitude, σ_a. From Eq. 7.2, it is seen that $\sigma_a = \sigma_m = \Delta\sigma/2$ for $R = 0$. The compressive surface stress arising from shot-peening lowers the mean stress in surface to a new value: $\sigma'_m = (\sigma_m - 550) = (\sigma_a - 550)$ MPa. Assuming that the modified Goodman approximation, Eq. 7.4, provides a reasonably accurate measure of the high-cycle fatigue life under nonzero mean stress, we see that

$$\frac{\sigma_a}{\sigma_e} + \frac{\sigma'_m}{\sigma_{TS}} = 1. \tag{7.14}$$

Substituting the appropriate values,

$$\frac{\sigma_a}{240} + \frac{\sigma_a - 550}{1500} = 1. \tag{7.15}$$

This equation gives: $\sigma_a = 283$ MPa. Thus, the leaf spring could be subjected to an applied stress amplitude of up to 283 MPa for infinite life. (Note that the possible beneficial effect of shot-peening in enhancing the fatigue endurance limit has not been taken into account in the numerical calculations. Thus, the stress amplitude estimate is likely to be very conservative.)

7.6.2 Case study: HCF in aircraft turbine engines

Fatigue cracking under low-amplitude, high-cycle loading is the dominant failure process in a number of engineering applications. In this section, which is largely predicated upon a review by Cowles (1996), we examine the origins and effects of high-cycle fatigue in advanced gas turbine jet engines used in military aircraft.

The principal cause of failure of components, from which a gas turbine jet engine of a modern military aircraft is made, is high-cycle fatigue. As illustrated in Fig. 7.9(*a*), fatigue failure accounts for 49% of all component damage in jet engines. High-cycle fatigue (HCF) is responsible for nearly half of all these failures, whilst low-cycle fatigue (LCF) and all other modes of fatigue lead to the remainder of fatigue failures in roughly equal proportions. Failure by HCF affects a variety of engine components, as shown in Fig. 7.9(*b*).

The origin of HCF in the gas turbine aircraft engine can be attributed to one or more of the following causes.

(1) Mechanical vibration arising from rotor imbalance (which affects plumbing, nonrotating structures and external members) and rub (which affects blade tips and gas path seals).

(2) Aerodynamic excitation occurring in upstream vanes, downstream struts and blades, whereby engine excitation frequencies and component response frequencies corresponding to different modes of vibration may overlap.

(3) Aeromechanical instability, primarily in blades, accompanying aerofoil flutter.

(4) Acoustic fatigue of sheet metal components in the combustor, nozzle and augmentor.

The above sources of low-amplitude, HCF are augmented by the following damage processes which create microscopic notches and other sites at which fatigue cracks can nucleate and advance subcritically to catastrophic proportions.

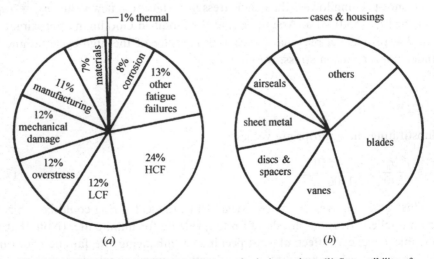

(*a*)   (*b*)

Fig. 7.9. (*a*) Distribution of different failure modes in jet engines. (*b*) Susceptibility of different components to HCF problems. (After Cowles, 1996.)

(1) *Foreign object damage* (FOD), which usually occurs in compressor and fan blades: FOD can induce micronotches, tears, dents and gouges that may vary in dimensions from tens of micrometers to tens of millimeters, depending on the size, nature and severity of impact of the foreign object. Sources of FOD are as diverse as sand particles and birds. As noted earlier, the leading edge of the fan blade, which is susceptible to FOD, is laser shock peened for improved fatigue resistance.

(2) *Domestic object damage* (DOD), which arises from a dislodged debris or component from another location of the engine.

(3) *Fretting fatigue*, which occurs at blade and disc attachment surfaces (dove-tail or fir-tree section), bolt flanges, and shrink-fit areas. See Chapter 13 for terminology definitions and detailed discussions pertaining to fretting fatigue.

(4) *Galling*, which occurs in the same regions as fretting, except that it involves greater displacements due to major engine throttle and speed changes.

(5) *HCF–LCF interactions*, where HCF is exacerbated by LCF as, for example, when creep and thermo-mechanical fatigue in hot sections (such as turbine blades) cause further reductions in fatigue life, over and above that due to vibrations.

Current methods to assess the HCF life of critical components in aircraft gas turbine engines entail the following general steps. (i) Appropriate stress analysis (largely based on the finite-element method) are performed to determine the mean stress level. (ii) Structural dynamics simulations are carried out to determine resonant frequencies and excitation modes. (iii) The design of the component is then carried out in such a way that (*a*) it meets the criteria for safe life for HCF with the appropriate mean stress level (using the modified Goodman diagram), and (*b*) no resonance-related problems arise. The parameters that serve as input to design are gathered from specimen testing and component testing, and the stress–life approach is empirically modified to allow for reductions in life due to FOD, DOD, fretting and galling.

7.7 Stress–life response of polymers

7.7.1 General characterization

The characterization of the fatigue life of polymeric materials on the basis of stress amplitudes is done in much the same way as for metals. Figure 7.10 is a typical Wöhler (S–N) curve for a polymer, where three distinct regions are seen in the variation of the stress amplitude σ_a as a function of the number of cycles to failure N_f.

The existence of *region I* and the slope of the $\Delta\sigma$–N_f plot in this region depend on whether crazes form at the high values of $\Delta\sigma$, and on whether the crazes cause microscopic cracks to nucleate. PS and PMMA, which are prone to craze formation, exhibit a well defined region I where the fatigue life depends strongly on $\Delta\sigma$. If the maximum tensile stress in the very first cycle is not

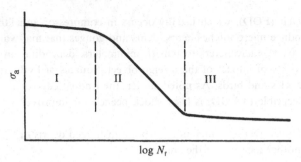

Fig. 7.10. Schematic representation of the typical variation of stress amplitude σ_a with the number of cycles to failure N_f for polymeric materials.

sufficiently large to form crazes, a distinct region I may not exist and the slope of the $\Delta\sigma$–N_f curve at the high $\Delta\sigma$ end will be a mere extrapolation of that in region II.

The dependence of total fatigue life on the stress amplitude in *region II* is a manifestation of the enhanced role of microscopic crack growth on fatigue fracture. At the higher σ_a end of region II, slow growth of crazes and their transformation into cracks are dominant mechanisms of failure. A slope of 14 MPa per decade of N_f seems to be characteristic of region II in a variety of polymeric materials fatigued at room temperature. As the test temperature is raised, there is a competition between shear banding and crazing in this region, as in the case of tensile deformation.

The high-cycle fatigue region represented in *region III* essentially forms the endurance limit for the polymer. The fatigue life here is controlled by the incubation time for the nucleation of microscopic flaws. The relative dominance of nucleation and growth of crazes and cracks constitutes the main distinction between regions II and III.

7.7.2 Mechanisms

The mechanisms controlling the total fatigue life of polymers also vary with many morphological, thermal, mechanical and environmental factors.

(1) When the cyclic loading involves high strain rates, the material is prone to hysteretic heating and to thermal softening. Consequently, increasing the test frequency (typically beyond 10 Hz for experiments conducted in the laboratory environment) results in a reduction in fatigue life. In an attempt to address these issues, the American Society for Testing and Materials (Philadelphia) developed standard methods in 1971 for the fatigue testing of polymers. These procedures are spelled out in Specification D-671-71. The standard calls for the measurement of temperature at fatigue failure unless it can be demonstrated that the heat rise is not significant. Furthermore, when thermal softening controls fatigue fracture, this standard defines the fatigue failure life operationally as the number of loading

cycles at a fixed stress amplitude which leads to an apparent reduction in modulus of 30% from the commencement of the fatigue test.

(2) The increase in the temperature of the polymer during high frequency fatigue loading also depends on the dimensions of the specimen. In thinner specimens, a greater fraction of the heat generated by damping is lost to the environment. Therefore, thicker fatigue test specimens, which retain a larger fraction of the temperature rise induced by hysteretic heating, exhibit a lower fatigue endurance limit (Riddell, Koo & O'Toole, 1967). In an analogous fashion, increases in test temperature (caused by external heating) also cause reductions in fatigue life.

(3) Most polymeric solids exhibit a longer fatigue life in inert environments than in chemically aggressive media. Such deleterious effects of aggressive environments are caused by complex synergistic interactions between the structure of the polymer and the surrounding medium. For example, the absorption of a nonsolvent can reduce the glass-transition temperature and modulus to the point that cavitation and crack initiation are promoted (Gent, 1970). Some liquid media, such as water, change the state of a polymer by processes involving plasticization or antiplasticization. Similarly, a small amount of acetone can transform polycarbonate from a glass to a crystalline solid. Specific combinations of material, environment and load can also improve the fatigue resistance (Manson & Sperling, 1976).

(4) Although current understanding of the effects of molecular properties on fatigue life is incomplete, it appears that the total fatigue life increases with increasing molecular weight (Sauer, Foden & Morrow, 1977) and crystallinity (Riddell, Koo & O'Toole, 1967), and with decreasing level of crosslinking (Sauer, 1978).

(5) As in the case of metallic materials, the total fatigue life derived from constant stress-amplitude fatigue decreases with increasing mean stress.

7.8 Fatigue of organic composites

The reinforcement of polymeric matrices with particles, whiskers and continuous fibers provides many possibilities for improving the strength, stiffness and resistance to creep and fatigue. The extent to which the improvements in the resistance to fatigue are realized is, of course, dictated by the specific combinations of matrix and reinforcement materials, processing methods, geometrical factors associated with the arrangement of the reinforcement phase in the matrix, and interfacial characteristics. Discontinuous and continuous fiber reinforcements of glass, boron, carbon, and high modulus organic materials such as aromatic polyarides are commonly used in the synthesis of polymeric composites. Polymers widely used as matrices for organic composites include polyesters, epoxies, Nylon 66, PC, polyphenylenesulphide, polyamideimide, PSF, PVC, and polyetherimide.

7.8.1 Discontinuously reinforced composites

The fatigue behavior of polymers reinforced with particulates and short fibers has been extensively reviewed in Hertzberg & Manson (1980, 1986). In this section, we present two examples to illustrate: (i) how the introduction of a strong second phase can enhance the resistance of a polymer to high cycle fatigue, and (ii) how such improvements are influenced by the specific choice of the reinforcement phase.

The stress–life curves for tension fatigue in several injection-molded composites with PSF matrices and with carbon or glass (short) fibers are presented in Fig. 7.11. Also included in this figure, for comparison purposes, are the stress–life characteristics of the unreinforced matrix material with a similar processing history. It is clear that the polymeric composites offer a superior (total) fatigue life than the unreinforced matrix, with the resistance to fatigue fracture increasing with increasing concentration of the glass fibers. The superiority of the reinforcement with carbon fibers to that of glass fibers is also evident. The high stiffness of the carbon whiskers contributes to the fatigue life by reducing the cyclic strains in the matrix at a given stress amplitude. Furthermore, the high thermal conductivity of carbon moderates the temperature rise due to hysteretic heating. It should, however, be noted that the beneficial effects of whisker reinforcements shown in Fig. 7.11 may be markedly reduced if the fatigue tests are conducted at higher cyclic frequencies where thermal softening begins to control the fatigue life.

7.8.2 Continuous-fiber composites

Cyclic loading of organic composites with continuous and unidirectional fibers can engender a more deleterious effect on service life than monotonic loading. The residual strength, toughness, and elastic response are all adversely affected by prolonged cyclic loading. This effect is clearly demonstrated in the stress–life data plotted in Fig. 7.12 for an organic composite with an epoxy resin (Bakelite ERL 2256) matrix and 0.6 volume fraction of unidirectional E-glass fibers. When the composite is subjected to tension fatigue at a stress ratio of 0.1 and a test frequency of 19 Hz in the direction of fiber orientation, an endurance limit of about 686 MPa results. This value is significantly lower than the static tensile strength of 1245 MPa. Also shown in Fig. 7.12 are the results of fatigue tests conducted for different orientations, ϕ, of the tensile axis with respect to the fiber direction. It is seen that an increase in ϕ leads to a monotonic reduction in the endurance limit, with the

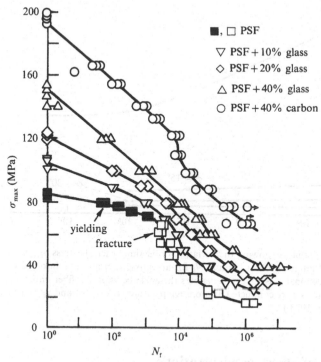

Fig. 7.11. A stress–life plot showing the variation of maximum tensile stress σ_{max} as a function of the number of fatigue cycles to failure N_f for injection-molded PSF which is reinforced with different amounts of short glass fibers and with 40% short carbon fibers. The results are based on room temperature fatigue tests conducted at $R = 0.1$, and $\nu_c = 5$–20 Hz. The small arrows denote that no failure occurred at the indicated number of cycles. (After Mandell *et al.*, 1983.)

matrix playing an increasingly dominant role in the fatigue process. Similar characterizations of fatigue failure have also been developed using strain-based approaches (e.g., Dharan, 1975; Talreja, 1987).

Nondestructive test methods based on acoustic emission, radiography, and modulus and damping measurements of fiber composites subjected to fatigue loads show fiber breakage, debonding, or matrix cracking fairly early in the fatigue process. The growth of such microscopic degradations continues to result in the lowering of the stress at which failure occurs with increasing numbers of fatigue cycles. In view of this progressive degradation, the useful fatigue life of many continuous-fiber composites is often defined as the number of cycles needed to cause a certain reduction in the elastic modulus. It is also common to characterize the distributed (microscopic) fatigue damage using Weibull statistics (see, for example, Talreja, 1987). At high cyclic frequencies, hysteretic heating also becomes an important factor. Furthermore, anomalous improvements in fatigue life may result in some cases due to the realignment of fibers, crack blunting, periodic overloads or frequency fluctuations.

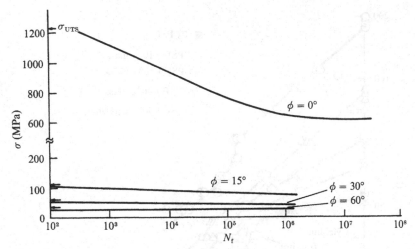

Fig. 7.12. Stress–life plots showing the variation of maximum tensile stress σ_{max} as a function of the number of fatigue cycles to failure N_f for an epoxy resin matrix reinforced with 60 volume% unidirectional glass fibers. ϕ denotes the angle between the fiber direction and the tensile axis. The tests were conducted at room temperature at $R = 0.1$ and $v_c = 19$ Hz. (After Hashin & Rotem, 1973.)

7.9 Effects of stress concentrations

The discussion up to this point has focused on nominally smooth-surfaced solids. However, engineering structures invariably contain stress concentrations which are the principal sites for the inception of fatigue flaws. The stress and deformation fields in the immediate vicinity of the stress concentration have a strong bearing on how the fatigue cracks nucleate and propagate. In this section, we examine continuum approaches based on stress–life concepts which deal with the issue of fatigue failure ahead of stress concentrations.

7.9.1 Fully reversed cyclic loading

The theoretical elastic stress concentration factor K_t relates the local stress ahead of the notch tip to far-field loading and is defined as the ratio of the maximum local stress σ_{max} to the nominal stress S. Under fatigue loading conditions, the elastic stress concentration factor is replaced by the so-called fatigue notch factor:

$$K_f = \frac{\text{unnotched bar endurance limit}}{\text{notched bar endurance limit}}. \qquad (7.16)$$

In general, fatigue experiments suggest that notches produce a less stress concentrating effect than predicted by theoretical elastic analysis such that $K_f < K_t$; $K_f \rightarrow K_t$ for large notch-root radii and for higher strength materials. The degree of agreement between theoretical predictions of elastic stress concentration and actual effects is often measured by the so-called notch sensitivity index which is defined as

$$q = \frac{K_f - 1}{K_t - 1}. \tag{7.17}$$

The parameter q varies from zero for no notch effect to unity for the full effect predicted by the elasticity theory. K_t is a function only of the component geometry and loading mode and is available in many handbooks (e.g., Peterson, 1959). However, K_f is determined from experimental measurements or empirical, engineering 'rules-of-thumb'. An example of such a measure of K_f is the well known Peterson equation for ferrous wrought alloys (Peterson, 1959):

$$K_f \approx 1 + \frac{(K_t - 1)}{\{1 + (A_n/\rho)\}}, \tag{7.18}$$

where A_n is a constant whose value depends on the strength and ductility of the material ($A_n \sim 0.25$ mm for annealed steels and $A_n \sim 0.025$ mm for steels of very high strength) and ρ is the notch-root radius.

The stress–life approach is employed for high-cycle fatigue failures ahead of stress concentrations by appropriately modifying the smooth specimen (unnotched) endurance limit σ_e. This involves either dividing the stress range $\Delta\sigma$ for all fatigue lives by the factor K_f (which often leads to very conservative results) or merely reducing the fatigue limit σ_e by a factor of K_f. This method is unsuitable for situations where considerable plastic deformation occurs ahead of the stress concentration.

7.9.2 Combined effects of notches and mean stresses

Now we consider a notched member which is subjected to cyclic loading with a nonzero mean stress in such a way that only elastic conditions prevail at all times throughout the member. If S, S_m, and S_a are the instantaneous value, the mean value, and the amplitude, respectively, of the nominal far-field cyclic stress imposed on the notched member (Fig. 7.13), the local stress amplitude and the mean stress at the tip of the notch can be computed from the fatigue notch factor K_f such that

$$\sigma_a = K_f S_a, \qquad \sigma_m = K_f S_m, \tag{7.19}$$

where the peak value of $K_f S$ is always smaller than the yield strength of the material σ_y in both tension and compression. In order to assess the effect of the mean stress on fatigue life, the modified Goodman equation, Eq. 7.4, or some other approach discussed in Section 7.2 may be employed. Using the modified Goodman diagram for the notched member, we see that

$$S_a = \frac{S_a|_{\sigma_m=0}}{K_f} \cdot \left\{1 - \frac{K_f \sigma_m}{\sigma_{TS}}\right\}, \quad \text{for} \quad K_f \cdot S|_{\max} < \sigma_y. \tag{7.20}$$

Figure 7.13 schematically shows the modified Goodman diagram for smooth and notched specimens which are subjected to cyclic loading under a nonzero mean stress. These specimens are subjected to an applied stress amplitude of S_a and $K_f S_a$, respectively, as indicated in the figure. Note that for the elastically deforming notched member, the intercepts in both the ordinate and the abscissa are reduced by a factor of K_f as compared to the unnotched fatigue specimen.

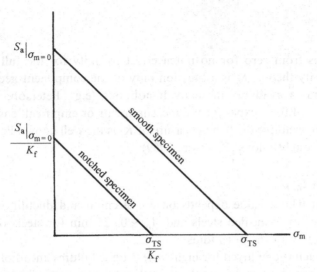

Fig. 7.13. The effect of mean stress on fatigue life as determined from the modified Goodman diagram for notched and unnotched fatigue specimens.

7.9.3 *Nonpropagating tensile fatigue cracks*

An intriguing phenomenon associated with tensile fatigue cracks growing ahead of stress concentrations involves the so-called *nonpropagating* flaws. It was first shown by Frost & Dugdale (1957) and Frost (1960) that fatigue cracks emanating from notches can arrest completely after growing some distance. A number of subsequent studies have unequivocally documented the existence of nonpropagating fatigue cracks ahead of stress concentrations in ductile solids (e.g., Smith & Miller, 1978; El Haddad *et al.* 1980; Tanaka & Nakai, 1983). Experimental measurements based on total life have shown that the nonpropagation or arrest of fatigue flaws occurs only ahead of sharp notches, above a certain critical value of K_t. Figure 7.14, where the fatigue limit is plotted as a function of K_t, schematically illustrates the conditions for the occurrence of nonpropagating fatigue flaws. Here, the threshold stress for crack initiation, expressed as the unnotched fatigue limit σ_e divided by K_f or K_t, is lower than the stress required to cause catastrophic fracture of nonpropagating cracks above the critical K_t. A discussion of the possible causes for the arrest of fatigue cracks ahead of stress concentrations is provided in Chapter 15 in connection with the topic of short fatigue cracks.

7.9.4 *Example problem: Effects of notches*

Problem:

A circumferentially notched cylindrical rod of a high-strength, low alloy steel is subjected to repeated bending at a nominal stress amplitude,

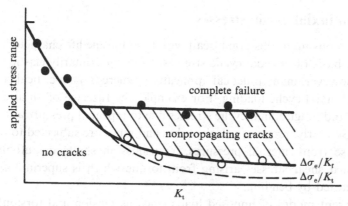

Fig. 7.14. Threshold stress range (fatigue limit) for crack initiation at a notch tip, characterized by the unnotched fatigue limit $\Delta\sigma_e$ divided by K_f or K_t and plotted as a function of K_t. (After Frost, Marsh & Pook, 1974.)

$S'_a = 200\,\text{MPa}$, and a mean stress, $S_m = 250\,\text{MPa}$. For the notch geometry of the rod, the elastic stress concentration factor, K_t, and the notch sensitivity index, q, are found from appropriate handbooks to be 2.3 and 0.92, respectively. Fully reversed cyclic stress tests conducted on smooth (unnotched) laboratory specimens give the following material parameters for the Basquin equation, Eq. 7.1: $\sigma'_f = 1000\,\text{MPa}$ and $b = -0.12$. The endurance limit was estimated from these tests to be 280 MPa. The yield and tensile strengths of the steel are $\sigma_y = 600\,\text{MPa}$ and $\sigma_{TS} = 1050\,\text{MPa}$. Estimate the high-cycle fatigue life of the notched rod.

Solution:
From Eq. 7.17, we find that the fatigue notch factor, $K_f = 1.2$. For the given loading conditions, i.e. $S_m = 250\,\text{MPa}$, $S_a = 200\,\text{MPa}$, we then find an equivalent value of fully reversed cyclic stress, $S_a|_{\sigma_m=0}$, for an unnotched specimen using Eq. 7.20. For this purpose, we rewrite Eq. 7.20 as:

$$S_a|_{\sigma_m=0} = \frac{S_a}{K_f} \cdot \left\{ 1 - \frac{K_f \sigma_m}{\sigma_{TS}} \right\}^{-1}. \tag{7.21}$$

Substituting the appropriate numerical values, it is seen that $S_a|_{\sigma_m=0} = 336\,\text{MPa}$. (Note that the requirement in Eq. 7.20 that $K_f \cdot S|_{max} < \sigma_y$ is satisfied.) The equivalent value of $S_a|_{\sigma_m=0} = 336\,\text{MPa}$ exceeds the fatigue endurance limit of 280 MPa. Thus, the specimen is expected to have a finite life. Using Eq. 7.1,

$$\sigma'_f(2N_f)^b = 1000(2N_f)^{-0.12}, \quad \Rightarrow \quad 2N_f = 8855. \tag{7.22}$$

Thus, the number of stress reversals to failure is 8855, or the number of cycles to failure is 4427.

7.10 Multiaxial cyclic stresses

The discussions up to this point dealt with total fatigue life characterization for loading on the basis of uniaxial cyclic stresses involving primarily elastic conditions. There are, however, many practical applications where fatigue-critical components undergo multiaxial cyclic loading. For example, the fuselage of an aircraft is subjected to hoop and longitudinal cyclic stresses due to repeated pressurization and depressurization. Similarly, pressure vessels, tubes and pipes are subjected to biaxial stresses which arise from internal pressures. Transmission shafts in automobiles experience combined shear stresses arising from torque which is superimposed on axial stresses generated by bending.

In general, different modes of imposed loads (such as tension and torsion) may occur either in a static mode (as, for example, due to a steady bending moment, mean pressure or steady torque) or they may fluctuate in phase or out of phase. The overall fatigue life under general multiaxial conditions is inevitably dictated by the complex phase relations between the different modes of loading. In this section, we consider stress-based approaches to total life under multiaxial loading conditions. The corresponding discussions of multiaxial strain-based approaches are taken up in the next chapter.

7.10.1 Proportional and nonproportional loading

Consider a fatigue specimen which is subjected to multiaxial cyclic loads. *Proportional loading* is considered to occur if, during changes and fluctuations in the imposed loads, the different components of the stress tensor vary in constant proportion to one another. At a given reference point in the material, proportional loading exists if the principal stresses at that point, σ_1, σ_2 and σ_3, vary in the following manner:

$$\frac{\sigma_2}{\sigma_1} = \lambda_1, \qquad \frac{\sigma_3}{\sigma_1} = \lambda_2. \tag{7.23}$$

Here λ_1 and λ_2 are scalar constants which may vary from point to point, but are constants for a given material point in the solid.

As an example, an increase in the internal pressure of a close-ended, thin-walled cylindrical tube causes the hoop stress σ_{hoop} and the axial stress σ_{axial} to vary in proportion to each other so that at all times, $\sigma_{\text{hoop}}/\sigma_{\text{axial}} = 2$. If the thin-walled cylinder is subjected to an axial tension P and a torque about the axis T, proportional loading occurs only if $P \propto T$. Any other mutual variation of P and T gives rise to nonproportional loading. Further discussions of nonproportional and out-of-phase loading are considered in the next chapter in the context of strain–life fatigue.

7.10.2 Effective stresses in multiaxial fatigue loading

We begin with a consideration of fully reversed cyclic loading (zero mean stress) where all cyclic loads are perfectly in phase or 180° out of phase with one another. Let σ_x, σ_y and σ_z denote the three normal components of stress applied to a fatigue specimen and let τ_{xy}, τ_{yz} and τ_{zx} denote the three shear stress components. Let σ_1, σ_2, σ_3 denote the principal stresses, with $\sigma_1 \geq \sigma_2 \geq \sigma_3$. The effective stress which characterizes the deformation of the material is then assumed to be that given by the von Mises criterion based on the octahedral shear stress,† τ_{oct}:

$$\sigma_e = \frac{3}{\sqrt{2}} \tau_{oct} = \frac{1}{\sqrt{2}} \sqrt{(\sigma_1 - \sigma_2)^2 + (\sigma_2 - \sigma_3)^2 + (\sigma_3 - \sigma_1)^2},$$

$$= \frac{1}{\sqrt{2}} \sqrt{(\sigma_x - \sigma_y)^2 + (\sigma_y - \sigma_z)^2 + (\sigma_z - \sigma_x)^2 + 6(\tau_{xy}^2 + \tau_{yz}^2 + \tau_{zx}^2)}.$$

$$(7.24)$$

Another measure of the intensity of the multiaxial stress state (and used in the Tresca yield criterion which was discussed in Section 1.4.3) is the maximum shear stress, τ_{max}, which acts on at most three mutually perpendicular sets of planes that intersect the principal stress axes at 45°:

$$\tau_{max} = \max \left| \frac{\sigma_1 - \sigma_3}{2} \right|. \qquad (7.25)$$

For fluctuating multiaxial stresses, Eq. 7.24 is written in terms of the amplitudes of principal stresses σ_{ia} ($i = 1, 2, 3$) and the amplitude of the effective stress $\sigma_{a,e}$ as

$$\sigma_{a,e} = \frac{1}{\sqrt{2}} \sqrt{(\sigma_{1a} - \sigma_{2a})^2 + (\sigma_{2a} - \sigma_{3a})^2 + (\sigma_{3a} - \sigma_{1a})^2}. \qquad (7.26)$$

In the most elementary stress–life analysis for multiaxial fatigue involving a zero mean stress, the Basquin exponent b and the fatigue strength coefficient σ_f' in Eq. 7.1, which are determined from experiments conducted on smooth specimens subjected to fully reversed *uniaxial* cyclic loads, are used in conjunction with the following stress–life equation for multiaxial stresses to determine the fatigue life, N_f:

$$\sigma_{a,e} = \sigma_f'(2N_f)^b. \qquad (7.27)$$

In some cases, a steady, nonfluctuating load in one mode (such as bending) may be superimposed on cyclic stresses in another mode (such as cyclic torsion). Here, the effects of mean stress on fatigue failure should be considered. If it is assumed that the controlling mean stress is related to the steady value of the hydrostatic stress, an effective value of this mean stress, $\overline{\sigma}_m$, may be defined in terms of the mean values of the principal stresses, σ_{im} ($i = 1, 2, 3$), as

$$\overline{\sigma}_m = \sigma_{1m} + \sigma_{2m} + \sigma_{3m} = \sigma_{xm} + \sigma_{ym} + \sigma_{zm}. \qquad (7.28)$$

---

† Recall from Chapter 1 that the octahedral shear stress is the resolved shear stress on the Π-plane, which is the plane oriented at equal angles to the three principal stress directions.

Alternatively, an effective mean stress may be defined based on the octahedral shear stress as

$$\sigma_{m,e} = \frac{1}{\sqrt{2}} \sqrt{(\sigma_{1m} - \sigma_{2m})^2 + (\sigma_{2m} - \sigma_{3m})^2 + (\sigma_{3m} - \sigma_{1m})^2}, \qquad (7.29)$$

The simplest way to invoke stress-based approaches to fatigue in the presence of multiaxial stresses which involve a nonzero mean stress is to combine the effective stress amplitude $\sigma_{a,e}$, determined from Eq. 7.26, and the effective mean stress as, for example, determined from Eq. 7.29, with the modified Goodman equation to obtain

$$\sigma_{a,e} = \sigma_{a,e}|_{\substack{\sigma_{1m}=0 \\ \sigma_{2m}=0 \\ \sigma_{3m}=0}} \left\{ 1 - \frac{\sigma_{m,e}}{\sigma_{TS}} \right\}. \qquad (7.30)$$

With this equation, the generalization of the modified Goodman equation to multi-axial fatigue then involves an *equivalent fully-reversed uniaxial cyclic stress*. The fatigue life under multiaxial cyclic stresses involving nonzero mean stresses is next determined using Eq. 7.27.

One of the major drawbacks of such effective stress approaches is that the differing effects of axial tension and compression mean stresses in a multiaxial fatigue test may not be accurately captured. In addition, the orientation of fatigue cracks with respect to the loading axes is not quantitatively determined from such criteria. Furthermore, a wide variety of experimental observations (to be discussed next) reveal that the normal stress also plays a critical role in influencing fatigue lives in multiaxial loading.

7.10.3 Stress–life approach for tension and torsion

For uniaxial tensile loading with a nonzero mean stress, the only nonzero components of stress are σ_{1a} and σ_{1m}. For this case, as anticipated, Eq. 7.30 reduces to Eq. 7.4. Now consider pure torsion loading only. The only nonzero components of stress in this case are the amplitude and mean value of the shear stress, $\tau_{xy,a}$ and $\tau_{xy,m}$, respectively. From Eqs. 7.24, 7.26 and 7.28, it is readily seen that

$$\sigma_{a,e} = \sqrt{3}\,\tau_{xy,a} \qquad \text{and} \qquad \bar{\sigma}_m = 0. \qquad (7.31)$$

An interesting point to observe here is that $\bar{\sigma}_m = 0$ even if the mean shear stress for torsional loading is nonzero. A variety of independent experimental studies (see, for example, the work dating back to Sines, 1959) have shown that *a superimposed mean static torsion has no effect on the fatigue limit of metals subjected to cyclic torsion*, although a superimposed static tension or bending stress has a marked effect on the fatigue life in normal cyclic loading.

The differing effects of torsion and tension loading on the microscopic and macroscopic modes of cracking and on the overall fatigue life are illustrated in Figs. 7.15(a) and (b), respectively, for an Inconel 718 alloy. Essentially throughout the fatigue life, failure occurs on planes of maximum shear. This damage process is classified as regime A in Fig. 7.15(a). In this regime, the cracks were confined to the planes of

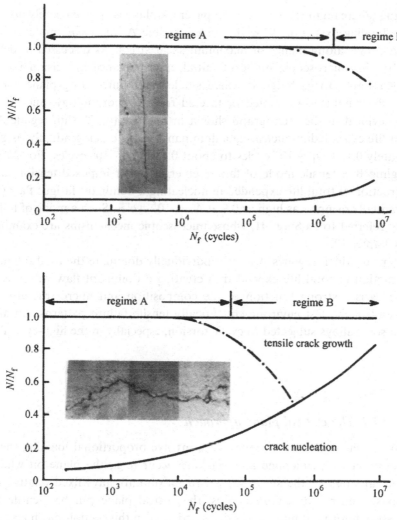

Fig. 7.15. The evolution of different regimes of fatigue failure during (*a*) torsion fatigue and (*b*) tension fatigue of an Inconel 718 alloy (yield strength, $\sigma_y = 1160$ MPa. (From Socie, 1993. Copyright American Society for Testing and Materials, Philadelphia. Reproduced with permission.)

maximum shear through the fatigue life. The fraction of life, N/N_f, expended in initiating a dominant shear crack is less than 0.1; the remaining 90% of life is spent in propagating this shear crack. For $N_f > 10^6$ cycles, the *local* mode of failure occurs on planes oriented normal to the local principal tensile stress, with the microscopic tensile cracks oriented at 45° to the shear cracks. This failure process is categorized as regime B in Fig. 7.15(*a*). The fraction of life expended in initiating the dominant crack rises to 0.2 at $N/N_f \sim 10^7$ cycles.

When the Inconel 718 alloy is subjected to axial tension fatigue, Fig. 7.15(*b*), a different failure pattern emerges throughout the life. In both the low-cycle and high-cycle fatigue failure regimes, the macroscopic crack plane is approximately normal to the tensile loading axis. For $N_f < 10^5$ cycles, regime A, the *local* mode of microscopic cracking is along planes of maximum shear stress. As discussed in detail in Chapter 10, this microscopic mode of initial crack advance in tension fatigue is commonly referred to as Stage I, where single-slip failure along planes of local maximum shear induces a serrated or faceted fracture morphology. Such features are clearly evident in the micrograph shown in Fig. 7.15(*b*). Within regime A, the fraction of life expended in nucleating a dominant fatigue crack gradually rises from approximately 0.1 at $N_f \sim 10^3$ cycles to about 0.4 at $N_f \sim 10^5$ cycles. For $N_f > 10^6$ cycles, regime B, a tensile mode of failure emerges under imposed tension fatigue, with the fraction of total life expended in nucleating a dominant fatigue flaw (1 mm in size) gradually rising to as high as 0.9 at $N_f \sim 10^7$ cycles. Such a mode of failure is commonly referred to as Stage II, whose microscopic mechanisms are examined in detail in Chapter 10.

The extent to which regimes A and B individually dominate the total fatigue life, and the fraction of total life expended in creating a dominant flaw within each of these regimes is a strong function of the composition and microstructure of the material, and of the test environment. Locally tensile failure patterns can also be induced in some alloys subjected to cyclic torsion, especially in the high-cycle fatigue regime.

7.10.4 *The critical plane approach*

When the multiaxial fatigue cycles involve proportional loading, the most critical plane on which damage and cracking occur (e.g., the plane on which the tensile normal stress or the shear stress is a maximum) stays fixed relative to the loading axis. Under such circumstances, the critical plane can be identified in a relatively straightforward manner, as discussed later in this section. Such a situation continues to hold even if the cyclic stresses are of variable amplitude. By contrast, when the components of the principal stress tensor do not vary in constant proportion to one another, the orientations of the critical planes in the specimen vary with time and loading path. As a result, an instantaneous definition of these critical planes derived from local stress analyses is needed. The critical plane approach to multiaxial fatigue life estimation then involves the following steps: (1) identification of the orientation of the plane which represents the critical site in the material for the onset of damage based on some criterion (such as the maximum normal/shear stress/strain criterion, or a combined normal stress/shear stress criterion) and (2) estimation of the number of cycles to initiate a fatigue crack or the total number of cycles to failure using experimental results on stabilized cyclic stress–strain curves and fatigue lifes, preferentially from uniaxial stress–life data obtained under fully reversed loading conditions (in conjunction with the modified Goodman diagram or some other approach to account of the effect of mean stresses; see later discussion in this chapter).

Figure 7.16(a) schematically shows a fatigue specimen subjected to bending–torsion cyclic loads. The reference coordinate axes, x, y and z, are marked on the figure. Let the orientation of a critical plane be defined with reference to this original coordinate system with the aid of two angles: the angle, γ, between the axis of the specimen (the x axis) and the line of intersection of the critical plane with the surface, and the angle, φ, between the normal to the critical plane, which is along the z' direction in Fig. 7.16(b), and the z axis. A new set of coordinate axes, x', y' and z', is defined in Fig. 7.16(b) to facilitate the visualization of the critical plane orientation. Using straightforward methods for coordinate transformation, the components of the stress and strain tensor in the x', y' and z' coordinate system are written as

$$\sigma'_{ij} = S_{ki}\sigma_{kl}S_{lj}, \qquad \epsilon'_{ij} = S_{ki}\epsilon_{kl}S_{lj},$$

$$\text{where} \quad \{S_{kl}\} = \begin{pmatrix} \cos\gamma & -\cos\varphi\sin\gamma & -\sin\varphi\sin\gamma \\ \sin\gamma & \cos\varphi\cos\gamma & \sin\varphi\cos\gamma \\ 0 & -\sin\varphi & -\cos\varphi \end{pmatrix}. \qquad (7.32)$$

Then the important components of the stress acting on the potentially critical plane are identified as: the normal stress ($\sigma'_{z'z'}$), the (in-plane) shear stress ($\sigma'_{y'z'}$) which produces a mode II type crack opening, and the (out-of-plane) shear stress which produces a mode III type crack opening (see Chapter 9 for further discussions of mode II and mode III).

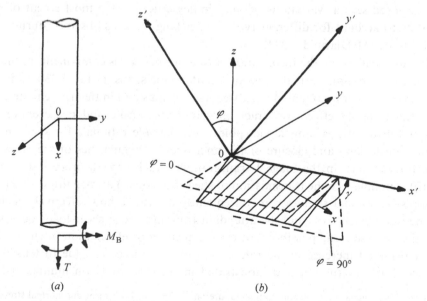

(a) (b)

Fig. 7.16. Schematic of a fatigue specimen subjected to bending–torsion cyclic loads (a) and the definition of the critical plane and the associate nomenclature (b).

In general, the identification of the orientations of the critical planes for general multiaxial loading with nonproportional stress fluctuations can become very cumbersome. However, as reviewed by Chu, Conley & Bonnen (1993), some simplifications can be extracted by noting that most fatigue cracking initiates at the surface. Given the state of plane stress at the surface elements of the specimen, criteria based on mode I (tensile failure) and mode III (torsional or out-of-plane shear failure) type damage lead to $\varphi = 90°$ as the critical orientation, whereas mode II type failure based on in-plane shear stresses lead to $\varphi = 45°$ as the critical orientation. With this fixed orientation of φ, only variations in the angle γ need to be calculated. If such approximations cannot be applied to a given material or loading situation, detailed computations based on the critical plane approach need to be undertaken.

7.10.4.1 Critical plane criteria invoking normal and shear stresses

On the basis of experimental observations of fatigue crack initiation patterns in steels and aluminum alloys, Findley (1959) postulated a criterion for multiaxial fatigue failure which is predicated upon the combined effects of normal stress and maximum shear stress. In his model, fatigue failure is deemed to occur when

$$\tau_a + \bar{\alpha}\sigma_{n,max} = \bar{\beta}. \tag{7.33}$$

Here, $\bar{\alpha}$ and $\bar{\beta}$ are material parameters determined experimentally, which are constants for a given life. An increase in the maximum normal stress $\sigma_{n,max}$ acting on the plane of the critical alternating shear stress causes a corresponding reduction in the permissible cyclic shear stress τ_a as per Eq. 7.33, for a fixed fatigue life. McDiarmid has proposed several variations of the Findley criterion, the most recent of which attempts to account for different types of cracking patterns observed during multiaxial fatigue (McDiarmid, 1994).

The justification for the incorporation of a normal stress component, in the form of either a maximum normal stress or a hydrostatic stress, in Eq. 7.33 and in other critical plane criteria presented in subsequent sections and in the next chapter, stems from the differing effects on crack growth from tensile and compressive mean stresses.[†] From a mechanistic viewpoint, one possible rationale for such an effect is that interlocking and closure of the microscopic irregularities of the crack faces seen, for example, in the Stage I transgranular crack growth process seen in Fig. 7.15(b), can influence the transmission of shear tractions. The magnitude of tensile or compressive normal stresses can consequently determine how effectively such contact between crack faces and the attendant transmission of shear loads occurs. One may then envision the possibility of a minimization or elimination of such contact, and a resultant increase in the rate of advance of the crack, due to tensile mean stresses. Such a situation is demonstrated in the case study on multiaxial fatigue

---

[†] Historically, criteria for multiaxial fatigue failure on the basis of both shear and normal stresses have also been considered in the earlier works of Gough, Pollard & Clenshaw (1951) and Stulen & Cummings (1954). A survey of historical developments in the area of multiaxial fatigue is given by Garud (1981).

presented in Chapter 8. A number of criteria have also been put forth to identify critical planes of failure on the basis of different combinations of normal and shear strains for proportional and nonproportional loading. These approaches are also discussed in Chapter 8.

7.10.4.2 Criteria invoking effective stresses

In order to develop a criterion which quantitatively captures the effects of superimposed normal loads on the observed fatigue limits, Crossland (1956) postulated that

$$\sqrt{J_{2a}} + \bar{\alpha}\sigma_{H,max} \leq \bar{\beta}, \tag{7.34}$$

where J_{2a} is the amplitude of the second invariant of the stress deviator defined in Section 1.4, $\sigma_{H,m} = \bar{\sigma}_m/3$ (see Eq. 7.28) is the mean value of the hydrostatic stress during a multiaxial fatigue loading cycle, and $\sigma_{H,max}$ is the maximum value of the hydrostatic stress during multiaxial loading; $\sigma_{H,max} = \sigma_{H,m} + \sigma_{H,a}$ (where $\sigma_{H,a}$ is the amplitude of the hydrostatic stress during a multiaxial fatigue loading cycle), and $\bar{\alpha}$ and $\bar{\beta}$ are material constants determined experimentally.

Sines (1959) modified the Crossland criterion in the following manner:

$$\sqrt{J_{2a}} + \bar{\alpha}\sigma_{H,m} \leq \bar{\beta}. \tag{7.35}$$

Let $\tau_{e,-1}$ be the endurance limit measured from a fully reversed cyclic torsion test (where $\sigma_{H,m} = 0$ and $J_{2a} = \tau_{e,-1}$) and let $\sigma_{B,0}$ be the endurance limit determined from a zero-to-tension ($R = 0$) uniaxial or bending cyclic load test (where $J_{2a} = \sigma_{B,0}/\sqrt{3}$ and $\sigma_{H,m} = \sigma_{B,0}/3$). Then, $\bar{\alpha} = [(3\tau_{e,-1}/\sigma_{B,0}) - \sqrt{3}]$ and $\bar{\beta} = \tau_{e,-1}$ in Eq. 7.35. The Sines criterion correctly captures the experimental observations that (a) the fatigue limit in torsion is independent of a mean shear stress and that (b) the endurance limit in bending varies linearly with a static normal stress. Following the same line of reasoning, the material parameters in Eq. 7.34 are: $\bar{\alpha} = [(3\tau_{e,-1}/\sigma_{B,-1}) - \sqrt{3}]$ and $\bar{\beta} = \tau_{e,-1}$, where $\sigma_{B,-1}$ is the endurance limit measured under fully reversed bending.

A number of subsequent variations of the Crossland (1956) and Sines (1959) criteria have been reported. For example, Dang Van (1973) has formulated a criterion where a linear combination of the shear stress acting on the plane of maximum shear stress and the instantaneous hydrostatic pressure, both of which are determined from *local* stress analyses, are used in an equation analogous to Eqs. 7.34 and 7.35. Such a criterion has also been suitably modified for damage and cracking predictions under in-phase and out-of-phase loading in multiaxial fatigue as well as contact fatigue. A variation of such an approach has also been suggested by Papadopoulos *et al.* (1997) who use volume-averaged stress quantities, ostensibly to simulate differing crystallographic orientations in a polycrystalline ensemble.

Experimental studies of proportional and nonproportional loading with fully reversed cycles in a variety of high-strength metals have shown that the ratio of the fatigue limit in torsion to that in normal/bend loading, $\tau_{e,-1}/\sigma_{B,-1} \approx 0.57\text{--}0.80$

(Zenner, Heidenreich & Richter, 1985; Froustey & Lasserre, 1989). For a phase difference of 0° between torsional and bending stress cycles, all of the foregoing theories give essentially the same result. For 90° out-of-phase loading, larger differences are seen among the different criteria. A quantitative comparison of the predictions of the foregoing theories with experimental results for in-phase and out-of-phase loading of high-strength metals is given in Papadopoulos *et al.* (1997).

Exercises

7.1   The S–N curve for an elastic material is characterized by the Basquin relationship, $\sigma_a = C \cdot N_f^b$, where C is a material constant, σ_a is the stress amplitude, N_f is the number of fully reversed stress cycles to failure and b is the Basquin exponent approximately equal to –0.09. When the stress amplitude is equal to the ultimate tensile strength of the material in this alloy, the fatigue life is 1/4 cycle. If a specimen spends 70% of its life subject to alternating stress levels equal to its fatigue endurance limit σ_e, 20% at $1.1\sigma_e$ and 10% at $1.2\sigma_e$, estimate its fatigue life using the Palmgren–Miner linear damage rule.

7.2   Explain why the modified Goodman diagram can be rewritten in terms of the endurance limit, σ_e, as

$$\sigma_e = \sigma_e|_{\sigma_m=0} \left\{ 1 - \frac{\sigma_m}{\sigma_{TS}} \right\},$$

where $\sigma_e|_{\sigma_m=0}$ is the endurance limit for zero mean stress cyclic loading.

7.3   A circular cylindrical rod with a uniform cross-sectional area of $20\,\mathrm{cm}^2$ is subjected to a mean axial force of $120\,\mathrm{kN}$. The fatigue strength of the material, $\sigma_a = \sigma_{fs}$, is $250\,\mathrm{MPa}$ after 10^6 cycles of fully reversed loading and $\sigma_{TS} = 500\,\mathrm{MPa}$. Using the different procedures discussed in this chapter, estimate the allowable amplitude of force for which the shaft should be designed to withstand at least one million fatigue cycles. State all your assumptions clearly.

7.4   A rotating bending machine produces a pure bending moment uniformly over the gage length of a fatigue specimen. Show that, in a bending specimen rotating at an angular velocity ω, the cyclic stress will be of the form $\Delta\sigma = A \sin \omega t$, where A is the peak stress amplitude and t is the time.

7.5   Why is tempered glass more resistant to tensile fracture than ordinary glass?

7.6   A cylindrical shaft of circular cross section is subjected to a bending moment of constant amplitude over its entire length. The deformation of the shaft material can be approximated by an elastic–perfectly plastic constitutive model and the yield strength in tension can be considered equal to that in compression. If the outer fibers of the shaft yield plastically over a depth of 1/4 of the diameter during the application of the bending moment, qualitatively discuss the development and distribution of mechanical residual stresses along the diameter of the shaft when the bending moment is fully removed.

7.7 Find the nonzero components of the principal values of the mean stress σ_{im} ($i = 1, 2, 3$) and the nonzero components of the principal stress amplitudes σ_{ia} ($i = 1, 2, 3$) for uniaxial cyclic loading with $R = 0.2$.

7.8 A thin-walled cylinder with closed ends is subjected to a steady twisting moment T, which is oriented parallel to the axis of the cylinder. The internal radius of the cylinder is r_i and its wall thickness is t.

(a) If the cylinder is now subjected additionally to an internal pressure p which fluctuates between zero and a maximum value, p_{max}, comment on the variation in the principal stress directions during a complete cycle in p.

(b) Under what conditions would you expect the principal stresses to be nearly aligned with the axial and hoop directions?

CHAPTER 8

Strain–life approach

The information derived from cyclic-stress-based continuum analysis mainly pertains to elastic and unconstrained deformation. In many practical applications, engineering components generally undergo a certain degree of structural constraint and localized plastic flow, particularly at locations of stress concentrations. In these situations, it is more appropriate to consider the strain–life approach to fatigue.

In this chapter, strain-based approaches to total fatigue life are discussed for smooth-surfaced and notched components subjected to constant amplitude and variable amplitude cyclic loading of metals and nonmetals. Also examined are available methods for fatigue life estimation for multiaxial cyclic loading.

8.1 Strain-based approach to total life

Coffin (1954) and Manson (1954), working independently on thermal fatigue problems, proposed a characterization of fatigue life based on the plastic strain amplitude. They noted that when the logarithm of the plastic strain amplitude, $\Delta\epsilon_p/2$, was plotted against the logarithm of the number of load reversals to failure, $2N_f$, a linear relationship resulted for metallic materials, i.e.

$$\frac{\Delta\epsilon_p}{2} = \epsilon_f'(2N_f)^c, \tag{8.1}$$

where ϵ_f' is the fatigue ductility coefficient and c is the fatigue ductility exponent. In general, ϵ_f' is approximately equal to the true fracture ductility ϵ_f in monotonic tension, and c is in the range of -0.5 to -0.7 for most metals. Typical values of ϵ_f' and c for a number of engineering alloys are listed in Table 8.1.

8.1.1 Separation of low-cycle and high-cycle fatigue lives

Since the total strain amplitude in a constant strain amplitude test, $\Delta\epsilon/2$, can be written as the sum of elastic strain amplitude, $\Delta\epsilon_e/2$, and plastic strain amplitude, $\Delta\epsilon_p/2$,

$$\frac{\Delta\epsilon}{2} = \frac{\Delta\epsilon_e}{2} + \frac{\Delta\epsilon_p}{2}, \tag{8.2}$$

Table 8.1. *Cyclic strain–life data for some engineering metals and alloys.*

Material	Condition	σ_y[‡] (MPa)	σ_f' (MPa)	ϵ_f'	b	c
Al alloys[†]						
1100	annealed	97	193	1.80	−0.106	−0.69
2014	T6	462	848	0.42	−0.106	−0.65
2024	T351	379	1103	0.22	−0.124	−0.59
5456	H311	234	724	0.46	−0.110	−0.67
7075	T6	469	1317	0.19	−0.126	−052
Steels[†]						
1015	aircooled	228	827	0.95	−0.110	−0.64
4340	tempered	1172	1655	0.73	−0.076	−0.62
Ti alloys[†]						
Ti–6Al–4V	solution-treated + aged	1185	2030	0.841	−0.104	−0.69
Ni-base alloys[†]						
Inconel X	annealed	700	2255	1.16	−0.117	−0.75

[†] *Source:* Osgood, 1982
[‡] Refers to the monotonic yield strength

the Coffin–Manson relationship, Eq. 8.1, provides a convenient engineering expression for characterizing the total fatigue life. Using the Basquin equation (Eq. 7.1) and noting that

$$\frac{\Delta\epsilon_e}{2} = \frac{\Delta\sigma}{2E} = \frac{\sigma_a}{E}, \tag{8.3}$$

where E is Young's modulus, it is found that

$$\frac{\Delta\epsilon_e}{2} = \frac{\sigma_f'}{E}(2N_f)^b. \tag{8.4}$$

Combining Eqs. 8.1, 8.2 and 8.4, one obtains

$$\frac{\Delta\epsilon}{2} = \frac{\sigma_f'}{E}(2N_f)^b + \epsilon_f'(2N_f)^c. \tag{8.5}$$

The first and second terms on the right hand side of Eq. 8.5 are the elastic and plastic components, respectively, of the total strain amplitude. Equation 8.5 forms the basis for the strain–life approach to fatigue design and has found widespread application in industrial practice. Table 8.1 provides a list of representative data for the stress–life and strain–life characterization of some commonly used engineering alloys.

8.1.2 Transition life

The variations of the elastic, plastic and total strain amplitudes are plotted in Fig. 8.1 as functions of the number of load reversals to failure, $2N_f$, from Eqs. 8.4,

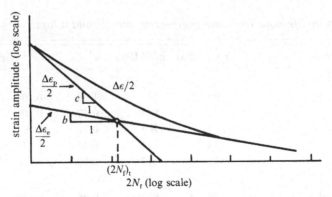

Fig. 8.1. The total strain amplitude versus life curve obtained from the superposition of the elastic and plastic strain amplitude versus life curves.

8.1, and 8.5, respectively. In order to examine the implications of Fig. 8.1 for 'short' and 'long' fatigue lives, it is useful to consider a transition life, which is defined as the number of reversals to failure $(2N_f)_t$ at which the elastic and plastic strain amplitudes are equal. From Eqs. 8.1 and 8.4,

$$(2N_f)_t = \left(\frac{\epsilon_f' E}{\sigma_f'}\right)^{1/(b-c)}. \tag{8.6}$$

At short fatigue lives, i.e. when $2N_f \ll (2N_f)_t$, plastic strain amplitude is more dominant than the elastic strain amplitude and the fatigue life of the material is controlled by *ductility*. At long fatigue lives, i.e. when $2N_f \gg (2N_f)_t$, the elastic strain amplitude is more significant than the plastic strain amplitude and the fatigue life is dictated by the *rupture strength*. Optimizing the overall fatigue properties thus inevitably requires a judicious balance between strength and ductility (e.g., Mitchell, 1978).

Mean stress effects have also been incorporated into the uniaxial strain-based characterization of fatigue life in a simple manner (Morrow, 1968). Assuming that a tensile mean stress reduces fatigue strength σ_f', such that (see Eq. 7.6):

$$\sigma_a = (\sigma_f' - \sigma_m)(2N_f)^b, \tag{8.7}$$

the strain–life relationship, Eq. 8.5, can be rewritten (see Section 7.2) as

$$\frac{\Delta\epsilon}{2} = \frac{\sigma_f' - \sigma_m}{E}(2N_f)^b + \epsilon_f'(2N_f)^c. \tag{8.8}$$

Cycle-dependent relaxation of the mean stress under strain-controlled fatigue loading is a counterpart of the cyclic creep mechanism discussed earlier (Section 3.9) in connection with stress-controlled fatigue. Consider a ductile solid which is subjected to a fixed amplitude of cyclic strains, where the mean strain is tensile, Fig. 8.2(*a*). If the material exhibits cyclic softening behavior, the flow stress at the

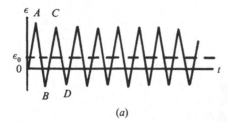

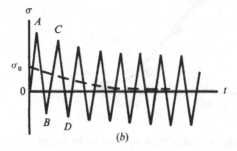

Fig. 8.2. Mean stress relaxation in a cyclically softening material subjected to strain-controlled fatigue.

imposed strain level C is expected to be lower than that at A. With a tensile mean strain level, the tendency for similar behavior in compression is not significant, and consequently, the shape of the hysteresis loop for the portion C to D will be roughly the same as that from A to B. This process results in a progressive reduction in the mean stress with increasing strain cycling, as shown in Fig. 8.2(b). The rate of decrease in mean stress progressively diminishes as the mean stress level approaches zero.

Mean stress relaxation can also occur in cyclically hardening materials (e.g., Sandor, 1972), although the mechanisms here are less obvious. Cyclic hardening reduces the plastic strain range and increases the stress range for a fixed total strain amplitude. With reference to Fig. 8.2(a), the material develops a higher flow stress at C than at A. However, with a tensile mean stress, the material yields more in tension than in compression. This preferential plastic straining alters the shape of the hysteresis loops in such a way that the stress at point D is lower than that at B, although C is at a higher stress level than at A. The net result is that a relaxation of mean stress occurs in a cyclically hardening material as well.

We conclude this section by noting that the low-cycle fatigue behavior of polymeric materials is also often characterized empirically using strain-based approaches. Figure 8.3 shows the variation of total strain amplitude $\Delta\epsilon$ (log scale) as a function of the number of load reversals to failure $2N_\mathrm{f}$ (log scale) for polycarbonate at 298 K. Note the similarity of this curve with the corresponding one for metals presented in Fig. 8.1.

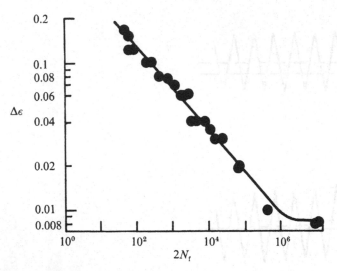

Fig. 8.3. Total strain amplitude $\Delta\epsilon$ plotted against the number of load reversals to failure $2N_f$ (log–log scale) for polycarbonate at 298 K. (After Beardmore and Rabinowitz, 1975.)

8.1.3 Example problem: Thermal fatigue life of a metal-matrix composite

Problem:

Refer to the example problem of a metal-matrix composite (Section 3.10.5) which is subjected to thermal cycling. Now consider situations under which the metallic matrix of the composite undergoes reversed plastic yielding during thermal cycling.

Let the matrix, a 2024 aluminum alloy, be reinforced with 25 volume% of spherical SiC particles. The composite is first cooled from a stress-free temperature of 400 °C to 25 °C. Subsequently, it is thermally cycled between these two temperature limits. It was determined from experiments and microscopic observations that damage arising from plastic deformation was more significant for thermal fatigue life than that from creep, for the particular testing conditions employed.

(i)   Compute the cyclic plastic zone size for one temperature reversal between 40 °C and 25 °C. Assume that the properties of the matrix alloy are the same as those given in the example problem in Section 3.10.5. In order to get analytical results, assume for a start that the mechanical properties of both phases of the composite are independent of temperature for the range of temperatures and heating/cooling rates employed.

(ii) Find the plastic strain accumulation per thermal cycle between 400 °C and 25 °C.

(iii)   If the fatigue ductility coefficient, ϵ'_f = 0.22, and the fatigue ductility
exponent, c = −0.59 (in Eq. 8.1), find the number of temperature reversals to failure. Identify the location at which failure initiates on the basis
of Coffin–Manson criterion.

(iv)   The results obtained in the above three parts are an oversimplification
in that they do not account for the temperature-dependence of material
properties. If the properties vary with temperature in the same manner
as in part (iv) of the example problem in Section 3.10.5, discuss the
effects of such temperature-dependence on your results in items (i)–(iii).

Solution:

It is given that: $f_p = (r_i/r_o)^3$ = 0.25, and T_{ini} = 400 °C, T_{room} = 25 °C,
and   the   applied   temperature   amplitude   (see   Fig.   3.16(b)),
$|\Delta T_a| = T_{ini} - T_{room}$ = 375 °C. It is noted that creep effects are less dominant
compared to plasticity effects. Denoting the properties of the matrix by the
subscript '1' and those of the particle by the subscript '2' and substituting the
appropriate values in Eq. 3.26, it is found that the elastic mismatch parameter
M_{el} = 0.35. Substituting the numerical values of the various terms in Eq. 3.28,
we find that the temperature change, from the initial stress-free temperature
(400 °C) at which plastic yielding begins is: $|\Delta T_1|$ = 174 °C, i.e. at 400 −
174 = 226 °C. The temperature change beyond which any reversal in tempera-
ture induces a reversed plastic zone is $|\Delta T_2| = 2|\Delta T_1|$ = 348 °C, i.e. at 400 −
174 = 52 °C. Since the first cooling is down to 25 °C, subsequent thermal
cycling would induce a reversed yield zone.

(i)   Substituting the numerical values of geometrical and material parameters
into Eq. 3.31, and solving iteratively for the reversed plastic zone r_c, we
see that that $(r_c/r_i)^3 \approx 1.06$.

(ii)   It can easily be shown, from the information provided in Section 3.10.3,
that the maximum plastic strains develop at the particle–matrix interface,
i.e. at $r = r_i$. The second term on the right hand side of Eq. 3.37 can be
used to estimate the rate of plastic strain accumulation per cycle.
Substituting $r = r_i$ into this term, we note that the rate of plastic strain
accumulation per cycle (for the given thermal amplitude ΔT_a), is

$$\Delta \epsilon^{pl} = \frac{2\sigma_y(1 - \nu_1)}{E_1}\left[\left(\frac{r_c}{r_i}\right)^3 - 1\right]. \tag{8.9}$$

Substituting the appropriate material properties and the result for r_c from
part (i), it is seen that $\Delta \epsilon^{pl} = 4.4 \times 10^{-4}$.

(iii)   The interface between the particle and the matrix develops the maximum
plastic strain which fully reverses during one complete thermal cycle.
From Eq. 8.1,

$$\frac{\Delta\epsilon^{pl}}{2} = \epsilon_f'(2N_f)^c, \qquad \Rightarrow \qquad 2N_f = \left(\frac{\Delta\epsilon^{pl}}{2\epsilon_f'}\right)^{1/c}. \tag{8.10}$$

Substituting the values of ϵ_f' and c from the given information, and noting that the plastic strain range during a thermal fatigue cycle is $\Delta\epsilon^{pl} = 4.4 \times 10^{-4}$ and that the plastic strain amplitude is $\Delta\epsilon^{pl}/2 = 2.2 \times 10^{-4}$, from part (ii), it is seen that $2N_f \approx 36\,000$.

(iv) As discussed in part (iv) of Section 3.10.5, a decrease in yield strength of the matrix with increasing temperature causes the monotonic plastic zone to increase. However, the cyclic plastic zone size is not expected to be affected to a significant extent because of the counteracting effect of an increasing yield strength during cooling. Therefore, the cyclic plastic zone size, the plastic strain accumulation per cycle and the total life would be expected to be affected only in a moderate manner. Precise estimates of the effects of changing properties during thermal cycling can only be quantified through numerical solutions of the relevant equations.

8.2 Local strain approach for notched members

The local strain approach relates deformation occurring in the immediate vicinity of a stress concentration to the remote stresses and strains using the constitutive response determined from fatigue tests on simple laboratory specimens. The analysis is divided into two steps:

(1) From a knowledge of the imposed loads on a notched component, the local stress and strain histories at the tip of the notch must be known.
(2) The fatigue life that can be expected for the local stress and strain histories must be determined.

For the first part, either simple analytical expressions or detailed finite element simulations of the notch tip deformation (using constitutive laws and hardening rules described in Section 3.3) are developed to relate the local stresses and strains to far-field loading. Alternatively, the notch tip deformation is experimentally monitored with the aid of strain gages or other displacement/strain measurement techniques. This is perhaps the most challenging aspect of design against fatigue fracture ahead of stress concentrations. For the second part, the damage accumulation from the local stress and strain histories must also be estimated so that the safe fatigue life of the component can be assessed based on the low cycle fatigue properties measured on smooth laboratory test specimens. For design purposes, it is often more convenient to relate notch tip fields to nominal (far-field) loading by simple engineering approximations. This section describes some such commonly adapted methods.

8.2.1 Neuber analysis

The stress and strain concentration factors are of the same value when only elastic deformation occurs at the tip of the notch. However, once the material yields at the notch tip, the stress and strain concentration factors take different values. (Note that the strain concentration factor K_ϵ is the ratio of the maximum local strain to the nominal strain.) Under conditions of plastic deformation, the theoretical elastic stress concentration factor is given approximately by the geometrical mean of the stress and strain concentration factors, K_σ and K_ϵ, respectively, as per the well known Neuber's rule (Neuber, 1961):

$$K_t = \sqrt{K_\sigma K_\epsilon}. \tag{8.11}$$

The prediction of fatigue lives for notched members has found increasing use of *local strain* approaches, which are modifications and applications of the Neuber's rule to fatigue loading conditions. In these approaches, the following expressions for fatigue notch factors are found to provide satisfactory predictions of the fatigue behavior in notched members of a wide variety of steels (Topper, Sandor & Morrow, 1969; Dowling, Brose & Wilson, 1977):

$$K_f = \sqrt{K_\sigma K_\epsilon} \tag{8.12}$$

for plastic deformation ahead of notches and

$$K_f = \frac{\sqrt{\Delta\sigma \Delta\epsilon E}}{\Delta\sigma^\infty} \tag{8.13}$$

for elastic deformation. In Eq. 8.13, $\Delta\sigma^\infty$, $\Delta\sigma$ and $\Delta\epsilon$ are the amplitudes of the fully reversed nominal stress, notch tip stress and notch tip strain, respectively. For fixed values of the imposed stress range $\Delta\sigma^\infty$, Eq. 8.13 is an equation of a rectangular hyperbola,

$$\Delta\sigma \Delta\epsilon = (K_f \Delta\sigma^\infty)^2/E = \text{constant}. \tag{8.14}$$

There is a family of curves, with different combinations of $\Delta\sigma$ and $\Delta\epsilon$, which satisfies this equation. K_f can be uniquely determined by simultaneously solving Eq. 8.13 with the cyclic stress–strain constitutive equation, Eq. 3.5, for the material. Multiplying both sides of Eq. 3.5 by $\Delta\sigma$, one obtains

$$\Delta\sigma \Delta\epsilon = \frac{(\Delta\sigma)^2}{E} + \Delta\sigma \left\{ \frac{\Delta\sigma}{K'} \right\}^{1/n_f}. \tag{8.15}$$

From Eqs. 8.14 and 8.15,

$$\frac{(\Delta\sigma)^2}{2E} + \Delta\sigma \left\{ \frac{\Delta\sigma}{2K'} \right\}^{1/n_f} = \frac{(K_f \Delta\sigma^\infty)^2}{2E}. \tag{8.16}$$

The factor '2' is introduced in the denominator on both sides of the equation because the stress amplitude is one-half of the total range, $\Delta\sigma^\infty$, for fully reversed loading.

Figure 8.4 is a schematic diagram of the approach outlined in Eqs. 8.11–8.16. Since the stress–strain response for the region at the tip of the notch must coincide

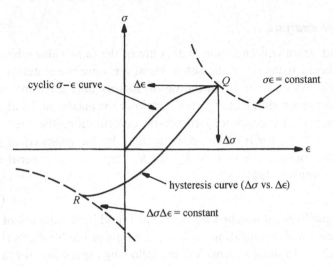

Fig. 8.4. Schematic illustration of the procedure used to determine the local stresses and strains at a notch tip.

with the characteristic cyclic stress–strain behavior of the material, the local stress σ_1 and the local strain ϵ_1 at the notch tip corresponding to a far-field tensile stress σ_1^∞ is determined by the intersection of two curves: (i) the Neuber hyperbola represented by the condition that $\sigma\epsilon = (K_f\sigma^\infty)^2/E =$ constant, and (ii) the cyclic stress–strain curve given by $\epsilon = \sigma/E + (\sigma/K')^{1/n_f}$. Point Q in Fig. 8.4 represents the local stress–strain coordinates at the notch tip corresponding to a far-field tensile stress σ_1^∞.

If the far-field stress is now reversed to a (compressive) value σ_2^∞, the stress range causing this reversal is $\Delta\sigma^\infty = \sigma_2^\infty - \sigma_1^\infty$. The corresponding local stress range and strain range values for the notch tip are $\Delta\sigma$ and $\Delta\epsilon$, respectively. To determine these values, the origin of the stress–strain coordinate system is now located at point Q. The stress–strain hysteresis curve obtained from the cyclic deformation tests is now used in conjunction with the Neuber rule, $\Delta\sigma\Delta\epsilon = (K_f\Delta\sigma^\infty)^2/E =$ constant, to locate the stress–strain coordinates for the notch tip for a far-field stress value of σ_2^∞. This is represented by Eq. 8.16 and by the point R in Fig. 8.4.

All subsequent reversals of loading employ Eqs. 8.14–8.16 to determine the local fields at the notch tip. This local strain approach can easily be implemented in a computer code.

In practical situations, particular attention has to be paid to the relationship between accumulated damage under variable amplitude loading and that measurable in a laboratory specimen under constant amplitude loading conditions. Cycle counting techniques have been developed to reduce complex fatigue loading histories to a series of discrete events so that cyclic damage could be properly accounted for. A number of counting techniques have been proposed to accomplish these goals: these include the so-called rainflow counting, range pair, level crossing and peak counting methods.

The extent to which any of the aforementioned methods for local strain analysis will provide successful predictions of fatigue life in a material depends on the fraction of fatigue life expended to initiate small flaws ahead of the notch and on the remaining fraction to propagate this flaw to failure. In addition to the local stress and strain calculations, fracture mechanics-based analyses of the stress and deformation fields ahead of notches are of considerable interest in developing a rational approach to this notch fatigue problem.

8.3 Variable amplitude cyclic strains and cycle counting

In variable amplitude fatigue, cycle counting methods are often employed to reduce the random load history into a series of discrete events which can be analyzed using the laboratory data obtained for constant amplitude fatigue loads. Cycle counting methods are commonly used in many life prediction models in the context of both local strain approach and defect-tolerant approach. A number of cycle counting methods have been developed over the years, on the basis of 'trial-and-error' approaches and 'educated guesswork'. Although empirical in their formulation, such counting methods have been tested (with varying degrees of success) on a wide variety fatigue-critical engineering components and have been found to provide useful guidelines for fatigue design. A detailed description of the various cycle counting techniques can be found in the reviews by Dowling (1972), Fuchs & Stephens (1980), Bannantine, Comer & Handrock (1990) and Dowling (1993). In this section, an example of the widely used rainflow counting procedure is presented to illustrate how complex loading histories imposed on a component can be reduced to a series of stress–strain hysteresis loops.

8.3.1 Example problem: Cycle counting

Problem:

Figure 8.5(a) shows an example of a strain–time loading sequence. The corresponding stress–time history (derived, for example, from uniaxial fatigue tests on smooth specimens) is shown in Fig. 8.5(b). (Note that a clear functional relationship does not exist between the stress–time and strain–time plots because of plastic deformation of the material. As an example, events 3–4 and 5–4' have identical values of mean strain and strain amplitude. However, the corresponding values of mean stress and stress amplitude are different.) Simplify the strain history using the rainflow counting method, and suggest strategies for fatigue life estimate for this variable amplitude cyclic straining.

Solution:

In order to simplify the strain history using the rainflow counting method, it is more convenient to replot Fig. 8.5(a) with the time axis oriented

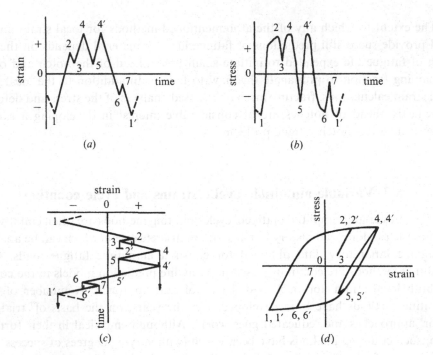

Fig. 8.5. (*a*) A random strain–time history imposed on a ductile solid. (*b*) The corresponding stress–time history. (*c*) Fig. (*a*) replotted with the time axis pointing downward, to illustrate the rainflow technique. (*d*) Stress–strain hysteresis loops extracted from the rainflow counting method. (After Landgraf & LaPointe, 1974.)

downward, as shown in Fig. 8.5(*c*). Imagine now that the lines connecting the strain peaks are a series of 'pagoda roofs' and that rain is dripping down these roofs (hence the name 'rainflow counting method'). Several rules are imposed on rain dripping down the roofs so that equivalent hysteresis loops can be extracted from the strain history (e.g., Dowling, 1972; Anzai & Endo, 1979). The following rules are imposed to define the flow of rain on the roofs: (i) The strain history is plotted such that the first and last peaks or valleys have the largest magnitude of strain. This ordering eliminates counting half cycles. (ii) Rainflow initiates at each peak (such as point 1) and is allowed to drip down continuously. However, the flow of rain from a peak must stop whenever it drips down a point which has a more positive strain value than the one from which it drips. For example, rain dripping down peak 2 must stop opposite peak 4 because the latter location has a more positive strain value than the former location. Similarly, the flow must stop when it comes opposite a minimum more negative than the minimum from which it is initiated. For example, flow from valley 5 must stop opposite valley 6 because the latter location is more negative than the former. (iii) Rainflow must stop if it encounters rain from the roof above. For example, during flow from points 3 to 4, rain dripping down from point 2 is encountered and hence flow must stop at point 2′. Note that every part of the strain–time history is counted once and only once.

Now apply the above rules to the strain history in Fig. 8.5(c). Rainflow begins on the outside at the highest peak strain and follows the pagoda roof down to the peak at 2. At this point, the flow drips down to location 2′ and continues down to point 4. From 4, rainflow continues down to location 4′ (which has the same strain magnitude as that at 4). From there, the flow path is along 4′ to 6, 6 to 7, and 7 to 1′. The stress–strain path from this sequence of events corresponds to the hysteresis loop defined by the circuit 1 → 4, 4′ → 1′, i.e. the outermost loop in Fig. 8.5(d).

Three additional hysteresis loops can be defined from the rainflow analysis of the remaining paths. These include paths 2 → 3 → 2′, 5 → 4′ → 5′, and 6 → 7 → 6′ in Fig. 5(c). Note that these hysteresis loops are not symmetrical about the origin of the strain axis.

If the random strain history shown in Fig. 8.5(a) were to be repeated m times, the rainflow method would characterize all of these random loading events in terms of the four hysteresis loops shown in Fig. 8.5(d), with each loop repeated m times. Constant amplitude, uniaxial fatigue data can now be generated on smooth laboratory test specimens using these hysteresis loops and the material constants b, c, σ_f' and ϵ_f' are determined experimentally. Mean stress effects on fatigue life are accounted for by recourse to Eq. 8.8, for example, where $\Delta\epsilon$ and σ_m appropriate for a particular strain cycle are substituted along with the foregoing material properties to obtain the number of cycles to failure, N_f, representative of that strain cycle. These results could be used in conjunction with the Palmgren–Miner rule to sum the fatigue damage. Specifically, this entails the summation of damage for each strain cycle to obtain

$$\sum_{i=1}^{M} \frac{m}{N_{fi}} = d,\qquad(8.17)$$

where M (= 4 in the example shown in Fig. 8.5) is the total number of key events (in the form of strain cycles) identified from the rainflow counting method and d is the accumulated fractional damage. Alternatively, the cycle counting method can be used in conjunction with the life prediction methods employing variable amplitude crack growth, which are discussed in Chapter 14.

Computer algorithms for cycle counting are available in the ASTM *Annual Book of Standards* section 3, vol. 03.01, 1986 (American Society for Testing and Materials, Philadelphia). Some simple algorithms have also been published by Downing & Socie (1982).

8.4 Multiaxial fatigue

Stress-based phenomenological approaches to total life under multiaxial fatigue were discussed in the preceding chapter. We now consider multiaxial loading under low-cycle fatigue conditions where combinations of different loading modes induce failure which is at least partly dictated by the critical values of different components of cyclic strains.

8.4.1 Measures of effective strain

In parallel with the definitions of effective stress quantities presented in the context of multiaxial stress–life approach in the preceding chapter, it is instructive to examine first the basic definitions of some effective strain quantities in the context of low-cycle fatigue. Following the discussion in Section 7.10, the effective intensity of the multiaxial stress/strain state is usually characterized in terms of the conjugate scalar pairs τ_{oct} and γ_{oct} (i.e. the octahedral shear stress and strain, respectively) in the von Mises distortional energy theory or in terms of τ_{max} and γ_{max} (i.e. the maximum shear stress and strain, respectively) in the Tresca yield theory. The effective strain in multiaxial loading is

$$\epsilon_e = \frac{\sqrt{(\epsilon_1 - \epsilon_2)^2 + (\epsilon_2 - \epsilon_3)^2 + (\epsilon_3 - \epsilon_1)^2}}{\sqrt{2}\,(1 + \nu)}, \tag{8.18}$$

where for the fully plastic state, the Poisson's ratio, $\nu = 0.5$ and for the elastic state, $\nu = 0.33$ for most metals and alloys. In Eq. 8.18, ϵ_1, ϵ_2 and ϵ_3 are the three principal strains, with $\epsilon_1 > \epsilon_2 > \epsilon_3$. The effective plastic strain based on the distortional energy theory is written as

$$\epsilon_e^p = \frac{1}{\sqrt{2}} \gamma_{oct}^p = \left\{ \frac{2}{3} \epsilon_{ij}^p \cdot \epsilon_{ij}^p \right\}^{1/2}, \tag{8.19}$$

where the superscript 'p' denotes plastic strains. The corresponding effective strain measures based on the maximum shear strain values are

$$\epsilon_e = \frac{\gamma_{max}}{1 + \nu} = \frac{\epsilon_1 - \epsilon_3}{1 + \nu}, \quad \text{and} \quad \epsilon_e^p = \frac{2}{3} \gamma_{max}^p = \frac{2}{3} (\epsilon_1^p - \epsilon_3^p). \tag{8.20}$$

The objective of the theories of multiaxial fatigue is to predict fatigue life under complex loading conditions in terms of laboratory data of strain–life curves gathered from uniaxial fatigue tests using simple criteria for failure. If the amplitude of the maximum principal strain, $\Delta\epsilon_1/2$, determines failure, Eq. 8.5 may be rewritten to obtain

$$\frac{\Delta\epsilon_1}{2} = \frac{\sigma_f'}{E} (2N_f)^b + \epsilon_f'(2N_f)^c. \tag{8.21}$$

In terms of the von Mises criterion, the strain-based expression for multiaxial fatigue life becomes

$$\frac{\Delta\epsilon_e}{2} = \frac{\sigma_f'}{E}(2N_f)^b + \epsilon_f'(2N_f)^c,$$ (8.22)

where the effective strain is defined in Eq. 8.18. Similarly, with the Tresca criterion, it is useful to note the correlation between axial strain and shear strain,

$$\frac{\Delta\gamma_e}{2} = \frac{(1+\nu)\Delta\epsilon_e}{2}.$$ (8.23)

Combining Eqs. 8.22 and 8.23 and taking $\nu = 0.3$ for elastic deformation and $\nu = 0.5$ for plastic deformation, it is seen that

$$\frac{\Delta\gamma_e}{2} = 1.3\frac{\sigma_f'}{E}(2N_f)^b + 1.5\epsilon_f'(2N_f)^c.$$ (8.24)

The main drawback of these effective strain measures is that they do not adequately capture the effects of mean stress on multiaxial fatigue life. In an attempt to overcome this limitation, Smith, Watson & Topper (1970) suggested a simple 'energy-based approach' to account for mean stress effects. In their approach, multiplying both sides of Eq. 8.5 by the maximum stress ($\sigma_{max} = \sigma_m + \{\Delta\sigma\}/2$) results in

$$\sigma_{max}\frac{\Delta\epsilon}{2} = \frac{(\sigma_f')^2}{E}(2N_f)^{2b} + \sigma_f'\epsilon_f'(2N_f)^{b+c}.$$ (8.25)

This model is predicated on the premise that no fatigue damage occurs when $\sigma_{max} < 0$, which can be at variance with experimental observations.

These rather strong limitations of the foregoing theories led to considerations of other approaches, such as the critical plane approach to multiaxial fatigue failure (Section 7.10), which is taken up in the following subsections specifically in the context of low-cycle fatigue.

8.4.2 Case study: Critical planes of failure

Conditions governing the identification of critical planes for the onset of fatigue damage and cracking were discussed in Section 7.10.4 by recourse to the stress–life approach to high-cycle multiaxial fatigue. In this section, we present a particular case study which illustrates the existence of critical planes of failure for strain-controlled, constant-amplitude cycling of an Inconel 718 alloy, which can be predicted on the basis of the maximum shear strain amplitude. In addition, this example, which is derived from the works of Socie & Shield (1984) and Socie (1993), clearly illustrates the pronounced effect of the normal stress on the rate of progression of damage and cracking on the critical plane.

Consider the three multiaxial loading paths schematically illustrated in Figs. 8.6(a)–(c). These figures signify three different experiments conducted with different combinations of axial strains and torsional shear strains on cylindrical tubes of Inconel 718 alloy. The multiaxial loading in these three experiments is designed such that they all have cyclic proportional straining with the same maximum shear strain amplitudes. Each specimen also experiences a static mean strain. The differences between the three experiments are that the normal stresses and strains across the plane of the maximum shear strain are

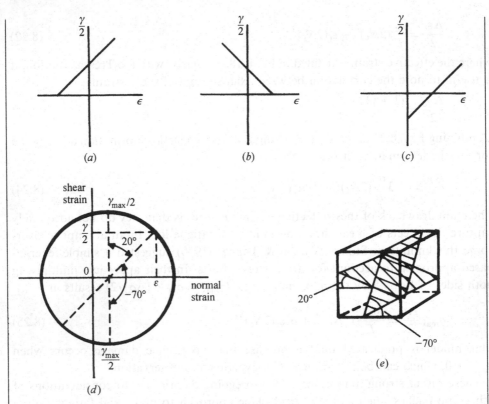

Fig. 8.6. (a)–(c) Three different proportional loading paths for the multiaxial straining experiments which have the same maximum shear strain amplitudes, but different normal stresses and strains across the planes of maximum shear strain. (d) Mohr's circle construction showing the orientations of the planes of maximum shear strain. (e) A schematic illustration of the orientations of the two planes of maximum shear strain. (After Socie, 1993.)

different. Figure 8.6(d) is the Mohr's circle of strain for these cases, where it is seen that two mutually perpendicular planes are subjected to the same maximum shear strain amplitudes. The orientations of these planes are schematically sketched in Fig. 8.6(e). Since the sign of the shear strain has no physical significance, these two planes would be expected to undergo the same extent of damage if shear strain amplitude alone determined the evolution of damage. Figures 8.7(a)–(c) show the orientations of the cracks formed for the three different loading paths sketched in Figs. 8.6(a)–(c), respectively. It is apparent that the cracking angles seen in Figs. 8.7(a)–(c) closely match the orientations of maximum shear strain planes predicted by the Mohr circle in Figs. 8.6(d) and (e). This also implies in this case that the maximum shear strain plane controls the evolution of damage and cracking in the Inconel 718 alloy. The loading paths shown in Figs. 8.6(a) and (c) give rise to crack planes oriented at $-70°$ and $+20°$, respectively, as seen from the angles of cracks in Figs. 8.7(a) and (c), respectively. The loading history shown in Fig. 8.6(b) promotes a crack angle of $-20°$, as seen in Fig. 8.7(b), because its loading direction is the reverse of that in the other two cases.

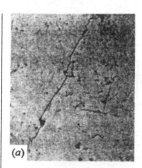
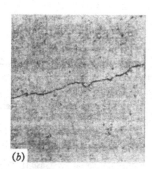
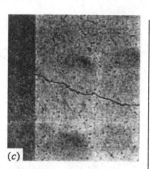

Fig. 8.7. (a)–(c) Observations of crack paths for the proportional straining histories shown in Figs. 8.6.(a)–(c), respectively. (From Socie (1993). Copyright the American Society for Testing and Materials. Reproduced with Permission.)

It is also apparent from Fig. 8.6(e), however, that the maximum normal strains on the two critical planes with the highest shear strain amplitude are different. Although the loading paths followed in the experiments entail proportional straining, the principal axes of stress and strain are not coincident, and consequently, the maximum normal stresses on the two planes of maximum shear strain amplitude are not the same. Observations of subcritical crack growth on the different planes clearly reveal that the maximum shear strain planes which have the highest tensile normal stress across them exhibit the highest crack propagation rates and the lowest fatigue lives. Compressive normal stresses inhibit the advance of cracks, while the tensile normal stresses facilitate crack growth.

The results discussed in this section thus clearly show how maximum shear strain amplitudes govern the initiation of fatigue cracking and how mean stresses on the planes of maximum shear influence fatigue crack growth and overall fatigue life. It is thus evident that the critical plane theories for multiaxial fatigue should incorporate the effects of both shear and normal components of multiaxial stresses/strains for life prediction. This case study also serves to provide a mechanistic justification for the various stress–life criteria described in Section 7.10 and for the additional criteria based on cyclic strains to be presented in the following sections.

8.4.3 Different cracking patterns in multiaxial fatigue

Brown & Miller (1973) suggest two different patterns for crack growth at surfaces of materials subjected to multiaxial fatigue loading on the basis of the orientations of the planes of maximum shear strain amplitude with the free surface. Consider the situation schematically sketched in Figs. 8.8(a) and (e), where a cubic element of a material is subjected to multiaxial strain cycles. The free surface plane is marked in these figures. Figures 8.8(b) and (c) show the planes of maximum shear strain amplitude wherein a small, stage I crack initiates by single shear process at the free surface and advances during subsequent cycling. In these figures, the shear stress acts parallel to the free surface (i.e. the front surface of the cube) and there is no shear stress acting normally to that surface. That is, the unit normal vectors to the

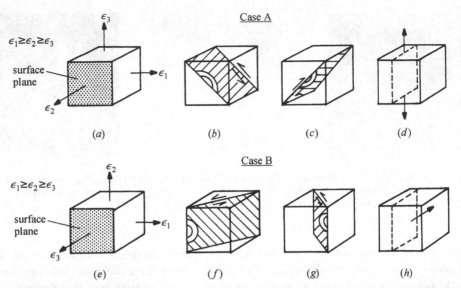

Fig. 8.8. Schematic illustrations of case A and case B fatigue cracking in multiaxial fatigue: (*a*) and (*e*) show multiaxial strains. (*b*), (*c*), (*f*) and (*g*) denote planes of maximum shear strain amplitude and the planes and directions of stage I crack growth for case A and case B. (*d*) and (*h*) show the planes and directions of stage II crack growth for case A and case B, respectively.

planes of maximum shear strain amplitude lie on the specimen surface plane. Under these circumstances, the cracks advance more in a direction parallel to the surface than normal to the surface, thereby increasing the aspect ratio of the crack. This pattern of cracking has been termed 'case A' by Brown & Miller. When the case A cracks become longer, i.e. when their critical dimensions span several grain diameters, stage II crack growth occurs as a result of simultaneous or alternating slip involving more than one slip system. At this time, the direction of crack advance and the plane on which it occurs are as shown in Fig. 8.8(*d*).

Now consider the possibility shown in Fig. 8.8(*e*) where the magnitudes of principal strains are such that the planes of maximum shear strain amplitude occur on the planes schematically sketched in Figs. 8.8(*f*) and (*g*). Here, the stage I cracks initiate at the surface and advance at 45° angles into the material, and this mode of cracking has been termed 'case B'. (The fatigue crack growth process along persistent slip bands (Chapter 4) for ductile single crystals represents case B cracks.) The direction of stage II crack growth for case B is also from the free surface into the material, as sketched in Fig. 8.8(*h*). Uniaxial tension fatigue leads to the same shear stress for case A and case B and hence it can facilitate either mode of failure. Mixed tension–torsion fatigue loading, however, invariably promotes case A cracks.

Brown & Miller (1973) postulate that the criterion for cracking for case A and case B follows the general relationships:

$$\frac{\Delta\gamma_{\max}}{2} = f_a\left\{\frac{\Delta\epsilon_n}{2}\right\}, \qquad \frac{\Delta\gamma_{\max}}{2} = f_b\left\{\frac{\Delta\epsilon_n}{2}\right\}, \tag{8.26}$$

where $\Delta\gamma_{max}/2$ and $\Delta\epsilon_n/2$ are maximum shear strain amplitude and normal strain amplitude, respectively, during the low-cycle fatigue loading, and f_a and f_b are different nonlinear functions of their arguments for case A and case B, respectively. The examples shown in the preceding subsection on Inconel 718 and the work of Fatemi & Socie (1988) demonstrate that the peak normal stress to the plane of maximum shear strain amplitude influences the propagation of stage I cracks under a variety of multiaxial loading conditions that induce case A and case B cracking.

8.4.4 Example problem: Critical planes of failure in multiaxial loading

Problem:

The cylindrical tube is subjected to different combinations of fluctuating torsional moment (ΔT) and static axial load (P) as well as static internal pressure (p). On the basis of the discussions presented in Sections 7.10.3, 7.10.4, 8.4.2 and 8.4.3, suggest possible orientations of (a) planes of maximum shear strain amplitude, (b) planes of maximum normal stress amplitude, (c) planes on which stage I shear cracks may form and (d) planes on which tensile cracks may form, and (e) the expected relative fatigue life, for the following four cases of loading.†

  (i) **Case 1:** The thin-walled cylindrical tube which is subjected to cyclic torsion loading only (i.e. ΔT is nonzero, $P = 0$ and $p = 0$).

 (ii) **Case 2:** Repeat the above problem for the situation where a static axial tensile load is imposed on the cylindrical tube which is subjected to cyclic torsional loading (i.e. ΔT is nonzero, $P > 0$ and $p = 0$).

(iii) **Case 3:** Repeat the above problem for the situation where a static axial compressive load is imposed on the cylindrical tube which is subjected to cyclic torsional loading (i.e. ΔT is nonzero, $P < 0$ and $p = 0$).

(iv) **Case 4:** Repeat the above problem for the situation where a static axial compressive load and a static internal pressure are imposed on the cylindrical tube which is subjected to cyclic torsional loading (i.e. ΔT is nonzero, $P < 0$ and p is nonzero).

Solution:

For all the cases considered in this problem, we indicate orientations of critical planes where the cyclic strains are marked by double-ended arrows and static stresses are indicated by single-ended arrows. Possible critical planes are denoted by dashed lines.

  (i) **Case 1:** For pure torsional cyclic loading, the maximum shear strains occur along the axial and hoop directions. These are also the planes

---

† This problem is adapted from a discussion in Socie (1993).

along which stage I, shear cracks (see Sections 7.10.3 and 7.10.4) are likely to form. The planes on which the tensile normal strains are maximum are oriented at 45° to the axis of the tube. Tensile cracking would be expected to occur on these planes. These results are schematically sketched in Fig. 8.9.

(ii) **Case 2:** For torsional cyclic loading with superimposed axial tension, there develops a superimposed tensile stress on only one of the two shear planes (i.e. the added tensile stress acts only parallel to the axial direction). Shear cracking due to stage I would then preferentially occur only on this plane (see Fig. 8.10), and the overall fatigue life would be expected to be smaller than that in part (i). Superimposed tension is not expected to alter the stresses on the vertical shear plane. The superimposed static tension would, however, raise the stresses equally on both the tension planes oriented at 45° to the tube axis. The resulting possibilities for failure are schematically shown in Fig. 8.10. Thus, the addition of an axial tensile load to a cyclically twisted cylindrical tube is expected to be detrimental to fatigue life, irrespective of whether the failure occurs by a shear mode or a tensile mode.

(iii) **Case 3:** For torsional loading with superimposed axial compression, one of the two shear planes (i.e. the plane normal to the applied compression axis) develops a normal compressive stress and hence is not expected to develop any cracks. On the other hand, the other shear plane which is parallel to the applied compression axis can easily develop a shear crack, and hence the overall fatigue life may not be higher compared to the shear failure for Case 1 or Case 2. The two 45° planes on which tensile stresses develop would both exhibit a reduced propensity for cracking as

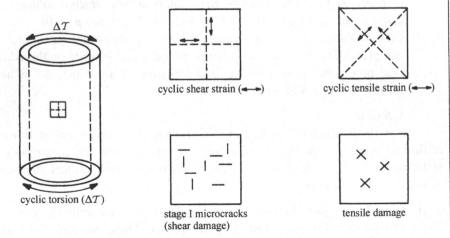

Fig. 8.9. A schematic illustration of the evolution of cyclic shear strains, normal strains, shear damage and tensile damage for cyclic torsional loading of the cylindrical tube, Case 1.

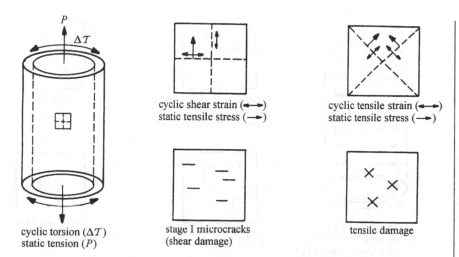

cyclic shear strain (◄—►)
static tensile stress (—►)

cyclic tensile strain (◄—►)
static tensile stress (—►)

cyclic torsion (ΔT)
static tension (P)

stage I microcracks
(shear damage)

tensile damage

Fig. 8.10. A schematic illustration of the evolution of cyclic shear strains, normal strains, shear damage and tensile damage for cyclic torsional loading of the cylindrical tube with a superimposed axial tensile load, Case 2.

a consequence of the applied compressive load. These possibilities are schematically sketched in Fig. 8.11. One may thus conclude that while the superimposed compressive load may offer a beneficial effect on fatigue life for materials which fail by tensile failure under cyclic torsion, it may not have any effect on fatigue life for materials which undergo a stage I shear failure.

(iv)   **Case 4:** For torsional loading with a superimposed static compression as well as a static internal pressure, a compressive normal stress is induced

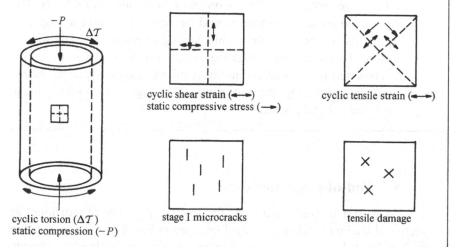

cyclic shear strain (◄—►)
static compressive stress (—►)

cyclic tensile strain (◄—►)

cyclic torsion (ΔT)
static compression (−P)

stage I microcracks

tensile damage

Fig. 8.11. A schematic illustration of the evolution of cyclic shear strains, normal strains, shear damage and tensile damage for cyclic torsional loading of the cylindrical tube with a superimposed axial compressive load, Case 3.

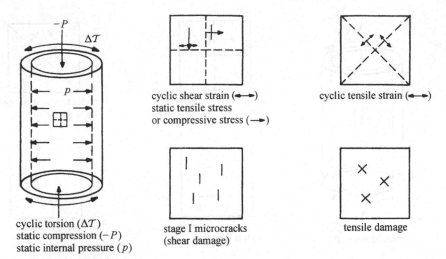

cyclic torsion (ΔT)
static compression ($-P$)
static internal pressure (p)

cyclic shear strain (◄─►)
static tensile stress
or compressive stress (─►)

cyclic tensile strain (◄─►)

stage I microcracks
(shear damage)

tensile damage

Fig. 8.12. A schematic illustration of the evolution of cyclic shear strains, normal strains, shear damage and tensile damage for cyclic torsional loading of the cylindrical tube with a superimposed axial compressive load and a static internal pressure, Case 4.

on the horizontal shear plane due to axial compression, and a tensile hoop stress develops on the vertical shear plane. Thus, shear failure, should it occur, would preferentially take place on the vertical shear plane (Fig. 8.12). Because of the superimposed tensile opening stress on this plane, materials which exhibit a shear mode of failure would be expected suffer a reduction in the fatigue life. On the two tension planes oriented at 45°, the stresses induced by the axial load would offset those stresses arising from the hoop stress; hence the overall tensile stress would remain the same as in Case 1 on these two planes (Fig. 8.12). Thus the superimposed static stresses would be expected to have a large influence on the shear mode of failure while having no effect on the tensile mode of failure. This situation is thus the opposite of that seen in Case 2.

The situations considered in this problem thus clearly show the need for carefully examining the evolution of damage and cracking on all the potential critical planes.

8.5 Out-of-phase loading

As a start to this discussion, the loading path for proportional and non-proportional loading is schematically illustrated in Fig. 8.13. A cylindrical specimen which is subjected to combined tension and torsion cyclic loads is shown in Fig. 8.13(*a*). The variations of axial strain ϵ and shear strain $\gamma/\sqrt{3}$ with time are illustrated in Fig. 8.13(*b*) for proportional loading. Here, the axial and shear strains vary

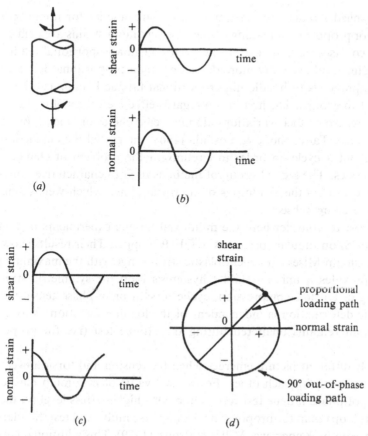

Fig. 8.13. (*a*) A cylindrical specimen subjected to combined axial and torsional fatigue loads. (*b*) Strain–time history for proportional loading where the axial and torsional strains are in phase. (*c*) Strain–time history for nonproportional loading where the axial and torsional strains are 90° out of phase. (*d*) The loading path for in-phase and 90° out-of-phase loading in strain space, for the strain histories considered here.

in constant proportion to each other and the phase angle between them is zero. Multiaxial fatigue involving 90° out-of-phase variation between the axial and shear strains is shown in Fig. 8.13(*c*). The loading path for proportional loading corresponds to a straight line in strain space, Fig. 8.13(*d*), where the axial strain is plotted against the shear strain. The 90° out-of-phase loading path is a circle in normalized strain space.

There exists experimental evidence which appears to suggest that in-phase cyclic straining is more damaging to fatigue life at low strain amplitudes, while out-of-phase cyclic straining is more damaging at high strain amplitudes, when the amplitudes of the applied tension (or bending) or torsion for proportional and nonproportional loading are comparable. It should, however, be noted that to obtain the same strain range, the applied normal and shear strains for nonproportional loading must be increased compared to those for proportional loading. This is because for comparable applied strains, the maximum strains are smaller for nonproportional loading than for proportional loading. When the maximum strains are held comparable for the two cases, it is almost always seen that nonproportional loading is at least as damaging, and generally more damaging, than proportional loading.

Available approaches to handle nonproportional fatigue loading can be broadly classified into two groups. The first group regards effective values of cyclic stress or strain without regard to their variations along specific planes or crack growth directions. For example, Taira, Inoue & Yoshida (1968) integrated the octahedral shear strains throughout a cycle, in order to circumvent the problem of changes in the direction of stresses. The second group of methods used to characterize nonproportional loading considers the conditions on a critical plane, which were discussed in detail in the preceding subsections.

It is of interest to consider here the multiaxial fatigue experiments of Lamba & Sidebottom (1978) on tubular specimens of OFHC copper. Their results show that a plot of the maximum Mises equivalent plastic stress range with the maximum plastic strain range provides a unique (stable) hysteresis loop for symmetric strain-controlled loading in tension–compression, cyclic torsion or in-phase tension–torsion. Here the cyclic deformation is independent of the loading direction as long as the loading direction remains unaltered during the fatigue test (i.e. for proportional loading).

A completely different picture emerges when the tension and torsion cyclic loads are 90° out of phase with each other. For a fixed value of maximum plastic strain range, the out-of-phase fatigue test results in a 40% higher stress level (in the stabilized hysteresis loop) than the proportional or in-phase multiaxial test. Similar trends have also been seen by Kanazawa, Miller & Brown (1979). This additional hardening occurs for out-of-phase loading as well as for any changes in cycling direction. These differences do not depend on whether the fatigue specimen is initially subjected to uniaxial loading or not. However, this additional hardening obtained in out-of-phase cyclic loading can be erased if uniaxial fatigue having the same maximum strain range is imposed after the multiaxial test.

Exercises

8.1   A metallic material is shot-peened in an attempt to improve its fatigue life. The shot-peening process results in a compressive residual stress of 250 MPa at the surface of the material. The material has the following monotonic and fatigue characteristics in the as-fabricated condition (before shot-peening): $E = 210$ GPa, $A' = 1000$ MPa, $\sigma'_f = 1100$ MPa, $\epsilon'_f = 1.0$, $n_f = 0.13$, $b = -0.08$, and $c = -0.63$. (See the discussions related to Eqs. 3.5, 7.1 and 8.5 for the definition of these variables.) On a log–log plot, show the variation of the total strain amplitude $\Delta\epsilon/2$ with the number of reversals to failure $2N_f$ for the material in both the as-fabricated and shot-peened conditions.

8.2   Consider a double-edge-notched plate of the following dimensions: width $W = 2.54$ cm, height $H = 3.05$ cm, thickness $B = 0.25$ cm, length of each notch $a_0 = 0.254$ cm and the net-section, theoretical stress concentration factor at the notch-tip $K_t = 2.4$. The monotonic and fatigue properties of the engineering alloy are as follows: $E = 210$ GPa, $A' = 1050$ MPa, $n_f = 0.1$, $\sigma'_f = 1150$ MPa, $\epsilon'_f = 1.15$, $b = -0.075$, and $c = -0.7$. (These variables are defined in connection with Eqs. 3.5 and 8.5.) The plate is subjected to a fully reversed ($R = -1$) tensile load of maximum value equal to 44.5 kN. Assuming that the plastic strains developed in the plate are small, determine the total fatigue life of the notched plate using the Neuber analysis.

8.3   In the example problem in Section 8.1.3, discuss the possible effects of cyclic strain hardening on the thermal fatigue life.

8.4   If the compressive axial load in Case 4 of the example problem in Section 8.4.4 were to be replaced by a tensile axial load, suggest possible orientations of (a) planes of maximum shear strain amplitude, (b) planes of maximum normal stress amplitude, (c) planes on which stage I shear cracks may form and (d) planes on which tensile cracks may form, and (e) the expected relative fatigue life.

Part three

DAMAGE-TOLERANT
APPROACH

CHAPTER 9

Fracture mechanics and its implications for fatigue

One of the most successful applications of the theory of fracture mechanics is in the characterization of fatigue crack propagation. An analysis of fatigue flaw growth based on fracture mechanics inevitably requires a thorough understanding of the assumptions, significance and limitations underlying the development of various crack tip parameters. An important part of such a study of fracture mechanics is the identification of the *regions of dominance* of the leading terms of asymptotic crack tip singular fields. The appropriate conditions for the dominance of critical fracture parameters are obtained from a knowledge of the accuracy of asymptotic continuum solutions and from the mechanistic understanding of microscopic deformation at the fatigue crack tip.

In this chapter, we present a focused discussion of the theories of linear elastic and nonlinear fracture mechanics that are relevant to applications in fatigue. Details of the mechanisms of fatigue crack propagation are examined in the following chapters.

9.1 Griffith fracture theory

Modern theories of fracture find their origin in the pioneering work of Griffith (1921) who formulated criteria for the unstable extension of a crack in a brittle solid in terms of a balance between changes in mechanical and surface energies. Consider a through-thickness crack of length $2a$ located at the center of a large brittle plate of uniform thickness B, which is subjected to a constant far-field tensile stress σ (Fig. 9.1). Griffith postulated that, for unit crack extension to occur under the influence of the applied stress, the decrease in potential energy of the system, by virtue of the displacement of the outer boundaries and the change in the stored elastic energy, must equal the increase in surface energy due to crack extension. Using the stress analysis of Inglis (1913) for an elliptical hole in an infinite elastic plate, Griffith deduced the net change in potential energy of the large plate, shown in Fig. 9.1, to be

$$W_{\mathrm{P}} = -\frac{\pi a^2 \sigma^2 B}{E'},$$

(9.1)

where, for plane strain and plane stress, respectively,

$$E' = \frac{E}{1 - \nu^2} \quad \text{and} \quad E' = E.$$

(9.2)

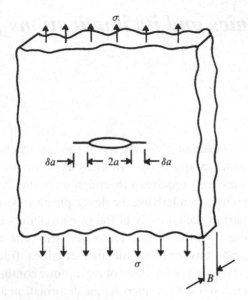

Fig. 9.1. A large plate of an elastic material containing a crack of length $2a$.

Here E is Young's modulus and v is Poisson's ratio. The surface energy of the crack system in Fig. 9.1 is

$$W_S = 4aB\gamma_s, \tag{9.3}$$

where γ_s is the free surface energy per unit surface area. The total system energy is then given by

$$U = W_P + W_S = -\frac{\pi a^2 \sigma^2 B}{E'} + 4aB\gamma_s. \tag{9.4}$$

Griffith noted that the critical condition for the onset of crack growth is:

$$\frac{dU}{dA} = \frac{dW_P}{dA} + \frac{dW_S}{dA} = -\frac{\pi a \sigma^2}{E'} + 2\gamma_s = 0, \tag{9.5}$$

where $A = 2aB$ is the crack area and dA denotes an incremental increase in the crack area. Note that the total surface area of the two crack faces is $2A$. The resulting critical stress for fracture initiation is

$$\sigma_f = \sqrt{\frac{2E'\gamma_s}{\pi a}}. \tag{9.6}$$

As the second derivative d^2U/da^2 is negative, the above equilibrium condition, Eq. 9.6, gives rise to unstable crack propagation.

Griffith's idealized model for brittle fracture considers a sharp crack for which the near-tip stresses exceed the cohesive strength of the material. In common engineering materials, nonlinear deformation processes are induced near the crack tip under the influence of the applied stress. Thus, although the Griffith concept laid the foundation for the physics of fracture, its energy balance considerations cannot be directly applied to most engineering solids. Orowan (1952) extended Griffith's brittle fracture concept to metals by simply supplementing the surface energy term in Eq. 9.6 with plastic energy dissipation. The resultant expression for fracture initiation is

$$\sigma_f = \sqrt{\frac{2E'(\gamma_s + \gamma_p)}{\pi a}}, \tag{9.7}$$

where γ_p is the plastic work per unit area of surface created. Note that γ_p is generally much larger than γ_s.

9.2 Energy release rate and crack driving force

Consider an elastic plate of uniform thickness B, as shown in Figs. 9.2 and 9.3. The plate contains an edge crack of length a. Let the plate be rigidly fixed at the upper end. The lower end of the plate is loaded with a force F and the displacement of the load point is u. (One may, in a broader sense, regard the variables F and u as a generalized force and a generalized displacement, respectively. F and u are work conjugate variables. For example, when F is taken to be the torque, u is the corresponding rotation.) The total mechanical potential energy of the cracked plate, W_P, is defined as

$$W_P = \Phi - W_F, \tag{9.8}$$

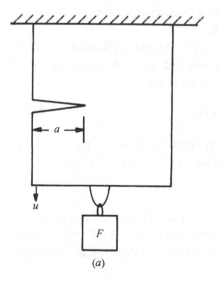

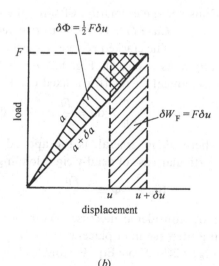

(a)   (b)

Fig. 9.2. (a) An elastic plate containing an edge crack subjected to dead weight loading. (b) Changes in the force–displacement curve and components of mechanical energy during incremental crack growth.

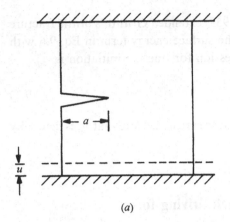
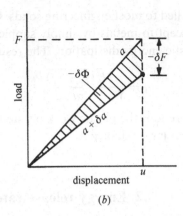

(a)   (b)

Fig. 9.3. (a) An elastic plate containing an edge crack subjected to displacement controlled loading. (b) Changes in the force–displacement curve and components of mechanical energy during incremental crack growth.

where Φ is the stored elastic strain energy and W_F is the work done by the external forces.

Irwin (1956) proposed an approach for the characterization of the driving force for fracture in cracked elastic bodies, which is conceptually equivalent to that of the Griffith model. Irwin introduced, for this purpose, the *energy release rate* $\mathcal{G}$ which is defined as

$$\mathcal{G} = -\frac{dW_p}{dA}. \tag{9.9}$$

Consider the estimation of $\mathcal{G}$ for the following two loading situations.

Case (1): Load-control or dead-weight loading

The cracked plate is subjected to a fixed force F by the application of a dead weight as shown in Fig. 9.2(a). In this load-controlled case, the components of mechanical energy (for a fixed crack length a) are written as

$$\Phi = \frac{1}{2}\int_0^u \mathcal{F}\,du = \frac{Fu}{2}, \quad \text{and} \quad W_F = Fu, \tag{9.10}$$

where $\mathcal{F}$, in general, is the applied load (which equals a fixed value, F, for the particular case of dead-weight loading). Combining Eqs. 9.8 and 9.10, it is seen that

$$W_P = -\Phi = -\frac{Fu}{2}. \tag{9.11}$$

Now consider an increase in crack length from a to $a + \delta a$. This causes a corresponding increase in displacement from u to $u + \delta u$ under the fixed force F, as shown in Fig. 9.2(b). From Eqs. 9.9 and 9.11, the energy release rate for dead-weight loading is written as

$$\mathcal{G} = -\frac{1}{B}\left[\frac{dW_p}{da}\right]_{F\text{ fixed}} = \frac{F}{2B}\left[\frac{\partial u}{\partial a}\right]_{F\text{ fixed}}. \tag{9.12}$$

As shown in Fig. 9.2(b), the advance of the crack by an increment δa (with F fixed) leads to a net increase in the stored strain energy by the amount

$$\delta\Phi\Big|_{F\ \text{fixed}} = -\frac{F\delta u}{2} + F\delta u = \frac{F\delta u}{2}. \tag{9.13}$$

Case (2): Displacement-controlled loading

Now consider the situation shown in Fig. 9.3, where the displacement u is controlled and the force F varies accordingly. When the crack advances by an increment δa under a fixed displacement u, the change in W_F is zero, and hence $\delta W_P = \delta\Phi$. From Eq. 9.9,

$$\delta\Phi\Big|_{u\ \text{fixed}} = -\mathcal{G}\delta\mathcal{A}, \quad \text{or} \quad \mathcal{G} = -\frac{1}{B}\left[\frac{\partial\Phi}{\partial a}\right]_{u\ \text{fixed}} = -\frac{u}{2B}\left[\frac{\partial F}{\partial a}\right]_{u\ \text{fixed}}. \tag{9.14}$$

As shown in Fig. 9.3(b), the advance of the crack by an increment δa (with u fixed) leads to a net decrease in the stored strain energy by the amount

$$\delta\Phi\Big|_{u\ \text{fixed}} = -\frac{u\delta F}{2}. \tag{9.15}$$

The compliance, C, of a cracked plate, which is the inverse of the stiffness, is defined as

$$C = \frac{u}{F}. \tag{9.16}$$

Combining Eq. 9.16 with Eqs. 9.12 and 9.14, it is seen that

$$\mathcal{G} = \frac{F^2}{2B}\cdot\frac{dC}{da}. \tag{9.17}$$

Equation 9.17 holds for both load control and displacement control, i.e. the energy release rate $\mathcal{G}$ is independent of the type of loading.[†] This result can also be rationalized by noting that the magnitudes of the change in stored energy under load control (Eq. 9.13) and displacement control (Eq. 9.15) differ only by $(\delta F \cdot \delta u)/2$, which is a negligible quantity. For crack advance by an increment δa with a given F and u, therefore,

$$\delta\Phi\Big|_{F\ \text{fixed}} = -\delta\Phi\Big|_{u\ \text{fixed}}. \tag{9.18}$$

It is noted that the definition of $\mathcal{G}$ given in Eq. 9.9 is valid for both linear and nonlinear elastic deformation of the body (see the Section 9.7.1). $\mathcal{G}$ is a function of the load (or displacement) and crack length, and is *independent* of the boundary conditions (i.e. type of loading) for the cracked body. The Griffith criterion for fracture initiation in an ideally brittle solid can be re-phrased in terms of $\mathcal{G}$ such that

$$\mathcal{G} = \frac{\pi\sigma^2 a}{E'} = 2\gamma_s. \tag{9.19}$$

[†] This is to be expected since, as shown later, the critical value of the energy release rate is related to the fracture toughness, which is a property of a material.

9.3 Linear elastic fracture mechanics

While the aforementioned theories regard fracture from an energy standpoint, the critical conditions for the growth of flaws can be formulated in more precise terms by means of linear elastic stress analyses. In this section, we provide a detailed discussion of linear elastic fracture mechanics.

9.3.1 Macroscopic modes of fracture

Before considering the variation of the stress and deformation fields in cracked bodies subjected to external loads, it is appropriate to examine the different modes of fracture. The crack surface displacements in the three basic modes of separation are schematically shown in Fig. 9.4. Mode I is the *tensile opening mode* in which the crack faces separate in a direction normal to the plane of the crack and the corresponding displacements of the crack walls are symmetric with respect to the $x–z$ and the $x–y$ planes. Mode II is the *in-plane sliding mode* in which the crack faces are mutually sheared in a direction normal to the crack front. Here the displacements of the crack walls are symmetric with respect to the $x–y$ plane and anti-symmetric with respect to the $x–z$ plane. Mode III is the *tearing or anti-plane shear mode* in which the crack faces are sheared parallel to the crack front. The displacements of the crack walls in this case are anti-symmetric with respect to the $x–y$ and $x–z$ planes. The crack face displacements in modes II and III find an analogy to the motion of edge dislocations and screw dislocations, respectively.

Irwin (1957), using the analytical methods of Westergaard (1939), quantified the near-tip fields for the linear elastic crack in terms of the stress intensity factor. Since the methods of characterizing the propagation of long fatigue cracks under a vast spectrum of microstructural, environmental and loading conditions are largely based on linear elastic fracture mechanics, we present here a detailed derivation of near-tip fields for mode I linear elastic fatigue cracks.

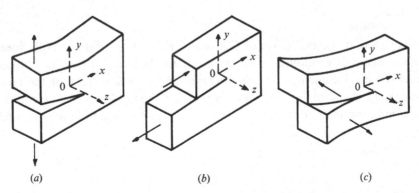

(a)   (b)   (c)

Fig. 9.4. The three basic modes of fracture. (*a*) Tensile opening (mode I). (*b*) In-plane sliding (mode II). (*c*) Anti-plane shear (mode III).

9.3.2 The plane problem

Consider a semi-infinite crack in an infinite plate of an isotropic and homogeneous solid (Fig. 9.5) in an attempt to develop crack solutions for plane strain and generalized plane stress. For the plane problem (i.e. for modes I and II), the in-plane displacements u_r and u_θ in the radial and angular coordinate directions, respectively, of a crack tip element are functions only of the polar coordinates r and θ, centered at the crack tip. In the absence of body forces, the equilibrium equations (see Section 1.4) in polar coordinates are

$$\frac{\partial \sigma_{rr}}{\partial r} + \frac{1}{r}\frac{\partial \sigma_{r\theta}}{\partial \theta} + \frac{\sigma_{rr} - \sigma_{\theta\theta}}{r} = 0,$$

$$\frac{\partial \sigma_{r\theta}}{\partial r} + \frac{1}{r}\frac{\partial \sigma_{\theta\theta}}{\partial \theta} + \frac{2\sigma_{r\theta}}{r} = 0, \qquad (9.20)$$

where r and θ are the polar coordinates shown in Fig. 9.5. The in-plane strain components are related to the in-plane radial and angular displacements according to

$$\epsilon_{rr} = \frac{\partial u_r}{\partial r},$$

$$\epsilon_{\theta\theta} = \frac{u_r}{r} + \frac{1}{r}\frac{\partial u_\theta}{\partial \theta},$$

$$\epsilon_{r\theta} = \left(\frac{1}{2}\right) \times \left\{ \frac{1}{r}\frac{\partial u_r}{\partial \theta} + \frac{\partial u_\theta}{\partial r} - \frac{u_\theta}{r} \right\}. \qquad (9.21)$$

The condition of strain compatibility (in polar coordinates) requires that

$$\frac{\partial^2 \epsilon_{\theta\theta}}{\partial r^2} + \frac{2}{r}\frac{\partial \epsilon_{\theta\theta}}{\partial r} - \frac{1}{r}\frac{\partial^2 \epsilon_{r\theta}}{\partial r \partial \theta} - \frac{1}{r^2}\frac{\partial \epsilon_{r\theta}}{\partial \theta} + \frac{1}{r^2}\frac{\partial^2 \epsilon_{rr}}{\partial \theta^2} - \frac{1}{r}\frac{\partial \epsilon_{rr}}{\partial r} = 0. \qquad (9.22)$$

The components of stress and strain in the r–θ plane are related by Hooke's law wherein, for plane stress ($\sigma_{zz} = 0$),

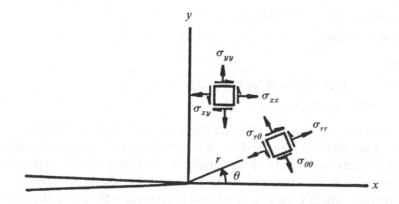

Fig. 9.5. Coordinate system and stresses in the near-tip region of a crack in a plate.

$$E\epsilon_{rr} = \sigma_{rr} - \nu\sigma_{\theta\theta},$$

$$E\epsilon_{\theta\theta} = \sigma_{\theta\theta} - \nu\sigma_{rr},$$

$$2\mu\epsilon_{r\theta} = \mu\gamma_{r\theta} = \sigma_{r\theta}, \qquad (9.23)$$

where μ is the shear modulus and, for plane strain ($\epsilon_{zz} = 0$),

$$2\mu\epsilon_{rr} = (1 - \nu)\sigma_{rr} - \nu\sigma_{\theta\theta},$$

$$2\mu\epsilon_{\theta\theta} = (1 - \nu)\sigma_{\theta\theta} - \nu\sigma_{rr},$$

$$2\mu\epsilon_{r\theta} = \sigma_{r\theta}. \qquad (9.24)$$

For the plane problem, the equations of equilibrium, Eqs. 9.20, are satisfied when the stress components are expressed by the Airy stress function χ through

$$\sigma_{rr} = \frac{1}{r}\frac{\partial\chi}{\partial r} + \frac{1}{r^2}\frac{\partial^2\chi}{\partial\theta^2}, \quad \sigma_{\theta\theta} = \frac{\partial^2\chi}{\partial r^2}, \quad \sigma_{r\theta} = -\frac{\partial}{\partial r}\left(\frac{1}{r}\frac{\partial\chi}{\partial\theta}\right). \qquad (9.25)$$

The compatibility condition, Eq. 9.22, when expressed in terms of the Airy stress function, satisfies the biharmonic equation,

$$\nabla^2(\nabla^2\chi) = 0, \quad \nabla^2 \equiv \frac{\partial^2}{\partial r^2} + \frac{1}{r}\frac{\partial}{\partial r} + \frac{1}{r^2}\frac{\partial^2}{\partial\theta^2}. \qquad (9.26)$$

The boundary conditions for the plane problem of the plate containing a traction-free crack are

$$\sigma_{\theta\theta} = \sigma_{r\theta} = 0 \quad \text{for} \quad \theta = \pm\pi. \qquad (9.27)$$

The choice of the Airy stress function for the present crack problem should be such that χ has a singularity at the crack tip and is single-valued. A possible form of χ which satisfies this requirement is

$$\chi = r^2 p(r, \theta) + q(r, \theta), \qquad (9.28)$$

where p and q are harmonic functions of r and θ which satisfy the Laplace equations $\nabla^2 p = 0$ and $\nabla^2 q = 0$.

Following the approach of Williams (1957), we consider solutions of separable form for the Airy stress function, $\chi = R(r)\Theta(\theta)$, based on

$$p = A_1 r^\lambda \cos\lambda\theta + A_2 r^\lambda \sin\lambda\theta,$$

$$q = B_1 r^{(\lambda+2)} \cos(\lambda+2)\theta + B_2 r^{(\lambda+2)} \sin(\lambda+2)\theta, \qquad (9.29)$$

which lead to

$$\chi = r^{(\lambda+2)}[A_1 \cos\lambda\theta + B_1 \cos(\lambda+2)\theta]$$

$$+ r^{(\lambda+2)}[A_2 \sin\lambda\theta + B_2 \sin(\lambda+2)\theta]. \qquad (9.30)$$

This equation consists of a symmetric part (the term within the first set of brackets on the right of the equality sign) and an anti-symmetric part (the term within the second set of brackets on the right of the equality sign). The symmetric part, which is an even function of θ, provides the mode I solution for crack tip fields and the antisymmetric part, which is an odd function of θ, provides the mode II solution.

Taking only the first term here to obtain the mode I fields,

$$\sigma_{\theta\theta} = \frac{\partial^2 \chi}{\partial r^2} = (\lambda + 2)(\lambda + 1)\, r^{\lambda}[A_1 \cos \lambda\theta + B_1 \cos(\lambda + 2)\theta],$$

$$\sigma_{r\theta} = -\frac{\partial}{\partial r}\left(\frac{1}{r}\frac{\partial \chi}{\partial \theta}\right)$$

$$= (\lambda + 1)\, r^{\lambda}[\lambda A_1 \sin \lambda\theta + (\lambda + 2)B_1 \sin(\lambda + 2)\theta]. \tag{9.31}$$

Now applying the boundary conditions, Eq. 9.27, one obtains

$$(A_1 + B_1)\cos \lambda\pi = 0,$$

$$[\lambda A_1 + (\lambda + 2)B_1]\sin \lambda\pi = 0. \tag{9.32}$$

The admissible cases are: (i) $\cos \lambda\pi = 0$ and hence

$$\lambda = \frac{2Z + 1}{2}, \tag{9.33}$$

where Z is an integer including zero, and

$$B_1 = -\frac{\lambda}{\lambda + 2}\, A_1, \tag{9.34}$$

or (ii) $\sin \lambda\pi = 0$ and hence

$$\lambda = Z \qquad \text{and} \qquad B_1 = -A_1. \tag{9.35}$$

Since the governing equations 9.20–9.24 are linear, any linear combination of the admissible solutions also provides a solution. Hence, from Eqs. 9.33–9.35,

$$\lambda = \frac{Z}{2}, \tag{9.36}$$

where Z is a positive or negative integer, including zero. Although, from a purely mathematical standpoint, there is no basis to reject any value of λ, the solution can be chosen to be of the lowest order singularity which is consistent with physical arguments. From Eqs. 9.31, it is seen that $\sigma_{ij} \sim r^{\lambda}$ and $\epsilon_{ij} \sim r^{\lambda}$. Therefore, the strain energy density is given by

$$\phi = \frac{1}{2}\, \sigma_{ij}\epsilon_{ij} \sim r^{2\lambda}. \tag{9.37}$$

The total strain energy within any annular area of inner and outer radii r_0 and R, respectively, centered at the crack tip, is

$$\Phi = \int_0^{2\pi}\int_{r_0}^{R} \frac{1}{2}\, \sigma_{ij}\epsilon_{ij}\, r\, dr\, d\theta \sim \int_0^{2\pi}\int_{r_0}^{R} r^{(2\lambda+1)}\, dr\, d\theta. \tag{9.38}$$

Invoking the argument that Φ should be bounded (i.e. $\Phi < \infty$) as $r_0 \to 0$, we see that $\lambda > -1$. ($\lambda = -1$ gives the trivial solution that $\sigma_{ij} = 0$.) Note that, since the displacements $u_i \sim r^{(\lambda+1)}$, the boundedness of displacements also requires that $\lambda > -1$. Thus, the physically admissible values of λ are

$$\lambda = -\frac{1}{2},\ 0,\ \frac{1}{2},\ 1,\ \frac{3}{2},\ 2,\ldots, \qquad \text{(or)} \qquad \lambda = \frac{Z}{2}, \tag{9.39}$$

where Z is -1, 0, or a positive integer. Taking the most dominant singular term represented by $\lambda = -1/2$ (and $B_1 = A_1/3$),

$$\chi = r^{3/2} A_1 \left[\cos\frac{\theta}{2} + \frac{1}{3} \cos\frac{3\theta}{2} \right] + O(r^2) + O(r^{5/2}) + \cdots,$$

$$\sigma_{ij} = A_1 r^{-1/2} \tilde{\sigma}_{ij}^{\mathrm{I}}(\theta) + O_{ij}(r^0) + O_{ij}(r^{1/2}) + \cdots . \tag{9.40}$$

The second term on the right hand side of Eq. 9.40, with an exponent of 0 for r, is a nonsingular, but nonvanishing, term. The higher order terms, with exponents greater than zero, vanish as $r \to 0$. Rewriting $A_1 = K_{\mathrm{I}}/\sqrt{2\pi}$,

$$\sigma_{ij} = \frac{K_{\mathrm{I}}}{\sqrt{2\pi r}} \tilde{\sigma}_{ij}^{\mathrm{I}}(\theta) + T\delta_{ix}\delta_{jx}$$

$$+ \text{ (terms which vanish at crack tip)}, \tag{9.41}$$

where there is no summation over x in the second term on the right hand side. K_{I} is known as the *stress intensity factor* for mode I loading. δ_{ij} is the Kronecker delta defined in Section 1.4.1. In terms of the in-plane stress components,

$$\begin{pmatrix} \sigma_{xx} & \sigma_{xy} \\ \sigma_{yx} & \sigma_{yy} \end{pmatrix} = \frac{K_{\mathrm{I}}}{\sqrt{2\pi r}} \begin{pmatrix} \tilde{\sigma}_{xx}^{\mathrm{I}}(\theta) & \tilde{\sigma}_{xy}^{\mathrm{I}}(\theta) \\ \tilde{\sigma}_{yx}^{\mathrm{I}}(\theta) & \tilde{\sigma}_{yy}^{\mathrm{I}}(\theta) \end{pmatrix} + \begin{pmatrix} T & 0 \\ 0 & 0 \end{pmatrix}$$

$$+ \text{ (terms which vanish at crack tip)}, \tag{9.42}$$

where the first term is the leading singular term for linear elastic mode I crack problems. The second term, generally referred to as the 'T term', contains the non-singular stress $\sigma_{xx} = T$ (Williams, 1957; Irwin, 1960; Larsson & Carlsson, 1973; Rice, 1974). For example, a brittle crack of length $2a$ lying on the x–z plane under remotely uniform biaxial stresses σ_{xx}^∞ and σ_{yy}^∞, is subjected to

$$K_{\mathrm{I}} = \sigma_{yy}^\infty \sqrt{\pi a} \quad \text{and} \quad T = \sigma_{xx}^\infty - \sigma_{yy}^\infty. \tag{9.43}$$

Although the leading singular term of the asymptotic solution, Eqs. 9.41–9.43, is adequate for characterizing most linear elastic fatigue crack growth problems, the omission of the T-stress can introduce significant errors in certain fatigue situations. Examples of such situations include: (i) short fatigue cracks, (ii) cracks subjected to mixed-mode loading where the in-plane shear stresses are substantially larger than the tensile stresses, and (iii) small cracks inclined at a small angle to the far-field tensile axis. Furthermore, different geometries of cracked specimens can influence the near-tip yield behavior in different ways because of the differences in the T-stress term. This effect and the attendant influence on fatigue crack closure are considered in Chapter 14.

For the plane problem, the leading terms for mode I stress fields in cartesian coordinates are

$$\begin{Bmatrix} \sigma_{xx} \\ \sigma_{yy} \\ \sigma_{xy} \end{Bmatrix} = \frac{K_{\mathrm{I}}}{\sqrt{2\pi r}} \cos\frac{\theta}{2} \begin{Bmatrix} 1 - \sin\frac{\theta}{2} \sin\frac{3\theta}{2} \\ 1 + \sin\frac{\theta}{2} \sin\frac{3\theta}{2} \\ \sin\frac{\theta}{2} \cos\frac{3\theta}{2} \end{Bmatrix}. \tag{9.44a}$$

When written in cylindrical coordinates, the stress fields for mode I have the follow-ing leading terms:

$$\left\{\begin{array}{c} \sigma_{rr} \\ \sigma_{\theta\theta} \\ \sigma_{r\theta} \end{array}\right\} = \frac{K_I}{\sqrt{2\pi r}} \cos\frac{\theta}{2} \left\{\begin{array}{c} 1 + \sin^2\frac{\theta}{2} \\ \cos^2\frac{\theta}{2} \\ \sin\frac{\theta}{2}\cos\frac{\theta}{2} \end{array}\right\},$$

$$\sigma_{zz} = v_1(\sigma_{xx} + \sigma_{yy}) = v_1(\sigma_{rr} + \sigma_{\theta\theta}),$$

$$\sigma_{xz} = \sigma_{yz} = \sigma_{rz} = \sigma_{\theta z} = 0. \tag{9.44b}$$

The corresponding displacements are

$$\left\{\begin{array}{c} u_x \\ u_y \end{array}\right\} = \frac{K_I}{2E}\sqrt{\frac{r}{2\pi}} \left\{\begin{array}{c} (1+v)\left[(2\kappa-1)\cos\frac{\theta}{2} - \cos\frac{3\theta}{2}\right] \\ (1+v)\left[(2\kappa+1)\sin\frac{\theta}{2} - \sin\frac{3\theta}{2}\right] \end{array}\right\},$$

$$\left\{\begin{array}{c} u_r \\ u_\theta \end{array}\right\} = \frac{K_I}{2E}\sqrt{\frac{r}{2\pi}} \left\{\begin{array}{c} (1+v)\left[(2\kappa-1)\cos\frac{\theta}{2} - \cos\frac{3\theta}{2}\right] \\ (1+v)\left[-(2\kappa-1)\sin\frac{\theta}{2} + \sin\frac{3\theta}{2}\right] \end{array}\right\},$$

$$u_z = -\left(\frac{v_2 z}{E}\right)(\sigma_{xx} + \sigma_{yy}) = -\left(\frac{v_2 z}{E}\right)(\sigma_{rr} + \sigma_{\theta\theta}). \tag{9.45}$$

For plane stress,

$$\kappa = \frac{(3-v)}{(1+v)}, \quad v_1 = 0, \quad v_2 = v,$$

and, for plane strain,

$$\kappa = (3-4v), \quad v_1 = v, \quad v_2 = 0. \tag{9.46}$$

The term K_I in Eqs. 9.44 and 9.45 is the mode I stress intensity factor which incor-porates the boundary conditions of the cracked body and is a function of loading, crack length and geometry. For plane problems, it is independent of the elastic constants.

The near-tip fields for mode II can be derived in a similar fashion by applying the boundary conditions, Eq. 9.27, to the antisymmetric part of the Airy stress function, Eq. 9.30. The resulting asymptotic solutions for mode II are:

$$\left\{\begin{array}{c} \sigma_{xx} \\ \sigma_{yy} \\ \sigma_{xy} \end{array}\right\} = \frac{K_{II}}{\sqrt{2\pi r}} \left\{\begin{array}{c} -\sin\frac{\theta}{2}\left(2 + \cos\frac{\theta}{2}\cos\frac{3\theta}{2}\right) \\ \sin\frac{\theta}{2}\cos\frac{\theta}{2}\sin\frac{3\theta}{2} \\ \cos\frac{\theta}{2}\left(1 - \sin\frac{\theta}{2}\sin\frac{3\theta}{2}\right) \end{array}\right\}, \tag{9.47a}$$

and, in cylindrical coordinates,

$$\begin{Bmatrix} \sigma_{rr} \\ \sigma_{\theta\theta} \\ \sigma_{r\theta} \end{Bmatrix} = \frac{K_{II}}{\sqrt{2\pi r}} \begin{Bmatrix} \sin\dfrac{\theta}{2}\left(1 - 3\sin^2\dfrac{\theta}{2}\right) \\[1mm] -3\sin\dfrac{\theta}{2}\cos^2\dfrac{\theta}{2} \\[1mm] \cos\dfrac{\theta}{2}\left(1 - 3\sin^2\dfrac{\theta}{2}\right) \end{Bmatrix},$$

$$\sigma_{zz} = \nu_1(\sigma_{xx} + \sigma_{yy}) = \nu_1(\sigma_{rr} + \sigma_{\theta\theta}),$$

$$\sigma_{xz} = \sigma_{yz} = \sigma_{rz} = \sigma_{\theta z} = 0. \tag{9.47b}$$

$$\begin{Bmatrix} u_x \\ u_y \end{Bmatrix} = \frac{K_{II}}{2E}\sqrt{\frac{r}{2\pi}} \begin{Bmatrix} (1+\nu)\left[(2\kappa+3)\sin\dfrac{\theta}{2} + \sin\dfrac{3\theta}{2}\right] \\[2mm] -(1+\nu)\left[(2\kappa-3)\cos\dfrac{\theta}{2} + \cos\dfrac{3\theta}{2}\right] \end{Bmatrix},$$

$$\begin{Bmatrix} u_r \\ u_\theta \end{Bmatrix} = \frac{K_{II}}{2E}\sqrt{\frac{r}{2\pi}} \begin{Bmatrix} (1+\nu)\left[-(2\kappa-1)\sin\dfrac{\theta}{2} + 3\sin\dfrac{3\theta}{2}\right] \\[2mm] (1+\nu)\left[-(2\kappa+1)\cos\dfrac{\theta}{2} + 3\cos\dfrac{3\theta}{2}\right] \end{Bmatrix},$$

$$u_z = -\left(\frac{\nu_2 z}{E}\right)(\sigma_{xx} + \sigma_{yy}) = -\left(\frac{\nu_2 z}{E}\right)(\sigma_{rr} + \sigma_{\theta\theta}). \tag{9.48}$$

For mode III, it can be shown (e.g., Anderson, 1995) that

$$\begin{Bmatrix} \sigma_{xz} \\ \sigma_{yz} \end{Bmatrix} = \frac{K_{III}}{\sqrt{2\pi r}} \begin{Bmatrix} -\sin\dfrac{\theta}{2} \\[1mm] \cos\dfrac{\theta}{2} \end{Bmatrix},$$

$$\begin{Bmatrix} \sigma_{rz} \\ \sigma_{\theta z} \end{Bmatrix} = \frac{K_{III}}{\sqrt{2\pi r}} \begin{Bmatrix} \sin\dfrac{\theta}{2} \\[1mm] \cos\dfrac{\theta}{2} \end{Bmatrix},$$

$$\sigma_{xx} = \sigma_{yy} = \sigma_{rr} = \sigma_{\theta\theta} = \sigma_{zz} = 0,$$

$$\sigma_{xy} = \sigma_{r\theta} = 0. \tag{9.49}$$

$$u_z = \frac{2K_{III}}{E}\sqrt{\frac{r}{2\pi}}\left\{2(1+\nu)\sin\dfrac{\theta}{2}\right\},$$

$$u_x = u_y = u_r = u_\theta = 0. \tag{9.50}$$

The above singular solutions for all three modes of fracture indicate that the stresses and displacements, respectively, are of the form

$$[\sigma_{ij}]_M = \frac{K_M}{\sqrt{2\pi r}}\,[\tilde{\sigma}_{ij}(\theta)]_M,$$

$$[u_i]_M = \frac{K_M}{2E}\sqrt{\frac{r}{2\pi}}\,[\tilde{u}_i(\theta)]_M, \tag{9.51}$$

where the subscript M refers to the modes of failure, I, II, and III. The appropriate stress intensity factor for each mode is defined as

$$K_I = \lim_{r \to 0} \left\{ \sqrt{2\pi r} \sigma_{yy} \Big|_{\theta=0} \right\},$$

$$K_{II} = \lim_{r \to 0} \left\{ \sqrt{2\pi r} \sigma_{xy} \Big|_{\theta=0} \right\},$$

$$K_{III} = \lim_{r \to 0} \left\{ \sqrt{2\pi r} \sigma_{yz} \Big|_{\theta=0} \right\}. \tag{9.52}$$

Stress intensity factor calibrations for a wide range of specimen and crack geometries, and loading conditions can be found in the Appendix.

9.3.3 Conditions of K-dominance

The stress intensity factors in Eqs. 9.51–9.52 are a measure of the *intensity* of the near-tip fields under linear elastic conditions. The radial component term $\sqrt{2\pi r}$ and the angular component term $\tilde{\sigma}_{ij}(\theta)$ in Eq. 9.51 depend only on the spatial coordinates. These terms determine the *distribution* of the near-tip fields.

There exists an annular zone ahead of the crack tip, known as 'the region of K-dominance', within which the stress intensity factor provides a unique measure of the intensity of stress, strain or deformation. The outer radius of the annular zone is determined by the radial distance at which the approximate, asymptotic singular solutions Eqs. 9.44–9.50 deviate significantly (say, by more than 10%) from the full *elasticity* solutions which include the higher order terms, e.g., the T-term and the nonsingular terms in Eqs. 9.40–9.43. The higher order terms need to be included beyond the annular zone because they begin to influence the overall accuracy of the solutions. The foregoing near-tip solutions were determined on the basis of the assumption of a sharp crack (of tip radius $\to 0$). However, the K-concept holds even when the crack is not sharp and when there is nonlinear deformation in a small zone near the crack tip. In ductile solids, the material at the crack tip yields when the near-tip stresses exceed the flow strength and the linear elastic solutions lose their validity within this plastic zone. Even if the crack tip plastic deformation zone is very small or nonexistent (as, for example, in the case of brittle ceramics), the foregoing continuum solutions are not expected to hold within the near-tip region of intense deformation (such as the zone of microcracking or phase transformation). Thus, the inner radius of the region of K-dominance is dictated by the size scale of microscopic failure processes (generally referred to as the 'process zone'). The near-tip fields within the plastic zone will be considered in Section 9.7.

The usefulness of the K-fields to characterize the onset or continuation of crack advance in materials that undergo inelastic deformation, such as plasticity, creep, microcracking or phase transformations, is predicated on conformity to the so-called 'small-scale yielding' condition. This condition requires that the crack tip zone of inelastic deformation, whatever its origin, be confined well inside the region of

K-dominance over which the asymptotic results, Eqs. 9.44–9.50, provide a reasonable approximation to the full solution. As alluded to earlier, an understanding of the conditions of K-dominance is essential for the characterization of fatigue fracture involving highly crystallographic crack growth, mixed-mode loading conditions or small fatigue flaws. A detailed discussion of each of these cases will be taken up in later chapters.

9.3.4 Fracture toughness

In linear elastic fracture mechanics, the initiation of crack advance under monotonic, quasi-static loading conditions is characterized by the critical value of the stress intensity factor, K_c. The value of K_c is a function of the mode of loading, the chemical environment, the material microstructure, the test temperature, strain rate, and the state of stress (i.e. plane stress or plane strain). The experimental test specimens used for the determination of critical stress intensity factor must conform to the requirements of small-scale yielding and other conditions of K-dominance which are spelled out in detail in the fracture test standard E-399 developed by the American Society for Testing and Materials (Philadelphia) in 1974. The critical value of the mode I stress intensity factor measured under plane strain conditions is commonly referred to as the *fracture toughness*, K_{Ic}, of the material at the particular test temperature. (Plane strain conditions are assumed to exist in the fracture toughness test specimen when the thickness of the test specimen is at least about 25 times the monotonic plastic zone size, which is defined in Section 9.5.) The corresponding fracture toughness values in the sliding and tearing modes are designated as K_{IIc} and K_{IIIc}, respectively.

9.3.5 Characterization of fatigue crack growth

Under cyclic loading conditions, the onset of crack growth from a preexisting flaw or defect can occur at (maximum) stress intensity values that are well below the quasi-static fracture toughness. For conditions of small-scale yielding, where the nonlinear zone at the crack tip is a mere perturbation in an otherwise elastic material, Paris, Gomez & Anderson (1961) and Paris & Erdogan (1963) postulated that the growth of a crack under *cyclic* loading should be governed by the 'law',

$$\frac{da}{dN} = C\Delta K^m, \tag{9.53}$$

where da/dN is the change in the length of the fatigue crack per load cycle (a is the crack length and N is the number of fatigue cycles) and ΔK is the stress intensity factor range defined as

$$\Delta K = K_{max} - K_{min}. \tag{9.54}$$

K_{max} and K_{min}, respectively, are the maximum and minimum stress intensity factors corresponding to the maximum load, P_{max} (or maximum nominal stress, σ_{max}) and the minimum load, P_{min} (or minimum nominal stress, σ_{min}). Recall that $K_{max} = Y\sigma_{max}\sqrt{\pi a}$ and $K_{min} = Y\sigma_{min}\sqrt{\pi a}$ for a center-cracked plate containing a crack of length $2a$ which is subjected to tensile fatigue with a far-field stress range, $\Delta\sigma = \sigma_{max} - \sigma_{min}$. Y is the finite size correction factor for the plate. The terms C and m in Eq. 9.53 are empirical constants which are functions of the material properties and microstructure, fatigue frequency, mean stress or load ratio, environment, loading mode, stress state and test temperature. The *empirical* crack growth law, Eq. 9.53, due to Paris *et al.* is the most widely used form of characterizing fatigue crack growth rates for a vast spectrum of materials and test conditions. Equation 9.53 also represents one of the most useful applications of the theory of linear elastic fracture mechanics. Further details of the fracture mechanics-based approach to characterizing fatigue will be considered in subsequent chapters.

9.4 Equivalence of $\mathcal{G}$ and K

The stress intensity factor approach to fracture has a direct equivalence to the energy approach. Consider the definition of energy release rate for the displacement boundary value problem involving the fixed-grip case, Eq. 9.17. Here, the change in strain energy (per unit thickness of the crackfront) as the length of the crack is increased from a to $a + \delta a$ is given by

$$\delta\Phi \approx -2\int_a^{a+\delta a} \frac{1}{2}\left(\sigma_{yy}u_y + \sigma_{xy}u_x + \sigma_{zy}u_z\right)dx. \tag{9.55}$$

The factor 2 before the integral sign appears because of the displacement of the two opposing crack surfaces, and the factor $1/2$ after the integral sign is introduced following the assumption of proportionality between the stresses and displacements. Substituting the appropriate stresses σ_{ij} for $r = x - a$ and $\theta = 0$ and displacements u_i for $r = a + da - x$ and $\theta = 0$ from the previous section into Eq. 9.55, it is readily seen, in the limit of $\delta a \to 0$, that

$$\mathcal{G} = -\left.\frac{\partial\Phi}{\partial a}\right|_u = \mathcal{G}_I + \mathcal{G}_{II} + \mathcal{G}_{III}. \tag{9.56}$$

The energy release rate $\mathcal{G}$ and the stress intensity factors K_I, K_{II} and K_{III} in the three modes of fracture are uniquely related. For the general three-dimensional case involving plane strain and anti-plane strain loading,

$$\mathcal{G} = \frac{(1 - \nu^2)}{E}\left(K_I^2 + K_{II}^2\right) + \frac{(1 + \nu)}{E}K_{III}^2 \tag{9.57}$$

and, for plane stress,

$$\mathcal{G} = \frac{1}{E}\left(K_I^2 + K_{II}^2\right). \tag{9.58}$$

Note that, when the crack advances in its own plane, i.e. for self-similar, coplanar crack growth, the energy release rates for the different modes of fracture are simply additive. This relationship also provides a means for developing *mode-invariant* criteria for the onset of failure under multiaxial loading conditions.

9.4.1 Example problem: $\mathcal{G}$ and K for the DCB specimen

Problem:

The double-cantilever beam (DCB) specimen is one of the widely used fracture test specimens for both brittle and ductile solids.

(i)   Using the concepts discussed in Section 9.2, calculate the energy release rate for the DCB specimen subjected to a mode I load as shown in Fig. 9.6.

(ii) Using the result from (i) and Eqs. 9.57 and 9.58, determine the stress intensity factor for the DCB specimen.

(iii) Discuss how stress intensity factor calibrations may be obtained experimentally for the DCB specimen using measurements of compliance changes.

Solution:

(i)   As illustrated in Fig. 9.6(a), the out-of-plane end deflection $u/2$ of a slender cantilever beam ($a \gg 2h$) of length a and height h subject to an end load F is given by beam theory:

$$\frac{u}{2} = \frac{Fa^3}{3EI}, \qquad I = \frac{Bh^3}{12}. \tag{9.59}$$

From Eq. 9.17 and Fig. 9.6(c),

$$\mathcal{G} = \frac{F^2}{2B}\frac{\mathrm{d}C(a)}{\mathrm{d}a} = \frac{F^2 a^2}{BEI} = \frac{12F^2 a^2}{EB^2 h^3}. \tag{9.60}$$

(ii) From Eqs. 9.57, 9.58 and 9.60, the mode I stress intensity factor for plane stress is derived as

$$\frac{K_{\mathrm{I}}^2}{E} = \mathcal{G} = \frac{12F^2 a^2}{EB^2 h^3}, \qquad K_{\mathrm{I}} = 2\sqrt{3}\cdot\frac{a}{h^{3/2}}\cdot\frac{F}{B}, \tag{9.61}$$

and, for plane strain, as

$$\frac{K_{\mathrm{I}}^2}{E}(1 - \nu^2) = \mathcal{G} = \frac{12F^2 a^2}{EB^2 h^3}, \qquad K_{\mathrm{I}} = \frac{2\sqrt{3}}{\sqrt{1 - \nu^2}}\cdot\frac{a}{h^{3/2}}\cdot\frac{F}{B}. \tag{9.62}$$

(iii) From Eq. 9.60, for plane strain,

$$\frac{K_{\mathrm{I}}^2}{E}(1 - \nu^2) = \mathcal{G} = \frac{1 - \nu^2}{E}k_{\mathrm{I}}^2(a)F^2 = \frac{1}{2}\frac{\mathrm{d}c(a)}{\mathrm{d}a}F^2. \tag{9.63}$$

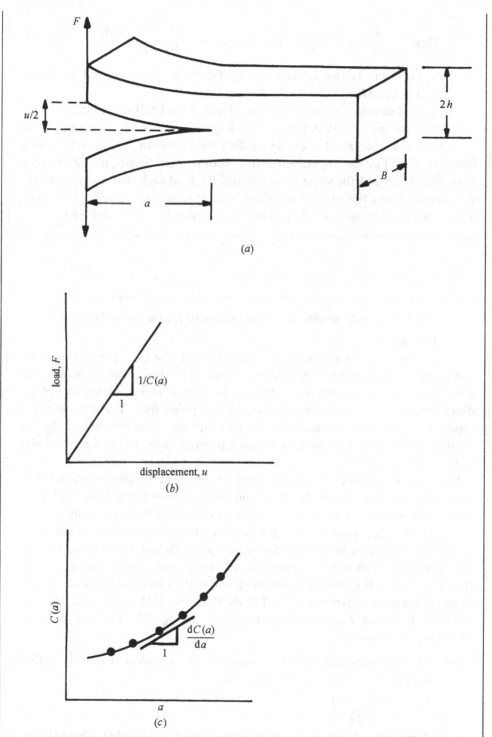

Fig. 9.6. (a) The double cantilever beam specimen. (b) A plot of the load versus the displacement showing the compliance. (c) Compliance change as a function of crack length.

Thus, $\dfrac{dc(a)}{da} = 2\dfrac{1-v^2}{E}k_f^2(a), \quad k_f(a) = \sqrt{\dfrac{E}{2(1-v^2)} \cdot \dfrac{dc(a)}{da}},$ (9.64)

where $k_f(a)$ is the stress intensity factor per unit applied load and $c(a) = C(a)/B$.

In order to calibrate the stress intensity factor for a DCB specimen, prepare a number of specimens with different crack lengths, a. Use a clip gage to measure the opening displacement u at the loading point, shown in Fig. 9.6(a), as a function of F. For each specimen with a known crack length, plot F versus u as in Fig. 9.6(c) and the slope $C(a)$. Control the load such that no crack extension occurs during this experiment. From similar results on multiple DCB specimens, plot $c(a)$ versus a. Find $k_f(a)$ and $K_I(a)$ using Eqs. 9.63 and 9.64.

9.4.2 Example problem: Stress intensity factor for a blister test

Problem:

The interface toughness of an adhesive is to be estimated on the basis of a proposed method which entails the bonding of a thin elastic disk with the adhesive on to a rigid substrate. Pressure is applied to the bonded side of the disk through a tiny hole in the substrate, by pumping a fluid. This pressurization results in the partial debonding of the disk and the formation of a 'bulge' or 'blister' in the disk. This method is hence referred to as the pressurized 'bulge test' or 'blister test'.

Consider a two-dimensional version of this test which is shown in Fig. 9.7. A thin beam is adhesively bonded to a rigid surface through which an incompressible fluid is forced resulting in a blister of half-width l. Assuming that all concepts of linear elastic fracture mechanics may be extended to this case, apply the compliance method to determine the tensile 'crack-tip' stress intensity factor in terms of the blister dimension l and the piston displacement q. Assume that $q = 0$ when the beam is undeflected, and that all length dimensions normal to the plane of the figure are unity. Let the thickness of the beam and its elastic modulus be h and E, respectively, as shown in Fig. 9.7. You are given the following information.

(i)   The transverse deflection of an elastic double cantilever beam of length $2l$ subjected to a uniform pressure p is

$$w(x) = \frac{p(l^2 - x^2)^2}{2Eh^3}.$$ (9.65)

(ii) A measure of compliance $c(l)$ may be defined as $q = c(l)Q$, where Q is the force acting on the piston per unit thickness as shown in the figure.

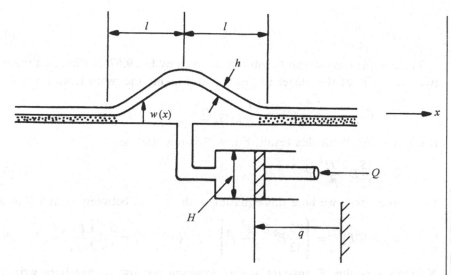

Fig. 9.7. A schematic of the pressurized bulge or blister test and the associated nomenclature.

(iii)

$$\int_{-l}^{l} w(x)\mathrm{d}x = qH. \tag{9.66}$$

(iv)   From the compliance interpretation of the energy release rate, $\mathcal{G}$, it is known that

$$\mathcal{G} = \frac{1}{2}\frac{\mathrm{d}c}{\mathrm{d}l}Q^2. \tag{9.67}$$

Solution:

The compliance of the system is defined as

$$q = c(l)Q. \tag{9.68}$$

Observe that (i) the pressurized liquid is incompressible, (ii) the pressure under the blister is equal to the pressure in the piston, and (iii) the pressure in the piston is directly related to Q by

$$p = \frac{Q}{H}. \tag{9.69}$$

Substituting Eq. 9.65 in Eq. 9.66, we obtain

$$p = \frac{15}{8}\frac{EqHh^3}{l^5}. \tag{9.70}$$

Now substitute Eq. 9.69 into Eq. 9.70 to get

$$q = \frac{8}{15}\frac{1}{E}\frac{l^5}{h^3 H^2}Q. \tag{9.71}$$

Arranging this equation into the form of Eq. 9.68, we note that

$$c(l) = \frac{8}{15} \frac{1}{E} \frac{l^5}{h^3 H^2}. \tag{9.72}$$

The compliance interpretation of $\mathcal{G}$ is given by Eq. 9.67. Noting that there are two crack tips at the blister in Fig. 9.7 and taking the result from Eq. 9.72,

$$\mathcal{G} = \mathcal{G}_{\text{per tip}} = \frac{1}{4} \frac{dc}{dl} Q^2 = \frac{4}{6} \frac{1}{E} \frac{l^4}{h^3 H^2} Q^2. \tag{9.73}$$

But $q = c(l)Q$. With this result, Eq. 9.73 is rewritten as

$$\mathcal{G} = \frac{75}{32} E \frac{h^3 H^2}{l^6} q^2. \tag{9.74}$$

For plane stress, we take this equation with the link between K_{I} and $\mathcal{G}$ to get

$$K_{\text{I}} = \{\mathcal{G}E\}^{1/2} = \left\{ \frac{75}{32} E^2 \frac{h^3 H^2}{l^6} q^2 \right\}^{1/2}, \quad \text{(or)} \quad K_{\text{I}} = \frac{5}{4} \sqrt{\frac{3}{2}} \frac{H h \sqrt{h}}{l^3} q E. \tag{9.75}$$

Young's modulus E appears in this equation because K_{I} has been written in terms of the displacement q. Instead, if use is made of Eq. 9.71, the stress intensity factor becomes

$$K_{\text{I}} = \sqrt{\frac{2}{3}} \frac{l^2}{H h \sqrt{h}} Q. \tag{9.76}$$

This equation gives us the desired result.

9.5 Plastic zone size in monotonic loading

9.5.1 The Irwin approximation

Estimates of the boundary of the plastic zone ahead of a crack in a ductile solid can be derived by considering the crack tip zone within which the von Mises equivalent stress (which is defined in Eq. (1.22) and calculated from the stress analysis of Section 9.3) exceeds the tensile flow stress σ_y (Irwin, 1960). The extent of the plastic zone r_p ahead of the crack tip ($\theta = 0$) is proportional to the square of the stress intensity factor. In mode I,

$$r_p = \frac{1}{3\pi} \left(\frac{K_{\text{I}}}{\sigma_y} \right)^2, \qquad \text{for plane strain,}$$

$$r_p = \frac{1}{\pi} \left(\frac{K_{\text{I}}}{\sigma_y} \right)^2, \qquad \text{for plane stress.} \tag{9.77}$$

Precise analyses of the plastic zone size and shape in modes I and II in strain-hardening solids are discussed in Section 9.10 for plane strain and plane stress.

Following similar arguments, the plastic zone size ahead of a mode III crack is found to be

$$r_{\rm p} = \frac{1}{\pi}\left(\frac{K_{\rm III}}{\tau_{\rm y}}\right)^2, \tag{9.78}$$

where τ_y is the shear yield stress of the material.

9.5.2 The Dugdale model

The size of the yield zone ahead of a mode I crack in a thin plate of an elastic–perfectly plastic solid (subject to plane stress deformation) was estimated by Dugdale (1960). In the Dugdale model, the plastic region is envisioned as a narrow strip (of near-zero height) which extends a distance $r_{\rm p}$ ahead of the crack tip and is loaded by the traction $\sigma_{yy} = \sigma_y$ over the length $r_{\rm p}$ (Fig. 9.8). If σ_{yy} were to be zero over the whole length $|x| < a + r_{\rm p},\ y = 0$, a far-field tensile stress σ^∞ would produce a positive stress intensity factor $K_{\rm I} = \sigma^\infty\sqrt{\pi(a + r_{\rm p})}$ (in an infinitely large plate). If the traction $\sigma_{yy} = \sigma_y$ were to be applied simultaneously along the length of the strip $a < |x| < a + r_{\rm p}$, it would superimpose a negative stress intensity factor $K_{\rm I}''$ on $K_{\rm I}'$ where

$$K_{\rm I}'' = -\sigma_y\sqrt{\pi(a + r_{\rm p})} + 2\sigma_y\sqrt{\frac{a + r_{\rm p}}{\pi}}\sin^{-1}\left(\frac{a}{a + r_{\rm p}}\right). \tag{9.79}$$

The requirement of bounded stresses at the point $x = a + r_{\rm p}$ provides the condition that $K_{\rm I}' + K_{\rm I}'' = 0$. Solving for $r_{\rm p}$, one finds that

$$\frac{r_{\rm p}}{a} = \sec\left(\frac{\pi\sigma^\infty}{2\sigma_y}\right) - 1. \tag{9.80}$$

For $\sigma^\infty \ll \sigma_y$ and hence for $r_{\rm p} \ll a$, this equation asymptotically leads to a plastic zone size

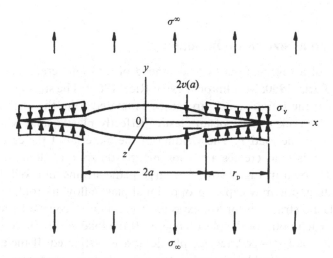

Fig. 9.8. A schematic representation of the Dugdale plastic zone model.

$$r_p = \frac{\pi}{8}\left(\frac{K_I}{\sigma_y}\right)^2. \tag{9.81}$$

This asymptotically exact result due to Dugdale compares well with the Irwin approximation, Eq. 9.77, for plane stress.

In the above model, one notes that an opening displacement $\delta = 2v(a)$ at $x = \pm a$ and $y = 0$ exists (which may be regarded as a consequence of necking ahead of the crack). It can be shown that the crack tip opening displacement takes the form

$$\delta_t = \frac{8\sigma_y a}{\pi E}\ln\left[\sec\left(\frac{\pi\sigma^\infty}{2\sigma_y}\right)\right], \tag{9.82}$$

or asymptotically, when $\sigma^\infty \ll \sigma_y$,

$$\delta_t = \frac{K_I^2}{\sigma_y E}. \tag{9.83}$$

9.5.3 The Barenblatt model

The strip zone model due to Barenblatt (1962) provides an analogue for ideally brittle materials of the Dugdale plastic yield zone analysis. For the perfectly brittle solid, one may envision, in Fig. 9.8, that the crack face traction $\sigma_{yy} = \sigma_{th}$, where σ_{th} is the theoretical bond rupture strength $\approx E/10$ (see, for example, Lawn, 1993). The critical condition for the fracture of the brittle solid may then be expressed in terms of the critical size of the cohesive zone at the crack tip, $r_p = r_{c0}$, or in terms of the critical crack opening displacement $\delta_c = 2v_c$ (Rice, 1968) such that

$$\mathcal{G}_c = 2\int_0^{v_c}\sigma_{yy}dv = \frac{8\sigma_{th}^2 r_{c0}}{\pi E} = 2\gamma_s. \tag{9.84}$$

9.6 Plastic zone size in cyclic loading

The existence of a reversed plastic zone ahead of a fatigue crack has long been recognized (e.g., Paris, 1960; McClintock, 1963; Rice, 1967). The stresses within the reverse yield zone at the fatigue crack tip can be estimated from the analysis of Rice (1967). Consider a cracked plate of an elastic–perfectly plastic solid which is subjected to a far-field tensile load P. The yielding of the material at the crack tip under the influence of this load creates a monotonic plastic zone of dimension r_p given by Eq. 9.66 and shown in Fig. 9.9(a). For large plane strains and well-developed plane stress yielding, one may expect proportional plastic flow to hold, i.e. the components of the plastic strain tensor are expected to remain in constant proportion to one another at each point of the plastic region. If the load P is reduced by an amount ΔP to a tensile load $P - \Delta P$, reverse plastic flow is instigated. If the extent of crack tip blunting is assumed to be negligible, the infinitely large stress concentra-

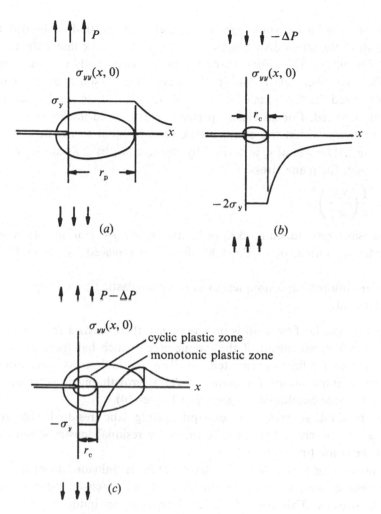

Fig. 9.9. Schematic representation of the development of cyclic plastic zone upon unloading. (After Rice, 1967.) (a) Monotonic plastic zone created by a far-field load P. (b) Stress distribution due to the reduction of the load by ΔP which, when superimposed with (a), gives the result in (c).

tion factor at the tip of the sharp crack leads to the formation of a reversed flow zone which is embedded within the monotonic plastic zone.

For proportional loading, the changes in the near-tip fields due to the reduction of the load are given by the solution derived earlier for monotonic loading (Section 9.3) with the exception that the loading parameter is replaced by the load range ΔP and that the yield stress and strain are replaced by *twice* their values corresponding to the load P. This modification is introduced to obtain the correct values of stresses in the reversed flow zone after subtracting the changes due to the load reduction ΔP, Fig. 9.9(b). If closure of the crack faces is not encountered, the superposition of the near-tip stresses in the fully loaded state, Fig. 9.9(a), and in the partially unloaded state, Fig. 9.9(b), leads to the stress distribution at the tip of the crack under the far-field load $P - \Delta P$, Fig. 9.9(c). Thus, for a crack which is only partially unloaded from a far-field tensile load, there exists within the monotonic plastic zone a region of reversed flow (termed the 'cyclic plastic zone') of size r_c in which *residual* compressive stresses are induced. For an elastic–perfectly plastic solid undergoing proportional flow, the stress within the cyclic plastic zone is equal to the flow stress in compression $(-\sigma_y)$. The size of r_c is derived by replacing K_I by ΔK_I and σ_y by $-2\sigma_y$ in Eq. 9.66 so that, for plane stress,

$$r_c \approx \frac{1}{\pi}\left(\frac{\Delta K_I}{2\sigma_y}\right)^2. \tag{9.85}$$

For zero–tension–zero loading, $\Delta K_I = K_I$ and $r_c = r_p/4$. For materials which cyclically harden or soften, σ_y in Eq. 9.85 should be replaced by the cyclic yield strength, σ_y'.

There are some interesting consequences of reversed plastic flow during unloading from a far-field load:

(1) Even after the far-field load is fully removed, there exists a zone of *residual* compressive stress ahead of the fatigue crack which has (previously) been subjected to far-field *cyclic* tension. This residual stress zone can have important implications for transient crack growth phenomena observed under variable amplitude fatigue (see Chapter 14).

(2) Since residual stresses are self-equilibrating, the residual compressive stresses at the crack tip must be offset by residual tensile stresses away from the crack tip.

(3) If a nonclosing flaw (such as a sharp notch) is unloaded from a far-field compressive load, reversed plastic flow at the notch-tip creates residual *tensile* stresses. This zone of residual tension can induce stable mode I fatigue crack growth in notched plates loaded in uniaxial cyclic compression (see Chapter 4).

(4) A *cyclic* variation in the stress intensity factor ΔK_I, from 0 to K_I, gives rise to a cyclic crack tip opening displacement $\Delta \delta_t$, which is one half of the total opening displacement δ under a monotonic stress intensity factor K_I, Eq. 9.83:

$$\Delta \delta_t \approx \frac{\Delta K_I^2}{2\sigma_y E}. \tag{9.86}$$

(5) In the absence of crack closure, the value of r_c and cyclic variations in stresses, strains and displacements depend *only* on ΔP and are independent of the maximum load P.

(6) When crack closure does not occur, the plastic superposition is valid up to the point when $r_c = r_p$, that is, for complete load reversal involving equal tension–compression fatigue.

Whereas the cyclic plastic zone size for a *stationary fatigue crack* in an elastic–perfectly plastic solid is one-quarter the size of the monotonic plastic zone, a plane stress analysis by Budiansky & Hutchinson (1978) shows that the reversed yield zone for an *extending fatigue crack* is less than 10% of the Dugdale monotonic yield zone size. This point is taken up for further discussion in Chapter 14. Analyses of cyclic damage zones ahead of fatigue cracks in ceramics are discussed in Chapter 11.

9.7 Elastic–plastic fracture mechanics

The stress intensity factor K provides a unique characterization of the near-tip fields under small-scale yielding conditions, while the corresponding loading parameter for the characterization of monotonic, nonlinear fracture in rate-independent materials is the J-integral proposed by Rice (1968). Although some of the features of this integral were embedded in the energy concepts derived by Eshelby (1956) and were discussed independently by Sanders (1960) and Cherepanov (1969), the particular form of this line integral proposed by Rice (1968) has led to the unifying theoretical basis for nonlinear fracture mechanics.

9.7.1 The J-integral

Consider a cracked body subjected to a monotonic load, Fig. 9.10. Assuming that the tractions **T** are independent of crack size and that the crack

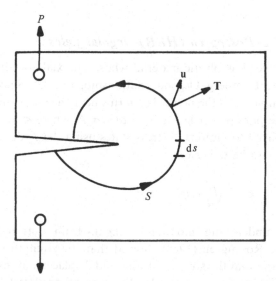

Fig. 9.10. A contour around a crack tip and the nomenclature used in the definition of the J integral.

faces are traction-free, the line integral J along any contour Γ which encircles the crack tip is given by

$$J = \int_{\Gamma} \left(w\,\mathrm{d}y - \mathbf{T} \cdot \frac{\partial \mathbf{u}}{\partial x}\,\mathrm{d}s \right), \tag{9.87}$$

where $\mathbf{u}$ is the displacement vector, y is the distance along the direction normal to the plane of the crack, s is the arc length along the contour, $\mathbf{T}$ is the traction vector and w is the strain energy density of the material. The stresses σ_{ij} are related to w by the relation $\sigma_{ij} = \partial w / \partial \epsilon_{ij}$. For a material which is characterized by linear or nonlinear elastic behavior (i.e. by deformation plasticity), J is *independent* of the path Γ taken to compute the integral.

Rice (1968) showed that J is the rate of change of potential energy (with respect to crack advance) for a nonlinear elastic solid and that J reduces to the energy release rate $\mathcal{G}$ for a linear elastic material:

$$J = \mathcal{G} \equiv -\frac{\partial(\mathrm{PE})}{\partial a}. \tag{9.88}$$

If Γ is regarded as the contour which just encircles the cohesive zone in the Barenblatt model, it is found that J is equal to the energy release rate given in Eq. 9.84: $J = \mathcal{G}_{\mathrm{c}} = 2\gamma_{\mathrm{s}}$. Furthermore, if the line integral is applied to the Dugdale model, the following relationship between J and the crack tip opening displacement δ_{t}, Eq. 9.83, is obtained:

$$J = \sigma_{\mathrm{y}}\delta_{\mathrm{t}}. \tag{9.89}$$

9.7.2 Hutchinson–Rice–Rosengren (HRR) singular fields

Consider an elastic–power law plastic material whose uniaxial constitutive response is characterized by the Ramberg–Osgood relationship, Eq. 3.4. Since the elastic strains are negligible compared to the power law terms near the crack tip, Eq. 3.4 can be approximated by the pure power law, $\epsilon/\epsilon_{\mathrm{y}} = \alpha(\sigma/\sigma_{\mathrm{y}})^{n}$, where $\epsilon_{\mathrm{y}} = \sigma_{\mathrm{y}}/E$. This relationship, when generalized to multiaxial stress states using the J_2 deformation theory (see Section 1.4.3), can be written as

$$\frac{\epsilon_{ij}}{\epsilon_{\mathrm{y}}} = \frac{3\alpha}{2}\left(\frac{\sigma_{\mathrm{e}}}{\sigma_{\mathrm{y}}}\right)^{(n-1)}\frac{s_{ij}}{\sigma_{\mathrm{y}}}, \qquad \sigma_{\mathrm{e}} = \sqrt{\frac{3}{2}s_{ij}s_{ij}}. \tag{9.90}$$

For nonlinear elastic solids undergoing monotonic, small-strain deformation, Hutchinson (1968) and Rice & Rosengren (1968) showed that the strength of the near-tip fields is the J-integral and that the stresses, strains and displacements exhibit $r^{-1/(n+1)}$, $r^{-n/(n+1)}$ and $r^{1/(n+1)}$ singularity, respectively. These so-called 'HRR fields' are written as

$$\sigma_{ij} = \sigma_y \left(\frac{J}{\alpha \sigma_y \epsilon_y I_n r} \right)^{1/(n+1)} \tilde{\sigma}_{ij}(\theta, n),$$

$$\epsilon_{ij} = \alpha \epsilon_y \left(\frac{J}{\alpha \sigma_y \epsilon_y I_n r} \right)^{n/(n+1)} \tilde{\epsilon}_{ij}(\theta, n),$$

$$u_i = \alpha \epsilon_y \left(\frac{J}{\alpha \sigma_y \epsilon_y I_n} \right)^{n/(n+1)} r^{1/(n+1)} \tilde{u}_i(\theta, n). \tag{9.91}$$

The universal functions $\tilde{\sigma}_{ij}(\theta, n)$, $\tilde{\epsilon}_{ij}(\theta, n)$, and $\tilde{u}_i(\theta, n)$ in Eqs. 9.91 vary with the polar angle θ, the strain hardening exponent n and the state of stress, i.e. plane stress or plane strain. The factor I_n depends mildly on the strain hardening exponent n.

Since J is a measure of the intensity of crack-tip fields, the onset of crack advance under quasi-static loads can be formulated on the basis of a critical value, $J = J_c$. When conditions of J-dominance (see Section 9.7.4) are satisfied in the plane strain test specimen, the measured critical value of plane strain fracture toughness is denoted by J_{Ic}. Detailed test procedures for the experimental measurement of J_{Ic} are spelled out in the test standard E-813 developed by the American Society for Testing and Materials (Philadelphia) in 1981. Under linear elastic, plane strain conditions,

$$J_{Ic} = \frac{K_{Ic}^2}{E} (1 - v^2). \tag{9.92}$$

This relation is used to infer an equivalent K_{Ic} value from J_{Ic} measurements in high toughness ductile solids in which valid K_{Ic} testing will require unreasonably large test specimens.

9.7.3 Crack tip opening displacement

Expressions for the crack tip opening displacement δ_t, Eqs. 9.83 and 9.89, were derived earlier using the Dugdale yield zone model. More precise estimates of δ_t can be obtained for strain-hardening materials from the solutions (Eqs. 9.91) for crack face displacements. The definition of δ_t is somewhat arbitrary because the distance between the crack faces, $\delta = u_y(x, 0^+) - u_y(x, 0^-)$, varies as $(-x)^{1/(n+1)}$ as the crack tip is approached. A commonly used operational definition of δ_t is based on the distance between two points on the upper and lower crack faces where two 45° lines drawn from the deformed crack tip intercept the crack faces (Fig. 9.11). With this definition,

$$\delta_t = d_n \frac{J}{\sigma_y}, \tag{9.93}$$

where d_n is a function of α, ϵ_y, and n. d_n ranges in value from about 0.3 to 0.8 as n is varied from 3 to 13 (Shih, 1981).

The crack tip opening displacement provides a measure of the size of the region at the crack tip where finite strain deformation is dominant. The condition for the onset of quasi-static fracture can also be stated as $\delta_t = \delta_{tc}$, where δ_{tc} is a critical crack tip

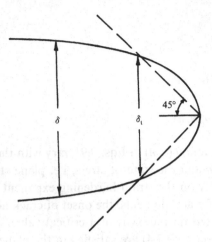

Fig. 9.11. Definition of crack tip opening displacement, δ_t.

opening displacement for the material under consideration. This approach to determining critical conditions for fracture initiation is sometimes appealing in that δ_t provides a physical length scale for fracture which is often needed for developing the vital link between microscopic failure processes and macroscopic fracture toughness.

The magnitude of δ_t varies continuously during fatigue due to the fluctuations in load. The effective range of δ_t in a given fatigue cycle is determined by the following factors: (i) the extent of reversed flow ahead of the crack tip, as seen in the development of Eq. 9.86, (ii) the roughness of the fracture surfaces (which is influenced by the microscopic fatigue mechanisms and the microstructural size scale), and (iii) the presence of any corrosion films or residual stretch of plastically deformed or transformed material on the fracture surfaces which cause premature closure of the crack even at a far-field tensile stress (see Chapter 14).

Noting the connection between J and K from Eqs. 9.57, 9.58 and 9.88, an expression for cyclic crack tip opening displacement can be derived. It is of the form given in Eq. 9.86.

9.7.4 Conditions of J-dominance

The J-integral provides a unique measure of the strength of singular fields in nonlinear fracture. However, some conditions must be ascertained before J can be used to characterize fracture in real ductile materials. The following list of requirements for the use of J is taken from a review article by Hutchinson (1983):

(1) The deformation theory of plasticity must be an adequate model of the small-strain behavior of real elastic–plastic materials under the monotonic loads being considered.

(2) The region in which finite strain effects dominate and the region in which microscopic failure processes occur must each be contained well within the region of the small-strain solution dominated by the singular fields, Eqs. 9.91.

The HRR fields exactly satisfy the first requirement for the use of deformation theory of plasticity if proportional loading occurs everywhere. Then, the singularity fields, Eqs. 9.91, based on the deformation theory assumptions are also solutions to the corresponding J_2 flow (incremental) theory equations (Section 1.4.3). Although this requirement for proportional loading is not exactly fulfilled in general in an elastic–power law plastic solid, the conditions encountered during the application of *monotonic*, uniaxial loads to *stationary* cracks do provide a reasonable justification for the use of deformation theory.

The aforementioned second requirement for the validity of J to characterize non-linear fracture provides the physical basis for determining the inner radius, r_0, of the annular zone of J-dominance. Let R denote the outer boundary of the zone of J-dominance which may be taken as the maximum distance ahead of the crack tip within which the singularity solutions, Eqs. 9.91, match (say, within 10% error) the full solutions (estimated using techniques such as the finite-element method) for the particular specimen geometry under consideration. Finite-element, flow theory model calculations of the crack-tip fields by McMeeking (1977) and McMeeking & Parks (1979), which take into account crack tip geometry changes, reveal that the finite strain effects are significant only over a distance of at most $3\delta_t$. This result provides a measure of r_0. From a microstructural standpoint, r_0 should be bigger than the size of the process zone; for example, the grain size for transgranular cleavage or intergranular fracture, and the mean spacing of void-nucleating particles for ductile failure by void growth. Numerical simulations of near-tip fields for *small-scale yielding* conditions in power law hardening materials show that the HRR singular solutions hold over a distance of 20–25% of the size of the plastic zone directly ahead of a mode I crack for essentially the entire range of strain hardening exponents found in ductile alloys.

Under *large-scale yielding*, however, the size of the region of J dominance is strongly dependent on specimen configuration. For these cases, where the entire uncracked ligament may be fully engulfed in a plastic zone, the size of the region of J dominance, R, is as small as 1% of the length of uncracked ligament for a center-cracked tension specimen or 7% of the length of the uncracked ligament for a deeply cracked bend bar or a compact tension specimen (McMeeking & Parks, 1979). This is the reason why standardized test procedures for determining the critical value of J in quasi-static loading, J_{Ic}, require the use of deeply cracked bend or compact tension specimens where the initial pre-crack length to the specimen width ratio is at least 0.5 (see ASTM Standard E-813 for J_{Ic} testing; American Society for Testing Materials, Philadelphia). With this requirement and the result that the zone of finite strains at the crack tip spans a distance of $3\delta_t$, it can be shown that the minimum uncracked ligament size, b, needed to obtain a valid J_{Ic} estimate is: $b = 25 J_{Ic}/\sigma_y$.

For J-controlled crack growth, Hutchinson & Paris (1979) have suggested that the regime of elastic unloading and nonproportional loading should be confined to well within the zone of J-dominance. In other words,

$$\frac{dJ}{da} \gg \frac{J}{R} \quad \text{and} \quad \Delta a \ll R. \tag{9.94}$$

The growth of cracks also causes changes in the near-tip fields as compared to those predicted by the HRR solutions for stationary cracks. Asymptotic analyses of near-tip stress fields during quasi-static crack growth in an elastic–perfectly plastic solid reveal that the stresses are generally unchanged from the Prandtl slip line fields representative of stationary crack tips (e.g., Rice, Drugan & Sham, 1980), except behind the crack tip where differences of about 10% emerge as a result of a wedge of elastic unloading for $112° < \theta < 162°$ (see Fig. 9.5 for the coordinate system). However, the near-tip strain distribution for a growing crack has a logarithmic singularity which is weaker than the $1/r$ strain singularity for a stationary crack.

9.7.5 Example problem: Specimen size requirements

Problem:

A low strength steel is to be used in an application which requires appreciable damage tolerance. It is known from prior testing that the alloy has a plane strain fracture initiation toughness, $K_{Ic} = 175 \text{ MPa}\sqrt{\text{m}}$. The yield strength of the steel is $\sigma_y = 350 \text{ MPa}$.

(i)   If the fatigue crack growth characteristics of this alloy in zero–tension cyclic loading are to be experimentally measured over the entire range of crack growth until final failure, determine the minimum specimen size requirements for a compact (tension) specimen which conform to ASTM Standard E-399. It is desired to have plane strain conditions during the entire range of crack growth. A schematic of the geometry of the compact specimen is given in Fig. A.2(a) in the Appendix. It may be assumed that the final failure occurs in this case when the the maximum stress intensity factor for the fatigue cycle approaches K_{Ic}.

(ii) For the same alloy and specimen geometry, determine the minimum specimen size requirements for a 'valid' J_{Ic} test which conforms to the specifications of ASTM Standard E-813.

Solution:

For the steel, the values of Young's modulus and Poisson ratio are taken as $E = 210 \text{ GPa}$ and $\nu = 0.33$.

(i) In order to carry out a 'valid' plane strain fracture initiation toughness test, small-scale yielding should be ensured. For the compact specimen shown in Fig. A.2(a), this implies that the length of the crack, a, the size of the uncracked ligament, $(W - a)$, and the specimen thickness, B, should be larger than at least 25 times the plane strain plastic zone size (Eq. 5.1), i.e.

$$a, (W - a), B \geq 25 \times \frac{1}{3\pi} \left(\frac{K_{Ic}}{\sigma_y} \right)^2. \tag{9.95}$$

It is thus seen that a, $(W - a)$, $B \geq 66.3$ cm (26.12 inches). This rather severe requirement for specimen size makes testing very difficult.

(ii) For J_{Ic} testing which conforms to ASTM specifications, start with a deeply cracked compact specimen whose initial crack length to width ratio $a/W > 0.5$. The geometrical requirement in this is that the uncracked ligament length be

$$(W - a) \geq 25 \frac{J_{Ic}}{\sigma_y}. \tag{9.96}$$

For small-scale yielding, J_{Ic} and K_{Ic} are related by Eq. 9.92 so that the above equation can be rewritten as

$$(W - a) \geq 25 \frac{K_{Ic}^2}{E \sigma_y} (1 - v^2). \tag{9.97}$$

Substitution of the appropriate numerical values gives the result that $(W - a) = 9.3$ mm. For materials with high fracture toughness and low yield strength values, the measurement of J_{Ic} thus provides an appealing test procedure for the estimation of fracture toughness.

9.7.6 Characterization of fatigue crack growth

Although conditions of nonproportional loading and the occurrence of elastic unloading would appear to violate the fundamental basis on which the application of J-integral to fracture problems is predicated, Dowling & Begley (1976) and Dowling (1977) have proposed a power law characterization of fatigue crack advance under elastic–plastic conditions based on the cyclic J-integral, J^c (also referred to as ΔJ), during a fatigue cycle, i.e. $da/dN \propto (J^c)^{m'}$, where m' is an exponent analogous to m in Eq. 9.53. This approach, despite its apparent shortcomings, provides suprisingly good characterization of the growth of short fatigue cracks of length comparable to the near-tip plastic zone size and longer fatigue flaws in nearly fully yielded specimens in some materials under certain cyclic loading conditions.

For the cyclic loading of a specimen under displacement control, the J-integral is usually determined in a straightforward manner by employing Eqs. 9.56 and 9.87

(Dowling & Begley, 1976). This procedure is schematically illustrated in Fig. 9.12(*a*). Here the rising part of the load–displacement hysteresis loops for two different crack lengths a_1 and a_2 are shifted to a common origin. The *J*-integral is then obtained from the potential energy difference corresponding to the shaded area and from Eqs. 9.56 and 9.88.

For the cyclic loading of a specimen under load control, there is some ambiguity in defining the proper limits of integration in the determination of the strain energy.

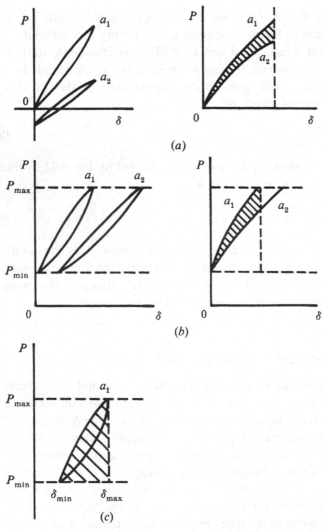

Fig. 9.12. Determination of *J*-integral with stabilized cyclic hysteresis loops. (*a*) Hysteresis loops for two different crack lengths in displacement-controlled fatigue and the translation of the rising part of the stabilized hysteresis loop to a common origin. (*b*) Similar method for load-controlled fatigue with the minimum load being employed as the reference point. (*c*) Determination of *J* using a single specimen.

Sadananda & Shahinian (1979) have suggested use of the minimum load of the fatigue cycle as a reference point for shifting the load–displacement curves, Fig. 9.12(b). Here the rising portions of the load–displacement curves are translated to a common origin at the minimum load. The shaded area gives the potential energy from which the J-integral for cyclic loading can be computed. Note that these methods require the measurement of at least two load-displacement curves corresponding to two different crack lengths. Approximate methods have been developed in an attempt to overcome this limitation and to determine J from a single specimen. Figure 9.12(c) schematically illustrates this method. If A_e is the total area under the load–displacement curve (shaded region), J for characterizing fatigue crack growth (for an edge-cracked or compact tension specimen) is determined from

$$J = \frac{2}{Bb}\{\alpha_1 A_e + \alpha_2 P(\delta_{\max} - \delta_{\min})\}, \tag{9.98}$$

where B is the thickness of the specimen, b is the length of the uncracked ligament, P is the load, $\delta_{\max}$ and $\delta_{\min}$ are the maximum and minimum displacements shown in Fig. 9.12(c), and α_1 and α_2 are correction coefficients which are functions of the crack length and have been published by Merkle & Corten (1974).

The justification for the above methods apparently rests on the argument that the phenomenological constitutive models for cyclic plasticity (Chapter 3) can be formulated in terms of *stable* hysteresis loops and that, if such loops can be mathematically translated to a common origin after each load reversal, the requirement for the stress to be proportional to the current plastic strain can be effectively satisfied. Although many researchers have employed the J-integral to characterize fatigue fracture at room and elevated temperatures, it should be noted that the cyclic J approach can *seriously violate* the basic assumptions leading to the development of the J-integral. Severe nonproportional loading and the rapid advance of the fatigue crack promote conditions where material descriptions based on the deformation theory of plasticity do not hold. At this point, experimental documentation of a reasonably good characterization of fatigue crack growth under some elastic–plastic conditions is the main justification that can be provided for the application of J-integral to cyclic loading. See Chapter 15 for a further discussion of this topic.

An alternative approach for characterizing fatigue crack growth under elastic–plastic conditions is often formulated in terms of crack opening displacements. Here, the fatigue crack growth rate is envisioned as being proportional to the cyclic crack tip opening displacement, $\Delta\delta_t$, defined in Eq. 9.86 such that $da/dN \propto \Delta\delta_t$. This type of analysis provides a size scale, $\Delta\delta_t$, for comparisons with striation spacing, residual crack wake stretch or fracture surface oxide thickness to correlate crack growth and crack closure (see Chapters 10 and 14). Furthermore, the crack opening displacement offers a convenient means for comparing the fatigue crack growth rates in different modes of fracture on a common scale. It can be seen from Eq. 9.93 that characterizations of crack advance based on J and on δ_t are essentially equivalent for proportional loading. The implications and limitations of the approaches for fatigue crack growth under nonlinear fracture conditions are examined in later chapters.

9.8 Two-parameter representation of crack-tip fields

Consider the small-strain, linear elastic solution, Eqs. 9.41–9.43, where the first two terms provide an adequate solution for the crack tip stress fields:

$$\sigma_{ij} = \frac{K_I}{\sqrt{2\pi r}}\,\tilde{\sigma}_{ij}^I(\theta) + T\delta_{1i}\delta_{1j}. \tag{9.99}$$

Under small-scale yielding, different levels of hydrostatic stresses induced at the crack tip in different specimen geometries can be characterized in terms of the non-dimensional parameter, T/σ_y, where σ_y is the yield strength (Bilby *et al.*, 1986; Harlin & Willis, 1988; Betagón & Hancock, 1991; Parks, 1992). The use of K_I and T, as two parameters which fully characterize the near-tip fields, becomes increasingly more invalid as the plastic zone size at the crack tip enlarges beyond the limits of small-scale yielding.

It was ostensibly first noted by McClintock (1971) that the states of near-tip stress fields in fully yielded cracked specimens require more than a single parameter, such as the J-integral, to adequately capture the triaxial fields. For example, center-cracked panels with large-scale yielding exhibit markedly reduced triaxial stresses at the crack tip than fully-yielded, deeply-cracked compact and bend specimens. In order to account for the role of differing states of triaxial stresses in influencing crack-tip fields and to overcome the inadequacy of the K_I–T two-parameter characterization under large-scale yielding, O'Dowd & Shih (1991, 1992) have proposed the so-called two-parameter J–Q theory. The salient features of this theory are outlined next.

Within the context of the small-strain deformation plasticity theory and an elastic–power law plastic material model, the stress fields at the crack tip take the form

$$\sigma_{ij} = \sigma_y\left(\frac{J}{\alpha\sigma_y\epsilon_y I_n r}\right)^{1/(n+1)}\tilde{\sigma}_{ij}(\theta, n) + \text{higher order terms.} \tag{9.100}$$

The first term on the right hand side of Eq. 9.100 may easily be recognized as the HRR singular stress field, $(\sigma_{ij})_{HRR}$, see Eq. 9.91. Higher order asymptotic analyses of crack-tip fields under large-scale yielding have been carried out by Li & Wang (1986), Sharma & Aravas (1991) and Xia, Wang & Shih (1993). These studies reveal that, in the region ahead of the crack tip where $|\theta| \leq \pi/2$ and $J/\sigma_y < r < 5J/\sigma_y$, Eq. 9.100 can be approximated in the following manner:

$$\sigma_{ij} = (\sigma_{ij})_{HRR} + Q\sigma_y\delta_{ij}, \qquad |\theta| \leq \frac{\pi}{2}, \tag{9.101}$$

where Q is a measure of the crack tip stress triaxiality, defined as

$$Q \equiv \frac{\sigma_{\theta\theta} - (\sigma_{\theta\theta})_{HRR}}{\sigma_y} \quad \text{at} \quad \theta = 0, \quad r = \frac{2J}{\sigma_y}. \tag{9.102}$$

That is, in this operational definition, Q represents the difference, normalized by the yield strength σ_y, between the actual hoop stress at the crack tip and that given by the HRR singular field at a fixed distance $2J/\sigma_y$ directly ahead of the crack tip. The distance $r = 2J/\sigma_y$ is chosen so as to lie just outside the blunting zone characterized by the finite strains; under such conditions, Q is found to be essentially independent of r.

Alternatively, the K-dominated small-scale yielding field, $(\sigma_{ij})_{SSY;T=0}$, can also serve as the reference solution. O'Dowd & Shih (1991, 1992) define Q as

$$Q \equiv \frac{\sigma_{\theta\theta} - (\sigma_{\theta\theta})_{SSY;T=0}}{\sigma_y} \quad \text{at} \quad \theta = 0, \quad r = \frac{2J}{\sigma_y}. \tag{9.103}$$

In terms of the hydrostatic or mean stresses, Q can also be defined as

$$Q_m \equiv \frac{\sigma_m - (\sigma_m)_{SSY;T=0}}{\sigma_y} \quad \text{at} \quad \theta = 0, \quad r = \frac{2J}{\sigma_y}. \tag{9.104}$$

Equations 9.102 and 9.103 can be interpreted in the following manner. Positive (negative) values of Q raise (reduce) the level of crack-tip hydrostatic stress from that given by the HRR solutions or the small-scale yielding solution.

The variation of hoop stress, $\sigma_{\theta\theta}$, as a function of the normalized distance ahead of the crack tip, $r/(J/\sigma_y)$, for plane strain and for $E/\sigma_y = 500$ and $\nu = 0.3$ are plotted in Fig. 9.13(a) and (b) for small strain (HRR solution) and finite strain, respectively, for different values of n and for $\theta = 0$. These pertain to the reference fields: $T = 0$ and $Q = 0$.

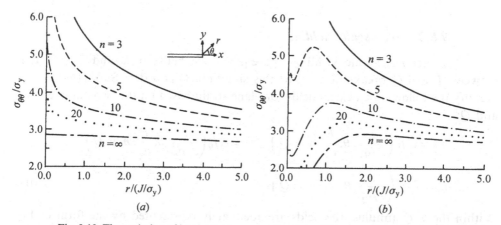

Fig. 9.13. The variation of hoop stress, $\sigma_{\theta\theta}$, as a function of the normalized distance ahead of the crack tip, $r/(J/\sigma_y)$, for plane strain and for $E/\sigma_y = 500$ and $\nu = 0.3$ for (a) small strain (HRR solution) and (b) finite strain, respectively, for different values of n and $\theta = 0$ for the reference fields: $T = 0$ and $Q = 0$. (After O'Dowd & Shih, 1991, 1992.)

O'Dowd & Shih have calculated Q and Q_m for several finite width geometries and find the difference between them to be always smaller than 0.1. Consequently, the definition of Q given in Eq. 9.103 has been recommended as the basis upon which crack-tip stress triaxiality is quantified.

9.8.1 Small-scale yielding

The two-parameter characterization of crack tip fields by K_I and T or by J and Q have equivalence under small-scale yielding. Recall from Section 9.7.1 that under plane strain and small-scale yielding,

$$J = \frac{K_I^2}{E} (1 - v^2).$$ (9.105)

Also, from dimensional considerations, it can be shown that

$$\sigma_{ij} = \sigma_y \tilde{\sigma}_{ij}^I \left(\frac{r}{J/\sigma_y}, \theta; \frac{T}{\sigma_y} \right).$$ (9.106)

Making use of Eq. 9.103, we write

$$\sigma_{ij} = (\sigma_{ij})_{\text{SSY};T=0} + Q\sigma_y \delta_{ij}.$$ (9.107)

For a power-law hardening solid characterized by the two-term solutions in Eqs. 9.99 and 9.101, a unique relationship can be defined between Q and T in the following form:

$$Q = F\left(\frac{T}{\sigma_y}; n \right)$$ (9.108)

for given E/σ_y and v. Figure 9.14 shows the variation of Q (based on the definition in Eq. 9.103) with T/σ_y for different values of n. These plots are essentially the same for both small strain and finite strain deformation at crack tip.

9.8.2 Large-scale yielding

Under large-scale yielding, the equivalence between K and J, and that between T and Q breaks down. For this situation, O'Dowd & Shih (1991, 1992) identify the following crack tip fields for plane strain based on the two parameters J and Q:

$$\sigma_{ij} = \sigma_y f_{ij}\left(\frac{r}{J/\sigma_y}, \theta, \frac{E}{\sigma_y}, n, v; Q \right), \quad \epsilon_{ij} = \epsilon_y f_{ij}\left(\frac{r}{J/\sigma_y}, \theta, \frac{E}{\sigma_y}, n, v; Q \right),$$

$$u_i = \frac{J}{\sigma_y} h_i\left(\frac{r}{J/\sigma_y}, \theta, \frac{E}{\sigma_y}, n, v; Q \right).$$ (9.109)

Within the J–Q annulus, the fields are adequately represented by the form of Eq. 9.107. Values of Q for a variety of specimen geometries and crack sizes (a/W) can be found in O'Dowd & Shih (1991, 1992). Implications of the two-parameter approach to fatigue crack growth characterization is discussed in Chapter 15.

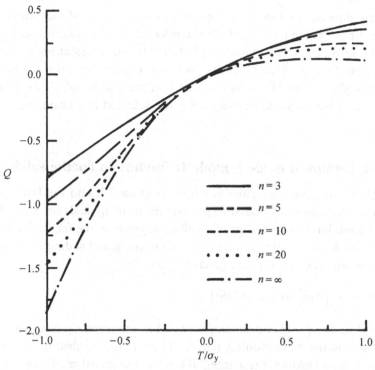

Fig. 9.14. The variation of Q with T/σ_y for different values of n (for $E/\sigma_y = 500$ and $\nu = 0.3$). (After O'Dowd & Shih, 1991, 1992.)

9.9 Mixed-mode fracture mechanics

The stress analyses discussed thus far have centered around the development of near-tip fields for stationary, mode I cracks. Fatigue cracks in structural materials, however, are generally subjected to combined-mode loading conditions. In order to develop proper design methodologies for combined-mode fatigue fracture, it is necessary to gain insights into the nature of near-tip fields under conditions of load mixture. Even when a component containing a fatigue crack is subjected to purely tensile far-field loading, mixed-mode conditions may prevail ahead of the crack tip if it is inclined at some arbitrary angle to the tensile axis, or microstructural and environmental factors promote a nonplanar fatigue crack growth. In composite materials consisting of two dissimilar components, a fatigue crack located at the interface between the two materials is subjected to *local* mixed-mode loading even when the far-field stresses are purely in mode I. The inducement of such mixed-mode conditions at the tip of a fatigue crack can lead to pronounced changes in the 'effective driving force' for crack advance, the size of the crack tip plastic zone, the rate of crack propagation, the extent of fatigue crack closure and the microscopic mechanisms of damage.

Among mixed-mode fracture problems, complete solutions of the near-tip fields are available for combined mode I–mode II cracks subjected to small-scale nonlinear deformation due to power law plasticity (Shih, 1974) and transient power law creep (Brockenbrough, Shih & Suresh, 1991). We first present here the elastic–plastic solutions for mode I–mode II cracks in strain-hardening materials. This is followed by a discussion of mixed-mode near-tip fields for deflected and branched cracks.

9.10 Combined mode I–mode II fracture in ductile solids

The linear elastic fracture mechanics solutions, Section 9.3, and the HRR singular fields, Section 9.7.2, can be extended to provide the near-tip fields for mode I–mode II crack problems. Under conditions of small-scale yielding, the far-field stress components for a crack subjected remotely to tensile opening and sliding stress intensity factors, K_I and K_{II}, respectively, are given by

$$\sigma_{ij} = \frac{1}{\sqrt{2\pi r}} \left[K_I \tilde{\sigma}_{ij}^I(\theta) + K_{II} \tilde{\sigma}_{ij}^{II}(\theta) \right], \tag{9.110}$$

where r and θ are the polar coordinates centered at the crack tip (see Fig. 9.5) and $\tilde{\sigma}_{ij}^I(\theta)$ and $\tilde{\sigma}_{ij}^{II}(\theta)$ are the dimensionless universal functions described in Eqs. 9.44 and 9.47. For small-scale yielding, the J-integral is related to mixed-mode stress intensity factors by

$$J = \frac{1}{E'} \left(K_I^2 + K_{II}^2 \right), \tag{9.111}$$

where E' is defined in Eq. 9.2. The relative strengths of K_I and K_{II} can be characterized in terms of an *elastic* mixity parameter, M^e, which is defined as (Shih, 1974)

$$M^e = \frac{2}{\pi} \tan^{-1} \left| \lim_{r \to 0} \frac{\sigma_{\theta\theta}(r, \theta = 0)}{\sigma_{r\theta}(r, \theta = 0)} \right| = \frac{2}{\pi} \tan^{-1} \left| \frac{K_I}{K_{II}} \right|. \tag{9.112}$$

In this characterization, $M^e = 0$ for pure mode II, $M^e = 1$ for pure mode I, and $0 < M^e < 1$ for different mixities of modes I and II.

The near-tip fields for the mixed-mode crack problem in an elastic–plastic solid whose constitutive response is represented by the nonlinear elastic (deformation) theory are analogous to the HRR fields for mode I described in Eq. 9.91 and are of the form

$$\sigma_{ij} = \sigma_y K_M^P \, r^{-1/(n+1)} \, \tilde{\sigma}_{ij}(\theta, M^P, n),$$

$$\sigma_e = \sigma_y K_M^P \, r^{-1/(n+1)} \, \tilde{\sigma}_e(\theta, M^P, n),$$

$$\epsilon_{ij}^P = \frac{\alpha \sigma_y}{E} (K_M^P)^n \, r^{-n/(n+1)} \, \tilde{\epsilon}_{ij}^P(\theta, M^P, n),$$

$$u_i = \frac{\alpha \sigma_y}{E} (K_M^P)^n \, r^{1/(n+1)} \, \tilde{u}_i(\theta, M^P, n). \tag{9.113}$$

The dimensionless functions $\tilde{\sigma}_{ij}$, $\tilde{\sigma}_e$, $\tilde{\epsilon}_{ij}^p$, and $\tilde{u}_i$ depend only on the polar angle θ, the strain hardening exponent n, and the near-tip *plastic* mixity parameter, M^P, which is defined similar to M^e as

$$M^P = \frac{2}{\pi} \tan^{-1} \left\{ \lim_{r \to 0} \frac{\sigma_{\theta\theta}(r, \theta = 0)}{\sigma_{r\theta}(r, \theta = 0)} \right\}$$
$$= \frac{2}{\pi} \tan^{-1} \left\{ \frac{\tilde{\sigma}_{\theta\theta}(\theta = 0, M^P)}{\tilde{\sigma}_{r\theta}(\theta = 0, M^P)} \right\}, \tag{9.114}$$

where M^P equals 0 for pure mode II, 1 for pure mode I, and $0 < M^P < 1$ for different mixities of modes I and II. The strength of the singular fields given in Eqs. 9.113 is the parameter K_M^P. The superscript P denotes that it is a plastic stress intensity factor and the subscript M refers to the mixed-mode condition. A definite meaning to this parameter has been given (Shih, 1974) by setting the maximum value of the θ-variation of the effective stress, $\tilde{\sigma}_e = \sqrt{\{(3/2)\tilde{s}_{ij}\tilde{s}_{ij}\}}$, to unity where $\tilde{s}_{ij} = \tilde{\sigma}_{ij} - (\tilde{\sigma}_{kk}/3)\delta_{ij}$. The strength of the dominant mixed-mode singularity, K_M^P, can be related to the J-integral via M^P:

$$J = \frac{(1 - \nu^2)}{E} \left(K_I^2 + K_{II}^2 \right) = \frac{\alpha \sigma_y^2}{E} I_n(M^P) (K_M^P)^{n+1}, \tag{9.115}$$

where $I_n(M^P)$ is a numerical constant which is a function of the strain hardening exponent n and the plastic mixity factor M^P. Thus, for combined-mode fracture, the two parameters, K^P and M^P, completely specify the near-tip fields for a given value of n, irrespective of whether small-scale yielding conditions prevail. A complete description of the near-tip fields in terms of J and M^e (or, equivalently, K_I and K_{II}) requires that the relationship between M^e and M^P be specified. By performing a detailed finite element analysis of mixed-mode loading under small-scale yielding, Shih (1974) has computed the relationship between M^e and M^P for values of $n = 1$–99. While $M^e = M^P$ for $n = 1$, $M^P > M^e$ for $n > 1$, with the difference between M^P and M^e increasing with increasing n.

The (circumferential) θ-variations of the universal dimensionless functions $\tilde{\sigma}_{ij}$, $\tilde{\epsilon}_{ij}$ and $\tilde{\sigma}_e$ are shown in Fig. 9.15 in order of increasing asymmetry. One observes a pronounced decrease in the hoop stress $\tilde{\sigma}_{\theta\theta}$ and in the hydrostatic stress, and a noticeable change in the distribution of plastic strains as the mode of loading is changed from pure tension (mode I) to pure sliding (mode II).

This effect of plasticity on the near-tip fields is more directly evident in Fig. 9.16, where the elastic–plastic boundaries are shown as functions of n and M^e for plane stress and plane strain. At the same amplitude of far-field loading (that is, at a fixed amplitude of J), the plastic zone size in mode II is up to five times bigger than that in mode I. This result implies that any deviation of the fatigue crack from the nominal mode I growth plane, as often found in crystallographic crack propagation, causes not only a change in near-tip stress intensity factors due to the inducement of mixed-mode conditions, but also leads to an increase in the size of the plastic zone ahead of the crack tip. The consequences of this effect are especially important in the study of

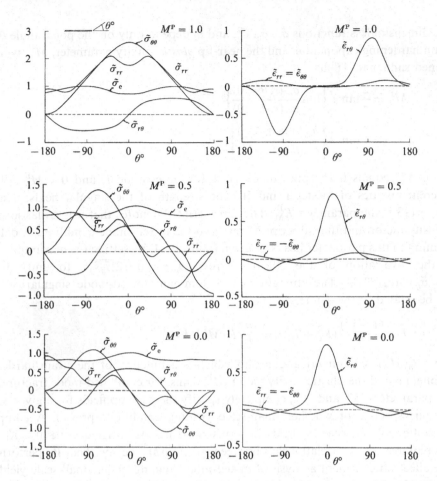

Fig. 9.15. Circumferential variations of stresses and strains shown in order of increasing load asymmetry for small-scale yielding and plane strain loading conditions. Strain hardening exponent, $n = 13$. (From Shih, 1974. Copyright American Society for Testing and Materials. Reprinted with permission.)

short fatigue cracks where plastic zone dimensions comparable in size to the crack length cause uncertainties in characterization.

9.11 Crack deflection

Tensile fatigue cracks in both brittle and ductile solids can deviate significantly from their normal mode I growth plane under the influence of far-field multiaxial stresses, interaction of the crack tip with microstructural inhomogeneities such as grain boundaries and interfaces, variable amplitude loading in the form of overloads, crystallographic separation, or the embrittling effect of an aggressive environment. As the local, near-tip stress intensity factor which is responsible for the

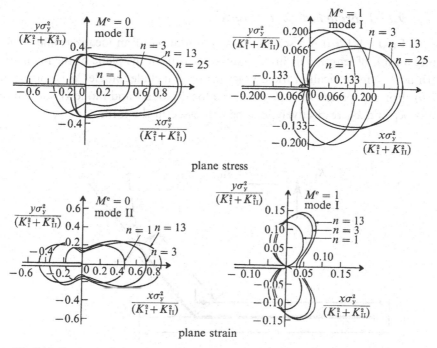

Fig. 9.16. Contours of plastic zones shown for different values of the strain hardening exponent n for mode I and mode II cracks subjected to plane stress and plane strain loading. (After Shih, 1973, 1974.)

advance of a deflected or branched crack (i.e. an 'effective driving force') can be considerably smaller than that of a straight crack of the same (projected) length, periodic changes in crack path lead to apparently beneficial (slower) fatigue crack growth rates (for a fixed amplitude of far-field loading). The deflection of a tensile crack from the nominal mode I plane induces mixed-mode near-tip conditions even if the far-field loading is purely mode I. Under cyclic loading conditions, the increase in the roughness of the fracture surface created by crack deflections can also cause an increase in the far-field stress at which the crack faces begin to close upon unloading from the maximum tensile stress. This enhanced closure level leads to further reductions in the near-tip 'driving force' for fatigue crack growth (see Chapter 14).

It should be recognized at the outset that microstructurally induced crack deflection in engineering materials is a problem for which it is very difficult to develop accurate continuum solutions. The mixed-mode fracture solutions are useful for deflected cracks only if the kink length greatly exceeds the size scale of the microstructural inhomogeneity responsible for crack deflection and the size scale of the near-tip plastic zone. Furthermore, if the entire crackfront does not deflect uniformly, multiaxial conditions involving modes I, II and III would be expected to develop. The discussions in this section are, therefore, intended to provide only a feel for the changes in 'effective driving force' arising as a consequence of simple deflections in the path of a linear elastic fatigue crack.

9.11.1 Branched elastic cracks

Figure 9.17(*a*) is a schematic representation of an idealized line crack containing a kink of length *b* inclined at an angle α from the plane of the main crack of length *a*. A symmetrically forked crack with an included angle 2α is shown in Fig. 9.17(*b*). For a pupative kink ($b \to 0$), the local mode I and mode II stress intensity factors, k_1 and k_2, respectively, can be expressed in the form

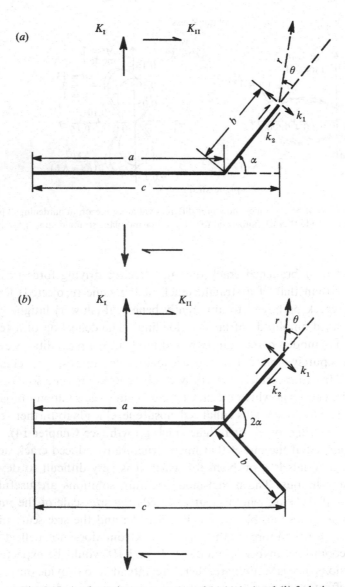

Fig. 9.17. A schematic representation of (*a*) kinked and (*b*) forked crack geometries and the associated nomenclature.

$$k_1 = a_{11}(\alpha)K_{\text{I}} + a_{12}(\alpha)K_{\text{II}},$$
$$k_2 = a_{21}(\alpha)K_{\text{I}} + a_{22}(\alpha)K_{\text{II}},\qquad(9.116)$$

where K_{I} and K_{II} denote the mode I and mode II stress intensity factors for the main crack in the absence of the kink or fork. To a first order approximation in α, the dimensionless factors for the infinitesimal kink are (Bilby, Cardew & Howard, 1977; Cotterell & Rice, 1980)

$$a_{11}(\alpha) = \frac{1}{4}\left(3\cos\frac{\alpha}{2} + \cos\frac{3\alpha}{2}\right),$$

$$a_{12}(\alpha) = -\frac{3}{4}\left(\sin\frac{\alpha}{2} + \sin\frac{3\alpha}{2}\right),$$

$$a_{21}(\alpha) = \frac{1}{4}\left(\sin\frac{\alpha}{2} + \sin\frac{3\alpha}{2}\right),$$

$$a_{22}(\alpha) = \frac{1}{4}\left(\cos\frac{\alpha}{2} + 3\cos\frac{3\alpha}{2}\right).\qquad(9.117)$$

Suresh & Shih (1986) have presented a summary of available stress intensity factor solutions for kinked and forked cracks as functions of the deflection angle and the length of the deflected part of the crack. The variations of the near-tip stress intensity factors, k_1 and k_2, for a line crack containing a kink of length $b = 0.1a$ and subjected to a far-field tensile stress intensity factor K_{I} are plotted in Fig. 9.18(a) as a function of the kink angle, α. Similar results for a symmetrically forked crack with a fork length $b = 0.1a$ are presented in Fig 9.18(b) as a function of the included fork angle, 2α (Kitagawa, Yuuki & Ohira, 1975; Lo, 1978). Note that k_2 vanishes at $2\alpha = 32°$ for $b/a = 0.1$. A similar observation was made by Bilby, Cardew & Howard (1977) who calculated the included angle 2α for $k_2 = 0$ to be $36°$ for $b/a = 0.025$.

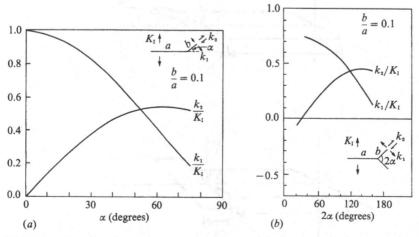

Fig. 9.18. Variation of normalized k_1 and k_2 for $b/a = 0.1$ as a function of (a) kink angle α and (b) fork angle 2α. (After Kitagawa, Yuuki & Ohira, 1975.)

The studies discussed above also show that for $b/a > 0.5$, k_1 and k_2 are independent of b/a. A similar trend is observed for symmetrically forked elastic cracks. This is consistent with the known result that the kinked and forked crack solutions for $b/a > 0.5$ approach those for a crack inclined at an angle β (radians) ($\beta = \pi/2 - \alpha$) to a remote tensile stress intensity K_I:

$$\frac{k_1}{K_I} = \sin^2 \beta, \qquad \frac{k_2}{K_I} = \sin \beta \cos \beta. \tag{9.118}$$

9.11.2 *Multiaxial fracture due to crack deflection*

For an elastic crack containing an infinitesimal kink ($b/a \to 0$) subjected to pure mode I far-field loading, the functions a_{11} and a_{12} in Eq. 9.116 can also be obtained to a reasonable degree of accuracy from coordinate transformations (Lawn, 1993). Here,

$$k_1 = K_I \tilde{\sigma}^I_{\theta\theta}, \qquad k_2 = K_I \tilde{\sigma}^I_{r\theta}, \tag{9.119}$$

where $\tilde{\sigma}^I_{\theta\theta}$ and $\tilde{\sigma}^I_{r\theta}$ are defined in Eq. 9.44.

The crack deflection processes discussed thus far pertain to the tilt configuration represented by the rotation of the crack plane about an axis parallel to the crack front, axis Oz in Fig. 19(a). Whereas this tilt deflection gives rise to near-tip mode I–mode II displacements for far-field mode I loads, the twist deflection represented by the rotation of the crack plane about an axis parallel to the initial direction of crack advance, axis Ox in Fig. 9.19(b), induces mode I and mode III displacements at the crack tip. For an infinitesimal twist angle ϕ,

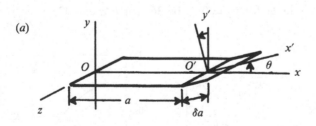

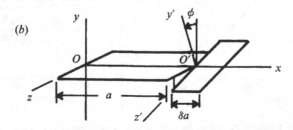

Fig. 9.19. Crack deflection leading to (a) tilting and (b) twisting of the crack front.

$$k_1 = K_I \tilde{\sigma}^I_{\phi\phi}, \qquad k_3 = K_I \tilde{\sigma}^I_{z'\phi}, \tag{9.120}$$

where the functions $\tilde{\sigma}^I_{\phi\phi}$ and $\tilde{\sigma}^I_{z'\phi}$ can be found listed in Lawn (1993). Equations 9.120 provide only a crude result since the twisting of a crack plane, Fig. 9.19(b), causes the formation of steps along the crack path; this may promote mode II displacements as well. Furthermore, the near-tip stress intensity factors may deviate considerably from the predictions of Eqs. 9.120 as a consequence of frictional sliding along the deflected segments of the crack.

The *elastic* solutions for k_1 and k_2 for kinked or forked cracks are strictly valid only when the plastic zone size is smaller than the zone of dominance of the k_1 and k_2 singular fields, which itself is a fraction of the kink or fork length b. From a knowledge of (i) the elastic solutions for k_1 and k_2 as a function of the deflected crack geometry, (ii) the universal mixed-mode plastic near-tip fields, and (iii) the numerically determined relationship between M^e and M^P as a function of the strain hardening exponent n for a given deflected crack geometry, Suresh & Shih (1986) have determined the near-tip fields ahead of a kinked or a forked crack for plane strain and small-scale yielding conditions. Their results show that the combined effects of crack tip plasticity and crack deflection can promote a more beneficial crack growth resistance than deflection alone.

9.12 Case study: Damage-tolerant design of aircraft fuselage

The fuselage of a passenger jet aircraft is one of the most complicated structures in the aircraft. It is a stiffened, thin-walled cylinder. Its diameter to wall thickness ratio is approximately 2000, somewhat analogous to a fully pressurized toy balloon. At cruising altitude, the differential pressure between the passenger cabin and the outside is approximately 55 kPa (8 psi), as discussed in Section 1.1.1. The cabin pressure causes a tensile stress in the hoop direction. The bending induced by routine flight loads causes tensile loads in the crown section (top part of the fuselage), shear loads on the sides of the fuselage, and compressive loads in the bottom of the fuselage (keel) in the longitudinal direction (e.g., Budiman, 1996).

Stiffeners are used in the following manner in order (i) to prevent bending, (ii) to avoid instability due to bending and (iii) to maintain the circular shape of the fuselage.

(1) Stiffeners in the circumferential direction, termed the *frames*. The spacing between neighboring frames in a medium-range passenger aircraft, such as the Boeing 737, is 500 mm (20 in.).

(2) Stiffeners in the longitudinal direction, termed the *longerons*. The spacing between neighboring longerons in a medium-range passenger aircraft, such as the Boeing 737, is 250 mm (10 in.).

Consider the following possibility for an accident. A turbine blade, dislodged from one of the jet engines of the aircraft during the course of a flight, penetrates the skin of the fuselage which is fully pressurized. The likely orientation for a through-thickness crack formed by this impact would be along the hoop direction. The hoop stress of the cylindrical fuselage with stiffeners, however, is nearly twice that of the longitudinal stress. Consequently, the worst situation here would be for the crack to be oriented along the axial direction. In order to arrest the advance of such a crack, tear straps or crack arresters are commonly introduced under the frames, midway between two neighboring frames, and under the longerons. The thickness of the tear straps is approximately the same as that of the fuselage skin.

The spacing between adjacent tear straps provides the basis for the maximum crack size in design. It means that, for the medium-range passenger transport aircraft, the fuselage has to be designed with provisions for the existence of a longitudinal crack as long as 500 mm.

Exercises

9.1   In the compliance calibration of an edge-cracked fracture toughness test-piece of an aluminum alloy, it was observed that a load of 100 kN produced a displacement between the loading pins of 0.3 mm when the crack length was 24.5 mm, and 0.3025 mm when the crack length was 25.5 mm. The fracture load of an identical testpiece containing a crack of length 25.0 mm was 158 kN. Calculate the critical value of the mechanical potential energy release rate G at fracture and hence the plane strain fracture toughness K_{Ic} of the alloy. All testpieces were 25 mm thick. For the alloy, Young's modulus, $E = 70$ GPa and Poisson's ratio, $v = 0.3$.

9.2   Starting with the unsymmetric part of the Airy stress function, Eq. 9.30, where

$$\chi = r^2\, p\,(r,\theta) + q\,(r,\theta),$$

and assuming that χ is separable, i.e. $\chi = R(r) \cdot \Theta(\theta)$, derive the leading term of the asymptotic singular solution for mode II,

$$\sigma_{ij} \approx \frac{K_{II}}{\sqrt{2\pi r}}\, \tilde{\sigma}_{ij}^{II}\,(\theta),$$

following the procedure discussed in Section 9.3.2 for Mode I. Write complete expressions for the different components of the stress field and compare your results with Eq. 9.47.

9.3   Use the result from the previous problem to show that the expression for the leading terms for the displacements is of the form

$$u_i = \frac{K_{II}}{2\mu}\,\sqrt{\frac{r}{2\pi}}\,\tilde{u}_i\,(\theta),$$

where μ is the shear modulus. Derive complete expressions for u_x and u_y and compare your results with those given in Eq. 9.48.

9.4   A piston (89 mm in diameter) is designed to increase the internal pressure in a cylinder from 0 to 55 MPa. The cylinder (closed at the other end!) is 200 mm long with internal diameter = 90 mm, outer diameter = 110 mm, and made of 7075–T6511 aluminum alloy (extruded bar with yield strength = 550 MPa, and K_{Ic} = 30 MPa$\sqrt{m}$). On one occasion, a malfunction in the system caused an unanticipated failure and the cylinder burst. Examination of the fracture surface revealed a metallurgical defect in the form of an elliptical flaw 4.5 mm long at the inner wall, 1.5 mm deep, and oriented normal to the hoop stress in the cylinder. Compute the magnitude of the pressure at which failure took place. (For the purpose of this problem, assume that the stress intensity factor for the elliptical flaw is $K = (1.12/\sqrt{Q})\sigma\sqrt{\pi a}$, where σ is the appropriate normal stress and Q is a shape factor. The calibration for Q as a function of the aspect ratio of the flaw is given in Section A.7.)

9.5 A steel used for an engineering application has a specified yield strength of 1000 MPa and a plane strain fracture toughness of 150 MPa$\sqrt{m}$.
   (a) Calculate the minimum dimensions required to carry out a 'valid' plane strain fracture toughness test.
   (b) Estimate the weight of the single edge cracked bend specimen and of the compact specimen of this steel which would have sufficient dimensions to provide valid plane strain fracture toughness. (See the Appendix section for K calibrations for different specimen geometries.) Assume that a/W = 0.45.
   (c) Estimate the load capacity of the mechanical testing machine you would need to carry out the fracture test.
   (d) If the available testing machine has a load capacity of 200 kN, do you need to alter the dimensions of the specimens to obtain a valid K_{Ic}? If so, what will you do?

9.6 The ASTM standard test method based on linear elastic fracture mechanics (standard E-399, 1974) for plane strain fracture toughness of metallic materials requires that the characteristic specimen dimensions (such as the crack size, the size of the uncracked ligament and specimen thickness) be greater than $2.5 \times (K_{Ic}/\sigma_y)^2$, where K_{Ic} is the mode I fracture toughness in plane strain and σ_y is the yield strength of the material. Similarly, the standard test method for the elastic–plastic fracture toughness J_{Ic} (standard E-813, 1981) requires that the size of the uncracked ligament and the specimen thickness be greater than $25J_{Ic}/\sigma_y$, where J_{Ic} is the critical value of the J-integral for fracture initiation.
   (a) Noting the relationship between J and K, Eqs. 9.57 and 9.8, derive an expression for the ratio b_K/b_J for a linear elastic material, where b_K is the uncracked ligament length for a K_{Ic} test specimen and b_J is that for a J_{Ic} test specimen. σ_y, E and v, respectively, are yield strength, Young's modulus and Poisson's ratio for the material.
   (b) Calculate the ratio b_K/b_J for a steel for which $K_{Ic} = 250$ MPa$\sqrt{m}$, σ_y = 350 MPa, E = 200 GPa and v = 0.33.

9.7   For plane strain and small-scale yielding conditions, the region of J-dominance spans a distance R of up to 25% of the size of the monotonic plastic zone at the crack tip. Assuming the requirement for J-dominance to be that $R > 3\delta_t$ (where δ_t is the crack tip opening displacement defined in Section 9.7.3) and noting that $\delta_t \approx 0.6 J/\sigma_y$ for a low hardening material, prove that the requirement for J-dominance is always satisfied for a low hardening steel with the following properties: $E/\sigma_y = 500$ and Poisson's ratio, $\nu = 0.33$.

9.8   Two engineering alloys, of the same overall yield strength (300 MPa) and elastic properties, are being considered as candidate materials for a particular structural application. Material 1 has a grain size of 10 μm, while the grain size of Material 2 is 220 μm. Both alloys fail in an intergranular fracture mode by the nucleation, growth and coalescence of voids at grain boundaries. The structural component, for which the alloys are being considered, is subjected to a tensile stress of 85 MPa. It contains a throughthickness single edge crack of length $a = 25$ mm. The thickness of the part containing the crack is 30 mm and its width is 100 mm. (The equations necessary to calculate the stress intensity factor for this problem can be found in the Appendix.)

       (a) Comment on the extent and validity of J-dominance for the two materials under consideration.

       (b) An examination of the ASTM standard (E813) for elastic–plastic fracture toughness testing of these two materials reveals the requirement that the depth of the initial crack (i.e. the notch length plus the length of the fatigue pre-crack) be at least one-half of the width of the specimen (e.g., compact tension, three-point bend specimen, or four-point bend specimen). Why?

9.9   Derive expressions for the near-tip mode I and mode II stress intensity factors, k_1 and k_2, respectively, for a center-cracked plate containing a crack which is inclined at an angle β to the far-field tensile load and subjected to a far-field mode I stress intensity factor, K_I. Check your answers with Eq. 9.118.

CHAPTER 10

Fatigue crack growth in ductile solids

As discussed in earlier chapters, the total life of a cyclically loaded component is composed of both the crack initiation and crack propagation stages. Modern *defect-tolerant* design approaches to fatigue are based on the premise that engineering structures are inherently flawed; the useful fatigue life then is the time or the number of cycles *to propagate* a dominant flaw of an assumed or measured initial size (or the largest undetected crack size estimated from the resolution of the nondestructive inspection method) to a critical dimension (which may be dictated by the fracture toughness, limit load, allowable strain or allowable compliance change). In most metallic materials, catastrophic failure is preceded by a substantial amount of stable crack propagation under cyclic loading conditions. The rates at which these cracks propagate for different combinations of applied stress, crack length and geometrical conditions of the cracked structure, and the mechanisms which influence the crack propagation rates under different combinations of mean stress, test frequency and environment, are topics of considerable scientific and practical interest.

In this chapter, we examine the mechanics and micromechanisms of stable crack propagation in ductile solids subjected to uniaxial and multiaxial cyclic loads of fixed amplitudes. Attention is focused on circumstances for which linear elastic fracture mechanics concepts are expected to be valid. Fatigue crack advance where considerable nonlinear deformation occurs ahead of a crack tip or notch tip is considered in Chapters 15 and 16. Situations involving crack growth under variable amplitude fatigue are examined in Chapter 14. It is also worth noting here that mechanisms of crack propagation to be discussed in this chapter pertain to those of 'long' fatigue cracks, typically several millimeters or more in all dimensions, for which growth rates are independent of the crack size. The corresponding discussions of small crack growth, including detailed definitions of what constitutes a 'small' crack, can be found in Chapter 15. Fatigue crack growth processes in brittle solids and semi-/noncrystalline solids are described in Chapters 11 and 12, respectively. Aspects of fatigue crack growth under some contact fatigue conditions are addressed in Chapter 13.

10.1 Characterization of crack growth

The rate of growth of a fatigue crack subjected to constant amplitude stress reversals is expressed in terms of the crack length increment per cycle, da/dN. Values of da/dN for different loading conditions are determined from experimentally determined changes in crack length over a number of elapsed fatigue cycles. When the applied stress range is held constant, the rate of growth of a fatigue crack generally increases with increasing number of fatigue cycles. (Exceptions to this trend are discussed in Chapter 15.) Figure 10.1 schematically illustrates a typical fatigue crack growth curve.

10.1.1 Fracture mechanics approach

One of the goals in fatigue design is to develop reliable methods for characterizing the crack growth rate in terms of an appropriate loading parameter which enables a quantification of the intrinsic resistance of the material to fatigue crack growth for different combinations of applied stresses, specimen geometry and crack geometry. When cyclic stresses applied to a component are so small that the zone of plastic deformation ahead of the advancing fatigue crack is a minor perturbation in an otherwise elastic field, linear elastic fracture mechanics solutions provide appropriate continuum descriptions for fatigue fracture. As noted in the preceding chapter, Paris, Gomez & Anderson (1961) and Paris & Erdogan (1963) suggested that for a cyclic variation of the imposed stress field, the linear elastic fracture mechanics characterization of the rate of fatigue crack growth should be based on the stress intensity factor range,

$$\Delta K = K_{max} - K_{min}. \tag{10.1}$$

K_{max} and K_{min} are the maximum and minimim values, respectively, of the stress intensity factor during a fatigue stress cycle. For an edge-cracked fatigue test specimen,

$$K_{max} = Y\sigma_{max} \sqrt{\pi a}, \quad K_{min} = Y\sigma_{min} \sqrt{\pi a},$$
$$\Delta K = Y\Delta\sigma \sqrt{\pi a}, \quad \Delta\sigma = \sigma_{max} - \sigma_{min}, \tag{10.2}$$

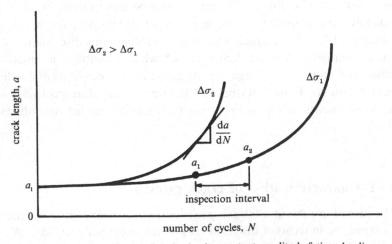

Fig. 10.1. Typical crack growth behavior in constant amplitude fatigue loading.

where Y is a geometrical factor which depends on the ratio of crack length a to the width of the specimen W, and σ_{max} and σ_{min} are the maximum and minimum values, respectively, of the fatigue stress cycle. Paris, Gomez & Anderson (1961) and Paris & Erdogan (1963) showed that the fatigue crack growth increment da/dN is related to the stress intensity factor range by the power law relationship

$$\frac{da}{dN} = C(\Delta K)^m,$$ (10.3)

where C and m are scaling constants. These constants are influenced by such variables as material microstructure, environment and temperature (both of which could promote a strong effect of cyclic load frequency and waveform), and load ratio, R. The load ratio is defined as

$$R = \frac{\sigma_{min}}{\sigma_{max}} = \frac{K_{min}}{K_{max}}.$$ (10.4)

The exponent m in Eq. 10.3 is typically between two and four for ductile metallic alloys. Equation 10.3 is applicable for a single mode of far-field loading and for a fixed value of R. For tensile fatigue, it is understood that ΔK refers to the range of mode I stress intensity factors during the stress cycle. Similarly, a stress intensity factor range ΔK_{II} or ΔK_{III} can also be used in Eq. 10.3 to characterize fatigue crack growth in mode II or mode III, respectively.[†]

Although the fracture mechanics approach to fatigue crack growth was a subject of some controversy in the early 1960s, experimental data gathered for a wide range of metallic materials confirmed a power law relationship of the form given in Eq. 10.3. Specifically, crack growth experiments on aluminum alloys with different combinations of stress range and crack length and with different specimen geometries (e.g., Paris & Erdogan, 1963) unambiguously established the validity of such a characterization. Although Eq. 10.3 is empirical, it has remained one of the most useful expressions in the analysis of fatigue crack growth for a vast spectrum of materials and fatigue test conditions. It is important to note here that stable fatigue crack growth occurs at stress intensity factor levels, $K_{max} = \Delta K/(1 - R)$, that are well below the quasi-static fracture toughness, K_{Ic}. In very ductile metallic materials with high fracture toughness, such as low strength steels, the onset of fatigue crack growth can occur at K_{max} values as low as one-hundredth of K_{Ic}.

As noted earlier, it has been known since the 1960s that the unique characterization of fatigue crack growth using Eq. 10.3 holds only for fixed values of the load ratio, R, and fixed environmental conditions. If environmental conditions, including temperature, have a strong effect on fatigue fracture, loading parameters such as cyclic freqency and waveform would also be expected to have a pronounced effect on crack growth rates (see Chapter 16). Changes in these test conditions can lead to changes in the empirical constants C and m and in the fatigue crack growth rates. For fixed environmental conditions, one would then expect the crack growth to be

---

[†] This definition of a 'driving force' for fatigue crack growth pertains to situations where the crack remains fully open during the entire fatigue cycle.

strongly influenced not only by ΔK, but also by the load ratio, R. This possibility then naturally gives rise to the notion that the general mechanical conditions governing fatigue crack growth are most accurately characterized by two parameters, i.e. ΔK and R, or equivalently, ΔK and K_{max}, for constant amplitude fatigue because $K_{max} = \Delta K/(1 - R)$:

$$\frac{da}{dN} = f(\Delta K, \ R) = g(\Delta K, \ K_{max}), \tag{10.5}$$

where f and g are functions which determine the numerical values of C and m in Eq. 10.3. The need for such a two-parameter characterization was ostensibly first demonstrated by Schmidt & Paris (1973) with experimental results for the threshold stress intensity factor range, ΔK_0, for fatigue crack growth in steels. They showed that the magnitude of the threshold for the onset of fatigue crack growth, ΔK_0, is intimately linked to the existence of a threshold value of applied ΔK *and* a critical value of R, or equivalently, to the threshold values of ΔK *and* K_{max}. Similar interpretations can also be gained from the experimentals of Döker & Marci (1983). Two-parameter characterizations involving ΔK and K_{max} have also been advanced to rationalize the subcritical fatigue crack growth characteristics of ceramics and organic composites by Kishimoto *et al.* (1987) and Hojo *et al.* (1987), respectively (see the next chapter for a discussion of these approaches for nonmetallic materials). On the basis of a survey of available experimental information, Vasudevan & Sadananda (1995) have shown how the conditions for the onset of fatigue crack growth are governed by the intrinsic threshold values of both ΔK and K_{max}. A more detailed discussion of these approaches is taken up in Sections 10.4 and 11.2.2.

10.1.2 Fatigue life calculations

The simplicity of the Paris power law relationship provides a means of estimating the useful life of a fatigued component for design or failure analysis. The fatigue life or the number of fatigue cycles to failure is calculated by integrating Eq. 10.3 from an assumed initial defect size a_0 to a critical crack size a_f. When the stress intensity factor is defined as in Eq. 10.2,

$$\frac{da}{dN} = C(Y\Delta\sigma\sqrt{\pi a})^m, \tag{10.6}$$

for a fixed value of the load ratio, R. If fatigue loading involves a constant amplitude of far-field stresses and a crack length change $(a_f - a_0)$ over which Y is roughly constant,

$$CY^m(\Delta\sigma)^m\pi^{m/2}\int_0^{N_f} dN = \int_{a_0}^{a_f} \frac{da}{a^{m/2}}. \tag{10.7}$$

The resulting fatigue life is

$$N_f = \frac{2}{(m-2)CY^m(\Delta\sigma)^m\pi^{m/2}}\left\{\frac{1}{(a_0)^{(m-2)/2}} - \frac{1}{(a_f)^{(m-2)/2}}\right\} \tag{10.8}$$

for $m \neq 2$ and

$$N_\mathrm{f} = \frac{1}{CY^2(\Delta\sigma)^2\pi}\ln\frac{a_\mathrm{f}}{a_0} \tag{10.9}$$

for $m = 2$. In general, Y varies with a and hence the integration is performed numerically. It is easily seen from Eq. 10.8 that when $a_0 \ll a_\mathrm{f}$ (which is usually the case for ductile alloys), N_f is not very sensitive to a_f, but is markedly influenced by a_0.

10.2 Microscopic stages of fatigue crack growth

The microscopic mode of fatigue crack growth is strongly affected by the slip characteristics of the material, characteristic microstructural dimensions, applied stress level and the extent of near-tip plasticity. In ductile solids, cyclic crack growth is envisioned as a process of intense localized deformation in macroscopic slip bands near the crack tip.

10.2.1 Stage I fatigue crack growth

When the crack and the zone of plastic deformation surrounding the crack tip are confined to within a few grain diameters, crack growth occurs predominantly by single shear, in the direction of the primary slip system, Figs. 10.2(a) and (b). This single slip mechanism leading to a zig-zag crack path has been termed *stage I* crack growth by Forsyth (1962). Stage I crack growth has also been observed in a variety of ferrous, aluminum and titanium alloys even for crack length values substantially larger than the grain size provided that the zone of near-tip plasticity is smaller than the grain dimensions (i.e. for very low ΔK levels). Further discussion of stage I crack growth is given in later sections.

10.2.2 Stage II crack growth and fatigue striations

At higher stress intensity range values, the plastic zone at the crack tip encompasses many grains. The attendant crack growth process involves simultaneous or alternating flow along two slip systems. This duplex slip mechanism, termed *stage II* by Forsyth (1962), results in a planar (mode I) crack path normal to the far-field tensile axis, Figs. 10.3(a) and (b). In single crystals, the transition from stage I to stage II growth causes the formation of dislocation cell structures and the breakdown of PSBs at the crack tip.

While the fracture surfaces created by stage I fatigue exhibit a serrated or faceted profile, stage II crack growth in many engineering alloys leads to the formation of fatigue 'striations'. These striations, first observed by Zappfe & Worden (1951), are ripples on the fracture surface. Figure 10.4 shows an example of fatigue striations formed in cyclic tension. (Additional examples are provided in Section 10.7 and

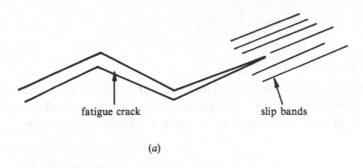

fatigue crack   slip bands

(a)

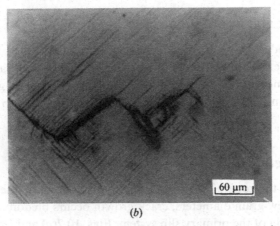

| 60 µm |

(b)

Fig. 10.2. (a) A schematic of stage I fatigue crack growth. (b) An example of stage I fatigue crack growth in a single crystal of Mar M-200 nickel-base superalloy. The tensile axis is vertical. The {100} plane is normal to the tensile axis. (Photograph from P.B. Aswath & S. Suresh.)

Chapter 14.) For cyclic loads in the Paris regime of fatigue crack advance, it has been found that the spacing between adjacent striations correlates with the experimentally measured average crack growth rate per cycle. Such correlations, first shown by Forsyth & Ryder (1960), have been most clearly documented for stage II crack growth in aluminum alloys and have led to the widespread use of striation spacing as an *a posteriori* indicator of crack growth history for failure analysis in both constant amplitude and variable amplitude fatigue.

It is important to note that not all engineering materials form striations during stage II fatigue. Striations are clearly seen in pure metals and many ductile alloys. Many engineering polymers also exhibit well-defined fatigue striations on the fracture surfaces (see Chapter 12). However, in steels they occur very infrequently and are often barely visible in cold-worked alloys. The possibility of striation development is also known to be strongly influenced by the value of ΔK, stress state, environment and alloy content (e.g., Pelloux, 1970; Lankford & Davidson, 1983; Roven, Langoy & Nes, 1987). Furthermore, as shown in later sections, microvoid coalescence and intergranular failure associated with 'static modes' are seen at high ΔK levels, whereas faceted crack advance has been documented for low ΔK levels.

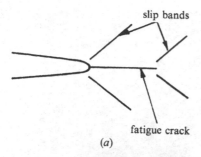

(a)

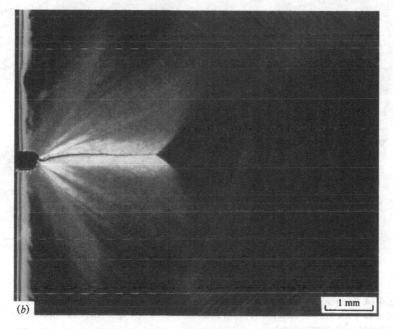

(b)

Fig. 10.3. (*a*) An idealization of stage II fatigue crack growth. (*b*) An example of stage II crack growth in a Cu single crystal. Two sets of {111} slip planes (white areas) are seen. The tensile axis is vertical. (From Neumann, 1969. Copyright Pergamon Press plc. Reprinted with permission.)

10.2.3 Models for striation formation

A number of conceptual models have been proposed over the years to rationalize the formation of fatigue striations and the planar growth of Stage II fatigue cracks. Of these, an appealing physical concept is the plastic blunting idealization of Laird (1967). In this model, the increment of crack extension per fatigue cycle is envisioned as occurring due to the plastic blunting of the crack tip. A crack tip amenable to duplex slip plastically blunts as shown in Fig. 10.5 upon the application of a tensile stress. The blunting process also effectively results in the extension of the crack by a distance of the order of the crack tip opening displacement. If the far-field stress is reversed into compression, the crack tip re-sharpens only to be blunted

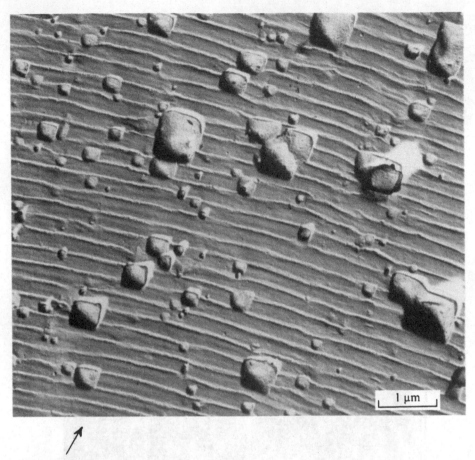

Fig. 10.4. Fatigue striations on the etched failure surface in 2024-T3 aluminum alloy. The arrow indicates crack growth direction. (From Pelloux, 1969. Copyright American Society for Metals. Reprinted with permission.)

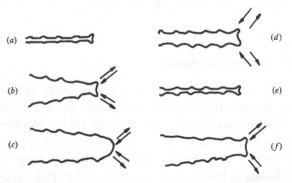

Fig. 10.5. An idealization of plastic blunting and re-sharpening which leads to stage II fatigue crack growth in fully-reversed fatigue. (*a*) Zero load, (*b*) small tensile load, (*c*) peak tensile load, (*d*) onset of load reversal, (*e*) peak compressive load, and (*f*) small tensile load in the subsequent tensile cycle. Arrows indicate slip direction. (After Laird, 1967.)

again during the next tension sequence (e.g., Laird & Smith, 1962). Since the closure of the crack during compression cannot fully negate the blunting and the attendant extension of the crack during the preceding tension load, *net* crack growth occurs during a fatigue cycle leading to the formation of striations, as depicted schematically in Fig. 10.5.

Laird's model applies to a wide variety of ductile materials, including polymers, because of its general description of the plastic blunting process. On the other hand, in ductile metal single crystals, shear on two slip planes which intersect the crack-front can be modeled as shown in Fig. 10.6. The work hardening of the primary slip plane (say, plane 1) leads to alternating shear on another slip plane (plane 2). Kinematic irreversibility of cyclic slip (see Chapter 4) causes net crack growth in a planar fashion along the line of intersection of the two slip planes. Many crystallographic details of this alternating slip process have been verified experimentally by Neumann (1969, 1974) on FCC crystals. For example, if an FCC monocrystal is loaded along the $\langle 100 \rangle$ direction, two $\{111\}$ planes are symmetrically loaded about the crack plane. The resulting crackfront should be perfectly planar and parallel to the intersection line of the two $\{111\}$ slip planes, which is the $\langle 011 \rangle$ direction. This is evident from the micrograph of Fig. 10.3(*b*).

The fatigue striations discussed up to this point are a consequence of duplex slip and they are known as *ductile striations*. The formation of *brittle striations*, which lie on crystallographic $\{100\}$ facets in high strength aluminum alloys, also occurs as a

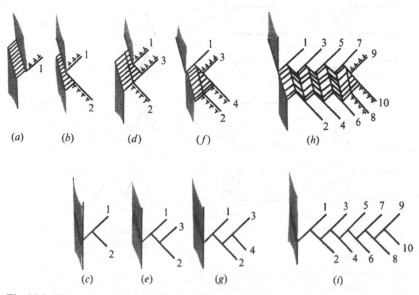

Fig. 10.6. Alternating slip model for stage II crack growth under fully reversed cyclic loads. The figures on the top row show tensile loading stages, and those on the bottom pertain to unloading stages. (After Neumann, 1969.)

consequence of environmental interactions (e.g., Forsyth & Ryder, 1960; Laird, 1967; Wanhill, 1975; Lynch, 1981). The term *cleavage striations* is also employed to describe these brittle striations (see Section 16.3.2).

10.2.4 Environmental effects on stage II fatigue

Some aspects of the effects of environment on the formation of fatigue striations can also be incorporated into the above 'unzipping' model of Neumann. It was first shown by Meyn (1968) that striation formation may be completely suppressed *in vacuo* in aluminum alloys which form well-defined striations in moist air. Similar findings were also reported by Pelloux (1969, 1970) for 2024-T3, 7075-T6 and Ti–6Al–4V alloys fatigued in a vacuum of 5×10^{-6} torr. Pelloux suggested that the alternating shear process is reversible unless an oxide film is formed on the slip steps created at the crack tip. This oxide layer impedes slip upon load reversal. Figure 10.7 illustrates the opening and closing of a crack during two fully reversed fatigue cycles in air and *in vacuo*. Pelloux postulated that the inhibition of reverse slip due to oxidation results in a larger *net* fatigue crack growth increment per cycle in air than *in vacuo*. Furthermore, the crack profiles associated with the opening and closing of the fatigue crack *in vacuo* do not produce any fatigue striations such as those found in air, Fig. 10.7.

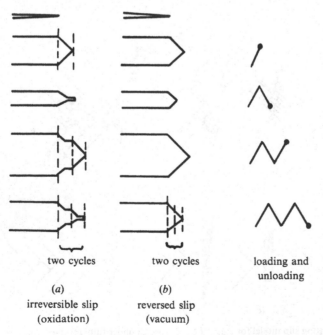

two cycles   two cycles   loading and
                                        unloading

(a)   (b)
irreversible slip   reversed slip
(oxidation)   (vacuum)

Fig. 10.7. Effects of environment on fatigue crack growth by alternating slip. See text for details. (After Pelloux, 1969.)

This idealization found experimental backing on several accounts: (i) Fatigue striations were not found *in vacuo* in a number of alloys which formed striations in air. (ii) The macroscopic rate of fatigue crack growth *in vacuo* was more than an order of magnitude slower than that in air. It has also been suggested (e.g., Neumann, 1983) that enhanced rubbing of the crack faces (due to the absence of any protective surface films on the freshly formed macroscopic fracture surfaces) leads to the obliteration of fatigue striations by fretting and causes a reduction in the rate of crack growth *in vacuo*.

10.3 Different regimes of fatigue crack growth

The Paris power law relationship, showing a linear variation of $\log da/dN$ with $\log \Delta K$, pertains to stable fatigue fracture over only a portion of the total crack growth resistance curve for an alloy. At extreme values of ΔK, both below and above that of the Paris regime, there is a precipitous rise in crack growth rates with increasing ΔK. For most engineering alloys, a plot of $\log da/dN$ against $\log \Delta K$ exhibits a sigmoidal variation, as shown schematically in Fig. 10.8. In this plot, three distinct regimes of crack growth can be identified: regime A, in which the *average* growth increment per cycle is smaller than a lattice spacing, is associated with the existence of a threshold stress intensity factor range ΔK_0. Below this threshold, cracks either remain dormant or grow at undetectable rates; above the threshold, there is a steep increase in da/dN with ΔK. Regime B, known as the Paris regime, exhibits a linear variation of $\log da/dN$ with $\log \Delta K$, Eq. 10.3. Regime C

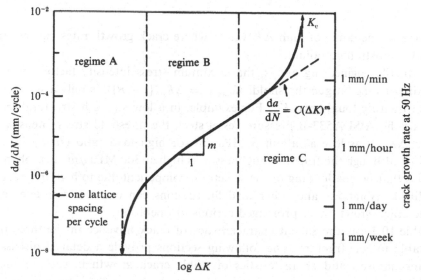

Fig. 10.8. Schematic illustration of the different regimes of stable fatigue crack propagation.

Table 10.1. *Characteristics of the three regimes of fatigue crack growth.*

Regime	A	B	C
Terminology	Slow-growth rate (near-threshold)	Mid-growth rate (Paris regime)	High-growth rate
Microscopic failure mode	Stage I, single shear	Stage II, (striations) and duplex slip	Additional static modes
Fracture surface features	Faceted or serrated	Planar, with ripples	Additional cleavage or microvoid coalescence
Crack closure levels	High	Low	—
Microstructural effects	Large	Small	Large
Load ratio effects	Large	Small	Large
Environmental effects	Large	*	Small
Stress state effects	—	Large	Large
Near-tip plasticity[†]	$r_c \leq d_g$	$r_c \geq d_g$	$r_c \gg d_g$

*Large influence on crack growth for certain combinations of environment, load ratio and frequency.

[†] r_c and d_g refer to the cyclic plastic zone size given in Eq. 9.85, and the grain size, respectively.

pertains to the range of high ΔK values where crack growth rates increase rapidly causing catastrophic failure.

In ductile engineering alloys, the maximum stress intensity factor value corresponding to the fatigue threshold, $K_{max,0}$ $(= \Delta K_0 / \{1 - R\})$, is only a small fraction of the fracture toughness, K_{Ic}. For example, in a low strength structural material, such as the AISI A533-B pressure vessel steel, the threshold stress intensity factor range ΔK_0 is as low as about 3 MPa$\sqrt{m}$ for high load ratio $(R = 0.75)$ fatigue loading, although the fracture toughness is as high as 300 MPa$\sqrt{m}$. This observation implies that fatigue loading of the structural component due to high frequency, low amplitude vibrations and other load fluctuations can cause undesirable levels of stable crack growth over prolonged periods of operation.

Table 10.1 lists the salient characteristics of crack advance in the three regimes of stable fatigue fracture. The following sections provide a detailed discussion of the mechanisms and characteristics of stable crack growth in each of the three regimes.

10.4 Near-threshold fatigue crack growth

Fatigue crack propagation in regime A deals primarily with noncontinuum failure processes, where the increment of average crack extension is less than 10^{-6} mm/cycle. In this region, the stress intensity factor range approaches the fatigue crack growth threshold ΔK_0. An operational definition for ΔK_0 is commonly used in terms of a maximum crack growth rate (typically 10^{-8} mm/cycle), based on the accuracy of the crack monitoring system and the number of elapsed cycles. For example, if the method used for crack length measurement is accurate to at least 0.1 mm, and if no crack advance is detected for at least 10^7 fatigue cycles, a crack growth threshold is considered to have been reached.

Structural engineers are often faced with the task of designing components which can withstand very high frequency, low amplitude loads over fatigue lifetimes as high as $10^{10} - 10^{12}$ load reversals. A high speed turbine rotor operating at 3000 revolutions per minute may be subjected to 10^{10} stress cycles over a typical life span of 20 years. If fatigue crack growth were to occur in this rotor at a seemingly insignificant rate of 3×10^{-9} mm/cycle (i.e. one-hundredth of an atomic spacing per cycle), this would still represent a total crack advance of 30 mm during the life of the rotor, which is sufficient to cause catastrophic failure. Although fatigue design based on ΔK_0 may prove to be overly conservative in very ductile alloys, information on fatigue thresholds is often included in the analysis of failure in turbine blades, turbine shafts, and alternator rotors, as well as in acoustic fatigue of welds in gas circuitry used in nuclear reactors. From both the scientific and practical viewpoints, near-threshold fatigue offers a challenging research topic in that large changes in the values of ΔK_0 (and hence in the total fatigue life) can be effected by alterations in the material microstructure, load ratio, environment and crack size (see Table 10.1).

There have been many research studies on the near-threshold fatigue of engineering alloys. Several international conferences (e.g., Davidson & Suresh, 1984) and a research monograph (Taylor, 1989) have been devoted exclusively to the discussion of fatigue thresholds. Figure 10.9 shows the typical range of ΔK_0 values as a function of load ratio, R, for a variety of Fe, Ni, Cu, Ti and Al alloys. Note that the threshold values for different microstructures of the same alloy system as well as for different alloys exhibit significantly less variation at high R values than at low R values.[†] Some mechanisms underlying this trend will become evident from the discussions to be presented in Chapter 14.

It is evident from Fig. 10.9 that the threshold ΔK_0 values for Fe and Ni alloys decrease with increasing R before reaching a constant plateau. This dependence of ΔK_0 on R was examined first in detail by Schmidt & Paris (1973), as discussed earlier in Section 10.1.1. They rationalized the occurrence of a fatigue threshold, for any load ratio R, in terms of two critical values: a constant value of the stress intensity factor

---

[†] From a comprehensive survey of published data, Liaw, Leax & Logsdon (1983) note that ΔK_0 values, when modified to exclude the effects of crack closure, vary in proportion to Young's modulus for a wide variety of alloy systems.

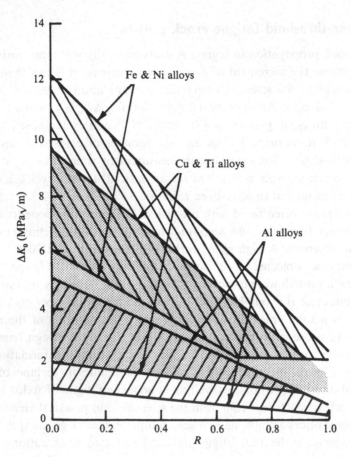

Fig. 10.9. Typical range of threshold ΔK_0 values measured in room temperature air and plotted as a function of the load ratio, R, for a variety of engineering alloys. (After Liaw, Leax & Logsdon, 1983.)

range, ΔK_0 and a critical value of the load ratio, R_{cr}. These two critical parameters to characterize the fatigue threshold can also be expressed in terms of ΔK_0 and $K_{max,0} = \Delta K_0/(1 - R)$, as shown in Fig. 10.10. In this figure, ΔK_{th} is the lowest value of ΔK_0 (at $R \geq R_{cr}$) below which a fatigue crack does not advance at any R.

Schmidt & Paris showed that when the fatigue crack growth threshold is characterized in terms of the amplitude of the stress intensity factor, ΔK_0, there exists a critical value of $R = R_{cr}$, below which a decrease in R causes an increase in the apparent ΔK_0. For $R \geq R_{cr}$, $\Delta K_0 = \Delta K_{th}$ is independent of R. Alternatively, when the fatigue crack growth threshold is characterized in terms of the maximum value of the stress intensity factor, $K_{max,0}$, it is found that $K_{max,0}$ is a constant for $R < R_{cr}$. For $R \geq R_{cr}$, $K_{max,0}$ increases with R, consistent with a constant value of ΔK_0 (= $K_{max,0} \times (1 - R)$). For any R, one would then expect that there exist two critical driving force parameters, ΔK_0 and $K_{max,0}$ which govern the threshold condition for

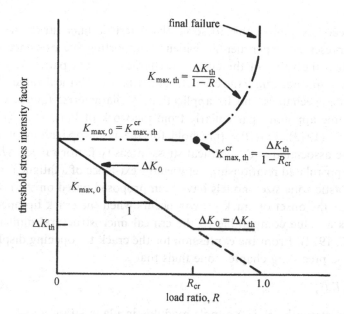

Fig. 10.10. Schematic variation of the amplitude or the maximum value of the threshold stress intensity factor as a function of load ratio.

crack growth. A constant value of the former one determines the threshold conditions at high R values while a constant value of the latter one determines the threshold at low R.[†]

Vasudevan & Sadananda (1995) have plotted the trends shown in Fig. 10.10 in terms of ΔK_0 versus $K_{\text{max},0}$ for a variety of available experimental data on fatigue thresholds. These plots show how fatigue cracks begin to advance when intrinsic values of both ΔK_{th} and $K_{\text{max,th}}^{\text{cr}}$ are exceeded for a given constant amplitude cyclic loading condition. Vasudevan and Sadananda thus demonstrate for a wide range of alloys that the threshold for fatigue crack growth is governed by critical values of both ΔK and K_{max}.

10.4.1 Models for fatigue thresholds

The existence of a fatigue crack growth threshold was first postulated by McClintock (1963). He used the criterion that fatigue crack growth occurs when a critical value of local strain or accumulated damage over certain characteristic distance ahead of the crack tip reaches a critical value. Using elastic–plastic analyses, McClintock argued that fatigue cracks may cease to advance when the extent of

---

[†] Schmidt & Paris (1973) rationalize the increase in ΔK_0 with decreasing R or the plateau in $K_{\text{max},0} = K_{\text{max,th}}^{\text{cr}} = \Delta K_{\text{th}}/(1 - R_{\text{cr}})$ versus R for $R \leq R_{\text{cr}}$ in terms of reduction to the effective 'driving force'; see Chapter 14.

near-tip plasticity becomes comparable to some characteristic microstructural size scale. Frost (1966) presented experimental evidence supporting the existence of a fatigue threshold when the value of the empirical characterization parameter, $\sigma_a^3 a$, where σ_a is the stress amplitude and a the crack length, attained a critical value. With the advent of fracture mechanics and its application to characterize fatigue crack growth, it soon became apparent, particularly from the work of Paris *et al.* (1972) and Schmidt & Paris (1973), that the threshold for the nonpropagation of long fatigue cracks can be associated with a critical stress intensity factor range, ΔK_0.

In addition to the postulated relationship between the existence of a fatigue threshold and the cyclic plastic zone size, models have been proposed based on the notion that the threshold for the onset of crack growth occurs when the crack tip opening displacement attains a value comparable to the critical microstructural dimension (e.g., Donahue *et al.*, 1972). From the expression for the crack tip opening displacement presented in the preceding chapter, one finds that

$$\Delta K_0 \propto \sqrt{\sigma_y E' l^*},\tag{10.10}$$

where σ_y is the yield strength, E' is Young's modulus in plane strain and l^* is the microstructural size scale such as grain size. The prediction that the threshold increases with increasing flow strength is generally inconsistent with experiment.

Sadananda & Shahinian (1977) proposed that the threshold for crack growth is reached when the shear stress, τ, required to nucleate and move a dislocation from the crack tip reaches a critical value. This criterion led to the result that

$$\Delta K_0 \propto \tau \sqrt{b},\tag{10.11}$$

where b is the magnitude of the Burgers vector. This model, as well as other arguments where a fatigue threshold is expected to be attained when a critical stress or strain is exceeded over a characteristic microstructural distance (Weiss & Lal, 1974), fail to capture the effects of various microstructural factors on fatigue thresholds.

10.4.2 Effects of microstructural size scale

The transition from the near-threshold regime to the intermediate stage of fatigue crack growth is also accompanied by a noticeable change from a microstructure-sensitive to a microstructure-insensitive fracture behavior. Experimental observations of such transitions and of the dependence of such transitions on metallurgical variables, particularly the grain size and slip characteristics, have now been reported for titanium alloys (e.g., Irving & Beevers, 1974; Lindigkeit *et al.*, 1979; Yoder, Cooley & Crooker, 1979), aluminum alloys (e.g., Asaro, Hermann & Baik, 1981; Carter *et al.*, 1984; Suresh, Vasudevan & Bretz, 1984), ferrous alloys (e.g., Masounave & Baïlon, 1976; Ritchie, 1979; Taira, Tanaka & Hoshina, 1979; Gray, Williams & Thompson, 1983) and nickel-base superalloys (King, 1982).

The transition in fatigue failure modes from the slow growth rate regime to the Paris regime typically occurs when the size of the cyclic plastic zone r_c becomes comparable to the characteristic microstructural dimension l^* of the alloy system. If the stress intensity factor range corresponding to this transition is termed ΔK_T,

$$r_c = \alpha \left(\frac{\Delta K_T}{\sigma_y'}\right)^2 \approx l^*. \tag{10.12}$$

Here σ_y' is the cyclic yield strength and $\alpha = 1/4\pi$ is a numerical factor from Eq. 9.85. Rearranging this equation, it is seen that

$$\Delta K_T \propto \sigma_y' \sqrt{l^*}. \tag{10.13}$$

Some of the clearest examples of a transition from a microstructure-sensitive to a microstructure-insensitive growth behavior at ΔK_T have been reported by Yoder, Cooley & Crooker (1979) for beta-annealed Ti–6Al–4V and Ti–8Al–1Mo–1V alloys with Widmanstätten structures. Here the size of the Widmanstätten packet size is the characteristic microstructural size scale, l^*. Yoder *et al.* found that the transition from a crystallographic, serrated mode of fracture to a striated mode of failure occurred at a value of ΔK_T which increased with the Widmanstätten packet size as per Eq. 10.13. Similar results have also been reported for ferritic–pearlitic steels by Yoder, Cooley & Crooker (1983). In this case, the threshold ΔK_0 (taken to be approximately equal to ΔK_T in view of the steep slope in crack growth curve in the near-threshold regime) is found to be of the form

$$\Delta K_0 = A + B\sqrt{d_g}, \tag{10.14}$$

where d_g is the ferrite grain size, and A and B are material constants. Taira, Tanaka & Hoshina (1979) also found a similar trend in low carbon steels. These latter authors postulated a criterion for predicting ΔK_0 ($\approx \Delta K_T$) based on a critical slip band zone size, which is analogous to the cyclic plastic zone size. Taira *et al.* assumed that a crack growth threshold is determined by whether the slip band near the crack tip propagates into the adjacent grain or not. When the size of the slip band zone at the crack tip, r_s, becomes greater than the grain size d_g, the slip band can traverse the grain boundary and propagate into the adjacent grain. For low carbon steels, Taira *et al.* found that $r_s \propto d_g$.

10.4.3 Effects of slip characteristics

Even when the size of the cyclic plastic zone at the tip of a near-threshold crack is the same for two different microstructures (with the same grain size), different fatigue thresholds may result because of the differences in slip characteristics. The influence of slip behavior on thresholds is best illustrated with the example of a precipitation-hardened alloy system. Consider the case of under-aged and over-aged microstructures of an aluminum alloy, where the two ageing conditions are designed to give the same macroscopic yield strength. In the case of the under-aged alloy, the matrix consists of shearable and coherent precipitates. These strengthening particles are amenable to be sheared by slip bands ahead of a fatigue crack.

It was suggested by Hornbogen & Zum Gahr (1976) that in such heterogeneously deforming microstructures, dislocation glide occurs within grains predominantly along a single slip system. The single shear failure promotes a highly serrated and tortuous crack path in the under-aged condition, with the wavelength of crack path tortuosity being of the order of the grain size. On the other hand, incoherent precipitates present in the over-aged alloy are bypassed, rather than sheared, by dislocations which form (Orowan) loops around the particles. The resulting deformation mode in the over-aged alloys is one of duplex slip and the near-threshold cracks exhibit a planar transgranular fracture profile.† Furthermore, the kinematic irreversibility of cyclic slip is more restricted in the over-aged alloy than in the under-aged condition. This difference is viewed as a factor contributing to the more beneficial crack growth characteristics of the under-aged aluminum alloys observed in the near-threshold regime. (It will be shown in Chapter 14 that, in addition to these intrinsic microstructural effects, crack deflection mechanisms associated with the more tortuous crack paths in the under-aged condition and the attendant possibility of contact between crack faces can promote lower crack growth rates than in the over-aged microstructure.)

The dramatic effect of slip characteristics on the path of a fatigue crack is evident in the optical micrographs, shown in Fig. 10.11, of near-threshold cracks in an Al–Cu–Li alloy commercially available under the trade name Weldalite 049. When this alloy is age-hardened to a T3 temper (under-aged condition), a tortuous, crystallographic crack path is observed, Fig. 10.11(a). However, over-aging (T8 temper) promotes a perfectly straight planar crack growth, Fig. 10.11(b), in accord with the expectations based on the above arguments.

The under-aged and over-aged microstructures of an aluminum alloy, with the same yield strength and grain structure, can also exhibit vastly different threshold ΔK_0 values in both vacuum and moist air, Fig. 10.12. For similar ageing treatments and flow strength values, increases in grain size result in an improvement in ΔK_0. As ΔK levels are increased to higher values corresponding to the Paris regime, the large plastic zones (of average dimensions greater than the grain size) promote duplex slip in both the ageing conditions, thereby minimizing the sensitivity of crack growth to the underlying microstructure.

The above results clearly point to the dilemma faced by materials scientists when designing engineering alloys with optimum fatigue resistance. In the near-threshold regime of fatigue crack growth, where a significant portion of stable crack growth life is expended, an increase in the grain size of the material or a decrease in yield strength generally results in a marked reduction in the near-threshold fatigue crack growth rates and an increase in ΔK_0. On the other hand, the resistance of a (smooth-surfaced) material to fatigue crack initiation, expressed in terms of an endurance

---

† Cleavage-like growth of fatigue cracks along grain boundaries and through grains have also been observed in the near-threshold region for a wide variety of alloy systems (e.g., Irving & Beevers, 1974). These failure modes, termed *cyclic cleavage*, do not represent cleavage fracture in the true sense because they exhibit slow, progressive failure rather than a brittle separation.

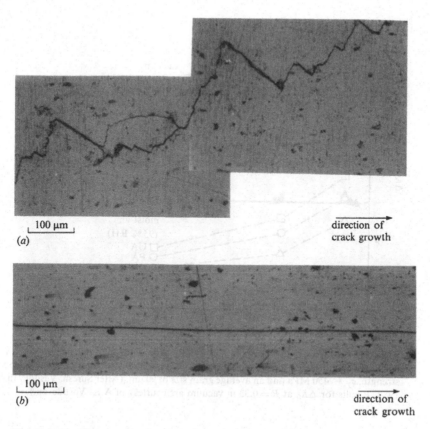

Fig. 10.11. Profiles of near-threshold fatigue cracks observed in (*a*) an under-aged condition (T3 temper) and (*b*) an over-aged condition (T8 temper) of an Al–Cu–Li alloy in laboratory air at a high load ratio, *R*. (From Blankenship & Starke, 1991. Copyright Pergamon Press plc. Reprinted with permission.)

limit, often increases with decreasing grain size and increasing strength of the material. Although these results may appear contradictory, they can be rationalized by noting that the former inference is valid solely for fatigue crack *growth*, especially at low stress intensity range levels, whereas the latter conclusion pertains primarily to crack *initiation*. If pre-existing flaws are not of a major concern, a finer grain size which promotes higher strength and higher endurance limit is favored. On the other hand, if a defect-tolerant design is adopted and the flawed component is expected to be subjected to high frequency, low stress amplitude loading over a prolonged time span, a coarser grained microstructure would be more appealing. Optimization of the microstructural dimensions for improved resistance to both crack initiation and crack growth would require a trade-off in the choice of grain size or even the development of a structure with a distribution of grain sizes.

It is necessary to emphasize here that one of the principal factors contributing to the dependence of fatigue thresholds on microstructural variables (such as grain size

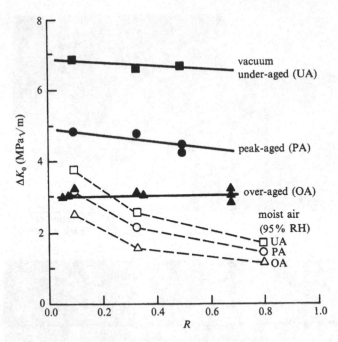

Fig. 10.12. Variation of the threshold ΔK_0 as a function of the load ratio, R, for under-aged and over-aged conditions of a 7075 aluminum alloy. Both microstructures have the same yield strength, $\sigma_y = 450$ MPa and an average grain size of 90 µm. (After Suresh, Vasudevan & Bretz, 1984. Results for ΔK_0 at $R = 0.33$ in vacuum are courtesy of A.K. Vasudevan.)

and yield strength), mechanical factors (such as load ratio) and environment is the development of a tortuous or deflected crack path (which is promoted by stage I mechanism) and the resultant possibility of crack face contact. These effects can alter the 'effective driving force' for fatigue crack growth. Consequently, the parameters used to characterize fatigue crack growth, Eqs. 10.1–10.3, should be appropriately modified to account for the changes in the near-tip stress intensity factor range. Failure to develop a quantitative description of the changes in driving force due to crack deflection and the attendant 'crack closure' effects has resulted in opposing viewpoints and controversy about the mechanisms of fatigue failure in many alloy systems. In Chapter 14, a detailed discussion is presented on the effects of crack deflection and crack face contact on the propagation of fatigue cracks, particularly in the near-threshold regime.

Since the rate of crack advance is slow in the near-threshold regime, embrittling environments can have a marked effect on fracture resistance. Environmental influence on apparent ΔK_0 values has been documented even when the maximum stress intensity factor value at threshold is appreciably smaller than the threshold for stress corrosion cracking K_{Iscc} under static loads. Mechanisms of corrosion fatigue in the near-threshold regime are discussed in Chapters 14 and 16.

10.4.4 Example problem: Issues of length scales

Problem:

Consider a low strength alloy steel which has the following mechanical properties: Young's modulus, $E = 210$ GPa, Poisson's ratio, $\nu = 0.33$, monotonic yield strength, $\sigma_y = 400$ MPa, stabilized cyclic yield strength, $\sigma_y' = 440$ MPa, plane strain fracture initiation toughness, $K_{Ic} = 100$ MPa$\sqrt{m}$, and nominal values of fatigue crack growth threshold stress intensity factor range, $\Delta K_0 = 7$ MPa$\sqrt{m}$ and 3 MPa$\sqrt{m}$ at R values of 0.0 and 0.75, respectively, at room temperature. The characteristic microstructural length scale for this alloy is the average diameter of an equiaxed grain which is approximately 50 μm. The near-threshold crack growth in this alloy leads to a fracture path which leads to the formation of an average asperity height of 3 μm at both R values. All the fatigue crack growth measurements in this alloy were made with specimens containing long cracks (crack length greater than at least 15 mm) where plane strain and small-scale yielding conditions were ensured at all times. Compute all the characteristic length scales associated with fatigue fracture in this alloy at ΔK and K_{max} values close to the fatigue threshold and compare them with the microstructural and other important dimensions. Speculate about the mechanisms which might possibly be associated with such length scales, and comment on the applicability of the various microscopic mechanisms to this alloy.

Solution:

The maximum plastic zone size at the fatigue crack growth threshold is:

$$r_p = \frac{1}{3\pi}\left(\frac{K_{max,0}}{\sigma_y}\right)^2 = \frac{1}{3\pi}\left(\frac{\Delta K_0}{(1-R)\sigma_y}\right)^2,$$

$$= \frac{1}{3\pi}\left(\frac{7}{400}\right)^2 \text{ m} = 32.5 \text{ } \mu\text{m, for } R = 0.0,$$

$$= \frac{1}{3\pi}\left(\frac{3}{0.25 \times 400}\right)^2 \text{ m} = 95.5 \text{ } \mu\text{m, for } R = 0.75. \tag{10.15}$$

Thus the monotonic plastic zone is fully confined within an average grain at $R = 0.0$, and it spans less than two grain diameters at $R = 0.75$. One would thus expect stage I mechanisms to be operational in this alloy at ΔK values close to the threshold.

The cyclic plastic zone size at threshold is:

$$r_c \approx \frac{1}{3\pi}\left(\frac{\Delta K_0}{2\sigma_y'}\right)^2$$

$$\approx \frac{1}{3\pi}\left(\frac{7}{2 \times 440}\right)^2 \text{ m} = 6.7 \text{ } \mu\text{m, for } R = 0.0,$$

$$= \frac{1}{3\pi}\left(\frac{3}{2 \times 440}\right)^2 \text{ m} = 1.2 \text{ } \mu\text{m, for } R = 0.75. \tag{10.16}$$

It is evident that the criterion for transition from stage I to stage II crack growth, which is implicit in Eq. 10.12, would not hold in this case because the critical crack length l^* is much smaller than the grain size; values of $\Delta K \gg \Delta K_0$ would be predicted in this case for such a transition.

The maximum crack tip opening displacement, Eq. 9.93, at threshold is:

$$\delta_t \approx 0.5 \left(\frac{K_{max,0}^2 (1 - \nu^2)}{2E\sigma_y'} \right) = 0.5 \times \left(\frac{(\Delta K_0)^2}{(1 - R)^2 E\sigma_y} \right),$$

$$\approx 0.5 \left(\frac{7^2 [1 - (0.33)^2]}{2 \times 210 \times 10^3 \times 400} \right) \quad m = 0.26 \ \mu m, \text{ for } R = 0.0,$$

$$\approx 0.5 \left(\frac{3^2 [1 - (0.33)^2]}{(0.25)^2 \times 2 \times 210 \times 10^3 \times 400} \right) \quad m = 0.76 \ \mu m, \text{ for } R = 0.75.$$

$$(10.17)$$

It is evident that the maximum extent of crack tip opening displacement is small compared to the characteristic microstructural dimension, and even smaller than the fracture surface asperity dimension. If oxidation of the fracture surfaces (as, for example, from corrosion of the alloy in a highly moist or aqueous environment) can lead to corrosion products whose dimensions within the crack are comparable to the crack opening displacement, it is possible to close the crack even under an imposed tensile cyclic stress. Similarly, if there is severe asperity locking between the crack faces during cyclic loading, the 'effective driving force' for fracture can be significantly smaller than the apparent or nominal ΔK in the near-threshold regime. As a result, the crack may prematurely arrest. (See Chapter 14 for further discussions of these effects.)

10.4.5 On the determination of fatigue thresholds

The experimental determination of threshold ΔK_0 and near-threshold crack growth rates is strongly influenced by the artifacts of the test methods. ΔK_0 values are generally measured using the so-called load-shedding technique. In this method, a fatigue pre-crack is initiated ahead of a starter notch in a compact tension (CT) or a center-notched specimen at ΔK values corresponding to the Paris regime of crack growth. The far-field stress range, and hence the value of ΔK, are successively reduced in stages, with the cyclic frequency and load ratio held constant. At each stage of constant far-field load range, the fatigue crack is allowed to propagate over distances of the order of a millimeter (or at least four times the maximum current plastic zone size) before the far-field ΔK is reduced by an amount not more than 10%. This step is repeated until the threshold, ΔK_0, for no crack growth or crack growth typically smaller than 10^{-8} mm/cycle is reached. With the ready availability of computer-controlled electro-servohydraulic fatigue test machines, load-shedding procedures for measuring fatigue thresholds can be carried out automatically without human intervention. For example, Saxena, *et al.* (1978) suggested that a particular K gradient can be achieved during load-shedding by controlling cyclic loads such that the following relationship is satisfied:

$$K_j = K_0 \exp\left[C(a_j - a_0)\right], \qquad (10.18)$$

where K_j and K_0 are the instantaneous and initial values of the stress intensity factor, C is a normalized gradient of K which is equal to $(dK/da)/K$, and a_j and a_0 are the instantaneous and initial values, respectively, of fatigue crack length. If C is assigned a pre-determined negative value, the imposed ΔK values and the resulting crack growth rates will decrease with the progression of the fatigue test.

If the far-field ΔK is dropped rapidly, the growth of the current crack tip through the greatest prior plastic zone induced at the previous higher load level will cause artificial retardation or even complete arrest of fatigue crack growth (see Chapter 14 for the mechanisms associated with such retardation effects during changing load amplitudes). However, if the ΔK values are reduced very slowly, the development of crack closure in some materials may once again result in premature closure of the crack faces and an apparently high value of ΔK_0 due to fracture surface oxidation in moist environments. The choice of a particular gradient of ΔK is somewhat arbitrary and can lead to considerable experimental uncertainties and errors in the measurement of fatigue thresholds. Furthermore, reductions in imposed ΔK with increasing crack length lead to the determination of a *crack growth arrest* threshold rather than the threshold for the *onset of crack growth*. It is this latter quantity which is of significance in most practical applications.

Extensive threshold testing programs involving a number of different metallic materials have resulted in the development of Standard E-647-86a by the American Society for Testing and Materials (Philadelphia) for cyclic load reductions to measure fatigue thresholds. However, the load-shedding methods, in general, have come under considerable criticism in recent years in view of the aforementioned experimental uncertainties. At least two alternative methods have been proposed: (i) Suresh (1985b) and Christman & Suresh (1986) proposed a technique whereby self-arresting fatigue pre-cracks are introduced in notched specimens under fully compressive far-field cyclic loads. Subsequently, the specimen is subjected to cyclic tensile loads beginning at stress intensity levels well below the anticipated threshold ΔK_0 value. If no crack growth is detected, the value of ΔK is raised by a small percentage. This procedure is repeated until the threshold ΔK_0 for the onset of crack growth is reached; beyond this point the fatigue test can be carried out at constant far-field cyclic load levels. (ii) Herman, Hertzberg & Jaccard (1988) have popularized a method of increasing R ratio to measure an *intrinsic* fatigue threshold. In this method, the imposed $K_{\max}$ is held fixed and ΔK is progressively reduced by raising R. The argument here is that when the maximum stress intensity factor of the fatigue cycle is held fixed, the size of the maximum plastic zone at the crack tip remains unaltered. Therefore, load reductions would not be expected to cause retardation effects typical of variable amplitude fatigue. Furthermore, closure of the crack faces is virtually eliminated for the high load ratio levels at which ΔK_0 is measured. This latter method, by its very design, cannot be used to measure the effects of load ratio on near-threshold crack growth rates.

It is perhaps worth noting that the use of widely different methods for the determination of fatigue thresholds has resulted in considerable differences in the degree of accuracy and meaning of the data reported in the literature. Caution should be exercised in the interpretation of crack growth mechanisms because the artifacts of the test method can often dominate the intrinsic effects of the microstructural or mechanical variables on fatigue thresholds. In addition to the various methods by which ΔK_0 is measured, the operational definition of ΔK_0, corresponding to the detectable limit of crack growth, itself is somewhat arbitrary. Fatigue tests conducted at conventional test frequencies (typically 0.1 to 100 Hz) take ΔK_0 to be the value of ΔK corresponding to a maximum growth rate of $10^{-7} - 10^{-8}$ mm/cycle. However, with the use of ultrasound methods capable of imposing frequencies as high as 20 000 Hz, it is feasible to measure average crack growth rates as low as 10^{-11} mm/cycle in about the same time period as it takes to measure ΔK_0 using conventional frequencies (e.g., Mitsche, Stanzl & Burkert, 1973; Weiss *et al.*, 1979). The threshold values measured with the ultrasound method are generally lower than those obtained at conventional test frequencies.

10.5 Intermediate region of crack growth

The intermediate or Paris region of fatigue crack growth in engineering alloys has been the subject of most extensive research. The microscopic mechanisms of fatigue crack growth (involving duplex slip and crack blunting and re-sharpening) in the intermediate regime are also generally well represented by the discussions and conceptual models described in detail in Section 10.2.

It is clear from the discussions up to this point that the rates of fatigue crack growth are influenced by several concurrent and mutually competitive mechanistic processes involving the microstructure, mechanical load variables and environment, as well as by crack closure effects. Therefore, it is not surprising to note that no single theory can successfully predict crack propagation rates for a wide variety of test conditions, even for a single alloy system, from fundamental principles. Despite this limitation, numerous attempts have been made in the past several decades to predict the rate of fatigue crack growth in the Paris regime from physical principles. These quantitative models of fatigue crack growth provide useful information for both scientific and practical purposes: (i) In many engineering applications, it is desirable to predict residual lives of components from a knowledge of stresses and crack lengths using models of crack growth. (ii) Theoretical analyses based on realistic mechanisms provide valuable parametric evaluations of the dependence of crack growth rates on different variables, which may not otherwise be done experimentally in a controlled manner. (iii) Crack growth models are sometimes useful for relative ranking of different microstructures in terms of their resistance to fatigue crack growth.

The multitude of theories proposed to predict constant amplitude fatigue crack growth in terms of ΔK can be broadly classified into the following groups.

(a) *Geometrical models* based on the crack tip displacement which, in the Paris regime of crack growth, is related to the crack length increment per cycle (Laird, 1967; Lardner, 1967; McClintock (1967) cited in Pelloux, 1969). The physical basis for these models evolves from the experimental correlations of striation spacings with crack growth rates and from the geometrical relationship between the striation spacing and the crack tip blunting process. Since the cyclic crack tip opening displacement $\Delta \delta_t$ is related to the stress intensity factor by Eq. 9.86,

$$\frac{\mathrm{d}a}{\mathrm{d}N} \approx \Delta \delta_t = \beta \frac{(\Delta K)^2}{\sigma'_y E'}, \tag{10.19}$$

where σ'_y is the cyclic yield strength, E' is Young's modulus in plane strain, and β is a function of yield strain, cyclic strain hardening exponent, the efficiency of blunting and re-sharpening in fatigue, and the numerical constant relating $\Delta \delta_t$ and ΔK in Eq. 9.86. Note that this equation predicts a Paris exponent m of 2, which is reasonable for many alloys. The expectation that $\mathrm{d}a/\mathrm{d}N$ is inversely proportional to E' is generally true for metallic materials. However, the prediction that $(\mathrm{d}a/\mathrm{d}N) \propto (1/\sigma'_y)$ is rarely observed. Furthermore, the quantitative prediction of $\mathrm{d}a/\mathrm{d}N$ based on Eq. 10.19 is good only in some small segments of the Paris regime.

(b) *Damage accumulation models* which invoke various criteria for fatigue fracture based on the critical value of accumulated strains or plastic work at the crack tip. The first such model was apparently due to McClintock (1963) who employed a Coffin–Manson type damage law. In this model, fatigue crack growth occurs when the following function of the local average plastic strain amplitude γ^p_a and the true plastic strain for monotonic fracture γ^p_f, integrated over the number of fatigue cycles N, reaches unity:

$$\int_0^N 4 \left(\frac{\gamma^p_a}{\gamma^p_f} \right)^k \mathrm{d}n = 1, \tag{10.20}$$

where k is an exponent taken to be 1 or 2. Using the near-tip strain fields for shear cracks, McClintock also characterized the growth rates of fatigue cracks in terms of the radius of the plastic zone. His damage theory predicts crack growth in proportion to the square of the plastic zone size, or to the fourth power of ΔK.

Weertman (1966) and Rice (1967) independently proposed essentially identical damage accumulation models which also predict that the Paris exponent m in Eq. 10.3 is 4. Both analyses are based on the plasticity model of a discrete surface of tensile yielding or slip ahead of the crack. Rice considered the deformation history of a particular point in an elastic–perfectly plastic solid from when that point was first engulfed by the plastic zone to when the crack tip reached that point. Crack growth was assumed to occur when the total absorbed hysteresis energy equaled a postulated critical value U^* per unit area of newly created fracture surface. If $\Delta u_y(x, 0)$ is the plastic displacement of tensile yielding per load reversal when the crack tip is at $x = 0$ and if crack growth rate is assumed to be constant for fracture over a distance of a cyclic plastic zone r_c, the increment of crack advance is given by

$$\frac{da}{dN} = \frac{4\sigma_y}{U^*} \int_0^{r_c} \Delta u_y(x, 0) dx, \tag{10.21}$$

where σ_y is the flow strength. Obtaining $\Delta u_y(x, 0)$ by doubling the yield strength and strain in the monotonic loading situation (see the discussion in Section 9.6), Eq. 10.21 becomes, for small-scale yielding,

$$\frac{da}{dN} = \frac{5\pi(1 - v^2)\epsilon_y \sigma_y}{96U^*} \left(\frac{\Delta K}{\sigma_y}\right)^4. \tag{10.22}$$

ϵ_y is the yield strain and v is the Poisson's ratio. For cyclically hardening or softening materials, σ_y and ϵ_y should be replaced by the corresponding values for cyclic loading.

Using the Bilby, Cottrell & Swinden (1963) model of infinitesimal dislocations distributed on a crack plane to calculate displacements in the crack tip yield zone, Weertman (1966) developed a damage accumulation theory similar to that of Rice. His analysis, as well as its subsequent extensions to work-hardening materials (e.g., Weertman, 1973; McCartney, 1976), revealed that

$$\frac{da}{dN} \propto \frac{(\Delta K)^4}{G\sigma_y^2 U^*}, \tag{10.23}$$

where G is the shear modulus. It is worth noting here that a feature common to all of the damage accumulation models is that they predict a Paris exponent of 4, in agreement with experimental observations for most metals.

Attempts have also been made to experimentally determine the magnitude of U^* during fatigue crack growth. Ikeda, Izumi & Fine (1977) measured the local stress–strain curves in the plastic zone around a propagating fatigue crack in a 7075 aluminum alloy using miniature strain gages. The areas in the stress–strain hysteresis loops were integrated over the plastic zone for a unit area of crack advance to give an approximate value of U^*. Davidson & Lankford (1977), on the other hand, calculated the energy dissipated by the advancing fatigue crack in Fe–0.05C steel using microstructural considerations. Their method utilized measurements of the distribution of subcell sizes in the vicinity of the crack plane using the channeling contrast technique in the scanning electron microscope. While Ikeda *et al.* estimated that $U^* \approx 6 \times 10^5$ J/m^2 for the 7075-T76 aluminum, Davidson & Lankford reported that U^* values were 17.3×10^5 and 5.7×10^5 J/m^2 for the Fe–0.05C steel in dry nitrogen and wet air, respectively. The values of U^* needed to match the predictions of the damage accumulation models with observed crack growth rates differed considerably from experimental measurements.

A number of damage accumulation models, which employ low cycle fatigue data to quantify damage at the crack tip, have also been developed (e.g., Majumdar & Morrow, 1974; Antolovich, Saxena & Chanani, 1975; Chakrabortty, 1979). Charkrabortty, for example, envisions fatigue crack growth as a consequence of the exhaustion of ductility due to the cyclic plastic strain in the reverse yield zone ahead of the crack tip. Exhaustion of ductility is described by the Coffin–Manson low cycle fatigue rule, Eq. 8.1. The main criticism of this approach is that the microscopic damage phenomena and the gradients in the field quantities ahead of a fatigue crack tip are not truly represented by low-cycle fatigue data obtained on smooth test specimens.

10.6 High growth rate regime

At very high ΔK values, the fatigue crack growth rates are significantly higher than those observed in the Paris regime. The sensitivity of crack growth to microstructure, load ratio and stress state (i.e. plane stress and plane strain loading) is also very pronounced. However, crack propagation rates are too rapid to be affected strongly by the test environment.

Figure 10.13 shows the variation of crack propagation rates as a function of ΔK and load ratio, R. The enhanced influence of load ratio in regime C is a consequence of the critical condition that the maximum stress intensity factor value for the fatigue cycle, K_{max}, approaches the fracture toughness of the material, K_c (or K_{Ic}, in plane strain):

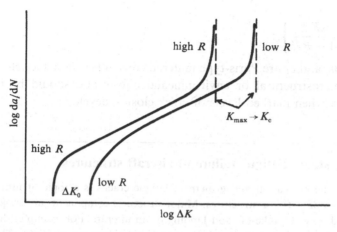

Fig. 10.13. A schematic representation of the effect of load ratio, R, on fatigue crack growth behavior.

$$K_{max} = \frac{\Delta K}{(1 - R)} \to K_c. \tag{10.24}$$

Since the ΔK values at which K_{max} begins to approach K_c are lower for high R ratios, the catastrophic fatigue failure occurs at lower ΔK values with an increase in load ratio.

As the fracture toughness is strongly influenced by the stress state, fatigue crack growth in regime C is also susceptible to the influence of constraint along the crack-front. Plane stress conditions and the possibility of a macroscopic shear mode of failure (for low constraint along the thickness direction of the specimen) promote apparently higher fracture toughness and high values of critical ΔK for catastrophic failure. It should be noted that at high values of ΔK in ductile solids, the plastic zone dimensions become large in typical laboratory test specimens; consequently, linear elastic fracture mechanics characterization of fatigue crack growth may become invalid as $K_{max} \to K_{Ic}$. At high ΔK levels approaching fast fatigue fracture, static fracture modes, such as cleavage, intergranular separation and fibrous failure, occur in addition to striation growth (Richards & Lindley, 1972; Ritchie & Knott, 1973). These *additional static modes* cause a marked sensitivity of propagation rates to microstructure.

Many semi-empirical and empirical models for fatigue crack growth have been proposed in the literature to account for the load ratio dependence of da/dN and the deviation of crack growth behavior in the near-threshold and final failure regimes from that described by the Paris law. Among these models, the most notable and widely used (for industrial design purposes) are the empirical approaches of Forman, Kearney & Engle (1967) and Walker (1970). These models are described, respectively, by the following two equations:

$$\frac{da}{dN} = C_a \left\{ \frac{\Delta K^{m_a}}{(1 - R)K_c - \Delta K} \right\} \tag{10.25}$$

and

$$\frac{da}{dN} = C_b \left\{ \frac{\Delta K^{m_b}}{(1 - R)^{c_1}} \right\}, \tag{10.26}$$

where C_a, C_b, m_a, m_b and c_1 are Paris-type material constants. These models do not account for any microstructural or environmental effects and should be further modified empirically when marked levels of crack closure develop.

10.7 Case study: Fatigue failure of aircraft structures

Fracture by the nucleation and growth of fatigue cracks is a topic of utmost concern in the design of airframe structures. Major structural components are subjected to repeated loads during the take-off and landing of an aircraft. For example, the fuselage is subjected to pressurization and depressurization during each flight. These variable loads may cause nucleation and growth of fatigue cracks which, if undetected, can lead to catastrophic failure. An example of such an event is an accident involving an Aloha Airlines Boeing 737 jet in which much of the upper half of a fuselage section was 'blown off' during a flight. The cause was fracture resulting from the undetected growth of multiple fatigue cracks (see *New York Times*, May 1, 1988, p. 1), which were initiated by the corrosion-induced delamination of the adhesive between the aluminum skin panels. In this section, we consider the case history of the crash of a Japan Airlines Boeing 747 in 1985 due to the fatigue failure of the rear pressure bulkhead.

On August 12, 1985, a Boeing 747SR (short range) owned by Japan Airlines, crashed into the mountains of Ueno village in Gumma Prefecture during a flight from Tokyo to Osaka. Of the 524 occupants of the aircraft, 520 persons (505 passengers and fifteen crew members) were killed; the four surviving passengers were seriously injured. This plane crash, resulting from the growth of fatigue cracks, is the worst air accident in the history of aviation involving a single aircraft. Fracture originated in the rear pressure bulkhead (Fig. 10.14) when the airplane reached an altitude of 7200 m (24 000 feet), about twelve minutes after take-off and, consequently, the vertical fin was blown off.

The rear pressure bulkhead has the shape of a hemispherical dome with an edge-to-edge diameter of 4.56 m, a radius of curvature of 2.56 m, and a height of 1.39 m. The main structure of this rear pressure bulkhead consists of eighteen webs which form the shape of the dome, four straps which form concentric circles and thirty-six edge-to-edge stiffeners which form radial lines. Furthermore, a short-length stiffener is placed at each middle point of the above arrangement; all the above parts are riveted. The upper webs, lower webs and straps are made of 2024-T42 clad aluminum alloy (of 0.8 mm, 0.9 mm

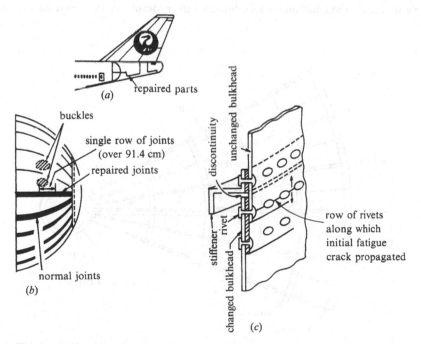

Fig. 10.14. Details of the geometry of the rear pressure bulkhead of the crashed Boeing 747 SR aircraft. (After Kobayashi, 1987.)

and 1.0 mm thick sections, respectively) and the edge-to-edge and short-length stiffeners are made of 2024-T42 aluminum alloy (of 2.4 mm and 1.0 mm thickness, respectively). The rivets are made of a 2017-T3 aluminum alloy.

This aircraft had an earlier accident in June, 1978, in which the rear pressure bulkhead was damaged. The aircraft was repaired by replacing one-half of the bulkhead structure, as shown in Figs. 10.14(*b*) and (*c*). Also included in this figure are the areas containing defective joints which spanned a distance of 91.4 cm. To compensate for local short edge margins in areas 2 and 3 (shown in Fig. 10.15), an additional splice plate, shown by the cross-hatched region in Fig. 10.14(*c*), was sandwiched between the old and new structural parts of the bulkhead. An error during final installation resulted in full load transfer across one row of rivets.

Figure 10.15 is a view of the left side of the rear bulkhead. The row of rivets along which tensile fatigue fracture occurred (denoted as row L18) is also indicated. The shaded areas marked 2 and 3 are regions adjacent to which only one row of rivets sustained all the loads produced by the fluctuations in the cabin pressure. The changes in cabin pressure due to each flight resulted in fatigue crack nucleation and growth from the rivet holes. The fatigue crack had propagated through a number of rivet holes during 12 319 flights (16 196 flight hours, with approximately six flights per day) since the 1978 repair. The dominant fracture occurred through the line connecting the rivet holes on the first row, starting at the lower edge of the upper web (see Figs. 10.14 and 10.15). Failure analysis indicated that fracture of the rear bulkhead originated from this crack as the cabin pressure increased during the ascent of the aircraft on its final flight. The cabin pressure vented, following failure of the rear bulkhead, into the vertical fin which was unable to sustain the additional loads. Since all the hydraulic systems were concentrated

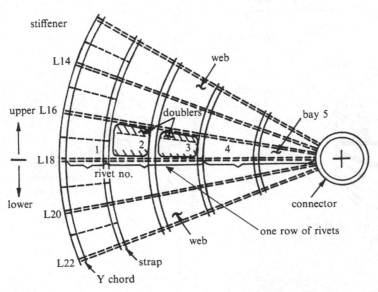

Fig. 10.15. Details of the left side of the rear pressure bulkhead. (After Kobayashi, 1987, and Kobayashi, Arai & Nakamura, 1987.)

in the base of the fin, hydraulic power was lost and the flight crew lost normal control. The airplane crashed into the mountains about thirty minutes later.

Post-crash analyses of the fracture zones revealed subcritical crack growth by fatigue. Figure 10.16 is a histogram showing the percentage of area that suffered fatigue damage between adjacent rivet holes versus the rivet hole (designation) number along the L18 row. Note that, in the region between rivets 35 and 55 of this row, most of the cross section suffered considerable fatigue damage. Evidence of subcritical crack growth by fatigue was also obtained with the aid of scanning electron fractography of the failed surfaces of the rear bulkhead. Figure 10.17(*a*) is a microscopic view of fatigue striations observed at a distance of 0.28 mm from rivet hole 34. The average striation spacing is 0.23 μm; this value, for the 2024 aluminum alloy, corresponds to a stress intensity factor range, $\Delta K \approx 10 - 15$ MPa$\sqrt{\mathrm{m}}$ (assuming that the striation spacing correlates with the crack growth increment per fatigue cycle). Figure 10.17(*b*) shows the fracture surface features of the same fatigue crack, taken at a distance of 3.05 mm from the same rivet hole. Note the significant increase in striation spacing from that seen at the shorter crack length. This observation implies a significant increase in fatigue crack growth rate with increasing numbers of flights. Figure 10.17(*b*) represents the fatigue crack growth regime just prior to the critical condition.

Laboratory simulations of fatigue crack growth are often conducted to gage the crack growth resistance of alloys used in fatigue-critical structures of commercial aircraft. Such studies are prompted by the identification of cracks nucleated at rivets, known as multi-site damage or MSD (Goranson and Miller, 1989). In order to analyze the rate of growth of a fatigue crack emanating from a rivet hole, it is necessary to estimate the stress intensity factor for the plate containing the rivet holes. For the particular case of a fatigue

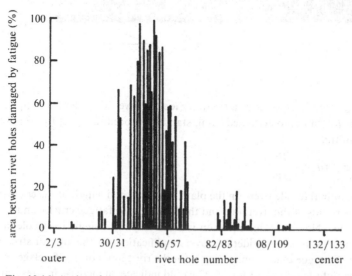

Fig. 10.16. A plot of the fractional area between rivet holes which had undergone failure by fatigue versus the designation number of the rivet holes along row L18. (After Kobayashi, 1987.)

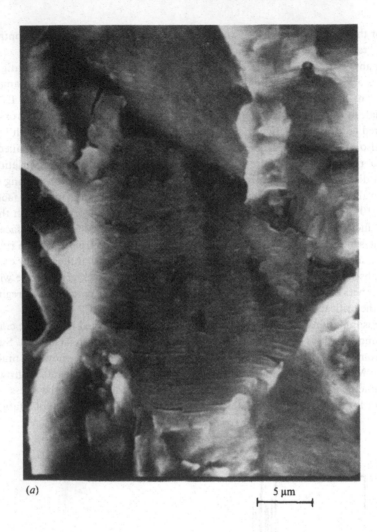

(a)   5 μm
                                                                        ├──────┤

crack emanating from both sides of a centrally located rivet hole, the crack tip stress
intensity factor, K_I, can be determined using standard procedures (see, for example,
Broek, 1986), so that

$$K_I = \frac{1}{2}\sigma\sqrt{\pi a} + \frac{\sigma W}{2\sqrt{\pi a}},$$ (10.27)

where σ is the nominal tensile stress on the plate, $2a$ is the total length of the crack (which
is the sum of the radius of the rivet hole and the length of the fatigue crack), and W is the
width of the plate. (Note that this equation is valid only for a single rivet hole in an
infinite plate and that it is a considerable oversimplification for the aircraft structures.)

   For the case of fatigue cracks emanating from the rivet holes on the fuselage skin, the
first term on the right hand side of Eq. 10.27 would indicate an increase in stress intensity
factor range with an increase in crack length. However, the second term causes a
reduction in stress intensity with an increase in crack length. These two factors result in a

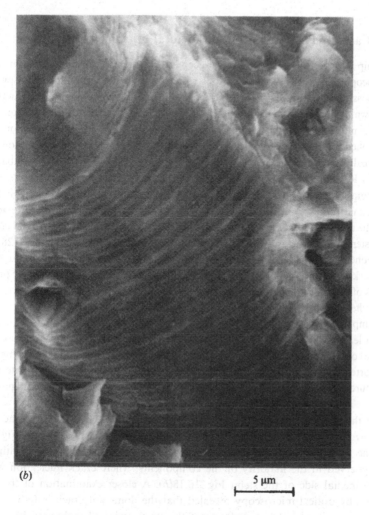

(b)

5 μm

Fig. 10.17. Scanning electron fractographs showing fatigue striation markings near rivet hole 34 along the L18 row. (*a*) Fatigue fracture surface at about 0.28 mm from the rivet hole. (*b*) Fatigue fracture surface at about 3.05 mm from the rivet hole. (From 'Final Report of the JAL Boeing Crash', The Japan Ministry of Transportation, Tokyo, 1987. Photos courtesy of H. Kobayashi, Tokyo Institute of Technology.)

nearly constant value of stress intensity factor range ($\Delta K \approx 15$ MPa$\sqrt{\text{m}}$ and $R \approx 0$) over several millimeters of fatigue crack growth from the rivet holes prior to catastrophic failure. For the 2024 aluminum alloy, this corresponds to an average crack growth rate of $0.1 - 0.2$ μm/cycle (Paris regime of crack growth). Striation spacing measurements made on the fatigue fracture surfaces, such as those seen in Fig. 10.17, typically reveal crack growth increments of this magnitude per fatigue cycle.

10.8 Case study: Fatigue failure of total hip components

Total hip replacement is a commonly accepted treatment for diseased or damaged hip joints in people. One of the complications resulting from the hip replacement procedure is the possibility of fatigue failure in the metallic femoral component. This fracture process is known to depend on such factors as the material and design of the component, the surgical technique, and the weight and activity level of the patient. A case history of such fracture phenomena in hip replacement components is presented in this section to illustrate the reality and complexity of fatigue in implant devices. This example is based on the work of Rimnac et al. (1986).

A metallic total hip component, known as the Trapezoidal-28^{TM}, was developed in the early 1970s in an attempt to improve mobility, stability and stem strength over previous hip designs. The name T-28 was given for this component because of the trapezoidal cross section of its stem and neck, and the 28 mm diameter of its femoral head. The T-28 femoral component was made of surgical grade 316L stainless steel in the wrought and lightly cold-worked condition. The trapezoidal shape of the stem was preferred over prior considerations of a diamond shape because it maximized the pressure of the acrylic grout in the femur when the prosthesis was inserted. Between 1973 and 1979, 805 patients received this implant at The Hospital for Special Surgery in New York City. Out of this population, at least twenty-one patients are known to have received medical treatment for the removal of a fractured femoral component. This failure rate was four times higher than that reported for other femoral component designs. In an effort to investigate the origin of fracture, a failure analysis was conducted on all twenty-one fractured femoral parts.

Eighteen of the twenty-one components failed by fracture of the stem. This fracture generally occurred within the proximal third of the stem portion, as shown in Fig. 10.18(a). In the remaining three cases, fracture occurred in the neck region. Multiple cracks were observed in the majority of the components, where crack nucleation was sighted on the medial side of the stem, Fig. 10.18(b). A closer examination of the fracture surface by optical microscopy revealed that the dominant crack initiation site was the posterior corner on the medial side of the stem, region A in Fig. 10.19. Near the central portion of the trapezoidal stem cross section, a step was usually observed on the fracture surface. This step was identified as the boundary between the medial crack and a crack which originated on the lateral side of the stem. The lateral crack exhibited clam shell markings (region B in Fig. 10.19), which are typically found on fatigue fracture surfaces.

On the basis of a number of observations, the origin of the femoral component failure was traced to fatigue loading. This conclusion was also supported by related results which showed that 316L stainless steel is resistant to stress corrosion cracking in the physiological environment. Microscopic observations in the transmission electron microscope (TEM), conducted by means of two-stage replication of the fracture surfaces, revealed isolated locations with fatigue striation markings that were associated with the medial cracks. On the lateral side, however, more uniform and greater numbers of striations were observed, Fig. 10.20; the striation spacing increased with increasing distance from the crack initiation site.

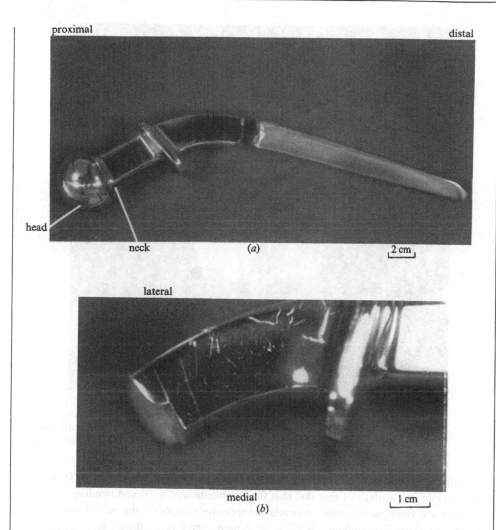

Fig. 10.18. (a) Photograph of the failure of the T-28 femoral component showing the proximal and distal sections and the femoral head and neck regions. (b) Micrograph showing multiple cracks which initiate on the medial side of the stem of the femoral component. (From Rimnac et al., 1986. Copyright American Society for Testing and Materials. Reprinted with permission.)

With the information derived from striation spacing measurements, the stress intensity factor solution for a corner crack, and the fatigue crack growth characteristics of 316L stainless steel, the stress intensity factor range, ΔK, and the stress range, $\Delta \sigma$, to which the components were subjected on the lateral side were estimated. Table 15.1 shows these results for the lateral side cracks in four of the failed femoral components. It is seen that the lateral side cracks were subjected to a tensile cyclic stress of 115 to 244 MPa. Note that this estimation procedure assumes that the striation spacing measured on the fracture surface corresponds to the fatigue crack growth rate per cycle which, in turn, is taken to be characterized by the nominal value of ΔK.

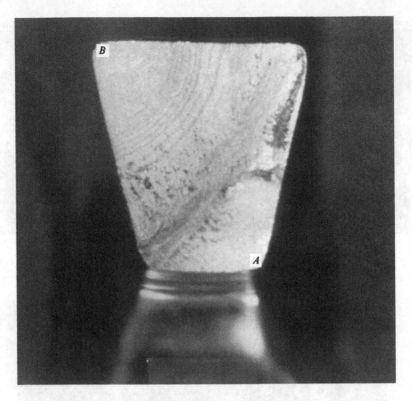

Fig. 10.19. Optical fractograph of the failed T-28 stem showing crack growth from the posteromedial corner, region *A*, and clam shell markings originating from the anterolateral corner, region *B*. (From Rimnac *et al.*, 1986. Copyright American Society for Testing and Materials. Reprinted with permission.)

Although crack growth from the medial side apparently exhibited a tensile fatigue mechanism, stress analysis revealed that the combination of axial and bending loads on the femoral component resulted in cyclic compressive stresses on the medial side where the cracks first initiated. Rimnac *et al.* used curved beam theory and the finite-element method to estimate the magnitude of stresses on the medial side; a typical sequence of loads was used in the simulation. This sequence consisted of an overload with a bending moment of 110 Nm and a compressive force of 4.3 kN, which corresponded to a load on the femoral head equivalent to seven times the body weight for a 91 kg (200 lb) individual. This caused the development of a compressive stress level of 604 MPa on the medial side and a tensile stress level of 556 MPa on the lateral side. Noting that the yield strength of the steel in (both compression and tension) is approximately 550 MPa, plastic yielding would be expected to occur. From the discussion of cyclic plasticity (Chapters 4 and 9), a residual tensile stress of at least 275 MPa is expected to be created on the medial side.

The overload was followed by normal cyclic loads which consisted of a bending moment of 55 Nm and a compressive force of 2.17 kN; these values correspond to a load on the femoral head which is equivalent to 3.5 times the normal body weight for a 91 kg individual. (It is well known in the biomechanics community that even the seemingly mild

Fig. 10.20. Fatigue striations observed on the fracture surface of a lateral crack in the femoral component. The arrow indicates the fatigue crack growth direction. (From Rimnac *et al.*, 1986. Copyright American Society for Testing and Materials. Reprinted with permission.)

activity of standing on one foot can translate into a load equivalent to 1.5 times the body weight on the implanted hip components.) This constant amplitude load sequence subjects the medial side to a maximum compressive stress of 207 MPa and the lateral side to a tensile stress of 289 MPa. The stresses estimated from the striation spacing measurements (Table 10.2) were in reasonable agreement with this independent prediction of cyclic stresses on the lateral side.

Fatigue failure of hip components is controlled by many variables and it is very difficult to identify and quantify *a posteriori* the various factors which led to the failure. As a result of these uncertainties, the mechanisms responsible for such failures are the subject of much controversy. However, on the basis of microscopic observations and stress analyses, the following sequence of events was proposed by Rimnac *et al.* for the fracture of the femoral stem. Residual tensile stresses induced by the large compressive overload on the medial side of the stem result in the growth of a fatigue crack prior to crack nucleation on the lateral side. With crack growth, there occurs a progressive reduction in the extent of residual tensile stresses (and an increase in the extent of crack closure) ahead of a fatigue crack subjected to cyclic compressive loads (see Chapter 14). Consequently, the medial cracks decelerate or arrest after a certain amount of crack advance. The nucleation of the medial crack enhances the stress levels to which the lateral side of the stem is exposed. Therefore, a crack originates on the lateral side and eventually links up with the medial crack. The coalescence of the two crackfronts, originated at opposite locations, produces a step on the fracture surface. This line of reasoning is

Table 10.2. *Estimation of the stress intensity factor range, ΔK, and cyclic stress range, $\Delta\sigma$, from the measurements of striation spacing, s, on four of the fractured femoral components.*

No.	Crack length, a (mm)	s (mm)	ΔK (MPa$\sqrt{m}$)	$\Delta\sigma$ (MPa)
1	3.0	2.15×10^{-4}	18.1	233
	5.0	3.00×10^{-4}	20.9	208
	9.0	4.15×10^{-4}	25.0	186
2	7.0	7.73×10^{-5}	13.7	115
	9.0	1.80×10^{-4}	17.0	127
3	3.0	1.35×10^{-4}	15.6	201
	7.5	2.75×10^{-4}	20.0	163
4	1.5	6.70×10^{-5}	13.4	244
	11.0	2.60×10^{-4}	19.6	132

Source: Rimnac *et al.*, 1986.

consistent with the experimental observations discussed earlier. Rimnac *et al.* concluded that the failure of the femoral stem was a consequence of the combined effects of the trapezoidal cross sectional design and of the material properties of the stainless steel used to make the component.

Surgical implants using the trapezoidal design were discontinued in 1978. Newer designs involve a rectangular cross section and a micro-grained (forged) Co alloy, which replaced the surgical grade 316L stainless steel.

In closing this section, we note that there is a continual search for nonmetallic materials for use in hip arthoplasty in view of the following drawbacks of a metallic hip prosthesis: (1) The metallic ions released as a result of wear of the prosthesis can lead to undesirable side effects, such as hypersensitivity, in patients with metallic hip components. (2) The stiffness of the metallic implant is much higher than that of the cortical bone. For example, the surgical grade 316L stainless steel and Ti-6Al-4V alloy have Young's modulus values which are approximately six and four times higher, respectively, than Young's modulus of the cortical bone which is the load-carrying member in the human femur. This elastic mismatch preferentially shields the stress in such a way that bone loosening and prosthesis fracture result. In view of these problems, composite materials such as carbon-fiber-reinforced poly(ether ether ketone), C/PEEK, are also considered candidates as implant materials.

10.9 Combined mode I–mode II fatigue crack growth

The discussion up to this point dealt with the growth of nominally mode I fatigue cracks in ductile solids. We now direct attention at mixed-mode crack growth.

10.9.1 Mixed-mode fatigue fracture envelopes

Gao *et al.* (1985) studied the growth of fatigue cracks in several ferrous and nonferrous alloys where biaxial loading was introduced in two different specimen geometries: an edge-notched specimen subjected to asymmetric four-point bend cyclic loading, and a plate with an inclined center crack subjected to biaxial tension. Figure 10.21 is the fatigue fracture envelope for combined mode I–mode II crack growth, where the *nominal value* of the in-plane shear stress intensity factor range ΔK_{II} is plotted against the nominal value of the mode I stress intensity factor range ΔK_I; both ΔK_{II} and ΔK_I are normalized by the threshold stress intensity factor range in mode I, ΔK_{I0}. The solid curve in this figure indicates the lower bound threshold condition below which no crack growth is detected at a load ratio, $R = 0.2$, and cyclic frequency, $v_c = 15$–80 Hz. The dashed line is the fracture locus outside of which predominantly mode I crack growth occurred on a plane normal to the maximum value of the near-tip tensile stress, $\sigma_{\theta\theta}$. Between these two curves, the characteristics of crack growth were strongly influenced by the frictional rubbing of the crack faces. The location of the upper bound curve (corresponding to

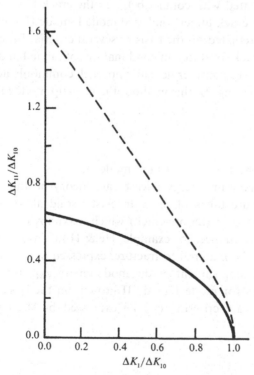

Fig. 10.21. Combined mode I–mode II fatigue fracture envelope for AISI 316 stainless steel showing the locus for the onset of fatigue crack growth (solid curve). The dashed line indicates the mixed-mode loading conditions at which fatigue fracture occurred on a plane normal to the maximum hoop stress, $\sigma_{\theta\theta}$. (After Gao *et al.*, 1985.)

the onset of mode I crack growth) was dependent on crack closure due to the interlocking of fracture surface asperities and on the formation of oxide debris. Such crack closure also promoted crack arrest in the region bounded by the two curves. When the amount of crack closure was reduced by increasing the load ratio of the stress cycle in tension, the upper bound curve began to approach the lower bound threshold curve. When slit cracks (where closure in the wake of the crack tip is not a consideration), instead of sharp fatigue cracks, were subjected to mixed-mode loading, the upper bound curve for the onset of mode I crack growth coincided with the threshold (lower bound) curve. These results clearly point to the significant role of frictional sliding in influencing mixed-mode fatigue crack growth.

Gao *et al.* also found that, for certain combinations of mixed-mode loading, the rate of crack growth was severely affected by the T-stress. It may be recalled from Section 9.3 that the T-stress is the nonsingular stress acting parallel to the crack plane. Crack growth rates increased with increasing values of T-stress at high ΔK_{I}–ΔK_{II} levels, although the threshold fracture envelope was unaffected by the T-stress.

10.9.2 *Path of the mixed-mode crack*

If frictional effects associated with contact between the crack faces are not significant, the path of the fatigue crack under combined mode I–mode II loading or in pure mode II loading can be predicted on the basis of several criteria. In order to quantify the path of the fatigue crack for different combinations of mode I and mode II load parameters, it is instructive to consider here an approach commonly adopted in conjunction with monotonic loading. In this method, the near-tip mode mixity is defined by a single phase angle,

$$\varphi_{\mathrm{M}} = \tan^{-1}\left\{\frac{K_{\mathrm{II}}}{K_{\mathrm{I}}}\right\}. \tag{10.28}$$

φ_{M} varies from zero for pure mode I to $\pi/2$ for pure mode II.

The Griffith theory or the maximum energy release rate theory postulates that combined mode I–mode II fracture takes place in an elastic solid along a plane (oriented at an angle $\hat{\theta}$ to the plane of the pre-crack) which maximizes the energy release rate. Implicit in this approach (see, for example, He & Hutchinson, 1989) is the notion that the resistance of the material to fracture, expressed in terms of the Griffith energy release rate $\mathcal{G}$ is independent of the mode-mixity φ_{M}. To a good approximation, the fracture locus for mode I–mode II growth on the basis of the maximum strain energy release rate criterion can be expressed by the equation (Palaniswamy & Knauss, 1978):

$$\left(\frac{K_{\mathrm{I}}}{K_{\mathrm{Ic}}}\right) + \frac{3}{2}\left(\frac{K_{\mathrm{II}}}{K_{\mathrm{Ic}}}\right)^2 = 1, \tag{10.29}$$

where K_{Ic} is the mode I fracture initiation toughness. For pure mode II fracture, this theory predicts that $K_{\mathrm{IIc}} = \sqrt{2/3}\,K_{\mathrm{Ic}}$ (where K_{IIc} is the mode II fracture initiation toughness) and that $\hat{\theta} = 77.4°$.

According to the maximum (tensile) hoop stress theory, the crack advances in a direction (oriented at an angle $\hat{\theta}$ to the pre-crack) in which the hoop stress at the crack tip is maximum (Erdogan & Sih, 1963). For brittle fracture, the crack advances when the hoop stress over a certain distance ahead of the crack tip reaches a critical value. The hoop stress under combined mode I–mode II loading is

$$\sigma_{\theta\theta} = \frac{K_{\mathrm{I}}}{\sqrt{2\pi r}} f_{\mathrm{I}}(\theta) + \frac{K_{\mathrm{II}}}{\sqrt{2\pi r}} f_{\mathrm{II}}(\theta), \tag{10.30}$$

where $f_{\mathrm{I}}(\theta)$ and $f_{\mathrm{II}}(\theta)$ are the universal dimensionless angular functions defined in Eqs. 9.44 and 9.47. For a given combination of mode I and mode II, the direction of the maximum hoop stress is determined from the condition $(\partial\sigma_{\theta\theta}/\partial\theta) = 0$, from which $\hat{\theta}$ is determined. Substituting this value of $\hat{\theta}$ in Eq. 10.30, one finds that

$$\frac{K_{\mathrm{I}}}{K_{\mathrm{Ic}}} f_{\mathrm{I}}(\hat{\theta}) + \frac{K_{\mathrm{II}}}{K_{\mathrm{Ic}}} f_{\mathrm{II}}(\hat{\theta}) = 1. \tag{10.31}$$

For pure mode II fracture, the maximum hoop stress theory predicts that $K_{\mathrm{IIc}} = 0.866 K_{\mathrm{Ic}}$ and that $\hat{\theta} = 70.5°$.

The values of $\hat{\theta}$ derived from the maximum strain energy release rate theory and from the maximum hoop stress theory for different mode mixities φ_{M} are plotted in Fig. 10.22. Also shown in this figure are the trends predicted by the criterion which states that the crack advances in a direction along which $K_{\mathrm{II}} = 0$ (He & Hutchinson, 1989). Note that little difference is found among the predictions of the different theories for brittle fracture. In addition to the above theories, it has been proposed

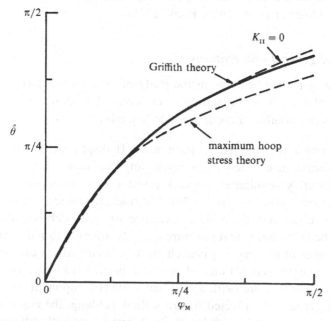

Fig. 10.22. Predictions of initial crack growth angle $\hat{\theta}$ in combined mode I–mode II loading as a function of the crack-tip mixity expressed by the phase angle φ_{M}.

by Sih (1974) that the mixed-mode crack advances in a direction along which the strain energy density S (which is a function of K_I and K_{II}) is the lowest. For pure mode II fracture, the strain energy density theory predicts that $K_{IIc} = 1.054K_{Ic}$ and that $\hat{\theta} = 79.2°$ (for Poisson's ratio, $v = 0.22$). Experimental studies (Tanaka, 1974; Kitagawa *et al.*, 1985) indicate that this approach, among the various criteria discussed here, generally leads to nonconservative estimates of critical conditions for mixed-mode fatigue. A review of results for mixed-mode fracture and fatigue is presented in Suresh *et al.* (1990) and Qian & Fatemi (1996).

When the load ratio R for the fatigue cycle is zero, the mixed-mode failure criteria can be applied by replacing K_I and K_{II} by K_{Imax} and K_{IImax}, respectively. For $R > 0$, Tanaka (1974) and Kitagawa *et al.* (1985) have employed nominal values of ΔK_I and ΔK_{II} in place of K_I and K_{II}, respectively, in an attempt to predict the mixed-mode fatigue thresholds (such as the lower bound threshold curve in Fig. 10.21). This choice was based on the earlier results of Frost & Dugdale (1957) who reported that the condition for the nonpropagation of fatigue cracks ahead of stress concentrations was insensitive to the mean level of the cyclic tensile loads. For combined mode I–mode II fatigue of aluminum at $R = 0.65$ (where crack closure is minimal), Tanaka and Kitagawa *et al.* compared their experimental results with the predictions of the various criteria listed above, and found that the maximum strain energy release rate theory and the maximum hoop stress theory provided conservative estimates of mixed-mode fatigue thresholds. It should, however, be noted that because of crack face contact in mixed modes I and II, the nominal use of any of the foregoing failure criteria is not likely to provide perfect agreement with experimental observations (Hourlier *et al.*, 1985; Pook, 1985).

10.9.3 Some general observations

One of the major complications in the study of mixed-mode fatigue crack growth is the frictional contact of the mating crack faces. The following trends are observed in fatigue under combined mode I–mode II loading.

(1) If a fatigue crack is subjected to pure mode II displacements, the crack flanks are always in contact. As a result, any coplanar crack growth in mode II invariably terminates in crack arrest (as in the region bounded by the two curves shown in Fig. 10.21). The fracture surfaces of specimens subjected to mode II fatigue show evidence of wear debris formation. Because of the frictional contact in pure mode II, fracture can also proceed by the development of a mode I branch crack. Closure effects can be minimized or even completely eliminated only if a tensile load is superimposed.

(2) Since mixed-mode fracture (with a stronger mode II component than mode I component) is severely affected by crack flank rubbing, the crack growth rates are a function of the crack length itself. Specimens with different pre-crack lengths exhibit different fatigue crack growth characteristics. If speci-

mens containing slits, rather than fatigue pre-cracks, are subjected to mixed-mode fatigue, lower mode II thresholds are often observed (despite the fact that the tip of the slit crack is blunter than that of the fatigue pre-crack).

(3)   In a study of mode II fatigue crack growth in structural steel, Smith & Smith (1988) identified three sources of crack growth attenuation: compressive residual stresses ahead of the fatigue pre-crack, interlocking of fracture surface asperities, and gross plastic deformation of the interlocking asperities. They estimated the effective stress intensity range in mode II to be less than one-half of the nominal value.

10.10 Combined mode I–mode III fatigue crack growth

Fatigue failure as a consequence of cyclic torsional loads is commonly encountered in many components used in automotive applications and in turbogenerator shafts in electric power generation and transmission systems. Failure under cyclic torsion may occur in anti-plane shear (mode III) along the transverse shear plane or along the longitudinal shear plane, or in mode I along ±45° to the axis of the shaft along planes of maximum principal tensile stresses (Fig. 10.23). A thumbnail crack in either of the two (longitudinal or transverse) shear planes will be subjected to mode II displacements where the thumbnail intersects the surface; however, mode III fracture will be induced as the thumbnail crack penetrates the interior of the rod (Hurd & Irving, 1982). With rare exceptions (e.g., Pook, 1985), laboratory simulations of mode III fatigue crack involve torsional loading of a circular cylinder with a pre-exising flaw (i.e. a circumferential notch plus a concentric fatigue pre-crack) introduced normal to the axis of the cylinder. At low stress values, typically less than 70% of yield, locally mode I growth dominates over mode III failure. As a result, torsional failure takes place either in a macroscopic tensile mode with a helical fracture surface, in a microscopic tensile mode with a 45° branch crack, or by a combination of these two modes.

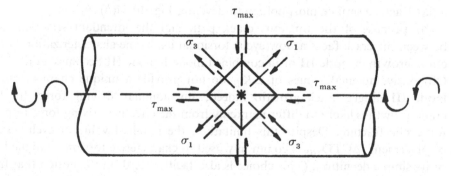

Fig. 10.23. The state of stress in a cylindrical rod subjected to cyclic torsion.

It is perhaps instructive to begin the discussion of mode III fatigue crack growth with a brief reconsideration of some fracture mechanics concepts discussed in Chapter 9. As shown in Eq. 9.78, the plastic zone at the tip of a mode III crack has a circular shape with a diameter

$$r_y = \frac{1}{\pi}\left(\frac{K_{III}^2}{\tau_y^2}\right), \tag{10.32}$$

where τ_y is the shear yield strength. (Note that the extent of plasticity ahead of the torsional crack is greater than that ahead of a tensile crack at comparable magnitudes of K_I and K_{III}. Because of this, it is difficult to maintain small-scale yielding conditions in torsional fracture specimens, of circular cylindrical rod shape and typically 15–30 mm in diameter, over the entire range of mode III fatigue crack growth rates.) The displacement of the crack tip in mode III is given by

$$(CTD)_{III} \approx \frac{2}{\pi}\frac{K_{III}^2}{\mu\tau_y}, \tag{10.33}$$

where μ is the shear modulus. The range of $(CTD)_{III}$ in fully reversed torsional fatigue ($R = -1$) is given by

$$(\Delta CTD)_{III} \approx \frac{2}{\pi}\frac{K_{III}^2}{2\mu\tau_y}, \tag{10.34}$$

analogous to the definition of the cyclic crack tip opening displacement, $(\Delta CTD)_I = \Delta\delta_t$, in mode I, Eq. 9.86.

10.10.1 Crack growth characteristics

Studies of mode III fatigue crack growth have identified the following basic characteristics in ductile solids (Hurd & Irving, 1982; Nayeb-Hashemi, McClintock & Ritchie, 1983a,b; Tschegg, 1983a,b,c; Gross, 1985; Pook, 1985).

(a) At low amplitudes of cyclic torsion, the fatigue cracks propagate locally in a tensile opening mode. This leads to the formation of the so-called *factory-roof* appearance on the fracture surface, as shown in Fig. 10.24(*a*). At high amplitudes of cyclic torsion, pure torsional crack growth (in the radial direction) occurs causing a flat fracture surface morphology to develop, Fig. 10.24(*b*).

(b) Because of the tortuous crack path and the attendant frictional contact between the crack faces at low cyclic torsional loads, the characterization of fatigue crack growth in mode III or in combined mode I–mode III becomes a difficult task. Obviously, nominal values of ΔK_{III} do not provide a unique characterization of mode III fatigue crack growth. Indeed, any nominal 'driving force' for fatigue crack growth is likely to differ markedly from the effective driving force responsible for cyclic fracture. Despite this limitation, the nominal value of cyclic crack tip displacement $(\Delta CTD)_{III}$ is commonly used to characterize torsional fatigue because of its simple definition. (This choice is also tacitly based on the notion that fracture occurs when the crack tip displacement attains a critical value; see Chapter 9 and

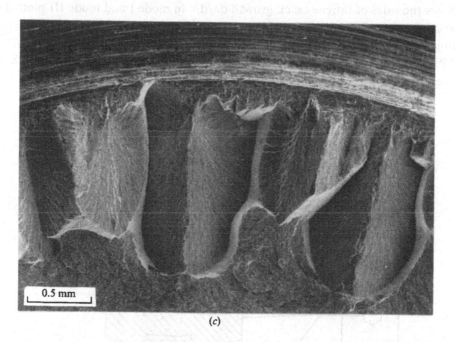

Fig. 10.24. (*a*) 'Factory-roof' fracture (radial mode III with 45° mode I branch cracks) at nominal $\Delta K_{III} = 12$ MPa$\sqrt{m}$, and (*b*) macroscopically flat, coplanar mode III fracture at $\Delta K_{III} = 45$ MPa$\sqrt{m}$ in AISI 4340 steel tempered at 650 °C (tensile flow strength, $\sigma_y = 956$ MPa) subjected to fully-reversed cyclic torsion. (From Tschegg & Stanzl, 1988. Copyright American Society for Testing and Materials. Reprinted with permission.) (*c*) A higher magnification view of the 'factory roof' fracture mode. (Courtesy of E.K. Tschegg, Technical University of Vienna.)

earlier sections of this chapter.) Furthermore, with the cyclic crack tip displacement as a characterizing parameter, the crack propagation rates in all three modes of failure (modes I, II and III) can be compared on a common scale. Nayeb-Hashemi, McClintock & Ritchie (1983a) have used a cyclic plastic strain intensity parameter, $\Delta\Gamma_{III}$, for characterizing mode III fatigue crack growth. This approach is essentially the same as the CTD approach in that $(\Delta CTD)_{III} \approx 2\Delta\Gamma_{III}$.

(c) A significant outcome of the factory-roof type fracture at low values of imposed ΔK_{III} is that the fracture surface asperities interlock during cyclic torsion. Figure 10.25 schematically illustrates possible mechanisms of this 'sliding mode crack closure' (Tschegg, 1983a,b). As a consequence of frictional contact between the crack face asperities, the apparent crack growth rates measured at such stress intensities are substantially lower than the corresponding mode I crack growth rates at identical magnitudes of *nominal* ΔK_I. This trend is evident in Fig. 10.26, which shows the rates of fatigue crack growth da/dN in mode I and mode III plotted as a function of $(\Delta CTD)_I$ and $(\Delta CTD)_{III}$, respectively, for 4340 steel (650 °C temper) subjected to fully reversed torsional loads. (Here, a small axial load, $K_I = 3$ MPa$\sqrt{m}$, was superimposed on cyclic torsion to enable accurate crack length measurements.)

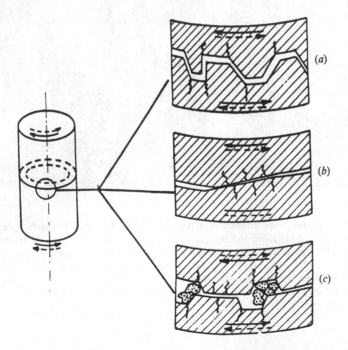

Fig. 10.25. Schematic illustration of the mechanisms of sliding mode crack closure showing (a) interlocking of asperities, (b) frictional contact, and (c) formation of fretting debris. (From Tschegg, 1983a. Copyright Chapman & Hall. Reprinted with permission.)

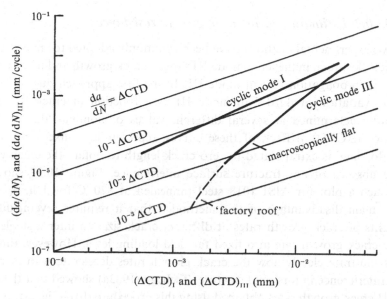

Fig. 10.26. Variation of mode I and mode III fatigue crack growth rates with nominal values of $(\Delta CTD)_I$ and $(\Delta CTD)_{III}$, respectively, in 4340 steel. (After Tschegg, 1983a.)

(d) Micromechanical models for torsional fatigue crack growth have been based on the assumption that crack growth in mode III takes place by the coalescence, in mode II shear, of voids parallel to, and in the vicinity of, the mode III crack front (Nayeb-Hashemi, McClintock & Ritchie, 1983a,b). Two methods which are similar to those discussed earlier in this chapter for mode I fatigue have been employed for macroscopic fatigue fracture: (i) the crack tip displacement model, in which growth rates in anti-plane shear are related geometrically to rates of mode II shear coalescence along the crackfront, on the basis of the assumption that the crack increment is proportional to the instantaneous ΔCTD, and (ii) the damage accumulation model, in which a relationship between mode III crack advance and mode II shear along the crackfront is developed by invoking the assumption that damage accumulation, of the Coffin–Manson type, governs the rate of shear coalescence. Controlled experiments in model microstructures are needed to substantiate the validity of these models.

(e) At low mode III fatigue crack growth rates, approaching the threshold, the fracture mode is in fact representative of combined mode I–mode III failure with typical factory-roof features. Although some investigators (e.g., Pook, 1985) define a threshold ΔK_{III} for the onset of torsional fatigue crack growth, others (e.g., Hurd & Irving, 1982; Tschegg, 1983a) merely associate the occurrence of faceted fracture with the existence of a fatigue threshold in mode III. This latter approach is somewhat more appealing than the former one because, near the fatigue threshold in torsion, pure shear fracture is not observed.

10.10.2 *Estimation of intrinsic growth resistance*

Two experimental methods have been commonly adopted to circumvent the issue of crack closure in influencing mode III fatigue crack growth and to estimate an *intrinsic* crack propagation rate in mode III. In the first approach due to Tschegg (1983a), the variation of da/dN in mode III as a function of crack length a is experimentally determined at several different values of nominal 'driving force', e.g., ΔK_{III} or $(\Delta CTD)_{III}$. Each of these curves, representing a fixed level of far-field cyclic loading, is extrapolated to zero crack length to obtain the crack growth rate in the absence of any fracture surface interference. Figure 10.27 shows the results of such a plot for AISI 1018 steel (annealed at 870 °C for 1 h, $\sigma_y = 260$ MPa). The main disadvantage of this method is that it requires several different measurements of crack growth rates at different crack sizes to infer a single value of intrinsic crack growth rate at a fixed far-field loading level. However, this technique does illustrate clearly how the crack growth rates decelerate with increasing crack face interference in torsional fracture. Tschegg (1983a) showed that the intrinsic mode III crack growth rates obtained using this procedure (from the extrapolated values at zero crack length) are somewhat higher than the corresponding mode I

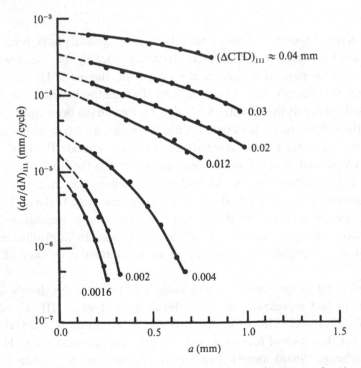

Fig. 10.27. Variation of mode III fatigue crack growth rates as a function of crack depth a (measured from the tip of the circumferential notch in the cylindrical specimen) in AISI 1018 steel. The numbers shown for each curve correspond approximately to fixed values of $(\Delta CTD)_{III}$ in millimeters. (After Tschegg, 1983b.)

fatigue crack growth rates at an identical range of CTD. This result implies that the principal reason for the apparently superior resistance to fatigue crack growth in mode III, as compared to the mode I growth, is the existence of crack closure. (This inference, however, is not fully substantiated since the effects of crack path tortuosity on mode III crack tip fields and the effects of crack closure on mode I growth rates were not considered.) It is of interest to note here that this extrapolation procedure to derive an intrinsic fracture resistance has also been employed by Tschegg & Suresh (1988) to estimate the effect of frictional sliding on mode III fracture toughness in 4340 steel (200 °C temper). It was found that the mode III fracture toughness increased by as much as a factor of two as the depth of the fatigue pre-crack in the circumferentially notched cylindrical specimen (19 mm diameter) was varied from 0.2 to 1.2 mm.

In the second approach aimed at minimizing the complications of fracture surface interference in torsional fatigue, a static tensile load is superimposed on the cyclic torsion loads in order to separate the interlocking fracture surface asperities. With this technique, the application of even a small static tensile load, under certain conditions, can markedly increase the rate of mode III fatigue crack growth by reducing the extent of crack closure (Nayeb-Hashemi, McClintock & Ritchie, 1983a; Tschegg, 1983a). However, the amount of tensile load needed to fully separate the fracture surfaces is a function of the material properties, the amplitude of the cyclic torsional loads, and the micromechanisms of crack growth. In high strength materials, a substantial tensile opening load is required to eliminate crack face contact. The superposition of such large tensile loads, however, induces strong mixed-mode conditions at the crack tip.

In agreement with the acceleration of cyclic mode III crack growth upon the superposition of a static mode I load, the works of Hourlier, McLean & Pineau (1978) and Akhurst, Lindley & Nix (1983) suggest that the superposition of static K_{III} on cyclic K_I causes the mode I crack to decelerate. This extent of reduction in growth rates increases with the increasing values of K_{III}.

Exercises

10.1 A very large plate in an engineering component, which is in continuous operation for 10 hours each day, is subjected to a mean tensile stress of 200 MPa during service. Superimposed on this operating stress are fluctuations arising from vibrations, which have an amplitude of 15 MPa (i.e. total range of 30 MPa) and a cyclic loading frequency of 50 Hz. Nondestructive crack detection techniques, which have a resolution of 0.2 mm in crack size, do not reveal any flaws in the component. If a crack were to exist at the critical location, the operating stress would be normal to the plane of the centrally located, through-thickness crack and small-scale yielding condi-

tions would prevail in the plate. The plane strain fracture initiation toughness of the material from which the plate is made is 100 MPa$\sqrt{m}$. The
fatigue threshold stress intensity factor range, ΔK_0, of the material is 3
MPa$\sqrt{m}$ for load ratio $R = 0$, and 1.5 MPa$\sqrt{m}$ for $R = 0.85$, and can
be assumed to be independent of the cyclic frequency.

   (a) Which of the following factors is/are most significant in the safe design
      of the component against catastrophic fracture: (i) the plane strain
      fracture toughness, K_{Ic}, (ii) the threshold stress intensity factor range,
      ΔK_0, or (iii) the rate of crack growth in the Paris regime? Why?

   (b) Suggest a procedure for the calculation of fatigue life in the plate.

10.2   Consider stage II crack growth by alternating slip in an FCC crystal. Prove
that

   (a) The line of intersection of two {111} slip planes in an FCC crystal
      subject to tension fatigue along the $\langle 100 \rangle$ direction lies along the $\langle 011 \rangle$
      direction.

   (b) Crack growth occurs on the {100} plane with the crack front parallel to
      the $\langle 011 \rangle$ direction.

10.3   Compare and contrast the anticipated effects of (*a*) stacking fault energy, (*b*)
grain size, and (*c*) load ratio on the fatigue life predicted on the basis of the
following three approaches to fatigue: (i) stress–life, (ii) strain–life, and (iii)
fracture mechanics.

10.4   Performing a fatigue threshold test using manual load controls in a servohydraulic testing machine can be an extremely time-consuming task.
Calculate the minimum number of days required to determine the threshold
stress intensity factor range, ΔK_0, of a material at a cyclic frequency of 1 Hz,
if ΔK_0 is defined as the threshold corresponding to a growth rate of no more
than 10^{-11} m/cycle and if the resolution of the crack detection technique is
0.05 mm.

10.5   Fatigue crack growth thresholds can be determined much more quickly
using ultrasonic loading devices than with conventional servohydraulic
test machines. If the fatigue test in the previous problem were to be conducted under ultrasonic loading at a frequency of 15 000 Hz, estimate the
amount of time required to determine the fatigue threshold. Discuss the
advantages and disadvantages of the ultrasonic test method over those of
the conventional test method.

10.6   A disgruntled employee in an engineering company is intent on destroying a
large steel plate used in a valuable engineering component. One evening,
after the normal working hours, he cuts a very sharp notch in the edge of the
plate. He orients the sharp flaw normal to the direction in which tensile
stresses are applied to the plate, with the hope that the flaw would eventually
result in catastrophic fracture of the plate. As he is committing the crime, a
bystander telephones the plant foreman at the foreman's home. The fore-

man, who leaves his home two minutes after the notification is received, travels to the scene of crime on his bicycle at an average speed of 15 km/hr. Would he have reached the plant in time to shut off the equipment before the plate fractures catastrophically? Here are some useful data for your analysis: (i) The plate is cyclically loaded uniformly from 5 to 85 kN at a frequency of 75 Hz. (ii) The plate is 18 cm wide and 0.5 cm thick. (iii) The yield strength of the steel is 1250 MPa and the plane strain fracture toughness is 48 MPa$\sqrt{m}$. (iv) The flaw introduced by the employee is 1 cm in length, through the thickness of the plate. (v) The bystander, who notified the foreman, reported to the police, by phone, that, in his judgement, the crack in the plate was propagating at a velocity proportional to the crack tip opening displacement corresponding to the peak load of each cycle. (vi) Post-failure fractographic examination revealed the presence of fatigue striations with a mean spacing of 2.5×10^{-4} mm where the crack was 2.5 mm long. (vii) The distance from the foreman's home to the factory is 5 km. (viii) The time interval between the two phone calls was 10 minutes. State all your assumptions clearly.

10.7 A piston, 9 cm in diameter, is used to increase the internal pressure in a cylinder from 0 to 50 MPa. The cylinder, which is made of peak-aged aluminum alloy (yield strength = 500 MPa, $K_{Ic} = 35$ MPa$\sqrt{m}$), has the following dimensions: length = 20 cm, inner diameter = 9 cm and outer diameter = 11 cm. The piston–cylinder arrangement suffered a malfunction which caused the cylinder to burst. Post-failure analyses showed the presence of a defect in the form of an elliptical flaw 4.4 mm long at the inner wall, 1.5 mm deep and oriented normal to the hoop stress in the cylinder. Compute the magnitude of the pressure at which the cylinder burst.

10.8 Consider a piston–cylinder arrangement, identical to the one in the previous problem, where the cylinder contains a semi-circular flaw 0.5 mm in radius instead of an elliptical flaw. If the fatigue crack growth behavior of the aluminum alloy is characterized by the relationship $da/dN = 2 \times 10^{-10}(\Delta K)^3$, where da/dN is in m/cycle and ΔK is in MPa$\sqrt{m}$, how many pressure cycles could the cylinder withstand before bursting?

10.9 The fuselage of a prototype commercial jet aircraft has a diameter of 2.6 m. The aircraft is used typically for an average of six short-haul passenger flights each day. During each flight, the cabin is pressurized to 52 kPa. The fuselage skin, 0.9 mm in thickness, is made of an aluminum alloy with Young's modulus, $E = 75$ GPa, yield strength, $\sigma_y \approx 400$ MPa, and plane strain fracture toughness, $K_{Ic} \approx 30$ MPa$\sqrt{m}$. In laboratory tests, the aluminum alloy exhibited the following fatigue crack growth characteristics. The threshold stress intensity factor range, $\Delta K_0 \approx 3.0$ MPa$\sqrt{m}$ for $R = 0$.

In the Paris regime, $da/dN = 10^{-6}$ mm/cycle and 10^{-5} mm/cycle at $\Delta K = 5$ and 9 MPa$\sqrt{m}$, respectively, for $R = 0$.

(a) When a routine maintenance check was conducted on the aircraft, the presence of a 0.4-mm long, through-thickness crack in the fuselage skin, oriented parallel to the longitudinal axis of the fuselage, was left undetected due to human error. If the aircraft fuselage is designed such that the cabin pressure could be sustained even with a longitudinal through-thickness crack of up to 10 cm length, find the safe fatigue lifetime of the aircraft from the time of the maintenance.

(b) What would be the safe lifetime in part (a) if the crack were to be oriented along the circumference of the fuselage skin?

CHAPTER 11

Fatigue crack growth in brittle solids

The mechanisms of fatigue crack growth in ductile metals and alloys, which were discussed in the preceding chapter, involve cyclic dislocation motion along one or two glide systems at the crack tip. The ensuing subcritical crack growth process in very ductile metals can typically cover a stress intensity factor range which spans a (threshold) value of 1–7 MPa$\sqrt{m}$ to one that corresponds to final failure of well over 100 MPa$\sqrt{m}$. Another noteworthy feature of fatigue cracking in ductile alloys is that there exist some distinct microscopic markings, such as striations, on the fracture surfaces from which the occurrence of failure unique to cyclic loading can be clearly identified in many situations.

Subcritical crack growth in brittle solids under cyclic loads, however, involves much more complex phenomena. In many brittle solids, there are no known differences between the micromechanisms of static and cyclic crack growth at low temperatures. Despite this apparent similarity of deformation and failure mechanisms, the superimposition of a cyclic load on a static mean stress can lead to noticeable differences in the lifetime and, in some cases, the rate of crack growth may become a strong function of cyclic frequency. Furthermore, some macroscopic crack growth phenomena may occur solely as a consequence of the imposition of cyclic loads as, for example, in the case of crack initiation ahead of stress concentrations under fully compressive cyclic loads (see Chapter 5).

This chapter deals with the growth of cracks in brittle solids subjected to fluctuating loads. The focus here is primarily on the mechanistic origins of the experimentally observed effects of cyclic loading on the growth of cracks in brittle solids. Consequently, no attempt will be made to review the vast amount of empirical data on fatigue crack growth reported for ceramics and other brittle solids. A survey of such observations can be found in Roebben et al. (1996). The reader is also advised to review the introduction to Chapter 5 before proceeding with the sections to follow in this chapter.

This chapter begins with a summary of various situations under which a 'cyclic effect' may occur in brittle solids. This is followed by a discussion of methods to characterize the rate of crack growth. The evolution of resistance curves (R-curves) in ceramics and the stability of subcritical crack growth are then addressed. Mechanisms of fatigue crack growth facilitated by crack-tip and crack-wake phenomena are considered next. The chapter concludes with a discussion of mechanisms of fatigue crack growth in brittle solids at elevated temperatures where particular attention is paid to the effects of intergranular amorphous films. Fatigue crack growth in brittle polymers and polymeric composites is considered in the next chapter.

11.1 Some general effects of cyclic loading on crack growth

Distinct effects of cyclic loading on the growth of cracks have been documented in a wide range of ceramics and ceramic composites. A close examination of these effects reveals the following mechanistic features.

(1) In ceramic materials containing long cracks (whose main dimensions are significantly larger than the characteristic microstructural size scale such as the grain size), cyclic loading is generally found to cause a pronounced increase in crack growth rate compared to static load experiments at room temperature. This deleterious effect, which is most clearly evident in ceramics and ceramic composites that exhibit a strong R-curve behavior (i.e. ceramics with an increasing resistance to quasi-static fracture with crack advance), primarily stems from the progressive breakdown of the bridging ligaments in the wake of the crack tip by repeated contact (Grathwohl, 1988; Evans, 1980; Horibe & Hirahara, 1991; Lathabai, Rödel & Lawn, 1991; Kishimoto *et al.*, 1994; Ramamurty, Hansson & Suresh, 1994). This deleterious effect of cyclic loading is more accentuated at lower stress intensity factor values and lower temperatures.

(2) Residual stresses arising from thermal and elastic anisotropy between the grains in a polycrystalline ensemble can promote spontaneous distributed intergranular microcracking during thermal excursions (e.g., during cooling from the processing temperature). Such microcracking can be exacerbated by mechanical loads involving axial compression as well as indentation (see Chapter 5). When a zone of microcracking is so formed ahead of a stress concentration or around the contact region of an indented surface, the release of the compressive mechanical load or the indenter engenders a zone of tensile residual stresses. A macroscopic crack then initiates and advances perpendicularly to this tensile residual field (Brockenbrough & Suresh, 1987; Lawn *et al.*, 1994). Repeated compressive cycling of notched ceramics or cyclic indentation can lead to a growth of the crack in a subcritical manner.

(3) At elevated temperatures, intergranular glassy films added during the processing of the ceramic become viscous. In addition, environmental interactions, such as oxidation, can lead to the *in-situ* formation of viscous glassy films at grain boundaries and interfaces in some ceramic materials. The viscous flow of these amorphous films during cyclic loading imparts a strong sensitivity to cyclic frequency, waveform and mean stress (Han & Suresh, 1989; Ewart & Suresh, 1992; Lin *et al.*, 1992; Dey, Socie & Hsia, 1995).

(4) In ceramics and ceramic composites with coarse grains or elongated grain structure arising from *in-situ* reinforcements, the bridging of crack faces can be promoted by both intact grains and glassy ligaments (Liu, Chen & Tien, 1994; Ramamurty, Hansson & Suresh, 1994).

(5) The mutual competition among the corrosive effects of environment, the viscous flow of *pre-existing* and *in-situ formed* intergranular and interfacial glassy films, and the breakdown of bridging ligaments due to repeated crack face contact determines whether cyclic loading is beneficial or detrimental to crack growth.

11.2 Characterization of crack growth in brittle solids

11.2.1 Crack growth under static loads

Many ceramics exhibit a tendency for subcritical crack growth under static/quasi-static loads owing to their susceptibility to stress corrosion by an aggressive environment, to the existence of an R-curve effect which stems from a variety of toughening mechanisms, or to creep cavitation at elevated temperatures. In these situations, the time rate of growth of a long crack, da/dt, is characterized on the basis of linear elastic stress intensity factor, K, such that

$$\frac{da}{dt} = A(K)^p, \tag{11.1}$$

where A and p are experimentally determined constants.[†] This characterization generally holds from a threshold value of K until K approaches the fracture toughness of the material.

If the ceramic is toughened, the 'driving force' for crack growth can be significantly altered as a result of mechanisms which are operational either at the crack tip or in the crack wake. These mechanisms include stress-induced martensitic transformations, crack deflection at grain boundaries and interfaces between the matrix and the reinforcements, and crack bridging by intact grains, fracture surface asperities, particles or fibers. (A discussion of such mechanisms is presented in Chapter 14.) A consequence of the toughening mechanisms is that the effective stress intensity factor, K_{eff}, at the crack tip is reduced from the nominal or applied K by the shielding component, K_s, such that

$$K_{\text{eff}} = K - K_s. \tag{11.2}$$

[†] As discussed in detail in Chapters 9 and 16, the conditions of validity of linear elastic fracture mechanics should first be ensured, especially at elevated temperatures. Methods to characterize high-temperature crack growth are discussed further in Chapter 16.

11.2.2 *Crack growth under cyclic loads*

It was shown in Chapter 10 that the cyclic crack growth characteristics of metals are generally characterized on the basis of the Paris law. Here the rate of crack advance per cycle, da/dN, is related to the stress intensity factor range $\Delta K = K_{max} - K_{min}$; K_{max} and K_{min} denote the maximum and minimum values of K, respectively. For a fatigue load cycle of constant-amplitude,

$$\frac{da}{dN} = C(\Delta K)^m, \tag{11.3}$$

where C and m are experimentally determined constants. The Paris exponent m for metallic materials is typically in the range 2–4. The fatigue crack growth rates, as shown in Chapter 10, are also strong functions of the stress ratio R, microstructure and environment.

Brittle materials, on the other hand, exhibit very limited subcritical crack growth at low temperatures. Within this restricted range of damage tolerance, it is of interest to know whether the cyclic component of the imposed loads induces subcritical fracture by a purely mechanical fatigue effect, or whether crack advance under the applied fluctuations in loads is merely a consequence of stable fracture facilitated by a static mean level of the applied load. One way to quantify this distinction is to estimate the rate of fatigue crack growth per cycle on the basis of experimentally determined static crack velocities, Eq. 11.1. Such an approach, commonly employed for the prediction of corrosion fatigue crack growth rates on the basis of stress corrosion crack velocities (see Chapter 16), was used by Evans & Fuller (1974) for ceramics. They estimated the time to failure under cyclic loading conditions on the basis of experimentally determined values of the material constants in Eq. 11.1. This approach implicitly invokes the following assumptions.

(1) The mechanisms of failure under cyclic loads are the same as those occurring under static loading conditions.

(2) Fluctuations in applied loads can be modeled as a series of small increments in fixed loads.

(3) Compressive stresses do not induce any crack growth; only the tensile portion of the fatigue cycle is considered in the calculation of crack growth rate.

(4) The crack propagation rate per fatigue cycle is obtained by integrating the crack velocity for static load, Eq. 11.1, over the time period of the fatigue cycle, $1/v_c$, where v_c is the cyclic frequency, such that

$$\frac{da}{dN} = \int_0^{1/v_c} A(K)^p \, dt. \tag{11.4}$$

With the above assumptions, the growth of a crack per unit period of time (i.e. for cyclic loads, the time period of a fatigue cycle) can be determined for any time-dependent stress $\sigma(t)$. Consider, for example, a cracked specimen for which the stress intensity factor is given by

$$K = \overline{Y}\sigma(t)\sqrt{a}, \tag{11.5}$$

where $\overline{Y}$ is a parameter which depends on specimen geometry, crack geometry, and crack length, a. Substituting this expression for K in Eq. 11.1 and integrating within the limits $t = 0$ and $t = t_f$ (where t_f is the time to failure) gives:

$$\int_0^{t_f} \{\sigma(t)\}^p dt = \left\{\frac{2}{(p-2)A\overline{Y}^p}\right\} \cdot \left\{\left(\frac{1}{a_0}\right)^{\frac{p-2}{2}} - \left(\frac{1}{a_{cr}}\right)^{\frac{p-2}{2}}\right\}, \quad p \neq 2. \tag{11.6}$$

Here a_0 is the initial crack size and a_{cr} the critical crack size, both of which are assumed to be the same for static and cyclic failures.

If σ_{static} is the static stress under which the crack advances and $\sigma(t)$ is the cyclic stress which fluctuates between a maximum of σ_{max} and a minimum of $\sigma_{min} = R\sigma_{max}$ (where R is the load ratio), the ratio of the time to failure under static stress to that under cyclic stress, obtained from Eq. 11.6, is

$$\frac{t_{f,static}}{t_{f,cyclic}} = \left[\nu_c \int_0^{1/\nu_c} \left\{\frac{2\sigma(t)}{\sigma_{max}(1-R)}\right\}^p dt\right] \cdot \left\{\frac{2\sigma_{static}}{\sigma_{max}(1+R)}\right\}^{-p}. \tag{11.7}$$

This ratio can be evaluated numerically for any waveform of the cyclic load.

The above approach of integrating the static crack growth response to derive the fatigue crack growth rates can also be used to assess if a pure mechanical fatigue effect or a synergistic effect involving cyclic loads and environmental interactions occurs. Let $(da/dN)_{exp}$ be the experimentally determined crack growth rate under cyclic loading conditions for a given K_{max} and R. If the difference

$$\left|\left(\frac{da}{dN}\right)_{exp} - \int_0^{1/\nu_c} A(K)^p dt\right| \neq 0, \tag{11.8}$$

it is inferred that the mechanisms of failure or crack-wake contact conditions under static loads may be different from those occurring under cycling loads. Whether the difference is greater than or less than zero indicates whether cyclic loading leads to a detrimental or beneficial effect, respectively, compared to static loading.

The isolation of the contributions to crack growth from static loads, as influenced by the K_{max}, and those from cyclic loads, as influenced by ΔK, can also be accomplished in a different manner. Kishimoto *et al.* (1987) carried out systematic experiments on sintered silicon nitride in an attempt to identify the contributions to fracture from K_{max} control and an effective ΔK control. Following the work of Hojo *et al.* (1987) on organic composites, they proposed a power-law characterization which involves both K_{max} and ΔK. This approach has been adopted for metals and nonmetals by others (e.g., Vasudevan & Sadananda, 1995; Jacobs & Chen, 1995) who present crack growth rates in terms of a modified Paris-type equation:

$$\frac{da}{dN} = B(K_{max})^g (\Delta K)^q. \tag{11.9}$$

In most monolithic ceramics subjected to constant-amplitude, tension–tension fatigue at room temperature, the static crack growth exponent g in Eq. 11.9 can be as much as a factor of ten or more greater than the cyclic crack growth exponent q.

11.3 Crack growth resistance and toughening of brittle solids

Before considering the crack growth characteristics under static and cyclic loads, it is instructive to examine the resistance of brittle solids to subcritical crack growth and the stability of the fracture process under a given loading configuration. Indeed, as alluded to earlier in this chapter and as shown in more detail in subsequent sections, the resistance of brittle solids to quasi-static crack growth has a strong bearing on their cyclic crack propagation characteristics and on the extent of fatigue damage tolerance. In addition, the stability of subcritical crack growth is a topic which has important implications for the fracture testing of brittle solids, as illustrated in the example problem which follows this section.

Consider the large plate with the central crack of initial length $2a_i$, Fig. 9.1. When a tensile stress σ is applied to this plate, the strain energy release rate is given by Eq. 9.19. First consider a plate material whose resistance to static fracture remains constant during crack growth. The resistance curve (R-curve) for this case, marked $\mathcal{G}_R$, is schematically shown in Fig. 11.1(a). Here it is evident that the critical value of the strain energy release rate during crack growth remains unaltered at $\mathcal{G} = \mathcal{G}_c$ in this material. Equivalently, the critical condition for crack growth can be formulated in terms of the mode I stress intensity factor K_c, since $\mathcal{G} = K^2/E'$, Eqs. 9.58 and 9.59. If the applied stress is σ_1, the initial value of the strain energy release rate is $\mathcal{G}_i \propto \sigma_1^2 a_i$. Since $\mathcal{G}_i < \mathcal{G}_c$, no crack growth occurs at the applied stress of σ_1. As the applied stress is raised progressively from σ_1 to σ_2 to σ_3, there comes a critical stress σ_3 at which $\mathcal{G}_i = \mathcal{G}_c$. Beyond this point, marked A in Fig. 11.1(a), the strain energy release rate (or the stress intensity factor) corresponding to the applied stress and the initial crack size begins to exceed the intrinsic fracture resistance of the material characterized by $\mathcal{G}_c$ or K_c, and causes the onset of unstable fracture. Thus, unstable crack growth occurs when

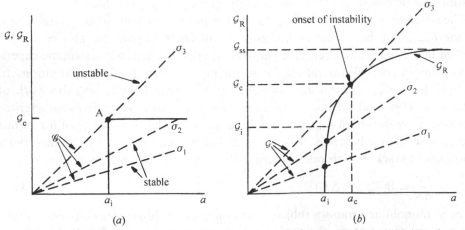

Fig. 11.1. Schematic variation of the driving force for fracture versus the material resistance to crack growth for a material with (a) a flat R-curve and (b) a rising R-curve.

$$\frac{d\mathcal{G}}{da} > \frac{d\mathcal{G}_R}{da}. \tag{11.10}$$

Now consider the cracked plate which is made of a brittle solid whose resistance to continued fracture is enhanced as a result of some toughening mechanism, Fig. 11.1(b). As described in Chapters 5 and 14, this mechanism may involve (i) the development of a dilatational transformation zone which is displaced to the wake of the crack-tip during crack growth, causing a net reduction in the effective crack opening displacement and 'driving force', (ii) the bridging of the crack faces by fibers, intact grains or fracture surface asperities, or (iii) periodic microscopic deflections in the path of the crack. Such a material exhibits a rising R-curve, marked $\mathcal{G}_R$ in Fig. 11.1(b). For applied stress levels σ_1 and σ_2, a small amount of crack extension occurs when $\mathcal{G}_i$ intersects the $\mathcal{G}_R$ curve. However, crack growth terminates quickly because the material resistance $\mathcal{G}_R$ rises faster with crack growth than the strain energy release rate due to the applied stress. The conditions for such stable fracture are that

$$\mathcal{G} = \mathcal{G}_R \quad \text{and} \quad \frac{d\mathcal{G}}{da} \le \frac{d\mathcal{G}_R}{da}. \tag{11.11}$$

When the applied stress σ_3 is of such a magnitude that at the point of tangency to the $\mathcal{G}_R$ curve, i.e. at the point where $\mathcal{G} = \mathcal{G}_c$, the rate of increase of the strain energy release rate due to the applied stress with crack growth exceeds the rate of increase of $\mathcal{G}_R$ with crack advance, unstable fracture ensues. The initial value of $\mathcal{G}$ at which crack growth begins and the steady-state value of $\mathcal{G} = \mathcal{G}_{ss}$ are known to differ by as much as a factor of ten or more for toughened ceramics.

Whereas the driving force for fracture, $\mathcal{G} \propto a$, for the centrally cracked wide plate, $\mathcal{G} \propto a^2$ for the double-cantilever beam specimen (see the worked example in Section 9.4.1). It is thus evident that the dependence of the variation of the driving force with crack extension is a strong function of the loading system, the crack geometry and the specimen geometry. Thus, while the critical value of the strain energy release rate $\mathcal{G}_c$ is expected to be unaffected by the loading configuration, the stability of crack growth is strongly influenced by it. Therefore, when performing static and cyclic fracture experiments on brittle solids such as ceramics, it is important to consider loading modes, such as displacement control and double-cantilever specimens, which facilitate stability of subcritical crack growth in brittle solids. These ideas are further illustrated with the following example.

11.3.1 Example problem: Fracture resistance and stability of crack growth

Problem:

In this worked example, we discuss the driving force for fracture in brittle solids and the issues of stability of subcritical crack growth by recourse to the concepts of strain energy release rate which were discussed in Section 9.1. Consider a cracked specimen which is loaded under compliant conditions in mode I, as shown in Fig. 11.2. The specimen has unit thickness ($B = 1$). The symbol C is the compliance (a function of crack length a) of the specimen and C_M is the compliance of the testing machine (which may be regarded as a spring) in which the specimen is loaded in tension. The testing machine is connected in series with the specimen. Let Δ_T be the total displacement which can be regarded as prescribed.

$$\Delta_T = \Delta + C_M P = \Delta + \left(\frac{C_M}{C}\right)\Delta. \tag{11.12}$$

The potential energy is given by

$$W_P = \Phi + \frac{1}{2}C_M P^2 = \frac{1}{2}\frac{\Delta^2}{C} + \frac{1}{2}\frac{C_M}{C^2}\Delta^2$$

$$= \frac{1}{2}C^{-1}\Delta^2 + \frac{1}{2}[C_M^{-1}(\Delta_T - \Delta)^2]. \tag{11.13}$$

(i)   Show that $C_M = 0$ corresponds to fixed grip loading (constant displacement), and that $C_M \to \infty$ corresponds to dead load.

(ii) Show that the mechanical strain energy release rate,

$$\mathcal{G} = -\left[\frac{\partial(W_P)}{\partial a}\right]_{\Delta_T} = \frac{1}{2}P^2\frac{dC}{da}, \tag{11.13}$$

and therefore, does not depend on the compliance of the testing machine.

(iii) Show that $\left[\frac{\partial \mathcal{G}}{\partial a}\right]_{\Delta_T}$ depends on the nature of the loading system.

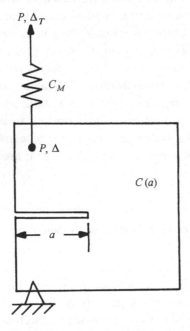

Fig. 11.2. A schematic of the cracked specimen and the loading arrangement.

(iv) For an ideally brittle solid, show that crack advance is stable if $\left[\frac{\partial \mathcal{G}}{\partial a}\right]_{\Delta_T} \leq 0$, and that it is unstable if $\left[\frac{\partial \mathcal{G}}{\partial a}\right]_{\Delta_T} > 0$.

Solution:

(i) For fixed grip loading, $\Delta_T = \Delta$. Substituting this in Eq. 11.12, we note that $C_M/C = 0$ and $C_M = 0$. For dead-weight loading, the rate of change of load P with time is zero, i.e.

$$\dot{P} = \frac{\dot{\Delta}}{C} = 0, \tag{11.15}$$

where $\dot{\Delta}$ is the rate of change of displacement. Combining with Eq. 11.12,

$$\dot{\Delta} = \frac{\dot{\Delta}_T}{1 + (C_M/C)} = 0. \tag{11.16}$$

From this, we note that $C_M \to \infty$ for dead-weight loading.

(ii) Using Eq. 9.9 and the information that $B = 1$, the mechanical strain energy release rate is

$$\mathcal{G} = -\frac{dW_p}{da}. \tag{11.17}$$

Combining Eqs. 11.12–11.14, we obtain

$$W_P = \frac{\Delta_T^2}{2(C + C_M)}, \quad \text{and}$$

$$\mathcal{G} = -\left[\frac{\partial(W_P)}{\partial a}\right]_{\Delta_T} = \frac{\Delta_T^2}{2(C + C_M)^2} \cdot \frac{dC}{da}. \tag{11.18}$$

Note that

$$\frac{\Delta_T}{C_T} = \frac{\Delta_M}{C_M} = \frac{\Delta}{C} = P, \quad \text{and} \quad \Delta_T = \Delta_M + \Delta. \tag{11.19}$$

From this equation, we note that $C_M + C = C_T$. Combining these results with Eq. 11.14, it is seen that

$$\mathcal{G} = \frac{P^2}{2} \cdot \frac{dC}{da}. \tag{11.20}$$

Note that this expression for $\mathcal{G}$ does not involve the compliance of the testing machine.

(iii) From part (ii),

$$\frac{\partial \mathcal{G}}{\partial a} = -\frac{\Delta_T^2}{(C + C_M)^3}\left(\frac{dC}{da}\right)^2 + \frac{1}{2}\frac{\Delta_T^2}{(C + C_M)^2}\frac{d^2C}{da^2}$$

$$= -\frac{P^2}{(C + C_M)}\left(\frac{dC}{da}\right)^2 + \frac{P^2}{2}\frac{d^2C}{da^2}. \tag{11.21}$$

Note that this equation involves the compliance of the testing machine, and hence the nature of the loading system.

(iv) For an ideally brittle solid, $d\mathcal{G}_R/da = 0$ (see Fig. 11.1). Combining Eq. 11.21 with Eqs. 11.10 and 11.11, it is readily seen that crack advance is stable if $\left[\frac{\partial \mathcal{G}}{\partial a}\right]_{\Delta_T} \leq 0$, and that it is unstable if $\left[\frac{\partial \mathcal{G}}{\partial a}\right]_{\Delta_T} > 0$.

11.4 Cyclic damage zone ahead of tensile fatigue crack

The phenomenon of crack growth under cyclic compression, which was discussed in Section 5.6, is a consequence of the development of a *cyclic damage zone* ahead of the stress concentration. Similar cyclic damage zones and residual stress fields also develop ahead of cracks in ceramic materials that are subjected to tensile fatigue loads. Although cyclic damage in tension fatigue of ceramics is qualitatively similar to that described in Section 9.6 for metals, a brief, focused discussion for brittle solids is perhaps appropriate here because of the vast differences in the constitutive responses of ductile and brittle materials.

As noted in the preceding section, cyclic damage zones evolve when permanent strains are retained within the deformation zone ahead of a flaw or a notch. The evolution of permanent strains in the damage zone is evident from the existence of *open* microcracks in the fully unloaded state in Fig. 5.1(a) (cyclic compression loading of Al_2O_3 at room temperature).

Suresh & Brockenbrough (1990) carried out finite-element analyses of the evolution of a reversed damage zone ahead of a tensile fatigue crack in coarse-grained alumina which undergoes stress-induced microcracking. By employing the constitutive model described in Section 5.3.2, they simulated the near-tip fields after one zero–tension–zero fatigue cycle for a brittle solid with a population of penny-shaped microcracks.

Figure 11.3(a) shows the normal stress ahead of the crack tip (corresponding to the peak far-field tensile stress) as a function of the distance x directly ahead of the crack tip (normalized by ρ). In this figure, $\sigma_{yy}\sqrt{\pi\rho}/K_{max}$ denotes the nondimensional value of the normal stress σ_{yy} on the plane of the main crack. For this condition, tensile stresses prevail everywhere in the edge-cracked solid. Figure 11.3(b) shows the unloaded state where the residual compressive stress zone which, for the conditions of the simulation, spans a distance of 9ρ ahead of the crack tip. Beyond this distance, tensile residual stresses exist so as to satisfy equilibrium requirements. Note that for the conditions of the analysis, the predicted size of the cyclic damage zone due to microcracking is significantly smaller than the size of the reverse plastic zone for the elastic–perfectly plastic solid (Section 9.6). Also indicated in Fig. 11.3 are the contours of constant microcrack density β drawn with x/ρ as the abscissa and y/ρ as the ordinate. This example serves to illustrate the origin of a reversed damage zone in a brittle solid which develops distributed microcracking ahead of a notch or a crack as a result of the release of residual stresses induced by processing, phase transformations or grain-boundary shear-induced tensile opening (see Section 5.3). This cyclic

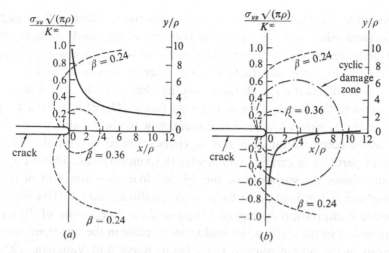

Fig. 11.3. Variation of normal stress as a function of distance directly ahead of the crack tip (solid curve) in a microcracking brittle solid (*a*) at the peak tensile stress and (*b*) upon unloading. The dashed lines denote the contours of stable damage zone shapes for the indicated values of β. (After Suresh & Brockenbrough, 1990.) See text for details.

damage zone is expected to have an effect on fatigue phenomena such as transient crack growth following tensile overloads.

11.5 Fatigue crack growth at low temperatures

The phenomenon of mode I crack growth under cyclic compression (Section 5.6) provides an example of failure where the macroscopic mode of crack growth under cyclic loads is distinctly different from the monotonic fracture mode in brittle solids. Even when the mechanisms and macroscopic modes of failure are similar in monotonic and cyclic loading, apparent differences may arise between the rates of crack growth in the two cases as a consequence of the mechanical contact of the crack faces and/or environmental interactions. The origins of such differences in crack growth in ceramic materials subjected to tension–tension or tension–compression fatigue at room temperature can be ascribed to a number of factors.

The thermal and elastic mismatch between adjacent grains in brittle solids with noncubic crystal structures can engender residual stresses upon cooling from the processing temperature (e.g., Evans, 1978; Ortiz & Suresh, 1993). In coarse grained ceramics, compressive residual stresses at grain boundaries can exert a 'clamping' action which leads to the attachment of intact grains to one or both crack faces. During the early fatigue cycling process and during short crack growth, the frictional contact between the pulled-out grains and the crack faces, and the associated crack bridging stresses promote an energy dissipation mechanism which serves to 'shield' the crack from the far-field loads. This beneficial effect on crack growth resistance is also manifested as a rising R-curve behavior in monotonic fracture. Repeated cyclic loading, in conjunction with the 'smashing together' of the crack surfaces, progressively leads to a degradation of these bridging tractions, and an ensuing elevation in the effective driving force for fracture. Consequently, the rates of fatigue crack growth progressively increase with increasing number of cycles of constant amplitude fluctuation in the nominal stress intensity factor. The effective crack opening displacement also registers a progressive increase with the breakdown of bridging ligaments which comprise intact grains in coarse-grained monolithic ceramics or reinforcement particles in ceramic composites (Kishimoto *et al.*, 1994). Figure 11.4 schematically shows the variation of the applied load as a function of the crack opening displacement at load point for a polycrystalline ceramic. Here the progression of tension–compression cycling and the attendant degradation of the frictional bridging ligaments in the crack wake lead to an increase in the apparent compliance and hysteresis in the tensile portion (e.g., Guiu, Reece and Vaughan, 1991). The extent of deterioration in crack shielding with cyclic loading can be a strong function of the grain size distribution, reinforcement particle size and distribution, maximum stress intensity factor level, mean stress, and the accumulation of debris particles released from the loss of bridging tractions within the crack faces (e.g., Grathwohl, 1988; Horibe & Hirahara, 1991; Lathabai, Rödel & Lawn, 1991; Ogawa, 1992; Dauskardt, 1993; Jacobs & Chen, 1995).

Figure 11.5 shows an example of the crack growth characteristics of a 99% pure alumina (grain size $\approx 10\,\mu\text{m}$) subjected to fully reversed cyclic loads ($R = -1$) and to static loads in the room temperature air environment. In this figure, the fatigue crack growth rates da/dN obtained in direct push–pull using a wedge-opening load specimen are plotted as a function of the maximum stress intensity factor of the fatigue

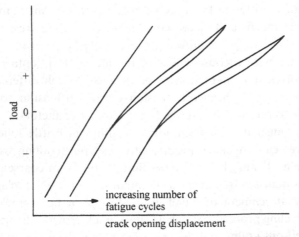

Fig. 11.4. A schematic of the cyclic variation of the applied load as a function of the load point displacement for a coarse-grained polycrystalline ceramic subjected to tension–compression fatigue at room temperature.

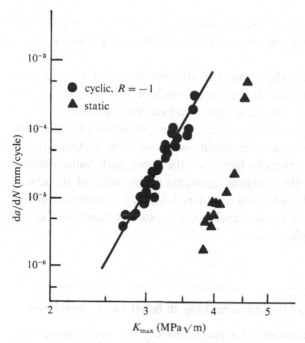

Fig. 11.5. Variation of fatigue crack growth rate da/dN as a function of the maximum stress intensity factor K_{max} under fully reversed cyclic loads ($v_c = 5$ Hz). Also indicated are the rates of crack growth per cycle derived from static load fracture data. (After Reece, Guiu & Sammur, 1989.)

cycle, K_{max}. Also shown in Fig. 11.5 are the static crack growth data for the alumina, where the crack velocity da/dt is converted to crack growth rate per cycle using the relationship $da/dN = (da/dt)/v_c$ and is plotted against the (fixed) stress intensity factor K_{max} at which the static test was performed. The crack growth rates in the cyclic loading case are significantly faster than those seen in static tension.

Transforming ceramics, which exhibit stable crack growth in monotonic loading, are also known to be susceptible to progressive degradation under fully or partially tensile cyclic loading. Subcritical fatigue crack growth also occurs under cyclic tension in transformation-toughened ceramics, such as polycrystalline ZrO_2, which is partially stabilized with MgO (Swain & Zelizko, 1988; Sylva & Suresh, 1989; Dauskardt, Marshall & Ritchie, 1990). Here, the exponent g in Eq. 11.19 is more than an order of magnitude higher than the exponent q, implying that the maximum value of the stress intensity factor, rather than its amplitude, determines the overall crack growth rates. Consistent with this trend, it is seen that despite the occurrence of an accelerated crack growth under cyclic loads, no differences exist between the mechanisms of monotonic and cyclic fracture. Furthermore, the bounds of the fatigue crack growth curve scale with the fracture toughness of the material; crack extension commences at a threshold value which is approximately 50% of the toughness (K_c) for all the heat treatments. The strains associated with phase transformations are essentially the same for both monotonic and cyclic loading conditions, and the crack propagation rates are influenced by the specimen geometry and crack size (Sylva & Suresh, 1989).

These results collectively show that the aforementioned crack-wake phenomena associated with contact between the surfaces, which in this particular case are accentuated by the translation of the transformed material (with an increase in volume) to the wake of the crack tip, most likely determine this apparent fatigue effect. The accelerated crack growth rates seen upon switching from a static stress intensity factor to a cyclic stress intensity factor of the same peak value ostensibly stem from the degradation of the bridging ligaments in the wake of the crack, even in this transformation-toughened ceramic material. Moisture-containing environments, such as humid air, lead to higher crack growth rates as compared to inert media under both static and cyclic loads.

11.6 Case study: Fatigue cracking in heart valve prostheses

The foregoing discussions of the characteristics of fracture in brittle solids under cyclic loading conditions deal primarily with mechanistic phenomena. There are many practical applications of such cyclic fractures in a wide variety of engineering structures, biomechanical devices and components used in microelectronic structures and packages. In this section, we consider a case study which highlights the possible role of cyclic loading in inducing failure in synthetic heart valves.†

Artificial heart valves implanted in humans continuously regulate blood flow. Components of the valve should exhibit mechanical endurance to fluctuating stresses, bending, contact fatigue and wear at articulating surfaces and resistance to stress corrosion cracking and cavitation erosion in the physiological environment.

The use of metallic components, made from such materials as titanium and Co–Cr alloys, for the complex-shaped housing of the heart valve necessitate such fabrication methods as welding and casting. These methods, however, are prone to severe structural damage arising from fatigue failure of the housing. Consequently, a brittle, ceramic-like solid made from pyrolytic carbon and/or pyrolytic carbon/graphite laminated composite has been used to fabricate heart valves. Over 600 000 of these have already been implanted in patients, and they constitute the majority of synthetic heart valves that are commercially produced in the United States.

The mechanism for regulating blood flow in artificial heart valves commonly involves two semi-circular leaflets (bileaflets) or a tilting disc, which open(s) and close(s) during heart beat. The leaflets (or the disc) are housed in a circular orifice, which is stiffened by a metallic restraining ring in some designs. In order to enable attachment to the cardiac tissue, the outer diameter is linked to a cloth sewing ring.

Damage and cracking under cyclic loads can be a significant form of failure for mechanical heart valves. The fatigue loading arises from the pumping of the human heart

---

† This case study is derived from a review by Ritchie (1996).

at a rate of 72 beats per minute (i.e. at a cyclic frequency of 1.2 Hz), which translates to approximately 38 million beats per year. The physiological environment in which this loading occurs is usually simulated by carrying out fatigue tests in Ringer's solution at 37 °C. The load ratio, R, of this cyclic loading pattern is zero, which is representative of the mean stress conditions of fatigue *in vivo*. Both total-life and fracture mechanics based methodologies for fatigue life estimation have been developed over the years for metallic and nonmetallic heart valves. However, since fatigue cracking in the heart valves can pose a life-critical situation, the regulations of the Food and Drug Administration (FDA)† in the United States require that defect-tolerant design principles be used to safeguard against the catastrophic failure of mechanical heart valves in the physiological environment.

In view of its durability, strength, resistance to wear, high thromboresistance, and cellular biocompatibility with human blood and soft tissue, the low-temperature isotropic (LTI) form of pyrolytic carbon has found application in the mechanical heart valve. Silicon-alloyed LTI pyrolytic carbon is used in mechanical heart valves either in the form of a coating, approximately 250-μm thick, on a polycrystalline graphite substrate or as a monolithic material. Up to 20 wt% Si is added to improve mechanical response of the carbon without any appreciable loss of biocompatibility. There exists a residual stress, σ_R, due to thermal mismatch between the carbon coating and the graphite substrate, which from stress measurements has been estimated to be in the range 20–48 MPa.

Figures 11.6(*a*) and (*b*) show examples of bileaflet and tilting disc valves, respectively, made of the LTI pyrolytic carbon. Figure 11.7(*a*) shows the inside of the seating lip of the cross section of a 29-mm mitral valve housing, where the highest dynamic stressing occurs during repeated heart beat. At this location, the maximum surface stress, estimated from finite-element simulations, is of the order of 5.5 MPa under static conditions and 30.1 MPa under dynamic conditions. This site is one critical location for the inception and growth of a fatigue flaw. Figure 11.7(*b*) shows another location on the outside surface of the housing at the upper corner of the stellite stiffening ring groove. At this site, the maximum surface stress is about 43 MPa for a corner root radius of 70 μm. In addition to these mechanical loads at these sites, the residual stresses, σ_R, remaining from the processing steps should be considered in the fatigue life calculations.

Laboratory experiments show that when cyclic loads with a load ratio $R \approx 0$ are applied, the rates of crack growth in the pyrolytic carbon valve material are significantly faster than those measured under a static load of the same peak value. This indicates a potential deterioration in the damage tolerance of this material due to stress fluctuations.

The chances of occurrence of a catastrophic *in-vivo* failure solely as a result of cyclic loads in an initially (nominally) flaw-free synthetic heart valve are generally expected to be very small. It is, however, acknowledged that should defects be introduced at a critical site, such as the one shown in Fig. 11.7(*a*) or (*b*), either as a result of poor fabrication and manufacturing quality or during prosthetic surgery, the subcritical advance of the defects during heart beat can lead to catastrophic failure well within the design life of the valve.

---

† *Replacement heart valve guidance document.* Food and Drug Administration, Division of Respiratory and Neurological Devices, U.S. Department of Health and Human Services, Washington, D.C., 1994.

(a)

(b)

Fig. 11.6. (a) Bileaflet and (b) tilting disc valves, respectively, made of the LTI pyrolytic carbon. (Photographs courtesy of R.O. Ritchie. Reprinted with permission.)

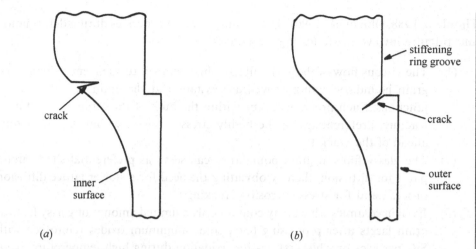

Fig. 11.7. (*a*) The inside of the seating lip of the cross section of a 29-mm mitral valve housing, where the highest dynamic stressing occurs during repeated heart beat. (*b*) Another location on the outside surface of the housing at the upper corner of the stellite stiffening ring groove. (After Ritchie, 1996.)

11.7 Fatigue crack growth at elevated temperatures

At temperatures, typically in excess of 1000–1200 °C, grain boundary diffusion, grain boundary sliding and stress corrosion cracking from environmental interactions become prominent in most ceramic materials. In addition, the glassy films at grain boundaries exert a dominant effect on deformation at these elevated temperatures, a consequence of which is enhanced cavitation and sliding at grain facets and possibly enhanced environmental interactions.

11.7.1 Micromechanisms of deformation and damage due to intergranular/interfacial glassy films

In most commercially synthesized ceramics, liquid-phase forming sintering additives are introduced during fabrication. In some structural ceramics such as silicon nitride, the presence of sintering additives, which invariably promote the formation of intergranular glassy films, is necessary in order to achieve full densification. At least a fraction of these additives invariably remains as intergranular glassy films in the material upon completion of all the processing steps. When the ceramic is exposed to a high temperature, the amorphous phase becomes viscous as the temperature surpasses the glass transition temperature of the phase. The flow of the glass phase (under the influence of an applied stress) along grain boundaries imparts a nonlinear (creep) deformation at the crack tip (e.g., Clarke, 1987;

Thouless, 1988; Ohji *et al.*, 1990). The ensuing damage processes during deformation and fracture involve the following mechanisms.

(1) The viscous flow of the glassy films at high temperatures promotes enhanced grain boundary sliding as well as cavitation at the grain boundary triple junctions. Such processes occur within the bulk of the ceramics, or during fracture, preferentially at the highly stressed region within the creep zone ahead of the crack tip.

(2) The glassy films at grain boundaries can serve as preferential short-circuit paths for diffusion, thereby obviating the need for a slower lattice diffusion process, and for stress-corrosion cracking.

(3) Even in ceramics which may contain only a limited amount of glassy films at grain facets after processing (e.g., some aluminum oxides reinforced with SiC particles or whiskers), *in-situ* oxidation during high-temperature exposure (e.g., the oxidation of SiC) can lead to the formation of glassy films. In unnotched or uncracked specimens, this oxidation occurs primarily at free surfaces. In cracked ceramics, however, the access of the environment to the highly-stressed region of the crack tip facilitates preferential formation of a high concentration of *in-situ*-formed glass at grain boundaries and interfaces immediately ahead of the crack tip. Figure 11.8(a) shows an example of a region of material (white in color) which has been oxidized around a growing crack in an Al_2O_3–SiC composite subjected to crack growth under cyclic loads at 1400 °C. During crack advance, this transformed region is displaced to the wake of the crack tip.

(4) During static or cyclic loading, the cavities which form in conjunction with the viscous flow of glassy films along grain triple points extend along grain facets (in monolithic ceramics) and along interfaces (in ceramic composites) to create microscopic flaws. Figure 11.8(b) is an example of such microcracking in the region immediately ahead of the crack tip in an alumina ceramic reinforced with SiC whiskers.

(5) As the population of such microcracks increases in the crack-tip region, a diffuse microcrack zone evolves ahead of the crack, Fig. 11.9(a). Note that the presence of open microcracks in the fully unloaded state is indicative of the development of permanent strains within the crack-tip damage zone.

(6) Bridging of the faces of the microcracks by the whiskers, Fig. 11.9(b), and frictional sliding along the debonded matrix–whisker interface in the damage zone ahead of the crack tip may modify the near-tip stress intensity factor. In addition, the crack faces may be bridged by glassy ligaments, e.g., Fig. 11.8(a).

(7) These mechanisms of crack-tip and crack-wake damage are essentially the same under static and cyclic loading conditions. However, the strain-rate sensitivity of the viscous flow of the amorphous film, in conjunction with the time-dependence of environmental interactions, renders the crack growth

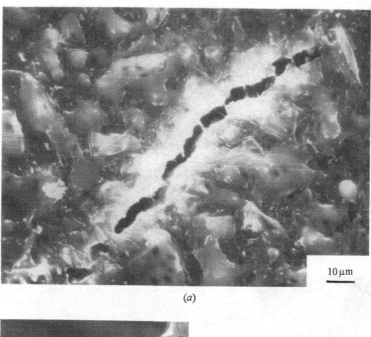

(a)

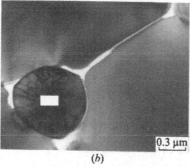

(b)

Fig. 11.8. Crack-tip and crack-wake damage processes facilitated by the glass phase in an Al₂O₃-matrix composite reinforced with 33 volume% SiC whiskers, which is subjected to tension fatigue (in a bend geometry) at $R = 0.15$ and $v_c = 0.1$ Hz at a temperature of 1400 °C in air. (a) A zone of material around the crack in which oxidation of SiC leads to the formation of glassy phase. (b) TEM observations of debonding along the whisker–matrix interface within the crack tip damage zone. (From L.X. Han and S. Suresh, Brown University, Providence, Rhode Island, 1990.)

process apparently frequency-sensitive under cyclic loads (see further discussions on this issue in subsequent sections).†

(8) The foregoing mechanistic processes associated with the effects of glassy films are also seen in unreinforced alumina which contains impurities and amorphous intergranular films that remain from the processing additives (e.g., Ewart & Suresh, 1992).

---

† Transmission electron microscopy has identified some differences between the high-temperature crack-tip damage processes under static and cyclic loads (Han & Suresh, 1989). Such differences are highlighted in Section 5.3.3 and Fig. 5.5.

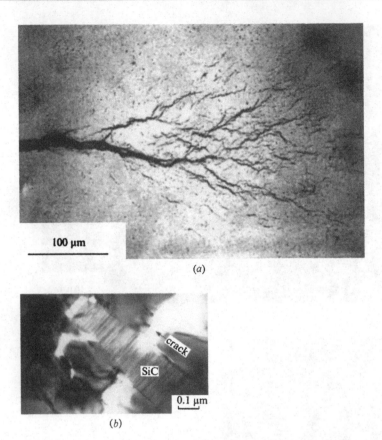

(a)

(b)

Fig. 11.9. (a) The development of a diffuse microcrack zone ahead of a crack in Al₂O₃–33 volume% SiC whisker composite subjected to zero–tension fatigue cycle at 1500 °C. (b) Debonding along the whisker–matrix interface at 1400 °C. (From Han & Suresh, 1989. Copyright American Ceramic Society. Reprinted with permission.)

(9) In ceramic systems, such as SiC-reinforced Si_3N_4, where heat treatments have been carried out to crystallize the amorphous intergranular/interfacial films, it is found that crystallization promotes an enhanced resistance to fatigue crack growth at high temperatures (Zhang, Edwards & Plumbridge, 1998).

11.7.2 Crack growth characteristics at high temperatures

Figure 11.10 shows the crack velocity da/dt for an Al_2O_3–33 volume% SiC whisker composite as a function of the stress intensity factor K_I under monotonic loads in the 1400 °C air environment. Also shown in this figure are the crack velocities $da/dt = da/dN \times v_c$ measured under cyclic loads at $R = 0.15$ and at frequencies $v_c = 0.1$ and 2 Hz in the same high temperature environment. In this figure, the

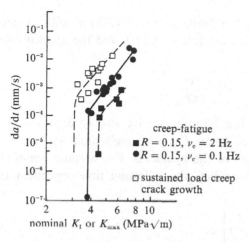

Fig. 11.10. Crack growth characteristics of Al_2O_3–33 volume% SiC whisker composite in 1400 °C air. Variation of fatigue crack velocity da/dt as a function of K_{max} under static and cyclic loading conditions. (After Han & Suresh, 1989.)

crack velocities under cyclic loads are *lower* than those under static loads. Furthermore, the cyclic crack growth response is apparently affected by the frequency of the stress cycle. The same trend has also been found for 90% pure alumina ceramic with intergranular glassy films during static and cyclic crack growth at 1050 °C (Ewart & Suresh, 1992). A rationale for this apparently beneficial effect of cyclic loading is considered in the next subsection.

It should be noted that the characterization of crack growth involving linear elastic fracture mechanics, as in Fig. 11.10, must necessarily involve small-scale creep conditions at the tip of the crack. For crack advance under both static and cyclic load, methods to ensure the validity of such characterization are outlined in Section 16.7.2.

11.7.3 Role of viscous films and ligaments

It was shown above that the presence of viscous films imparts an apparent ductility to the brittle solid at elevated temperature whereby enhanced levels of subcritical crack growth are facilitated as a consequence of grain boundary and interfacial sliding, cavitation ahead of the crack tip and bridging of the crack faces by viscous ligaments. This process, which is mechanistically nearly the same under static and cyclic loading conditions, is influenced by the loading rate, cyclic frequency and waveform.

Quantitative models have been developed in an attempt to rationalize the apparently enhanced resistance to crack growth under cyclic loads, especially at higher cyclic frequencies and maximum stress intensity values. By invoking an analogy between cavitation in the glassy film ahead of the crack tip and bridging of the faces of a linear elastic crack by viscous films, Thouless (1988) postulated an expression linking the crack-tip stress intensity factor, $K_{tip}(t)$, and the applied (nominal) stress intensity factor, $K(t)$:

$$\frac{dK_{tip}(t)}{dt} + \frac{K_{tip}(t)}{t_1^*} = \frac{K(t)}{t_1^*}, \qquad t_1^* = \frac{8\eta}{\pi E_m} \qquad (11.22)$$

where t_1^* is the characteristic time for the rate-sensitive viscous deformation of the glassy ligaments bridging the crack faces, E_m is Young's modulus of the matrix material, η is the effective viscosity of the ligaments. For a static nominal stress intensity factor, K_{static}, Eq. 11.22 predicts the following time-dependent near-tip stress intensity factor:

$$K_{tip}(t) = K_{static}\left\{1 - \exp\left(\frac{-t}{t_1^*}\right)\right\}. \qquad (11.23)$$

The effectiveness of viscous glassy ligaments in bridging the crack and altering the 'driving force' for fracture was quantitatively assessed by Lin *et al.* (1992) for polycrystalline alumina subjected to static and cyclic loads. They numerically integrated Eq. 11.22 for high-temperature cyclic loading tests conducted at a frequency of 2 Hz and load ratio of 0.1 to obtain the effective near-tip stress intensity factor for different creep times. They compared cyclic loading whose peak stress intensity factor K_{max} was the same as that of the static crack growth test, and forced t_1^* to match experiments to extract the effective viscosity η. Figure 11.11 shows the variation of the near-tip stress intensity factor, normalized by the maximum applied stress intensity factor, as a function of time normalized by the characteristic time, t_1^*. It is seen that the effective stress intensity factor under cyclic loads is only about one-half of the maximum stress intensity factor K_{max} which is the imposed driving force for

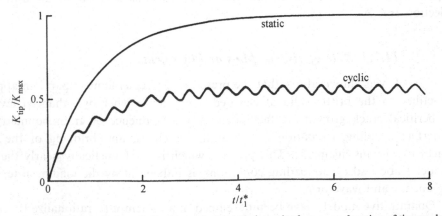

Fig. 11.11. The evolution of effective crack-tip stress intensity factor as a function of time under static and cyclic loads for polycrystalline alumina subjected to fatigue at $v_c = 2$ Hz and $R = 0.1$ at 1200 °C. The imposed K_{max} for the fatigue loading was the same as the stress intensity factor for static loading. (After Lin *et al.*, 1992.)

static crack growth. This simple calculation rationalizes the beneficial effect of cyclic load and hence the apparently lower rate of crack growth compared to the static load fracture experiment. For the conditions of the experiments simulated in Fig. 11.11, the time period of one fatigue cycle is 0.5 s (i.e. the inverse of 2 Hz) whereas the $t_1^* = 1.25$ s.

Some general trends, which are consistent with experimental observations, emerge from these simulations:

(1) The rates of fatigue crack growth typically decrease with increasing cyclic frequency.
(2) For fixed values of the maximum stress intensity factor and cyclic frequency, an increase in fatigue crack growth rates occurs in the following sequence of waveforms: triangular $\rightarrow$ sinusoidal $\rightarrow$ trapezoidal or square.
(3) These results imply that the time spent at the maximum stress intensity primarily governs the crack growth response, and that the attendant crack growth process is primarily time-dependent rather than cycle-dependent.

Different types of ceramics (e.g., silicon carbide, silicon nitride and aluminum oxide) exhibit different sensitivity to subcritical crack growth at elevated temperature. For example, silicon carbide ceramics can be synthesized in such a way that they contain little or no intergranular glassy films. At elevated temperatures, they exhibit essentially no difference in crack growth rates between static and cyclic loads. Polycrystalline alumina containing large amounts of impurities and glassy phase (e.g., 5–10 volume%) typically shows static and cyclic crack growth characteristics similar to those illustrated in Fig. 11.10 for alumina-based composites. However, when the extent of glassy films is significantly reduced in the polycrystalline alumina, the propensity for subcritical crack growth is also markedly suppressed (Ewart & Suresh, 1992). In this case, the crack growth curve essentially follows the limits for subcritical fracture seen in the R-curve for the ceramic.

In silicon-nitride ceramics with elongated grains introduced for toughening purposes, both grain bridging and viscous ligament bridging are known to occur during static and cyclic crack growth at elevated temperatures (Liu, Chen & Tien, 1994; Ramamurty, Hansson & Suresh, 1994). Whether crack growth under cyclic stresses is slower or faster compared to that under static stresses of the same peak values is determined by whether crack bridging by glassy ligaments or pulled-out grains, respectively, dominates crack-wake contact. As a result, beneficial or detrimental effects of cyclic loads on high temperature fracture (*vis-à-vis* static load fracture) are possible in silicon-nitride ceramics.

We conclude this section by noting that the preferential evolution of damage at grain boundaries and interfaces at elevated temperatures in brittle solids containing glassy films can also dominate over other types of deformation, such as matrix plasticity within the grains by dislocation motion. This trend has been seen in moly-disilicide-based intermetallics containing intergranular and interfacial glassy films where cavitation induced by the amorphous films is the dominant crack-tip damage process, although the test temperature (above the brittle-to-ductile transition temperature) is capable of promoting slip within the grains. The overall effects of the introduction of glassy films to a brittle solid are that: (1) the threshold stress intensity factor at which stable crack growth commences is reduced under both static and cyclic loading conditions (which constitutes a detrimental effect) and (2) the range of stress intensity factors over which subcritical crack growth occurs is increased (which constitutes a beneficial effect from the viewpoint of damage tolerance). Thus, allowance for the existence of glassy films in brittle solids should be made such that these competing trends are judiciously balanced for optimal damage tolerance as well as for optimizing other mechanical properties at high temperatures.

Exercises

11.1   Toughening of brittle solids by crack deflection, by the addition of a second phase, has received considerable attention. Consider the combined mode I–mode II loading experienced at the tip of a deflected fatigue crack in a brittle solid which is loaded remotely in pure mode I to a stress intensity factor K_I. If the angle of deflection (kink) is α and the deflected segment (length b in Fig. 9.17a) is very much smaller than the straight segment (length a in Fig. 9.17a), show that the near-tip mode I and mode II stress intensity factors, k_1 and k_2 are related to K_I by the angular functions, $a_{11}(\alpha)$ and $a_{21}(\alpha)$, respectively, in Eq. 9.117. (Hint: See Eq. 9.118.)

11.2   In the previous problem, if the remote loading includes both mode I and mode II, i.e. K_I and K_{II}, derive expressions relating k_1 and k_2 to K_I and K_{II} in terms of the tilt angle, α.

11.3   Consider the elevated temperature crack growth characteristics of ceramics in which glassy films at grain boundaries undergo viscous deformation at high temperature. Provide a physical reason for the argument in Section 11.7.3 that, for fixed values of the maximum stress intensity factor and cyclic frequency, an increase in fatigue crack growth rates occurs in the following sequence of waveforms: triangular → sinusoidal → trapezoidal or square.

11.4   Consider the crack growth results shown in Fig. 11.10 for the alumina–silicon carbide composite at 1400 °C. Using the procedure outlined in Section 11.2.2 predict the crack growth characteristics under cyclic loads on the basis of the static crack growth results by assuming that the mechanisms of static and cyclic fracture are the same at the elevated temperature. Compare your predictions with the measured fatigue crack growth rates for both frequencies and discuss possible reasons for the discrepancies, if any, between your predictions and the experimentally determined growth rates.

11.5 The K_{Ic} value of a ceramic is $3\,\text{MPa}\sqrt{\text{m}}$. Its tensile rupture strength and Young's modulus are 250 MPa and 375 GPa, respectively. If a circumferentially cracked cylindrical rod of this ceramic is fractured in cyclic torsion, the average height of an asperity on the fracture surface is 0.75 mm due to locally mode I fracture at 45° to the axis of the rod. With appropriate calculations, show that the fracture surface asperities of the ceramic in mode III can never be fully separated by the superposition of a mode I load and that the apparent fracture or fatigue cracking resistance of this ceramic in mode III is always likely to be higher than that in mode I.

CHAPTER 12

Fatigue crack growth in noncrystalline solids

The crack growth characteristics of ductile and brittle crystalline solids were the topics of attention in the two preceding chapters. We now direct focus on semi-crystalline and noncrystalline materials including metallic glasses and a wide variety of polymers. The main purpose here, as in Chapter 6, is to compare and contrast the mechanisms of fatigue fracture in crystalline and amorphous solids, and to highlight those failure features which are specific to semi-crystalline and noncrystalline materials. Further information on the topics discussed here, including more details on experimental observations, can be found in the relevant literature cited throughout this chapter.

12.1 Fatigue crack growth characteristics

Early attempts to characterize the rate of fatigue crack growth in rubbery solids invoked an energy approach along the lines of the Griffith concept (Section 9.1). Rivlin & Thomas (1953) extended the surface energy concept of Griffith to situations involving large strains and nonlinear elastic deformation, such as that found in rubber, and characterized the critical condition for the onset of fracture in terms of the surface work parameter, Υ. This parameter is a function of strain energy density and crack length. For the case of a single edge-notched panel, the surface work parameter is defined as

$$\Upsilon = m_1 a U, \tag{12.1}$$

where m_1 is a constant which is mildly sensitive to the material properties, a is crack length, and U is strain energy density. For a linear elastic solid subjected to uniform far-field stresses, $U = \sigma^2 / 2E$. With $m_1 = \pi$, one finds that Υ in Eq. 12.1 is equivalent to the strain energy release rate $\mathcal{G}$ introduced in Chapter 9.

A vast amount of empirical work has shown that, for time-independent fatigue fracture, the rate of crack growth can be characterized in terms of the surface work parameter, such that

$$\frac{da}{dN} = \overline{C} \Upsilon^{\overline{m}}. \tag{12.2}$$

Here $\overline{C}$ is a material constant and $\overline{m}$ is an exponent with a value roughly equal to two for natural rubber. This equation is strictly applicable for zero–tension loading, where the minimum value of Υ is $\Upsilon_{min} = 0$ (e.g., Andrews, 1974). Although Υ is sometimes replaced by its amplitude $\Delta\Upsilon$ for nonzero mean stress, there is no convincing experimental evidence that da/dN is uniquely related to $\Delta \Upsilon$, except when $\Upsilon_{min} = 0$.

A more common method of characterizing fatigue crack growth in polymers under conditions of small-scale damage ahead of the crack tip involves use of the Paris law:

$$\frac{da}{dN} = C(\Delta K)^m. \tag{12.3}$$

This approach is merely an extension of the application of linear elastic fracture mechanics, widely adopted in metal fatigue, to the fatigue of polymers as well. It is now known (e.g., Hertzberg & Manson, 1980, 1986) that the variation of fatigue crack growth in a wide variety of amorphous and semi-crystalline polymers can be characterized in terms of the stress intensity factor range, ΔK. Figure 12.1 shows examples of this linear elastic fracture mechanics characterization in PMMA, PS, PVC, PA, PDVF and Nylon 66. Also indicated in this figure, for comparison purposes, are the fatigue crack growth characteristics of 7075-T651 aluminum alloy and a $2\frac{1}{4}$Cr–1Mo steel (bainitic microstructure with yield strength, $\sigma_y = 500$ MPa). Several observations can be drawn from these results.

(1) When characterized in terms of the nominal values of ΔK, polymers exhibit a markedly inferior resistance to fatigue crack growth than metal alloys.

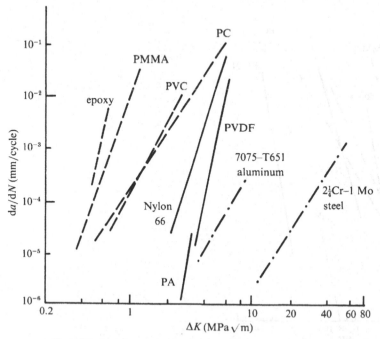

Fig. 12.1. Fatigue crack growth characteristics of several amorphous and semi-crystalline polymers represented by dashed and solid lines, respectively. Also shown for comparison purposes are the fatigue crack growth rates of an aluminum alloy and a steel. (After Hertzberg, Nordberg & Manson, 1970; Hertzberg, Skibo & Manson, 1979.)

(2)   There is a correlation between fatigue crack growth rates and fracture toughness in that polymers with a higher toughness generally exhibit a lower fatigue crack growth rate. This is similar to the trend exhibited by transformation-toughened ceramics (Section 11.4), but is in sharp contrast to the behavior of ductile alloys where no unique correlation exists between fatigue crack growth rates and fracture toughness in the near-threshold and Paris regimes.

(3)   Since the materials for which the fatigue characteristics are plotted in Fig. 12.1 have very different elastic moduli, it is instructive to replot the crack growth rates from this figure in terms of $\Delta K/E$ where E is the relevant Young's modulus. This normalized plot is shown in Fig. 12.2. As discussed in Section 10.4, the differences in fatigue crack growth characteristics of various metallic materials are, in general, substantially diminished when ΔK is normalized by Young's modulus. In the case of polymers, however, this normalization does not eliminate the vast differences in crack growth response.

(4)   Similar to the trends noted in connection with fatigue crack initiation (Chapter 6), semi-crystalline polymers exhibit a greater resistance to fatigue crack growth than amorphous polymers.

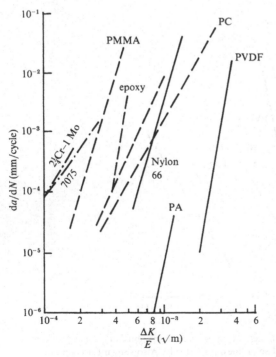

Fig. 12.2. Fatigue crack growth rates shown in Fig. 12.1 replotted as a function of $\Delta K/E$. (After Hertzberg, Skibo & Manson, 1979; Manson, Hertzberg & Bretz, 1981.)

Although Figs. 12.1 and 12.2 show fatigue crack growth only in the Paris regime, the fatigue crack growth curves for many polymeric materials, with log (da/dN) plotted as a function of log (ΔK), exhibit a sigmoidal variation consisting of three distinct regimes (near-threshold, intermediate growth rate and high growth rate), analogous to the behavior shown in Fig. 10.8 for ductile solids. An example of this sigmoidal growth behavior is presented in a later section.

12.2 Mechanisms of fatigue crack growth
12.2.1 Fatigue striations

The discussions of fatigue crack growth in metals (Chapter 10) dealt with the mechanisms of striation formation and the usefulness of fatigue striations in the analysis of failure. Under some loading conditions, many polymers also exhibit clear striation markings on the fracture surfaces. The term *striation* in metal fatigue usually refers to a growth band on the fatigue fracture surface, irrespective of whether the striation spacing correlates with the crack growth rate per cycle or not. In the polymer fatigue literature, the term striation is exclusively used to denote growth bands on the fracture surface whose spacing equals the rate of crack growth in each stress cycle. This restriction is in view of the fact that other types of dis-continuous growth marks (where the fatigue crack does not advance in each stress cycle), which apparently resemble striations, are also formed during the fatigue fracture of polymers (see the next subsection).

Figures 12.3(a) and (b) show examples of fatigue striations formed in PSF and PMMA, respectively. These striation markings are oriented normal to the crack growth direction. Skibo, Hertzberg & Manson (1976) and Hertzberg, Skibo &

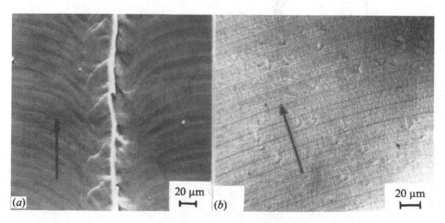

Fig. 12.3. Fatigue striations formed in (a) PSF and (b) PMMA. The arrow indicates the crack growth direction. (From Hertzberg, Skibo & Manson, 1979. Copyright American Society for Testing and Materials. Reprinted with permission.)

Manson (1979) conducted a comprehensive study of striation formation in semi-crystalline and amorphous polymers, and showed that the striation spacing could be related to the macroscopic crack growth rates. The variation of fatigue crack growth rates da/dN with the stress intensity factor range ΔK is plotted in Fig. 12.4 for two polymeric materials: PSF and PMMA. The data shown in this figure were obtained at a fixed load ratio, $R = 0.1$, at room temperature from macroscopic measurements of crack growth rates made during the fatigue tests, as well as from post-failure measurements of striation spacing in the scanning electron microscope. The results of this figure support the argument that the crack growth rate per cycle corresponds to the striation spacing for the materials and loading conditions under consideration.

The conditions under which striations form in polymeric materials are somewhat different from those found in metals.

(1) Polymers exhibit striated growth at high ΔK values (typically $da/dN \geq 5 \times 10^{-4}$ mm/cycle). On the other hand, the one-to-one correspondence between da/dN and s generally occurs in metals at lower growth rates in the Paris regime.

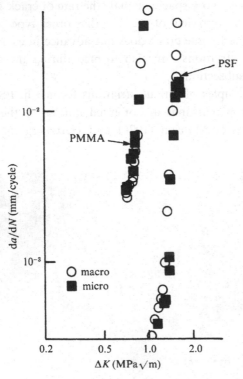

Fig. 12.4. Macroscopic fatigue crack growth rates da/dN (open symbols) and post-failure measurements of striation spacing $s = da/dN$ (filled symbols) plotted against the stress intensity factor range ΔK for PSF and PMMA. (After Hertzberg, Skibo & Manson, 1979.)

(2) The most popular interpretation of striation formation in metals relates the striation spacing to the crack-tip opening displacement. As shown in Section 10.2.3, this viewpoint invariably leads to the prediction that the Paris exponent m in Eq. 12.3 is 2. Such a trend is in distinct contrast to the results obtained for polymers where the exponent m is neither equal to 2 nor constant. Instead, m for polymers exhibiting striated fatigue crack growth varies from about 4 to 20. For this reason, the development of models based on the crack tip opening displacement to rationalize striation formation in polymers is not likely to produce realistic quantitative results.

(3) In metallic materials, post-failure analysis of striation spacing can be exploited, in conjunction with Eq. 10.19, to infer *a posteriori* the ΔK conditions that prevailed at the time of fatigue fracture for certain ranges of crack growth rates. The case studies discussed in Chapter 10 provide some examples of this approach. In polymeric solids, however, only empirical correlations exist between striation spacing and macroscopic crack growth rates. Therefore, no quantitative estimates of ΔK can be derived without recourse to independent experimental measurements of crack growth rates.

12.2.2 Discontinuous growth bands

Another fatigue fracture surface feature in polymers is the so-called *discontinuous growth band* (DGB). These growth bands corresponding to a single burst of fatigue crack advance after every several hundred fatigue cycles have been found in polyvinylchloride (PVC). These bands apparently resemble striations, although their spacing is considerably larger than the crack length increment per cycle. The broad generality of the discontinuous growth process to many amorphous and semi-crystalline polymers was established in the work of Skibo *et al.* (1977). These authors coupled *in situ* cinematographic observations with acoustic emission signals of audible 'clicks' during bursts of fatigue crack growth to document the conditions for the formation of DGBs in PVC. It is now known that the DGBs form predominantly at low ΔK levels. The possibility of formation of DGBs and the spacing between them is strongly dependent upon the chemistry and molecular weight of the polymer as well on upon the ΔK level, test temperature and loading frequency. Figure 12.5 shows discontinuous growth bands oriented perpendicularly to the fatigue crack growth direction in PVC.

Interpretations of the formation of discontinuous growth bands center around the notion that the accumulation of damage ahead of the fatigue crack over many cycles causes the crack to jump suddenly. At low ΔK levels approaching the threshold for fatigue crack growth, the microscopic deformation at the crack tip involves crazing for many polymers. Drawing an analogy between the craze zone at the crack tip in a polymer and the plastic zone at the crack tip in a metal, Elinck, Bauwens & Homés

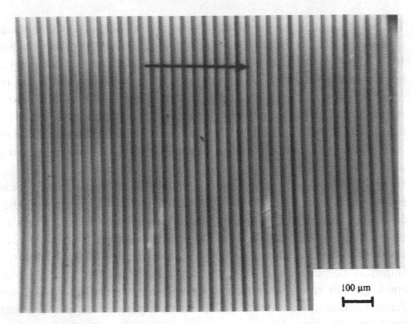

Fig. 12.5. Discontinuous fatigue crack growth bands seen on the fracture surface of PVC. The arrow indicates the crack growth direction. (From Skibo *et al.*, 1977. Courtesy of R.W. Hertzberg, Lehigh University.)

(1971) used the Dugdale strip yield model (Section 9.5.2) to estimate the maximum craze length to be

$$r_{cz} = \frac{\pi}{8} \frac{K_{max}^2}{\sigma_{cz}^2}, \tag{12.4}$$

where σ_{cz} is the stress at which crazing occurs (which is assumed constant over the length of the craze) and K_{max} is the maximum value of the stress intensity factor during a stress cycle. Hertzberg, Skibo & Manson (1979) measured the craze length r_{cz} at different K_{max} levels in several polymers and estimated σ_{cz} using Eq. 12.4. These calculated values of σ_{cz} agreed well with the values reported in the literature for tensile fracture. Furthermore, the value of craze stress estimated by this procedure was roughly a constant over the entire range of ΔK values where discontinuous crack growth was observed. The value of σ_{cz} was also comparable to the tensile yield strength for many polymers. A notable outcome of these investigations is the realization that the discontinuous crack increment occurs within the craze zone which develops at the crack tip.[†] The burst of fatigue crack growth is illustrated by the micrographs shown in Fig. 12.6.

Figure 12.7 shows another example of the type of damage that evolves ahead of a fatigue crack tip in a crazable polymer. This figure is a transmission electron micro-

---

[†] Fatigue fracture profiles resulting from crack growth through a single craze or a bundle of crazes are termed 'mirror' and 'mist' regions, respectively, by Skibo, Hertzberg & Manson (1976).

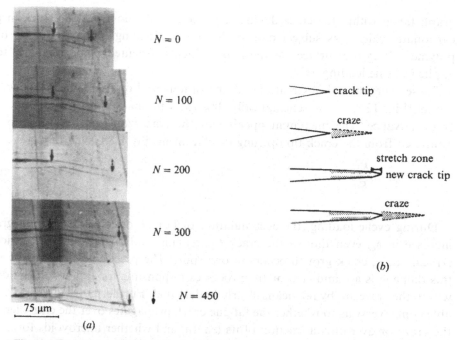

$N = 0$

$N = 100$

$N = 200$

$N = 300$

$N = 450$

75 μm

(a)

crack tip

craze

stretch zone
new crack tip

craze

(b)

Fig. 12.6. (*a*) Micrographs revealing the positions of the crack tip (arrows on the left) and of the tip of the craze (arrows on the right) in fatigued PVC. The number of fatigue cycles N corresponding to the stress cycles imposed between each observation is indicated. (From Hertzberg & Manson, 1973. Photo courtesy of R.W. Hertzberg, Lehigh University.) A schematic representation of the crack tip craze zone is also provided in (*b*).

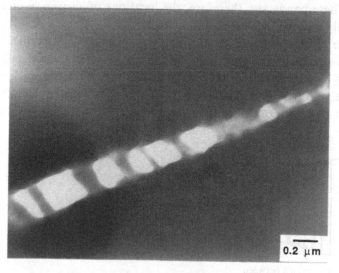

0.2 μm

Fig. 12.7. Transmission electron micrograph of a craze within the cyclic damage zone ahead of a fatigue crack in polycarbonate which was subjected to tension–tension fatigue at room temperature. (From Pruitt & Suresh, 1993. Reprinted with permission.)

graph taken within the craze damage zone ahead of a long fatigue crack in polycarbonate which was subjected to cyclic tension loading. Note the fibrils of the
polymer filling the surfaces of the craze which is oriented perpendicularly to the
applied tensile loading axis.

These experimental observations can be rationalized on the basis of a conceptual
model, Fig. 12.8, which schematically illustrates the mechanisms of discontinuous
crack advance. The maximum opening of the craze zone at the crack tip δ_{cz} is
estimated from the crack tip opening displacement, Eq. 9.83, such that

$$\delta_{cz} = \frac{K_{max}^2}{E\sigma_{cz}}. \tag{12.5}$$

During cyclic loading, the accumulation of fatigue damage results in a gradual
increase in δ_{cz} even though the crack tip remains stationary. When δ_{cz} reaches a
critical value, crack growth occurs in one spurt. The precise mechanisms by which
this damage is accumulated or the processes responsible for fatigue crack advance
within the craze are by no means clearly understood. Furthermore, there is considerable controversy as to whether the fatigue crack propagates over the entire length of
the craze or over only a fraction of its length, and whether microvoids form in the
craze midribs (e.g., Hertzberg, Skibo & Manson, 1979; Mackay, Teng & Schultz,
1979; Mills & Walker, 1980). Despite these uncertainties, it has been hypothesized
(e.g., Könczöl, Schincker & Döll, 1984; Takemori, 1984) that the crack grows initially along the craze-matrix interface and subsequently along the fibril midrib until
crack arrest. The nonplanar microscopic path is sketched in Fig. 12.8. This process
repeats itself causing dark bands of discontinuous growth to appear on the fatigue
fracture surface. These bands are, therefore, the locations of successive crack tips
which are repeatedly blunted, advanced and arrested.

It is appropriate at this juncture to examine the quasi-static subcritical crack
growth process involving crazes, which has been investigated in some detail in the
theoretical study of Kramer & Hart (1984). Their analysis shows that a crack extending with a constant velocity v and having a coplanar craze of constant length ahead
of its tip is shielded by the craze from the far-field loading. The fracture of the craze

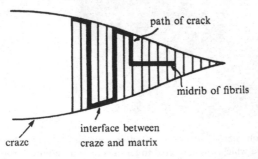

Fig. 12.8. An idealization of the path of a fatigue crack within the craze zone.

occurs by a local creep process. The fracture time t_f of the craze matter is related to the local stress σ_1 ($= \lambda\sigma$, where λ is the craze fibril extension ratio and σ is the local craze traction) by the expression

$$t_f = \frac{1}{\dot{\epsilon}_1} \left(\frac{\bar{\sigma}_1}{\sigma_1} \right)^{m_c}, \tag{12.6}$$

where $1/\dot{\epsilon}_1$ is the craze fibril failure time under a local craze true stress $\bar{\sigma}_1$ and m_c is an empirically determined exponent ($m_c = 12$ for PMMA). The dependence of the crack velocity da/dt on the net crack tip stress intensity factor K is given by

$$\frac{da}{dt} = \frac{2\dot{\epsilon}_1 D_0}{(m_c - 2)} \left(\frac{K\lambda}{\bar{\sigma}_l \sqrt{2\pi D_0}} \right)^{m_c}, \tag{12.7}$$

where D_0 is the craze fibril spacing and $\sigma(D_0) = K/\sqrt{2\pi D_0}$. This quasi-static creep process can have some applicability to the fatigue fracture of crazable polymers, although repeated cyclic loading may additionally exacerbate craze fracture by aiding in the breakdown of craze matter.

12.2.3 Combined effects of crazing and shear flow

The foregoing discussion of discontinuous crack growth focused on fatigue fracture which was promoted by crazing ahead of the crack tip. In some polymeric materials subjected to low ΔK levels, discontinuous crack growth occurs as a consequence of both crazing and shear banding ahead of the crack tip. The resultant crack tip process zone has the shape of the Greek letter ϵ (epsilon), and hence this fracture process is commonly referred to as *epsilon discontinuous crack growth*.

Mills & Walker (1980) and Takemori & Kambour (1981) made the experimental observation that fatigue cracks in smooth specimens of polycarbonate exhibited an unusually shaped crack tip plastic zone which consisted of a pair of shear bands and a single craze. The shear bands arch forward asymptotically at 45° above and below the crack plane, whereas the craze is formed directly ahead of the crack tip. In subsequent studies, Takemori (1982) observed that this combined craze–shear feature of the short surface cracks was distinctly different from the manner in which longer, through-thickness cracks advanced. The longer cracks typically exhibited discontinuous crack growth due to the presence of a single craze or a bundle of crazes at the tip (as discussed in the preceding section) or advanced purely by the shear banding process. This difference in the lateral profile was evident even when the nominal ΔK values for the surface flaws (as wide as 10 mm) and for the corresponding longer flaws emanating from notches were identical. In this respect, the epsilon discontinuous crack growth process may be regarded as the polymer-version of the well known short fatigue crack anomaly discussed in Chapter 15 in the context of metal fatigue. The development of epsilon plastic zones for short cracks has been identified in at least four polymeric materials: PC, PSF, a polyester carbonate copolymer, and a polyarylate block copolymer (Mills & Walker, 1980; Takemori, 1982; Rimnac, Hertzberg & Manson, 1983).

In the epsilon-shaped plastic zone, the shear bands originate at the crack tip and arch forward asymptotically along planes oriented at 45° to the mode I crack plane. The plane of the craze, which bisects the shear bands, coincides with the crack plane. Figure 12.9 shows a sequence of optical micrographs in which the discontinuous crack growth bands associated with the epsilon plastic zone are evident.

The growth behavior of (physically) short and long fatigue cracks in PC is presented in Fig. 12.10, where the discontinuous growth associated with the short part-through surface cracks is indicated by the closed symbols. The remainder of the data points were collected using a compact tension specimen containing a long, through-

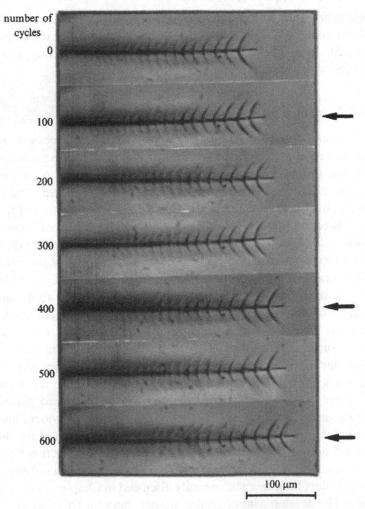

Fig. 12.9. A sequence of optical photographs taken after every 100 stress cycles of loading which show discontinuous crack growth in PC by the formation of an epsilon-shaped deformation zone. The arrows indicate the occurrence of sudden jumps of the crackfront. (From Takemori, 1982. Reprinted with permission.)

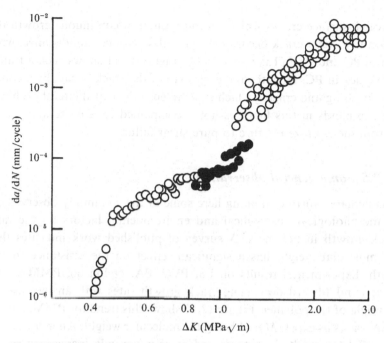

Fig. 12.10. Fatigue crack growth rate data for PC at room temperature. The filled symbols denote discontinuous cracking in initially smooth specimens due to the development of an epsilon-shaped process zone. The open symbols represent growth data obtained from compact tension specimens. (After Takemori, 1984.)

thickness crack. Note that the overall shape of this plot is similar to that shown in Fig. 10.8 for ductile metallic materials.

In the regime of ΔK where surface cracks exhibit epsilon discontinuous growth, the propagation rates are generally less than expected on the basis of discontinuous crack growth by crazing alone. Furthermore, the growth rates corresponding to the epsilon fracture mode are less than those predicted by the extrapolation of crack growth from higher ΔK levels (where the epsilon plastic zones do not develop). This apparently beneficial and anamolous behavior associated with the epsilon fracture process has been attributed to the larger amount of work expended in creating the shear bands and crazes (than that needed for craze formation alone) and to the shielding of the crack tip by the plastic zone.

12.2.4 Shear bands

Experimental results indicate that increases in the applied ΔK or test temperature generally cause the microscopic fracture mode in polymers (which are prone to shear deformation) to change in the following sequence: discontinuous growth bands formed by crack tip crazing → epsilon discontinuous growth bands formed by the combined effects of crazing and shear banding → formation of shear bands. For the part-through surface cracks which exhibit epsilon discontinuous growth bands, this transition to a shear crack occurs after growth over a certain distance, which is up to 1 mm in PC and PSF (Takemori, 1984). Figure 12.11 shows such a transition for a fatigue crack in PC. Here the initial growth of the crack occurs by a combination of shear banding and crazing which leads to epsilon-shaped lateral profiles. The last pair of slip bands in this figure is not accompanied by a craze and there is a transition from shear–craze fracture to pure shear failure.

12.2.5 *Some general observations*

It is perhaps worth examining here some of the commonly observed effects of various morphological, mechanical and environmental factors on the rates of fatigue crack growth in polymers. A survey of published work indicates that an increase in molecular weight has a significant effect on the resistance to fatigue crack growth. Experimental results on PS, PVC, PA, Nylon 66, PMMA and PE show a clear trend toward decreasing crack growth rates with an increase in the molecular weight of the polymer. Figure 12.12 shows this trend for PMMA and PVC where da/dN varies as $\exp(1/M)$, M being the molecular weight. An increase in $1/M$ by a factor of three results in a three orders of magnitude increase in the crack growth rate at a fixed value of ΔK.

On the basis of experimental information available in the literature, Hertzberg & Manson (1980) suggest that only molecules longer than some characteristic size scale

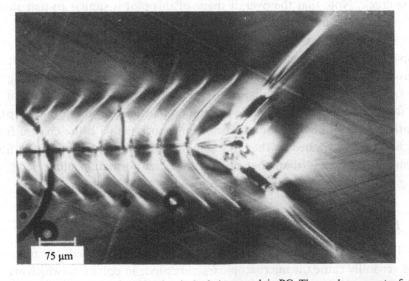

Fig. 12.11. Shear band cracks ahead of a fatigue crack in PC. The crack propagates from a free surface by a combined shear–craze process which leads to an epsilon-shaped lateral profile. The last pair of shear bands is not accompanied by a craze; it denotes pure shear failure. (From Takemori, 1982. Reprinted with permission.)

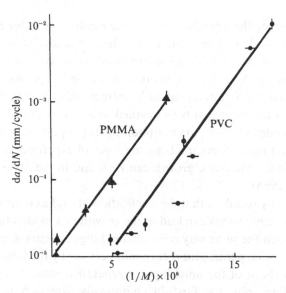

Fig. 12.12. Variation of fatigue crack propagation rates, at fixed values of ΔK (= 0.6 MPa$\sqrt{\text{m}}$) and test frequency v_c (= 10 Hz), as a function of reciprocal molecular weight for PMMA and PVC. (After Kim *et al.*, 1977.)

(related to the width of the crazes) form an entanglement network which is capable of resisting crack growth. In other words, the craze is populated with an entanglement network consisting of several molecules, rather than individual molecular chains. Hertzberg & Manson suggest that the networks are entangled by fatigue loading. A reduced possibility of disentanglement as well as enhanced orientation hardening at higher molecular weights may then promote an increased resistance to fatigue crack growth. It is noted that the effect of M on fatigue crack growth is much stronger than that on the glass transition temperature, yield strength and elastic modulus. These latter (monotonic) properties are governed by the bulk behavior in that they are a reflection of some weighted average of the weak and strong molecular entanglements of widely differing chain lengths. Therefore, above a certain critical M, the monotonic properties are not significantly enhanced by M provided that there is a finite number of weakly entangled molecular chains. By contrast, in fatigue fracture involving crazes, the core of the craze fibrils is made up of strongly entangled molecular chains; the craze possesses a relatively high M with a relatively narrow distribution of molecular weight. Hertzberg & Manson postulate that the rate of fatigue crack growth should decrease with increasing M because only the highest M entanglements are left in the craze fibrils. This line of reasoning also appears to find experimental backing in the effect of molecular weight distribution on fatigue crack growth.

An increase in the degree of crystallinity also causes a similar reduction in the rate of fatigue crack growth (Fig. 12.1). On the other hand, increases in test temperature and the degree of cross-linking, or the introduction of a chemically aggressive environment generally promote a weakened resistance to fatigue fracture. It is often difficult to derive conclusions on the individual contributions to the overall fatigue crack growth rate from each of these factors because controlled experiments in simple polymeric systems have been attempted only infrequently. Furthermore, only in relatively few instances has attention been devoted to a careful examination of the microscopic fracture modes *vis-à-vis* the morphology of the polymer and the macroscopic crack propagation rates. A comprehensive survey of experimental data on various factors influencing fatigue crack growth can be found in the reviews by Hertzberg & Manson (1980, 1986).

We conclude this discussion by recalling that the application of cyclic compressive loads to notched plates of polymeric solids can lead to the growth of a mode I fatigue crack from the notch tip in much the same way compression fatigue occurs in ductile metals (Chapters 4–6). In polymers, permanent deformation caused by the formation of crazes and/or shear bands ahead of the notch tip can generate a zone of residual tensile stresses upon unloading from the far-field compressive stress. A mode I fatigue crack propagates at a progressively slower rate under the influence of the localized residual tensile field at the notch tip.

12.2.6 Example problem: Fatigue crack growth in epoxy adhesive

Problem:

Two plastic beams are glued together using an epoxy adhesive as shown in Fig. 12.13. The air bubbles trapped in the adhesive during the joining process lead to the formation of several flat, circular, disk-like cracks in the adhesive; these cracks are aligned parallel to the interface between the two plastic beams which are joined together. The dominant flaw is usually a semi-circular (thumbnail) crack located along the interface in the adhesive near the outer surface of the composite beam. The average radius of such a crack is $a = 1.2$ mm. The bilayer beam, of dimensions shown in Fig. 12.13 is loaded in three-point bending.

  (i) If the fracture toughness of the epoxy is 0.8 MPa$\sqrt{\text{m}}$, calculate the maximum load P which the beam can support.

 (ii) During service, the load P fluctuates between zero and one-half of the maximum value calculated in part (i), and during each fluctuation the crack in the adhesive advances according to the relationship:

$$\frac{da}{dN} = C(\Delta K)^8, \tag{12.8}$$

where $C = 6 \times 10^{-6}$ (MPa)$^{-8}$·m^{-3}. Find the number of load fluctuations before final failure occurs.

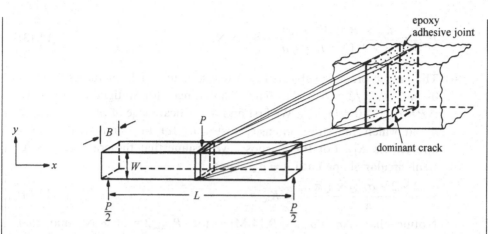

$L = 1.7$ m, $W = 10$ cm, $B = 7.5$ cm

Fig. 12.13. Crack in the joint between two beams which are glued together and subjected to three point-bending.

Solution:

(i) The maximum bending moment M_{max} in the simply supported beam occurs at the point of application of the load P:

$$M_{max} = \frac{P}{2} \cdot \frac{L}{2} = \frac{PL}{4}. \qquad (12.9)$$

The tensile opening stress acting perpendicularly to the interface cracks is

$$\sigma_{xx} = \frac{My}{I}, \qquad \text{where } I = \frac{1}{12}BW^3. \qquad (12.10)$$

The maximum value of this stress, σ_{max} occurs at the outer fiber where $y = W/2$, such that

$$\sigma_{max} = \frac{M_{max}}{I} \cdot \frac{W}{2} = \frac{PL}{4} \cdot \frac{12}{BW^3} \cdot \frac{W}{2} = \frac{3}{2} \cdot \frac{PL}{BW^2}. \qquad (12.11)$$

Assume that the worst-case scenario is one where the semi-circular, thumb-nail crack is located near the outer fiber on the tension side of the beam, and that the crack is located entirely within the adhesive (so that the complications arising from interface fracture can be ignored). For this geometry, the stress intensity factor is given by Eq. A.14 in the Appendix which is:

$$K_I = \frac{1.12\sigma_{max}\sqrt{\pi a}}{\sqrt{Q}}. \qquad (12.12)$$

From Eq. A.12 in the Appendix, we see that $\sqrt{Q} \approx \pi/2$. Substituting this value of Q and the given information that $L = 1.7$ m, $W = 100$ mm, $B = 75$ mm, $K_I = K_{Ic} = 0.8$ MPa$\sqrt{m}$, and $a = 1.2$ mm into Eqs. 12.11 and 12.12, and solving for P, the maximum load which the beam can support is found to be

$$P_{max} \approx \frac{K_{Ic} \times B \times W^2 \times \sqrt{\pi}}{3 \times 1.12 \times L \times \sqrt{a}} \approx 5375 \text{ N.} \tag{12.13}$$

(ii) The fatigue crack is subjected to a tensile load which fluctuates between 0 and $P_{max}/2 = 2688$ N. The load range for fatigue loading is: $\Delta P = P_{max} - 0 = 2688$ N. We first find the critical value of the crack length a_{cr} at which the stress intensity factor K_I for the semi-circular crack approaches K_{Ic}. For this purpose, we assume that the crack retains its semi-circular shape until $a = a_{cr}$. Thus,

$$\frac{1.12 \times 2 \times \sigma_{max} \times \sqrt{\pi a_{cr}}}{\pi} = K_{Ic}. \tag{12.14}$$

Noting that $\Delta\sigma = \sigma_{max} = 9.14$ MPa for $P_{max}/2 = 2688$ N, and that $K_{Ic} = 0.8$ MPa$\sqrt{m}$, this equation gives $a_{cr} = 4.8$ mm. Next, the crack growth equation is integrated to obtain the fatigue life:

$$\frac{da}{dN} = C(\Delta K)^8 = C\left\{\frac{1.12 \times 2 \times \sigma_{max} \times \sqrt{\pi a}}{\pi}\right\}^8$$

$$\int_0^{N_f} dN = \int_{a_0}^{a_{cr}} \frac{a^{-4} da}{6.5 \times C \times (\Delta\sigma)^8}, \tag{12.15}$$

where N_f is the number of fatigue cycles to failure, $a_0 = 1.2$ mm is the initial crack radius, $C = 6 \times 10^{-6}$ (MPa)$^{-8}$ m^{-3}, $\Delta\sigma = 9.14$ MPa, and $a_{cr} = 4.8$ mm. Integrating, we find that $N_f = 99\,698$ load cycles.

12.3 Fatigue of metallic glasses

When molten metals are cooled extremely rapidly, at rates of the order of 10^6 K/s, an amorphous microstructure develops. These *metallic glasses* with a noncrystalline structure have found commercial applications in power transformers, magnetic sensors, magnetic switches, and recording heads for video and audio equipment. Metallic glasses are generally fabricated in the shape of thin ribbons by rapid solidification processes when the molten metal is cooled upon coming into contact with the surface of a rotating drum. Some examples of metallic glasses are $Pd_{80}Si_{20}$, $Pd_{77}Cu_6Si_{17}$, $Pt_{64}Ni_{16}P_{20}$, $Fe_{78}B_{13}Si_9$, and $Co_{66}Fe_4Ni_1B_{14}Si_{15}$, wherein the subscripted number for each element indicates the concentration of that element in atomic per cent.

Metallic glasses exhibit some unusual mechanical properties and deformation response. They have very high tensile strengths (of the order of 2% of Young's modulus), which approach the theoretical strength values. At low temperature and low applied stresses, metallic glasses undergo linear elastic deformation. An increase in temperature promotes viscoelastic response at low stresses. Elevating the applied stress level induces plastic deformation prior to fracture. Viscous flow of the material at high temperature occurs uniformly through the bulk of the material. On the other hand, plastic deformation is restricted to narrow bands of shear within which strains in excess of unity are produced. In compression, deformation proceeds by multiple shear banding with essentially no work hardening. Deformation in tension is generally concentrated along a singularly intense shear band, and tensile failure occurs coincident with yielding.

Only a limited amount of work on the fatigue of metallic glasses has been reported in the literature (e.g., Davis, 1975; Ogura, Fukushima & Masumoto, 1975; Chaki & Li, 1984). Most of these studies have been performed using thin specimens (i.e. under predominantly plane stress conditions) because of the fabrication of these materials in the form of ribbons that are only tens of micrometers in thickness. An examination of the results reveals the fatigue crack growth behavior of metallic glasses is similar to that of metals in that three distinct regimes of growth (threshold regime, Paris regime and fast fracture regime) are observed.

Experimental results obtained by Ogura *et al.* (1975) and Chaki & Li (1984) for $Co_{72}Fe_6Mo_2B_{15}Si_5$ and $Pd_{80}Si_{20}$, respectively, indicate threshold ΔK_0 values in the range 7–9 MPa$\sqrt{m}$, Paris exponents m in the range 2–4, and maximum stress intensity factor at final failure, $K_{max} = \Delta K/(1 - R)$, in the range 10–20 MPa$\sqrt{m}$ at $R = 0.1$. Permanent deformation ahead of the fatigue crack occurs by the formation of shear bands with a clearly visible plastic zone, Fig. 12.14, somewhat analogous to that seen in metals. Ogura *et al.* report that the plastic zone dimensions observed in $Pd_{80}Si_{20}$ compare well with those predicted by linear elastic fracture mechanics. Under some conditions of fatigue loading, metallic glasses also exhibit striations and equiaxed vein patterns indicative of localized shear rupture (Davis, 1975). Because of their susceptibility to plastic deformation, metallic glasses also exhibit transient retardation characteristics similar to those found in metals (Chapter 14), although the extent of such retardation is significantly less pronounced than in metals (Chaki & Li, 1984).

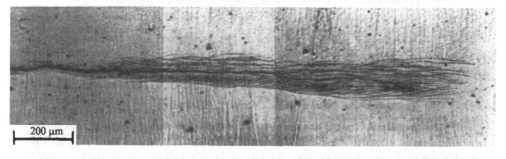

Fig. 12.14. Slip bands around a fatigue crack in $Pd_{80}Si_{20}$ metallic glass subjected to zero–tension cyclic loads at room temperature. (From Ogura, Fukushima & Masumoto, 1975. Copyright Pergamon Press plc. Reprinted with permission.)

12.4 Case study: Fatigue fracture in rubber-toughened epoxy

It was shown in Chapter 7 how the introduction of a strong second phase can enhance the fatigue life of a polymer, and how such improvements are influenced by the specific choice of the reinforcement phase. In this section, we demonstrate similar beneficial effects associated with fatigue crack propagation in two different polymeric matrices reinforced with different particles, fibers or fillers.

Owing to their creep and solvent resistance, which is promoted by a cross-linked structure, epoxies have been broadly used as matrices for structural components in automotive and aircraft applications. The cross-linked structure of the epoxy, however, promotes a very weak resistance to fatigue crack propagation. In order to circumvent the difficulties with the poor damage tolerance, epoxy matrices are reinforced in one of several ways: (a) compliant rubbery particles are added to the matrix, or (b) the matrix is reinforced with rigid inorganic fillers or thermoplastic particles. In the former case, apparent toughening and enhancement in damage tolerance stems from a reduction in effective 'driving force' for fracture by such mechanisms as cavitation, plastic void growth and shear banding in the vicinity of the crack tip. In the latter case, such processes as bowing and deflection of the crack tip around the reinforcement phase and the bridging of the crack faces by the reinforcement apparently improve the crack growth resistance.†

In this section, we consider an example of the beneficial effects of rubber-particle and glass-filler additions to epoxy matrices on the overall fatigue crack growth behavior. These results are taken from the work of Azimi (1994) and Azimi, Pearson & Hertzberg (1995).

Figure 12.15 shows the fatigue crack growth behavior of an epoxy matrix (comprising a diglycidyl ether of bisphenol A) with and without rubber modifiers. Two different types of rubber modifiers are considered in this figure: (i) a liquid carboxyl terminated butadiene-acrylonitrile (CTBN) in volume fractions of 0.01 or 0.10 with a particle size of approximately $1.5\,\mu m$, and (ii) structured core-shell PMMA particles, of volume fraction 0.10 with a particle size of $0.2\,\mu m$, comprising a methacrylated butadiene-styrene copolymer (MBS) with a small amount of carboxyl groups (COOH) included in the PMMA shell (this reinforcement is hereafter referred to as MBS–COOH). It is evident from Fig. 12.15 that even a small concentration of CTBN particles (volume fraction as low as 0.01) is sufficient to enhance crack growth resistance at high ΔK levels. An increase in the volume fraction of the particles to 0.10 significantly diminishes the rate of fatigue crack growth as well as the slope m of the Paris regime of fatigue fracture. It is also evident that the smaller particles of the MBS–COOH reinforcement are more effective in reducing crack growth rates than the CTBN, at a fixed volume fraction. In this last case, the beneficial effect of the reinforcement is evident through the entire range of fatigue crack growth.

The mechanistic origins of these beneficial effects of rubber toughening have been interpreted by recourse to the fractographs taken in different regimes of fatigue failure. On the basis of these observations, Azimi (1994) has concluded that the bridging of the crack faces by intact rubber particles plays a dominant role in enhancing the resistance to fatigue failure at low ΔK levels. In the Paris regime as well as in the final failure regime,

---

† Detailed discussions of such mechanisms are presented in Chapter 14.

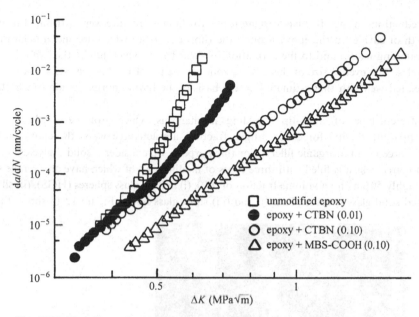

Fig. 12.15. The effects of rubber particle concentration and type on the fatigue crack growth characteristics of the epoxy matrix. Test conditions: $R = 0.1$, frequency = 10 Hz, and temperature = 20 °C. (After Azimi, 1994.)

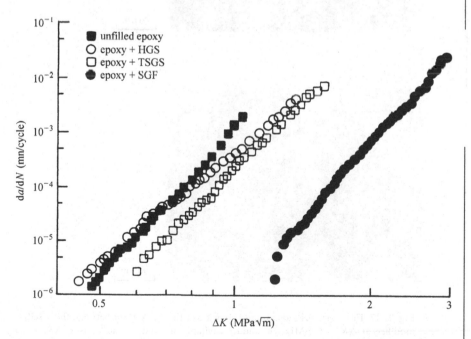

Fig. 12.16. The effects of inorganic filler concentration and type on the fatigue crack growth characteristics of the epoxy matrix. Test conditions: $R = 0.1$, frequency = 10 Hz, and temperature = 20 °C. (After Azimi, 1994.)

the reductions in the effective 'driving force' for fatigue fracture were attributed to the growth of cavities in the epoxy around the rubber particles which debonded from the surrounding matrix and to the formation of shear bands which linked the rubber particles. The overall size of the 'plastic zone' at the tip of the fatigue crack also correlated well with that estimated on the basis of the Irwin approximation (see Section 9.5.1).

Different types of crack-tip 'shielding' mechanisms, which promote a reduction in the apparent 'driving force' for fatigue fracture, are activated when the epoxy matrix is reinforced with inorganic fillers. For this purpose, we consider a solid diglycidyl ether resin matrix which is filled with three types of modifiers all of which have an average size of roughly 50 μm and a volume fraction of 0.10: (i) hollow glass spheres (HGS), (ii) silane treated solid glass spheres (TSGS), and (iii) short glass fibers. Figure 12.16 shows the

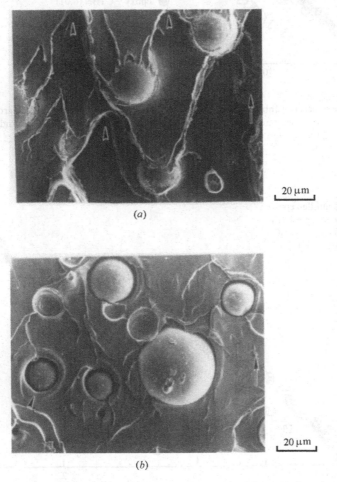

$20\,\mu\text{m}$

(*a*)

$20\,\mu\text{m}$

(*b*)

Fig. 12.17. Fractographs indicating (*a*) bowing of the crack front between the TSGS modifiers at $\Delta K = 0.6$ MPa$\sqrt{\text{m}}$ and (*b*) cavitation around the modifiers at $\Delta K = 1.5$ MPa$\sqrt{\text{m}}$. Test conditions: $R = 0.1$, frequency $= 10$ Hz, and temperature $= 20\,°$C. (From Azimi, 1994. Photographs courtesy of H. Azimi. Intel Corporation, Chandler, Arizona. Reprinted with permission.)

fatigue crack growth response of the unfilled and filled epoxy matrix. Hollow glass spheres do not appear to be very effective as fillers, especially at low ΔK levels, from the viewpoint of fatigue crack growth resistance. While the solid glass spheres provide a tangible improvement in fatigue damage tolerance, the short glass fibers facilitate a marked enhancement in crack growth resistance. Figures 12.17(a) and (b) point to possible causes of such beneficial effects in the TSGS-filled epoxy at ΔK values of 0.6 and 1.5 MPa√m, respectively. The larger arrow in Fig. 12.17(a) shows the direction of fatigue crack advance at the low ΔK, and the smaller arrows indicate the bowing of the crack between the filler particles. It is evident that the crack front bends so as to circumvent the intact particles; this process leads to a reduction in the effective 'driving force' for fracture. At the higher ΔK, however, debonding of the fillers from the surrounding matrix is apparent, Fig. 12.17(b), although some evidence exists of the crack bowing mechanism between particles.

The presence of short glass fibers in the epoxy matrix promotes significant levels of crack-tip deflection and fiber-bridging of the crack faces. These mechanisms engender a million-fold reduction in the crack growth rate compared to the unfilled epoxy matrix. Quantitative effects of these processes on the rates of fatigue crack propagation will be considered in Chapter 14.

We conclude this section by illustrating that similar enhancements of fatigue damage resistance can also be activated in other polymeric matrices. Consider the results of Fig. 12.18 for polyethersulfone which is reinforced with glass fibers. The plot of da/dN versus

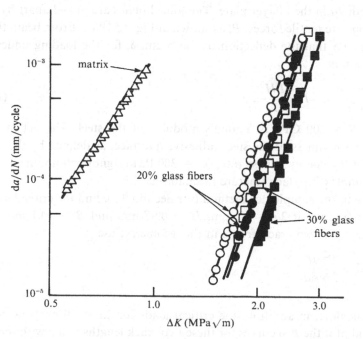

Fig. 12.18. Fatigue crack growth characteristics of polyethersulfone with and without reinforcements of short fibers of glass at $v_c = 5$ Hz and $R = 0.2$. The open and closed symbols refer to fracture parallel to and normal to the mold-fill direction, respectively. (After Voss & Walter, 1985.)

nominal ΔK shifts to higher values of ΔK as the concentration of the fibers is increased. The values of threshold ΔK_0 for fatigue crack growth, the Paris exponent m, and the overall toughness are all higher for the composites than for the unreinforced matrix. Note that the fatigue crack growth resistance in this material is not isotropic; resistance to tensile fatigue fracture normal to the mold-fill direction is superior to that along the mold-fill direction.

Exercises

12.1   Explain how you would distinguish between fatigue striations and discontinuous growth bands on the fracture surfaces of polymers.

12.2   Repeat the example problem in Section 12.2.6 for the case where the dominant flaw is a through-thickness edge crack of length 1.2 mm (instead of a thumb-nail crack).

12.3   A laminated composite is made by bonding two long thin strips of steel with an adhesive epoxy, as shown in Fig. 12.19(*a*). A patch of the adhesive was intentionally left out in the central section in order to create a central crack of length $2a$ in the bilayer plate. The joined plates are pulled apart by equal and opposite tensile forces, P, as shown in Fig. 12.19(*a*). From beam theory, it is known that the deflection of the beam, δ, for the loading under consideration is

$$\frac{\delta}{2} = \frac{Pa^3}{24EI}, \qquad I = \frac{BH^3}{12}, \tag{E12.1}$$

where $E = 200$ GPa is Young's modulus of the steel. The delamination fracture toughness of the steel–adhesive interface, as defined by the critical value of the energy release rate, $\mathcal{G}_c = 300$ Pa m. Ignore all nonlinear effects and complex interface fracture mechanisms.

(a) Using the compliance method (see Section 9.2), find the critical load P_{cr} for fracture if $2a = 60$ mm, $H = 0.97$ mm and $B = 10.1$ mm. Since there are two crack fronts in this geometry, use

$$2\mathcal{G} = \frac{P^2}{2B}\frac{\mathrm{d}C}{\mathrm{d}a}. \tag{E.12.2}$$

(b) Calculate, in addition, the critical loads for $2a = 70$ mm and 80 mm, and plot the P–δ curves for these two crack lengths along with the result from part (a). For each crack length, mark the location where fracture occurs by the symbol, 'x'. Connect all the 'x's to obtain the 'toughness locus', as shown schematically in Fig. 12.19(*b*).

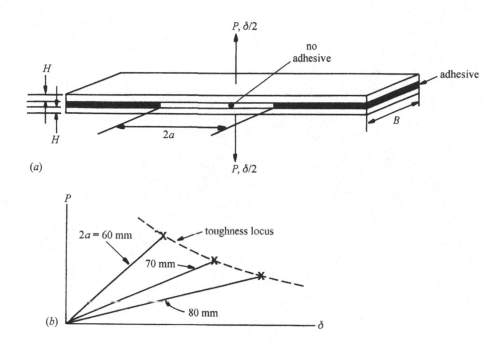

Fig. 12.19. (*a*) Geometry and loading of the joined plates. (*b*) Procedure to obtain the toughness locus.

(c) Compute the area bounded by the P–δ curves for $2a = 70$ mm and 80 mm, and the toughness locus. Using this area, estimate the critical energy release rate $\mathcal{G}_c$, and show that it is approximately equal to 300 Pa m.

12.4 Discuss possible mechanisms which improve the fatigue crack growth resistance of rubber-toughened epoxy as compared to that of the untoughened matrix, as shown in Fig. 12.15.

12.5 Discuss possible mechanisms which improve the fatigue crack growth resistance of epoxy containing inorganic fillers as compared to that of the unfilled matrix, as shown in Fig. 12.16.

Part four

ADVANCED TOPICS

CHAPTER 13

Contact fatigue: sliding, rolling and fretting

The term *contact fatigue* broadly refers to the surface damage process that leads to pitting, wear debris formation and fatigue cracking when the surfaces of two bodies repeatedly touch each other. The relative motion between the bodies may involve global/partial slip or rolling, or a combination of these modes of contact. Usually either or both surfaces in contact may also be subjected to fluctuating stresses from vibration or other mechanical loads.

In this chapter, attention is directed at the mechanics and mechanisms of different contact fatigue phenomena. Table 13.1 provides the definitions of key terminology encountered in contact fatigue, along with examples of practical situations where such phenomena apply. These processes will be discussed in detail in various sections of this chapter.

13.1 Basic terminology and definitions

Figure 13.1 provides a general frame of reference with which the various parameters of interest in contact mechanics are defined. In this figure, adapted from Johnson (1985), two surfaces in nonconforming contact are shown whose shapes prior to contact deformation are characterized by the functions, $z_1(x, y)$ and $z_2(x, y)$. The separation between the surfaces is $h(x, y) = z_1 + z_2$. Let $\mathbf{V}_1$ and $\mathbf{V}_2$ denote the linear velocity of surfaces 1 and 2, respectively, and let $\mathbf{\Omega}_1$ and $\mathbf{\Omega}_2$ be their respective angular velocities. The frame of reference, centered at the instantaneous contact point O, moves with linear and angular velocities $\mathbf{V}_O$ and $\mathbf{\Omega}_O$, respectively, so as to preserve its orientation to the indicated common tangent plane and the common normal at O.

The linear velocity $\mathbf{v}$ and angular velocity ω of the two bodies, with respect to the reference frame, are:

$$\mathbf{v}_1 = \mathbf{V}_1 - \mathbf{V}_O, \quad \mathbf{v}_2 = \mathbf{V}_2 - \mathbf{V}_O, \quad \omega_1 = \mathbf{\Omega}_1 - \mathbf{\Omega}_O, \quad \omega_2 = \mathbf{\Omega}_2 - \mathbf{\Omega}_O.$$

$$(13.1)$$

In order that the two surfaces be in continuous contact without either separating or interpenetrating, the contact should be continuous, i.e. the cartesian components of the velocity of the two surfaces along the common normal must be equal, with $V_{z1} = V_{z2} = V_{zO}$ or $v_{z1} = v_{z2} = 0$. With this reference frame, the three basic types of motion relevant to contact fatigue can now be defined.

Table 13.1. *Some basic definitions in contact problems.*

Terminology	Definition	Example
Conforming contact	Contact between surfaces whose profiles fit exactly without deformation	Journal bearings, flat slider bearings
Nonconforming contact	Contact between surfaces whose profiles do not match exactly	Gear teeth, ball and roller bearings
Sliding contact	Contact which arises due to a relative linear displacement	Skidding of a locked wheel of vehicle on a track, some aspects of contact between gear teeth
Fretting	Degradation of properties due to a repeated sliding between surfaces over small relative displacements	Riveted joints, dove-tail section of jet engines, interwire contacts in cables, shaft, keyways; also see Table 13.3
Free rolling contact	Contact which arises due to a relative angular displacement between surfaces about an axis located in their common tangent plane	Contact along the pitch line of gears, wheel on a railwaytrack (in the absence of a skid); $V_1 = V_2$
Tractive rolling contact	Rolling accompanied by tangential tractions between contacting surfaces	Contact above and below the pitch line of gears
Spin	Contact which arises due to a relative angular displacement between surfaces about a common normal to tangent plane	Rotational motion induced in a wheel while turning a corner

(1) The relative linear velocity, v_r, between the two surfaces at the instantaneous contact point characterizes sliding: $v_r = v_1 - v_2 = V_1 - V_2$. The cartesian components of the sliding velocity on the common tangent plane are: $v_{rx} = v_{x1} - v_{x2}$ and $v_{ry} = v_{y1} - v_{y2}$. Here one may envision *sliding contact fatigue* to occur in one of several ways. (a) The contact area between two surfaces is displaced in one direction, as a result of global sliding, and then the surfaces are disengaged and moved back to the original point of first contact; this process repeats itself at some frequency. The sliding displacement amplitude then is the total distance of linear relative motion between the contacting surfaces from the point of first contact to the point of complete loss of contact. (b) Complete relative sliding between the two surfaces occurs in one direction, followed by sliding in the reverse direction without any loss of contact.

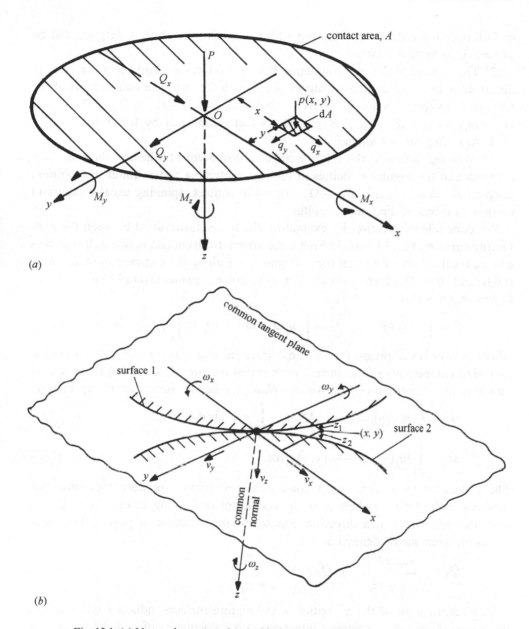

Fig. 13.1. (*a*) Nomenclature associated with the contact of two nonconforming surfaces. (*b*) An enlarged view of the contact area around O showing the various forces and moments.

Oscillatory slipping can also take place by small relative displacements (typically with a cyclic amplitude of 5–50 μm) only over a portion (near the outer periphery) of the contact area. Here, the two surfaces are 'stuck together' in the central portion of the contact region. This process is commonly referred to as *fretting*. The small-amplitude cyclic slip, in conjunction with superimposed cyclic loads in one or both contacting members, leads to *fretting fatigue*. Descriptions of fretting fatigue, including a more detailed definition of what constitutes fretting fatigue, will be presented in Section 13.6.

(2) The motion of the two contacting bodies produces a relative angular velocity about an axis on the common tangent plane, with the cartesian components of the relative angular velocity given by: $\omega_{rx} = \omega_{x1} - \omega_{x2} = \Omega_{x1} - \Omega_{x2}$ and $\omega_{ry} = \omega_{y1} - \omega_{y2} = \Omega_{y1} - \Omega_{y2}$. Repeated contact established by this rolling action leads to *rolling contact fatigue*.

(3) *Spinning* refers to the relative angular displacement whereby there arises a difference in the angular velocities of the two contacting bodies about the common normal: $\omega_{rz} = \omega_{z1} - \omega_{z2} = \Omega_{z1} - \Omega_{z2}$. In cyclic contact, spinning usually occurs in conjunction with sliding and/or rolling.

We conclude this section by examining the forces transmitted between the contacting surfaces, Fig. 13.1(*b*). The net force arising from contact between the surfaces can be resolved into a normal force P which acts along the common normal, and a tangential force Q which is on the common tangent plane. Overall equilibrium of forces requires that

$$P = \int_A p(x, y)\mathrm{d}A, \quad Q_x = \int_A q_x(x, y)\mathrm{d}A, \quad Q_y = \int_A q_y(x, y)\mathrm{d}A, \tag{13.2}$$

where $p(x, y)$ is the pressure or normal traction, and $q_x(x, y)$ and $q_y(x, y)$ are the cartesian components of the shear or tangential tractions.† Assuming that the contact area lies almost entirely on the x–y plane, the equilibrium of moments leads to:

$$M_x = \int_A p(x, y)y\mathrm{d}A, \qquad M_y = -\int_A p(x, y)x\,\mathrm{d}A,$$

$$M_z = \int_A \{q_y(x, y)x - q_x(x, y)y\}\mathrm{d}A. \tag{13.3}$$

The tangential force arises as a consequence of friction between the contacting surfaces with $Q \leq \mu P$, where μ is the coefficient of limiting kinetic friction. This frictional force acts in a direction counter to the direction of pure sliding, with cartesian components defined as

$$Q_x = \frac{-v_{rx}\mu P}{\sqrt{v_{rx}^2 + v_{ry}^2}}, \qquad Q_y = \frac{-v_{ry}\mu P}{\sqrt{v_{rx}^2 + v_{ry}^2}}. \tag{13.4}$$

The deformation of the asperities of the mating surfaces induces contact over a finite area instead of at a single contact point. A consequence of such deformation is that both forces and moments are transmitted across the contact area. The rolling moments, M_x and M_y in Eq. 13.3, oppose rolling motion by providing frictional resistance. The spin component of the moment, M_z, arises from friction within the contact area.

Contact fatigue usually occurs by the conjoint action of one or more of the foregoing contact phenomena. Consider, for example, a deep-groove ball bearing. When

---

† The discrete nature of asperity contact may lead to the transmission of forces and moments non-uniformly and discontinuously over the contact area A.

the ball rolls on the race support, pure rolling contact fatigue occurs only along the two line contacts schematically shown in Fig. 13.2. All other contact points between the ball and the race undergo both repeated tangential motion with friction (sliding contact fatigue) and rolling contact fatigue. Similarly, mating gear teeth are generally subject to both sliding and rolling contact fatigue, with the point of contact moving along with the rotation of the gear teeth.

13.2 Mechanics of stationary contact under normal loading

To set the scene for basic analyses and to motivate interpretations of the mechanistic processes relevant to contact fatigue, we begin with some classical results of normal indentation. Attention is confined to the simple cases of a sphere and a cylinder contacting a planar surface of the same material. These two cases provide insights into the contact fatigue situations encountered in a number of practical situations. These analyses are followed by discussions of the mechanics and micro-mechanisms of different types of relative motion between the contacting surfaces. The discussion is necessarily brief and is confined only to those issues with which connection can be made to fatigue damage processes. More details of contact mechanics can be found in Gladwell (1980), Johnson (1985) and Hills & Nowell (1994).

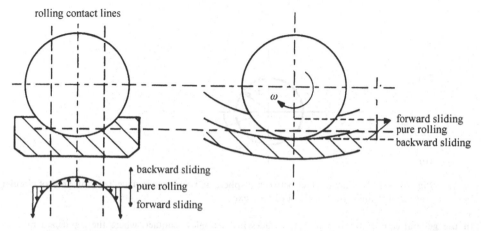

Fig. 13.2. Combined sliding and rolling contact in a deep-groove roller bearing and the local variation of surface velocity.

13.2.1 *Elastic indentation of a planar surface*

13.2.1.1 *Sphere indenting a planar surface*

Let a sphere, of radius R,[†] be pressed against a flat surface by a normal force P, as illustrated in Fig. 13.3(a). Assume that the contact between the surfaces is nonconforming and frictionless, involving small strains, and that the dimensions of the two bodies in contact are significantly larger than the contact area. For this Hertzian contact (Hertz, 1882), the radius of the circular contact zone is

$$a = \left\{\frac{3PR}{4E^*}\right\}^{1/3}, \qquad \frac{1}{E^*} = \frac{1 - v_1^2}{E_1} + \frac{1 - v_2^2}{E_2}, \tag{13.5}$$

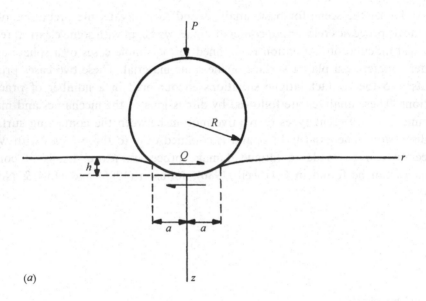

(a)

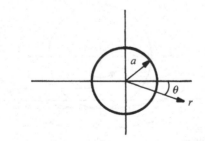

(b)

Fig. 13.3. (a) Normal contact between a sphere and a planar surface. (b) Plan view of circular contact area and cylindrical coordinate axes.

[†] In the general case of two nonplanar surfaces in frictionless contact, where the significant radii of curvature of the indenter and the substrate are R_1 and R_2, respectively, a relative radius of curvature is defined: $R \equiv \{(1/R_1) + (1/R_2)\}^{-1}$.

where E and v are Young's modulus and Poisson's ratio, respectively, the subscripts '1' and '2' refer to the sphere and the flat surface, respectively, and E^* is a composite modulus of the indenter–substrate system. When the two contacting bodies are made of the same material with $E_1 = E_2 = E$ and $v_1 = v_2 = v$ (which will be the case in all subsequent discussions), Eq. 13.5 becomes

$$a = \left\{ \frac{3PR}{2E}(1 - v^2) \right\}^{1/3}. \tag{13.6}$$

The pressure, $p(r)$, within the contact area $(0 \leq r \leq a)$ is

$$p(r) = \frac{p_{max}}{a}\sqrt{a^2 - r^2} = \frac{3p_m}{2a}\sqrt{a^2 - r^2} = \frac{3P}{2\pi a^2}\left(1 - \frac{r^2}{a^2}\right)^{1/2}. \tag{13.7}$$

The maximum pressure in the contact area, p_{max}, is 1.5 times the mean pressure, p_m.[†] The mutual approach of remote points in the two bodies is represented by the distance h (see Fig. 13.3):

$$h = \frac{a^2}{R} = \left\{ \frac{9P^2(1 - v^2)^2}{4RE^2} \right\}^{1/3}. \tag{13.8}$$

In the idealized case of frictionless contact with only a normal force P, the following general results are obtained from the Hertz analysis:

(1) The maximum value of the von Mises effective stress occurs in the substrate along the loading axis (i.e. the z-axis in Fig. 13.3) at a depth $z = 0.48a$ beneath the reference planar surface, for $v = 0.3$.

(2) The maximum tensile normal stress, $\sigma_r = (1 - 2v)p_{max}/3$, occurs on the surface $(z = 0)$ at the contact perimeter $(r = a)$.

(3) The maximum shear stress, $\tau_{max} = 0.31p_{max}$, exists beneath the surface at $r = 0$ and $z = 0.48a$.

(4) The foregoing results imply that a brittle mode of failure is likely to initiate near the contact perimeter *at the surface*, whilst the onset plastic yielding is favored *beneath the surface*.

13.2.1.2 Cylinder indenting a planar surface

It was seen in Fig. 13.2 that the roller of the roller bearing makes a line contact with the supporting race. In general, line contact is produced when the shapes of two bodies are such that conforming contact arises in one direction while a nonconforming contact exists in a perpendicular direction. Such a situation occurs when two cylinders, with parallel axes, are pressed against each other.[‡]

[†] If the contact area is an ellipse whose major and minor radii are a and b, respectively, the maximum Hertzian pressure is $p_{max} = 3P/(2\pi ab)$.

[‡] The *macroscopic* contact between the blade and the compressor rotor in a turbine jet engine could be likened approximately to a two-dimensional line contact, whilst the indentation of rough surface asperities between the contacting surfaces of the rotor and the blade may be modeled on a *microscopic* scale by the Hertzian contact of a sphere indenting a flat surface.

Consider the two-dimensional case of a circular cylinder, of radius R, which is pressed against a planar surface in such a way that its axis is parallel to the surface. In the notation used in Fig. 13.1, the axis of the cylinder is parallel to the y-axis, and contact occurs over a strip of width $2a$ along the y-axis. In the absence of interface friction, the distribution of normal tractions is given by the Hertz theory:

$$p(x) = \frac{2P}{\pi a^2}\sqrt{a^2 - x^2}, \qquad a = \left\{\frac{4PR}{\pi E}\right\}^{1/2}, \tag{13.9}$$

where P is the applied normal load per unit length of the cylinder. The maximum contact pressure, $p_{max} = 2P/(\pi a) = (4/\pi)p_m$. The maximum shear stress, $\tau_{max} = 0.3p_{max}$ exists beneath the contact interface at $x = 0$ and $z = 0.78a$. At the contact interface, $\sigma_x = \sigma_y = -p(x)$. All stresses vanish at the surface outside the contact interface.

13.2.2 *Plastic deformation*

We now present a brief summary, following Johnson (1983), of the plastic deformation of the substrate indented by a normal load. The onset of plastic yielding for normal indentation by a sphere occurs at a depth of approximately $0.5a$ beneath the indented surface when the applied load P reaches a value P_y signifying the onset of yield. For the spherical indenter, the maximum Hertzian pressure at the commencement of plastic flow is:

$$p_{max} = \frac{3P}{2\pi a^2} \approx 1.6\sigma_y, \tag{13.10}$$

where σ_y is the yield strength of the indented material in compression. Essentially the same maximum pressure develops for a general elliptical contact zone (with any combination of a and b) at $P = P_y$.

The nascent plastic zone formed at $P = P_y$ below the indenter is engulfed by elastic material. Further increases in P lead to an enlargement of the plastic zone as the material is compressed along the loading axis and stretched in the radial direction parallel to the indented surface.

At the other limiting case of fully plastic deformation, slip-line field theory solutions are available, particularly for rigid straight-sided indenters such as flat punches and cones (Shield, 1955). A wide variety of analyses, corroborated by experiments, have shown that, for a rigid indenter pressed against a rigid–perfectly-plastic substrate,

$$p_m \approx 3\sigma_y, \tag{13.11}$$

for essentially all geometries of general interest of the two contacting bodies and for all shapes of contact zones. The contact pressure is also approximately the same everywhere.

For strain-hardening materials, Eq. 13.11 could be extended by replacing σ_y with $\bar{\sigma}_y$, where $\bar{\sigma}_y$ is the stress required to induce a representative strain ϵ_R in a uniaxial compression test. For a spherical indenter, $\epsilon_R \approx 0.2a/R$ (Tabor, 1951). For a strain-hardening engineering alloy whose plastic deformation may be approximated by a pure power-law function, $\epsilon/\epsilon_y = (\sigma/\sigma_y)^n$, the mean contact pressure and the pressure distribution take the form (Mathews, 1980):

$$p_m = \frac{6n\sigma_y}{2n+1}\left\{\frac{8a}{9\pi R\epsilon_y}\right\}^{1/n}, \quad p(r) = p_m\frac{2n+1}{2n}\left\{1-\frac{r^2}{a^2}\right\}^{1/(2n)}. \tag{13.12}$$

For $n = 1$ (a linear elastic solid with $E = \sigma_y/\epsilon_y$), Eq. 13.12 reduces to the Hertzian elastic result, Eq. 13.7. For the other extreme case of $n \to \infty$ (an ideally plastic solid), $p_m \approx 3\sigma_y$ becomes uniformly the same everywhere (i.e. independent of r). This result is the same as that of Eq. 13.11. Equation 13.12 may be rewritten in terms of $\bar\sigma_y$ and $\bar\epsilon_y$ as

$$\bar\sigma_y = \frac{p_m}{3} = \frac{2n\sigma_y}{2n+1}\left\{\frac{8a}{9\pi R\epsilon_y}\right\}^{1/n}, \quad \bar\epsilon_y = \frac{8a}{9\pi R}\left\{\frac{2n}{2n+1}\right\}^n. \tag{13.13}$$

Note that $\bar\epsilon_y$ ranges from $0.188a/R$ to $0.171a/R$ as n is changed from 1 to ∞ in Eq. 13.13. This narrow range of $\bar\epsilon_y$ is close to the empirical value of $0.2a/R$ proposed by Tabor (1951) for the spherical indenter.

13.2.3 Residual stresses during unloading

When a zone of permanent inelastic deformation is induced during indentation, releasing the applied load causes a zone of residual stresses to be generated in the deformed region. This process is analogous to the creation of a residual stress field and a cyclic plastic zone at the tip of a crack or a notch which is subjected to loading and unloading (see Sections 4.11 and 9.6). The residual stresses which remain in the material after the release of the indenter can play a significant role in influencing the fatigue response as well as brittle failure processes.

For an ideally plastic solid, the stresses directly beneath the indenter (i.e. along the z axis, $r = 0$) evolve in such a way that the Tresca yield criterion, Eq. 1.23, gives the following result:

$$\sigma_z - \sigma_r = -\sigma_y, \tag{13.14}$$

where σ_z and σ_r are the normal and the radial stresses (which are also the principal stresses and hence the shear stresses are zero along the z axis). Elastic unloading causes an effective superposition of a uniform tension (negative pressure), approximately equal to $3\sigma_y$ (as shown in Eq. 13.11), at the contact interface. A uniform tension of $3\sigma_y$ acting on the contact area gives (Timoshenko & Goodier, 1951), upon unloading,

$$(\sigma_z - \sigma_r)|_{max} = 0.66 \times (3\sigma_y) \qquad \text{at } z = 0.64a. \tag{13.15}$$

Using superposition, we combine Eqs. 13.14 and 13.15, to obtain the residual stress upon unloading the indenter (as indicated by the subscript 'R'), at $z = 0.64a$ (which is the location at which both the maximum Tresca and von Mises effective stresses develop)

$$(\sigma_z - \sigma_r)|_R = [(3 \times 0.66) - 1]\sigma_y = 0.98\sigma_y. \tag{13.16}$$

Thus, at the location of maximum effective stress at $z = 0.64a$, the material is nearly at a point of reverse plastic flow when the normal load is released.

At the surface outside the contact area, the stress state during elastic–plastic loading involves radial compression and a small circumferential tension. If, during unloading, a uniform tensile stress of $3\sigma_y$ is superimposed, there occurs a net increase in the radial compression and circumferential tension by:

$$\sigma_\theta = -\sigma_r = \frac{1}{3}(1 - 2v)3\sigma_y \frac{a^2}{r^2}, \qquad r > a. \tag{13.17}$$

This result is simply the circumferential and radial stress distribution corresponding to a tensile point force of $3\sigma_y \pi a^2$ (as seen from far away from the contact area) according to the classical Boussinesq solution (Boussinesq, 1885). Plastic flow is thus assisted by the uniform stress generated during unloading. As noted by Johnson (1983), this result provides a rationale for the observation that semi-brittle materials, such as PMMA and some ceramics which are capable of plastic flow or microcracking within the hydrostatic stress field under the indenter, form radial cracks upon release of the indenter (since the principal circumferential stress is tensile).

13.2.4 Example problem: Beneficial effects of surface compressive stresses

Problem:

Shot peening and cold rolling are palliatives that are commonly used to suppress the formation of a fatigue crack on surfaces that are subjected to sliding, fretting and rolling contact fatigue. These processes produce a layer of equi-biaxial compressive residual stresses ($\sigma_x|_R = \sigma_y|_R$).[†]

Let a sphere be normally pressed against a flat surface. Assume that the contact is frictionless. Let the indented surface contain an equi-biaxial compressive residual stress, $\sigma_x|_R = \sigma_y|_R \to -\sigma_y$; $\sigma_z|_R = 0$ on planes parallel to the indented surface (e.g., from shot peening). Show how much higher normal force P can be applied without plastically yielding the surface with a residual stress induced by shot peening as compared to the untreated surface with no residual stress. Assume the indented surface to be elastic–ideally plastic.

Solution:

The equi-biaxial residual compressive stress is added to the radial stress to give a net radial stress of $[\sigma_r + \sigma_x|_R]$. When the sphere is indented on to the flat surface with a load P, plastic yielding is suppressed if, as per the Tresca yield condition,

$$|[\sigma_r + \sigma_x|_R] - \sigma_z| \le \sigma_y. \tag{13.18}$$

[†] The simple example problem presented here is derived from Hills & Ashelby (1982).

We consider situations where the residual surface compressive stresses are approaching the yield strength, i.e.

$$|\sigma_x|_R| = |\sigma_y|_R| \to \sigma_y. \tag{13.19}$$

Combining Eqs. 13.18 and 13.19,

$$\sigma_r - \sigma_z \leq 2\sigma_y. \tag{13.20}$$

As seen from general result (3) in Section 13.2.1.1, however, elastic indentation of the surface by a sphere gives rise to

$$(\sigma_r - \sigma_z)|_{max} = 2\tau_{max} = 0.62 p_{max} \qquad \text{at } z = 0.48a. \tag{13.21}$$

Combining Eqs. 13.20 and 13.21, it is found that plastic yielding of the indented surface is suppressed during loading (as well as during unloading, as seen from Eq. 13.16) when

$$p_{max} \leq \frac{2}{0.62}\sigma_y \qquad \text{(or)} \qquad p_{max} \leq 3.2\sigma_y. \tag{13.22}$$

A comparison of Eq. 13.10 (which gives the maximum pressure for the spherical indentation of a flat surface without any residual stress) with Eq. 13.22 readily shows that the maximum Hertzian contact pressure which is imposed by the indenter on the surface is elevated by a factor of two when a state of equi-biaxial compressive residual stress, approaching the yield strength in compression, is introduced. Since p_{max} scales with $P^{1/3}$ (see Eqs. 13.6 and 13.7), the surface with the residual compressive stress is capable of bearing an indentation load which is up to eight times (i.e. 2^3) larger without undergoing plastic flow. Thus, surface modification techniques such as shot peening can lead to significant enhancements in contact fatigue resistance (see Fig. 13.16).

13.3 Mechanics of sliding contact fatigue

Any propensity for sliding to occur engenders a tangential frictional force, Q, during frictional contact. This force acts counter to the direction of motion on each surface. So long as there exists a steady sliding motion between the bodies, the tangential force Q may be envisioned as a force of kinetic friction between the surfaces. Before assessing the effects of tangential forces on the stress fields and displacements, recall that the two surfaces 'conform' to each other when their elastic properties are the same. The size and shape of the contact area as well as the contact pressure are then controlled only by the profiles of the contacting surfaces and by P, and not by Q. Consequently, the interactions between the normal pressure and the tangential friction can be neglected, if the two materials are the same, and the tangential stress, $q(x, y)$ and the contact pressure $p(x, y)$ on the sliding surfaces can be superimposed to determine the overall elastic stresses in the bodies. This situation will be assumed in all subsequent analyses.

13.3.1 Sliding of a sphere on a planar surface

If μ is the coefficient of kinetic friction for relative motion between the surfaces, the relationship between $q(x, y)$ and $p(x, y)$ is usually postulated to follow the Amonton–Coulomb friction law:

$$\frac{|\, q(x, y)\, |}{p(x, y)} = \frac{|\, Q\, |}{P} = \mu. \tag{13.23}$$

Combining Eqs. 13.5–13.7 with this equation, we obtain the radial variation of the interface tangential stress within the contact area:

$$q(r) = \frac{3\mu P}{2\pi a^3}\sqrt{a^2 - r^2}, \tag{13.24}$$

which always acts in a direction counter to the direction of sliding (i.e. along the negative or positive x-axis).

From the surface tractions, Eqs. 13.7 and 13.24, the stress fields on the contact surface (x–y plane) and in the symmetry plane (y–z plane) have been extracted by Hamilton & Goodman (1966) for $\mu = 0.25$ and 0.50 and for $\nu = 0.3$ by recourse to a method proposed by Green (1949) and to the work of Mindlin (1949). Hamilton (1983) and Sackfield & Hills (1983) extended these results by reporting explicit expressions for the stresses at any point beneath the sliding contact for Hertzian loading. The following general trends can be captured from the above studies.

(1) The maximum value of the von Mises effective stress, which occurs along the loading axis ($z = 0.48a$ and $r = 0$) for frictionless Hertzian contact ($\mu = 0$), is raised to the sliding surface when the friction coefficient $\mu \geq 0.3$. In this latter case, the second invariant of the deviatoric stress tensor, J_2 (defined in Eq. 1.22) becomes:

$$J_2 = \frac{p_{max}^2}{3}\left\{\frac{(1 - 2\nu)^2}{3} + \frac{(1 - 2\nu)(2 - \nu)\mu\pi}{4} + \frac{(16 - 4\nu + 7\nu^2)\mu^2\pi^2}{64}\right\}. \tag{13.25}$$

Note that J_2 can be identified with τ_y^2, where τ_y is the shear yield strength of the material.

(2) The maximum tensile normal stress, $\sigma_{rr} = (1 - 2\nu)p_{max}/3$, occurs on the surface ($z = 0$) at the contact perimeter ($r = a$) for $\mu = 0$. The tangential friction force elevates the local compressive stress at the leading edge of the sliding contact ($x = +a$, $z = 0$), while the tensile stress intensifies at the trailing edge ($x = -a$, $z = 0$).

(3) For frictional sliding with $\mu \geq 0.3$, the maximum tensile normal stress occurs at the surface at ($x = -a$, $z = 0$):

$$(\sigma_1)_{max} = \sigma_{rr} = p_{max}\left\{\frac{1 - 2\nu}{3} + \frac{4 + \nu}{8}\mu\pi\right\}. \tag{13.26}$$

13.3.2 Partial slip and complete sliding of a cylinder on a planar surface

If friction is introduced along the contact interface between the cylinder and a flat surface, which was considered in Section 13.2.1.2, the distribution of the tangential tractions is of the form:

$$q(x) = \frac{Q}{\pi\sqrt{a^2 - x^2}}, \qquad (13.27)$$

whereas the normal traction distribution, $p(x)$, is given in Eq. 13.9. An examination of this equation readily reveals that the shear tractions approach infinity at the edges of the contact ($x = \pm a$). In reality, such high tractions can not be accommodated by finite values of μ. Instead, the high shear stresses at the contact rim are relieved by small relative slip between the bodies, termed *microslip*, which initiates at the contact perimeter and spreads radially inward. Thus, the contact region comprises a region of 'stick' in the center, which is surrounded by an annular zone of microslip extending from the periphery of the stick region to the outer boundary of contact. Note that all points within the microslip region move in the same direction. This process of *partial slip* exists as long as the magnitude of the tangential frictional force Q is less than the limiting force for the onset of sliding, μP. Figure 13.4 schematically illustrates this partial slip behavior.

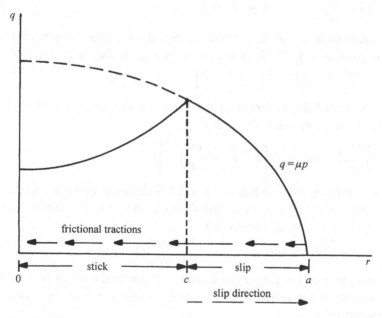

Fig. 13.4. Schematic illustration of the variation of tangential stresses at the contact interface during partial slip. Equation 13.27 describes the curve for no slip, while Eq. 13.29 describes the curve for partial slip.

Quantitative analyses for partial slip were presented by Cattaneo (1938), Mindlin (1949) and Deresiewicz (1954). The radius of the central stick region, c, which is concentric with the contact circle, is

$$\frac{c}{a} = \left[1 - \frac{Q}{\mu P}\right]^{1/2}; \qquad 0 \le c/a \le 1. \tag{13.28}$$

In the stick region, the tangential tractions are

$$q(x) = \frac{\mu p_{max}}{a}\left\{\sqrt{a^2 - x^2} - \sqrt{c^2 - x^2}\right\}; \qquad -c \le |x| \le c. \tag{13.29}$$

At the edge of the stick region ($c \le |x| \le a$), $q(x) = \mu p(x)$.

The conditions governing the occurrence of partial slip (which, as shown in Section 13.6, plays a major role in fretting fatigue) can now be formulated from the above equations. If we hold P fixed and slowly increase Q, partial slip initiates at the edges of contact and spreads inward. Eventually, as $Q \to \mu P$, $c \to 0$, i.e. the stick zone shrinks down to a line at $x = 0$. Any propensity for Q to increase further would cause the entire cylinder to slide on the flat surface.

13.3.3 *Partial slip of a sphere on a planar surface*

Equation 13.24 provides the tangential traction distribution at the onset of complete sliding of a sphere on a planar surface. If only partial slip occurs, the radius of the central stick region, c, which is concentric with the contact circle, is:

$$\frac{c}{a} = \left\{1 - \frac{Q}{\mu P}\right\}^{1/3}; \qquad 0 \le c/a \le 1. \tag{13.30}$$

Within the region of microslip at the contact rim, the resultant tangential tractions, $q(r)$, are also given by Eq. 13.24. Inside the stick region, the shear tractions are

$$q(r) = \frac{\mu p_{max}}{a}\left\{\sqrt{a^2 - r^2} - \sqrt{c^2 - r^2}\right\}; \qquad 0 \le r \le c. \tag{13.31}$$

The relative tangential displacement of the substrate with respect to the sphere (in the direction of Q), as shown in Fig. 13.5, is

$$\delta \approx \frac{3\mu P}{4Ea}(2 - v)(1 + v)\left\{1 - \left(1 - \frac{Q}{\mu P}\right)^{2/3}\right\}. \tag{13.32}$$

Obviously, $\delta = 0$ for $Q = 0$. When $Q \to \mu P$, full sliding is initiated. At this point, $dQ/d\delta = 0$, and a limiting state of global sliding with $\delta = \delta_{sl}$ is reached. The initial slope of the Q–δ curve and δ_{sl} are given by

$$\frac{dQ}{d\delta}\Big|_{T=0} = \frac{32a}{E}(2 - v)(1 + v), \qquad \delta_{sl} \approx \frac{3\mu P}{8Ea}(2 - v)(1 + v). \tag{13.33}$$

Once full sliding begins, the tangential stiffness of spherical contact is lost, and any subsequent sliding of the sphere on the planar surface can only be facilitated by displacement control.†

---

† This situation is termed *slide*, whereas the term *slip* is reserved for relative tangential displacement over only a portion of the contiguous surfaces.

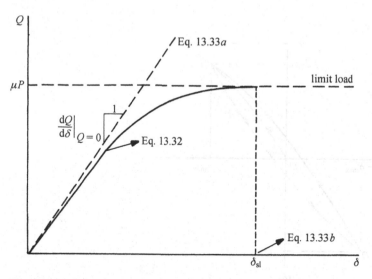

Fig. 13.5. Variation of frictional force with the relative tangential displacement during the partial slip and complete sliding of a sphere on a planar surface.

13.3.4 Cyclic variations in tangential force

Consider an elastic sphere indenting a planar surface of the same material under a fixed normal load $\overline{P}$ and a cyclic tangential load Q which fluctuates between the constant limits $+Q_{max}$ and $-Q_{max}$. Partial slip conditions are considered here such that $Q_{max} \leq \mu \overline{P}$, and that the deformation is elastic.

During the first increase in Q from 0 to $+Q_{max}$, i.e. loading path OA in Fig. 13.6(a), a partial slip zone develops as described in the preceding section. Figure 13.6(b) shows the radial variations of the tangential force Q (for a fixed normal load $\overline{P}$) corresponding to this situation. For this point, $\delta = \delta_{max}$, as determined from Eq. 13.32.

When the tangential load is reduced from Q_{max}, slip of opposite sign develops near the outer periphery of the contact region ($b \leq r \leq a$). The size of this zone of reversed slip at point B in Fig. 13.6(a) is found to be

$$\frac{b}{a} = \left\{ 1 - \frac{Q_{max} - Q}{2\mu \overline{P}} \right\}^{1/3}.$$
(13.34)

Upon unloading further to point C in Fig. 13.6(a), where $Q = 0$, the radial variation of the tangential tractions become

$$q = -\frac{3\mu \overline{P}}{2\pi a^3} \sqrt{a^2 - r^2}, \qquad b \leq r \leq a,$$

$$q = -\frac{3\mu \overline{P}}{2\pi a^3} \left\{ \sqrt{a^2 - r^2} - 2\sqrt{b^2 - r^2} \right\}, \qquad c \leq r \leq b,$$

$$q = -\frac{3\mu \overline{P}}{2\pi a^3} \left\{ \sqrt{a^2 - r^2} - 2\sqrt{b^2 - r^2} + \sqrt{c^2 - r^2} \right\}, \quad r \leq c.$$
(13.35)

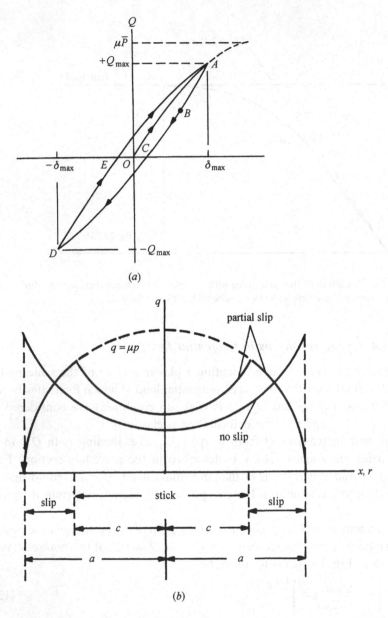

(a)

(b)

Fig. 13.6. (*a*) Cyclic variation in the shear force as a function of the tangential displacement. (*b*) The evolution of circular contact area during the application of a constant normal force $\bar{P}$ and a cyclic shear force fluctuating between the limits $+Q_{max}$ and $-Q_{max}$.

The curve corresponding to point C in Fig. 13.6(*b*) shows the radial variation of the tangential force.

Upon further increases in Q in the reverse direction until $Q = -Q_{max}$, point D is reached in Fig. 13.6(*a*). At this point, $\delta = -\delta_{max}$, and reverse sliding extends as much as forward sliding displacement with $b = c$. During the unloading path $ABCD$ in

Fig. 13.6(a), the variation of the shear traction with displacement follows the relationship

$$\delta = \frac{3\mu\overline{P}(2-v)(1+v)}{4Ea}\left\{2\left(1-\frac{Q_{max}-Q}{2\mu\overline{P}}\right)^{2/3}-\left(1-\frac{Q_{max}}{\mu\overline{P}}\right)^{2/3}-1\right\}. \quad (13.36)$$

The situation at point D is the complete reversal of that at A. The next reversal of Q from $-Q_{max}$ to $+Q_{max}$ (i.e. from point D to point E to point A) leads to a situation that is similar to the unloading path $ABCD$ but with an opposite sign. Thus, the $Q-\delta$ loop becomes closed and this trend repeats itself for constant-amplitude partial cyclic slip between $-Q_{max}$ and $+Q_{max}$.

The area enclosed by the hysteresis loop $ABCDEA$ in Fig. 13.6(a) is indicative of the irreversible energy, ΔW, dissipated by the frictional microslip in the regime $c \le r \le a$. From the analysis of Mindlin *et al.* (1952), it is known that

$$\Delta W = \frac{18(\mu\overline{P})^2}{5Ea}(2-v)(1+v)$$

$$\times\left\{1-\left(1-\frac{Q_{max}}{\mu\overline{P}}\right)^{5/3}-\frac{5Q_{max}}{6\mu\overline{P}}\left[1-\left(1-\frac{Q_{max}}{\mu\overline{P}}\right)^{2/3}\right]\right\}. \quad (13.37)$$

13.4 Rolling contact fatigue

During rolling contact fatigue, which was introduced in Section 13.1, elastic deformation from normal loading generates a contact area whose features are described by the Hertz theory. As seen in Section 13.1, relative sliding between the two bodies occurs during rolling if the tangential velocities V_1 and V_2 are not equal, and there exists a tangential force Q at the contact interface. Following the discussions in Section 13.3.3, it is to be expected that during tractive rolling, a fraction of the contact area undergoes microslip while the surfaces stick together in the remaining fraction. The difference between the circumferential strains in the two bodies in the stick region of the contact area ostensibly leads to a small slip displacement which is widely known as *creep*.

The foundations for analyzing rolling contact fatigue are generally predicated upon the mechanics of contact presented in earlier sections of this chapter. A detailed discussion of the mechanics of rolling contact fatigue, which is strongly influenced by the specific geometries, material response and boundary conditions pertinent to different applications, will not be undertaken here. The interested reader is referred to the monograph by Johnson (1985) for the theory of rolling contact.

In this section, we begin with a consideration of energy loss during repeated rolling contact. This is followed by a discussion, in the next section, of the shakedown limits for sliding and rolling contact fatigue. The microscopic mechanisms of contact fatigue are considered next. The discussion of rolling contact fatigue is then concluded with a case study of crack initiation in gear teeth in a ground vehicle application.

13.4.1 Hysteretic energy dissipation in rolling contact fatigue

Figure 13.7 schematically shows the variation of stress state with location on a stationary, planar substrate on which a cylinder of radius R and unit length is rolled with an angular velocity ω. The interface is assumed to have negligible friction; the normal load is P. Assuming that Hertzian contact conditions prevail, the rate of elastic strain energy storage per unit time period is found to be

$$\dot{W} = \frac{2Pa\omega}{3\pi}, \tag{13.38}$$

where a is shown in Fig. 13.7. If the imperfect elastic response of the material leads to the dissipation of stored energy, as for example due to the microscopically inelastic deformation arising from asperity contact, a fraction of the stored energy is irreversibly dissipated. Let this fraction be denoted as h_{hys}. The moment M_y on the rolling cylinder required to maintain motion is then found by setting the net work done equal to the energy dissipated, so that

$$M_y\omega = h_{hys}\dot{W} = h_{hys}\frac{2Pa\omega}{3\pi}, \quad \rightarrow \quad \mu_R \equiv \frac{M_y}{PR} = h_{hys}\frac{2a\omega}{3\pi R}, \tag{13.39}$$

where μ_R is the coefficient of rolling friction (Tabor, 1955). For an elliptical or circular contact, $\mu_R = (3h_{hys}a)/16R$. This simple analysis, where the loss factor h_{hys} is simply chosen as a fraction of the stored energy, does not account for the variation of hysteresis energy loss with strain, nor is it amenable to be generally correlated with uniaxial properties of materials. For metallic materials, $h_{hys} \sim 0.01$, whereas for polymers and rubber h_{hys} has much higher values.

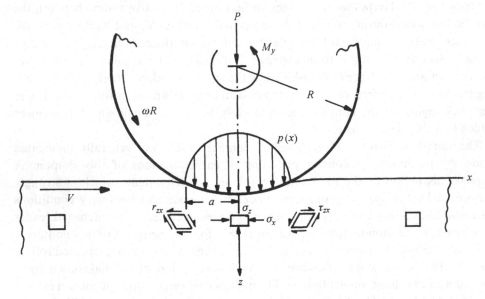

Fig. 13.7. Rolling of a cylinder on a planar surface and the associated nomenclature and stress fields. (After Johnson, 1985.)

13.4.2 Shakedown limits for rolling and sliding contact fatigue

The generation of residual stresses upon unloading the indenter can suppress subsequent plastic yielding during cyclic indentation, as shown in Section 13.2.3. For the case of frictionless normal cyclic indentation, the conditions governing the onset of reversed yielding, i.e. the shakedown limits, are uniquely prescribed by analyses such as the ones presented in Section 13.2.3. For cyclic indentation with friction, however, the shakedown limits fall between two bounds. The upper bound and lower bound theorems, discussed in Section 3.7 for this shakedown phenomenon, have been applied to contact fatigue problems. In this section, we examine the shakedown limits for sliding and rolling contact fatigue for two-dimensional line contact (between a cylinder and a planar surface) and for three-dimensional point contact (between a sphere and a planar surface).

13.4.2.1 Shakedown limits for two-dimensional contact

The mechanics of frictionless normal indentation of a cylinder on a planar surface was examined in Section 13.2.1.2. The cylinder is placed on the flat surface with its axis parallel to the y-axis and the friction force acts along the x-axis in Fig. 13.1. It may be recalled from that discussion that the maximum contact pressure, $p_{max} = 2P/(\pi a) = (4/\pi)p_m$ (where P is the applied load per unit length of the cylinder), that the maximum shear stress, $\tau_{max} = 0.3p_{max}$ exists beneath the contact interface at $x = 0$ and $z = 0.78a$, and that $\sigma_x = \sigma_y = -p(x)$ at the contact interface.

Consider the situation where the long cylinder, oriented parallel to the y axis, rolls freely without friction in the x direction on the planar surface (x–y plane), as shown for example in Fig. 13.7. For this case, the statically admissible field of residual stresses that satisfy symmetry and equilibrium requirements includes only the components $\sigma_x|_R$ and $\sigma_y|_R$ (where, as shown in Section 13.2.3.1, the subscript 'R' denotes a residual stress). These stresses are functions only of z and not of x and y.

In plane strain (applicable to the plane of symmetry, the x–z plane), the von Mises yield condition is written as

$$\{[\sigma_x + \sigma_x|_R] - \sigma_z\}^2 + \{\sigma_z - [\sigma_y + \sigma_y|_R]\}^2 + 6\tau_{zx}^2 \leq 2\sigma_y, \tag{13.40}$$

where σ_y is the yield strength of the indented material in simple compression. Johnson (1962) applied Melan's lower bound theorem (see Section 3.7) to this case to derive the shakedown limits in pure rolling. For this purpose, the terms within the braces {.......} in Eq. 13.40 can be made to vanish by appropriately choosing $\sigma_x|_R$ and $\sigma_y|_R$ to give

$$\tau_{zx}|_{max} = \frac{\sigma_y}{\sqrt{3}}, \tag{13.41}$$

where $\tau_{zx}|_{max}$ is the maximum value of the shear stress on the x–z plane which contains the direction of sliding. From the Hertzian analysis, it is known (e.g.,

Johnson, 1985) that $\tau_{zx}|_{max} = 0.25\, p_{max}$, occurs at $x = \sqrt{3}a/2$ and $z = a/2$. Thus, the shakedown limit for free rolling of the cylinder occurs when

$$\frac{p_{max}}{\sigma_y} = \frac{4}{\sqrt{3}} = 2.309. \tag{13.42}$$

Similarly, in terms of the Tresca yield condition, the magnitude of the maximum principal shear stress is equated to the yield strength k in simple shear to determine the maximum contact pressure p_{max} for first yield and shakedown. Note that $k = \sigma_y/\sqrt{3}$. Combining this result with Eq. 13.42, it is seen that the shakedown limit for pure rolling is:

$$\frac{p_{max}}{k} = 4. \tag{13.43}$$

For tractive rolling involving both rolling and frictional sliding with line contact, Johnson & Jefferis (1963) applied Melan's theorem in conjunction with both Tresca and von Mises yield criteria. They derived the shakedown limits as a function of the traction coefficient, $\mu = Q/P$, as plotted in Fig. 13.8. When the friction coefficient is low ($\mu \leq 0.3$), yielding occurs beneath the surface. For higher values of friction, the yield point moves to the surface. For tractive rolling of a cylinder which involves combined rolling and sliding with an interface friction coefficient $\mu > 0.3$, the maximum principal shear stress in the plane of deformation is found to be μp_{max}. Applying the Tresca yield criterion, the shakedown limit for tractive rolling with high friction then satisfies the equation

$$\mu p_{max} = k, \qquad \text{or} \qquad \frac{p_{max}}{k} = \frac{1}{\mu}. \tag{13.44}$$

Thus, pure rolling has a shakedown limit which is 4μ times larger than the shakedown limit for tractive rolling with $\mu > 0.25$. This behavior is plotted graphically in Fig. 13.8. At loads above the shakedown limit, the surface layers of the substrate are subjected to a progressive, tangential shear displacement with respect to the material underneath. This shearing causes ratchetting and incremental collapse.†

13.4.2.2 Shakedown limits for three-dimensional contact

The shakedown limits for three-dimensional point contact under sliding and rolling are available from the work of Ponter, Hearle & Johnson (1985) who employed Koiter's kinematical upper bound theorem (see Section 3.7). Figure 13.9 shows their results for the steady-state response of a circular contact area subjected to repeated normal and tangential loads. The evolution of plastic yielding is divided into different regimes in this figure.

(1) When the maximum normal contact pressure, p_{max}/k, for any value of the traction coefficient $\mu = Q/P$, is below the line labeled 'elastic limit', no plastic flow occurs for any number of contact fatigue cycles.

---

† Note that the analyses here do not take into account the strain hardening of the substrate material.

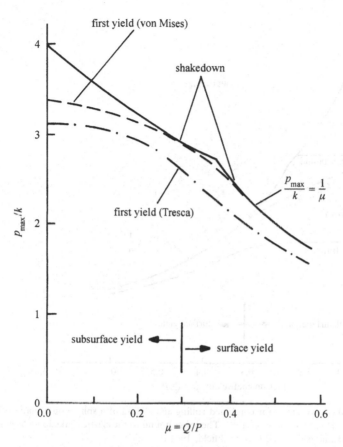

Fig. 13.8. Shakedown limits for combined rolling and sliding of a cylinder on a planar surface. (After Johnson & Jefferis, 1963.)

(2) In the regime marked 'elastic shakedown', plastic flow occurs during the first load. However, the development of residual stresses suppresses further plastic flow and a steady state of elastic deformation occurs during repeated tractive rolling. No cyclic plasticity occurs in this regime. Both lower bound and upper bound to this elastic shakedown are plotted in Fig. 13.9.

(3) Both upper bound and lower bound estimates for elastic shakedown show that $p_{max}/k = 1/\mu$ for $\mu > 0.3$. This result is the same as that derived in the preceding subsection for the shakedown limit for the tractive rolling of a cylinder under high friction.

(4) In the region marked 'cyclic plasticity' or 'plastic shakedown', a closed cycle of plastic straining is developed in a region of the substrate beneath the free surface. In this material, which is fully surrounded by undeforming material, progressive accumulation of plastic strains is suppressed. Such a zone of plastic shakedown is specific to the three-dimensional point contact, and is

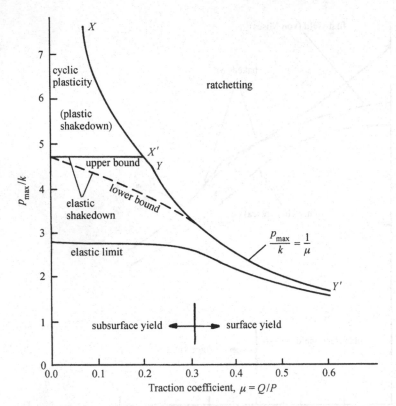

Fig. 13.9 Shakedown limits for combined rolling and sliding of a sphere on a planar surface. (After Ponter, Hearle & Johnson, 1985. The lower bound to the elastic shakedown limit marked by the dashed line is from Hills & Sackfield, 1984.)

not seen in the two-dimensional line contact analyzed in the preceding sub-section.

(5) Above the upper bound for the plastic shakedown, marked by the path $XX'YY'$, plastic strain accumulation takes place causing ratchetting and incremental collapse.

The foregoing analyses deal with contact situations where the substrate is modeled as elastic–ideally plastic with no strain hardening. Dang Van & Maitournam (1994) have reported a steady-state elastoplastic formulation to calculate shakedown limits in sliding with repeated contact reversals where von Mises associative plasticity was used along with linear kinematic hardening. (The hardening modulus was taken to be 8% of the elastic modulus.) The sliding was assumed to extend several times the contact diameter and the elastic sliding tractions were used in the calculations. The shakedown limits obtained by these authors are consistent with the the results of Hills & Asbelby (1982) and of Ponter, Hearle & Johnson (1985) for $\mu > 0.3$. For $0 \le \mu < 0.3$, Dang Van & Maitournam predict an almost linear variation of the shakedown limit between 4 ($\mu = 0$) and 3 ($\mu = 0.3$).

13.5 Mechanisms of contact fatigue damage

13.5.1 Types of microscopic damage

We now direct attention at microscopic failure processes in contact fatigue. The nature of specific failure mechanism is strongly influenced by such factors as the geometry and properties of the contacting bodies, and of the lubricant, if any, between the surfaces, the topology of the surfaces of the contacting bodies, the mechanical loading conditions, and the environment. While it is not feasible to generalize the failure modes for all contact fatigue situations, it is possible to identify a set of prominent failure modes for different contact conditions. Table 13.2 provides the terminology and definitions of the prominent damage mechanisms commonly encountered during rolling and sliding contact fatigue. The case study presented in the next subsection illustrates how damage evolves during rolling and sliding contact fatigue.

13.5.2 Case study: Contact fatigue cracking in gears

We examine in this section the mechanics of contact fatigue crack initiation in case-hardened gear teeth. This case study, dealing with cracking in gears used in a heavy ground vehicle, is based on the work of Olsson (1998).

Figure 13.10(*a*) is an optical micrograph of a surface spall, which has the typical shape of a sea shell, formed during the repeated contact occurring on the tooth of an involute pinion. The pitch line of the gear is visible in the center of the figure (region marked A). The positive radial direction of the gear is towards the top of the figure in the vertical direction, which is also the direction of motion for tooth-to-tooth contact. Figure 13.10(*b*) is a view of the cross section made along the spalling symmetry plane, where also the 'wing cracks' at the bottom of the spalling crater are visible. The main cracks, that create the spall, advance at an angle, θ_{init}, to the surface. As the crack advances, its orientation to the surface changes. The irregular profile of the bottom of the spall crater, as well as the presence of 'wing cracks', which kink out of the main crack path, are also observed in this figure.

As shown schematically in Fig. 13.11(*a*), the point of contact in the driving gear commences at the root of the gear tooth and progressively moves to the outer tip of the tooth during operation. For the gear which is the follower, the reverse situation occurs, i.e. the point of contact moves from the outer tip to the root of the tooth. The corresponding gear teeth contact forces are sketched in Fig. 13.11(*a*), and the relative sliding velocity as a function of contact position is shown in Fig. 13.11(*b*) for the driver. As seen in this figure and in Fig. 13.2, pure rolling occurs along the pitch line. Above the pitch line, the friction force acts in the positive radial direction causing positive slip (i.e. tangential force directed toward the tip of the gear tooth), whereas below the pitch line it is oriented in the negative radial direction causing negative slip (toward the root of the gear tooth), Fig. 13.11(*a*). Regions experiencing negative slip are known to be more susceptible to cracking than the positive slip region.

Table 13.2. *Some prominent damage processes in contact fatigue.*

Terminology	Description
Galling	Transfer of material in macroscopic patches from one contacting surface to the other, and possibly back to the first surface. This process occurs under high tractions acting over multi-asperity dimensions. Consequences include: increased vibrations, positioning errors and lubricant contamination.
Surface distress	Micro-scale spalling fatigue of rolling contact surfaces due mainly to asperity-level normal contact loads. This damage process is aided by burnished surface areas, asperity-level microcracks, and micro-spall craters. Can cause spalling fatigue.
Spalling	Macro-scale Hertzian contact fatigue leading to the formation of macroscopic craters in the contact region. Can cause fractures and seizures.
Pitting	Formation of microscopic surface notches and cavities due to the combined action of contact fatigue and environmental interactions. See Fig. 4.14 for an example of pitting and fatigue crack initiation from the pit due to the combined action of fretting fatigue and environment.

Figure 13.12(*a*) shows the microcracks which initiate first on the gear teeth in the region where large slip (i.e. away from the pitch line) occurs. These cracks, typically 10–30 μm in depth, are oriented roughly at 45° to the contacting surfaces; they are numerous in the region of negative slip and very much less frequent at locations of positive slip. Closer to the pitch line, where essentially pure rolling takes place with little or no slip, the crack initiation angle with respect to the free surface is much smaller, as shown in Fig. 13.12(*b*). The vast majority of these surface microcracks in the same region of the gear teeth have roughly the same orientation angle and length. One may, therefore, infer that their initiation conditions are governed primarily by the main loading mode, which is Hertzian spherical contact (which simulates asperity contact) with either small or large frictional slip (depending on the location with respect to the pitch line). The secondary mode of failure is the development of wing cracks which branch away from the main crack and lead to spallation.

Many factors, in general, govern the onset and advance of the spall cracks during the contact fatigue of gear teeth (e.g., Murakami, Sakae & Ichimaru, 1994; Olsson, 1998). These factors include:

(1) The local geometry of the tooth,
(2) Contact pressure,
(3) The extent of positive or negative slip (i.e. the relative velocity between the mating gear teeth), which increases linearly from the pitch line to the outer tip or to the root, respectively,
(4) The nature of elastohydrodynamic lubrication, including the pressure dependent viscosity, thickness and contamination of the lubricant,

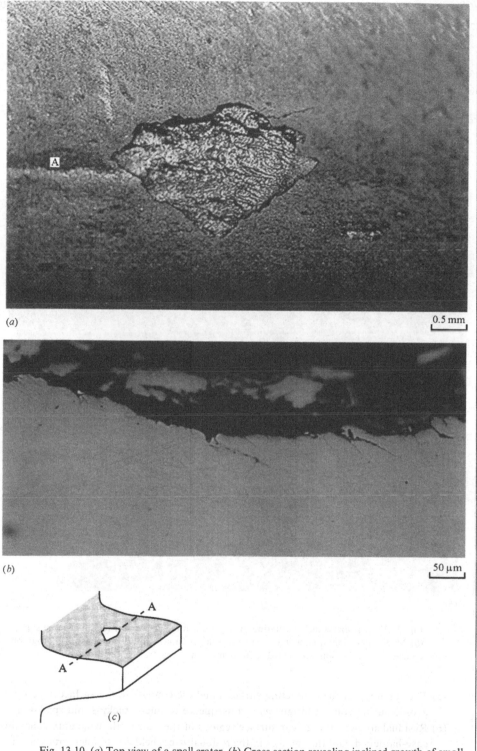

Fig. 13.10. (*a*) Top view of a spall crater. (*b*) Cross section revealing inclined growth of small cracks and wing cracks branching out of the main crack. (*c*) A schematic showing the location and orientation of the spall crater in (*a*) on the gear tooth. (Photographs courtesy of M. Olsson. Reprinted with permission.)

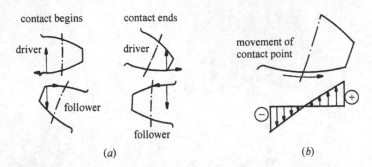

(a)   (b)

Fig. 13.11. (a) Contact forces on the gear teeth on either side of the pitch circle. (b) The relative velocity of the contact point of the driver.

(a)   10 μm

(b)   10 μm

Fig. 13.12. (a) Microcracks initiating at the contacting surface in a region of large slip. (b) Microcracks initiating at the contacting surface in a region of small slip. (Photographs courtesy of M. Olsson. Reprinted with permission.)

(5) The topology of the contacting surfaces and the changes in the surface asperity profile during contact fatigue as a consequence of microcracking and spallation,

(6) Residual stresses in the near-surface regions of the gear teeth, which are generally case-hardened, and the changes in the residual stress field due to microcracking and spallation,

(7) Embrittling effects of corrosive environments, if any, including pitting, and

(8) Debris particles trapped within the gear teeth which promote abrasive wear.

While each of these factors can, in general, play an important role in influencing the contact fatigue life of gear teeth, a simple analysis of the mechanics of crack initiation and of the initial direction of cracking is considered here to provide a rationale for the observations shown in Fig. 13.10. For this purpose, we begin with a consideration of Hertzian spherical contact with interfacial friction, where cracking is postulated to be governed primarily by the trajectory of the principal tensile stress. Here spherical contact is envisioned as that occurring between the asperities on one gear tooth (with approximately spherical tip) and the microscopically flat surface of the mating gear tooth. Figure 13.13(a) schematically shows this geometrical idealization along with the orientation angle ϕ of the maximum principal tensile stress plane (on which the principal stress σ_1 acts) with respect to the vertical axis z. From simple Mohr's circles construction, the orientation angle of the principal stress plane ϕ is found, in terms of the shear stress σ_{rz} and the normal stresses σ_z and σ_r, to be

$$\tan 2\phi = \frac{2 \tan \phi}{1 - \tan^2 \phi} = \frac{2\sigma_{rz}}{\sigma_z - \sigma_r}. \tag{13.45}$$

Solving this equation for $\tan \phi$, it is found that

$$\tan \phi = \frac{dr}{dz} = \left[\left\{ \frac{\sigma_z - \sigma_r}{2\sigma_{rz}} \right\} \pm \sqrt{1 + \left\{ \frac{\sigma_z - \sigma_r}{2\sigma_{rz}} \right\}^2} \right]^{-1}. \tag{13.46}$$

For $r \gg z$, $\mathcal{R} = \sqrt{(r^2 + z^2)} \approx r$ and $(1/\tan 2\phi) \gg 1$, and $dr/dz \approx (\tan 2\phi)/2$.

For a start, we view the contact loading from outside the contact area and approximate the contact loads by a concentrated normal force P and a concentrated tangential force μP where μ is the friction coefficient, Fig. 13.13(b). It is then postulated that a fatigue crack initiates at a radial distance r_0 (which is of the order of the contact radius, a). We then invoke the criterion that as the fatigue crack penetrates the substrate, its orientation, defined by the angle θ_{init} in Fig. 13.13(b), is normal to the local maximum principal tensile stress. The stresses in the substrate, i.e. σ_r, σ_z and σ_{rz}, are approximated

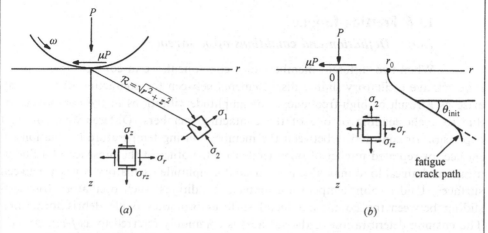

Fig. 13.13. (a) Schematic diagram showing the stresses due to asperity contact on a gear tooth during tractive rolling. (b) The postulated path of a fatigue crack and the associated nomenclature.

by superposing the well-known Boussinesq (1885) solution (for the concentrated normal load, P) and the Cerruti (1882) solution (for the concentration tangential force, μP). These stresses are then substituted in Eq. 13.46 to obtain

$$\frac{dr}{dz} \approx \frac{z}{r}\left[\left\{\frac{1-2\nu}{3\mu}+\frac{2(1+\nu)}{3}\right\}\right]^{-1}. \tag{13.47}$$

Integrating this equation with respect to both r and z, the function $z(r)$ is found. Invoking the limiting case, $z \to \infty$, the asymptotic value of the initial orientation of the fatigue crack with respect to the free surface is obtained:

$$\theta_{\text{init}} = \frac{dz}{dr} \approx \tan^{-1}\left[\left\{\frac{1-2\nu}{3\mu}+\frac{2(1+\nu)}{3}\right\}^{1/2}\right]. \tag{13.48}$$

For steel–steel contact, $\nu = 0.3$ and $0.3 \leq \mu \leq 1$ in the region of large slip. These values give: $45° \leq \theta_{\text{init}} \leq 49°$. This estimation of θ_{init} is close to the initial crack angle seen in Fig. 13.12(a) for large slip.

Along or in the immediate vicinity of the pitch line of the gear, where essentially pure rolling conditions prevail with little slip, Eq. 13.46 can be solved for the appropriate rolling contact conditions without friction using only the Boussinesq solution. This gives the result

$$\theta_{\text{init}} \approx \tan^{-1}\left[\left\{\frac{1-2\nu}{3}\right\}^{1/3}\right]. \tag{13.49}$$

For $\nu = 0.3$, $\theta_{\text{init}} = 27°$ for pure rolling.† This prediction is in agreement with the crack initiation angle seen in Fig. 13.12(b) for rolling with small slip. This case study thus serves to illustrate how approximate quantitative guidelines can be developed to rationalize experimentally observed cracking patterns and spallation processes in gear teeth subjected to contact fatigue.

13.6 Fretting fatigue

13.6.1 Definition and conditions of occurrence

When two different members of a component are in contact, small-amplitude relative oscillatory sliding displacements between the contacting surfaces may arise as a result of high-frequency, low-amplitude vibrations in the component or due to cyclic loading of one of the contacting members. Differences in thermal expansion or contraction between the members during temperature fluctuations in service, or repeated impact of one member on the other in the presence of a fluctuating mechanical load may also lead to small-amplitude oscillatory contact between surfaces. Under room-temperature ambient conditions, such oscillatory frictional sliding between the contacting metal surfaces promotes oxide debris formation. The ensuing deterioration of the surfaces is commonly referred to as *fretting wear*.

---

† Observe that Eq. 13.49 can not be obtained by simply setting $\mu = 0$ in Eq. 13.48 because of the manner in which the limit $z \to \infty$ influences the final result.

In the presence of an aggressive environment, such degradation is termed *fretting corrosion*. When the repeated tangential displacement also facilitates a reduction in the fatigue endurance limit and promotes an earlier nucleation and/or growth of fatigue cracks, the resulting damage is referred to as *fretting fatigue*. Examples of some prominent engineering applications where fretting fatigue is of concern are listed in Table 13.3.

As noted in Section 1.1, early studies of fretting (Eden, Rose & Cunningham, 1911; Tomlinson, 1927) and fretting fatigue (Warlow-Davies, 1941; McDowell, 1953) documented the deleterious effects of oscillatory sliding contact on cyclically loaded structures. The large body of experimental work carried out since these early studies has established that the deterioration in the fatigue resistance of a material due to fretting occurs by complex synergistic interactions from such factors as (see, for reviews, Waterhouse, 1992; Hills & Nowell, 1994; Waterhouse & Lindley, 1994; Fouvry *et al.*, 1996; Lindley, 1997):

- the amplitude of the cyclic slip displacement (typically 5–50 μm),
- the mismatch in the elastic and plastic properties between the contacting surfaces,
- cyclic frequency, waveform and hold periods,
- the normal contact pressure,
- the coefficient of friction between the fretted surfaces which is strongly influenced by the roughness of the surface asperities,
- environment and temperature,
- residual stresses induced by surface modification techniques such as shot-peening or coating, or by heat treatments, welding and other joining operations,
- mechanical loads imposed on one or both members engaged in fretting contact, and
- microstructural changes and phase transformations, if any, produced by the local temperature rise in the vicinity of the fretted surfaces.

13.6.2 Fretting fatigue damage

Laboratory simulations of fretting fatigue usually involve a gage section of a specimen which is subjected to uniaxial fluctuating stresses under load or displacement control, with a portion of the gage length of the specimen subjected to a transverse normal contact via a bridge-type pad, a sphere or a rounded-tip punch.† Generally, two contact points transverse to the specimen axis are used to ensure symmetry of the loading system. The frictional forces at the contact surfaces should be monitored in order to derive an independent measure of the friction coefficient and its variation as a function of the number of fatigue loading cycles. The extent of

---

† Fretting contact arising from a sphere or a rounded-tip punch results in finite stresses at the interface, whereas a flat-ended punch induces a stress singularity at the sharp corners.

Table 13.3. *Examples of practical applications where fretting fatigue failures are known to occur.*

Application	Location of fretting fatigue failure
Steam or gas turbine engines	• The seating of the turbine disk on the drive shaft in land-based turbines • The dove-tail or fir-tree section of aircraft jet engines where the blade is attached to the disk • The flanged joint between the disk/bevelled gear and the drive shaft in gas turbine transmission system in helicopters • Rotor-tooth contact land area in the vicinity of the gaps between the stiffness compensation wedges in axially slotted turbogenerator rotors (see Section 13.7 for a case study).
Cabes and ropes	• Fretting and rubbing in interwire contacts, with fluctuating loads arising from wind or water flow Examples include: • Electric power transmission lines • Ropes used for cable cars • Hawsers and mooring ropes in marine environments
Joints in fatigue-loaded structures	• Rivets in wing/fuselage structures of aircraft • Splines and keyways in shafts in couplings • Shrink-fitted components
Bio-implant devices	• Fracture on the underside of the screwhead and the countersink of the hole in the base plates that are screwed into the bone • Fretting fatigue at the femoral stem in total hip replacements where the lesser trochometer bone is wired around the stem of the femoral component

partial slip displacement can be easily controlled by adjusting the specimen geometry, the contact pad geometry, and the normal load associated with the contact. A review of the experimental and theoretical methods employed to simulate fretting fatigue damage can be found in Hills & Nowell (1994), Waterhouse & Lindley (1994) and Szolwinski & Farris (1996).

The conditions typical of fretting fatigue are such that partial slip develops at the periphery of contact area; the mechanics of microslip is completely described by the analyses presented in Sections 13.3.2 and 13.3.3. Repeated sliding in the partial slip region which surrounds the stick regime of the contact area generally leads to oxidation of the fretted surface. This results in wear debris formation and cracking. In addition, microplasticity may arise in the partial slip region as a result of asperity contact or when high friction conditions prevail at the contact interface. (Recall from Section 13.3 that the site of maximum effective stress during sliding moves to the surface when the friction coefficient is high.) The region of microslip, therefore, has a distinctly different 'worn out' appearance when compared to the remainder of the contact area.

Figure 13.14(a) schematically shows the annular region of the fretting scars which develops within the partial slip regime at the outer rim of the contact area. For a fixed normal force and cyclic variations in the tangential tractions between the fixed limits, $-Q_{max}$ and $+Q_{max}$, the partial slip displacements for fretting contact between a sphere and a planar surface are given by Eq. 13.36. Figure 13.14(b) is a plan view of a fretting scar produced in a 6061–T6 aluminum alloy subjected to a contact by a sphere of the same alloy (of tip radius 12.7 mm) under a fixed normal force (15 N) and cyclic tangential stress (± 13 MPa) with a friction coefficient $\mu \approx 0.6$ after

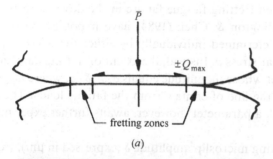

(a)

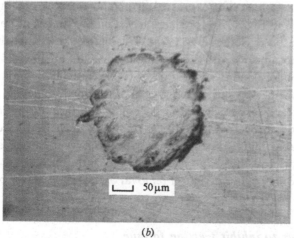

(b)

Fig. 13.14. (a) A schematic representation of fretting scars produced under a fixed normal load $\overline{P}$ and fluctuating tangential force $-Q_{max} \leq Q \leq +Q_{max}$. (b) A plan view of the fretting annulus observed in a 6061–T6 aluminum alloy. (Photograph courtesy of P.B. Birch, J. Dominguez & B. Wittkowsky, Massachusetts Institute of Technology.)

1.7×10^5 fatigue cycles; the aluminum alloy substrate was simultaneously subjected to a fully reversed cyclic stress of ± 13 MPa.

The energy dissipated within the region of partial slip is a source of vibration damping. During fretting fatigue, the maximum amplitude of the tangential force, $Q_{max} \ll \mu \bar{P}$. Under these conditions, the energy loss associated with frictional microslip, Eq. 13.37 reduces to

$$[\Delta W]_{\text{fretting}} = \left\{ \frac{(2 - v)(1 + v)}{9\mu \bar{P} a E} \right\} Q_{max}^3. \tag{13.50}$$

In other words, the energy loss associated with frictional microslip is proportional to the cube of the cyclic tangential force amplitude, Q_{max}. A comparison of Eqs. 13.50 and 13.36 reveals that the fretting microslip displacement is nearly proportional to the square of Q_{max}.

While the energy loss, Eq. 13.50, can potentially be used to develop a measure of fretting fatigue damage, other ad-hoc formulations that are amenable for easier implementation into component design have been postulated. For example, on the basis of their studies of simulated fretting fatigue failure in the dovetail sections of aircraft jet engines, Ruiz, Boddington & Chen (1984) have hypothesized that the overall fretting damage is not determined individually by either the micro-slip displacement δ, or the interface shear stress q. Instead, the extent of fretting damage was taken to be given by a parameter whose magnitude equals $\sigma_{\text{fat}} \cdot \delta \cdot q$, where σ_{fat} is the surface stress introduced along the line of contact from the fatigue loads. The clear physical basis for the use of such a parameter, however, awaits further experimentation.

The relationships among fretting microslip amplitude δ (expressed in μm), fretting wear volume loss, and the fatigue life (expressed in number of cycles to initiate a critical flaw, N_i) are plotted in Fig. 13.15 for steels, based on the studies of Sproles & Duquette (1978) and Vingsbo & Soderberg (1988). This figure shows that the wear volume loss increases monotonically with the slip amplitude, with the rate of increase being more pronounced under conditions of gross slip. The fatigue life, however, is most severely affected in the partial slip region typical of fretting fatigue (i.e. for $\delta \sim$ 8–20 μm), where fatigue cracks are nucleated at the contact surface during repeated sliding. (See the case study in Section 13.7 for examples of such fretting cracks.) During full sliding, the rapid removal of material by fretting wear can cause these fatigue cracks to disappear, and thus can ostensibly result in an increase in fatigue life, since newer cracks should be initiated after removal of the surface layer by wear.

13.6.3 Palliatives to inhibit fretting fatigue

Several approaches are commonly employed in industrial practice to combat damage and failure arising from fretting fatigue. These can be broadly classified into two groups:

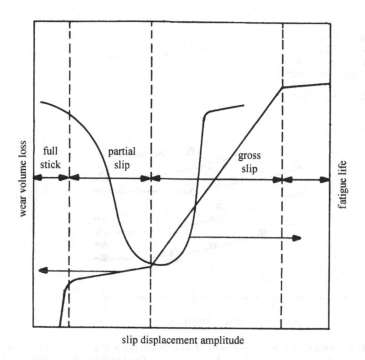

full stick partial slip gross slip

wear volume loss

fatigue life

slip displacement amplitude

Fig. 13.15. A schematic representation of the dependence of fretting wear and fretting fatigue life on the cyclic slip amplitude.

(1) Methods by which a layer of compressive residual stresses is induced in the vicinity of the surface which undergoes fretting fatigue. These methods include: shot-peening, cold working, and laser shock peening.

(2) Methods by which the coefficient of friction between the fretted surfaces is reduced by introducing lubricants such as polytetrafluoroethylene (PTFE or Teflon), molybdenum disulphide (MoS_2).†

Figure 13.16 shows the beneficial effects of different palliative treatments on the total fatigue life (characterized by the S–N curves) of a 2014 aluminum alloy fretted with a 3.5%NiCrMoV steel bridge pad. Here, the use of both shot-peening and lubrication leads to a much better improvement in total fatigue life than either peening or lubrication only. It is, however, clear that the total fatigue life in the absence of fretting is still superior to that with fretting where both peening and lubrication are used as palliatives.

---

† Caution should be exercised in deciding to introduce lubricants as palliatives, for the following reasons. (a) Repeated fretting over tens of thousands or more cycles inevitably leads to the removal or disappearance of the lubricant in most practical situations, thereby eventually enhancing the interfacial friction. (b) The slip displacement amplitude depends in a complex manner on the friction coefficient (Eq. 13.36). Experiments show (e.g., Lindley & Nix, 1992) that for a fixed amplitude of fatigue loading, the slip amplitude increases with increasing effectiveness of lubricant. Such an increase in slip amplitude may, in fact, introduce conditions of partial slip whereby the fretting fatigue damage is exacerbated (see Fig. 13.15).

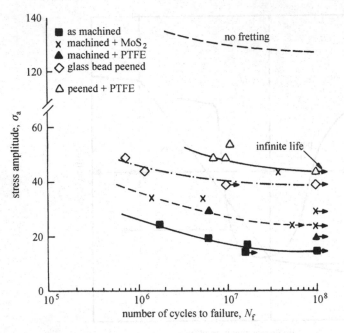

Fig. 13.16. The effects of different palliative treatments on the S–N curves for a 2014 aluminum alloy fretted with a 3.5%NiCrMoV steel bridge pad. (After Lindley & Nix, 1992.)

A three-dimensional computational simulation of fretting fatigue under conditions of partial slip has been carried out by Giannakopoulos & Suresh (1998). They examine the effects of superimposed compressive stresses on the evolution of stress fields in the vicinity of the contact area, in an attempt to rationalize the beneficial effects of shot-peening in fretting and to develop guidelines for the introduction of compressive residual stresses. It is found that when a zero mean stress or a tensile mean stress is superimposed on the fatigue specimen subjected to cyclic stresses and fretting contact, the location of the maximum principal tensile stresses occurs in the wake of the sliding contact. When a compressive mean stress is imposed, however, the location of the principal tensile stress moves to the front of the contact region. Thus, the onset of cracking is preferentially moved to the leading edge of fretting contact in the presence of compressive residual stresses, a phenomenon experimentally observed by Kuno *et al.* (1989). Giannakopoulos & Suresh also note that the compressive mean stresses diffuse the effective stresses along the slip direction, thereby reducing the depth to which these stresses penetrate into the substrate under fretting.

The characterization of the detrimental effects of fretting fatigue is made in Fig. 13.16 by recourse to total life approaches. We note, however, that fracture mechanics method can also be applied to the study of cracking induced in fretting fatigue. Such defect tolerant approaches are considered in the example problem and the case study to be presented in the following sections.

13.6.4 Example problem: Fracture mechanics methodology for fretting fatigue fracture

Problem:†

The singular stresses which develop at the edges of contact between a sharp-ended pad and a planar surface can be determined from the asymptotic crack-tip fields for equivalent geometries of cracked bodies. This asymptotic matching technique, described in detail in Giannakopoulos, Lindley & Suresh (1998), can be used in some cases to derive a life prediction methodology for fretting fatigue. Conditions under which such a 'crack analogue' is applicable include the contact between a rigid punch and a flat surface, or situations where the combination of elastic properties of the sharp-edged punch and the planar surface lead to square-root singularities.

Consider the two-dimensional contact between a flat-ended, straight-sided punch which is fretted against a planar substrate with large outer dimensions, Fig. 13.17(a). The normal contact force per unit contact length (which is normal to the plane of the figure) is P and the tangential contact force per unit contact length is Q. Assuming essentially incompressible deformation, the distribution of normal contact pressure, $p(x)$, and the tangential traction, $q(x)$, for this problem are known to be (e.g., Johnson, 1985)

$$p(x) = \frac{-P}{\pi\sqrt{a^2 - x^2}}, \qquad q(x) = \frac{Q}{\pi\sqrt{a^2 - x^2}}. \tag{13.51}$$

The loading conditions in Fig. 13.17(a) are such that P is constant while Q cyclically varies in the range $-Q_{max} \leq Q \leq +Q_{max}$. The width of the punch is $2a$. It may be assumed that the plane-strain contact fatigue problem considered here can be treated primarily as linear elastic and that any plastic zones which might develop at the contact boundaries are much smaller than the contact width $2a$.

(i) The asymptotic singular fields which develop at the boundaries of contact in Fig. 13.17(a) can be determined by recourse to the asymptotic stress fields at the tip of a double edge-cracked specimen which is schematically shown in Fig. A.1(c). For this purpose, the width of the contact zone $2a$ in Fig. 13.17(a) is made equal to the uncracked ligament length ($W - 2a$) in Fig. 13.17(b), and the crack length a is assumed to be infinitely large in Fig. A.1(c) because of the large dimensions of the substrate in Fig. 17(a). (If W is the width of the fretted substrate in Fig. 13.17(a), then the crack length a in Fig. A.1(c) is identified with ($W - 2a$)/2 in Fig. 13.17(a).) This analogy is illustrated schematically as the double edge-cracked plate in Fig. 13.17(b) which is loaded with a compressive normal (mode I) load P (per unit length) and an in-plane (mode II) shear load Q (per unit length).

† This example problem is derived from Giannakopoulos, Lindley & Suresh (1998).

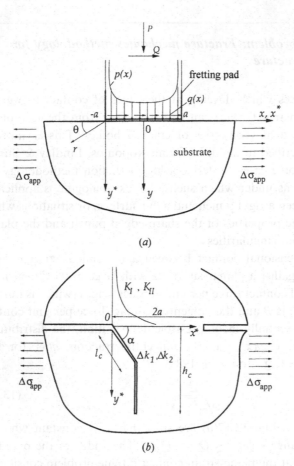

Fig. 13.17. (*a*) Schematic of the contact between a flat-ended punch and a planar surface of the same material, and the associated nomenclature. (*b*) Equivalent geometry of a double edge-cracked plate and the definitions of various geometrical and loading parameters associated with fretting fatigue fracture.

By asymptotically matching the stress fields at the boundaries of the contact in Fig. 13.17(*a*) with the corresponding crack-tip stress fields in the equivalent crack geometry shown in Fig. 13.17(*b*), find the stress intensity factors K_I and K_{II} for the cracked plate in Fig. 13.17(*b*).

(ii)   The contact conditions are such that $P = 39.4$ kN/m, while Q cyclically varies in the range $-Q_{max} \leq Q \leq +Q_{max}$, with $Q_{max} = 16$ kN/m, and $2a = 12.7$ mm. It is known from fatigue crack growth experiments that the threshold stress intensity factor range, ΔK_0, for small fatigue cracks (of characteristic length typically smaller than 1.5 mm) varies with load ratio R in the following manner: $\Delta K_0 = (2.7 - 1.9R)$ MPa$\sqrt{m}$. It is found experimentally that a fatigue crack of length, $l_c = 0.64$ mm in Fig. 13.17(*b*), initiates at the left contact edge at an angle α. If it can be assumed that the continued growth of this fatigue crack occurs along a direction where $\Delta k_2 = 0$, and that linear elastic fracture mechanics can be applied to characterize fatigue crack growth, find (a) the angle α and (b) whether the crack would arrest or continue to advance.

(iii) Now consider the situation where the contact loads are significantly increased: $P = 176.5$ kN/m, while Q cyclically varies in the range $-Q_{max} \leq Q \leq +Q_{max}$, with $Q_{max} = 236.2$ kN/m. The contact width $2a$ remains the same as before. The fretted substrate is now additionally subjected to a uniform cyclic tensile stress which fluctuates in the range 110 MPa $\leq \sigma_{app} \leq 140$ MPa. In this case, the fatigue crack would initially start from the contact edge at an angle α under the influence of P and ΔQ. However, once it propagates over some critical length l_c, its continued growth is expected to be governed solely by the applied uniform fatigue stress $\Delta \sigma_{app}$. Consequently, the crack would be expected to deflect in such a way that it advances perpendicularly to the direction along which $\Delta \sigma_{app}$ is applied, as shown in Fig. 13.17(b). (a) For these conditions, find α and l_c. (b) Does the crack arrest in this case?

(iv) For the loading conditions in part (iii), determine the total fretting fatigue life if the fracture toughness of the substrate, $K_{Ic} = 95$ MPa$\sqrt{m}$ and if its crack growth rate (over the entire range of crack advance considered here for any R ratio) can be characterized by the Paris law, $dl/dN = C(\Delta k_1)^m$, where $C = 10.8 \times 10^{-12}$ (MPa)$^{-3}$ m$^{-1/2}$, $m = 3$ and l is the appropriate crack length dimension.

Solution:

(i) In order to carry out asymptotic matching between Figs. 13.17(a) and (b), move the coordinate system from the center of contact to the edge of contact in Fig. 13.17(a). In this new coordinate system, $x^* = x - a$, and $y^* = y$. The x^*–y^* axes are also located at the crack-tip in Fig. 13.17(b). Now consider the asymptotic stress fields for the traction distributions given in Eq. 13.51 in the limit of $x^* \to 0$, from which it is seen that

$$\sigma_{yy} = \sigma_{y^*y^*} \to \frac{-P}{\pi\sqrt{2ax^*}}, \qquad \text{for mode I,}$$

$$\sigma_{xy} = \sigma_{x^*y^*} \to \frac{Q}{\pi\sqrt{2ax^*}}, \qquad \text{for mode II.} \qquad (13.52)$$

Using Eq. 9.52, we note that

$$\sigma_{yy} = \frac{K_I}{\sqrt{2\pi x^*}}, \qquad \sigma_{xy} = \frac{K_{II}}{\sqrt{2\pi x^*}}, \qquad \text{as } x^* \to 0. \qquad (13.53)$$

Comparing Eqs. 13.52 and 13.53, it is apparent that

$$K_I = -\frac{P}{\sqrt{\pi a}}, \qquad K_{II} = \frac{Q}{\sqrt{\pi a}}. \qquad (13.54)$$

These stress intensity factor results are exactly the same as those listed in stress intensity factor handbooks (e.g., Tada, Paris & Irwin, 1973) for the double edge-cracked plate geometry shown in Fig. 13.17(*b*). Note also that the above mode I stress intensity factor is the limiting value of Eq. A.4 for $a \to \infty$. (The negative value of K_I in Eq. 13.54 is a consequence of P being compressive. Implicit in this stress intensity calculation is the assumption that the crack is nonclosing under compression.)

Thus we have shown that there exists an equivalence, in the asymptotic limit, between the singular stress fields at the contact boundaries and those at the tips of cracks in analogous geometries. Consequently, known fracture mechanics solutions can be applied to contact mechanics problems or vice versa, under appropriate geometrical and loading conditions. It should be emphasized, however, that this equivalence of fields holds only in the asymptotic limit.

(ii) It is stated that the crack advances in a plane in which $\Delta k_2 = 0$. The local stress intensity factors for the deflected crack geometry can be determined from the solutions given in Eqs. 9.116 and 9.117.

(a) Specifically, use the expression for k_2 in Eq. 9.116, in conjunction with Eqs. (1) (where $P = 39.4$ kN/m, $\Delta Q = \pm 16$ kN/m, and $a = 6.35$ mm is one-half of the contact width), and solve for α by invoking the condition that $\Delta k_2 = 0$. Numerical solution yields that $\alpha = 36°$.

(b) Note from Eq. 9.116 that although P (and hence K_I) is fixed, k_1 fluctuates because K_{II} fluctuates. Using Eqs. 9.116 and 9.117 with $\alpha = 36°$, we find that, for the given loading conditions, $\Delta k_1 \approx 0.4$ MPa$\sqrt{\mathrm{m}}$. This value of the local mode I stress intensity factor range is much smaller than the threshold stress intensity factor range, ΔK_0, for any R ratio. Hence, we conclude that the small fatigue crack inclined at an angle of 36° to the surface would arrest.

Comment: Fretting fatigue experiments (Lindley & Nix, 1991) conducted on a rotor steel for the same geometrical and loading conditions as in the present example have shown that the small fatigue crack initiates at an angle of 30° to the fretted surface, and that it arrests after growth over a distance, $l_c \approx 0.6$ mm. See the case study in the next section for further details.

(iii) Use the expression for k_2 in Eq. 9.116, in conjunction with Eqs. 13.54 (where $P = 176.5$ kN/m, $\Delta Q = \pm 236.2$ kN/m, and $a = 6.35$ mm is one-half of the contact width), and solve for α by invoking the condition that $\Delta k_2 = 0$. Numerical solution yields that $\alpha = 57°$. Substituting this value of α in Eq. 9.116, it is seen that $\Delta k_1 = 7.2$ MPa$\sqrt{\mathrm{m}}$. Since this value of the mode I stress intensity factor range is significantly larger than ΔK_0, the fatigue crack is expected not to arrest.

As shown in Fig. 13.17(b), the fatigue crack which initiates at an angle of 57° to the surface would first propagate under the influence of P and ΔQ over a distance of l_c, before deflecting again to advance normally to $\Delta \sigma_{app}$. This transition in crack growth occurs when the local 'driving force' for fatigue crack growth due solely to $\Delta \sigma_{app}$ begins to exceed that due solely to P and ΔQ. At the critical length l_c, the local Δk_1 due to $\Delta \sigma_{app}$ is:

$$\Delta k_1 = 1.12 \Delta \sigma_{app} \left\{ \pi \cdot (l_c \sin \alpha) \right\}^{1/2} \geq 7.2 \text{ MPa} \sqrt{m}, \tag{13.55}$$

where $(l_c \sin \alpha)$ is the projected length of the fatigue crack normal to the uniform cyclic tensile stress. Using $\Delta \sigma_{app} = 125$ MPa, $\alpha = 57°$, we find that $l_c = 1$ mm.

Comment: Experiments by Lindley & Nix (1991) show that for the conditions of the calculations carried out in this part, $\alpha \approx 60°$.

(iv) The total number of fatigue cycles, N, prior to catastrophic failure comprises two parts: (1) the number of cycles, N_1, of fatigue crack growth along the initial angle α, and (2) the number of cycles, N_2, of fatigue crack growth perpendicular to $\Delta \sigma_{app}$, such that $N = N_1 + N_2$.

Integrating the Paris equation for the initial crack growth along the angle α, with $\Delta k_1 = 7.2$ MPa$\sqrt{m}$, $l_c = 1$ mm along with the given values of C and m, it is seen that

$$N_1 = \int_0^{l_c} \frac{dl}{C(\Delta k_1)^m} \approx \frac{l_c}{C(\Delta k_1)^m} = 2.48 \times 10^5 \text{ cycles.} \tag{13.56}$$

The fatigue crack would propagate perpendicularly to $\Delta \sigma_{app}$ until a crack length of h_c, as shown in Fig. 13.17(b), when the maximum stress intensity factor of the uniform tensile loading approaches the fracture toughness of the material, i.e. until

$$1.12 \sigma_{app,max} \sqrt{\pi h_c} \rightarrow K_{Ic}. \tag{13.57}$$

Substituting $\sigma_{app,max} = 140$ MPa and $K_{Ic} = 95$ MPa$\sqrt{m}$ into this equation, we find that $h_c = 117$ mm.

The number of cycles N_2 is found from

$$N_2 = \int_{l_c \sin \alpha}^{h_c} \frac{dl}{C(1.12 \Delta \sigma_{app} \sqrt{\pi l})^m}$$

$$= \frac{1}{C(1.12 \Delta \sigma_{app} \sqrt{\pi})^m} \cdot \frac{2}{m-2} \left\{ (l_c \sin \alpha)^{(2-m)/2} - h_c^{(2-m)/2} \right\}$$

$$= 2.42 \times 10^5 \text{ cycles.} \tag{13.58}$$

The total fatigue life then is $N = N_1 + N_2 = 4.9 \times 10^5$ cycles.

13.7 Case study: Fretting fatigue in a turbogenerator rotor

This case study deals with the failure of a 660 MW turbogenerator rotor in England during the 1970s as a result of fatigue cracking. The crack initiated and advanced by fretting fatigue in the rotor tooth contact land in the vicinity of the gaps between the longitudinal stiffness compensation wedges. This case study is predicated on the failure analysis by Lindley & Nix (1991).

13.7.1 Design details and geometry

The difference in flexural stiffness between the winding and pole axes of generator rotors is generally compensated by one of two methods. In one approach, transverse slots are machined in each pole face (the so-called 'cross-gashing' technique). In the other method, relevant to the present case study, longitudinal slots are machined along the entire length of the pole faces to facilitate stiffness compensation, Fig. 13.18(a). A consequence of this procedure is that the flexural stiffness is reduced. However, since high magnetic flux densities occur in the pole, steel filler blocks and retaining wedges, Fig. 13.18(b), are used to replace most of the machined away steel in order to minimize magnetic saturation. The rotor shaft is made of an AISI 3.5%NiCrMoV steel, while the wedges are made of an AISI 1%CrMo steel and the filler block, of mild steel.

Figure 13.19(a) shows the overall detail of the pole face slot and the location of the wedges and filler blocks. The cross section and elevation views of the wedge geometry are illustrated in Figs. 13.19(b) and (c), respectively. The gaps between the stiffness compensation slot wedges in the rotor tooth contact land were the sites for the initiation of fatigue cracks. There is no restriction, in principle, on the precise location of the wedges in the slots and on the gap between adjacent wedges. The wedge ends, however, were chamfered at the contact land, as shown in Fig. 13.19(c), to ensure a minimum wedge gap of approximately 1 mm. Figure 13.19(d) shows the likely distribution of frictional forces at the maximum load, and the location where a fretting fatigue crack initiated.

13.7.2 Service loads and damage occurrence

Most high-cycle fatigue problems, including fretting fatigue, in the rotors of turbogenerators arise from self-weight bending. Mean stresses vary with rotational speed due to centrifugal forces and to the changes in the level of shrink fit. The normal contact pressure varies with rotational speed; as the speed increases, the stresses due to shrink fit generally decline whereas the stresses due to centrifugal loading increase. This trend results in different fretting fatigue conditions at any one critical location during operation at the following three typical running speeds: (1) barring which is a low speed operation intended to avoid thermal distortion due to sagging which arises from self-weight bending;[†] (2) normal operation at 3000 rpm; and (3) overspeed testing at approximately

---

[†] With the relatively high thermal efficiency of the 660 MW turbogenerator, barring would be kept to a minimum. But, during maintenance periods, such as those arising during the repair of a boiler problem, the barring operation would be carried out.

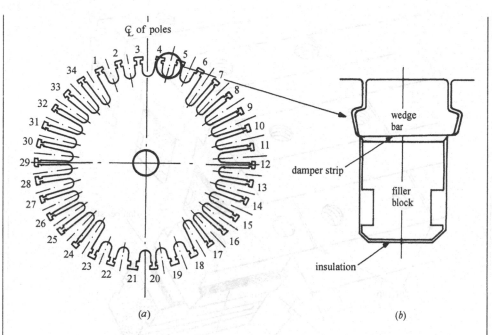

Fig. 13.18. (*a*) Schematic of a section through the longitudinally slotted pole face of a turbogenerator rotor. The outer diameter of the pole face is 1.12 m. (*b*) Details of the pole face slot showing the wedge and the filler block. After Lindley & Nix (1991).

3300 rpm. The contact pressure between the wedge and the rotor tooth contact land (Figs. 13.19(*a*) and (*b*)) during barring was estimated to be 30 MPa, while that during normal operation was 300 MPa.

The occurrence of a marked increase in the vibration levels of the generator rotor prompted ultrasonic inspection which confirmed the presence of several major cracks near the mid-span of the rotor, Fig. 13.20. Subsequently, two large 660 MW generators were taken out of service and were subjected to metallographic and fractographic analyses. It was discovered that the relative movement between the retaining wedge and the rotor teeth contact lands had led to the initiation of cracks by fretting fatigue, under the influence of the cyclic self-weight bending stresses.

Under typical service conditions at 3000 rpm, the retaining wedges effectively stick to the rotor contact lands under the high contact pressure ($\sim$ 300 MPa) arising from centrifugal loading. Given this 'stick contact', the free edge at the gap between the wedge and the rotor tooth (see Fig. 13.19(*d*)) becomes a site of stress concentration at which a small fretting fatigue crack (typically 0.25 mm deep) begins to advance into the rotor tooth. This is illustrated as stage 2 in Fig. 13.20(*b*). The bending of the tooth head, as a consequence of the centrifugal forces applied by the wedges and filler blocks, induces high tensile stresses locally around the tooth neck fillet radius. These tensile stresses cause continued growth of the fretting fatigue crack, which is marked as stage 3 in Fig. 13.20(*b*). The presence of high tensile residual stresses introduced by slot machining caused crack growth in stage 4 both across and down the tooth neck. Residual stress measurements were made both by employing the X-ray method and by hole-drilling at the center of a strain gage rosette to obtain the three-dimensional profile. These measurements indicated

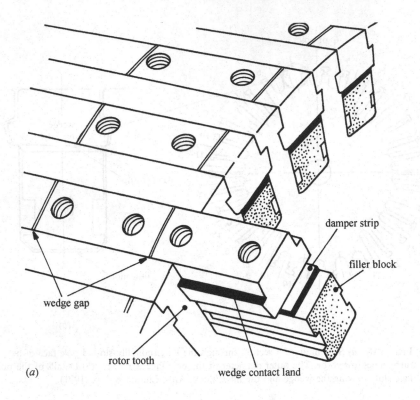

damper strip

filler block

wedge gap

rotor tooth

wedge contact land

(a)

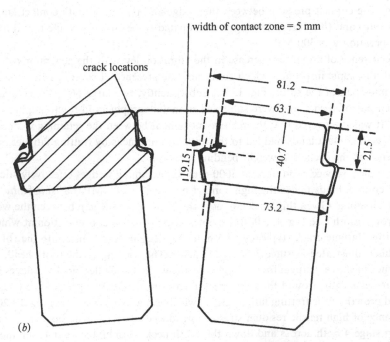

width of contact zone = 5 mm

crack locations

81.2

63.1

119.15

40.7

21.5

73.2

(b)

Fig. 13.19. Details of the fretting fatigue location. (a) Overall view of the pole face slot. (b) Cross section view.

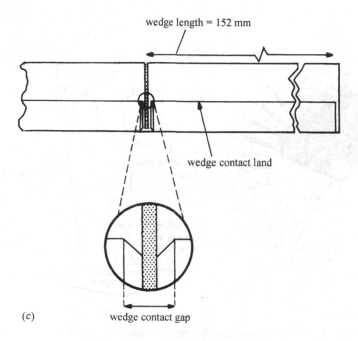

wedge length = 152 mm

wedge contact land

(c)

wedge contact gap

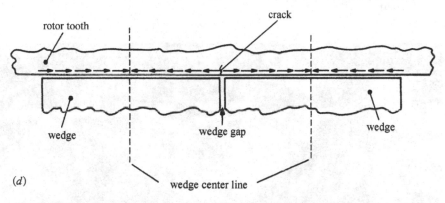

rotor tooth

crack

wedge

wedge gap

wedge

(d)

wedge center line

Fig. 13.19 *(continued)* (c) elevation view of the wedge geometry. (d) Likely friction force distribution in the rotor tooth and crack initiation site. All dimensions are in mm. (Courtesy of T.C. Lindley.)

a peak tensile stress of 370 MPa, decaying to zero at a depth of 3 mm beneath the surface. (Had these tensile stresses not been present, the crack initiated by fretting fatigue would have arrested in stage 4 in the tooth neck region, as inferred from fracture mechanics analyses. See the example problem in the preceding section.) Once the crack reached the bottom of the pole teeth in Stage 5, it was evident from both fracture analysis and beach mark observations that the final growth to the rotor bore had taken place rapidly, within a matter of a few thousand hours. The photographs in Fig. 13.21 show the actual fretting fatigue failure processs described above.

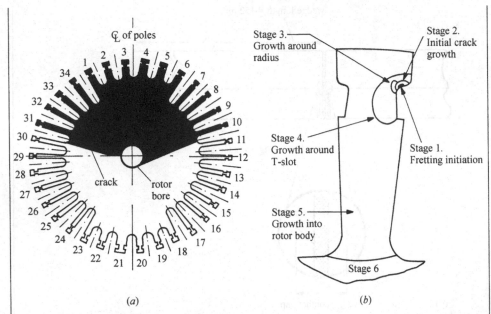

Fig. 13.20. (a) Regions of major cracking in the rotor mid-span position where the maximum bending moment occurs. (b) Different stages of crack development. After Lindley & Nix (1991).

Fig. 13.21. (a) An overall view of the fatigue failure in the pole face slot near the rotor mid-span.

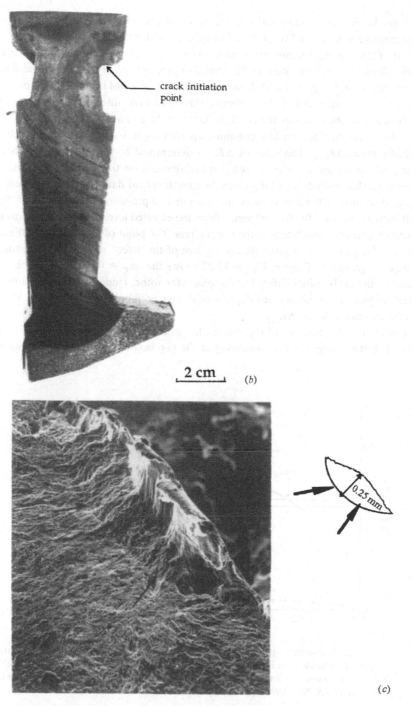

Fig. 13. 21 (*continued*) (*b*) Crack development from the tooth–wedge contact land. (*c*) The initiation of a small fretting fatigue crack, very similar to that found in the cracked rotor, in a laboratory simulation of the failure. The arrows denote the crack tip. (Photographs courtesy of T.C. Lindley.)

On the basis of computational stress analyses of the rotor components, under representative service conditions, and from the estimation of local effective stress intensity factors using the numerical methods available at the time of failure analysis, Lindley & Nix (1991) estimated the likelihood of propagation of fretting fatigue cracks in the generator rotor. Figure 13.22 shows the estimated values of the effective applied stress intensity factor range, ΔK_{app}, and effective threshold stress intensity factor range, ΔK_{th}, as a function of the depth of the crack initiated at the contact land (Fig. 13.19(d)), with and without fretting, for loading conditions pertaining to barring. Note that fretting markedly raises ΔK_{app}. The value of ΔK_{th} is determined by the material, and the local stress fields which influence the effective mean stress; these threshold stress intensity values were determined from long crack fatigue threshold data for the rotor steels subjected to mode I tension–tension and tension–compression fatigue loading.† The small increase in ΔK_{th} for fretting stems from the elevated level of compressive stress due to contact pressure, which lowers the mean stress. The point of intersection of the ΔK_{app} vs a and ΔK_{th} vs a curves gives the critical size of the defect, a_{cr}, which is capable of subsequent growth by fatigue. Figure 13.22 shows that $a_{cr} \approx 0.25$ mm, a crack size similar to the likely initial defect for the generator rotor. Under nonfretting conditions, several millimeters of initial crack depth would have been necessary for the crack to grow by fatigue, since $\Delta K_{th} > \Delta K_{app}$.

In summary, the analyses of fracture in the generator rotor indicated that the growth of small fretting fatigue cracks occurring at the gap between the longitudinal stiffness

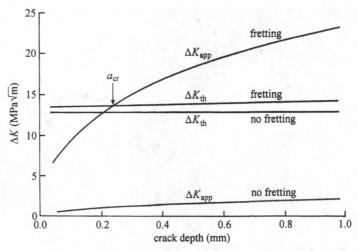

Fig. 13.22. Estimates of the effective applied stress intensity factor range, ΔK_{app}, and effective threshold stress intensity factor range, ΔK_{th}, as a function of the depth of the crack initiated at the contact land. The plots show the crack driving force with and without fretting. After Lindley & Nix (1991).

† Although these estimates could be improved somewhat by considering small crack threshold data and mixed-mode effective threshold stress intensities representative of local fretting conditions, the simple criteria adopted in the original failure analyses appear to have captured the salient features of the actual failure processes reasonably accurately.

compensation wedges was likely to occur under barring conditions. The critical defect size, $a_{cr} \approx 0.25$ mm, for the continued advance of the crack during barring was comparable to the size of the fretting defects likely to be present at the rotor tooth contact land, due to cyclic slip (fretting) between the wedge and the contact land during barring when the contact pressure was low. The analyses thus predicted that the rotor would be under a high risk of fretting fatigue failure at this location, consistent with the actual occurrence during the service life of the generator rotor.

As a consequence of the above failure in the generator rotor arising from fretting fatigue at the ends of short wedge bars, the design was changed to continous wedge bars, with no gaps, made from a high strength aluminum alloy to avoid further failure occurrence.

Exercises

13.1   Frictionless normal indentation of a surface by a sphere was considered in Section 13.2.1.1.

(a) Prove the statement in that section that the maximum tensile normal stress is

$$\sigma_r = \frac{1-2v}{3}p_{max}. \qquad (\text{E13.1})$$

(b) Show that it occurs at the periphery of contact, $r = a$.

(c) Discuss the consequences of this maximum stress for fatigue cracking in (a) polycrystalline aluminum oxide, (b) PMMA and (c) commercially pure aluminum during cyclic indentation.

(d) Describe qualitatively how a layer of residual compressive stress at the surface would affect the magnitude and location of this maximum stress.

13.2   The surfaces of metallic materials are usually shot-peened for the purpose of suppressing crack formation due to contact fatigue.

(a) Suggest at least three methods for enhancing the contact fatigue resistance of ceramic surfaces.

(b) List the advantages and disadvantages of each method.

13.3   Consider the example problem in Section 13.2.3.1 where a planar surface with a residual compressive stress is subjected to cyclic normal indentation without friction. If the equi-biaxial residual compressive stresses at the indented surface were to be $\sigma_x|_R = \sigma_y|_R = -0.5\sigma_y$, where σ_y is the yield strength of the material, calculate how much higher normal indentation force P can be applied without plastically yielding the surface as compared to the case where the surface has no residual compressive stresses.

13.4   Compare the shakedown limits for pure rolling (no friction) with those for tractive rolling (with a high friction coefficient). Which is higher? Why?

13.5   List at least four practical applications where each of the following contact fatigue damage processes is possible: (a) galling, (b) spallation, and (c) pitting.

13.6   Consider the example problem in Section 13.6.4. If the substrate which is contact-loaded with a punch were to be of finite width $W = 6a$ (where $2a$ is the width of the contact zone), instead of being infinitely wide, repeat part (i) of that problem and find the mode I and mode II stress intensity factors for the equivalent crack analogue. You may assume that the depth of the substrate beneath the punch is infinitely large.

13.7   Consider a circular cylinder which is pressed 'end on' against a flat surface of the same material.

(a) Find an equivalent geometry of a cracked body where the asymptotic crack-tip fields are the same as those around the contact zone boundary between the cylinder and the flat surface.

(b) Determine the distribution of crack-tip fields using fracture mechanics solutions and compare them with the solutions given in any monograph on contact mechanics. Show that they are equal in the asymptotic limit.

CHAPTER 14

Retardation and transients in fatigue crack growth

The discussions presented in Chapters 9–12 focused on constant amplitude cyclic loading situations where the nominal stress intensity factor amplitude (for fixed load ratio and environmental conditions) and/or the maximum stress intensity factor uniquely govern the rates of crack advance in ductile and brittle solids. There are, however, a variety of situations where the local or effective stress intensity factor range or peak value at the crack tip, which is responsible for fatigue crack growth, can be markedly different from the nominal imposed value. These differences between the apparent and actual 'driving force' for fatigue fracture may stem from such effects as (i) premature closure of the crack faces even under fully tensile far-field cyclic loads, (ii) periodic deflections in the path of the crack due to mcirostructural impediments to fracture or changes in local stress state and mode mixity, (iii) shielding of the crack tip from the far-field, applied loads by the residual stress fields generated within the cyclic plastic zone or stress-induced phase transformations, and (iv) by the bridging of the faces of the crack by fibers, particles, intact grains or corrosion products. These processes, many of which are applicable to crystalline and noncrystalline as well as brittle and ductile solids, can lead to an apparent retardation of the fatigue crack growth and hence can possibly enhance the damage-tolerance characteristics of fatigue-prone materials and structures. It is, therefore, not surprising to note that considerable research effort has been devoted to the study of these 'extrinsic' mechanisms of fatigure fracture whereby the *apparent* crack propagation rates can be markedly altered even when the intrinsic resistance of the material to fracture remains unaltered. The study of crack retardation mechanisms under constant amplitude fatigue loading is essential for developing accurate life prediction models and for improving microstructural design for enhanced damage tolerance.

While the study of constant amplitude fatigue provides valuable insights into the mechanistic processes by which fatigue failure occurs, structural components used in practical applications are invariably subjected to variable amplitude fatigue loads. The extrapolation of scientific knowledge gathered on the constant amplitude fatigue of materials to the more realistic situations involving varying amplitudes of cyclic stresses or strains is a challenge for engineers and scientists alike. This problem is further compounded by the seemingly different effects induced by variable amplitude fatigue loads on the resistance of a material to crack initiation and crack growth. Traditional models based on the damage accumulation concepts (see Chapter 7) imply that the *total fatigue life* of a (nominally defect-free) component is decreased by the application of periodic overloads. However, experimental observations on variable amplitude fatigue *crack growth* in a wide variety of metallic and non-metallic materials have unambiguously established that the application of peridoic overloads can significantly decelerate the rate of fatigute fracture. As discussed in Chapter 1, this apparent contradiction is merely a reflection of the degree to which the crack initiation stage of fatigue is quantitatively accounted for in the life prediction models.

In this chapter, we begin with a discussion of various mechanisms by which crack growth retardation occurs in materials even under constant amplitude cyclic loading. These mechanisms include different types of crack closure processes, crack deflection, phase transformations and crack bridging. We then direct attention at crack growth retardation or acceleration under variable amplitude fatigue where the effects of tensile and compressive overloads, block overloads and underloads, as well as load sequence effects are examined. It is demonstrated that variable amplitude and spectrum loads can have a distinctly different effect on the characteristics of fatigue crack growth than on fatigue crack initiation (which was addressed in Chapters 7 and 8).

14.1 Fatigue crack closure

The possibility that a fatigue crack can close even at a far-field tensile load was first rationalized by Elber (1970, 1971) on the basis of experimental observations. By monitoring changes in the compliance of thin sheets of cracked 2024-T3 aluminum alloy, Elber argued that a zone of residual tensile deformation is left in the *wake* of a fatigue crack tip. The attendant reduction in crack opening displacement gives rise to premature contact between the faces of the crack and causes a reduction in the apparent 'driving force' for fatigue crack advance.

Additional mechanisms of crack closure have been identified which have broadened the application of crack closure to rationalize a significantly greater number of fatigue crack growth characteirstics than what was originally anticipated from Elber's work. Experimental observations published in the later 1970s and early 1980s established that Elber's mechanism was not the sole cause of closure, but that other types of closure phenomena also influence the rate of fatigue crack advance. On the basis of their own results and of the work of other researchers, Ritchie, Suresh & Moss (1980), Suresh, Zamiski & Ritchie (1981), and Suresh & Ritchie (1982a, 1984a) categorized the various forms of fatigue crack closure that are induced by a variety of mechanical, microstructural and environmental factors, and coined the expression 'plasicity-induced crack closure' for Elber's closure due to residual plastic stretch at crack wake. The additional sources of closure arise as a result of: (i) corrosion layers formed within a fatigue crack (oxide-induced crack closure), (ii) microscopic crack closure), (iii) viscous fluids penetrated inside the crack (viscous fluid-induced crack closure), and (iv) stress- or strain-induced phase transformations at the crack tip (transformation-induced crack closure).

While concepts of fatigue crack retardation were developed over the years mainly in the contact of fatigue in conventional metals, expanding interest in the fatigue of advanced metallic systems, nonmetallic materials and composites has led to a surge in research on several other mechanisms which impede the growth of (constant amplitude) fatigue crack growth. These mechanisms include: (i) crack deflection, (ii) crack-bridging or trapping, and (iii) crack-shielding due to microcracking, crazing and phase transformations. Figure 14.1 schematically illustrates the various mechanisms by which the growth of a fatigue crack can be retarded. These processes are considered in this chapter in the context of constant amplitude and variable amplitude fatigue.†

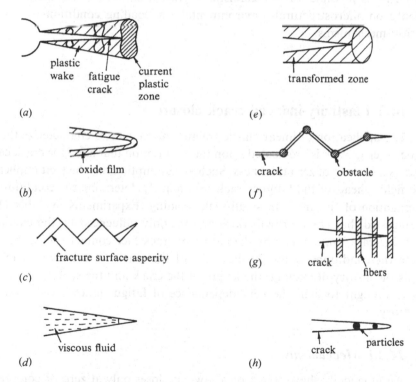

Fig. 14.1. A schematic illustration of the mechanisms which promote retardation of fatigue crack growth in constant amplitude fatigue. (*a*) plasticity-induced crack closure; (*b*) oxide-induced crack closure; (*c*) roughness-induced crack closure; (*d*) fluid-induced crack closure; (*e*) transformation-induced crack closure; (*f*) crack deflection; (*g*) crack-bridging by fibers; (*h*) crack-bridging (trapping) by particles.

---

† Quantitative analyses of the mechanics of these processes were addressed in Chapters 5, 6, 9, 11 and 12.

It is noted at the outset that the evolution of different crack closure and retardation mechanisms during both constant amplitude and variable amplitude fatigue is a process that can not be quantified accurately. These processes can be strongly influenced by even small variations in the path of the crack, environmental conditions, loading conditions and testing methods. Different retardation mechanisms assume different levels of significance in different microstructures even in the same class of materials. In addition a number of these mechanisms can simultaneously influence the overall crack growth behavior such that it is often impossible to identify the individual contributions to the overall crack growth rates from each of these retardation or shielding mechanisms. It is, therefore, not surprising that there exists considerable controversy and difference of opinion on the applicability and significance of different retardation mechanisms to the fatigue crack propagation. Since it is not possible to draw universal conclusions on the influence of each mechanism in a broad range of fatigue situations, we present in this chapter, in as quantitative and general terms as possible, the mechanistic origins, conditions of applicability, and dependence on microstructural, environmental and loading conditions of various retardation mechanisms.

14.2 Plasticity-induced crack closure

The application of linear elastic fracture mechanics to characterize the rate of fatigue crack growth is predicated upon the assumption that a fatigue crack can be idealized as a saw-cut of zero thickness. Such an assumption has the tacit implication that the fields ahead of the fatigue crack tip uniquely determine the conditions for the propagation of the crack in small-scale yielding. Experiments by Elber (1970, 1971) suggested that crack growth rates are not only influenced by the conditions *ahead* of the crack tip, but also by the nature of crack face contact *behind* the crack tip. Since the conditions extant in the wake of the crack tip are a result of such factors as the history of loading, the length of the crack and the stress state, Elber's work also brought to light the very dependence of fatigue crack growth rates on prior history.

14.2.1 Mechanisms

An atomically sharp notch or a saw-cut closes only at zero or compressive loads. However, the propagation of a fatigue crack gives rise to a wake of material that has previously been deformed plastically. Figure 14.2 shows the development of a plastic wake for three different crack lengths for a fatigue crack which propagates under a constant amplitude of cyclic tensile stresses (i.e. an increasing ΔK value). During one cycle of crack growth, residual tensile strains are left in the material behind the advancing crack front, as only elastic recovery occurs after the creation of

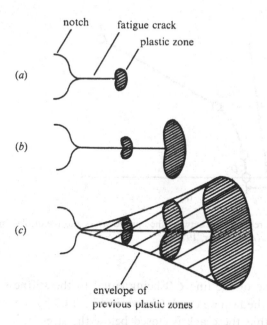

Fig. 14.2. The development of an envelope of prior plastic zones around an advancing fatigue crack. (After Elber, 1970, 1971.)

the fracture surfaces. With an increase in the stress intensity factor and the size of the plastic zone due to crack advance, the material which has previously been deformed permanently within the plastic zone now forms an envelope of plastic zones in the wake of the crack front.

By mounting strain gages above and below the plane of the crack, about 2 mm behind the crack tip, along the side surfaces of a center-cracked aluminum alloy specimen, Elber measured the far-field tensile load at which the two fracture surfaces opened completely during the fatigue cycle (Fig. 14.3). If one monitors the variation of the applied stress σ or load P as a function of the displacement of the crack faces δ (or the displacement of two reference points located just above and below the crack plane) at some distance behind the crack tip, the stress in the fatigue cycle at which the crack becomes fully open can be inferred. A schematic of the typical variation between σ and δ, as the specimen is unloaded fully from the far-field stress, is provided in Fig. 14.3.

In the regions between points A and B, the σ–δ plot exhibits a constant slope which is equal to the measured stiffness of an identical sheet with a saw-cut of the same length as the fatigue crack. This implies that the fatigue crack is fully open until the nominal stress is reduced from σ_{max} to σ_{op}. As unloading continues from point B to point C, the second derivative of the stress–displacement curve, $d^2\sigma/d\delta^2$ becomes negative. Closure of the crack faces is the only possible mechanism which can induce a change in $d^2\sigma/d\delta^2$ as shown in the region between points B and C in Fig. 14.3. In the final stages of unloading beyond point C, the stress–displacement curve exhibits

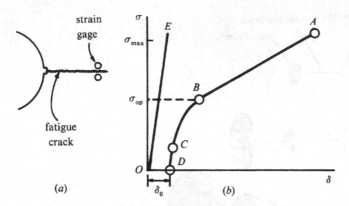

Fig. 14.3. A schematic of the relationship between the applied stress and the displacement measured by the strain gages. (After Elber, 1970.)

a linear variation, with the slope of the line CD being equal to the stiffness of an identical notched plate without the fatigue crack (shown in Fig. 14.3 by the straight line OE). This result indicates that the crack is closed below the stress level corresponding to point C. Under zero far-field stress, δ_0 is the residual crack opening displacement which is equal to the difference in the δ values between the saw-cut and the fatigue crack at the crack length where the reference strain gages are mounted. The compliance plot in Fig. 14.3 can also be presented in terms of the stress intensity factor K rather than the stress.

Figure 14.4(a) schematically shows a compliance curve typical of many alloys where the variation of the stress intensity factor during both the loading and unload-

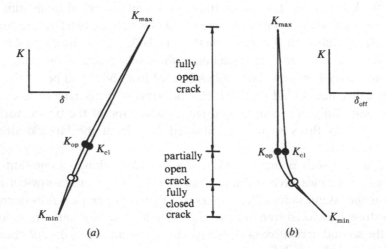

Fig. 14.4. (a) A plot of the stress intensity factor K as a function of the crack opening displacement during a complete fatigue cycle. (b) Modification of the compliance plot using the offset procedure.

ing portions of the fatigue cycle is plotted against the displacement measured on the side or back faces of the specimen (by strain gages) or at the crack mouth (by a clip gage). As the cracked specimen is loaded in tension, it initially remains closed until a certain stress intensity factor (denoted by the open circle in Fig. 14.4(a)) is reached. At this point, the crack gradually begins to 'peel open'. The crack opens fully at the stress intensity factor K_{op}. Upon unloading, the first contact between the crack faces occurs at the closure stress intensity K_{cl}. Below this loading point, the crack gradually closes. Complete crack closure occurs at the tensile stress intensity factor denoted by the open circle.

An offset procedure is commonly used to identify the crack opening stress intensity from the compliance curve. In this method, an offset displacement δ_{off} is computed by subtracting a displacement corresponding to the fully open configuration of the crack. Between K_{op} and K_{max}, the displacement $\delta = \bar{\alpha}K$, where $\bar{\alpha}$ is a proportionality constant. The offset procedure for closure measurement involves the subtraction of a proportion of the load signal from the displacement transducer signal using a simple summing amplifier. The resulting signal may be amplified electronically by a factor of up to 1000 (e.g., Fleck & Smith, 1982). The offset displacement is defined as $\delta_{off} = G_1(\delta - \bar{\alpha}K)$, where G_1 is the gain of the electronic circuit. When the crack is fully open, $\delta_{off} = 0$. When the crack begins to close, the δ_{off} trace takes a nonzero value. Figure 14.4(b) illustrates the variation of the stress intensity factor with the offset displacement during a complete fatigue cycle.

The following points should be noted in the context of crack closure measurements: (i) Since the crack closes gradually during a reduction in the far-field stress intensity factor, there is no unique definition for the closure stress intensity factor. Sometimes an average closure stress intensity factor, which is the mean value of the stress intensity factors corresponding to the points of complete opening and complete closure, is used. (ii) The stress intensity level for the first contact between the crack faces during unloading and the stress intensity level required for the complete separation of the fracture surfaces during loading are generally different because of compressive crushing induced by crack closure (see the next subsection). Such deformation and frictional contact between the mating fracture surface asperities also introduces a hysteresis in the K–δ plot, as shown in Fig. 14.4.

Elber argued that the crack can propagate only during that fraction of the fatigue loading cycle in which the crack faces are separated. The effective stress range, $\Delta\sigma_{eff}$, and the corresponding effective stress intensity factor range, ΔK_{eff}, which are responsible for crack growth are then given by

$$\Delta\sigma_{eff} = \sigma_{max} - \sigma_{op} = U\Delta\sigma, \qquad \Delta K_{eff} = K_{max} - K_{op} = U\Delta K, \qquad (14.1)$$

where $\Delta\sigma$ and ΔK are the applied stress range and the stress intensity factor range, respectively. The corresponding characterization of fatigue crack growth rates based on LEFM becomes

$$\frac{da}{dN} = C(\Delta K_{eff})^m = C(U\Delta K)^m. \qquad (14.2)$$

For the 2024-T3 aluminum alloy sheet specimen fatigue-tested (under predominantly plane stress conditions) over a ΔK range of 13 to 40 MPa$\sqrt{\text{m}}$ at test frequencies of 1 Hz and 30 Hz, and for variations in load ratio, R, of -0.1 to 0.7, Elber determined that

$$\frac{K_{op}}{K_{max}} = 0.5 + 0.1R + 0.4R^2. \tag{14.3}$$

While a plot of da/dN versus nominal ΔK (in the Paris regime) varied noticeably with the load ratio R in Elber's experiments, a reformulation of the same set of data in terms of the effective stress intensity factor range ΔK_{eff} using Eq. 14.1 collapsed the results for different R values into a single curve. Numerous subsequent studies have provided empirical modifications to Eq. 14.3, which are specific to the particular choice of material and testing conditions. It is now recognized that, in addition to its dependence on R, U is strongly influenced by the specimen geometry, the stress state, the stress intensity factor range and environment.

14.2.2 Analytical models

An analysis providing a justification for the use of ΔK_{eff} to characterize fatigue crack growth was developed by Budiansky & Hutchinson (1978). Their model is presented in this section.

The main assumptions of this model are: (i) The (long) fatigue crack is subjected to steady-state growth under plane stress and small-scale yielding conditions and is propagated under constant K_{max}. (ii) The crack opening displacements estimated from the Dugdale analysis describe the opening profile of the fatigue crack at K_{max} with the exception that a residual stretch of previously yielded material is attached to the crack faces. (iii) Upon unloading, this residual stretch leads to contact over the entire length of the fatigue crack. (iv) In the immediate vicinity of the crack tip, the residual displacements are determined by the reversed flow zone. (v) In the region ahead of the crack tip, beyond the cyclic plastic zone, the plastic stretch created at $K = K_{max}$ is left unaltered upon unloading.

The Dugdale expressions (see Section 9.5.2) for the plastic zone size and crack tip opening displacement, respectively, are

$$r_p = \frac{\pi}{8}\left(\frac{K}{\sigma_y}\right)^2, \quad \text{and} \quad \delta_t = \frac{K^2}{\sigma_y E}. \tag{14.4}$$

With the origin of the cartesian coordinate system located at the crack tip, the variation of plastic stretch over the distance $(0, r_p)$ is given by

$$\frac{\delta}{\delta t} = g\left(\frac{x}{r_p}\right), \quad g(\xi) = \sqrt{1-\xi} - \frac{\xi}{2}\log\left|\frac{1+\sqrt{1-\xi}}{1-\sqrt{1-\xi}}\right|. \tag{14.5}$$

Given the possibility of compressive yielding over the distance, $r_c = r_p/4$, during unloading from K_{max} (Section 9.6), the residual plastic stretch at $K_I = K_{min} = 0$ over the interval $(0, r_p/4)$ becomes

$$\frac{\delta}{\delta_t} = g\left(\frac{x}{r_p}\right) - \frac{1}{2} g\left(\frac{4x}{r_p}\right), \tag{14.6}$$

where δ_t is the crack tip opening displacement at K_{max}. In the interval (r_c, r_p), the plastic stretch equals that given by Eq. 14.7. The plastic stretch at $K_{min} = 0$ is one-half the value at K_{max}.

Budiansky & Hutchinson assume that for steady-state fatigue crack growth, a plastic stretch of magnitude $\delta_R/2$, which leads to crack closure all along the crack length at $K_{min} = 0$, is appended to the upper and lower crack faces. Using Mushkelishvili's complex potentials formulation, it can be shown that this approximation, in conjunction with the assumptions listed earlier, leads to the following expressions for plastic stretch:

$$\frac{\delta_R}{\delta_t} = \frac{\pi^2 \alpha}{4} + \int_\alpha^1 \sqrt{\frac{\xi - \alpha}{\xi}} h(\xi)\, d\xi, \quad \frac{\delta_R}{\delta_t} = -\frac{\pi^2 \alpha}{4} + \int_\alpha^1 \sqrt{\frac{\xi}{\xi - \alpha}} h(\xi)\, d\xi, \tag{14.7}$$

where

$$h(\xi) = -g'(\xi) = \frac{1}{2} \log\left|\frac{1 + \sqrt{1 - \xi}}{1 - \sqrt{1 - \xi}}\right|, \tag{14.8}$$

and α is the ratio of cyclic plastic zone size r_c for the growing crack to the monotonic plastic zone size (at K_{max}) for the stationary fatigue crack. Equations 14.7 and 14.8 reduce to

$$\int_\alpha^1 \frac{h(\xi) d\xi}{\sqrt{\xi(\xi - \alpha)}} = \frac{\pi^2}{2}. \tag{14.9}$$

The numerical solution of Eq. 14.9 gives $\alpha = 0.092\,86$ and $\delta_R/\delta_t = 0.8562$. In other words, the reverse plastic zone size for a growing fatigue crack is less than 10% the size of the maximum monotonic plastic zone and is less than 40% the size of the cyclic plastic zone for a stationary fatigue crack in an ideally plastic solid. Furthermore, the residual stretch left behind the advancing fatigue crack is as high as 86% of the crack tip opening displacement at K_{max}.

When the far-field stress intensity factor is reduced from K_{max}, the crack opening displacement along $x < 0$ varies with K as

$$\frac{\delta}{\delta_t} = g\left(\frac{x}{r_p}\right) - \left(\frac{2\bar{r}_c}{r_p}\right) g\left(\frac{x}{\bar{r}_c}\right), \tag{14.10}$$

where

$$\bar{r}_c = \frac{r_p}{4}\left(1 - \frac{K}{K_{max}}\right)^2 \tag{14.11}$$

is the instantaneous size of the cyclic plastic zone at any point in the unloading sequence. Noting that Eq. 14.10 has a local minimum in the interval $(-\infty, 0)$, the values of x and $\bar{r}_c$ at which the first contact between the fracture surfaces occurs (at $K_{min} < K = K_{cont} < K_{max}$) during unloading can be determined from the conditions,

$$g\left(\frac{x}{r_p}\right) - 2\left(\frac{\bar{r}_c}{r_p}\right)g\left(\frac{x}{\bar{r}_c}\right) = \frac{\delta_R}{\delta_t}, \quad g'\left(\frac{x}{r_p}\right) - 2g'\left(\frac{x}{\bar{r}_c}\right) = 0. \tag{14.12}$$

The solution of Eqs. 14.12, in conjunction with the previously cited finding that $\delta_R/\delta_t = 0.8562$ and that $\alpha = 0.092\,86$, leads to the result

$$\frac{K_{cont}}{K_{max}} = 1 - \sqrt{1 - \left(\frac{\delta_R}{\delta_t}\right)^2} = 0.483. \tag{14.13}$$

Figure 14.5(*a*) shows the process of crack closure as the far-field stress intensity factor is reduced from a peak value of K_{max} to the minimum value $K_{min} = 0$. Note the development of crack closure as the value of K/K_{max} is decreased below 0.48.

Budiansky & Hutchinson further extend their analysis to predict the opening stress intensity levels during reloading to $K = K_{max}$. As the fatigue crack is reloaded, the crack faces open by a gradual peeling process whereby the boundary of the contact region shifts toward the crackfront. $K \geq K_{op}$ is the point of complete separation of the crack faces. Figure 14.5(*b*) shows the crack opening profile and the development of crack tip plasticity. Note the opening of the crack faces for K/K_{max} values greater than 0.557. As the crack attempts to retrieve its original configuration at $K = K_{max}$, note that, for $a/r_p < x < 1$, the yield stress σ_y is barely re-established and that no further stretching occurs. It is also seen that, for plane stress plasticity-induced crack closure, K_{op}/K_{max} is only a function of load ratio R and that the predicted relationship is quite close to that reported by Elber. Furthermore, the model leads to the result that

$$\delta_t - \delta_R \approx 0.73\frac{(K_{max} - K_{op})^2}{E\sigma_y}, \quad \delta_t - \delta_R \approx 0.54\frac{(K_{max} - K_{cont})^2}{E\sigma_y}. \tag{14.14}$$

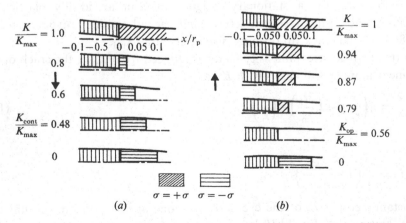

Fig. 14.5. (*a*) The process of plasticity-induced crack closure as the fatigue crack is unloaded from $K = K_{max}$ to $K = K_{min} = 0$. (*b*) The process of crack opening as the fatigue crack is reloaded to $K = K_{max}$. (After Budiansky & Hutchinson, 1978.)

When compared with Eq. 9.86, this result implies that ΔK_{eff} is a valid characterizing parameter for fatigue crack advance under small-scale yielding conditions. Budiansky & Hutchinson also explored the effects of strain hardening on plasticity-induced closure. It was found that cyclic strain hardening enhanced the effects of crack closure, whereas cyclic softening reduced closure levels.

The closure model by Budiansky & Hutchinson (1978) provides a comprehensive theoretical treatment of the phenomenon of crack closure. The strength of the model rests on the following features: (i) The analyses provide a theoretical justification for the use of ΔK_{eff} to characterize fatigue crack advance. (ii) The results rationalize the effect of R ratio on crack closure in the Paris regime of fatigue and are consistent with the experimental observations of Elber (1971) and of subsequent researchers for aluminum alloys. (iii) The calculations provide valuable insights into the development of monotonic and reversed flow plastic zones and plastic stretch during steady-state fatigue crack growth and into the effects of strain hardening on closure. The limitations of the model are: (i) The analysis pertains to plane stress deformation whereas predominantly plane strain conditions exist over a wide range of fatigue crack growth rates for most metals, especially in the near-threshold regime where the strongest effect of closure is usually observed. (ii) The approach does not invoke any mechanisms of fatigue crack growth. (iii) The model addresses just one aspect of fatigue crack closure, while, as discussed in the sections to follow, other microstructurally and/or environmentally enhanced crack closure effects may dominate over plasticity-induced crack closure.

14.2.3 Numerical models

One of the earliest finite-element models of plasticity-induced crack closure in plane stress was published by Ohji, Ogura & Yoshiji (1975). In their analysis, an incremental plasticity model incorporating kinematic hardening was used and crack growth was simulated by extending the fatigue flaw in each stress cycle by a prescribed length which was equal to the finite-element mesh size. For a crack growth distance of $0.5\,\text{mm}$ (ten times the mesh size), the opening stress ratio U in Eq. 14.1 was found to be as low as 0.5 for a fatigue crack emanating from a sharp notch. Although Ohji *et al.* did not explore the validity of ΔK_{eff} to characterize fatigue fracture over a wide range of R ratio or nominal ΔK values, their results indicated that the strain amplitude in the vicinity of the crack tip scaled with ΔK_{eff}.

In a parallel study, Newman (1976) also performed a two-dimensional finite-element analysis for plane stress using the incremental theory of plasticity. The model simulated crack growth by releasing the crack tip nodes and accounted for changing boundary conditions associated with crack extension and intermittent contact of the crack faces, without invoking any criterion for crack advance. The predictions of the dependence of crack closure on R ratio were found to be consistent with the experimental observations of Elber (1970) for aluminum alloys. These analyses have also been extended to study constant and variable amplitude fatigue in plane stress and plane strain using the finite element method (e.g., Newman, 1976; Blom & Holm, 1985; Fleck & Newman, 1988; McClung & Sehitoglu, 1989) and the finite difference method (Llorca & Sánchez Gálvez, 1990).

There appears to be a consistency among the finite-element results in that they all indicate a markedly reduced level of crack closure in plane strain than in plane stress. This trend is also reflected in the three-dimensional elastic–plastic finite-element analyses of fatigue crack growth in 'thick' center-cracked plates (Chermahini *et al.*, 1989). These authors found that the crack closure and opening stress varied through the thickness of the cracked plate. On the specimen side surface and in the mid-thickness plane, the crack opening stress levels approached the two-dimensional solutions for plane stress and plane strain, respectively. There also exists an apparent effect of specimen geometry on plastcity-induced crack closure in plane strain. This effect can be rationalized in terms of the 'T-stress', which is the nonsingular term in the asymptotic expansion for the in-plane stresses in the linear elastic crack analysis, Eq. 9.42. For a range of specimen geometries, the crack tip displacement and the plastic zone size for stationary cracks in elastic-ideally plastic materials are uniquely defined by the stress intensity factor K and the nonsingular stress T (Fleck & Newman, 1988). While a variation in T-stress (induced by a variation in crack geometry) has no significant effect on closure in plane stress, its role in modifying plane strain closure for different crack geometries can be pronounced.

The numerically predicted trends are also in qualitative agreement with the experimental measurements of closure by Fleck & Smith (1982) at the surface and mid-thickness sections of a 24 mm thick compact tension specimen of a low strength steel (at $\Delta K = 25\,\mathrm{MPa}\sqrt{\mathrm{m}}$ and $R = 0.05$) using a strain gage and a push-rod compliance gage, respectively.† These experiments showed a reduced level of closure in plane strain than in plane stress.

14.2.4 *Effects of load ratio on fatigue thresholds*

The effect of load ratio R on the fatigue threshold was discussed in Figs. 10.9 and 10.10. Schmidt & Paris (1973) employed crack closure concepts to rationalize this influence of R ratio on fatigue threshold stress intensity factor ΔK_0, although plasticity effects on crack closure in plane strain are not dominant. For a 2024-T3 aluminum alloy, they found a monotonic reduction in ΔK_0 with increasing R up to a critical value (R_{cr}), beyond which ΔK_0 was load ratio-independent

---

† In this method, closure measurements are made by interrupting the fatigue test and drilling two parallel holes, 1.5 mm in diameter, behind the crackfront and along the mid-thickness section of the specimen. From the top surface of the specimen, one hole is drilled to a depth of 1 mm below the fracture plane, while the other one terminates at about 1 mm above the fracture plane. A push-rod assembly is then attached to the specimen and the test is re-started. During fatigue loading, the relative displacements of the bottoms of the holes are measured with a double cantilever clip gage. The point of change in slope in the load–displacement plot is then used to measure the mid-thickness, plane strain closure level.

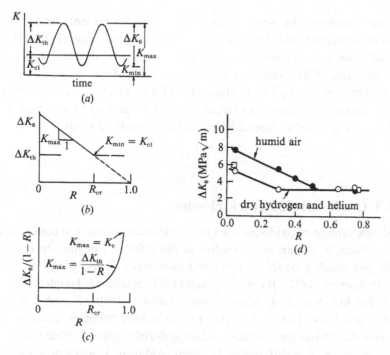

Fig. 14.6. A model, due to Schmidt & Paris (1973), for the dependence of ΔK_0 on R. (a) Definition of nominal and effective stress intensity factors (b) and (c) Predicted variation of ΔK_0 and $K_{0,max}$, respectively, with R. (d) Experimentally determined variation of ΔK_0 and $K_{0,max}$ with R for a $2\frac{1}{4}$Cr–1Mo (AISI A542 Class 3) steel (bainitic microstructure, monotonic tensile yield strength = 500 MPa) in laboratory air (23 °C, 30% relative humidity) and in dehumidified hydrogen and helium environments. (After Suresh, Zamiski & Ritchie, 1981, and Suresh & Ritchie, 1982b, 1984a.)

(Fig. 14.6). Assuming that (i) fatigue crack growth begins at a constant *effective* threshold stress intensity factor range ΔK_{th} for a given material, and that (ii) the closure stress intensity factor K_{cl} at the threshold is independent of R, Schmidt & Paris provided a rationale for the two distinct regions in the ΔK_0 versus R plot. For $R < R_{cr}$,

$$K_{0,max} = \frac{\Delta K_0}{1-R} = K_{cl} + \Delta K_{th}, \quad K_{min} < K_{cl}, \tag{14.15}$$

whereas for $R \geq R_{cr}$,

$$\Delta K_0 = \Delta K_{th}, \quad K_{min} \geq K_{cl}. \tag{14.16}$$

As $R \to R_{cr}$, $K_{min} \to K_{cl}$ and $\Delta K_0 \to \Delta K_{th}$.

This model provides an explanation for the load ratio dependence of fatigue thresholds and is consistent with experimental data for aluminum alloys as well as ferritic–pearlitic, bainitic and martensitic steels, e.g., Fig. 14.6(d). However, this line of reasoning did not offer mechanistic descriptions of closure, nor did it explain the pronounced effects of environment, e.g., Fig. 14.6(d), and microstructure on fatigue thresholds. Furthermore, the Schmidt and Paris model does not justify the high levels of closure observed under predominantly plane strain, near-threshold conditions, a result which is apparently counter to many experimental observations (Lindley & Richards, 1974; McEvily, 1977) and to the aforementioned numerical results. It is, therefore, evident that plasticity-induced crack closure alone cannot explain the influence of various mechanical, microstructural and environmental factors on fatigue crack growth, especially in the near-threshold regime.

14.3 Oxide-induced crack closure

The mechanism of oxide-induced crack closure evolved as a consequence of attempts to rationalize apparent anomalies in the effects of environment on near-threshold fatigue crack growth in steels and aluminum alloys. Many researchers, particularly Paris *et al.* (1972), Ryder & Lynch (1977), Skelton & Haigh (1978), Tu & Seth (1978), Ritchie, Suresh & Moss (1980), Stewart (1980), Suresh, Zamiski & Ritchie (1981), and Liaw, Leax & Logsdon (1983), made references to the possibility of crack closure due to fracture surface oxidation during fatigue at both ambient and elevated temperatures. The differences between oxidation kinetics near the fatigue threshold and at higher growth rates were also qualitatively described by Benoit, Namdar-Irani & Tixier (1980) for austenitic stainless steels. Suresh, Zamiski & Ritchie (1981) reported the first quantification of the influence of oxide layers formed within fatigue cracks on the threshold fatigue behavior in ferritic–pearlitic, bainitic and martensitic steels. Using scanning Auger spectroscopy, they estimated the thickness of oxide layers on fracture surfaces over a wide range of fatigue crack growth rates and found the oxide thickness within the fatigue crack to be comparable to the scale of the crack tip opening displacement near the threshold. In a parallel study, Vasudevan & Suresh (1982), employing secondary ion mass spectroscopy to measure the oxide layer thickness, noted the significance of oxide-induced closure to near-threshold fatigue fracture in 2XXX and 7XXX series aluminum alloys. Oxide-induced crack closure is also known to have a decisive effect on the near-threshold crack propagation response of copper (Liaw *et al.*, 1982) and nickel-base superalloys (King, 1982; Yuen, Roy & Nix, 1984) fatigued in oxidizing media.

14.3.1 Mechanism

During the propagation of a fatigue crack, the presence of a moist atmosphere leads to oxidation of the freshly formed fracture surfaces. At low amplitudes of cyclic crack tip opening displacements, i.e. at near-threshold ΔK levels and low R ratios, the possibility of repeated crack face contact during tensile fatigue is enhanced as a consequence of locally mixed-mode crack opening, microscopic roughness of the fracture surfaces and some plasticity-induced closure. There occurs, at low ΔK levels, a continual breaking and reforming of the oxide scale behind the crack tip. The attendant 'fretting' mechanism can lead to the build-up of oxide layers which are up to $0.2\,\mu m$ thick at threshold ΔK_0 in lower strength steels fatigue-tested in moist (40% relative humidity) laboratory air at $R = 0.05$. This oxide layer is about twenty times thicker than that formed on a freshly prepared surface which is exposed to the same moist environment for the same length of time as the fatigue test specimen. Noting that the maximum crack tip opening displacement at threshold for most lower strength steels (at low R ratios) is also of the order of a small fraction of a micrometer, one may envision the possiblity of complete wedging of the crack by the oxide film. It can be shown by means of simple calculations involving density and molecular weights of iron oxides that 1 cc of Fe oxidizes to 1.76 cc of FeO, 2.07 cc of Fe_3O_4 and 2.13 cc of Fe_2O_3. If it is assumed that the oxide emerges only at the free surface of the crack (which is consistent with experimental observations), the oxide thickness measurement made on any one fracture surface represents roughly the *excess* oxide thickness inside the crack (Suresh, Zamiski & Ritchie, 1981).

At high R ratios, where the possibility of crack face contact is minimized because of the larger crack tip opening displacements at any ΔK level, or at high ΔK levels where the rate of crack advance is generally too rapid to promote copious oxidation of any R value, the fretting oxidation mechanism is not expected to play a significant role in influencing closure. In Fig. 14.7(a), regions of corrosion deposits are clearly visible on the fatigue fracture surface (at growth rates that fall in the near-threshold regime) in a $2\frac{1}{4}$Cr–1Mo steel fatigue-tested at a low R ratio of 0.05. In the high R ratio test, however, significantly less oxidation takes place even in the near-threshold regime. This is more clearly seen in the corresponding quantitative estimates of the oxide thickness (based on argon ion sputter methods in the scanning Auger spectrometer), Fig. 14.7(b). Here the oxide thickness is plotted as a function of crack length and crack growth rates for the same specimens fatigue tested at low and high R levels.

14.3.2 Implications for environmental effects

The excess oxide layer thickness on the fracture surface varies considerably along the crack wake, Fig. 14.7(b), and the extent of oxidation is strongly dictated by specific conditions of microstructure, environment, ΔK, and load ratio. It is, therefore, a formidable task to formulate predictive models for oxide-induced crack closure and for its effects on crack growth rates. One may, however, develop an estimate of the effect of fracture surface corrosion on near-threshold crack 'driving force' by considering the simple analogy of a rigid wedge of constant thickness d_0 inside a linear elastic fatigue crack. If the wedge extends all along the length of the fatigue crack and its edge terminates at a distance $2l$ behind the crack tip, the sole mechanical effect of the rigid barrier to the closure of the fatigue crack yields a closure stress intensity factor of

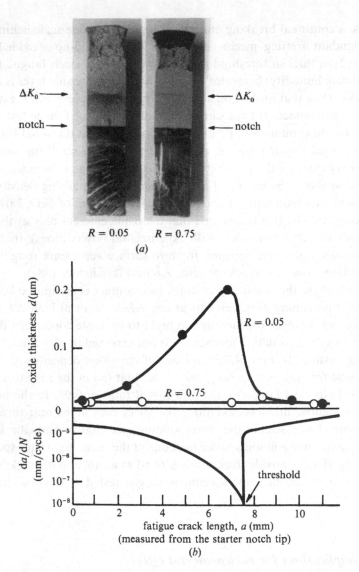

Fig. 14.7. (*a*) Macroscopic appearance of oxide deposits on near-threshold fatigue fracture surfaces in a $2\frac{1}{4}$Cr–1Mo bainitic steel fatigue tested in moist air at $R = 0.05$ and 0.75. (*b*) Corresponding Auger measurements of oxide layer thickness as a function of crack length and crack growth rates. (From Suresh, Zamiski & Ritchie, 1981. Copyright *Metallurgical Transactions*. Reprinted with permission.)

$$K_{cl} = \frac{d_0 E}{4\sqrt{\pi l}(1 - \nu^2)}, \tag{14.17}$$

where E and ν are Young's modulus and Poisson's ratio of the cracking material, respectively. Taking typical values of $l = 0.2\,\mu m$ and $d_0 = 0.2\,\mu m$ from experiments on lower strength steels (Suresh, Parks & Ritchie, 1982), it is found that $K_{cl} \approx 2.3\,MPa\sqrt{m}$. This estimate of the closure K due to wedge contact, although too simplistic to account for all the effects associated with the complex process of oxide-induced closure, provides a justification for the experimentally observed differences in the ΔK_0 values of $2\frac{1}{4}$Cr–1Mo steels in dry and moist environments at low load ratios. Figure 14.8 shows the fatigue crack growth characteristics of a martensitic $2\frac{1}{4}$Cr–1Mo steel (commercial designation AISI A542 Class 2) in moist laboratory air and in dehumidified, ultra high purity hydrogen and helium (138 kPa pressure) at $R = 0.05$ and 0.75. The near-threshold crack propagation in both the dry gaseous environments are up to two orders of magnitude higher and the ΔK_0 about 50% lower than in air at the lower R ratio (also see Fig. 14.6). The mechanism underlying this trend can be explained using the following line of reasoning.

Corrosion deposit formation, leading to an oxide thickness comparable in size to the crack opening displacements, is promoted in the moist environment at near-threshold growth rates and at low load ratios. However, when the advancing near-threshold crack is exposed to dehumidified environments, fracture surface oxidation and the attendant closure are suppressed due to the paucity of moisture. Because of the reduced level of oxide-induced closure, the ΔK_0 values of the steel in the dehumidified hydrogen and helium environments is lower than that in moist air (Ritchie, Suresh & Moss, 1980; Stewart, 1980; Suresh, Zamiski & Ritchie, 1981).

Suresh *et al.* found that the addition of moisture to hydrogen or helium led to crack growth retardation at low ΔK levels and R ratios due to oxide formation, while the dehumidification of the moist laboratory air promoted a marked accelera-

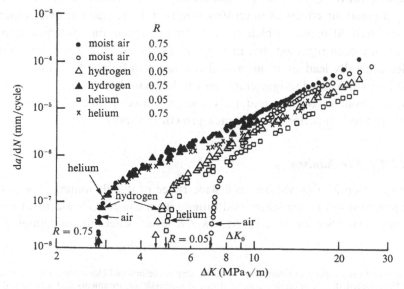

Fig. 14.8. Effects of dry and moist environments on near-threshold fatigue crack growth in $2\frac{1}{4}$Cr–1Mo steel (AISI A542 Class 2, martensitic microstructure, monotonic tensile yield strength = 769 MPa) at low and high load ratios. (After Suresh, Zamiski & Ritchie, 1981.)

tion in the near-threshold crack growth rates. At the high R ratio of 0.75, oxide-induced crack closure is essentially nonexistent in both wet and dry media because of the large minimum crack opening displacement (at K_{min}). Consequently, the ΔK_0 values are about the same in moist air, dry helium and dry hydrogen. Although these results are counter to the expectations based on conventional corrosion fatigue processes, they are fully consistent with the concept of oxide-induced crack closure.†

Oxide-induced crack closure is promoted by moisture-containing environments, elevated temperatures, low load ratios, low ΔK levels, high cyclic frequencies, and lower strength and coarser-grained microstructures where fretting and rubbing along fracture surface asperities become favorable. It is also known that, for a fixed alloy composition and yield strength, the variation in precipitation, induced by ageing treatments, can have a decisive effect on fracture surface oxidation during the fatigue of aluminum alloys (Vasudevan & Suresh, 1982). Although oxide-induced crack closure, in conjunction with plasticity-induced closure, provides a mechanistic rationale for many experimentally observed effects of environment and load ratio on crack propagation in certain regimes of crack growth in a number of materials, it does not account for microstructural effects on fatigue crack advance.

14.4 Roughness-induced crack closure

Roughness-induced crack closure has come to be recognized as one of the mechanisms by which some of the apparent microstructural effects on fatigue crack growth can be rationalized. This phenomenon provides an explanation for many apparently anomalous effects of microstructure on fatigue crack growth, especially in the near-threshold regime, which are counter to conventional interpretations. It has traditionally been regarded that finer-grained materials and wavy slip deformation modes generally lead to an improved resistance to fatigue fracture. However, research on near-threshold fatigue crack growth behavior has revealed that, in many alloy systems, coarser-grained materials and enhanced planar slip deformation modes give rise to a superior fatigue crack growth threshold.

14.4.1 Mechanism

Experimental observations in a wide range of ductile materials reveal that crack propagation in the near-threshold fatigue regime occurs by means of a single slip mechanism (see Section 10.2.1). At low ΔK levels, where the maximum plastic

---

† The *beneficial* effects of moist environments in promoting oxide-induced closure are specific to certain regimes of crack growth and to certain combinations of materials, environments and mechanical loads. It is emphasized that classical corrosion fatigue processes involving metal dissolution and hydrogen embrittlement are known to cause significant *deterioration* in fatigue crack growth resistance in a wide variety of situations. These deleterious effects of environments on fatigue are discussed in Chapter 16.

zone size, on the average, is typically smaller than a characteristic microstructural dimension such as a grain size, a crystallographic fracture process is promoted. This stage I growth mechanism leads to a highly serrated or faceted fracture morphology and an elevation in the crack closure stress. It is also known that permanent plastic deformation ahead of the crack tip, as well as the possibility of slip irreversibility during unloading from the peak stress, cause mis-match between the fracture surface asperities. *In situ* observations in the scanning electron microscope of the opening and closing of fatigue cracks in aluminum alloys have documented a strong mode II component associated with the fatigue crack opening displacements and the occurence of premature contact between the crack face asperities (Nowack *et al.*, 1979; Halliday & Beevers, 1981; Minakawa & McEvily, 1981; Ritchie and Suresh, 1981). The tortuous crack path promoted by the crystallographic growth mechanisms, in conjunction with the occurrence of mixed-mode sliding of crack faces and the mis-match between the crack face asperities, immediately provides a mechanism for enhanced crack closure. The maximum crack opening displacement at threshold (for low R ratios) for most metallic materials is of the order of a fraction of a micrometer. Given the development of rough fracture surface asperities, whose height can be tens of micrometers in coarser-grained materials, one may envision the ease with which crack closure can develop at low load ratios and growth rates.

14.4.2 *Implications for microstructural effects on threshold fatigue*

The role of coarser grains and rough fracture surfaces in promoting slower near-threshfold crack propagation rates at low load ratios as a consequence of roughness-induced crack closure is clearly documented in the study of Gray, Williams & Thompson (1983). In a fully pearlitic eutectoid steel, these authors examined the effects of prior austenite grain size on fatigue crack growth at low and high load ratios using well-controlled heat treatments whereby the pearlitie interlamellar spacing and colony size as well as the flow strength of the alloy were roughly kept constant. (The pearlite colony size $= 9$–$10\,\mu m$, pearlite interlamellar spacing $= 0.24$–$0.245\,\mu m$, and the 0.2% offset yield strength $= 406$–$11\,MPa$ for the microstructures investigated.) the coarser-grained material shows a significantly higher fatigue threshold at low load ratios, Fig. 14.9. However, grain size has little effect on near-threshold fatigue crack growth at $R = 0.7$ where even the minimum crack opening displacement of the fatigue cycle is well above zero. By repeating the experiments in a dry helium environment and conducting crack closure measurements, Gray *et al.* established that the apparently higher fatigue crack growth resistance observed in the coarser-grained material at $R = 0.1$ was due to roughness-induced crack closure. The differences between the ΔK_0 values at $R = 0.1$ were slightly lower in helium than in moist air for the two grain sizes, although the overall behavior at high load ratios was the same for the two environments.

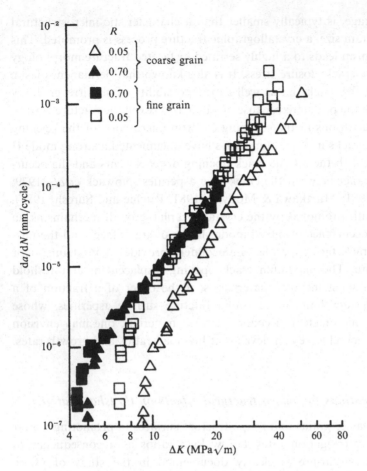

Fig. 14.9. The effect of prior austenite grain size on fatigue crack growth in a pearlitic eutectoid steel at $R = 0.1$ and 0.7. (After Gray, Williams & Thompson, 1983.)

Roughness-induced crack closure is promoted by: (i) low stress intensity factor levels where plastic zone dimensions at the crack tip are typically smaller than an average grain dimension, (ii) small crack tip opening displacements (at low ΔK levels and low R ratios) which are of a size scale comparable to the average height of the fracture surface asperities (see the example problem in Section 10.4), (iii) microstructures consisting of coarser grains and shearable and coherent precipitates which generally engender highly planar, crystallographic slip, (iv) periodic deflections in crack path induced by grain boundaries, second phase particles, composite reinforcements or abrupt load changes, and (v) enhanced slip irreversibility, especially due to slip step oxidation in moist environments. The most dramatic effects of roughness-induced crack closure are often observed whenever microstructurally induced crack deflections generate premature contact between fatigue fracture surfaces.

Examples of such effects as well as analytical models for the joint influence of crack deflection and roughness-induced crack closure are presented in Section 14.9.

14.5 Viscous fluid-induced crack closure

Problems of rolling contact fatigue in lubricated ball bearings have generated considerable research interest in the effects of oil environments on fatigue life (e.g., Endo, Okada & Hariya, 1972; Ryder, Martin & Abdullah, 1977). In the significant majority of these studies, the principal role of an oil medium is viewed as one of suppressing, or occasionally exacerbating, the effect of environment on fatigue life. For example, noncorrosive oils, such as mineral or silicone oils, are known to promote slower stage II fatigue crack growth rates, compared to air, in both ferrous and nonferrous alloys. However, when water is present in such oils, pitting corrosion fatigue processes occur as a result of which the surface fatigue life in lubricated bearing steels is severely diminished. Similarly, the presence of 4700 ppm (saturation level) concentration of H_2S in sour crude oil is known to increase stage II fatigue crack growth rates by as much as twenty times in high strength low alloy (HSLA) pipeline steels (Vosikovsky, 1976).

14.5.1 Mechanism

The mechanisms by which viscous fluids penetrating within growing fatigue cracks influence propagation rates have also been the subject of widely differing interpretations. Way (1935), for example, proposed that the wedge formed within the crack by a penetrating viscous fluid would *accelerate* crack growth rates, especially in the case of lower viscosity fluids which penetrate the crack more easily. On the other hand, Endo, Okada & Hariya (1972) differently interpreted the same mechanism by arguing that the hydrodynamic oil pressure generated by the *complete penetration* of the crack by the fluid would *decelerate* the crack by reducing the effective stress range. Since the oil pressure (for full penetration) is directly proportional to the kinematic viscosity, higher viscosity oils are expected to *decelerate* fatigue cracks, as per the model of Endo *et al.*

Tzou, Suresh & Ritchie (1985) conducted an experimental study of the influence of dehumidified silicone and paraffin oils with different kinematic viscosities on fatigue crack propagation in low strength bainitic steel. They identified three distinct, but mutually competitive, mechanisms for the apparent effects of *dry* viscous environments on fatigue crack propagation: (i) the suppression of hydrogen embrittlement or metal dissolution, primarily in the Paris regime of fatigue crack growth, (ii) the minimization of oxide-induced crack closure in the near-threshold regime of crack growth, and (iii) the development of crack closure due to the hydrodynamic wedging action of the oil inside the crack.

The net effect of oil environments in influencing crack closure is strongly dependent upon several competing factors such as suppression of environmental embrittlement, minimization of oxide-induced crack closure, penetration of fluids within cracks and the hydrodynamic wedging action. Therefore, it is difficult to extract from the analysis general trends pertaining to viscous fluid-induced closure. However, both experiments and calculations appear to indicate that crack closure promoted by viscous fluids may not be as significant as that arising from theroughness or oxidation of the fracture surfaces.

14.6 Phase transformation-induced crack closure

It has long been recognized that phase transformations at the tip of a fatigue crack can lead to a retardation in crack growth rates (e.g., Pineau & Pelloux, 1974; Hornbogen, 1978a). This phenomenon is commonly referred to as the TRIP (*TR*ansformation-*I*nduced *P*lasticity) effect in the metals literature (Zackay, Parker, Fahr & Busch, 1967; Chanani, Antolovich & Gerberich, 1972). The TRIP effect is characterized by phase changes at the tip of a crack leading to a net increase in the volume of the transforming region. As the enlarged material in the transformed zone is left behind the advancing fatigue crack tip, a net reduction in the crack opening displacement ensues. With reference to fatigue crack closure, crack tip phase transformation is analogous to crack tip plasticity in the sense that compressive residual stresses are induced within the nonlinear zone during cyclic tension. Furthermore, in both cases, the residual displacements left in the wake of the extending fatigue crack act to close the crack prematurely at a far-field tensile stress.

Pineau & Pelloux (1974) and Hornbogen (1978a) studied fatigue crack growth in metastable austenitic stainless steels and showed that strain-induced martensitic transformation generally resulted in reduced crack growth rates at low to mid-ΔK levels. The crack propagation rates increased in the following sequence of microstructural variations: metastable austenite $\rightarrow$ stable austenite $\rightarrow$ martensite of the same composition. Direct measurements of crack closure due solely to strain-induced martensitic transformation in metals have not been attempted. However, experiments suggest that such transformation-induced closure is promoted by phase changes which lead to a dilation (about 4% due to martensitic transformation in ferrous alloys) of the region undergoing transformation, and by conditions which promote phase changes, viz., higher strain rates, lower temperatures and increased metastability of the transforming phase. As in the case of plasticity-induced closure, the constraint of the surrounding elastic material on the deformed region at the crack tip is necessary to promote transformation-induced closure since the transformation of the entire specimen would not result in appreciable crack face contact. For this reason, transformation-induced closure is strongly influenced by the size and geometry of the test specimen and of the fatigue crack.

As discussed in detail in Chapters 5 and 11, transformation toughening has been the subject of considerable research in the ceramics field since the discovery of stress-induced martensitic transformations in zirconia-containing ceramics.† For purely dilatational transformations and for monotonic loading, McMeeking & Evans (1982) and Budiansky, Hutchinson & Lambropoulos (1983) calculated the reduction in stress intensity factor due to transformation-induced closure to be

$$K_I - K_{eff} = \frac{0.22}{1 - \nu} V_f \epsilon_T E \sqrt{h_T}, \qquad (14.18)$$

where V_f and h_T are the volume and height, respectively, of the transformed zone, ϵ_T is the transformation strain, and E and ν are Young's modulus and Poisson's ratio, respectively.

14.7 Some basic features of fatigue crack closure

It is evident from the previous five sections that crack closure is a phenomenon whose influence on fatigue crack growth rates is strongly dictated by microstructural and environmental factors, and mechanical loading parameters. However, there are some basic characteristics and trends which are common to various types of crack closure and to a wide variety of materials.

(1) Crack closure is generally more dominant at lower ΔK levels and at lower R ratios because of the smaller minimum crack opening displacements of the fatigue cycle. For plasticity-induced or transformation-induced crack closure, the possibility of enhanced crack closure may also increase as a result of the larger crack wake stretch or transformed zone size at higher ΔK levels and R ratios; however, this increased propensity for closure is offset by the larger minimum crack opening displacements.

(2) There is a characteristic size scale associated with each closure process, such as the height of the residual plastic crack wake for plasticity-induced crack closure, the thickness of the fracture surface oxide layer for oxide-induced crack closure, the height of the fracture surface asperities (and the extent of mode II displacements) for crack closure due to roughness, and the height of the transformation zone for closure arising from phase changes. When the size of this characteristic 'closure dimension' becomes comparable to the crack opening displacement, premature crack face contact has a marked effect on the rate of fatigue crack growth.

† Fatigue-induced transformation to a covalently bonded grey tin is also known to occur in metallic tin cycled below room temperature (Löhberg & Mustofon, 1976). Furthermore, among polymers, a transformation leading to a dilation of about 1% has been observed in polytetrafluoroethylene (PTFE) at 19 °C (Hornbogen, 1978b).

(3) As a fatigue crack emerges from a free surface or a stress concentration, the extent of crack closure generally increases with an increase in crack length up to a saturation crack length, beyond which closure is normally crack length-independent.

(4) Closure is produced by mechanisms which are operative both at the tip of the fatigue crack, such as plastic deformation or phase transformations, as well as phenomena which occur in the wake of the fatigue crack tip, such as fracture surface oxidation. For the former two crack tip processes, inelastic deformation should be constrained; if the entire specimen deforms plastically or undergoes martensitic transformation, closure would not influence crack growth.

(5) No unique conclusion can be reached about the effect of stress state on the extent of crack closure. While it is generally seen that plasticity-induced crack closure under *cyclic tension* is more dominant in plane stress than in plane strain (Section 14.1), the reverse situation is known to occur for crack growth under *cyclic compression* (see Section 4.11).

14.8 Issues and difficulties in the quantification of crack closure

There are a number of complex issues which pose a formidable challenge to experimental and theoretical attempts aimed at quantifying the effect of crack closure on fatigue behavior in engineering materials:

(1) Crack closure is often very specific to the conditions of the experiment and its magnitude is influenced by the synergistic effects of material microstructure, test environment, and stress state.

(2) Even small fluctuations in microstructural path of crack advance (induced, for example, by grain boundaries, impurities or other inhomogeneities in *local* microstructure) can cause large fluctuations in crack propagation rates, especially near the fatigue threshold where the crack opening displacements are small. Similarly, small differences in the moisture content of the test environment can cause marked variations in the extent of oxide-induced crack closure in some materials.

(3) Standard procedures, such as the 'load-shedding method', for the measurement of fatigue crack propagation rates at near-threshold stress intensity levels often lead to artificially high values of crack closure. In most metallic materials, if the ΔK values are decreased 'too rapidly', premature crack arrest may occur as a consequence of overload effects. On the other hand, if the rate of load reduction is 'too slow', an artificially high threshold may still occur due to oxide-induced crack closure. The 'correct' rate of load reduction is often arbitrarily chosen.

(4) Even when microstructural and environmental effects are set aside, crack closure is still strongly influenced by the specimen size and geometry, crack size, stress state, prior load history and the location, with respect to the crack tip, where closure is measured. Therefore, different methods for detecting crack closure can provide different estimates of closure.

(5) Fatigue crack closure is a phenomenon which leads to the breakdown of the similitude concept of fracture mechanics. Therefore, the application of nominally identical values of ΔK may lead to different levels of crack closure (and hence apparently different crack growth rates) in the same material for different crack sizes (see Chapter 15).

(6) Bulk measurements of crack closure (such as those based on compliance changes that are estimated from strain gages, clip gages and push-rod gages, the ultrasonic or acoustic emission methods, or changes in electrical potential) do not distinguish between the different mechanisms of crack closure discussed earlier. These bulk measurement techniques do not provide insights into the fundamental closure phenomena because multiple closure mechanisms may simultaneously influence the rate of fatigue crack growth.

14.9 Fatigue crack deflection

Methods by which the path of a crack can be periodically deflected from its nominal growth plane offer one possible way of enhancing the apparent resistance to fatigue crack growth (Suresh, 1983a, 1985b). Crack deflection is viewed as one of the mechanisms for the toughening of brittle and ductile matrix composites, wherein the obstacles in the path of the crack (reinforcements) may cause apparently beneficial resistance to crack growth by tilting or twisting the crackfront (e.g., Faber & Evans, 1983).

Improving the resistance of engineering materials to fatigue crack growth *via* crack deflection processes inevitably requires an understanding of the following issues:

(1) How should the micro- and macrostructure of a material be designed by optimizing the crack deflection characteristics?
(2) How does the structure of the material affect the fatigue crack path?
(3) What is the *sole* effect of crack path on the overall resistance to fatigue crack growth?

The answers to the first two questions are specific to the material microstructure and thermomechanical processing conditions, and have been thoroughly investigated only for a limited number of material systems. However, an understanding of the last issue can be achieved by theoretically exploring the geometrical consequences of crack deflection to the overall crack growth resistance. Such analyses are also essential for identifying the differences between the *intrinsic* and *apparent* crack growth resistance of a material under cyclic loading conditions.

In this section, we discuss the implications of the geometrical effects of crack deflection and of the combined influence of deflection and roughness-induced crack closure on the growth of fatigue flaws. This is followed by examples which illustrate the feasibility of improving fatigue crack growth resistance by crack deflection processes.

14.9.1 Linear elastic analyses

Deflections in the path of a nominally mode I fatigue crack are generally ignored in the characterization of fatigue behavior because of the difficulties in estimating the effective near-tip driving force. However, it is now established (see, for example, Section 9.11) that even small deflections in the path of a fatigue crack (compared to the length of the crack) can lead to a reduction in crack growth rates by several orders of magnitude, especially in the near-threshold fatigue regime. Although the realistic issue of three-dimensional deflections and the attendant fatigue crack closure effects are virtually impossible to track by way of analytical formulations, a quantitative understanding can be gained by examining the case of periodic two-dimensional deflections in the path of a fatigue crack.

Consider first the idealized case of a fatigue crack which consists of segments of periodic deflections in its path such as the one shown in Fig. 14.10(*a*). In this segment, θ is the kink angle, *D* is the distance over which the tilted crack advances along the kink, and *S* is the distance over which the plane of the growing crack is normal to

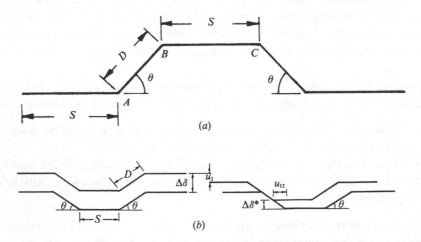

Fig. 14.10. (*a*) Idealization of a segment of a deflected crack with periodic tilts. (*b*) Schematic representation of a deflected crack in the fully opened condition at the peak load of the fatigue cycle (on the left) and relative mismatch between the crack faces leading to a premature crack closure during unloading. (After Suresh, 1983a, 1985b.)

the far-field tensile axis. The dimensions D and S are considerably smaller than the total length of the fatigue crack, a. The degree of tilting deflections, expressed in terms of the tilt angle θ and the extent of tilt, $\hat{D} = D/(D + S)$, are the same for all the segments, although the deflections are induced in opposite directions in any two adjoining segments. A first order estimate of the role of deflection on fatigue crack growth can be formulated by assuming that (i) the size of the plastic zone at the tip of the fatigue crack is much smaller than D and S (which is representative of near-threshold crack growth in some materials) and (ii) the sharp corners, such as A and B in fig. 14.10(a) do not have a significant effect on crack growth along the segments AB and BC, respectively.

These periodic deflections in the path of a fatigue crack affect the overall crack growth rates by means of the following three effects:

(1) When the crack tip is deflected from its nominal mode I growth direction, the effective driving force for crack growth is typically smaller than that of a straight crack of the same total (projected) length which is subjected to the same far-field ΔK. Therefore, the propagation of a deflected crack at the same rate as the corresponding straight crack requires an apparently larger driving force.

(2) If crack length is measured along the mode I growth direction, a deflected crack propagates at an *apparently slower* rate than a straight crack subjected to the same *effective* driving force.

(3) During the closing of the fatigue crack, even a small mismatch between the mating crack faces can lead to premature contact at far-field tensile stresses, Fig. 14.10(b). This crack closure effect would further amplify the apparent driving force required for the propagation of a deflected crack at the same rate as a corresponding straight crack.

For the elastic solutions for kinked cracks, Eqs. 9.116 and 9.117, the local mode I and mode II stress intensity factors, k_1 and k_2, respectively, for growth along the kink, of span D in Fig. 14.10(a), are

$$k_1 \approx a_{11}K_{\mathrm{I}} = \cos^3\left(\frac{\theta}{2}\right) K_{\mathrm{I}}, \quad k_2 \approx a_{21}K_{\mathrm{I}} = \sin\left(\frac{\theta}{2}\right) \cos^2\left(\frac{\theta}{2}\right) K_{\mathrm{I}}, \qquad (14.19)$$

where K_{I} is the instantaneous value of nominal mode I stress intensity factor for the fatigue cycle. The effective stress intensity factor, k, for *coplanar growth* along the deflected segment D is provided by the energy release rate criterion, Eq. 9.57 or 9.58, under *elastic conditions*, to be

$$k = \sqrt{k_1^2 + k_2^2}. \qquad (14.20)$$

The effective stress intensity factor range for crack growth along the deflected span D and the straight span S is given, respectively, by

$$\Delta k = k_{\max} - k_{\min} \quad \text{and} \quad \Delta k = \Delta K_{\mathrm{I}}, \qquad (14.21)$$

where k_{max} and k_{min} represent the values of k at the maximum and minimum loads of the fatigue cycle, respectively, and ΔK_I is the nominal far-field stress intensity factor range. Combining Eqs. 14.20–14.21, the weighted average of effective stress intensity factor range, $\overline{\Delta k}$, for the periodically deflected crack is written as

$$\overline{\Delta k} = \left\{ \frac{D \cos^2\left(\frac{\theta}{2}\right) + S}{D + S} \right\} \Delta K_I = \left\{ \hat{D} \cos^2\left(\frac{\theta}{2}\right) + (1 - \hat{D}) \right\} \Delta K_I. \qquad (14.22)$$

If the crack length is always measured along the mode I growth direction, the *apparent* crack propagation rate for a periodically deflected crack along the projected mode I plane will be

$$\left(\frac{da}{dN} \right) = \{ \hat{D} \cos\theta + (1 - \hat{D}) \} \left(\frac{da}{dN} \right)_L, \qquad (14.23)$$

where $(da/dN)_L$ is the growth rate of a straight crack subjected to the same effective stress intensity factor range.

Equations 14.22 and 14.23 describe the modifications to crack driving force and growth rates due *solely* to deflection. However, as discussed in Section 14.4, premature contact is promoted between the asperities of mating fatigue fracture surfaces because of the *net* mismatch between them (which arises from a variety of irreverisble deformation mechanisms). This mismatch causes roughness-induced crack closure and further reduces the effective driving force for fatigue crack growth in Eq. 14.22.

Figure 14.10(b) shows the opening profile of a segment of a fatigue crack (located just behind the crack tip) at the maximum far-field tensile stress; the opening displacement between the crack faces is $\Delta\delta$. As the crack unloads in both mode I and mode II, it is assumed that a *net* mismatch u_{II} develops above the K_{min} for the fatigue cycle. Note that U_{II} is not the mode II displacement due to the local stress intensity factor, k_2. At the instance of *first contact* due to the lateral shift u_{II}, the cyclic crack opening displacement is $\Delta\delta^* = \Delta\delta - u_I = u_{II} \tan\theta$, where u_I is the mode I displacement from the peak of the fatigue cycle to the point of first contact. (This approximation is strictly valid for the case when $R = 0$ and $\Delta K = K_{max}$.) Representing the extent of mismatch by the nondimensional parameter, $\chi = U_{II}/u_I$, the magnitude of the closure effect is deduced as $\Delta K_{cl}/\Delta K_I = \sqrt{\Delta\delta^*/\Delta\delta}$. The resulting ratio of the apparent to effective stress intensity factor range is

$$\frac{\Delta K_I}{\Delta k_{eff}} \approx \left\{ \frac{D \cos^2(\theta/2) + S}{D + S} \right\}^{-1} \left\{ 1 - \sqrt{\frac{\chi \tan\theta}{1 + \chi \tan\theta}} \right\}^{-1}. \qquad (14.24)$$

Thus, the *nominal* stress intensity factor ragne ΔK_I required to propagate a deflected crack at the same rate as a straight crack is always greater than that for the latter (undeflected) crack (where $\Delta k_{eff} = \Delta K_I$) by a factor equal to the right hand side of Eq. 14.24. (This inference is based on the assumption that the effective stress intensity factor range for the straight crack and the deflected crack are the same and that crack deflection and roughness-induced closure are the only processes affecting the rate of crack growth.) Equation 14.24 represents the closure level at any R where fracture surface contact is geometrically possible as a result of mismatch, i.e. for $\delta_{min} < D \sin \theta$, where δ_{min} is the crack opening displacement at $K_I = K_{min}$.

14.9.2 Experimental observations

It is known from many independent studies on 7xxx series aluminum alloys that the over-aged (T7351) heat treatment condition leads to a predominantly straight (undeflected) fatigue crack profile over a wide range of growth rates (10^{-8}–10^{-4} mm/cycle in vacuum). If one examines the under-aged microstructure (of roughly the same yield strength level) of the same alloy under identical test conditions in vacuum, it is found that a highly serrated and crystallographic, transgranular crack path is promoted. The fracture surface roughness in the under-aged alloy is increased with increasing grain size. The differences in the path of the crack for the two ageing conditions are promoted by the differences in their precipitation characteristics. Whereas the coherent strengthening precipitates in the under-aged condition are easily sheared by dislocations, the large incoherent precipitates in the over-aged microstructure form obstacles to dislocation motion with the result that the dislocations loop around the particles and bypass them (see Section 10.4.3).

Figure 14.11 shows the fatigue crack growth rates for two 7475 aluminum alloys, of grain sizes $18 \, \mu m$ and $80 \, \mu m$, whose under-aged and over-aged (T7351) microstructures were fatigue-tested in vacuum at $R = 0.1$. The yield strength values for the under-aged and over-aged microstructures were 505 MPa and 455 MPa, respectively, for the $18 \, \mu m$ grain size material, and 451 MPa and 445 MPa, respectively, for the $80 \, \mu m$ grain size material (Carter *et al.*, 1984). The ratio of the true length of the fatigue crack measured along the actual crack path to the projected length for the under-aged and over-aged tempers was 1.13 and 1.05, respectively, for the $18 \, \mu m$ grain size material, and 1.67 and 1.09, respectively, for the $80 \, \mu m$ grain size material. Figures 14.12(*a*) and (*b*), respectively, show examples of the near-threshold fatigue crack profile for the under-aged and T7351 tempers of the 7475 aluminum alloy ($80 \, \mu m$ grain size) in vacuum at $R = 0.1$. Figure 14.12(*c*) shows a micrograph of the closing of a near-threshold fatigue crack in an another aluminum alloy. The mismatch χ in this case is 0.25–0.50.

The following inferences can be drawn from Figs. 14.11 and 14.12: (i) The under-aged tempers lead to a substantially more tortuous crack path, whose roughness increases with increasing grain size. (ii) The under-aged tempers exhibit a markedly lower fatigue crack growth rate in the near-threshold regime than the over-aged alloys, with the threshold ΔK_0 increasing with increasing grain size. These trends point to the possibility that the principal cause of the extrinsic differences in the fatigue crack growth response of the two microstructures stems from crack deflection and the attendant roughness-induced closure. (Note that oxide-induced crack closure

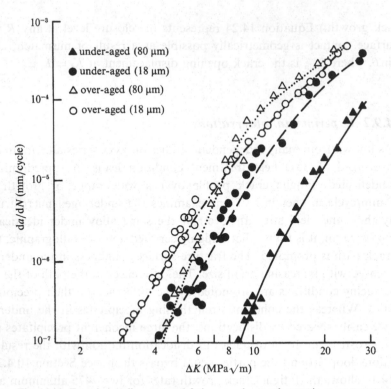

Fig. 14.11. Experimentally measured fatigue crack growth rates in room temperature vacuum for under-aged and over-aged (T7351) conditions of a 7475 aluminum alloy, of grain sizes 18 μm and 80 μm, at $R = 0.1$ and frequency, $v_c = 30$ Hz. (After Carter *et al.*, 1984.)

is essentially suppressed in vacuum, and that closure induced by plasticity is comparable for the two ageing conditions because of their similar yield strength levels.)

14.9.3 Example problem: Possible benefits of deflection

Problem:

The solid line in Fig. 14.13 indicates the typical near-threshold fatigue crack growth curve for the over-aged (T7351) condition of 7x75 (7075 or 7475) aluminum alloy tested *in vacuo* at room temperature at $R = 0.1$. This material exhibits a straight crack path, with $\hat{D} \approx 0$ and $\theta \approx 0°$ in the deflection model. An examination of the under-aged (T351) microstructure of the same alloy shows periodic deflections in crack path.

Invoking the assumption that the differences in the fatigute crack growth rates between the under-aged and the over-aged alloy stem primarily from the differences in their crack path, predict the fatigue crack growth rates for the under-aged alloy (using the deflection analysis given earlier in this section) for different values of the mismatch parameter χ, on the basis of the fatigue crack

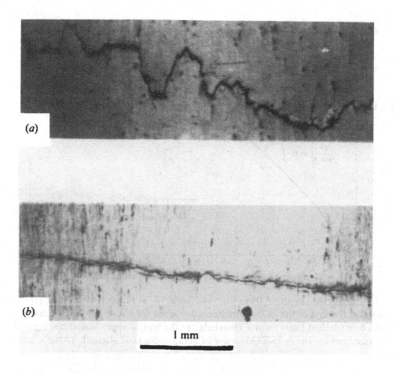

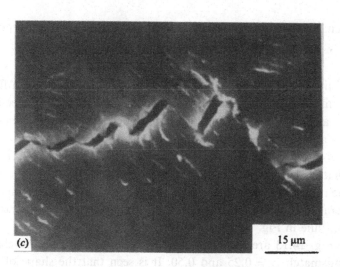

Fig. 14.12. (*a*) Typical near-threshold fatigue crack growth profiles of (*a*) under-aged and (*b*) over-aged (T7351 temper) of a 7475 aluminum alloy (grain size = 80 μm) tested in vacuum at *R* = 0.1. (from Carter *et al.*, 1984. Copyright *Metallurgical Transactions*. Reprinted with permission.) (*c*) A micrograph of the closing of the crack in a peak-aged Al–Li–Mg alloy showing the mismatch between the fracture surface asperities. (From Starke & Williams, 1989. Copyright American Society for Testing and Materials. Reprinted with permission.)

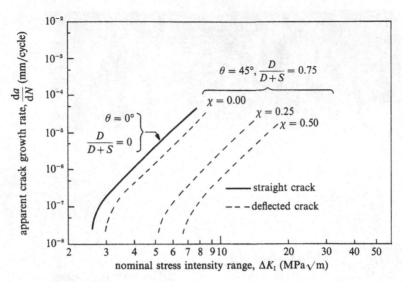

Fig. 14.13. Predicted variation, from Eqs. 14.23 and 14.24, of the fatigue crack growth rates for under-aged alloys of the 7x75 aluminum alloy series for $\theta = 45°$, $\hat{D} = 0.75$, and $\chi = 0.00$, 0.25 and 0.50 (dashed lines) from a knowledge of the typical upper bound crack growth rate curve (solid line) for the 7x75-T7351 alloy of comparable yield strength. (After Carter *et al.*, 1984.)

growth rates given for the over-aged alloy in Fig. 14.13. For this purpose, assume that $= 45°$, $\hat{D} = 0.75$, and $\chi = 0.00$, 0.25 and 0.50.

Solution:
    The dashed lines in Fig. 14.13 shows the predicted variations, from Eqs. 14.23 and 14.24, in the rates of fatigue crack growth due to crack deflection alone for $\theta = 45°$ and $\hat{D} = 0.75$ (which are typical deflection parameters for under-aged 7x75 aluminum alloys in vacuum) and $\chi = 0$ (no closure). Specifically, Eq. 14.24 gives the rise the nominal ΔK_I for the deflected crack, as compared to that of the straight crack (i.e. the extent of shift of the solid line in Fig. 14.13 to the right). Equation 14.23 gives the drop in the crack growth rate for the deflected crack, as compared to that of the straight crack (i.e. the extent of shift of the solid line in Fig. 14.13 to the bottom).
    Also shown in Fig. 14.13 are the growth rate predictions for high levels of fracture surface mismatch, $\chi = 0.25$ and 0.50. It is seen that the shape of the fatigue curves and the improvements in the fatigue behavior predicted for the deflected cracks are qualitatively similar to those measured experimentally (see Figs. 14.11 and 14.12).
    It is emphasized here that this example does not deal with any *intrinsic* improvements in the resistance of the material to fatigue fracture. Crack deflections induced by microstructural modifications merely reduce the apparent driving force for fracture. It is feasible that some microstructural design methods

which enhance crack deflection processes could also markedly reduce the intrinsic resistance of the material to both fatigue crack initiation and growth in which case the overall crack growth rates may in fact be increased.

14.10 Additional retardation mechanisms

14.10.1 Crack-bridging and trapping in composite materials

In composite materials with unidirectional, continuous fiber reinforcements, different failure mechanisms occur under far-field tension. If the fiber strength is sufficiently high, the tensile cracks extend completely through the matrix and the crack faces are bridged by the fibers. This type of failure is known to occur in brittle ceramics and glasses reinforced with SiC and carbon fibers (e.g., Marshall, Cox & Evans, 1985) as well as in titanium alloys reinforced with SiC fibers (Sensmeier & Wright, 1990). On the other hand, if the fiber strength is lower than a certain critical value, matrix failure results in complete failure of the composite since the fibers break in the wake of the advancing crack. There have been numerous theoretical studies of quasi-static toughening improvements associated with crack-bridging (e.g., Aveston & Kelly, 1973; Marshall, Cox & Evans, 1985; McCartney, 1987; Marshall & Cox, 1987).

Marshall *et al.* and Marshall & Cox developed a fracture mechanics analysis for fiber-reinforced composites where brittle matrix-cracking precedes fiber failure and where only frictional bonding exists between the fibers and the matrix. Consider the crack configuration schematically shown in Fig. 14.14(a) where continuous fibers aligned along the tensile loading direction bridge the crack over a distance b in the wake of the crack tip; a is the total length of the crack. such a crack-bridging process has, in fact, been observed in titanium-matrix composites, e.g., Fig. 14.14(b). The (SCS-6) SiC fibers bridging the crack over the distance b are subjected to a fiber pull-out force which is a function of the crack opening displacement, the interfacial sliding stress, the properties of the constituents of the composite and the diameter of the fiber. The effect of the frictional sliding force of the fibers bridging the crack faces, $\sigma_f(x)$ in Fig. 14.14(a), is modeled by imposing crack closure pressure, $p(x)$, where x is the position on the crack surface. Following the analysis of Marshall *et al.* for penny cracks, the closure pressure, for situations where the crack size a is significantly larger than the fiber spacing, is given by

$$p(x) = f\sigma_f(x), \tag{14.25}$$

where f is the volume fraction of the fibers. The effect of the far-field tensile stress σ^∞ on the crack tip is reduced by the closure pressure such that the near-tip stress intensity factor for the composite becomes

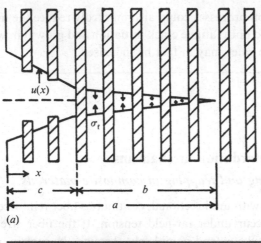

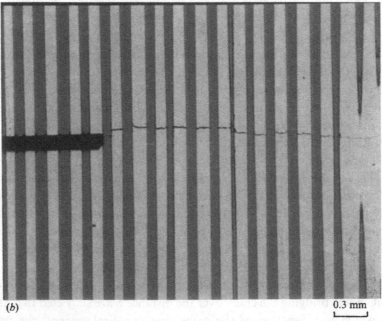

(b)   0.3 mm

Fig. 14.14. (*a*) A schematic representation of crack bridging. (*b*) Bridging of a fatigue crack by SiC fibers (32 volume%) in a Ti–6Al–4V alloy. (From Sensmeier & Wright, 1990. Copyright The Minerals, Metals and Materials Society. Reprinted with permission.)

$$K = 2\sqrt{\frac{a}{\pi}} \int_0^1 \frac{[\sigma^\infty - p(X)]X\,\mathrm{d}X}{\sqrt{1 - X^2}}, \tag{14.26}$$

where $X = x/a$. The closure pressure is related to the crack opening displacement by the relationship

$$p(x) = \alpha\sqrt{u(x)} \quad \text{for} \quad x > c, \quad p(x) = 0 \quad \text{for} \quad x < c, \tag{14.27}$$

where

$$\alpha = \sqrt{\frac{4\tau f^2 E_f E_c}{RE_m(1-f)}}. \tag{14.28}$$

Here, R is the radius of the fibers, E_f, E_m and E_c are Young's moduli of the fiber, matrix and composite, respectively, and τ is the frictional stress at the fiber–matrix interface. Equation 14.28 is valid when the interfacial shear stresses are neglected beyond the region of fiber/matrix slip. The crack opening displacement, $u(x)$, in the wake of the crack tip is obtained from the distribution of surface tractions such that

$$u(X) = \frac{4(1-\nu^2)a}{\pi E_c} \int_X^1 \frac{1}{\sqrt{s^2-X^2}} \left(\int_0^s \frac{[\sigma^\infty - p(\xi)]\xi}{\sqrt{s^2-\xi^2}} \, d\xi \right) ds, \tag{14.29}$$

where s and ξ are normalized position coordinates and ν is Poisson's ratio for the composite. Numerical solution of Eqs. 14.27 and 14.29 followed by the evaluation of the integral in Eq. 14.26 gives the effective stress intensity for the bridged crack.

Experimental studies of crack bridging during fatigue of titanium alloy–SiC fiber composites show that the rate of crack growth reaches a steady-state value as a consequence of reductions in near-tip ΔK induced by pressure on the crack faces by the bridging fibers. While the value of da/dN increases monotonically for (long) fatigue crack growth in a monolithic material subjected to a fixed value of far-field stress range, $\Delta\sigma^\infty$, the growth rate for the composite attains a ΔK-independent steady-state value, as shown in Fig. 14.15. Further descriptions of fatigue in fiber-

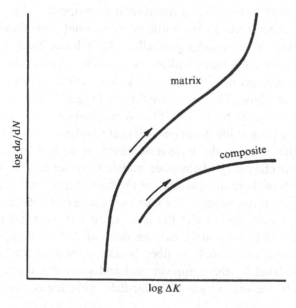

Fig. 14.15. Schematic representation of fatigue crack growth rate, da/dN, as a function of the nominal ΔK for a composite in which the fibers bridge the crack faces.

reinforced metal-matrix and ceramic-matrix composites can be found in McMeeking & Evans (1990), Evans, Zok & McMeeking (1995) and Cox (1996).

While the phenomenon of crack bridging promotes apparent improvements in the resistance to fatigue crack growth in composites with continuous fiber reinforcements, the interaction of the crack tip with discontinuous particles is also known to have a strong effect on the geometry on the crackfront and the rate of crack growth. The influence of particles on fatigue fracture can come about by one of two mechanisms: (i) The particles, if impenetrable, deflect the crack tip and cause a reduction in the effective ΔK. (ii) The particles, dispersed in the ductile matrix, *trap* (bridge) the crackfront (see Fig. 14.1(h)); the attendant changes in the geometry of the crackfront can lead to apparent improvements in fatigue crack growth resistance, depending on the size, shape and distribution of the particles and on whether the particles are tougher or more brittle than the matrix. Analyses of crack trapping have been published by Rice (1985) and extended to fatigue by Bower & Ortiz (1990).

14.10.2 On crack retardation in advanced metallic systems

The various retardation mechanisms discussed in this chapter offer possibilities for improving the crack growth resistance of engineering alloys by compositional and/or microstructural modifications, and by the addition of a reinforcement phase. Research in the areas of metal–ceramic composites provides some illustrative examples of such enhancements in the resistance to fatigue crack growth.

The need to exercise caution in the interpretation of fatigue crack growth resistance becomes abundantly clear in the case of metal-matrix composites. For example, ductile-matrix composites reinforced with brittle unidirectional fibers (which are oriented normal to the plane of the crack) generally exhibit lower fatigue crack growth rates than the unreinforced matrix alloy as a result of crack bridging. However, such beneficial fatigue properties cannot be guaranteed if the fiber orientation is parallel to the crack plane. This directionality of fatigue behavior is also strongly reflected in the stress–life (S–N) data for fiber-reinforced metal-matrix composites. In this latter case, the fatigue life decreases markedly as the loading direction changes from the axis of the fibers to the transverse direction normal to the fiber orientation.† In addition, the chemistry of the fiber–matrix interface and the differences between the coefficients of thermal expansion of the fiber and the matrix play a strong role in determining the mechanisms of fatigue damage. The far-field fatigue loading must be capable of overcoming the thermal residual stresses before the interface between the fiber and the matrix can be debonded. The competition among different failure mechanisms, such as fiber breakage, matrix cracking or interfacial debonding, is dictated by the compositional and geometrical properties of the composite as well as by the magnitude of the far-field cyclic loads *vis-à-vis* the

---

† The trend here is similar to that described in Fig. 7.12 for organic composites.

magnitude of the thermal residual stresses. Note that the composite is also highly susceptible to damage by pure thermal fatigue as a consequence of the vast differences in properties between the constituent phases.

An example of the role of the microscopic damage mechanism in influencing the fatigue properties is also found in the case of metal-matrix composites discontinuously reinforced with brittle particles. Aluminum alloys reinforced with particles or whiskers of SiC generally exhibit better near-threshold fatigue crack growth characteristics than the unreinforced matrix alloy if fatigue failure occurs predominantly within the ductile matrix (Christman & Suresh, 1989; Shang & Ritchie, 1989). Deflection of the fatigue crack by the brittle particles, enhanced crack closure and crack trapping contribute to this improved fatigue resistance in the composite by lowering the effective crack-tip opening displacement. Similar trends have also been seen in *in-situ* reinforced Ti-matrix composites with TiB whiskers (Sobeyejo, Ledrich & Sastry, 1995). However, if the particular combination of processing conditions, particle size, particle concentration and ageing treatment promotes particle fracture or interfacial separation, the fatigue crack growth resistance of the composite can be significantly lower than that of the matrix alloy (e.g., Sugimura & Suresh, 1992). Similarly, the low-cycle fatigue properties of the composite can also be inferior to those of the matrix (Bonnen, Allison & Jones, 1991).

14.11 Case study: Variable amplitude spectrum loads

The topics discussed thus far in this chapter have centered around fatigue crack growth in engineering materials subjected to a constant amplitude of cyclic loads. We now direct attention at concepts of variable amplitude fatigue crack growth. Particular attention is focused on the phenomenology and micromechanical processes associated with transient crack growth effects following single, periodic or block overloads in tension or compression.

There is a multitude of examples of variable amplitude loading in fatigue-critical structural components. Blades in gas turbine engines experience low amplitude, high frequency vibration during operation, superimposed on a relatively smaller number of cycles of fatigue loading due to start-up and shut-down. Railway tracks are subjected to random loading depending on the frequency and loading conditions associated with the passage of trains. The rotors and bearings of a turbogenerator are subjected to an overload during every start-up.

As an example of spectrum loading, the typical stress history to which the lower wing skin of a transport aircraft is subjected is shown in Fig. 14.16, where σ_m is the mean stress corresponding to an undisturbed flight. On the ground, the lower wing skin of the aircraft is loaded in compression. During flight, variable loads due to gust are superimposed on a mean tensile load corresponding to an undisturbed flight. The transition from a compressive mean stress on the ground to a tensile stress during flight is an important load cycle in itself and is usually referred to as the ground–air–ground cycle. Standard air spectra have also been developed to simulate the load sequence for transport aircraft

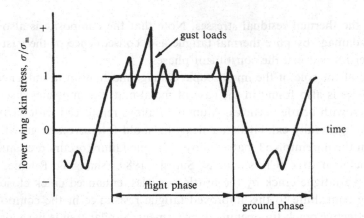

Fig. 14.16. Typical stress–time plot for the lower wing skin of a transport aircraft. (After de Jonge & Nederveen, 1980.)

wing tension skins near the main landing gear attachment. This program, known as Transport WIng STandard (TWIST), can account for ten different types of gust loading ranging from 'stormy' to 'tranquil' conditions. TWIST includes approximately 100 stress cycles per flight.

14.12 Retardation following tensile overloads

During the growth of a fatigue crack, load excursions in the form of single tensile overloads or high amplitude–low amplitude block loading sequences can result in the retardation of crack advance or even in the complete arrest of the crack. Figure 14.17 schematically illustrates the typical crack growth behavior following a single tensile overload.

During fatigue crack growth at a baseline stress intensity factor range ΔK_B, the material exhibits a characteristic crack growth rate $(da/dN)_B$. If a tensile overload is now applied, the ductile solid generally exhibits a small amount of temporarily accelerated growth which mainly occurs during the application of the overload (e.g., Corlby & Packman, 1973). The marking on the fracture surface corresponding to the rapid extension of the crack during the overload cycle is sometimes referred to as the *stretch zone*. These stretch zone markings are commonly observed when $K_{2\,max}$ is significantly greater than $K_{1\,max}$. If the overload causes a change in the microscopic or macroscopic fracture mode, accelerated crack growth may persist over many post-overload cycles.

The burst of accelerated crack advance, shown in Figs. 14.17(*b*) and (*c*), is generally followed by a prolonged period of decelerated crack growth, when the post-overload fatigue crack is subjected to the same baseline stress intensity factor range ΔK_B as in the pre-overload regime. This progressive reduction in crack velocity

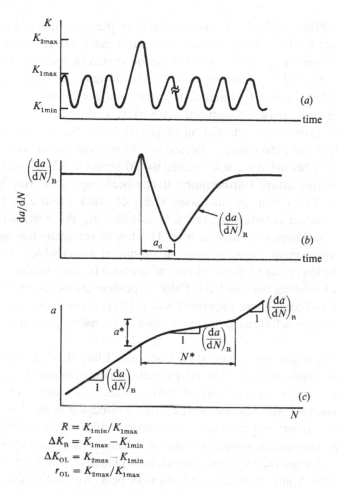

Fig. 14.17. Definitions of different parameters used to describe transient crack growth effects following single tensile overloads.

continues over a crack growth distance a_d, which is commonly known as the *delay distance*. After reaching a minimum, the crack growth rates begin to increase and eventually catch up with the pre-overload value $(da/dN)_B$. The total crack growth distance a^* and the total number of post-overload cycles N^*, over which the transient effects of the single tensile overload affect crack growth, are a strong function of ΔK_B, the material microstructure and environment, as well as such factors as the nominal load ratio R, and the overload ratio r_{OL}, which are defined in Fig. 14.17.

14.12.1 Plasticity-induced crack closure

The discussions presented in Section 14.1 demonstrated that permanent residual tensile displacements formed due to plastic deformation at the crack tip result in contact between the fracture surfaces as the crack tip advances through the prior plastic zone. Elber (1970, 1971) argued that such plasticity-induced crack closure can also account for the transient retardation phenomena due to overloads. The application of an overload produces a larger plastic stretch in the wake of the fatigue crack tip (as compared to the baseline ΔK), as the fatigue crack propagates through the overload plastic zone. The attendant fracture surface contact causes an enhancement in the level of plasticity-induced crack closure in the post-overload regime which, in turn, promotes a retardation of growth rates. Since some crack growth through the overload plastic zone is needed for the residual plastic stretch to induce premature closure behind the crack tip, one would expect a delayed retardation based on this interpretation. Furthermore, it has been suggested (von Euw, Hertzberg & Roberts, 1972) that the increased levels of crack closure and the increase in frictional contact between the crack faces following the overload can obliterate any surface markings, such as striations. This line of reasoning was ostensibly consistent with experimental observations on aluminum alloys which showed crack growth with striations prior to the overload and abraded fracture surfaces with no striation markings following the overload. Direct experimental measurements of closure using the push-rod compliance gage (see Section 14.1) also show higher levels of crack closure in plane stress following the application of a tensile overload (Fleck, 1988).

Although there is a variety of experimental evidence (e.g., Elber, 1970; Himmelein & Hillberry, 1976) supporting the role of plasticity-induced crack closure in influencing retardation effects, many observations on delay effects in a number of alloy systems are also inconsistent with this mechanism. (i) As discussed in Section 14.1, plasticity-induced crack closure generally leads to a greater reduction in effective ΔK in plane stress than in plane strain. However, more enhanced levels of post-overload retardation have been documented in some materials in plane strain than in plane stress (Suresh, 1983b). (ii) Many characteristic features of post-overload retardation, such as changes in the mode of failure following overloads and in the effects of microstructure on crack growth in many aluminum alloys, are inconsistent with the descriptions of plasticity-induced crack closure concepts (Knott & Pickard, 1977; Bucci et al., 1980; Suresh, 1983b; also see later sections).

14.12.2 Crack tip blunting

It has been argued that the blunting of the crack tip by the overload cycle, which can persist even during post-overload crack growth, can lead to crack growth retardation (Christensen, 1959; Rice, 1967). The underlying mechanism for retardation is that the blunted crack tip behaves like a notch with a less severe stress concentration than the originally sharp crack tip. Blunting of overloaded fatigue cracks is also observed in polymeric materials such as polycarbonate (Banasiak, Grandt & Montulli, 1977) and in a number of metal alloys (e.g., Lankford & Davidson, 1981); see Fig. 14.18 for an example of crack tip blunting following an overload. Although crack tip blunting does influence the rate of post-overload crack growth and provides a basis for interpreting certain load sequence effects (see Section 14.13), it cannot account for the existence of delayed retardation and cannnot quantitatively rationalize the experimentally observed reductions in post-overload crack growth.

14.12.3 Residual compressive stresses

Reverse yielding ahead of a fatigue crack loaded in cyclic tension produces residual compressive stresses (Section 9.6). When a tensile overload is applied, the size of the zone of residual compression is increased. X-ray measurements of stress fields ahead of fatigue cracks in steels (Allison, 1979; Taira & Tanaka, 1979) also indicate an enlarged residual compressive zone following tensile overloads (as compared to the pre-overload case). It has been suggested (Willenborg, Engle & Wood, 1971; Wheeler, 1972) that these residual compressive stresses can retard post-overload crack growth. As discussed in later sections, some of the most widely used life prediction models tacitly employ the concept of retardation due to residual compressive stresses. However, one encounters some fundamental difficulties with rationalizing a number of retardation events solely on the basis of residual compressive stresses: (i) The largest residual compressive stresses exist in the immediate vicinity of the crack tip (see Fig. 9.9). Hence, instantaneous retardation or crack arrest, rather than delayed retardation, is predicted contrary to experimental observations. (ii) Experiments show (e.g., Suresh, 1983b) that retardation can persist even when the post-overload crack has traversed through the (predicted) zone of residual compressive stresses.

14.12.4 Deflection or bifurcation of the crack

In some engineering alloys with a propensity for planar slip deformation, and in certain loading situations that involve plane stress conditions, the application of a tensile overload promotes deflections of the crack tip from the nominal mode I crack growth plane (Schijve, 1974; Lankford & Davidson, 1981; Suresh, 1983b). Figure 14.18 shows an example of crack deflection in an underaged Al–Li–Cu–Zr alloy in which the application of a tensile overload resulted in a severe bifurcation of the crack tip throughout the thickness of the edge-notched (compact) tensile specimen. Note that the fatigue crack remains blunted even after complete unloading from the far-field tensile stress. The enhanced planarity of slip due to the presence of δ' (Al$_3$Li) precipitates in this material has been identified as a major factor in promoting such bifurcated crack profile in the intense strain region ahead of the fatigue crack tip.

The growth of the fatigue crack along a deflected path implies a reduced crack propagation resistance away from the mode I growth plane. Thus, temporary accel-

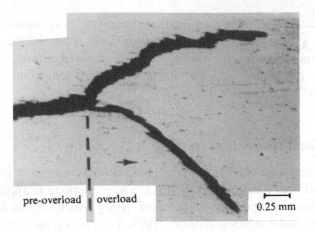

pre-overload ¦ overload

0.25 mm

Fig. 14.18. Branching of a fatigue crack upon application of a tensile overload in an under-aged Al–2.1Li–2.9Cu–0.12Zr alloy. Arrow indicates the crack growth direction. (From Suresh *et al.*, 1987. Copyright Pergamon Press plc. Reprinted with permission.)

eration in crack growth rates can occur during the application of the overload. However, from a continuum standpoint, a bifurcated crack tip profile is subjected to a lower effective ΔK than a straight crack of the same projected length subjected to an identical value of far-field ΔK (Section 9.11). On the basis of experimental results obtained on aluminum alloys and of linear elastic analyses of branched cracks, it has been demonstrated that a reduction in effective ΔK of approximately 25% can be realized in the post-overload regime solely from the crack deflection process (Suresh, 1983b). Thus, the geometrical effects associated with the bifurcation of the fatigue crack tip by tensile overloads can, in certain cases, contribute markedly to retardation effects. The development of changes in crack tip geometry during variable amplitude fatigue is a strong function of microstructure, stress state and baseline ΔK levels. It should be noted that if crack deflection occurs preferentially along a path of low fracture resistance (such as a weak interface) in some materials (such as composites), it is likely to accelerate rather than retard post-overload growth rates.

14.12.5 Near-threshold mechanisms

When a fatigue crack is subjected to tensile overloads, the development of enhanced levels of plasticity-induced crack closure, residual compressive stresses or crack deflection results in a post-overload stress intensity factor range whose effective value is much lower than the nominal baseline value ΔK_B. Therefore, even when the post-overload ΔK_B is well into the Paris regime of fatigue crack growth, the corresponding effective ΔK may fall in the near-threshold fatigue regime for the material. In such cases, *additional* mechanisms of fatigue crack growth retardation, which are of significance to constant amplitude near-threshold fatigue, are also

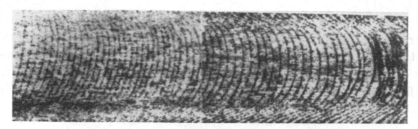

Fig. 14.19. Bands of oxide formed on the fracture surface in a 7010-T7 aluminum alloy fatigued in moist air with 8000 constant stress amplitude fatigue cycles separating tensile overloads (overload ratio = 1.8). Spacing between oxide bands ≈ 0.3–0.4 mm. (From Bucci *et al.*, 1980. Copyright American Society for Testing and Materials. Reprinted with permission.)

activated in the post-overload regime (Suresh, 1983b). For many alloys, these additional retardation mechanisms typically involve stage I crack growth as well as oxide-induced and roughness-induced crack closure. Figure 14.19 shows an example of oxide layer formation on the fracture surface of a 7010-T7 aluminum alloy which was subjected to a single tensile overload ($K_{2\,\mathrm{max}} = 1.8K_{1\,\mathrm{max}}$) after every 8000 cycles of constant amplitude fatigue crack growth at $\Delta K_{\mathrm{B}} = 7 - 10\,\mathrm{MPa\sqrt{m}}$ and $R = 0.33$. (This is a regime of growth in which oxide-induced closure is not significant during constant amplitude fatigue in the moist air environment.) Each dark band of oxide observed in this micrograph corresponds to retarded crack growth after one tensile overload.

The activation of near-threshold mechanisms in the post-overload regime is not the primary cause of retardation; it is a process which prolongs retardation. This interpretation is consistent with the observations of delayed retardation because the near-tip ΔK should be reduced first to near-threshold levels by other processes before the typical near-threshold growth and closure mechanisms for the materials begin to influence post-overload failure. (Oxide-induced and roughness-induced closure are referred to as near-threshold mechanisms in the present discussion primarily because their influence on crack growth is stronger at lower ΔK levels.)

Experimental evidence of the signficance of near-threshold mechanisms to post-overload crack growth is found in a number of aluminum alloys (Suresh, 1983b). Studies of constant amplitude fatigue in under-aged, peak-aged and over-aged aluminum alloys of 2xxx and 7xxx series indicate that the threshold stress intensity factor range ΔK_0 decreases and that the near-threshold growth rates increase with an increase in the degree of ageing (see the examples provided in Section 10.4). In the Paris regime of crack growth, however, the opposite trend is observed for most aluminum alloys, with the over-aged microstructure exhibiting the lowest crack propagation rates. If periodic overloads are superimposed on the constant amplitude stress cycles in the Paris regime, the ranking of different microstructures in terms of resistance to crack growth is switched. The pre-overload crack growth resistance at a fixed baseline stress intensity factor range, ΔK_{B}, may increase in the microstructural sequence: under-aged → peak-aged → over-aged condition. However, the post-overload crack growth resistance at the same baseline stress intensity range increases in the opposite sequence: over-aged → peak-aged → under-aged (Knott & Pickard, 1977; Bucci *et al.*, 1980). These interpretations of the role of threshold mechanisms in variable amplitude fatigue also find supporting evidence in the earlier data reported in the literature on aluminum and titanium alloys (Hertzberg & Mills, 1976; Hopkins *et al.*, 1976; Nowack *et al.*, 1979).

14.13 Transient effects following compressive overloads

The application of fracture mechanics concepts to fatigue crack growth characterization is generally based on the premise that crack advance occurs only during that portion of the fatigue cycle where the crack faces do not touch. Implicit in this approach is the notion that compressive stresses do not cause any significant changes in crack growth rates because (long) fatigue cracks remain closed during the compressive loading (Hudson & Scardina, 1969). However, there has been a growing body of evidence which unequivocally shows that, for both long and short fatigue cracks, the application of compressive overloads can lead to an acceleration in crack growth rates.

It was shown in Chapters 4–6 that fully compressive cyclic loads of both constant and variable stress amplitudes result in the nucleation and growth of fatigue cracks ahead of stress concentrations. The mechanisms underlying this effect involve the generation of residual tensile stresses upon unloading. The harmful effects of compressive loads on the growth of long tensile fatigue cracks has also been well documented. Crooker (1971) presented the results of a study which showed that the compression portion of a fully reversed, tension–compression cycle can contribute substantially (up to 50%) to fatigue crack growth rates in medium to high strength steels and in titanium alloys.

Some of the clearest evidence of the deleterious effects of compressive overloads on crack growth resistance is also found in the work of Topper & Yu (1985). These authors measured the propagation rates of fatigue cracks, 2 to 5 mm in length, in center-cracked plates of 2024-T351 aluminum alloy which were subjected to periodic compressive overloads superimposed on zero–tension fatigue. Figure 14.20 shows the variation of fatigue crack growth rates, da/dN, plotted as a function of the maximum stress intensity of the fatigue spectra, $K_{max} = \Delta K_B$, for different frequencies of application of a compressive overload of fixed amplitude (which was several times the magnitude of the maximum tensile stress of the fatigue cycle). The results given in this figure indicate that the near-threshold fatigue crack growth rates measured using a loading spectrum with periodic compressive overloads are always higher than those of constant amplitude, zero-tension fatigue ($N_c \rightarrow \infty$); this observation is true for all the values of N_c defined in Fig. 10.6. Similar results have also been reported by other workers. For example, de Jonge & Nederveen (1980) show

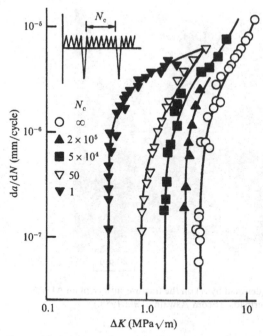

Fig. 14.20. Influence of compressive overloads on the rates of fatigue crack growth in a 2024-T351 aluminum alloy. (After Topper & Yu, 1985.) The growth rates are characterized in terms of the maximum stress intensity factor range. The inset shows the normalized stress intensity factor for the loading sequence as a function of time. N_c denotes the number of baseline tensile fatigue stress cycles in-between two overloads.

that when the ground–air–ground cycles are removed from the lower wing skin loading spectrum, Fig. 14.16, the fatigue life improves by as much as a factor of 3.3. A nonpropagating (compression) fatigue crack ahead of a stress concentration is also found to reinitiate upon the application of compressive overloads (Aswath *et al.*, 1988).

Interpretations of the plausible reasons for crack growth acceleration due to compressive overloads have centered around crack closure concepts. It has been postulated (Topper & Yu, 1985) that compressive overloads lead to the flattening of fracture surface asperities. Since most aluminum alloys are known to be susceptible to high levels of crack closure due to the premature contact between fracture surface asperities, this line of reasoning provides an appealing rationale for the lowering of crack growth resistance in the presence of compressive overloads. Observations of fracture surface abrasion induced by compressive overloads have also been reported for some alloys. Figure 14.21 is a scanning electron micrograph of the fracture surface in an AISI A542 Class 3 steel (of bainitic microstructure and with tensile yield strength = 500 MPa). This photograph shows compression-induced abrasion marks (marked by the letter 'A') in an area immediately behind the crack tip. In this case,

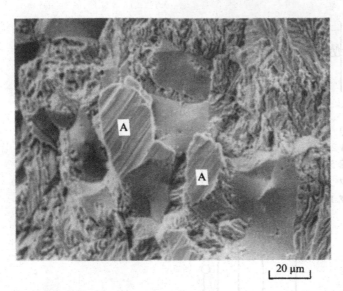

Fig. 14.21. Abrasion marks (denoted by *A*) on the fracture surface of an AISI A542 Class 3 steel subjected to compressive overloads. (From Aswath *et al.*, 1988.)

the crack was propagated ahead of a stress concentration under fully compressive far-field cyclic loads until crack arrest occurred. Subsequent application of compressive overloads reinitiated the fatigue crack. The abrasion marks correspond to the crack growth region where the compressive overloads were applied. Note that the abrasion lines on two different grains on the fracture surface are parallel. Had these marks been a consequence of cyclic slip traces, their orientations in different grains would have been different.

The influence of compressive overloads on the growth of long fatigue cracks is strongly dependent on the micromechanisms of crack growth, in particular on prior development of crack surface morphology and roughness-induced crack closure. These factors, in turn, depend on such microstructural dimensions as the grain size. Since roughness-induced crack closure has a more dominant effect near the threshold than in the Paris regime, the role of abrasion in affecting fatigue crack growth is also expected to be significant for baseline ΔK values which fall in the near-threshold regime. Therefore, compressive overloads are likely to have a more dominant effect on near-threshold fatigue crack growth than in the higher growth rate regimes of fatigue. Similarly, the generation of residual tensile stresses upon unloading from the peak compressive stress is less likely for a long fatigue crack. However, such effects can become important for short cracks emanating from stress concentrations (Chapters 8 and 15) and for long cracks previously blunted by a tensile overload (see next section).

14.13.1 Compressive overloads applied to notched materials

Aswath *et al.* (1988) conducted a series of systematic experiments to investigate the role of stress amplitude of the very first compression cycle on subsequent compression fatigue crack growth, while maintaining the stress state, specimen geometry, notch geometry and the rest of the loading sequence fixed. It was found that whereas stable crack growth (prior to crack arrest) occurs from the root of the notch over hundreds of thousands of compression cycles, the amplitude of the very first compression cycle played the most decisive role in determining the rate and total distance of crack advance. Specifically, a three-fold increase in the stress amplitude during the first compression cycle alone led to a ten-fold increase in the total distance of crack growth a^*, even though the subsequent million or more fatigue cycles were kept identical. Figure 14.22 shows the remarkable increase in the crack growth distance due to the increase in the amplitude of the first compression cycle. Numerical simulations rationalize this increase in a^* on the basis of the increase in the extent of the tensile residual stress zone generated by the compressive overload in the first cycle (Aswath *et al.*, 1988). These results clearly demonstrate the significant role of compressive overloads on crack initiation and growth ahead of stress concentrations. This mechanism is expected to have a profound effect on the fatigue life of structural components where cyclic compression loading occurs during service.

14.14 Load sequence effects

The preceding discussion has focused on the role of single overloads on crack growth. In this section, some examples of how different combinations of cyclic loads influence fatigue fracture are taken up.

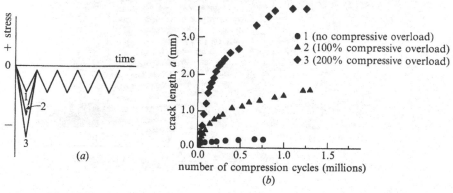

Fig. 14.22. (*a*) Schematic showing constant amplitude loading (test 1) and variable amplitude loading where a 100% (test 2) and 200% (test 3) overload was applied only in the first cycle. The applied stress amplitude of the remaining cycles was held fixed. (*b*) The effect of the stress amplitude of the very first compression cycle on crack growth behavior from a notch tip in an AISI A542-3 steel (tensile yield strength = 500 MPa) under plane stress. (After Aswath *et al.*, 1988.)

14.14.1 Block tensile load sequences

One of the most important load sequences in the study of variable amplitude fatigue fracture is that involving blocks of different cyclic stress amplitudes and/or load ratios. Let us now consider some examples of transient effects induced by block loads. Figure 14.23 is a scanning electron fractograph showing fatigue striations in a 2024-T3 aluminum alloy subjected to a high–low block loading sequence, which is

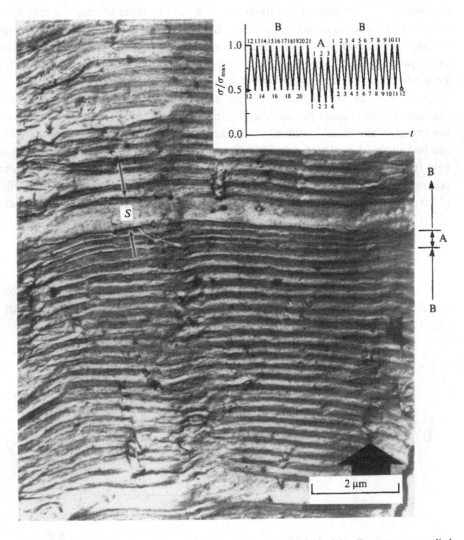

Fig. 14.23. Typical fractography resulting from a high–low–high block loading sequence applied to a 2024-T3 aluminum alloy. Crack growth corresponding to each block and the crack growth direction (large arrow) are indicated. The fatigue loading sequence is shown where the applied cyclic stress, normalized by its maximum value σ_{max}, is plotted as a function of time t. (From McMillan & Pelloux, 1967. Copyright American Society for Testing and Materials. Reprinted with permission.)

shown along with the micrograph. In this case, the amplitude of the cyclic loads in all three blocks is the same. Block B consists of 21 load cycles, whereas block A consists of four load cycles. Block B leads to the formation of well-defined striation markings during each load cycle, with the striation spacing corresponding to the rate of crack extension per cycle. Upon switching from block B to block A (i.e. from peak B21 to valley A1), crack growth continues at a slower rate with a corresponding reduction in striation spacing. (The four narrowly spaced striations just prior to the stretch zone marked *S* denote crack growth in load block A.) If the loading is switched again from block A to block B (i.e. from valley A4 to peak B1), a stretch zone (denoted *S*) is created as a consequence of the abrupt increase in the peak tensile stress. This sudden burst of crack advance is followed by uniform fatigue crack growth rates characteristic of the stress amplitude in block B.

Figure 14.24 shows another example of tensile block loading where the peak tensile stresses for the blocks are the same and the stress amplitude in each block is different. Blocks B and C cause crack growth, with the rate of crack advance per cycle corresponding to the respective striation spacings seen in the figure. The stress

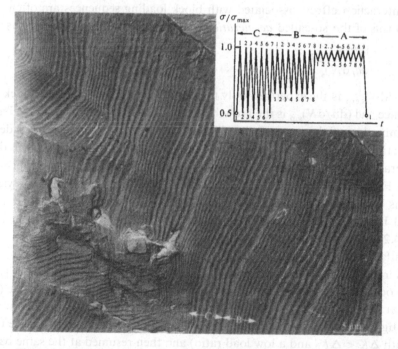

Fig. 14.24. Typical fracture surface due to block loading in a 2024-T3 aluminum alloy. Regions of crack growth corresponding to blocks B and C are marked. No crack growth occurred due to the application of block A. The fatigue loading sequence is shown in the inset where the applied cyclic stress, normalized by its maximum value σ_{max}, is plotted as a function of time t. (From McMillan & Pelloux, 1967. Copyright American Society for Testing and Materials. Reprinted with permission.)

amplitude of block A was too small to cause any crack growth. Note the absence of any stretch zones or temporary acceleration in this loading sequence with a constant value of maximum stress intensity factor. (Although the examples in Figs. 14.23 and 14.24 show clear striations, it should be noted that, as discussed in Chapter 10, the striation spacing does not always correlate with crack growth rates, and that not all materials exhibit well marked striations.)

The following transient effects are commonly observed when high–low or low–high block loading sequences are applied: (i) The application of a low–high block loading sequence can cause a stretch zone or a bifurcation crack tip profile, depending on the relative amplitudes and load ratios of the blocks. (ii) A transient acceleration in crack growth may be induced during the initial stages of ahigh amplitude block which follows a low amplitude block, if the blunting of the crack tip by the high block reduces the pre-existing level of fatigue crack closure (i.e. increases ΔK_{eff}). (iii) Sudden reductions in K_{max} corresponding to a high–low sequence with a reduced R ratio or ΔK can cause pronounced retardation or even complete crack arrest. (iv) Significant changes can occur in the mechanisms of crack growth upon changing from one block to another; the attendant transient effects are dictated by the specific microstructural properties of the material.

Load interaction effects associated with block loading sequences are often quantified in terms of the so-called *acceleration factor*, γ. This factor is defined as

$$\gamma = \frac{(da/dN)_{s,exp}}{(da/dN)_{s,p}},$$ (14.30)

where $(da/dN)_{s,exp}$ is the experimentally measured crack growth rate per block loading sequence and $(da/dN)_{s,p}$ is the growth rate per sequence predicted on the basis of linear summation of the constant amplitude crack growth response. $\gamma = 1$ denotes no interaction effects, whereas $\gamma > 1$ indicates acceleration in crack growth due to load interactions; $\gamma < 1$ indicates retardation. For the block loading programs shown in Figs. 14.23 and 14.24, $\gamma \approx 2$ (for crack growth where small-scale yielding conditions prevailed). Many researchers (e.g., Nisitani & Takao, 1978), who employed block tensile overloads and underloads (similar to the sequence shown in Fig. 14.24) to study interaction effects in steels, also reported acceleration factors significantly greater than unity.

In materials prone to certain types of crack closure, growth retardation ($\gamma < 1$) may also occur as a consequence of load interaction effects. Suresh & Ritchie (1981) found that crack growth retardation occurred in a bainitic steel, if near-threshold tensile fatigue crack growth was interrupted with millions of cycles of tensile underloads (with $\Delta K < \Delta K_0$ and a low load ratio) and then resumed at the same baseline ΔK. A similar trend was also reported by Kobayashi *et al.* (1984) for AISI A508-3 steel. This coaxing-like retardation effect is believed to be a result of enhanced oxide formation on the crack faces during the application of underloads.

14.14.2 Tension–compression load sequences

The rate of fatigue crack growth is also known to depend strongly on the order in which tensile and compressive overloads are applied. Figure 14.25 schematically shows typical effects of tension–compression sequence effects on variable amplitude fatigue fracture. Also shown, for comparison purposes, are the typical crack growth characteristics of a ductile solid under constant amplitude loads and under the influence of tensile overloads. As seen earlier, the application of a tensile overload can severely blunt the crack tip. If a compressive overload is applied immediately after a tensile overload, the crack tip which is blunted by the tensile overload behaves like a notch. As shown in Chapter 8, the application of a cyclic compressive load to a blunt notch results in the formation of residual tensile stresses which easily nucleate a fatigue crack. These residual tensile stresses can partly or fully nullify the beneficial delay effects associated with the application of a tensile overload. It has been experimentally established (Marissen, Trautmann & Nowack, 1984) that the application of a compressive overload immediately following a tensile overload gives rise to rates of crack growth which are significantly faster than those observed in the presence of the tensile overload alone.

If a compressive overload precedes a tensile overload, the rate of fatigue crack may increase or remain the same (as compared to the loading sequence involving the tensile overload alone) depending on the loading conditions. For long fatigue cracks in wavy slip materials and for baseline stress intensity range ΔK_B, which are well into the Paris regime, the application of a compressive overload prior to a tensile overload may not have any significant effect on the delayed retardation produced by the overload.

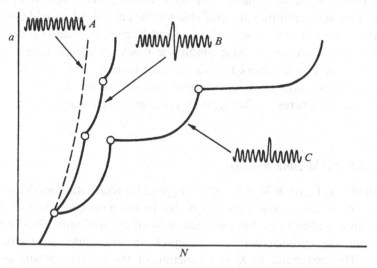

Fig. 14.25. A schematic illustration of transient crack growth during constant amplitude fatigue (*A*) and during variable amplitude loading involving single tensile overloads (*C*) or tensile-compressive overload sequences (*B*). The open circles represent the crack length locations at which each variable amplitude sequence is applied.

However, for low ΔK_B values and in planar slip materials prone to marked levels of roughness-induced closure, a compressive overload preceding a tensile overload may cause some reduction in the extent of delayed retardation and hence an increase in crack growth rate as compared to the application of the tensile overload alone.

14.15 Life prediction models

The ultimate aim of fatigue research is to develop quantitative models for the prediction of useful life in cyclically loaded structures. In view of the complexity of the mechanisms underlying variable amplitude fatigue, it is not surprising that only semi-empirical approaches are capable of providing reasonable estimates of service life in engineering components subjected to complex loading histories.

Available life prediction methods can be broadly classified into two categories: the cycle-by-cycle approach and the characteristic approach. In the cycle-by-cycle approach, the extent of cyclic damage or crack advance associated with each cycle is added to predict the overall fatigue life. The most well known cycle-by-cycle model is the Palmgren–Miner linear damage rule, Eq. 7.8, in which load interaction effects are not considered. The cycle-by-cycle summation concept can also be extended to fatigue crack growth. Here, the crack growth increment per cycle may be assumed to equal the crack growth rate associated with constant amplitude loading of the same magnitude; alternatively, it may also be deduced from a load interaction model (see below).

One of the critical issues in the use of the cycle-by-cycle method is to define what is meant by a fatigue cycle in variable amplitude loading. As noted in Chapter 8, several cycle counting techniques are available for the analysis of variable amplitude loads in conjunction with the low-cycle fatigue and fracture mechanics concepts.† In finite element simulations of variable amplitude crack growth, a fatigue cycle is represented by some pre-determined crack advance (usually one mesh dimension), rather than one fluctuation in stress or stress intensity factor. In this section, we examine the salient features of the cycle-by-cycle approach and the characteristic approach.

14.15.1 Yield zone models

Willenborg, Engle & Wood (1971) suggested a retardation model, the basic premise of which is that delayed growth is due to the increased levels of residual compressive stresses ahead of a fatigue crack induced by overloads. These compressive residual stresses cause a reduction in crack tip stress intensity factor by an amount K_{red}. This reduction in K is a function of the forward plastic zone size associated with the current load cycle of a random load spectrum and of the max-

---

† See Chapter 8 for an example of a cycle counting method.

imum plastic zone size created by the overloads. Crack growth retardation is predicted by this model if the minimum stress intensity factor K_{min} of the current loading cycle is less than K_{red}.

Although widely used in the aerospace industry, some of the basic features of this model run counter to current mechanistic understanding of crack closure. For example, the model implies that the delay in crack growth is due to a reduction in K_{max}, rather than a reduction in ΔK_{eff}.

Wheeler (1972) introduced a simple retardation model using similar assumptions. In this approach, the variable amplitude crack growth rate, $(da/dN)_{VA}$, in a fatigue cycle is related to the equivalent constant amplitude growth rate at the same nominal ΔK, $(da/dN)_{CA}$, by a retardation factor, β, so that

$$\left(\frac{da}{dN}\right)_{VA} = \beta \left(\frac{da}{dN}\right)_{CA}, \quad \beta = \left(\frac{r_{pi}}{r_{max}}\right)^k. \tag{14.31}$$

Here, r_{pi} is the plastic zone size associated with the ith (post-overload) stress cycle and r_{max} is the distance from the current crack tip to the greatest prior elastic–plastic boundary created by the overload. k is an exponent which is empirically chosen to obtain the best overall crack growth predictions for a given material and loading spectrum. Implicit in the Wheeler model is the notion that the main cause of retardation is the zone of residual compressive stresses at the crack tip created by the overload.

14.15.2 Numerical models of crack closure

On the basis of Elber's work (1970) on crack closure, many attempts have been made to develop models for variable amplitude fatigue. Experiments indicate that the crack closure stress varies in every cycle when the block of random loading history consists of many cycles. On the other hand, when the block length is short, the crack closure stresses remain roughly constant. The former (long) loading history is termed *nonstationary* and it denotes a loading sequence which repeats itself after a crack growth increment of greater than the largest prior plastic zone. The latter (short) loading waveform is termed *stationary* and it represents the loading sequence which is repeated regularly and exactly.

Crack closure concepts have also been developed into predictive models using detailed finite-element programs where iterative solution procedures are invoked for a cycle-by-cycle calculation of closure (e.g., Newman, 1983). Newman's finite-element formulation for plasticity-induced crack closure predicts a change in crack opening load after a fixed amount of crack growth rather than after each load cycle. It can, therefore, be used for both stationary and nonstationary loading histories. The main drawback of these models is that they are based on formulations which do not incorporate crack closure effects due to oxidation or fracture surface roughness on variable amplitude fatigue life. Other developments in the numerical modeling of

crack closure in variable amplitude fatigue can be found in the volumes edited by
Newman & Elber (1988) and Petit *et al.* (1988).

14.15.3 Engineering approaches

Many researchers (e.g., Socie, 1977; Schijve, 1980; Kikukawa, Jono &
Mikami, 1982; Sunder, Seetharam & Bhaskaran, 1984) have employed the crack
closure concept in conjunction with various cycle counting methods to predict
crack growth life in variable amplitude fatigue involving random and block program
loads. These approaches, which predict fatigue lives to within an accuracy of a factor
of two, involve the following assumptions. (i) The crack opening stress intensity
factor K_{op} is constant during each block of variable amplitude loading. K_{op} is
fixed by the maximum stress intensity factor $(K_{max})_{VA}$ and the minimum stress
intensity factor $(K_{min})_{VA}$ in each program block. In addition, the crack opening
load is assumed equal to that in a constant amplitude fatigue test with an equivalent
ΔK, i.e. $\delta K = K_{max} - K_{min} = (K_{max})_{VA} - (K_{min})_{VA}$. (ii) Crack growth is integrated on
a cycle-by-cycle basis. This involves the modified Paris rule,

$$\frac{da}{dN} = C_1(\Delta K_{eff})^{m_1},$$

where $\quad \Delta K_{eff} = K_{max} - K_{op}, \quad$ if $\quad K_{min} < K_{op},$

$\qquad\qquad\qquad = K_{max} - K_{min}, \quad$ if $\quad K_{min} \geq K_{op}.$ $\qquad$ (14.32)

Here da/dN is the crack growth rate due to an effective stress intensity factor range,
ΔK_{eff}, and C_1 and m_1 are empirical constants obtained from constant amplitude
data based on da/dN versus ΔK_{eff} plots.

(iii) Cycles are counted using the rainflow, range-pair or other methods (see
Chapter 8). The main appeal of this approach is that one can deduce the crack
opneing behavior in variable amplitude loading on the basis of fatigue crack growth
response measured using constant amplitude loading.

14.15.4 The characteristic approach

This approach was first proposed by Paris (1960) for random loading. The
basic hypothesis here is that, because of similarity, the random variation of the crack
tip fields are describable in terms of the root-mean-square value of the stress intensity
factor range, ΔK_{rms}. The variable amplitude crack growth rates are then given by the
Paris-type relationship

$$\frac{da}{dN} = C(\Delta K_{rms})^m,$$ $\qquad$ (14.33)

where C and m are, as discussed in Chapter 10, material constants and

$$\Delta K_{\text{rms}} = \frac{\sqrt{\left(\sum_{i=1}^{n} \Delta K_i^2\right)}}{n}. \tag{14.34}$$

Here, ΔK_i is the stress intensity factor range in the ith cycle in a sequence consisting of n stress cycles. Note that $\Delta K_{\text{rms}} = \Delta K$ for constant amplitude fatigue. Therefore, the characteristic approach suggests that the average crack growth rate in spectrum fatigue can be predicted from constant amplitude fatigue data using Eq. 14.34. Although empirical, this approach has been widely used in a number of fatigue-critical applications; examples include variable amplitude fatigue of steel bridges (Barsom, 1976). However, in some applications, such as flight simulation of gust loading spectra for aircraft, random loading and its statistically equivalent program loading are known to provide different results (Schijve, 1973).

Exercises

14.1 Discuss the advantages and disadvantages of each of the following methods for measuring (i) crack length and (ii) crack closure loads:
   (a) Crack mouth opening displacement gage.
   (b) Strain gages mounted in the wake of the crack, above and below the plane of the crack.
   (c) Strain gages mounted immediately ahead of the crack tip.
   (d) Back-face strain gage.
   (e) DC electrical potential drop.
   (f) AC electrical potential drop.
   (g) Ultrasonics.
   (h) Acoustic emission.
   (i) Travelling microscope providing a direct optical image of the crack on the specimen surface.

   Also discuss the experimental procedures you would use for the calibration of crack length and crack closure, where appropriate, for each of the above methods.

14.2 For each of the following cases, plot the load (P) versus the crack opening displacement (δ) curve for a metallic material subjected to loading and unloading phases in zero–tension–zero fatigue:
   (a) Plastic deformation at crack tip; no change in crack configuration (i.e. crack opening and crack length); loading phase.
   (b) Gradual opening of the crack during the loading phase and plastic deformation at crack tip.
   (c) Elastic behavior at constant crack configuration during the loading phase.

(d) Effect of plastic behavior = effect of configuration change on the P–δ plot during the unloading phase.

(e) Effect of plastic behavior > effect of configuration change on the P–δ plot during the unloading phase.

(f) Effect of plastic behavior < effect of configuration change on the P–δ plot during the unloading phase.

(g) Configuration change during the unloading, with negligible plastic deformation at crack tip.

14.3   Derive Eq. 14.24 following the discussion in Section 14.9.

14.4   The nominal variation of the fatigue crack growth rate, da/dN, as a function of the nominal stress intensity factor range, ΔK, for a hypothetical, fine-grained alloy is the same as that shown by the solid line in Fig. 14.12. This crack growth response was obtained at room temperature at a load ratio of zero. Optical microscopy of the crack profiles reveals a highly planar (straight) fracture path normal to the tension fatigue loading direction. Measurements of changes in compliance during loading and unloading do not show any crack closure over the entire range of ΔK. This material was given a proprietary heat treatment which results in a drastic change in the fracture mode, without altering the overall stress–strain characteristics of the alloy or the grain size. The heat-treated alloy exhibited a highly tortuous crack path, with periodic deflections of 45° from the nominal mode I growth plane and a deflected segment ratio $(D/\{D + S\})$ of 0.5. For $R = 0$,

(a) Predict the variation of da/dN with ΔK for the heat-treated alloy, by accounting for the combined effects of (i) changes in crack-tip stress intensity factor due to crack deflection, (ii) changes in apparent crack growth rate due to crack deflection, and (iii) no mismatch between fracture surface asperities.

(b) Repeat the calculations in (a) by considering crack closure due to fracture surface mismatch with the mismatch factor, $\chi = 0.1$ and 0.25.

Assume small-scale yielding and that crack deflection and the associated contact between the crack faces are the primary source of the change in crack propagation behavior.

14.5   Discuss how possible effects of crack tip plasticity (under conditions of plane strain and small-scale yielding) would modify the predictions of the model for fatigue crack deflection, Eqs. 14.23 and 14.24. (Hint: See the discussion in Section 9.11.2.)

14.6   Discuss a set of conditions under which the application of a tensile overload to a component can result in (i) an improvement, (ii) a reduction, or (iii) no change in fatigue life. Provide at least one practical example for each situation.

14.7   Repeat the previous problem for the case of a compressive overload.

14.8 Consider the engineering component with a stress concentration which is schematically sketched in Fig. 1.2. The component is subjected to variable amplitude fatigue which consists of constant amplitude stress cycles mixed with periodic overloads. The component is made of a high strength material and the extent of plastic deformation is small compared to the root radius of the notch in the component.

(a) Outline a procedure for the determination of the number of cycles to initiate and propagate the crack ahead of the stress concentration.

(b) Discuss the beneficial or detrimental effects of the overloads on the fatigue life.

14.9 A fatigue crack in a structural component made of an aluminum alloy, of yield strength $\sigma_y = 300$ MPa, propagates at a constant amplitude stress intensity factor range of 6 MPa $\sqrt{m}$ with a load ratio of zero. A malfunction in the operation of the component results in a single tensile overload with a peak stress which is 50% higher than the peak stress in the previous constant amplitude stress intensity fluctuation. After the overload, the component returns to normal operation.

(a) If the Wheeler model for retardation is used with $k = 1$ in Eq. 14.31, how much does the post-overload crack growth rate differ from the pre-overload crack growth rate immediately upon resumption of normal operation after the overload? Plane stress conditions exist throughout the range of fatigue crack growth.

(b) Calculate the distance that the crack has to advance after the application of the overload to resume pre-overload crack growth rate.

(c) Calculate the retardation factor β when the crack has extended a distance of $25 \, \mu m$ after the application of the overload.

(d) Discuss the limitations of this approach.

14.10 A fatigue crack, propagating at a constant amplitude stress intensity factor ragne $\Delta K = 3$ MPa$\sqrt{m}$ and load ratio $R = 0.2$, is subjected to a tensile overload with $\Delta K_{OL} = 6$ MPa$\sqrt{m}$. Following the single overload, the crack is subjected to a constant amplitude $\Delta K = 3$ MPa$\sqrt{m}$ and load ratio, $R = 0.1$. The application of the overload results in crack bifurcation, with the branched crack having a forked crack-tip which is symmetric about the mode I growth plane (i.e. the crack develops a geometry similar to that shown in Fig. 9.17(b); $\alpha = 45°$ and $b/a = 0.1$). The pre-overload crack growth rate followed the relationship,

$$\frac{da}{dN} = 2.5 \times 10^{-12}(\Delta K)^4,$$

where da/dN is in units of m/cycle and ΔK is in MPa$\sqrt{m}$. Assuming that (i) b is much larger than the crack-tip plastic zone size, (ii) the crack advances along the arms of the fork, and (iii) the changes induced in the geometry of the crack are the principal factors influencing growth retardation, calculate the crack propagation rate in the post-overload regime.

14.11   The effect of the amplitude of the first compression cycle on crack growth from notches under far-field cyclic compression was discussed in Section 14.13.1. Speculate about the effects of the following loading patterns on mode I fatigue crack growth from notches under cyclic compression:

(a) The fatigue crack is propagated ahead of a stress concentration under a constant amplitude of imposed cyclic compressive loads. The cracks arrest naturally after advancing a distance of ten times the notch-tip radius. At this point, the notched specimen is subjected to a 100% compressive overload. Will the application of the compressive overload result in the reinitiation of the crack? Why?

(b) The same situation as in the previous problem, except that the distance of total crack growth under constant amplitude cyclic compression is one-half of the notch-tip radius (instead of ten times the notch-tip radius). At this point, the notched specimen is subjected to a 100% compressive overload. Will the application of the compressive overload result in the reinitiation of the crack? If so, is the extent of crack growth following the overload greater than or less than that in part (a)? Explain.

CHAPTER 15

Small fatigue cracks

The practice of characterizing the growth of fatigue cracks on the basis of fracture mechanics primarily relies on laboratory fatigue tests on specimens containing 'long' flaws which are typically tens of millimeters in length. There are, however, a number of fatigue-critical engineering components, such as turbine discs and blades, whose design requires an understanding of the propagation of fatigue cracks of significantly smaller dimensions. In the majority of investigations where continuum approaches have been adopted for the characterization of small fatigue flaws (of size range from a fraction of a millimeter to several millimeters), it has been shown that the growth rates of small flaws can be significantly greater than the corresponding rates of long flaws when characterized in terms of the same *nominal* driving force (see Fig. 15.1). The direct application of laboratory data (derived from experiments on long fatigue cracks) to design against the failure of safety-critical components containing short flaws can, therefore, lead to dangerous overestimates of fatigue lives. Research effort in this area has led to an awareness of the *apparently anomalous* behavior of short fatigue cracks and has provided possible ways in which many seemingly conflicting viewpoints of total-life and defect-tolerant fatigue approaches can be rationalized in a unified fashion.

To focus attention on the practical significance of the 'short crack problem', consider the effect of crack growth characterization on the estimated fatigue life of a commercial alloy, Fig. 15.2. This figure contains experimental data on the number of fatigue cycles to failure for cracks of different initial dimensions in a nickel-base superalloy, commercially known as Astroloy. This alloy is used as a disc material for aircraft gas turbine engines. Current design methodology based on linear elastic fracture mechanics (LEFM) provides accurate estimates of fatigue life in this material when the initial size of the fatigue flaw is larger than 0.3 mm. However, when the material contains defects that are smaller than this size, life predictions nominally based on LEFM may give nonconservative values (even if small-scale yielding conditions prevail). This complication arises because the actual growth characteristics, crack path tortuosity or closure processes for small flaws can be different from those of longer flaws. The problem is further exacerbated by the fact that the limit for the detection of flaw size with existing nondestructive inspection methods is about 0.3 mm.

In this chapter, we present a discussion of the different types of small fatigue cracks which exhibit apparently anomalous growth behavior compared to long fatigue flaws. This is followed by an examination of the mechanisms of the growth of short fatigue cracks, models for the initiation and growth of short flaws,

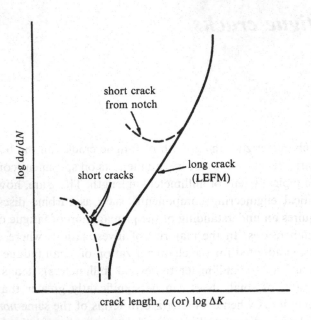

Fig. 15.1. A schematic of the typical fatigue crack growth behavior of long and short cracks at constant values of imposed cyclic range and load ratio.

methods of characterization as well as conditions for the breakdown of fracture mechanics.

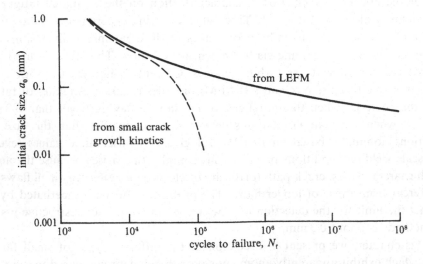

Fig. 15.2. The significance of the short crack problem is illustrated for an Astroloy. The plot shows the number of fatigue cycles to failure, estimated using LEFM and small crack growth kinetics, as a function of the initial flaw size. (After Hudak *et al.*, 1988.)

15.1 Definitions of small cracks

There are different size scales below which the growth rates of a fatigue crack may be found to exhibit a crack size dependence. Suresh & Ritchie (1984b) suggested the following definitions by which short cracks can be broadly classified:†

(1) Fatigue cracks for which the crack size is comparable to the scale of the characteristic microstructural dimension such as the grain size for monolithic materials and the interparticle spacing for particulate-reinforced composites; these are referred to as *microstructurally small* flaws.

(2) Small fatigue cracks in smooth specimens for which the near-tip plasticity is comparable to the crack size, or cracks which are engulfed by the plastic strain field of a notch; these are referred to as *mechanically small* flaws.

(3) Fatigue flaws which are significantly larger than the characteristic microstructural dimension and the scale of local plasticity, but are merely *physically small* with length typically smaller than a millimeter or two.

(4) Fatigue cracks which are nominally amenable to linear elastic fracture mechanics analyses, but exhibit apparent anomalies in propagation rates below a certain crack size as a consequence of the dependence of environmental stress corrosion fatigue effects on crack dimensions; these flaws are sometimes referred to as *chemically small* cracks.

A chief reason for the interest in the study of short fatigue cracks can be traced to the experimental findings that cracks which conform to any one of the foregoing definitions of smallness often propagate at rates which are not amenable to an unambiguous characterization based on continuum mechanics or linear elastic fracture mechanics parameters.

15.2 Similitude

The characterization of fatigue crack advance on the basis of linear elastic fracture mechanics principles is predicated upon the concept of similitude. This concept merely implies the fact that for a particular material–environment system subject to some particular loading history, identical near-tip conditions will prevail in specimens of different sizes containing different crack geometries if the magnitude of the characterizing parameter (e.g., K_I in small-scale yielding) is the same. Since LEFM uniquely characterizes the growth of a linear elastic fatigue crack (in the absence of the crack retardation mechanisms discussed in the previous chapter), the crack propagation mechanisms in these different specimens would be expected to be the same. However, when far-field loading conditions differ markedly from the

---

† Although the expressions 'small cracks' and 'short cracks' are often used interchangeably, sometimes a distinction is made between the two cases in that the former definition is employed for flaws which are small in all three dimensions whereas the latter type are taken to denote through-thickness flaws which are small in all but one dimension (e.g., Ritchie & Lankford, 1986).

near-tip conditions (see the case study at the end of this chapter) and when the size of the fatigue crack appreciably contributes to such differences, the similitude concept based on nominal values of the characterizing parameter is no longer applicable. Thus, the small crack problem is essentially an outcome of the inapplicability of the fracture mechanics parameters to uniquely characterize the growth of fatigue cracks independent of the crack size.

15.3 Microstructural aspects of small flaw growth

The first reported observations of the accelerated growth of short fatigue cracks are apparently due to Pearson (1975) who examined the effect of crack size on propagation rates in a precipitation-hardened aluminum alloy. He found that short surface flaws, 0.006 to 0.5 mm deep, grew up to 100 times faster than longer flaws, tens of millimeters in size, which were subjected to the same nominal ΔK. This investigation also indicated the possibility of the advance of short fatigue flaws at nominal stress intensity levels below the threshold ΔK_0 for long cracks. Subsequent fatigue studies on 7075 aluminum alloy (Lankford, 1982), silicon iron (Tanaka, Hojo & Nakai, 1983), and on a number of other alloy systems have documented several transient acceleration and retardation characteristics associated with the subcritical growth of short fatigue cracks (see, for example, conference proceedings edited by Miller & de los Rios, 1986; Ritchie & Lankford, 1986; Kitagawa & Tanaka, 1990).

Figure 15.3 is a schematic of the short crack growth behavior for a peak-aged 7075 aluminum alloy. This figure shows a marked reduction in the rate of growth of the microstructurally short crack with increasing crack length; the growth rate of the retarded short crack subsequently increases with crack length until it merges with the long crack growth data characterized by LEFM. The retardation of the crack occurs when the crack tip reaches a grain boundary in the material. Periodic retardations of the short flaw have also been reported whenever the crack tip encounters a grain boundary in aluminum alloys (Morris, 1979, 1980), low strength mild steels (Tanaka, Nakai & Yamashita, 1981), titanium alloys (Wagner *et al.*, 1986) and nickel-base superalloys (Sheldon *et al.*, 1981).

Note that the retardation of a short flaw, at least in the context of LEFM, is somewhat anomalous in that an increase in crack length, at fixed far-field cyclic stresses, is expected to increase the nominal ΔK for crack growth and the attendant crack growth rates. However, such transient effects have been rationalized by several mechanisms associated with short flaw growth:[†]

[†] Small flaws in transformation-toughened ceramics are also known to exhibit periodic growth retardation and arrest as a consequence of crack tip–grain boundary interactions and the development of closure as the transformed material is left in the wake of the advancing crack tip (Sylva & Suresh, 1989).

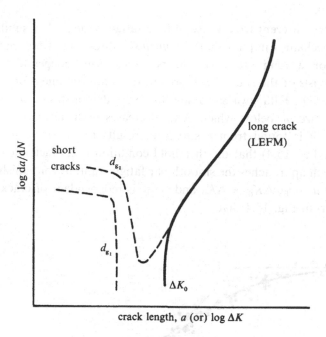

Fig. 15.3. A schematic of sub-threshold growth and transient retardation characteristics of a microstructurally small fatigue crack (dashed lines). $d_{g2} > d_{g1}$.

(1) Crack tip–grain boundary interactions which have the following consequences: (i) Cessation of crack growth occurs until a sizeable plastic zone is established in the neighboring grain (James & Morris, 1983), (ii) Slip bands emanating from the crack tip are pinned at the grain boundary (Tanaka, Nakai & Yamashita, 1981). (iii) A change in crack tip driving force is caused by crack deflection associated with the crystallographic reorientation of the crack tip as it traverses through the grain boundary into the adjacent grain (Suresh, 1983a). Experiments indicate that small crack retardation generally occurs when the crack size a is comparable to or is a multiple of the grain size, d_g. The depth of the retardation well also depends on the crystallographic orientations of the neighboring grains (Lankford, 1982).

(2) An increase in crack closure with an increase in crack length (Morris, 1980) and with crack deflection (Suresh, 1983a).

15.4 Threshold conditions for small flaws

15.4.1 Transition crack size

The predominantly stage I growth of microstructurally small flaws and the attendant retardation and arrest behavior point to the existence of threshold conditions which may be very different from those of long fatigue cracks. The similitude concept of fracture mechanics implies that the *intrinsic* threshold stress intensity factor range ΔK_{th} for a long crack should be crack size independent with $\Delta K_{th} = \Delta K_0$. On the basis of the short crack growth rate data obtained for a wide variety of ductile materials, Kitagawa & Takahashi (1976) demonstrated that there exists a critical crack size a_0 below which ΔK_{th} decreases with decreasing crack length, a. For $a < a_0$, it is found from a survey of results for a wide variety of engineering alloys (see Fig. 15.4) that the threshold condition is characterized by a critical stress $\Delta\sigma_{th}$ which approaches the smooth bar fatigue limit σ_e for vanishingly small crack sizes. For $a > a_0$, $\Delta K_{th} = \Delta K_0$ and ΔK_{th} is independent of crack size. Thus, one may infer from Fig. 15.4 that

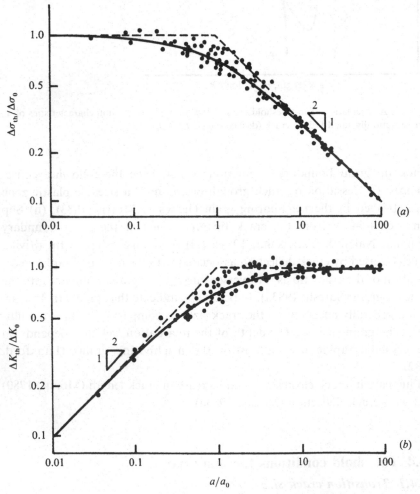

Fig. 15.4. The effect of crack size on (*a*) threshold stress and (*b*) threshold stress intensity factor range for a wide variety of engineering alloys with yield strength values ranging from 30 to 770 MPa. (After Tanaka, Nakai & Yamashita, 1981.)

$$a_0 = \frac{1}{\pi}\left(\frac{\Delta K_0}{\sigma_e}\right)^2,$$

$$\Delta K_{th} = \Delta K_0 \quad \text{for} \quad a > a_0,$$

$$\Delta\sigma_{th} = \sigma_e \quad \text{for} \quad a < a_0. \tag{15.1}$$

Attempts have been made to establish a link between the transition crack size a_0 and the characteristic microstructural dimensions in a vast spectrum of materials (e.g., Kitagawa & Takahashi, 1976; Tanaka, Nakai & Yamashita, 1981; Taylor & Knott, 1981; Lankford, 1982, 1983; Brown & Hicks, 1983; Wagner *et al.*, 1986). For very high strength steels, of monotonic yield strength σ_y up to 2000 MPa, $a_0 \approx 1 - 10\,\mu m$ and for very low strength steels with σ_y as low as 200 MPa, $a_0 \approx 100 - 1000\,\mu m$. Although there is no unique definition of a_0 which is representative of all materials, it is worth noting that a measure of a_0 can be extracted from continuum level measurements involving the fatigue limit σ_e (from S–N curves obtained from fully reversed cyclic stressing) and the long crack fatigue threshold ΔK_0 (from fracture mechanics approach to fatigue). In this sense, the field of a short fatigue crack promotes a marriage of several apparently contradicting philosophies associated with the total-life (Chapters 7 and 8) and the defect-tolerant (Chapters 9–12) approaches to fatigue.

The transition crack size a_0 has also been employed as an empirical parameter to account for the differences in growth rates between long and short fatigue cracks. El Haddad, Topper & Smith (1979) suggested that when a fictitious or intrinsic crack length a_0, defined in Eq. 15.1, is added to the fatigue crack length a, the resulting characterization based on LEFM provides crack propagation rates that are independent of crack size. Here,

$$\Delta K = Q\Delta\sigma^\infty\sqrt{\pi(a + a_0)}, \tag{15.2}$$

where $\Delta\sigma^\infty$ is the far-field stress range and Q is the finite size correction factor. Equation 15.2 implies that a_0 is the smallest crack size which is amenable to an LEFM characterization. El Haddad *et al.* demonstrated that the differences in the fracture mechanics-based growth characterization for long and short flaws vanish when the fictitious crack size a_0 is used by recourse to Eq. 15.2. The major drawback of this empirical approach is that attempts to provide a physical reasoning for the validity of Eq. 15.2 have so far remained unconvincing.

15.4.2 Critical size of cyclic plastic zone

The threshold conditions based on the transition crack size (Fig. 15.4) involve a critical stress criterion for short flaws and a critical stress intensity criterion for long cracks. Ohuchida, Usami & Nishioka (1975), Usami & Shida (1979), and Usami (1982) suggested a single criterion for fatigue thresholds which rationalized the overall shape of the threshold plots in Fig. 15.4. These researchers postulated that the fatigue limit is reached when a critical value of the cyclic plastic zone size, r_c^*, is reached. Using the Dugdale plastic zone estimate, Eq. 9.81, to derive the cyclic plastic zone size for positive load ratios ($R \geq 0$), it was shown that

$$r_c^* = a\left(\sec\frac{\pi\Delta\sigma_{th}}{4\sigma_y} - 1\right). \tag{15.3}$$

Noting that r_c^* is a constant which is independent of crack size, it is seen that Eq. 15.3 simulates the shape of the $\Delta\sigma_{th}$ versus a plot shown in Fig. 15.4(*a*). Usami & Shida have also extended this approach to include the effects of negative load ratios on short crack growth. Despite the apparent success of this hypothesis to predict the overall variation of the threshold stress with crack size, the basic premise of the model, namely that r_c^* is a constant for different crack sizes, remains to be established by experiments or by physical reasoning.

15.4.3 Slip band models

Alternative interpretations for the threshold conditions for small fatigue flaws have been formulated in terms of (crack tip) slip band–grain boundary inter-action. Tanaka, Nakai & Yamashita (1981) postulate that the threshold for short cracks is dictated by whether the grain boundaries obstruct or are penetrated by the slip bands emanating from crack tips. An analysis of the condition for the blockage of slip bands (see Fig. 15.5) by grain boundaries (see Tanaka, Nakai & Yamashita, 1981, for further details) provides the following results for σ_{th} and K_{th}:

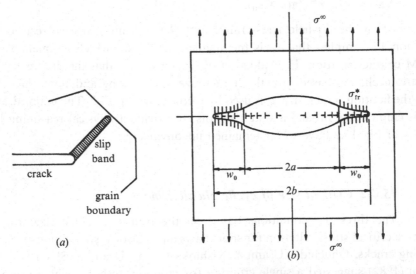

Fig. 15.5. Schematic representation of the slip band model. (*a*) A slip band emanating from a crack tip is blocked by a grain boundary. (*b*) Nomenclature associated with the coplanar slip band for an isolated crack. (After Tanaka, Nakai & Yamashita, 1981.)

$$\sigma_{th} = \frac{K_c^m}{\sqrt{\pi b}} + \frac{2}{\pi}\sigma_{fr}^* \cos^{-1}\left(\frac{a}{b}\right),$$

$$K_{th} = \sigma_{th}\sqrt{\pi a} = K_c^m\sqrt{\frac{a}{b}} + 2\sqrt{\frac{a}{\pi}}\sigma_{fr}^* \cos^{-1}\left(\frac{a}{b}\right). \tag{15.4}$$

In Eqs. 15.4, b is the sum of the crack length a and the width of the blocked slip band zone w_0, σ_{fr}^* is the friction stress for dislocation motion in the band, and K_c^m is the microscopic stress intensity factor at the tip of the slip band. For long cracks, wherein $w_0 \ll a$, the threshold stress intensity is given by

$$K_0 = K_c^m + 2\sqrt{\frac{2}{\pi}}\sigma_{fr}^*\sqrt{w_0}. \tag{15.5}$$

For vanishingly small fatigue flaws, the fatigue limit σ_e is obtained by letting $a = 0$ in Eq. 15.4 such that

$$\sigma_e = \sigma_{fr}^* + \frac{K_c^m}{\sqrt{\pi w_0}}. \tag{15.6}$$

There are several interesting implications of this analysis: (i) If one takes $\sigma_{fr}^* = 0$, $K_0 = K_c^m$ and $w_0 = a_0$ in Eqs. 15.4, it is found that

$$\sigma_{th} = \frac{K_0}{\sqrt{\pi(a + a_0)}} \quad \text{and} \quad K_{th} = \frac{K_0\sqrt{a}}{\sqrt{(a + a_0)}}. \tag{15.7}$$

This result is the same as the characterization based on the fictitious crack length a_0 assumption, Eq. 15.1. (ii) Similarly, the critical r_c hypothesis, Eq. 15.3, is found to be a special case of the blocked slip band model by setting $K_c^m = 0$ in Eqs. 15.4. (iii) Since one may envision the slip band to be confined within a single grain (at the fatigue limit), w_0 may be assumed to be of the order of one half the grain size, d_g. With this assumption, Eqs. 15.6 and 15.7 imply that the threshold stress intensity for long cracks ΔK_0 increases with increasing grain size while the threshold stress for small flaws ($\sigma_{th} \to \sigma_e$) decreases with increasing grain size.

This prediction has the consequence that microstructural modifications, such as increased yield strength and reduced grain size, which offer an enhanced resistance to crack initiation and the growth of small fatigue flaws may in fact have a detrimental effect on the growth of long fatigue cracks. These expectations are borne out by experiments in a number of different alloy systems: 300M maraging steels of different strengths (Ritchie, 1979), maraging steel and cast iron (Usami & Shida, 1979), and dual-phase steels (Suzuki & McEvily, 1979; Dutta, Suresh & Ritchie, 1984). Figure 15.6, for example, illustrates this effect where the threshold conditions are plotted as a function of the surface roughness (to simulate crack size) for a maraging steel and a cast iron. The high strength maraging steel has a superior resistance to crack initiation and short crack growth as compared to cast iron; the reverse situation is found for the growth of long flaws.

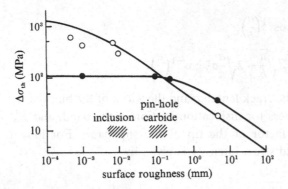

Fig. 15.6. Thresholds for the growth of long and short fatigue flaws in cast iron (filled circles, σ_y = 113 MPa) and maraging steel (open circles, σ_y = 1906 MPa). The surface roughness is used to simulate crack size. The roughness scale associated with inclusions and pin-hole carbides is also illustrated. (After Usami & Shida, 1979.)

15.5 Fracture mechanics for small cracks at notches

Linear elastic fracture mechanics solutions of the near-tip fields for sharp cracks (of tip radius, $\rho \rightarrow 0$) were derived in Chapter 9. If instead of a crack, one considers a notch where $\rho > 0$, similar solutions can be derived to characterize the fields at the tip of the notch. Consider the case of a sharp notch in a body subjected remotely to tensile stresses in a direction normal to the plane of the notch (Fig. 15.7). Linear elastic analyses of Creager & Paris (1967) show that elastic fields in the vicinity of the notch take the form:

$$\left\{ \begin{array}{c} \sigma_{xx} \\ \sigma_{yy} \\ \sigma_{xy} \end{array} \right\} = \frac{K_I}{\sqrt{2\pi r}} \cos\frac{\theta}{2} \left\{ \begin{array}{c} 1 - \sin\frac{\theta}{2}\sin\frac{3\theta}{2} \\ 1 + \sin\frac{\theta}{2}\sin\frac{3\theta}{2} \\ \sin\frac{\theta}{2}\cos\frac{3\theta}{2} \end{array} \right\}$$
$$+ \frac{K_I}{\sqrt{2\pi r}}\left(\frac{\rho}{2r}\right) \left\{ \begin{array}{c} -\cos\frac{3\theta}{2} \\ \cos\frac{3\theta}{2} \\ -\sin\frac{3\theta}{2} \end{array} \right\}, \qquad (15.8)$$

where the cylindrical coordinates r and θ and the notch-root radius ρ are defined in Fig. 15.7. In Eq. 15.8, K_I is the stress intensity factor at the tip of a sharp crack which is subjected to the same loading conditions as the notch and is of the same length as the notch in Fig. 15.7. The first term on the right hand side of Eq. 15.8 is identical to the near-tip fields for a sharp crack, Eqs. 9.44. The second term reflects the modification to the sharp crack solutions due to the bluntness of its tip. Note that, on the

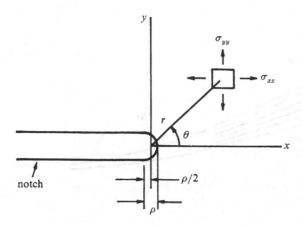

Fig. 15.7. Reference coordinates and stresses in the near-tip region of a notch in a plate.

plane of the notch ($y = 0$ in Fig. 15.7), the stress singularities for sharp notches of low ρ values is centered at a distance of $\rho/2$ *behind* the notch tip, which is the origin of the coordinate system.

15.5.1 Threshold for crack nucleation

During the application of a far-field cyclic load of amplitude $\Delta\sigma^{\infty}$, the maximum stress intensification occurs at the notch tip ($x = \rho/2$). The stress amplitude at the notch tip is given (by taking $\theta = 0$ and examining only the σ_{yy} component for mode I from Eqs. 15.8) by

$$\Delta\sigma_{\max} = K_{t}(\Delta\sigma^{\infty}) = \frac{2}{\sqrt{\pi}}\frac{\Delta K_{I}}{\sqrt{\rho}}, \tag{15.9}$$

where ΔK_{I} is the stress intensity factor range at the tip of a fatigue crack, of the same length as the notch, subjected to the same far-field cyclic load $\Delta\sigma^{\infty}$. Equation 15.9 holds true strictly for the case of a vanishingly small notch-root radius. However, it has been found that Eq. 15.9 is accurate to within 10% of the results obtained from more detailed finite-element calculations of notch tip fields under small-scale yielding conditions for notch tip radii $\rho < 4.5\,\text{mm}$, and for a notch length significantly greater than the root-radius. (e.g., Barsom & Rolfe, 1987).

The particular form of Eq. 15.9 suggests that the number of fatigue cycles to initiate a fatigue crack ahead of a stress concentration can possibly be correlated equivalently with either $\Delta\sigma_{\max}$ or $\Delta K_{I}/\sqrt{\rho}$, following the stress–life approach to fatigue, Chapter 7. Barsom & McNicol (1974) studied the initiation of fatigue cracks under uniaxial zero–tension cyclic loads in double edge-notched plates of HY-130 steel where the notch-root radius was systematically varied. They found that the initiation of a fatigue crack was characterized by a constant threshold value of the parameter $\Delta K_{I}/\sqrt{\rho}$ for the different notch tip geometries (Fig. 15.8), in accordance

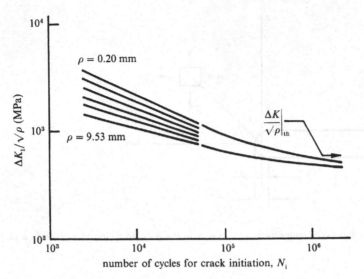

Fig. 15.8. Fatigue life plots correlating the number of cycles to initiate an engineering-sized crack (N_i) with the parameter $\Delta K_I / \sqrt{\rho}$ for double edge-notched specimens of HY-130 steel ($\sigma_y = 1000$ MPa) with notch-root radii $\rho = 0.2$–9.5 mm. (After Barsom & McNicol, 1974.)

with the predictions of Eq. 15.9. From a comparison of Figs. 7.1 and 15.8, and from Eq. 15.9, it is readily seen that this threshold value is essentially an endurance limit for notched bars expressed in terms of linear elastic fracture mechanics.

15.5.2 Example problem: Crack growth from notches

As a linear elastic fatigue crack begins to propagate ahead of a notch, its 'driving force' is affected by the elastic stress concentration at the notch tip. Once the length of the crack exceeds a certain fraction of the notch tip radius, the geometry of the notch tip region no longer has a bearing on the near-tip fields for the advancing fatigue crack. Issues pertaining to the role of stress concentrations in influencing the growth of small flaws that emanate from them and to the validity of the use of standard stress intensity calibrations for sharp fatigue cracks are clearly illustrated in this example problem, taken from Dowling (1979).

Problem:

Consider a large plate containing a central hole of radius c, as shown by the inset in Fig. 15.9. A fatigue crack emanates from the tip of the hole in response to the application of a far-field zero–tension fatigue cycle of amplitude σ^∞. As a fatigue crack initiates at the stress concentration and advances perpendicularly to the applied load, analytically determine transition crack size at which linear elastic fracture mechanics begins to provide a valid characterization

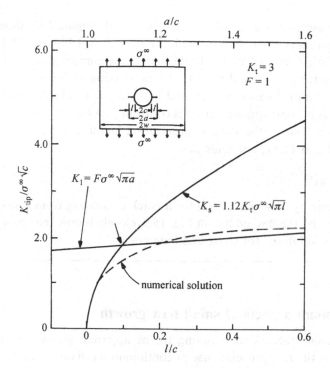

Fig. 15.9. Variation of stress intensity factor $K_{tip}/\sigma^{\infty}\sqrt{c}$ as a function of the fatigue crack length l (normalized by the radius of the circle c). (After Dowling, 1979.) The numerical values for the ordinate are given for $K_t = 3$ and $F = 1$. Limiting stress intensity solutions, $K_{tip} = K_s$ and $K_{tip} = K_l$, from Eqs. 15.10 and 15.11, respectively, are plotted along with the numerical predictions of Newman (1971).

of the 'driving force' for fracture. Ignore any nonlinear deformation processes at the root of the stress concentration.

Solution:

When the length of the fatigue crack l is small compared to the radius of the circle, the limiting analytical stress intensity factor solution for the short fatigue crack (assuming negligible plasticity in the vicinity of the hole and the crack) at the maximum imposed stress of the fatigue cycle is given by

$$K_s = 1.12 K_t \sigma^{\infty} \sqrt{\pi l}. \tag{15.10}$$

Here K_t is the theoretical elastic stress concentration factor (equal to three for a circle), the pre-multiplier 1.12 on the right hand side is the free surface correction factor, and the subscript S on the left hand side refers to the fact that Eq. 15.10 is the stress intensity solution for a 'short' crack ahead of a stress concentration. As the length of the fatigue crack l increases and $a = c + l$ becomes large compared to c, the stress intensity factor for the 'long' fatigue crack is given by

$$K_l = F\sigma^{\infty}\sqrt{\pi a}. \tag{15.11}$$

For a large plate whose semi-width is $W \gg (c + l)$, $F = 1$. Figure 15.9 shows the variation of the stress intensity factor, predicted from Eqs. 15.10 and 15.11, as a function of the fatigue crack length l. In this figure, the numerical values for the ordinate are given for $K_t = 3$ and $F = 1$. The intersection of the two curves corresponding to the limiting cases of short and long fatigue cracks provides a transition crack size l_0 beyond which fracture is controlled by the nominal stress intensity factor rather than by the stress concentration factor at the notch tip. Equating Eqs. 15.10 and 15.11, one finds that

$$l_0 = \frac{c}{\left[(1.12K_t/F)^2 - 1\right]}. \tag{15.12}$$

Numerical calculations by Newman (1971) for a crack emanating from a stress concentration, shown by the dashed line in Fig. 15.9, closely follow the limiting stress intensity factor solutions for short and long cracks.

15.6 Continuum aspects of small flaw growth

One of the major factors contributing to the apparent anomalies in the growth of small flaws is the inappropriate use of continuum mechanics and fracture mechanics methodologies for the characterization of fatigue cracks. While the limitations associated with the interpretation of microstructurally small fatigue crack growth in terms of continuum stress analysis are somewhat obvious, it has also become evident in recent years that there are some serious restrictions implicit in the nominal use of both linear elastic and nonlinear fracture mechanics for different types of short fatigue flaws.

15.6.1 Two-parameter characterization of short fatigue cracks

The leading singular term of the asymptotic expansion for the linear elastic crack problem, Eq. 9.44, is generally adequate for the complete description of near-tip fields for uniaxially loaded fatigue cracks under small-scale yielding conditions. For long fatigue cracks, the size of the zone of K-dominance is typically 10% of the characteristic dimension (i.e. crack size). As the length of the fatigue crack becomes smaller, the zone of K-dominance is also reduced. Furthermore, when the crack size is small, higher order terms including the 'T-stress' (see Section 9.3) and the terms which vanish at the crack tip should be incorporated into the expressions for the near-tip fields. These higher order terms not only serve to modify the stress and deformation fields ahead of the small fatigue crack, but they also influence the size of the plastic zone and the attendant plasticity-induced crack closure in its wake (Allen & Sinclair, 1982; Suresh & Ritchie, 1984b; Fleck & Newman, 1988). The near-tip stress distribution and the hydrostatic stress level at the crack tip can be significantly affected by the T-stress even if the strength of the singularity (i.e. J) is unaffected. In addition, these effects of the higher order terms are expected to be further enhanced when crack deflections associated with the stage I growth of small flaws promote strong mode II displacements.

It was shown in Section 9.8 that different levels of hydrostatic stresses induced at the crack tip in different specimen and crack geometries can be uniquely characterized, under small-scale yielding conditions, by the stress intensity factor K_I and the T-stress. It was also shown in that discussion that under large-scale yielding conditions, the corresponding two-parameter characterization involves the J-integral and the triaxiality measure Q. The need for a two-parameter characterization of the near-tip fields for small fatigue cracks can become significant on account of the following factors.

(1) The fatigue crack, which generally initiates at a free surface (i.e. essentially under plane stress conditions), advances into a region of increasing constraint as the crack tip penetrates into the subsurface material. There is a need to characterize, in a unique manner, the changes to the near-tip fields which involve differing levels of near-tip triaxiality as the small crack advances.

(2) The growth of short fatigue cracks (of dimensions of the order of several grain diameters) is markedly more sensitive to the elastic and plastic mismatch between adjacent grains. It is generally known that when the crack tip approches an interface (such as a grain boundary) between an elastically or plastically weaker material to one that is elastically or plastically stronger, the energy release rate at the tip of the crack is 'shielded' (i.e. reduced from the applied, remote value) as the crack tip approaches the interface from the weaker material (see the case study presented later in this chapter). Conversely, if the crack tip approaches the interface from the elastically or plastically stronger material, the energy release rate is 'amplified' locally. The extent of such shielding or amplification of the local energy release rate can be a strong function of the extent of constraint, as quantified by the T-stress or Q (Sugimura *et al.*, 1995; Kim, Besson & Pineau, 1998). Similarly, small fatigue cracks which initiate in the surface coatings and advance into the bond-coat or the substrate can experience a strong effect of T or Q as they approach the interface.

(3) Small cracks (of microstructural size scale) usually propagate along crystallographic directions. The ensuing local mixed-mode loading also promotes a stronger effect of T-stress or Q (see, for example, Eq. 9.43) than the growth conditions for longer flaws.

(4) If plasticity-induced crack closure plays a significant role in influencing crack growth, its magnitude and influence on fatigue fracture are also dependent on the extent of triaxiality (see Section 14.2).

(5) During fretting fatigue, the small crack which initiates at the fretting contact region experiences an effective T-stress as it advances deeper into the fretted

substrate, when the substrate is subjected to a uniform (static or cyclic) tensile or compressive load (see the example problem in Section 13.6.4).

15.6.2 Near-tip plasticity

When the (self-generated) plastic zone size of a small fatigue crack becomes comparable to its length, the use of LEFM becomes invalid and elastic–plastic characterization offers a possible means of accounting, at least partially, for the apparent differences in the growth behavior of long and short fatigue cracks. Dowling (1977) has suggested that the cyclic J integral, J^c ($\equiv \Delta J$), provides a measure of the driving force for elastic–plastic fatigue crack growth (see the discussion in Chapter 9). By measuring the amplitude of the cyclic stress range $\Delta\sigma$ and the plastic strain range $\Delta\epsilon_p$ from stable hysteresis loops (Fig. 15.10a), Dowling has suggested the following expression for J^c to characterize the growth of small semi-circular flaws in initially smooth low cycle fatigue specimens:

$$J^c \approx 3.2\Delta W_e a + 5.0\Delta W_p a = \frac{1.6\Delta\sigma^2 a}{E} + \frac{5.0\Delta\sigma\Delta\epsilon_p a}{n_f + 1}, \tag{15.13}$$

where ΔW_e and ΔW_p are the elastic and plastic components of the nominal strain energy density, a is the depth of the surface flaw which is assumed to be half its surface length $2c$, and $(1/n_f)$ is the fatigue hardening exponent.

Dowling argued for the effectiveness of J^c characterization, computed from Eq. 15.13, for small flaws based on his experimental results for long and short fatigue crack growth rate data on AISI A533B nuclear pressure vesssel steel of yield strength, $\sigma_y = 480$ MPa (Fig. 15.10b). A closer fit was obtained between the fatigue behavior of long flaws (crack size > 25 mm, filled symbols) and short flaws (crack size ≤ 0.18 mm) using this method. Dowling's method provides a noticeable improvement over other types of characterization techniques for elastic–plastic short flaws. Despite this good experimental correlation, the general applicability of the cyclic J integral to characterize rapidly growing small fatigue cracks remains questionable (see Chapter 9).

15.6.3 Notch-tip plasticity

Methods for the analysis of the growth and fatigue lives of flaws in notched components were discussed in detail in the preceding chapter. Here we focus our attention on the issue of the growth of a small fatigue flaw typically shorter than the inelastic strain field zone size of the notch.

When fatigue cracks emanate from notches in brittle solids, the initial growth of the nascent cracks is influenced by the plastic deformation occurring at the root of the notch. One of the commonly used definitions of a short crack pertains to flaws which emanate from stress concentrations and which are submerged within the strain field of the notch (see Section 15.1). Figure 15.11(a) schematically shows an example

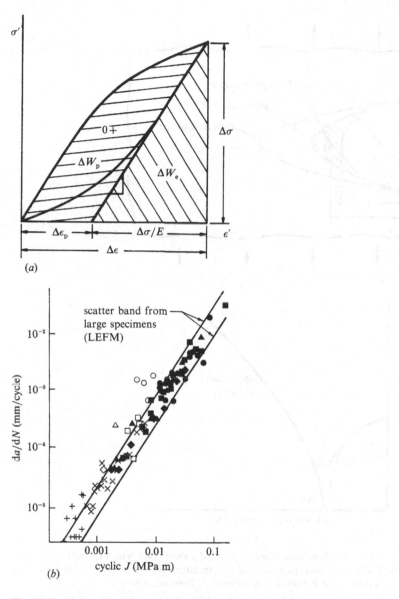

Fig. 15.10. (a) A method for estimating cyclic J from stable hysteresis loops for the characterization of small cracks propagating under elastic–plastic conditions. (b) Variation of crack growth rates da/dN as a function of $J^c(\Delta J)$ for long and short fatigue cracks in AISI A533B steel. The symbols, open squares, open circles, + and ×, refer to a crack length smaller than 0.18 mm and the filled symbols refer to cracks longer than 25 mm. The data pertain to different values of imposed strain range which varied from 0.005 to 0.04. (After Dowling, 1977.)

of a crack of length l propagating from the tip of a notch of depth c and root radius ρ. These small flaws often exhibit transient crack growth effects until the tip of the small flaw is relieved of the strain field associated with notch tip deformation during fatigue. A behavior often seen in such cases is the temporary retardation (and some-

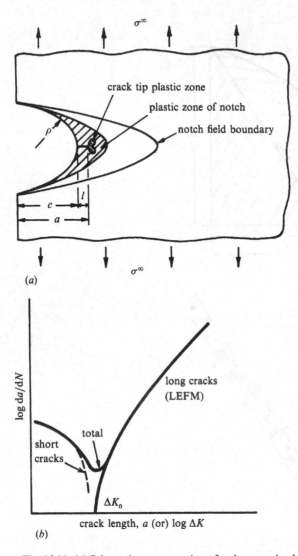

(a)

(b)

Fig. 15.11. (a) Schematic representation of a short crack which is submerged within the strain
field of a notch and the associated nomenclature. (b) Elastic–plastic and linear elastic crack
growth behavior for small flaws emanating from notches.

times even complete crack arrest, Section 8.3) as the small fatigue flaw advances
ahead of the notch tip (see Fig. 15.11b).

Both fracture mechanics-based models and empirical damage models for notch
fatigue have been proposed for the prediction of such deceleration of short crack
growth in notch fatigue. These arguments center around previously described procedures based on: (i) elastic–plastic analyses based on the variation of J integral, (ii)
the fictitious crack length (a_0) concept (El Haddad, Topper & Smith, 1979), and (iii)
the constant cyclic plastic zone size (r_c^*) criterion (Usami, 1982). Such approaches,

however, do not provide convincing descriptions of notch-field plasticity effects for small flaws for the following reasons: (i) none of the foregoing viewpoints allows for the possibility of a decrease in the effective driving force for fatigue fracture with an increase in crack length without recourse to crack closure arguments and (ii) the growth behavior of the small flaw initiated from a notch (Fig. 15.11) has a strong similarity to that of a microstructurally small crack in a smooth specimen (Fig. 15.3), where notch tip plasticity effects are nonexistent. The only continuum level mechanisms which could account for both of these observations involve reduction in effective crack tip driving force as a consequence of the development of crack closure with an increase in crack length and/or the deflection of the fatigue crack. The influence of these retardation mechanisms is examined in detail in the next section.

15.7 Effects of physical smallness of fatigue flaws
15.7.1 Mechanical effects

One of the most troublesome aspects of the short fatigue crack problem, as viewed from an engineering design standpoint, involves the so-called physically short flaws. These fatigue cracks, typically of the order of 0.5–2.0 mm in length, are in many cases significantly longer than both the scale of the microstructure and the size of the near-tip yield zone. Consequently, they are amenable, at least in principle, to an unambiguous characterization on the basis of linear elastic fracture mechanics. However, numerous investigators (e.g., McCarver & Ritchie, 1982; James & Morris, 1983; Suresh, 1983a, 1985b; Tanaka & Nakai, 1983) have demonstrated that physically small flaws can also propagate appreciably faster than long cracks subjected to the same *nominal* value of the stress intensity factor range.

As discussed in the previous chapter, the phenomenon of crack closure plays a major role in promoting marked differences between the effective (near-tip) ΔK and the nominal (far-field) ΔK. Since crack closure mechanisms arise as a result of premature contact between the crack faces behind the advancing crack tip, and since, by definition, a short crack has a limited wake, crack closure effects are less pronounced for a short flaw propagating in a smooth specimen or ahead of a notch tip. A clear illustration of this closure-induced difference between long and short fatigue cracks is found in the work of James & Morris (1983), Fig. 15.12. These researchers measured the nil-load crack tip opening displacement δ_t, after complete unloading following a fatigue test, as a function of the crack size in a Ti–6Al–4Zn–2Sn–6Mo alloy ($\sigma_y = 1140$ MPa, primary α grain size $\approx 4\,\mu$m and β grain size $\approx 12\,\mu$m). With an increase in crack length from $50\,\mu$m to $150\,\mu$m, there was an increase in δ_t, reflecting an enhanced level of closure primarily attributable to fracture surface roughness. This measure of the extent to which the fracture surfaces were 'propped-open' leveled off at a crack length of $200\,\mu$m, beyond which closure was crack size-independent. This closure effect, in conjunction with the other (crack

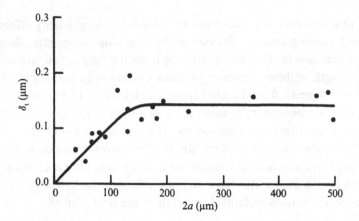

Fig. 15.12. The variation of the crack tip opening displacement δ_t at zero far-field load as a function of crack length $2a$ in a Ti–Al–Zn–Sn–Mo alloy. (After James & Morris, 1983.)

wake) retardation mechanisms, provides a rationale for the apparently higher growth rates found in the case of physically small flaws.

Finite-element simulations of plasticity-induced closure for (physically) small flaws also show that the development of crack closure in the wake of the advancing small crack can lead to a reduction in the effective driving force as the length of the crack increases. Finite-element computations by Newman (1983) of the evolution of closure for small fatigue cracks thus provide a numerical justification of the transient retardation behavior observed experimentally. However, it is worth noting that such numerical models are highly idealized and that they do not take into consideration the important effects of microstructural impediments to short crack growth or of crack deflection and fracture surface roughness.

The role of lower crack closure levels in enhancing the nominal growth rates of small flaws is now documented from a variety of experimental methods: (i) Tanaka & Nakai (1983) measured the closure loads for long and short fatigue cracks in a structural mild steel using compliance methods. The large differences between the long and short flaws, observed when the growth rates were characterized by nominal ΔK, vanished when the closure-corrected effective driving force ΔK_{eff} was used. (ii) Crack closure and crack tip opening displacements in aluminum and titanium–aluminum alloys based on highly precise laser interferometric techniques (Lee & Sharpe, 1986; Larsen *et al.*, 1986) and stereoimaging and selected area electron channeling methods (Davidson & Lankford, 1986) have unequivocally shown that closure stresses for short cracks are substantially smaller than those of longer cracks (in plane stress) at the same nominal ΔK. (iii) When the wake of a long fatigue crack (tens of millimeters in length) with fully developed closure stresses is removed either by drilling behind the crack tip or by machining away the crack wake leaving only a flaw of sub-millimeter length, it is found that crack propagation rates increase by several orders of magnitude upon resumption of fatigue loading at the pre-removal ΔK level (e.g., Minakawa, Newman & McEvily, 1983; Pineau, 1986). (iv) When through-thickness short fatigue cracks, a fraction of a millimeter in length, are introduced by cyclic compression in notched plate specimens, subsequent tensile fatigue loading leads to much higher growth rates and lower thresholds than those observed for long fatigue cracks (Suresh, 1985a; Christman & Suresh, 1986).

15.7.2 Environmental effects

Environmental interactions promote another situation involving physically small fatigue cracks where the smallness of the crack is responsible for anomalously faster crack propagation rates despite the validity of LEFM characterization as well as the absence of mechanical closure effects. This effect of the chemically short crack (definition (4) in Section 15.1) is a consequence of the strong dependence of corrosion fatigue phenomena on crack size. The significance of the chemically short crack problem was first realized from the work of Gangloff (1981) whose results for an AISI 4140 steel ($\sigma_y = 1300$ MPa) are plotted in Fig. 15.13. Gangloff showed that, in the ambient temperature, laboratory air environment, long fatigue cracks ($a \sim 50$ mm) and small fatigue cracks ($a \sim 0.1 - 0.8$ mm), both of which are amenable to a valid LEFM characterization, exhibit essentially the same growth rates in the Paris regime. Thus, in the nonaggressive environment of laboratory air, similitude based on LEFM is found to hold. However, in the embrittling environment of 3% NaCl, the short fatigue cracks propagated up to two orders of magnitude faster than the long cracks.

The chemical effects on short crack growth are not confined to the sodium chloride–steel system, but are known to occur in ferrous and aluminum alloys fatigued in hydrogen-containing or hydrogen-producing gaseous or aqueous media (Lankford, 1983; Saxena *et al.*, 1985). Quantitative modeling of the environmental effects based on crack size is still in its infancy. Existing experimental evidence appears to suggest an enhanced electrochemical production of hydrogen within small cracks as a result of lowering of the pH level (increased acidification) and restricted oxygen reduction (which consumes H^+). Gangloff & Wei (1986) identify the following crack size dependent mechanisms which influence corrosion fatigue in some ferrous alloys: (i) oxide- and roughness-induced crack closure in the wake of the small fatigue crack; (ii) hydrogen-enhanced crack tip straining; and (iii) mass transport/reaction to produce hydrogen within occluded cracks. The crack surface area to crack solution volume ratio affects the acidity and chemistry of the crack tip environment below a certain critical crack size which is specific to the combination of microstructure, environment, applied stress range and specimen geometry.

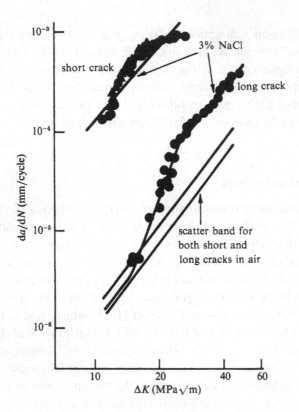

Fig. 15.13. The growth of long ($a \approx 50$ mm) and physically short ($a \approx 0.1 - 0.8$ mm) linear elastic fatigue cracks in laboratory air and 3% NaCl at room temperature. (After Gangloff, 1981.)

15.8 On the origins of 'short crack problem'

In this chapter, we have presented a variety of reasons for the apparent discrepancies associated with the growth of small fatigue cracks. These issues, which promote a lack of similitude, can be summarized as follows:

(1)   Local microscopic discontinuities, such as grain boundaries, inclusions, pre-cipitates and reinforcements, can affect the path and the rate of advance of small fatigue cracks because their size scale may be comparable to the depth of an incipient, microstructurally small flaw. These geometric and structural discontinuities as well as the crystallographic orientations of the grains have a substantially diminished impact on the growth of a long fatigue crack; in this case, the crack tip samples many grains and the fracture behavior is dictated principally by the bulk properties.

(2)   Microscopic growth mechanisms of small fatigue flaws may be vastly dif-ferent from those of long cracks at nominally identical values of ΔK because of the differences in the constraint imposed by the surrounding elastic

material. The restraint exerted on a small fatigue crack emanating at a free surface in a smooth specimen is different from that experienced by the tip of a through-thickness long crack in the same material (Schijve, 1982). The crystallographic, stage I growth of the near-surface small flaw also promotes severe mixed-mode loading at the crack tip.

(3) Because of the possible differences in the extension mechanisms of microstructurally short and long fatigue cracks, the modifications to effective driving force arising from crack deflection processes (see Chapters 9 and 14) also differ considerably. Deformation of small flaws submerged within individual grains is dictated by crystal plasticity and the crystallographic growth of these flaws promotes strong mixed-mode conditions at the crack tips. Consequently, the combined effects of crack deflection and crack tip plasticity for small flaws are also different from those of long fatigue cracks.

(4) Even when the loading conditions and the crack size satisfy the requirements for LEFM characterization, the physical smallness of the fatigue flaw generates faster crack propagation rates. In other words, a crack with a long wake benefits more substantially from premature contact due to a variety of closure mechanisms.

(5) In some corrosive media, the similitude normally expected for a fatigue crack with minimal or saturated levels of crack closure is violated as a consequence of corrosion fatigue mechanisms that are crack size-dependent.

Since the small crack problem is associated with some limiting situations involving the violation of continuum approximations or of similitude, which is vital to the successful use of fracture mechanics, no simple solutions are available for engineers faced with developing design guidelines for fatigue-critical components containing small flaws. However, recent advances and the tools available for further research in this area should provide help in tackling several of the issues connected with the small fatigue crack effect: (i) Crystal plasticity models for polycrystalline ensembles have not received enough attention from researchers in the field of fatigue. However, such crystalline models offer hope for at least partial success in handling the crack tip fields and growth behavior of microstructurally small fatigue flaws. (ii) Advances in the experimental capabilities (such as laser interferometry) for monitoring the near-tip opening displacements for small flaws may lead to a solution for characterizing small flaws whose dimensions are comparable to the scale of near-tip plasticity. (iii) Progress in the technology for the processing of monolithic and composite materials comprising both ductile and brittle matrices offer numerous possibilities for the improved control of fatigue crack path deflection, crack tip shielding or even complete crack arrest. Thus judicious processing techniques may, in the not-too-distant-future, lead to microstructural design procedures whereby problems associated with the paucity of a reliable characterization methodology for short fatigue cracks can be circumvented.

15.9 Case study: Small fatigue cracks in surface coatings

Fatigue crack initiation and the growth of small fatigue flaws are topics of considerable interest for the life assessment of structural coatings. Surface coatings, produced by such processes as physical vapor deposition (PVD), chemical vapor deposition (CVD) or plasma spray (e.g., PVD-processed CrN, CVD-processed TiC and TiN, and plasma-sprayed Cr_2O_3 and Al_2O_3–ZrO_2) are applied to such tribologically-critical and fatigue-critical automotive components as crank shafts, cam lobes, valve stems, piston rings, and turbocharger bearings. In turbine engines for aircraft, thermal-barrier coatings (e.g., plasma-sprayed 7 wt% Y_2O_3-stabilized ZrO_2 over an underlying plasma-sprayed NiCoCrAlY bond coat) are deposited on to the combustor liner, stator vanes and turbine blades.

In this section, we present a case study dealing with some issues related to the fatigue design of ceramic surface coatings on metallic substrates. This case study addresses some methods by which the fatigue life of wear- and contact-resistant coatings could be improved.

15.9.1 Theoretical background for cracks approaching interfaces perpendicularly

Fatigue cracks generally initiating at the surface coating advance through the coating in a direction approximately perpendicular to the interface between the coating and the substrate. Theoretical studies of the conditions governing the continued advance of a crack across an interface between two materials of different elastic and/or plastic properties provide helpful guidelines for rationalizing the fatigue crack growth response in such surface coatings. We, therefore, begin with some relevant results from such studies.

Consider first the idealized case of a linear elastic crack which perpendicularly approaches a 'sharp', well-bonded interface between two isotropic materials that have the same elastic properties but very different yield strengths. Begin with the situation where the crack is propagating normally toward to the interface entirely within the plastically weaker material. The crack tip is still within this material, at some distance from the interface, but the plastic zone just begins to spread across the interface. Finite-element simulations (Sugimura *et al.*, 1995; Kim, Suresh & Shih, 1997) have shown that the J-integral at the crack-tip, J_{tip} (computed from the contour integral taken entirely within the weaker material) becomes smaller than the remotely applied J-integral, J_{app} ($=$ $K_I^2(1 - v^2)/E$, where K_I is the applied stress intensity factor, and v and E are the Poisson ratio and Young's modulus, respectively, of both the materials on either side of the interface). In other words, the crack tip is partially shielded from the applied loads as the plastically stronger material further ahead of the crack tip bears a greater magnitude of the opening stress than the weaker material directly in front of the crack tip. Figure 15.14(a) shows this trend where the numerically computed ratio J_{tip}/J_{app} is plotted as a function of $K_I/(\sigma_y\sqrt{L})$. Here σ_y is the yield strength of the plastically weaker material and L is the distance from the crack-tip to the interface, as shown schematically in Fig. 15.14(b). When the crack approaches the interface from the plastically stronger solid,

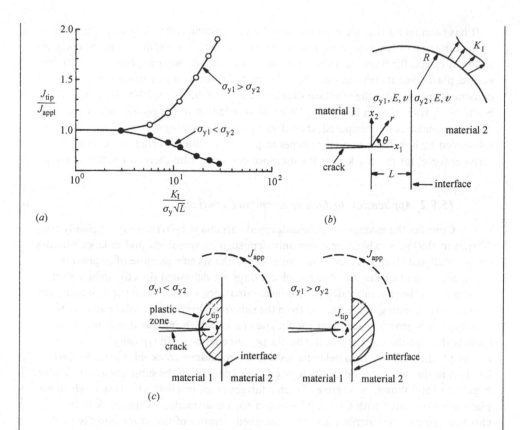

Fig. 15.14. (*a*) Numerically predicted variation of J_{tip}/J_{app} as a function of $K_I/(\sigma_y\sqrt{L})$ for a crack approaching an interface perpendicularly. σ_{y1} and σ_{y2} are the yield strengths of materials 1 and 2, respectively, with the crack always approaching the interface from material 1. σ_y refers to the lower of the two yield strengths. (*b*) The interlayer geometry is schematically sketched along with the associated nomenclature. (*c*) Definitions of J_{tip} and J_{app}. (After Sugimura *et al.*, 1995.)

however, the near-tip energy release rate, as quantified by the *J*-integral, is amplified, i.e. $J_{tip}/J_{app} > 1$, Fig. 15.14(*a*). (Recall that for a homogeneous material, $J_{tip} = J_{app}$ as the *J*-integral is path-independent.)

These predictions for the possible shielding or amplification of local energy release rates for cracks that approach interfaces perpendicularly find some backing in the experiments of Suresh, Sugimura & Tschegg (1992). Such experiments reveal that the conditions governing the growth or arrest of a fatigue crack at a normally oriented, well-bonded interface between a ferritic steel and an austenitic steel (with essentially the same elastic properties) are determined by whether it advanced toward the interface from the plastically weaker (ferritic) steel or the plastically stronger (austenitic) steel. In addition, experiments of fatigue crack growth across interfaces oriented at 30, 45 and 60° to the plane of the fatigue crack in this steel bimaterial also show trends which are consistent with the shielding and amplification processes discussed above (Sugimura, Grondin & Suresh, 1995).

It has been found that the extent of shielding and amplification is generally a function of the direction of crack advance relative to the interface, the distance from the crack tip to the interface, the finite thickness of the interface, the gradients in properties, including elastic, plastic and thermal mismatch, of the material across the interface and the level of constraint imposed by the interface (Sugimura *et al.*, 1995; Kim, 1996; Kim, Suresh & Shih, 1997; Kim, Besson & Pineau, 1998). If, in addition to the plastic mismatch, an elastic mismatch is superimposed, the extent of overall shielding and amplification can be influenced by how these two mismatches in properties differ. Further changes in the 'driving force' for the crack near the interface can arise from thermal residual stresses.

15.9.2 Application to fatigue at surface coatings

Consider the example of plasma-sprayed chromia (Cr_2O_3) coatings, typically 100–500 μm in thickness, which are commonly deposited inexpensively and in large volumes on to small and large structural components, for the specific purpose of improving resistance to contact fatigue. When such coatings are deposited directly upon a steel substrate, the thermal mismatch stresses generated upon cooling from the processing can lead to the debonding of the coating from the substrate. Even if the coating adheres to the substrate, it is generally seen that the fatigue cracks, which initiate at the surface run quickly through the thickness of the brittle, porous coating (with typically 8–15 volume% porosity); the interface then delaminates and a new fatigue crack initiates in a nearby location in the substrate and causes long cracks to develop in the substrate due to fatigue. Figure 15.15(*a*) shows an example of such a fatigue crack in a 0.45 wt% C steel which was plasma-spray-coated with Cr_2O_3. The reason for the attraction of the crack in the chromia layer to the interface and the continued advance of this crack into the steel

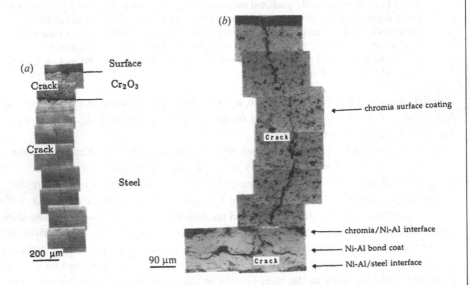

Fig. 15.15. Optical micrographs showing small fatigue crack growth in a chromia coated steel (*a*) with and (*b*) without a Ni–Al bondcoat. (From Suresh, Sugimura & Ogawa, 1993. Copyright Elsevier, Ltd. Reprinted with permission.)

substrate (even if the interface remains bonded) can be qualitatively justified from the trends shown in Fig. 15.14(*a*) for the case of the crack approaching the interface from the stronger material (where $J_{tip}/J_{app} > 1$).

A plasma-sprayed interlayer or bond coat of Ni–5 wt% Al (typically 50–100 μm in thickness) is usually deposited between the chromia coating and the steel substrate. If the thickness of this bond coat is at least several times larger than the plastic zone when the small fatigue crack approaches the interface between the chromia and the bond coat, one would expect the propensity for the crack to jump across the bond coat to be suppressed. Under these conditions, the results shown in Fig. 15.14(*a*) suggests that the crack would penetrate the chromia/Ni–Al interface and continue to advance through the bond coat. However, when the crack-tip plastic zone overlaps with the interface between the (soft) bond coat and the (relatively much harder) steel substrate (both of which have nearly the same Young's modulus and Poisson ratio), while the fatigue crack tip is still in the bond coat, the results shown in Fig. 15.14(*a*) suggest that the crack-tip would be shielded from the imposed fatigue loads. It would be expected to arrest before penetrating the Ni–Al/ steel interface (since $J_{tip}/J_{app} < 1$). This is indeed seen in the micrograph of the fatigue crack taken on a cross-section plane normal to the interface and shown in Fig. 15.15(*b*). The fatigue crack, in an attempt to avoid the interface, bifurcates and continues to advance well within the soft bond coat, along a plane parallel to the Ni–Al/steel interface, where final failure eventually occurs. The arrest of the fatigue crack significantly enhances the fatigue crack growth resistance of the material.

The overall benefits of the surface coating and the bond coat can be illustrated in a concise manner by recourse to the stress–life (S–N) plot shown in Fig. 15.16. This plot shows the stress amplitude (for fully reversed cyclic stressing) versus the number of fatigue cycles to failure, obtained from rotating bending fatigue tests conducted on the uncoated 0.45 wt% C steel and on that coated with chromia as well as with both chromia and Ni–Al. Compared to the uncoated steel, the steel deposited only with the chromia surface coating exhibits improved S–N fatigue response at all lives because a hard surface

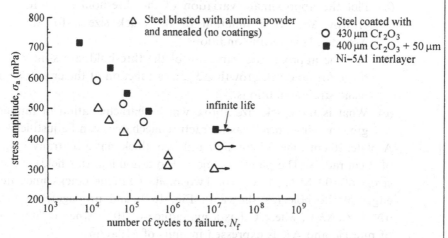

Fig. 15.16. S–N fatigue curves for the 0.45 wt% C steel without any coating, with only the chromia coating, and with the chromia coating and Ni–Al bondcoat. (After Suresh, Sugimura & Ogawa, 1993.)

layer enhances the resistance to fatigue crack initiation. When the bond coat is introduced between the chromia and the steel substrate, the resistance to fatigue crack initiation (by virtue of a hard outer surface) and the resistance to fatigue crack propagation (by virtue of the bifurcation of the small fatigue crack before it penetrates the Ni–Al/steel interface) are both improved. As a result, the coated steel, with an appropriately designed bond coat, enhances the highest life at any cyclic stress amplitude level compared to the other two cases.

The design of a bond coat is commonly predicated upon considerations of interfacial strength and protection against environmentally induced degradation (as, for example, in the case of the NiCoCrAlY bond coat used between the Ni-based superalloy substrate and the zirconia thermal-barrier coating used in aircraft jet engine components). The case study presented here demonstrates the possibility that, in addition to such considerations, interlayers and bond coats could be designed for the specific purpose of enhancing thermomechanical fatigue resistance in thermal-barrier and wear-resistant surface coatings. With advances in the processing and thermo-mechanical modeling of functional (structural) materials with stepwise or continuous gradients in the composition, microstructure and properties, there exist further possibilities for enhancing the damage-tolerance of coatings.

Exercises

15.1   Explain the concept of similitude in fracture mechanics. Under what conditions is it likely to fail in (*a*) quasi-static fracture and (*b*) cyclic fracture.

15.2   A steel has the following monotonic and cyclic properties: yield strength $\sigma_y = 1000$ MPa, smooth bar fatigue limit, $\sigma_e = 600$ MPa, threshold stress intensity factor range, $\Delta K_0 = 6$ MPa$\sqrt{\text{m}}$.

   (a) Plot the approximate variation of the threshold stress for no crack growth $\Delta\sigma_{\text{th}}$ as a function of the crack size a for this material. Assume plane strain conditions.

   (b) Plot the approximate variation of the threshold stress intensity factor range for no crack growth ΔK_{th} as a function of the crack size a under plane strain conditions.

   (c) What is the crack size above which characterization of fatigue crack growth using linear elastic fracture mechanics can be justified?

15.3   A plate, 10 cm wide, 12 cm high and 1 cm thick, has a centrally located hole of 2 cm radius. The plate is subjected to a nominal (far-field) cyclic tensile stress of 100 MPa at $R = 0$. Two mode I fatigue cracks nucleate at the edges of the hole (as shown in Fig. 15.9) and propagate at a rate da/d$N = C(\Delta K)^3$, where C has a magnitude of 10^{-11} when da/dN is in units of m/cycle and ΔK is expressed in units of MPa$\sqrt{\text{m}}$.

   (a) What is the transition crack size, l_t, below which the growth of the fatigue crack is affected by the elastic strain field of the notch?

Assume small-scale yielding conditions at the tip of the fatigue crack and neglect the possibility of plastic deformation in the vicinity of the hole.

(b) If the crack initiates during the first stress cycle, calculate the number of fatigue cycles necessary to propagate the crack a distance of 1.5 cm on either side of the circular hole.

15.4 Comment on the effects of T-stress on the growth of a fatigue crack which initiates at the free surface of a metal and propagates into the interior of the material. Specifically, discuss in general qualitative terms the possible role of constraint and crack length in influencing the T-stress and the crack growth rates.

15.5 A large plate of a brittle ceramic material contains a single edge notch. The plate is subjected to a far-field tensile load of 5 kN. The plate dimensions are: width $W = 5.0$ cm, height $H = 6.0$ cm and thickness $B = 1.2$ cm. The notch length a_0 is 2 cm and the notch-root radius $\rho = 0.5$ mm.

(a) Estimate the minimum length of the pre-crack one has to introduce at the tip of the notch to avoid possible effects of the strain field of the notch on subsequent fracture behavior.

(b) Calculate the stress intensity factors if through-thickness pre-cracks of the following lengths are introduced at the notch-tip: (i) 10 μm, (ii) 0.5 mm and (iii) 3 mm.

15.6 Discuss possible effects of thermal mismatch stresses on the propensity for crack growth and crack arrest in the coated steel in the case study presented in this chapter. Assume that the substrate is heated to a temperature of 200 °C during the plasma spray process.

CHAPTER 16

Environmental interactions: corrosion-fatigue and creep-fatigue

The effects of environments on the nucleation and growth of fatigue cracks were highlighted in previous chapters. These discussions touched upon the influence of oxygen-containing media on the kinematic irreversibility of cyclic slip (Chapter 4), crack initiation at corrosion pits (Chapter 4), the formation of brittle fatigue striations (Chapter 10), the role of fracture surface oxidation in promoting crack closure in constant amplitude and variable amplitude fatigue (Chapter 14), and the apparently anomalous growth of 'chemically short' fatigue cracks (Chapter 15). It is clear from these earlier descriptions that the usually deleterious (and occasionally beneficial) effects of environment must constitute an integral part of any complete mechanistic theory or design methodology for fatigue fracture.

This chapter deals specifically with the effects of environment on fatigue behavior. The discussions are presented in two parts: corrosion-fatigue effects and creep-fatigue effects. The first part begins with a survey of the micromechanisms of corrosion-fatigue. This is followed by examples which illustrate the effects of gaseous and aqueous media on fatigue failure for different mechanical conditions of cyclic loading. This section is concluded with a brief examination of models of corrosion-fatigue. In the second part, factors influencing the creep-fatigue behavior of engineering alloys are considered and many life prediction models are discussed. Particular attention is devoted to the mechanisms of high temperature fatigue deformation and to the issues pertaining to the characterization of creep crack growth using different fracture mechanics parameters.

16.1 Mechanisms of corrosion-fatigue

The deterioration of fatigue properties in engineering materials can be caused by an external medium in the form of a solid, liquid or gas. *Metal embrittlement* is the weakening of a higher melting point metal when it is in contact with certain lower melting point metals. *Liquid metal embrittlement* is a special case of metal embrittlement where the embrittling medium is a liquid metal. The embrittlement of alloys resulting from aqueous solutions is commonly referred to as *stress corrosion cracking* (SCC). In hydrogenous gases (such as H_2, H_2S, or water vapor) and in aqueous media, hydrogen can be introduced into the metal by the dissociation of hydrogen molecules into atomic hydrogen or by the release of hydrogen by metal dissolution, respectively. The resulting damage to mechanical response is known as *hydrogen embrittlement.* Under cyclic loading conditions, the embrittling environment can accelerate the initiation of a surface flaw in an initially crack-free material and then propagate the flaw to certain critical size. *Corrosion-fatigue* is a term which is commonly used to denote the damage and failure of a material under the combined action of cyclic stresses and any embrittling medium, although its most widespread adaptation is in the context of aqueous environments.

16.1.1 *Hydrogenous gases*

In metals exposed to hydrogenous gases, the key steps in the inducement of hydrogen embrittlement involve:

(1) the physisorption of hydrogen on the freshly created slip step or fatigue crack surface (e.g., Henaff, Marchal & Petit, 1995),

(2) the dissociation of the hydrogen molecule into atomic hydrogen,

(3) the transport of hydrogen by random walk diffusion through the lattice, by pipe diffusion along interfaces such as grain boundaries, or by transport along with mobile dislocations in plastically deforming regions to critical sites in the lattice or to regions of high triaxiality ahead of the crack tip (Fig. 16.1).

Once atomic hydrogen enters the subsurface regions, embrittlement of the material is promoted by one or a combination of the following processes:

(1) decohesion of atomic bonds or interfaces (e.g., Morlet, Johnson & Troiano, 1958; Oriani & Josephic, 1974; Gerberich & Chen, 1975),

(2) formation of molecular hydrogen at internal voids or defects by the recombination of transported hydrogen atoms and the attendant generation of large internal pressure (Zappfe & Sims, 1941),

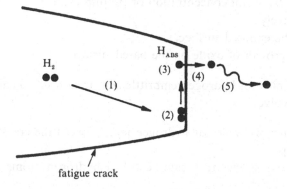

Fig. 16.1. An idealization of the critical steps involved in hydrogen embrittlement in hydrogenous gas media. (1) Transport of gaseous hydrogen. (2) Surface adsorption. (3) Dissociation into hydrogen atoms. (4) Hydrogen intake into the lattice. (5) Transport in material to regions of high tensile hydrostatic stresses.

(3)   lowering of surface energy due to adsorbed hydrogen (Petch, 1956),

(4)   hydride precipitation (in some alloys of Ti, Zr, V and Nb).

Although a single theory is unlikely to account for all the effects of hydrogen seen in metallic materials under monotonic and cyclic loading conditions, decohesion theories appear to offer at least a partial explanation of hydrogen embrittlement in high strength steels. In these approaches, the reduction in cohesive strength is assumed to be proportional to the local concentration of hydrogen c_h, where

$$c_h = c_0 \exp(\sigma_H \bar{v}/R_0 T). \tag{16.1}$$

c_0 is the equilibrium concentration of hydrogen in the unstressed ferrous lattice, $\bar{v}$ is the partial molar volume of hydrogen in iron, σ_H is the tensile hydrostatic stress, R_0 is the universal gas constant, and T is the absolute temperature. The fracture of a metal in the presence of hydrogen generally occurs in one of two ways: (i) the rate of failure is accelerated by the presence of hydrogen although there is no change in the microscopic mode of fracture, or (ii) a transition from a normally ductile mode of failure to a brittle mode of failure in the presence of hydrogen.

16.1.2 Aqueous media

For ductile alloy/aqueous environment systems subjected to both mono-tonic and cyclic loads, interpretations of embrittlement involve the mechanism of electrochemical reaction at freshly formed slip steps or at the crack tip. This process includes two principal candidate mechanisms of environmental damage: *anodic slip dissolution* or *hydrogen embrittlement*. The advance of fracture by slip dissolution (see Fig. 16.2a) occurs by:

(1)   the diffusion of the active species (such as water molecules or halide anions),

(2)   the rupture of the protective oxide film at a slip step or in the immediate wake of a crack tip by strain concentration or fretting contact between the crack faces, respectively,

(3)   the dissolution of the exposed surface,

(4)   the nucleation and growth of oxide on the bared surface.

For the alternative mechanism of hydrogen embrittlement in aqueous media, the critical steps (Fig. 16.2b) involve:

(1)   the diffusion of water molecules or hydrogen ions between the crack walls toward the crack tip,

(2)   the reduction of these species to create adsorbed hydrogen atoms at the crack tip surface,

(3)   the surface diffusion of adsorbed atoms to preferential surface locations,

(4)   the absorption of the adatoms to a critical location (e.g., a grain boundary, the region of high triaxiality ahead of a crack tip, or a void).

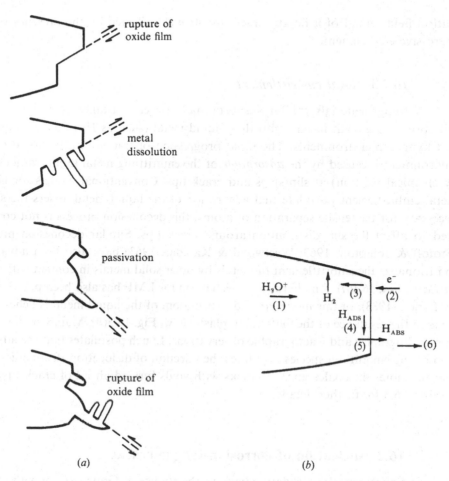

Fig. 16.2. A schematic illustration of (*a*) slip dissolution and (*b*) hydrogen embrittlement in aqueous media. (1) Liquid diffusion. (2) Discharge and reduction. (3) Hydrogen adatom recombination. (4) Adatom surface diffusion. (5) Hydrogen absorption in metal. (6) Diffusion of absorbed hydrogen. (After Ford & Silverman, 1979.)

In situations involving the formation of a passivating oxide film on the freshly created surface, the reaction at the crack tip which influences dissolution, hydrogen-ion reduction, or crack growth is controlled (at a given electrochemical potential) by three factors: the oxide rupture rate, the solution renewal rate, and the passivation rate.

Under cyclic loading conditions, fretting contact between the mating crack faces, pumping of the aqueous environments to the crack tip by the crack walls, and continual blunting and resharpening of the crack tip by the reversed loads will strongly influence the rate of anodic dissolution. Therefore, both cyclic frequency and stress waveform are known to strongly influence crack growth. It is also known from the crack tip strain measurements of Davidson & Lankford (1981) that deformation fields ahead of a fatigue crack are strongly affected by the presence of an aggressive environment.

16.1.3 Metal embrittlement

Many materials exhibit a substantially lower resistance to fatigue crack initiation and growth in an embrittling liquid metal (such as Hg) than in aqueous and hydrogen environments. The rapid progression of fracture in the liquid metal environment is caused by the *adsorption* of the embrittling metal atom (rather than by chemical reaction) at slip steps and crack tips. Conventional models for liquid metal embrittlement postulate that adsorption of the liquid metal lowers the stress necessary for the tensile separation of atoms; this decohesion process is not considered to affect the slip distribution around crack tips. Similar decohesion models (Stoloff & Johnston, 1963; Westwood & Kamdar, 1963) have also been advanced to rationalize the embrittlement of metals by other solid metals in contact with their surfaces, Fig. 16.3(*a*). An alternative mechanism for LME has also been popularized by Lynch (1988). In this model, the adsorbed atom of the liquid metal is believed to weaken the resistance of the material to plastic flow, Fig. 16.3(*b*). Mainly on the basis of metallographic and fractographic observations, Lynch postulates that the adsorption of the embrittling species facilitates the injection of dislocations from crack tips and promotes the coalescence of cracks with voids formed ahead of crack tips (see Section 16.3 for further details).

16.2 Nucleation of corrosion-fatigue cracks

Experimental work dating back to the studies of Gough & Sopwith (1932) and Thompson, Wadsworth & Louat (1956) shows that oxygen-containing media, and moist and aqueous environments, generally reduce the fatigue lives of ductile solids. There is ample experimental evidence in a wide range of engineering alloys

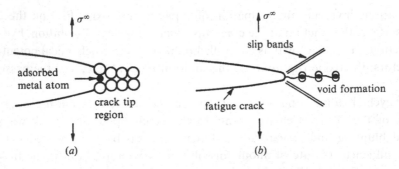

Fig. 16.3. Idealization of (*a*) solid or liquid metal embrittlement due to decohesion, and (*b*) liquid metal embrittlement due to increased local plastic flow and void growth.

indicating that inert atmospheres or vacuum enhance the fatigue limit and/or the fatigue life above the fatigue limit.

16.2.1 Gaseous environments

It is well established from experimental work conducted over many decades that noninert gaseous surroundings can reduce the fatigue lives of ductile solids. There are two different mechanistic interpretations to rationalize the phenomena underlying such deleterious effects of aggressive gaseous media. In the first type, which is based on the work of Thompson, Wadsworth & Louat (1956), the gaseous environment is believed to strongly affect the *growth* of the fatigue crack, but not the conditions for crack nucleation. In this model (Fig. 4.8), oxygen dissolved at the freshly formed slip steps is deemed responsible for enhanced slip irreversibility under cyclic loading and for the preclusion of the rewelding of nascent crack surfaces. In the second type, the mechanistic models (e.g., Shen, Podlaseck & Kramer, 1966) are based on the argument that the surfaces of the ductile alloys are strengthened with an oxide film. Under cyclic loads, the accumulation of dislocations in the surface regions leads to the formation of cavities and voids which eventually grow into cracks. Implicit in this interpretation is the assumption that, in the absence of oxygen, dislocations readily escape from the surface, thereby retarding crack nucleation.

Water vapor is also known to be a major contributing factor to the nucleation of fatigue cracks in precipitation-hardened aluminum alloys. Here the environmental influence may arise from enhanced slip irreversibility due to the oxidation of slip steps or from hydrogen embrittlement (see Fig. 16.3).

16.2.2 Aqueous environments

Figure 16.4 schematically illustrates the commonly observed effects of aqueous environment on the fatigue life of most metallic alloys. Models for the initiation of fatigue cracks in aqueous media have generally been based on the following mechanisms:

(1) Stress concentration at the roots of corrosion pits which serve as crack nucleation sites (e.g., Fig. 4.14).

(2) Enhanced slip irreversibility due to the oxidation of slip steps.

(3) Preferential electrochemical attack at regions (such as the PSBs) where plastic deformation is localized, with the relatively less deformed surrounding regions serving as cathodes.

(4) Preferential electrochemical attack at locations on the surfaces of a fatigued metal where the protective oxide film is ruptured.

(5) A reduction in the surface energy of the alloy as a consequence of the adsorption of the environmental species, and the increased rates of microcrack growth.

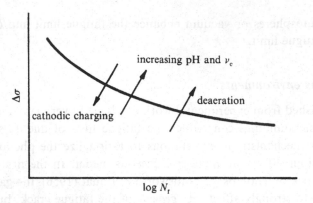

Fig. 16.4. A schematic of the effects of corrosive media on the number of cycles to failure, N_f, under a fixed cyclic stress range, $\Delta\sigma$.

The rates of environmental attack and their influence on fatigue lives are influenced by the electrochemistry of the medium as well as by the specific mechanical loading conditions. Therefore, universal conclusions cannot be derived on the mechanisms of corrosion-fatigue. Nevertheless, the following examples and general guidelines (see Duquette, 1979, for a detailed review) illustrate the role of various electrochemical, mechanical and microstructural factors in influencing corrosion-fatigue crack growth behavior.

(1)   For a fixed number of stress cycles, a lower value of cyclic frequency generally leads to lower fatigue life. However, in many cases, higher cyclic frequencies produce a greater level of fatigue damage for a fixed amount of time.

(2)   Dissolved oxygen is essential to corrosion-fatigue in neutral pH solutions. Deaeration of the solution generally causes enhancements in fatigue lives and aids in the development of a well-defined fatigue limit (in solutions which, in the aerated condition, do not exhibit a fatigue limit).

(3)   An increase in the imposed anodic current density can lead to a reduction in fatigue lives in low carbon steels. For example, Duquette & Uhlig (1968) show that increasing the anodic current density from $0.2\,\mu\mathrm{A/cm^2}$ to $40\,\mu\mathrm{A/cm^2}$ results in an order of magnitude reduction in the number of cycles to failure at fixed cyclic stress levels ($\Delta\sigma = 207$ MPa). Beyond a current density value of $40\,\mu\mathrm{A/cm^2}$, no change is observed in the fatigue life. The deleterious effect of anodic current density is more pronounced at higher imposed stress levels.

(4)   Cathodic charging of high strength aluminum alloys promotes a reduction in fatigue resistance.

(5)   Fatigue tests on aluminum alloys indicate that the reduced fatigue life in water vapor is due to hydrogen embrittlement. Specifically, low-cycle fatigue

experiments on Al–Zn–Mg alloy reveal that pre-exposure to humid air is just as detrimental to fatigue life as the presence of water vapor in the test environment. However, when the test specimens are stored in vacuum long enough to remove hydrogen (by allowing sufficient time for bulk diffusion with an assumed diffusion coefficient of 10^{-13} m^2/s), the pre-exposure embrittlement effect is reversed (Ricker & Duquette, 1988).

(6)   In high strength aluminum alloys which are susceptible to hydrogen embrittlement, the degree of embrittlement is affected by the extent of stress triaxiality ahead of the crack tip; see Eq. 16.1. For torsional loading, the triaxial component of stress at the crack tip is nominally zero. There is some experimental evidence indicating that environmental effects under conditions of mode III (anti-plane strain) loading are somewhat less pronounced than under mode I loading (where significant hydrostatic stresses are induced ahead of the crack tip) (Duquette, 1979).

(7)   Grain boundary precipitates can influence the fatigue behavior of aluminum alloys exposed to aqueous media by three different mechanisms (Vasudevan, Liu & Ricker, 1987): (i) The precipitates influence the rate of electrochemical reactions by aiding in the removal of material from the crack tip and by affecting the rate of hydrogen evolution. (ii) The grain boundary precipitates can modify the chemistry of the crack tip solution, thereby influencing the stability of the passivating layer at the crack tip. (iii) The size, shape and distribution of the grain boundary precipitates can alter the fracture stress at the crack tip which, in turn, can influence the manner in which the environment interacts with the advancing fatigue crack.

16.3 Growth of corrosion-fatigue cracks

It is convenient to characterize the effects of environment on the rates of fatigue crack growth by considering different combinations of crack growth rates measured under purely mechanical fatigue and under stress corrosion conditions. Figure 16.5(a) schematically illustrates (following the detailed discussions of Chapter 6) the sigmoidal variation of fatigue crack growth as a function of stress intensity factor range (on log–log scale) under purely mechanical cyclic loading conditions (i.e. in inert environments). The typical variation in crack velocity da/dt as a function of the applied stress intensity factor K is plotted (on log–log scale) in Fig. 16.5(b) for sustained load crack growth of metallic materials in the presence of an environment. In this latter figure, the environment has no effect on the fracture behavior of the material below a static stress intensity factor value K_{Iscc}, where the subscript denotes stress corrosion cracking in mode I. Above K_{Iscc}, the crack velocity exhibits a precipitous increase with increasing stress intensity factor K (region I). This is followed by a region of growth (region II) in which the crack increment per

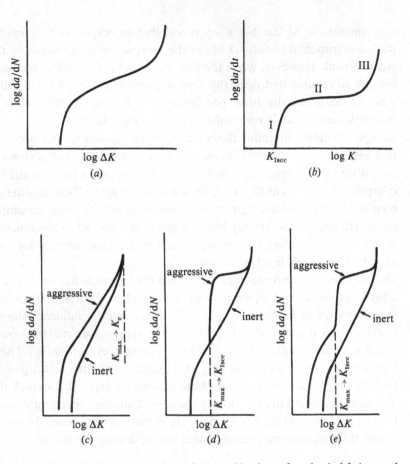

Fig. 16.5. Schematic representations of the combinations of mechanical fatigue and environmentally assisted crack growth. (*a*) Fatigue crack growth behavior in inert environments. (*b*) Stress corrosion crack growth under sustained loads. (*c*) True corrosion-fatigue arising from synergistic effects of cyclic loads and aggressive environment. (*d*) Stress corrosion-fatigue behavior obtained from a superposition of mechanical fatigue (*a*) and stress corrosion cracking (*b*). (*e*) Mixed corrosion behavior obtained from a combination of (*c*) and (*d*). (After McEvily & Wei, 1972.)

unit time is essentially independent of the applied K. As the stress intensity factor values approach the fracture toughness of the material (region III), there is a steep rise in crack velocity.

It is evident from Fig. 16.5(*b*) that subcritical crack growth can occur in the presence of an environment at stress intensity values that are significantly lower than the critical stress intensity for quasi-static fracture initiation K_c (in inert media). For example, AISI 4340 steel with a monotonic yield strength, $\sigma_y = 1690$ MPa, has a plane strain fracture toughness K_{Ic} of 56 MPa$\sqrt{m}$, while its K_{Iscc} value in seawater is as low as 17 MPa$\sqrt{m}$ (Hertzberg, 1995). For a low strength $2\frac{1}{4}$Cr–1 Mo steel (AISI A542 Class 3, $\sigma_y = 500$ MPa) with $K_{Ic} = 295$ MPa$\sqrt{m}$, the value of K_{Iscc} in dry gaseous hydrogen (138 kPa pressure) is approximately 85 MPa$\sqrt{m}$.

16.3.1 Types of corrosion-fatigue crack growth

Corrosion-fatigue crack growth in metallic materials can be schematically represented in three different ways (McEvily & Wei, 1972). Figure 16.5(c) illustrates the so-called *true corrosion-fatigue* behavior where the environment alone (in the absence of cyclic loads) is taken to have no effect on crack growth behavior. Instead, the environment accelerates crack growth rates, only under cyclic loading conditions, by embrittling the material by time-dependent, temperature-dependent, or electro-chemical processes, such as those examined in Section 16.1. True corrosion-fatigue influences cyclic fracture even at maximum stress intensity factor values K_{max} ($= \Delta K/\{1 - R\}$) in fatigue that are less than K_{Iscc}. Figure 12.5(d) shows the so-called *stress corrosion-fatigue* process which is a simple superposition of mechanical fatigue (Fig. 16.5a) and stress corrosion cracking (Fig. 16.5b). Note that stress corrosion-fatigue occurs only when $K_{max} > K_{Iscc}$. The combination of true corrosion-fatigue and stress corrosion-fatigue results in a *mixed corrosion behavior* which is schematically depicted in Fig. 16.5(e).

Figure 16.6 shows typical fatigue crack growth characteristics of high strength steels where the effects of gaseous and aqueous environments are illustrated. In these high strength materials, the presence of an aggressive environment accelerates crack growth rates by several orders of magnitude when $K_{max} > K_{Iscc}$. This acceleration of fatigue fracture is more pronounced at lower cyclic frequencies (where there is more time per stress cycle for the environment to interact with the crack tip). Furthermore, the threshold ΔK value, ΔK_T, at which stress corrosion-fatigue effects begin to influence crack growth rates, decreases with increasing load ratio because K_{max}

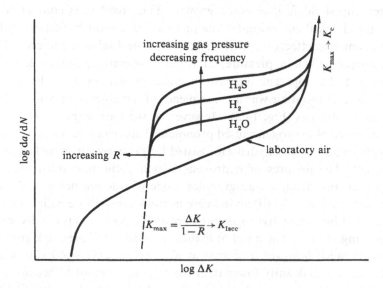

Fig. 16.6. Schematic illustration of the effects of environment, test frequency and R ratio on fatigue crack growth in high strength steels.

($= \Delta K / \{1 - R\}$) begins to approach K_{Iscc} at lower ΔK levels. The superposition of mechanical fatigue and stress corrosion cracking, therefore, provides a reasonably accurate description of corrosion fatigue in many high strength materials in certain environments (see Section 16.3.4 for further details).

Lower strength materials, however, exhibit corrosion-fatigue characteristics which are not amenable to such superposition of mechanical and environmental cracking modes. The definition of K_{Iscc} and its incorporation into corrosion-fatigue modeling for lower strength materials is less reliable because of the following factors:

(1) Low strength materials exhibit crack tip blunting under the sustained loading conditions used to measure K_{Iscc}. However, under repeated cyclic loading, the crack tip may be continually resharpened.

(2) Fretting fatigue contact in the immediate wake of the crack tip under cyclic loading can rupture the protective oxide film. The attendant exposure of fresh surface to the environment leads to the modification of material–environment interactions.

(3) The deformation of lower strength materials exhibits a higher sensitivity to variations in strain rate. Thus environmental interactions with the highly strained crack tip region are likely to be different under fatigue loading than under monotonic loading.

The classification of corrosion-fatigue processes illustrated in Fig. 16.5 does not account for the increase in crack closure due to the formation of corrosion deposits within fatigue cracks, especially at near-threshold ΔK levels. As discussed in Section 14.3, the enhanced levels of crack closure promoted by corrosion deposits can reduce the rate of near-threshold fatigue crack growth. This trend runs counter to the predictions of the classical corrosion-fatigue processes discussed in Section 16.1. In a study of environmental effects on AISI A542 Class 3 steel subjected to cyclic loads in gaseous hydrogen (138 kPa pressure) at room temperature, Suresh & Ritchie (1982b) found that the apparent increase in crack growth rates in dry gaseous hydrogen over those in moist air was a consequence of the absence of oxide-induced crack closure in the dry gas (Fig. 14.8). However, in the Paris regime, at $\Delta K \geq 20$ MPa$\sqrt{\text{m}}$, the presence of hydrogen caused pronounced accelerations in crack growth rates, where the mechanisms of hydrogen assisted fatigue were attributed to hydrogen embrittlement. The features of hydrogen assisted cyclic fracture in this steel included: (i) a transition from a transgranular mode of fracture below ΔK_T to an intergranular mode above ΔK_T, (ii) an increase in crack growth rate with decreasing cyclic frequency and increasing hydrogen pressure above ΔK_T, and (iii) a decrease in ΔK_T with increasing R ratio. The onset of hydrogen-assisted fatigue crack growth in the lower strength steel, however, occurred at K_{max} ($= \Delta K_T / \{1 - R\}$) values of 20 MPa$\sqrt{\text{m}}$ which were significantly lower than the K_{Iscc} value of 80 MPa$\sqrt{\text{m}}$.

As noted earlier, many experimental studies have shown that the susceptibility of a cyclically loaded material to gaseous environments is dependent upon the mate-

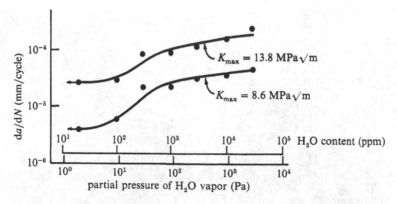

Fig. 16.7. The influence of water vapor content on crack growth rates in 7075-T6 (Alclad) aluminum alloy fatigued at 57 Hz at room temperature. (After Hartman *et al.*, 1967, and Wei, 1970.)

rial–environment combination and the composition and pressure of the gas. For example, the presence of water vapor within a fatigue crack in steels, aluminum and titanium alloys adversely affects its growth rates. Furthermore, fatigue crack propagation rates in steels and titanium alloys are known to be accelerated in the presence of hydrogen gas. Aluminum alloys, however, do not generally exhibit accelerated hydrogen-assisted crack growth in gaseous hydrogen, presumably because of the protective oxide film which readily forms on fresh fracture surfaces. On the other hand, water vapor has a marked effect on corrosion-fatigue crack growth in aluminum alloys. Figure 16.7 shows the variation of fatigue crack growth rate da/dN of a 7075-T6 aluminum alloy as a function of the partial pressure of water vapor at two different values of imposed ΔK. For both ΔK values, a partial pressure of 10 Pa (or 100 ppm) is sufficient to cause an order of magnitude increase in fatigue crack propagation rates. The role of moisture in enhancing crack growth rates in aluminum alloys has also become evident from experiments conducted in dry and wet environments of argon, hydrogen and oxygen (e.g., Hartman, 1965; Wei, 1970).

16.3.2 Formation of brittle striations

The formation of fatigue striations and the spacing between adjacent striations is affected by the presence of an environment. Figure 16.8 is an example of brittle striations formed in aggressive environments and of ductile striations formed in the presence of an inert gas in a Ni single crystal (with specimen axis close to the ⟨100⟩ orientation) which has been fatigued at room temperature at a frequency of 1 Hz. In mercury and gaseous hydrogen (100 kPa pressure), brittle striations form on {100} planes and their spacing is roughly the same in the liquid metal and in H_2 gas. No difference in the mode of fatigue crack growth is evident for the two environ-

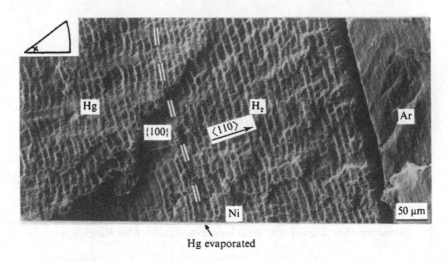

Hg evaporated

Fig. 16.8. Scanning electron micrograph of fatigue fracture surface of Ni single crystal produced by cyclic loading at 1 Hz and 20 °C first in mercury, then in dry hydrogen after evaporating mercury and then in dry argon. The symbol 'x' in the inset shows the orientation of the specimen axis with respect to the standard triangle of the stereographic projection. (From Lynch, 1981. Copyright The Metallurgical Society. Reprinted with permission.)

ments. However, when fatigue tests are resumed in dry argon atmosphere (100 kPa pressure), ductile striations are formed whose spacing is about one-fifth of that seen in Hg or H_2. In all three environments, crack growth occurred on {100} planes in ⟨110⟩ directions. Similar modes of crack growth were also observed for Al–Zn–Mg alloy subjected to the same three environments.

For the liquid metal environment, the adsorption of the embrittling atom is considered responsible for the enhancement in crack growth rates. However, for gaseous H_2, different theories have been proposed which postulate that either *dissolved* H_2 or *adsorbed* H_2 is the cause of embrittlement. On the basis of fractographic evidence, Lynch (1981, 1988) has proposed that adsorbed (rather than dissolved) hydrogen is responsible for accelerated crack growth in hydrogen-bearing environments. His hypothesis rests on the following experimental observations: (i) The crystallographic aspects of crack growth as well as slip line traces at crack tips in hydrogen-containing environments are very similar to those produced by adsorption-induced liquid metal embrittlement. (ii) Hydrogen charging prior to the fatigue test in inert gases has no effect on crack growth. (iii) Brittle crack growth in the environment can occur at velocities that are too high to permit the diffusion of hydrogen directly ahead of the crack tip. Lynch argues that adsorption of hydrogen facilitates localized plastic flow which, depending on the environment, causes apparent differences in the formation of fatigue striation spacings. It should, however, be recognized that the strain rate-sensitivity of liquid metal embrittlement in many alloys is significantly less than that in gaseous or aqueous media.

16.3.3 Effects of mechanical variables

As noted in several preceding discussions, mechanical load variables such as test frequency, stress waveform and load ratio have a marked effect on the rate of corrosion-fatigue crack growth. In this section, examples of such mechanical load effects are presented. Figure 16.9 shows the fatigue crack growth behavior of a mill-annealed Ti–8Al–1Mo–1V alloy in 3.5% NaCl at cyclic frequencies, $\nu_c = 0.5$, 2.5, 5.0, 15.0, and 50.0 Hz at $R = 0.05$ at room temperature. Also indicated in this figure are the fatigue crack growth curves for this material in dry argon at $R = 0.05$ and at

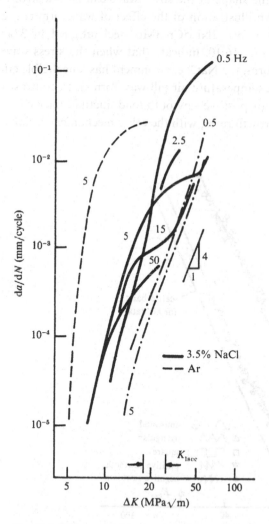

Fig. 16.9. Influence of frequency on the rate of fatigue crack growth in mill-annealed Ti–8Al–1Mo–1V alloy in 3.5% NaCl and in argon gas at $R = 0.05$ and sinusoidal cyclic load variation at room temperature. The dashed curve at left shows the fatigue crack propagation behavior in 3.5% NaCl at $R = 0.75$ and at 5 Hz frequency. (After Bucci, 1970.)

0.5 and 5.0 Hz. Although a change in frequency in the inert environment has no major influence on fatigue crack propagation, lowering the frequency (i.e. increasing the time for environmental interactions per stress cycle) in the NaCl environment causes an appreciable increase in the crack growth rate. Note that the environmental effects are pronounced at $K_{max} \geq K_{Iscc}$ ($\approx$ 20–30 MPa$\sqrt{m}$) in NaCl. At high growth rates, where the crack velocities are too rapid to be influenced by chemical effects, the inert gas and NaCl lead to similar crack extension behavior. Increasing the load ratio in the NaCl environment promotes a lowering of ΔK_T as per the description presented in connection with Fig. 16.6.

For a fixed cyclic frequency, the shape of the stress wave can have a strong effect on corrosion-fatigue fracture. An illustration of the effect of waveform is provided by the results of Barsom (1972) for 12Ni–5Cr–3Mo steel fatigued in 3% NaCl solution at 0.1 Hz. His results, Fig. 16.10, indicate that when the stress wave is of a square or negative sawtooth form, the NaCl environment has a negligible effect on crack growth over that of room temperature air (all waveforms). In contrast to this behavior, triangular, sinusoidal or positive sawtooth load fluctuations cause noticeable increases in fatigue crack growth rates, with the other mechanical variables held

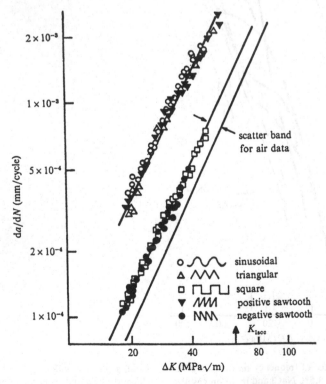

Fig. 16.10. The effect of stress waveform on fatigue crack growth in 12Ni–5Cr–3Mo steel in 3% NaCl solution at 0.1 Hz at room temperature. (After Barsom, 1972.)

fixed. Barsom rationalized this waveform effect by noting that, when the rising portion of the stress cycle is slow enough to permit sufficient time for environmental interactions, environmentally assisted crack growth results. For the square and negative sawtooth waves, loading occurs in one step, and consequently no environmental effects are evident. Note that for this lower strength steel, environmentally assisted crack growth occurs at K_{max} values that are well below K_{Iscc} in NaCl. This result is probably an outcome of the continual rupture of the protective oxide film in the vicinity of the crack tip by repeated cyclic loads.

16.3.4 Models of corrosion-fatigue

Corrosion-fatigue is a process which is an outcome of synergistic interactions among the environment, material microstructure, and cyclic loads. Since the mechanisms of corrosion-fatigue are specific to a particular material–environment–load system, it is not surprising that quantitative modeling of this complex process is still in its infancy. There does not exist a single model which is capable of quantitatively predicting the essential features of environmentally assisted fatigue for broad classes of materials and environments.

A simple approach to modeling corrosion-fatigue involves the so-called *superposition* approach (Bucci, 1970; Wei & Landes, 1970). This model assumes that the overall crack growth rates (da/dN) of a material under corrosion-fatigue conditions can be determined by a linear superposition of the crack extension rates due solely to mechanical fatigue (da/dN)$_F$ (determined in inert environments) and the stress corrosion crack growth rate (da/dN)$_{SC}$ (derived from the sustained load crack velocity da/dt), such that

$$\left(\frac{da}{dt}\right) = \left(\frac{da}{dt}\right)_F + \left(\frac{da}{dt}\right)_{SC}. \tag{16.2}$$

The second term on the right hand side is obtained by integrating the environmentally assisted crack growth rate da/dt, which is influenced by the stress intensity factor K, over the time period of one fatigue cycle, such that

$$\left(\frac{da}{dt}\right)_{SC} = v_c \int_0^{1/v_c} \frac{da}{dt}(K) \, dt. \tag{16.3}$$

Note that the effects of cyclic frequency, waveform, and load ratio can be incorporated into this model. The model provides a reasonable prediction of corrosion-fatigue behavior in a limited number of systems such as high strength steels. However, this linear superposition model implies that there is no synergistic interaction between the mechanical and environmental components of corrosion-fatigue. The model also invokes the tacit assumption that the mechanisms of (sustained load) stress corrosion cracking and corrosion-fatigue are the same. These two assumptions of the model run counter to experimental observations available for many engineering alloys, especially lower strength materials.

A consequence of the above superposition model is that corrosion-fatigue effects are predicted only when $K_{max} > K_{Iscc}$. This prediction is also not in accord with the experimental results available in the literature for many lower strength materials (see Fig. 16.10, for example). To overcome this limitation, Weir *et al.* (1980) and Wei & Simmons (1981) proposed an extension of Eq. 16.2, where

$$\left(\frac{da}{dt}\right) = \left(\frac{da}{dt}\right)_F + \left(\frac{da}{dt}\right)_{SC} + \left(\frac{da}{dt}\right)_{CF}. \tag{16.4}$$

The last term on the right hand side of this equation represents the true corrosion crack growth rate (at $K_{max} < K_{Iscc}$) which is a synergistic effect arising from the interaction of cyclic loading and environment. Wei & Simmons hypothesize that this cycle-dependent contribution arises from the reaction of the environment with fresh crack surfaces produced by fatigue, and that it is a function of surface reaction during one loading cycle. For material–environment combinations which lead to high reaction levels, this contribution also depends on the rate of transport of the aggressive environment to the crack tip.

An alternative approach to modeling corrosion-fatigue has been postulated by Austen & McIntyre (1979). These workers suggest that the processes of fatigue and stress corrosion cracking are mutually competitive. According to this *process competition model*, a crack will grow at the fastest available rate pertinent to the prevailing stress intensity factor.

The limitations of the foregoing models become apparent upon reexamination of the results discussed in Section 15.7.2. It may be recalled from this section that corrosion-fatigue crack growth rates are strongly influenced not only by the specific combinations of cyclic loads, material and environment, but also by the crack size. These results imply that critical steps in corrosion processes which depend on crack size, such as the rate of transport of the embrittling species to the crack tip, the rate of pumping the environment by the opening and closing of the crack walls, and the development of crack closure and fretting contact along the crack flanks, must be incorporated into any complete model for environmentally assisted fatigue. Considerable collaborative research among materials scientists, mechanical engineers, chemists and surface scientists is necessary before such quantitative models can be developed.

16.4 Case study: Fatigue design of exhaust valves for cars

Exhaust valves in automobile engines provide an example of an engineering component where the effects of fatigue, high-temperature creep, creep-fatigue interactions, corrosive environments, galling and wear pose a considerable challenge for successful design. We consider here some critical issues in the design of internal combusion engine exhaust valves for automobiles. This case study is based on the work of Dowling *et al.* (1993).

The material from which the valve is made should have an optimum combination of the following features:

(1) *Low density*. Light-weight valves lead to improved fuel efficiency, reduced noise and vibration, and better overall performance. They also facilitate operation at higher engine speeds and enable faster opening and closing. These benefits stem from the fact that

    (a) a lower density leads to lower valve-spring loads to maintain valve toss speeds (for a given performance goal),

    (b) a reduced valve-spring load lowers mechanical friction which must be overcome for the operation of the valve train.

    The extent of the overall benefit from low density, however, is very specific to the engine and the valve train mechanism.

(2) *High thermal conductivity and low thermal expansion coefficient*. Valve temperature and thermal gradients are reduced as a result of high thermal conductivity. This, in conjuction with a low coefficient of thermal expansion, reduces thermal stresses and valve-induced knock limitations. Consequently, the operating temperature of the valve can be increased.

(3) *Stiffness*. The valve should have a Young's modulus which is high enough for desirable valve train dynamics while, at the same time, being low enough to promote valve sealing on a distorted seat.

(4) *Manufacturability and low cost*. The large volume production of exhaust valves inevitably requires that the composition and microstructure of the material from which it is made be controllable within a narrow range for predictable performance, while the cost of manufacture is kept low. Consider, for example, a typical engine manufactured by the Ford Motor Co. in the 1990s. One to two million exhaust valves are required per annum for each engine type. With a mass of 30 g per valve, the material needed for valves is 80 000 kg/year, assuming a 75% material efficiency.

Figure 16.11 schematically shows the failure modes commonly encountered in exhaust valves of spark-ignited gasoline engines. The impingement of exhaust gases causes the

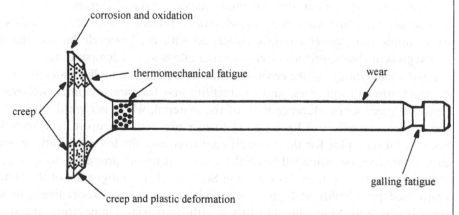

Fig. 16.11. Possible failure modes in the automobile exhaust valve. (After Dowling *et al.*, 1993.)

highest temperatures, typically in the range 650–800 °C, to develop at the valve stem blend and underhead radius; the maximum temperature, however, is very specific to the type of engine. The highest stresses, typically 35–70 MPa, also occur in the valve underhead radius and the seat area; again, the magnitudes of these stresses depend on such factors as the valve geometry, alignment and combusion gas pressure. These sites are also regions where creep deformation can occur. Corrosion and oxidation are also potential damage mechanisms in the valve seat area.

Mechanical fatigue and creep-fatigue cracking can occur at the site of the stem/blend radius, where the stresses are typically less than 20 MPa if the valve is properly aligned with the valve seat. Should misalignment occur here due, for example, to seat distortion, the bending stresses can become as high as 250 MPa. The valve stem and tip regions typically undergo wear, galling and mechanical fatigue.

Conventional austenitic stainless steels, such as 21-2N, are commonly used for exhaust valves for maximum temperatures in the range 650–800 °C. Inconel 751 is a commercial alloy which is widely used for high-performance exhaust valves in cars. The need for lower valve mass and higher valve temperatures, in order to move towards higher engine speeds and near-stoichiometric fuel to air ratios, has led to a continual search for newer materials for automobile exhaust valves. In view of their high-temperature capabilities and proven record in aircraft jet engine components, Ni-base superalloys are possible considerations despite their high density and high cost. The lower densities of conventional Ti alloys, TiAl intermetallics and ceramics also make them possible candidates for low-weight valves. The most critical obstacle to the use of such advanced materials in exhaust valves is cost-effective processing and manufacturing.

16.5 Fatigue at low temperatures

The initiation and growth of fatigue cracks at sub-zero temperatures is a topic of considerable interest in many practical applications. Some examples of these situations include: transportation vehicles used in cold climates such as the arctic environments, equipment used for the transportation and storage of liquefied fuels, and superconducting machinery and electrical transmission systems. In this section, we examine some general trends associated with the low-cycle fatigue and fatigue crack growth characteristics of engineering alloys at low temperatures.

As the temperature of the environment surrounding a metallic material is lowered, its yield strength increases, and its ductility and fracture toughness decrease. The (smooth specimen) endurance limit of the material, which is typically 0.3–0.5 times the tensile strength, will be expected to increase with a drop in temperature. The cyclic strain-life plot for the material characterizing the low-cycle fatigue behavior can, in general, be estimated from the change in tensile properties as a function of temperature. For example, as shown in Section 8.1, the fatigue life of the material is controlled by ductility at high imposed plastic strains and short lives (the second term in Eq. 8.5). Since ductility drops with decreasing temperature, the short life fatigue resistance also deteriorates with decreasing temperature. At long lives, where

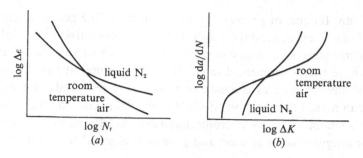

Fig. 16.12. Schematic illustration of the effect of low temperatures on (a) strain-based total fatigue life, and (b) fatigue crack growth in metallic materials.

the fatigue life is controlled by strength (the first term in Eq. 8.5), lower temperatures will be expected to promote a higher number of cycles to failure. These effects of temperature on the overall strain-life curve are schematically sketched in Fig. 16.12(a).

The reduction in cyclic plasticity accompanying low temperature fatigue also causes pronounced variations in the rates of fatigue crack growth from that seen at room temperature. Since the fracture toughness of the material decreases with decreasing temperature, regime III of (catastrophic) fatigue failure will occur at lower nominal ΔK values than at room temperature, as shown in Fig. 16.12(b). The reduced levels of cyclic plasticity at cryogenic temperatures also make it difficult for fatigue cracks to advance at low ΔK, with the result that the apparent threshold stress intensity range values are elevated at lower temperatures (e.g., Tschegg & Stanzl, 1981).

16.6 Damage and crack initiation at high temperatures

Failure-critical components in aircraft engines and power generating plants are subjected to cyclic loads at elevated temperatures. Consider, for example, some components used in a gas turbine aeroengine. The typical design life of the engine is 25 000 h involving some 30 000 civilian flights (Coles, 1980). In this engine, the compressor disc and casing, made of Ni-base alloy and age-hardened stainless steel, respectively, are subjected to high strain fatigue at 400 °C. The combustor liner, made of Co-base alloy, is subjected to thermal fatigue at 982 °C. The turbine blades, fabricated from Ni-base superalloy, undergo thermomechanical fatigue, high-cycle fatigue and/or stress rupture, at temperatures of up to 982 °C. The turbine discs, made of Ni-base alloy, are susceptible to high strain fatigue and creep in the temperature range of 371–649 °C. It is readily seen from this example that the initiation and growth of cracks at elevated temperatures is a topic of major concern. This subject constitutes one of the most complex topics in materials research because elevated temperature fatigue, in general, is a consequence of the conjoint action of creep, cyclic loads and environmental corrosion. Although our current knowledge of these complex interactions is far from complete, there is a rich variety of mechanistic processes which have been documented in elevated temperature fatigue. A comprehensive survey of this subject area is available in several reviews and volumes (e.g., Sadananda & Shahinian, 1981; Skelton, 1983; Riedel, 1987). This and the following sections provide a summary of the micromechanisms, models and characterization methodology for fatigue crack initiation and growth at elevated temperatures.

16.6.1 Micromechanisms of damage

Fatigue crack nucleation in an alloy at elevated temperatures is influenced by a number of different microscopic processes which depend on the temperature, applied stress level and environment. The mechanisms responsible for the inception of fatigue flaws can be generally grouped into the following categories: (i) cracking induced by cyclic slip, (ii) grain boundary cavitation, (iii) grain boundary sliding and the attendant development of wedge cracks, (iv) nucleation and growth of voids at inclusions and precipitates, and (v) oxidation and corrosion.

16.6.1.1 Cyclic slip

The nature of cyclic slip is affected by temperature. The following observations provide an illustration of the role of temperature in cyclic slip.

(1) At temperatures in excess of one-half of the homologous temperature, most materials begin to exhibit wavy slip characteristics. In solid solution alloys, the stacking fault energy usually increases with increasing temperature (e.g., Rémy, Pineau & Thomas, 1978). Thermal activation also permits dislocation climb and cross slip.

(2) The kinematic irreversibility of cyclic slip is enhanced by oxidation and/or gas diffusion at elevated temperatures.

(3) The tendency for wavy slip may be opposed, in some materials, by dynamic strain ageing which favors slip planarity.

(4) The time-dependence of dislocation climb also renders elevated temperature deformation strain rate-sensitive.

(5) The stability of microstructure in most engineering materials breaks down at elevated temperatures. Thermally activated microstructural changes may arise as a result of strain-ageing (e.g., carbon steels in the temperature range of 230–370 °C), recovery (e.g., AISI 316 stainless steel at about 800 °C; Challenger & Moteff, 1972) or precipitation (e.g., $M_{23}C_6$ carbide formation in AISI 316 stainless steel; Weiss & Stickler, 1972). These metallurgical changes can be beneficial or detrimental to fatigue resistance; in either case, they considerably modify the cyclic stress–strain characteristics.

16.6.1.2 Creep cavitation

At temperatures spanning one-third to two-thirds of the melting point and at low imposed stresses, most metals, alloys and ceramics exhibit creep deformation by grain boundary cavitation. These cavities nucleate at grain boundary triple point junctions, grain boundary particles or ledges on the intergranular facets. In many engineering alloys, creep cavities nucleate well below a nominal stress level of 100 MPa, and their nucleation occurs continuously over a substantial fraction of the creep life (Riedel, 1987). Shear sliding of the grain boundary facets also aids in the nucleation of cavities. A particularly interesting aspect of creep cavitation under cyclic loads is that the extent of cavitation is strongly affected by the shape of the stress wave. Slow tension-going and fast compression-going strain rates, i.e. loading wave shapes with long hold times at tensile stresses, exhibit a greater propensity for cavitation than balanced load cycles where the magnitude of the tensile loading part varies in an identical manner to the magnitude of the compression phase (Fig. 16.13a), e.g., Majumdar & Maiya (1980) and Baik & Raj (1982).

Riedel (1987) considered several different theories for creep cavitation for slow–fast fatigue loading and presented a mechanism map (Fig. 16.13b) for fatigue life as a function of temperature, for fixed values of strain range $\Delta\epsilon$ and test frequency. At

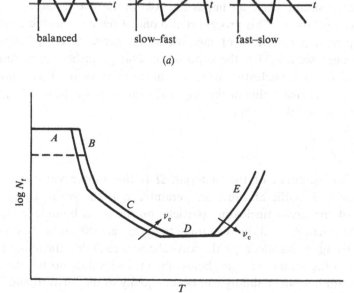

Fig. 16.13. (a) A schematic of balanced, slow–fast and fast–slow fatigue loading. (b) Mechanism map for fatigue life as a function of temperature (at fixed values of nonlinear strain range and frequency) for slow–fast creep-fatigue loading. (After Riedel, 1987.)

intermediate temperatures represented by regime C in Fig. 16.13(b), diffusive cavity growth occurs, where cavities nucleate continuously throughout the creep-fatigue test. At low values of accumulated strains $2\Delta\epsilon N_f$, the effects of frequency ν_c and temperature T in regime C are described by the relation, $N_f \propto (\nu_c T/\delta_g D_b)^{2/5}$, where δ_g is the thickness of the grain boundary, and D_b is the grain boundary diffusion coefficient. At larger values of N_f, regime B, the accumulated strain is so large that the cavity density saturates early in the fatigue life. For this case, $N_f \propto \nu_c T \delta D_b$. At lower temperatures represented by region A, plastic hole growth is postulated to be the dominant mechanism. This process is independent of frequency if the formation of cavity surfaces is fully irreversible.

At higher temperatures, region D, the cavities grow so rapidly that the imposed strain rate (which is assumed constant within the framework of Fig. 16.13b) can no longer accommodate the cavity volume. Consequently, cavity growth in the slow–fast loading mode becomes constrained. At still higher temperatures, region E, constrained cavity growth occurs in tension and shrinkage of cavities occurs in the fast compression segment. Since the diffusion rates are high at these temperatures, cavity growth begins to be constrained in both tension and compression. This causes essentially zero net growth of cavities and hence, improved fatigue life.

16.6.1.3 Grain boundary sliding

Many experimental studies (Driver, 1971; Min & Raj, 1979) have shown that grain boundary sliding plays an important role in the nucleation of intergranular cavities in creep-fatigue. This process is also one of the mechanistic considerations in the empirical partitioning of inelastic strain range in creep-fatigue life calculations (see next section). On the supposition that grain boundary sliding is the rate-limiting step in the nucleation of triple-junction cracks and that it does not occur above a certain critical value of the imposed strain rate $\dot\epsilon_{cr}$, Min & Raj (1979) developed a model which shows that

$$\dot\epsilon_{cr} = \frac{8\sigma_y \Omega \delta_g D_b}{kTd_g f_b d_p^2}. \tag{16.5}$$

Here σ_y is the yield strength of the material, Ω is the atomic volume, k is the Boltzmann constant, T is the absolute temperature, d_g is the grain size, f_b is the volume fraction of the cavity nucleating particles on the grain boundary, d_p is the average size of the particles, and the remaining terms are as defined in the preceding subsection. According to this theory, if the wave shape is such that the tension-going and compression-going strain rates are below the critical value, no net damage is produced since grain boundary sliding will occur equally in the forward and reverse loading directions. With slow–fast loading and with the strain rate in the compression phase greater than $\dot\epsilon_{cr}$, maximum damage is produced. Although this model is an oversimplification of the grain boundary sliding process, it offers insights into the effects of wave shape on high temperature fatigue in many alloys.

Grain boundary cavitation is also known to be affected by the relative variations of temperature and stress with time. Experiments by Fujino & Taira (1980) on low carbon steel and AISI 304 stainless steel show that in-phase cyclic variations in temperature and strain (i.e. tension-going strain phase accompanied by increasing temperature) produces more fatigue damage than out-of-phase thermomechanical fatigue loading. Furthermore, out-of-phase thermomechanical cyclic loading produces a lower fatigue life than isothermal mechanical fatigue. This variation of fatigue resistance for different thermomechanical loading conditions appears to exhibit a correlation with the extent of grain boundary sliding.

16.6.1.4 Inclusions and precipitates

Subsurface nucleation of fatigue cracks can occur at elevated temperature by the decohesion of inclusions and second phase particles from the surrounding matrix (see Fig. 4.13 for an example of this mechanism). In some aluminum alloys, void nucleation around grain boundary precipitates causes intergranular cavitation even at low temperatures (Vasudevan & Doherty, 1987). Furthermore, enhanced precipitation during elevated temperature fatigue provides a greater number of void nucleation sites.

16.6.1.5 Oxidation and corrosion

The presence of an oxidizing or aggressive environment can influence, in many ways, the mechanisms of high temperature fatigue crack nucleation.

(1) The presence of an environment may preclude the sintering or closure of cavities at elevated temperature.

(2) If a grain boundary near a free surface is not protected by an oxide layer, oxygen gas or other embrittling species may diffuse along the boundary and react with grain boundary precipitates. Under the influence of an applied stress, cavities initiate around the grain boundary precipitates (e.g., Wells, 1979). Furthermore, the precipitation of residual gases may provide the nuclei for cavities, with the internal gas pressure accelerating the growth of cavities.

(3) The combination of surface diffusion and slip step oxidation promotes enhanced kinematic irreversibility of cyclic slip (see Section 4.2), which causes fatigue cracks to nucleate in the absence of other mechanisms (Fig. 16.14). This process is commonly known as the *Fujita mechanism* (Fujita, 1963).

(4) Preferential oxidation at certain microstructural sites, such as at the intersection of a grain boundary with a free surface, causes microscopic stress concentrations (notches) to develop. The micronotches elevate the local stresses and promote crack nucleation (Duquette, 1979).

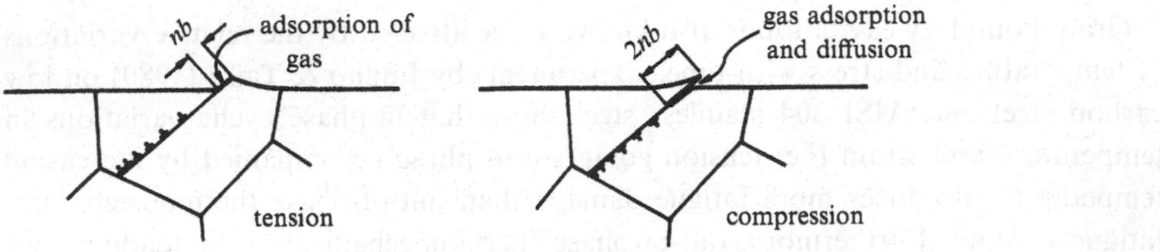

nb   adsorption of
gas

tension

2nb   gas adsorption
and diffusion

compression

Fig. 16.14. A schematic of the Fujita mechanism for fatigue crack nucleation.

(5)   Repeated fatigue loads rupture the protective oxide film on the surface of
the material. The absence of a protective film provides an active path for
chemical attack. Furthermore, the cracking of the brittle oxide film, by
itself, may lead to the growth of a catastrophic fatigue flaw (Wells, 1979).

For many materials, the deterioration of fatigue resistance at elevated temperature
is merely a consequence of environmental interactions. The experimental basis for
this inference is illustrated by the results of Coffin (1973) for AISI A286 ferrous
alloy, Fig. 16.15. Here the variation of plastic strain range $\Delta\epsilon_p$ with the number of
cycles to failure N_f is unaffected by either the temperature or the frequency *in vacuo*.
However, lowering the test frequency severely reduces the low-cycle fatigue life in
elevated temperature air.

16.6.2 Life prediction models

Numerous approaches have been developed over the years to predict the
total failure life of engineering materials subjected to elevated temperature fatigue.

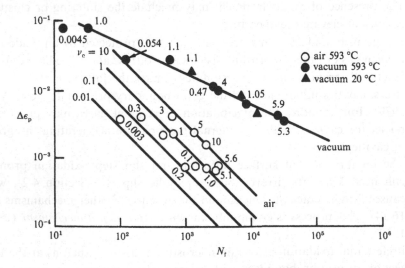

Fig. 16.15. A plot of $\Delta\epsilon_p$ versus N_f for AISI A286 alloy in laboratory air and vacuum at 593 °C.
Numbers adjacent to the data points denote the frequency in cycles per minute. The solid lines
represent regression analysis of data. (After Coffin, 1973.)

Comprehensive reviews of the available life prediction models can be found in Coffin (1974), Tomkins & Wareing (1977), and Batte (1983). The most widely used methods for life prediction can be classified into the following three groups: (i) damage accumulation models, (ii) frequency-modified Coffin–Manson equations, and (iii) strain range partitioning. A brief summary of each of these methods is presented in this section.

16.6.2.1 Damage accumulation models

The simplest approach to predicting creep-fatigue life involves the linear damage summation model. In this method, the damage accumulated by mechanical fatigue and by creep are linearly superposed, such that

$$\zeta_f + \zeta_c = 1, \tag{16.6}$$

where ζ_f and ζ_c are the fractional damage due to fatigue and creep, respectively. If the material is subjected to n_i fatigue cycles at a stress amplitude of $\Delta\sigma_i$ and the number of cycles to failure at this stress amplitude is N_{fi}, the fraction of fatigue damage is given (see the Palmgren–Miner rule in Section 7.3) by the expression $\sum n_i/N_{fi}$, where the summation is carried out over the total number of stress blocks with different amplitudes. Similarly, if t_j is the average time under an average imposed stress σ_j and t_{Rj} is the time to rupture at that stress level, then $\sum t_j/t_{Rj}$ denotes the amount of damage accumulated by creep. For variable amplitude loading at elevated temperatures, the damage summation rule is written as

$$\sum \frac{n_i}{N_{fi}} + \sum \frac{t_j}{t_{Rj}} = d, \tag{16.7}$$

where d is the accumulated damage fraction which is often taken to be unity.

This approach is widely used in conjunction with strain-controlled fatigue tests with dwell periods under load. As noted by Batte (1983), one encounters at least three problems in the implementation of the linear damage summation concept: (i) Under fixed strain dwell periods, the stresses continue to relax. (ii) Cyclic hardening or softening leads to different levels of stress relaxation during continued fatigue straining. (iii) Stress rupture data used to predict life in Eq. 16.7 are commonly derived from monotonic load tests and they are not, in general, representative of creep-fatigue behavior.

16.6.2.2 Frequency-modified Coffin–Manson equation

The frequency-modified strain range–life relationship was proposed by Coffin (1973) to account for the effect of cyclic frequency on elevated temperature cyclic response (such as that shown in Fig. 16.15). This relationship, which phenomenologically incorporates the frequency term ν_c in the Coffin–Manson relationship, Eq. 8.1, takes the form

$$\Delta\epsilon_p = c_f \left(N_f v_c^{k-1} \right)^{-\beta},$$ (16.8)

where the constant c_f and the exponents k and β are specific to the temperature–material–environment system. The behavior of many steels is satisfactorily described in an alternate form of Eq. 16.8 where the total strain range is separated into elastic and plastic components, similar to Eq. 8.5:

$$\Delta\epsilon_t = \Delta\epsilon_e + \Delta\epsilon_p = c_{f1} \left(N_f v_c^{k_1-1} \right)^{-\beta_1} + c_{f2} \left(N_f v_c^{k_2-1} \right)^{-\beta_2}.$$ (16.9)

It should be noted that the application of Eqs. 16.8 and 16.9 to predict creep-fatigue life requires a knowledge of the stress–strain hysteresis loops.

16.6.2.3 Strain range partitioning

The strain range partitioning methods, proposed by Halford, Hirschberg & Manson (1973), isolate the overall damage due to thermomechanical fatigue interactions into mechanical and time-dependent components of strain range. On a microscopic scale, cyclic slip is envisioned as the mechanism for time-independent mechanical fatigue, and grain boundary cavitation and sliding are considered the primary mechanisms influencing creep-fatigue damage. The strain range partitioning methods also take into consideration the reversal of strains associated with fatigue and creep phenomena. In this approach, any fully reversed inelastic strain range is divided into four components (see Fig. 16.16): (i) $\Delta\epsilon_{pp}$, which denotes tensile plastic

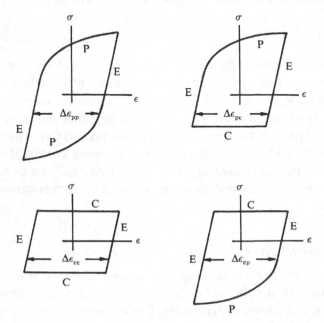

Fig. 16.16. Schematic representation of strain range partitioning into four components of inelastic strain range. E, P and C denote elastic, plastic and creep deformation, respectively.

deformation reversed by compressive plastic deformation; (ii) $\Delta\epsilon_{cp}$, which denotes tensile creep reversed by compressive plastic deformation; (iii) $\Delta\epsilon_{pc}$, which denotes tensile plastic deformation reversed by compressive creep; and (iv) $\Delta\epsilon_{cc}$, which denotes tensile creep reversed by compressive creep.

The cyclic endurance is related to the four components of strain range by Coffin–Manson type power law relationships which take the form

$$\Delta\epsilon_{pp} = c_{pp}N_{pp}^{-\beta_1}, \qquad \Delta\epsilon_{cp} = c_{cp}N_{cp}^{-\beta_2}, \quad \text{etc.} \tag{16.10}$$

Figure 16.17 shows the fatigue–life relationships for the partitioned strain ranges. The total fatigue life is then determined by the summation of the damage fractions from the four modes of deformation such that

$$\frac{1}{N_f} = \frac{1}{N_{pp}} + \frac{1}{N_{cp}} + \frac{1}{N_{pc}} + \frac{1}{N_{cc}}, \tag{16.11}$$

where the subscripts denote the same modes as those of the strain ranges. Sometimes the damage fractions are also weighted in proportion to the magnitude of the strain ranges in each of the four modes. Many empirical modifications of this approach are also available for incorporating the effects of slow–fast fatigue loading (including the effects of different tension-going and compression-going parts of the cycle) and of creep-fatigue dwell periods into the fatigue life calculation (see, for a review, Batte, 1983).

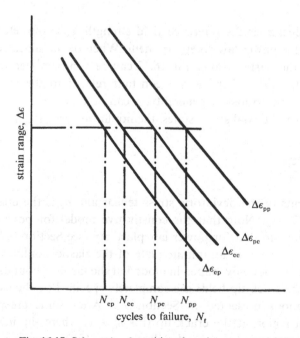

Fig. 16.17. Schematic of partitioned strain ranges and fatigue life relationships.

16.7 Fatigue crack growth at high temperatures

At temperatures that are significant fractions of the homologous temperature, the mechanisms responsible for fracture become time-dependent rather than cycle-dependent. Time-dependent processes may be controlled by creep and/or environment, both of which are, in general, thermally activated. For time-dependent fatigue crack propagation, the stress intensity factor range is still the controlling parameter if the zone of inelastic deformation at the crack tip is small in size compared to crack length and the size of the uncracked ligament. However, at very high temperatures and low test frequencies, the limitations of linear elastic fracture mechanics become apparent as the inelastic zone becomes an appreciable fraction of the characteristic specimen dimensions. Furthermore, the decrease in flow strength of the material at elevated temperatures causes a greater extent of nonlinear deformation to develop. In this section, different fracture parameters used to characterize elevated temperature crack growth are examined, and the conditions for the validity of each parameter are pinpointed. This is followed by discussions of elevated temperature crack growth in engineering alloys.

16.7.1 Fracture mechanics characterization

Consider an elastic–nonlinear viscous material for which the total strain rate in uniaxial tension is given by

$$\dot{\epsilon} = \frac{\dot{\sigma}}{E} + \dot{\epsilon}_y \left(\frac{\sigma}{\sigma_y} \right)^{n_c}, \tag{16.12}$$

where E is Young's modulus, σ_y is the reference yield strength, $\dot{\epsilon}_y$ is the reference creep strain rate, and n_c is the power law creep exponent. When the nonlinear creep strain rate dominates over the elastic strains, Eq. 16.12 can be approximated by the pure power law, $\dot{\epsilon}/\dot{\epsilon}_y = \alpha(\sigma/\sigma_y)^{n_c}$, where α is a constant related to the material parameters in Eq. 16.12. (With no loss of generality, α can be taken to be unity in the present discussion.) For multiaxial stress states, the nonlinear creep strain rate is given by

$$\dot{\epsilon}_{ij}^c = \frac{3}{2} \dot{\epsilon}_y \left(\frac{\sigma_e}{\sigma_y} \right)^{(n_c-1)} \frac{s_{ij}}{\sigma_y}, \tag{16.13}$$

where s_{ij} are the components of the deviatoric stress tensor and σ_e is the effective stress defined by $\sigma_e^2 = (3/2)s_{ij}s_{ij}$. Note that the constitutive model for power law creep, Eq. 16.13, is analogous to that for power law plasticity (see Section 9.7.2).

Consider a stationary edge crack in an infinite plate of the elastic–nonlinear viscous material. Let the plate be suddenly loaded in mode I at the far-field boundary at time $t = 0$. At this point, the crack tip fields are completely specified by the asymptotic elastic singular solutions presented in Section 9.3. After some creep has occurred, there develops a region at the crack tip (for $n_c > 1$) where the inelastic part of deformation is more dominant than the elastic part. If this region is very small, *small-scale creep* conditions prevail, and fatigue crack growth at the elevated temperature is still amenable to a characterization based on ΔK. For contours far from the crack tip (i.e. in regions where elastic strains prevail), the value of the J-integral is

$$J_\mathrm{o} = \frac{1 - \nu^2}{E} K^2 \tag{16.14}$$

for plane strain conditions. At time $t > 0$ and at regions close to the crack tip, the near-tip stresses σ_{ij}, creep strain rates $\dot{\epsilon}_{ij}^\mathrm{c}$, and the displacement rates $\dot{u}_i$ are given by the universal fields of the HRR-type discussed in Section 9.7.2, Eqs. 9.91. This is a consequence of the similarity between Eq. 16.13 and the pure power-law model described in Section 9.7.2. With respect to the polar coordinates centered at the crack tip,

$$\sigma_{ij} = \sigma_\mathrm{y} \left(\frac{C(t)}{\alpha \sigma_\mathrm{y} \dot{\epsilon}_\mathrm{y} I_{n_\mathrm{c}} r} \right)^{1/(n_\mathrm{c}+1)} \tilde{\sigma}_{ij}(\theta, n_\mathrm{c}),$$

$$\dot{\epsilon}_{ij}^\mathrm{c} = \alpha \dot{\epsilon}_\mathrm{y} \left(\frac{C(t)}{\alpha \sigma_\mathrm{y} \dot{\epsilon}_\mathrm{y} I_{n_\mathrm{c}} r} \right)^{n_\mathrm{c}/(n_\mathrm{c}+1)} \tilde{\epsilon}_{ij}(\theta, n_\mathrm{c}),$$

$$\dot{u}_i = \alpha \dot{\epsilon}_\mathrm{y} \left(\frac{C(t)}{\alpha \sigma_\mathrm{y} \dot{\epsilon}_\mathrm{y} I_{n_\mathrm{c}}} \right)^{n_\mathrm{c}/(n_\mathrm{c}+1)} r^{1/(n_\mathrm{c}+1)} \tilde{u}_i(\theta, n_\mathrm{c}). \tag{16.15}$$

The amplitude of the singularity is $C(t)$, which depends on the elapsed time t, the magnitude of the far-field loads, the crack geometry, and the material constants given in Eq. 16.12. The near-tip fields in creeping solids scale with $C(t)$ in the same way the rate-independent plastic fields scale with J in an elastic–plastic solid. Therefore, $C(t)$ can be expressed as a line integral on any counterclockwise path Γ around the crack tip which lies within the region where the creep strains are well in excess of the elastic strains (Bassani & McClintock, 1981):

$$C(t) = \int_\Gamma \left\{ \left(\frac{n_\mathrm{c}}{n_\mathrm{c} + 1} \right) \sigma_{ij} \dot{\epsilon}_{ij} \mathrm{d}y - \sigma_{ij} n_j \frac{\partial \dot{u}_i}{\partial x} \mathrm{d}s \right\}, \tag{16.16}$$

Here n_j is the j-component of the unit outward normal along Γ. Note the similarity of Eq. 16.16 to Eq. 9.87. The $C(t)$ integral is path-independent in the region where the creep strain rates are significantly greater than the elastic strain rates.

The extent of the creep zone is defined by the boundary ahead of the crack tip where the creep strain measure, $\bar{\epsilon}^\mathrm{c} = \sqrt{(2/3)\epsilon_{ij}^\mathrm{c}\epsilon_{ij}^\mathrm{c}}$, is equal to the elastic strain measure, $\bar{\epsilon}^\mathrm{e} = \sqrt{(2/3)\epsilon_{ij}^{\mathrm{e}'}\epsilon_{ij}^{\mathrm{e}'}}$; $\epsilon^{\mathrm{e}'}$ is the deviatoric elastic strain determined from the initial elastic K-field. Using this definition, Riedel & Rice (1980) argued that the total strain well within the creep zone depends uniquely on the stress. The self-similar nature of the inner fields leads to the following relation between J and $C(t)$:

$$J_\mathrm{i} = C(t)(n_\mathrm{c} + 1)\, t, \tag{16.17}$$

where the subscript for J denotes the inner field. Invoking the assumption that J_i equals J_o, Eq. 16.14, Riedel & Rice estimated that

$$C_{RR} = \frac{1 - v^2}{(n_c + 1)Et} K^2. \tag{16.18}$$

C_{RR} denotes the Riedel–Rice estimate of $C(t)$ based on the assumption that $J_i = J_o$. For short creep times, $C_{RR} = C(t)$, and the amplitude of the near-tip fields decays as $1/t$. At long times where extensive creep develops, the situation is analogous to the fully plastic state. For this case, steady-state conditions are attained and $C(t)$ approaches a value C^* which is path-independent everywhere:

$$C^* = \dot{\epsilon}_y \sigma_y a \left(\frac{P}{P_0}\right)^{n_c+1} \text{function} (n_c, \text{geometry}), \tag{16.19}$$

where P and P_0 are the applied load and reference load, respectively, and a is the crack length. It is apparent that, for the same specimen and crack configurations, values of C^* are obtained from solutions developed for J. Expressions for C^* for different specimen geometries and loading conditions have been published by Kumar, German & Shih (1981). The amplitude factor $C(t)$ for any creep condition is given by

$$C(t) = \frac{1 - v^2}{(n_c + 1)Et} K^2 + C^*. \tag{16.20}$$

The time for transition, t_T, from small-scale creep to extensive creep is obtained by equating the two terms on the right hand side of Eq. 16.20 and solving for t:

$$t_T = \frac{(1 - v^2)K^2}{(n_c + 1)EC^*}. \tag{16.21}$$

C^* and $C(t)$ have been suggested as candidate parameters for characterizing elevated temperature crack growth under monotonic loads (Landes & Begley, 1976; Riedel & Rice, 1980; Ohji, Ogura & Kubo, 1980).

Under small-scale creep conditions, Riedel & Rice (1980) estimate the size of the creep zone to be

$$r_{cr} = \frac{1}{2\pi} K^2 \{EB_1 t\}^{2/(n_c-1)} F_c(\theta), \tag{16.22}$$

where B_1 is a constant related to the material parameters in Eq. 16.12 and F_c is a function of the polar angle θ. In a finite element study of the dominance of near-tip fields for stationary cracks, Li, Needleman & Shih (1988) found that, for $n_c = 5$, the HRR-type fields, Eqs. 16.7.1, described near-tip stresses and deformation over the inner one-fifth of the creep zone; finite strain effects dominated over a region comparable to the extent of the crack tip opening displacement. It is also of interest to note that microstructurally induced changes in crack path, which promote mixed-mode fracture, can have a marked effect on the rate of growth of the creep zone. The numerical analysis of Brockenbrough, Shih & Suresh (1991) shows that, for $n_c = 5$, the creep zone grows about seven times faster in mode II than in mode I, with a corresponding decrease in the transition time, t_T, from small-scale to extensive creep.

Landes & Begley (1976) have also presented an interpretation of C^* in terms of the stress-power dissipation rate in cracked bodies. This interpretation provides a useful (and experimentally measureable) value of C^* at the loading pins for fracture specimens in which mainly steady-state creep deformation exists.

The fracture mechanics parameter $C(t)$ for nonsteady-state conditions is a near-tip amplitude factor which cannot be measured at the loading pins in a fracture test specimen. Its value is only calculated using Eq. 16.16. To overcome this limitation, Gieseke & Saxena (1989) suggest the use of another crack tip parameter, C_t, which is based on the stress–power dissipation rate definition of C^*:

$$C_t = -\frac{1}{B}\frac{\partial U^*}{\partial a},$$

(16.23)

where U^* is the instantaneous power release rate and B is the thickness of the cracked specimen. (Note the analogy between Eqs. 16.23 and 9.88.) Bassani, Hawk & Saxena (1986) showed that the parameter C_t is uniquely related to the rate of expansion of the creep zone under small-scale creep by the equation

$$C_t = 2(1 - v^2)\frac{K^2}{EW}\frac{D_1'}{D_1}b_1\dot{r}_{cr}.$$

(16.24)

Here W is the width of the cracked specimen, D_1 is the stress intensity calibration factor defined as $D_1 = (K/P)B\sqrt{W}$, $D_1' = dD_1/d(a/W)$, and b_1 is a scaling factor which is determined numerically.

The discussions presented thus far in this section pertain to stationary creep cracks. Certain aspects of near-tip fields for growing creep cracks have been reported by Hart (1980) and Hui & Riedel (1981). The latter authors conclude that, for $n_c > 3$, the stress singularity varies as $r^{-1/(n_c-1)}$ and that the scalar multiplier (i.e. the amplitude of the singularity) depends only on the crack growth rate (and not the applied load). For crack growth rates greater than some minimum value, no steady-state conditions appear to exist.

To summarize this section, it should be noted that there are several different parameters which have been proposed for the characterization of creep crack growth. The choice of a particular parameter is dictated by the temperature, the material constants in the constitutive equation for creep, and the size of the creep zone in relation to the characteristic dimensions of the test specimen. The above considerations are exclusively based on continuum analyses and they do not account for microstructural aspects of failure, the stability of microstructure, or environmental interactions.

16.7.2 Characterization of creep-fatigue crack growth

Under cyclic loading conditions at elevated temperature, the conditions governing the choice of an appropriate characterizing parameter can be estimated by comparing the cycle time t_c (i.e. the duration of the fatigue cycle) with the transition time t_T (Riedel, 1983). Small-scale creep conditions occur when $t_c \ll t_T$ and extensive creep conditions prevail when $t_c \gg t_T$. Based on this comparison, three distinct regions of fatigue crack growth can be identified. At lower temperatures and higher cyclic frequencies (i.e. low t_c), fatigue crack growth is essentially cycle-dependent and it is reasonably well characterized by ΔK. At very high temperatures and low frequencies (i.e. high t_c), crack growth is completely controlled by time-dependent processes. For this purpose, C^*, $C(t)$ and C_t have been used as candidate parameters depending on the material, environment and loading conditions. In the region lying between these two extremes, fatigue crack growth is an outcome of the synergistic interactions between cycle-dependent and time-dependent processes. For this creep-fatigue situation, two different approaches have emerged (e.g., Gieseke & Saxena, 1989): superposition of crack growth rates and superposition of crack driving forces.

In the first approach, the mechanical fatigue and time-dependent creep components of crack growth are linearly superposed to derive an overall crack extension rate (e.g., Saxena, 1988). The total fatigue crack growth rate is then given by

$$\left(\frac{da}{dN}\right) = \left(\frac{da}{dN}\right)_F + \left(\frac{da}{dN}\right)_{CR}, \tag{16.25}$$

where the subscripts F and CR denote the contributions from fatigue and creep, respectively. Note the similarity of this equation with Eq. 16.2 for corrosion-fatigue. Similar partitioning methods and their various empirical adaptations have also been employed in the context of thermomechanical fatigue where both the loads and temperatures vary cyclically (e.g., Nicholas, Heil & Haritos, 1989). The fatigue component of high temperature crack growth is then characterized by

$$\left(\frac{da}{dN}\right)_F = C(\Delta K)^m \quad \text{or} \quad \left(\frac{da}{dN}\right)_F = C_1(\Delta K_{eff})^{m_1} \tag{16.26}$$

under small-scale inelastic conditions, where ΔK and ΔK_{eff} can be taken as the nominal and effective values of the stress intensity factor range, respectively, and C, C_1, m and m_1 are the usual empirical constants in the Paris equation. Under large-scale plasticity, attempts have been made to characterize the mechanical fatigue component in terms of J^c ($\equiv \Delta J$), where $(da/dN)_F = C_2(J^c)^{m_2}$; C_2 and m_2 are empirical correlation constants (e.g., Dowling & Begley, 1976). Similarly, the time-dependent component $(da/dN)_{CR}$ is characterized by

$$\left(\frac{da}{dt}\right)_{CR} = C_3\{P_{CR}\}^{m_3}, \tag{16.27}$$

where C_3 and m_3 are material constants. P_{CR} is the creep fracture parameter which is chosen to be C^* by Landes & Begley (1976), $C(t)$ by Saxena (1988) and C_t by Bassani, Hawk & Saxena (1986). One of the methods for converting da/dt into $(da/dN)_{CR}$ is to use the relationship (e.g., Saxena, 1988)

$$\left(\frac{da}{dN}\right)_{CR} = \int_0^{1/v_c} \frac{da}{dt}(P_{CR})dt. \tag{16.28}$$

This approach, in principle, can also account for the effect of test frequency and waveform on the rates of high temperature crack growth.

Some researchers (e.g., Okazaki & Koizumi, 1983; Riedel, 1987) have proposed characterization for combined elastoplastic–creep under cyclic loads using a partitioning of J^c into an elastic component ($\Delta J_e = (1 - \nu^2)\Delta K^2/E$) and an inelastic component which is taken as the sum of the contributions from plastic and creep deformation. (The cyclic J-integral has also been termed the Z integral by Wüthrich, 1982, and Riedel, 1987.) Although success has been claimed for the characterization of inelastic creep-fatigue interactions using such approaches, there is doubt about the validity of cyclic J-based concepts under creep-fatigue conditions involving severely nonproportional loading. The main justification at this time is that these concepts appear to provide somewhat reasonable characterization for limited material–environment–loading combinations.

Finally, it is worth noting that under extensive creep in some materials, the crack growth rates have been assumed proportional to the net section stress and crack length. Furthermore, numerous micromechanical models have been proposed for fatigue crack growth based on the mechanisms (described in Section 6) of diffusion, grain boundary sliding and intergranular cavitation. A review of these models can be found in Sadananda & Shahinian (1981) and Riedel (1987).

16.7.3 Summary and some general observations

Although the use of linear elastic fracture mechanics to characterize room temperature fatigue crack growth has found widespread appeal, similar applications for elevated temperature fatigue failure must be approached with caution. For mixed fatigue–creep crack growth involving extensive near-tip deformation, there is no consensus at present on the validity of any one fracture mechanics parameter. Furthermore, the linear superposition models, albeit applicable in some cases, do not generally account for the experimentally observed mechanistic dissimilarities between time-dependent and cycle-dependent failure. The possibility of premature closure in the wake of the advancing fatigue crack, nonproportional deformation at the crack tip and rapid crack growth rates would obviously rule out the scientific basis for the nominal use of nonlinear fracture parameters for high temperature fatigue.

Increases in temperature and decreases in cyclic frequency generally lead to an increase in the rate of fatigue crack growth in most alloys (e.g., James, 1972) as a result of the environmental effects and damage processes discussed in Section 16.6. However, this trend may not be observed in situations where corrosion deposits formed at elevated temperatures block the faces of the crack (e.g., Skelton & Haigh, 1978), when cyclic waveforms and dwell periods cause changes in deformation mechanisms (e.g., Sadananda & Shahinian, 1981) or when changes in test temperature or frequency cause a change in the micromechanisms of damage ahead of the crack tip (e.g., Pineau, 1983). It is also emphasized that the effects on fatigue crack growth of most test variables are specific to the particular combination of material, environment, test temperature and mechanical loads.

16.8 Case study: Creep-fatigue in steam-power generators

Approaches to characterize creep-fatigue were discussed earlier in this chapter wherein the complexities associated with choosing an appropriate characterization methodology were highlighted. In this section, we present a case study of the assessment of residual life of a material which was tested after thirty years of service in the high-temperature turbine casing of a steam-power generation plant in France. This case study is based on the work of Mudry & Pineau (1983) and Moliné, Piques & Pineau (1991).

In view of their hardenability and the amenability to undergo large changes in mechanical properties in response to heat treatments and precipitation, Cr–Mo–V steels are used for such components as boilers, steam pipes, forged turbine rotors and cast turbine casings. During service, many components of the steam power plant are prone to crack initiation and growth due to the combined effects of cyclic loading and creep. In the case of thick sections of turbine casings, in particular, fatigue cracks initiate as a result of thermal transients generated during start-up and shut-down. The most damaging transients are introduced between 150 and 200 °C at the locations between the steam admission valves. Once initiated by fatigue, these cracks then advance subcritically by creep at the operating temperature of 540 °C under steady loads. Figure 16.18(a) shows an enlarged view of such a crack in the turbine casing. Figure 16.18(b) is the view of the cracked region after weld repair.

A research effort was sponsored in the 1980s by Electricité de France to assess the remnant creep life of a 1Cr–1Mo–0.25V steel which was taken out of a large turbine casing of a 125 MW thermal power station. Over a period of thirty years, this steel had been subjected to approximately 2000 start-up/shut-down fatigue cycles and 150 000 hours of operation.

The material that was exposed to long-term service at temperature in the power plant was examined for changes in composition, microstructure and mechanical properties. Auger spectroscopy studies of the steel identified the segregation of P to grain boundaries. Tensile as well as Charpy-impact experiments showed that during service, long-term aging at 540 °C for 150 000 hours had produced only small variations in tensile strength, a modest shift (by approximately 20 °C) in the ductile–brittle transition temperature (which was ascribed to the grain boundary segregation of P), a small change in creep ductility, and a reduction in time to creep rupture by as much as a factor of 3 to 5 (as compared to material taken from a cold section of the turbine).

Creep crack initiation and growth experiments were carried out on the specimen taken from the hot section of the turbine casing (i.e. the aged specimen) and the one taken from the cold section (unaged). Compact geometry test specimens were used for this purpose (see Section A.4 in the Appendix where the mode I stress intensity factor K_I for this geometry is given). The value of C^* was estimated on the basis of the experimental measurements in the following way:

$$C_{exp}^* = \frac{n_c}{n_c + 1} \cdot F_1\left(\frac{a}{W}, n_c\right) \cdot \frac{P \dot{\Delta}_{exp}}{B_{vgn}(W - a)}. \tag{16.29}$$

Here n_c is the stress exponent for secondary creep law, $F_1(a/W)$ (≈ 2 for the plane strain compact specimen) is a nondimensional function of the crack length to specimen width ratio (a/W) and n_c, P is the applied load, $\dot{\Delta} = (d\Delta/dt)$ is the load-line displacement rate

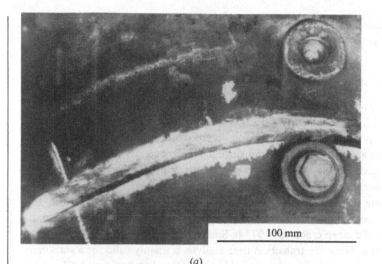

(a)

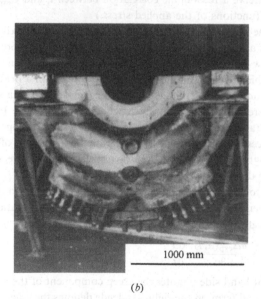

(b)

Fig. 16.18. (a) An enlarged view of the crack (after grinding) in the thick-section turbine casing of a 125 MW steam power plant. (b) View of the initially cracked region which was repaired by welding. Compare the region between the bolts in Figs. (a) and (b). (From Mudry & Pineau (1983). Photographs courtesy of A. Pineau, Ecole des Mines, Evry, France. Reprinted with permission.)

(where t is the time), B_{vgn} is the minimum thickness of the specimen at the v-groove notch, a is the crack length, $(W - a)$ is the width of the uncracked ligament and the subscript 'exp' denotes that the subscripted quantity is estimated on the basis of experimental measurements.

Experiments revealed that, for both the hot section and cold-section specimens, the time to initiate the growth of a creep crack, t_i, exhibited the following power-law dependence on C^*_{exp}:

$$t_i \cdot \{C^*_{exp}\}^{0.85} = \text{constant}. \tag{16.30}$$

This result indicates that there exists a unique correlation between the creep crack initiation time and C^*_{exp}, which reduces to C^* (see Eq. 16.19) under conditions of large scale power-law creep. Since the effect of aging on creep strain rate was included in the experimentally-based assessment of C^*_{exp} and since the creep ductility of the material was not strongly influenced by long exposures to high temperature, the correlation given in Eq. 16.30 was independent of the initial condition of the material.

The transition time from small-scale creep to extensive creep, Eq. 16.21, for the steel taken from the hot-section, $t_T \approx 12$ h, while the experimentally determined time to initiate the advance of a creep crack, $t_i \approx 975$ h. Similarly, for the cold-section steel, $t_T \approx 23$ h, while $t_i \approx 1000$ h. Since the transition time analysis is mainly valid for a stationary crack, it is not surprising to observe a reasonable correlation between t_i and C^*_{exp}, Eq. 16.30. (Note that t_T and t_i are functions of the applied stress.)

The creep fracture of both the hot-section and cold-section steel specimens exhibited a two-stage crack growth process at high temperatures: (1) Stage I, which corresponded to a threshold value of C^*_{exp}, below which a pre-existing crack did not advance under a steady creep load and above which even a small change in C^*_{exp} resulted in a precipitous rise in the crack velocity, $\dot{a} = da/dt$, and (2) Stage II, which corresponded to a steady creep crack growth regime wherein $\log \dot{a}$ had a linear dependence on $\log C^*_{exp}$. In this latter regime, both $\dot{a}$ and $\dot{\Delta}_{exp}$ (and hence, C^*_{exp}) advance precipitously by creep crack growth. The crack growth process involves primarily intergranular separation of the material, as shown in the micrograph of Fig. 16.19. Let the transition time between Stage I and Stage II crack growth be denoted as t_{stage}.

Once the creep crack begins to advance rapidly, i.e. when $t > t_{stage}$, the experimentally determined load-line displacement $\Delta_{exp}(a, t)$ changes because of creep deformation as well as an increase in crack length. Thus,

$$\frac{d\Delta_{exp}(a, t)}{dt} = \frac{\partial\Delta_{exp}(a, t)}{\partial t} + \frac{\partial\Delta_{exp}(a, t)}{\partial a} \cdot \frac{da}{dt}, \tag{16.31}$$

where the first term on the right hand side denotes the creep component of the load-line displacement rate, while the second term on the right-hand side denotes the crack-growth component. By setting the magnitudes of these two components equal, one can define a transition time, t^{cal}_{stage}. From the experimental measurements of the change in load-line displacement rate during the creep fracture experiments, t^{cal}_{stage} can be computed using a reference stress concept (which is outlined in Moliné et al. 1991). An experimental assessment of this transition time, t^{exp}_{stage} can also be made from transition from Stage I to Stage II during creep crack growth. A good agreement was found between such calculated and experimentally determined transition times between Stage I and Stage II crack growth for both the hot-section and cold-section steels. These correlations were also in agreement with transition time estimates derived from finite-element analyses.

The foregoing results reveal that in the steam power plant turbine casing, fatigue cracks, initiated due to the start-up and shut-down cycles, advance under a steady creep load. The conditions for the initiation of growth of such creep cracks was shown to be

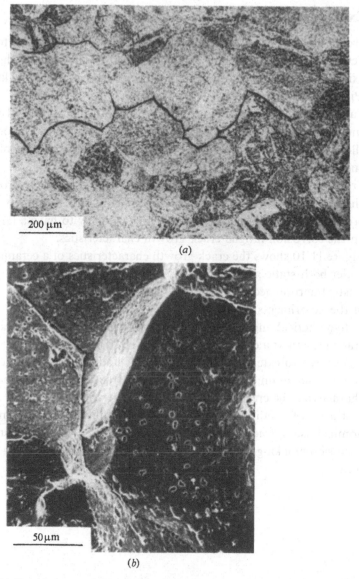

Fig. 16.19. (*a*) Intergranular creep crack growth in the Cr-Mo-V steel. Note also the presence of secondary cracks in this polished and etched specimen surface. (*b*) A higher magnification micrograph showing creep cavities along grain boundaries. (From Mudry & Pineau (1983). Photographs courtesy of A. Pineau, Ecole des Mines, Evry, France. Reprinted with permission.)

directly related to C^*_{exp}. Once rapid creep crack growth begins at the end of Stage I at $t > t_{stage}$, the commencement of Stage II creep fracture could also be correlated with C^*_{exp} on the basis of the transition time from Stage I to Stage II. This case study thus serves to illustrate how fracture mechanics principles for creep failure can be applied, with suitable modifications, to industrial components.

Exercises

16.1   Prove the statement in Section 14.3 that 1 cc of Fe oxidizes to 1.76 cc of FeO, 2.07 cc of Fe_3O_4 and 2.13 cc of Fe_2O_3. (The ratio of the volume of oxide produced to the volume of metal consumed is commonly known as the 'Pilling–Bedworth' ratio.)

16.2   Calculate the Pilling–Bedworth ratio for the formation of Al_2O_3 from Al and compare the efficacy of oxide-induced crack closure for aluminum alloys with that for steels solely on the basis of the volume of oxide produced on the fatigue fracture surface.

16.3   From the results presented in Eq. 9.49, show that the hydrostatic stress ahead of a mode III fatigue crack is nominally zero. List a few mechanistic processes for which the hydrostatic stress is likely to have a large influence on the corrosion-fatigue crack growth characteristics.

16.4   Figure 11.10 shows the crack growth characteristics of a ceramic composite under both static and cyclic loading conditions in 1400 °C air. See if linear elastic fracture mechanics provides valid characterization for the conditions of the experiments.

16.5   A hypothetical engineering alloy is subjected to high temperature fatigue crack growth at a cyclic frequency of 5 Hz. The transition time from small-scale creep to extensive creep, Eq. 16.21, under those testing conditions is 50 s. Comment on the validity of different fracture mechanics parameters to characterize the creep-fatigue fracture of the alloy.

16.6   List a set of conditions under which the similitude concept implicit in the nominal use of fracture mechanics is likely to lose validity during (*a*) stress-corrosion cracking, (*b*) corrosion-fatigue fracture and (*c*) creep-fatigue fracture.

Appendix: Stress intensity factors for some common crack geometries

This section presents a summary of stress intensity factors for some commonly used fatigue test specimens and crack configurations. These results are compiled from stress intensity factor handbooks (e.g., Rooke & Cartwright, 1976; Sih, 1973; Tada, Paris & Irwin, 1973) and monographs on fracture mechanics (e.g., Broek, 1986; Hellan, 1984; Kanninen & Popelar, 1985), where further details and derivations can be found.

A.1 Single-edge-cracked tension specimen

The stress intensity factor at the tip of a through-thickness, single edge crack of length a in a rectangular plate of width W (and height $H \geq 2W$) subjected to a tensile stress σ (Fig. A.1a) is

$$K_\mathrm{I} = \sigma \sqrt{a} \, f\left(\frac{a}{W}\right), \tag{A.1}$$

$$f\left(\frac{a}{W}\right) = 1.99 - 0.41\frac{a}{W} + 18.7\left(\frac{a}{W}\right)^2 - 38.48\left(\frac{a}{W}\right)^3 + 53.85\left(\frac{a}{W}\right)^4. \tag{A.2}$$

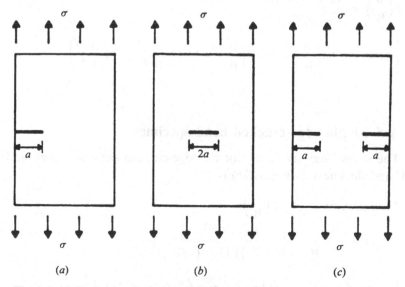

Fig. A.1. (a) Single-edge-cracked tension specimen. (b) Center-cracked tension specimen. (c) Double-edge-cracked tension specimen. The width of the plate is W in all cases.

A.2 Center-cracked tension specimen

The stress intensity factor for a center crack of length $2a$ in a plate (width W) which is subjected to a remote tensile stress σ (Fig. A.1b) is

$$K_{\mathrm{I}} = \sigma\sqrt{\pi a}\left(\sec\frac{\pi a}{W}\right)^{1/2}. \tag{A.3}$$

A.3 Double-edge-cracked tension specimen

The stress intensity factor of a plate of width W which contains two edge cracks each of length a (Fig. A.1c) is

$$K_{\mathrm{I}} = \sigma\sqrt{a}\, f\left(\frac{a}{W}\right), \tag{A.4}$$

$$f\left(\frac{a}{W}\right) = 1.99 + 0.76\left(\frac{a}{W}\right) - 8.48\left(\frac{a}{W}\right)^2 + 27.36\left(\frac{a}{W}\right)^3. \tag{A.5}$$

A.4 Compact specimen

The mode I stress intensity factor for a compact specimen (Fig. A.2a) subjected to a tensile load P is

$$K_{\mathrm{I}} = \frac{P}{B\sqrt{W}}\, f\left(\frac{a}{W}\right), \tag{A.6}$$

$$f\left(\frac{a}{W}\right) = \frac{(2 + \{a/W\})}{(1 - \{a/W\})^{3/2}} \times$$

$$\left[0.886 + 4.64\left(\frac{a}{W}\right) - 13.32\left(\frac{a}{W}\right)^2 + 14.72\left(\frac{a}{W}\right)^3 - 5.6\left(\frac{a}{W}\right)^4\right]. \tag{A.7}$$

A.5 Single-edge-cracked bend specimen

The stress intensity factor for an edge-cracked bend specimen of span S, width W and thickness B (Fig. A.2b) is

$$K_{\mathrm{I}} = \frac{PS}{BW^{3/2}}\, f\left(\frac{a}{W}\right), \tag{A.8}$$

$$f\left(\frac{a}{W}\right) = \frac{3(a/W)^{1/2}}{2(1 + 2\{a/W\})\,(1 - \{a/W\})^{3/2}} \times$$

$$\left[1.99 - \left(\frac{a}{W}\right)\left(1 - \frac{a}{W}\right)\left\{2.15 - 3.93\left(\frac{a}{W}\right) + 2.7\left(\frac{a}{W}\right)^2\right\}\right]. \tag{A.9}$$

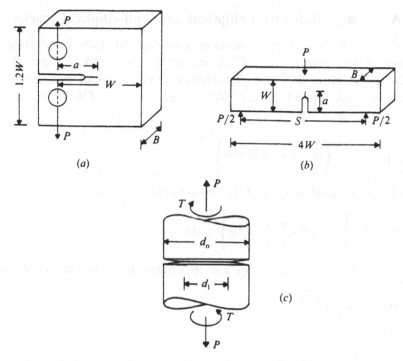

Fig. A.2. (*a*) Compact specimen. (*b*) Single-edge-cracked bend specimen. (*c*) Circumferentially cracked cylindrical specimen.

A.6 Circumferentially cracked cylindrical specimen subjected to tension and/or torsion

The mode I and mode III stress intensity factors for a circumferentially cracked cylindrical specimen (uncracked ligament diameter d_i, outer diameter d_o, and $D = \{d_i/d_o\}$) subjected to tension and torsion (Fig. A.2c) is

$$K_{\mathrm{I}} = \frac{2P}{\pi d_i^2} \left\{ \frac{\pi d_i}{2}(1-D) \right\}^{1/2} \times$$

$$\{1 + 0.5D + 0.375D^2 - 0.363D^3 + 0.731D^4\}, \qquad (A.10)$$

$$K_{\mathrm{III}} = \frac{6T}{\pi d_i^3} \left\{ \frac{\pi d_i}{2}(1-D) \right\}^{1/2} \times$$

$$\{1 + 0.5D + 0.375D^2 + 0.3125D^3 + 0.273D^4 + 0.208D^5\}. \qquad (A.11)$$

A.7 Large plates with elliptical and semi-elliptical cracks

Consider a plate specimen containing an elliptical crack. Let the major and minor axes of the ellipse be $2c$ and $2a$, respectively, as shown in Figure A.3. The stress intensity factor for the embedded elliptical flaw varies along the crackfront as a function of the angle ϕ (see Fig. A.3b). When the dimensions of the cracked body are much larger than a and c,

$$K_I = \frac{\sigma\sqrt{\pi a}}{\Psi}\left(\sin^2\phi + \frac{a^2}{c^2}\cos^2\phi\right)^{1/4}, \tag{A.12a}$$

where Ψ is the elliptical integral of the second kind, which is given by

$$\Psi = \int_0^{\pi/2}\left\{1 - \left(1 - \frac{a^2}{c^2}\right)\sin^2\phi\right\}^{1/2}d\phi. \tag{A.12b}$$

K_I is maximum when $\phi = 90°$. Using a series expansion for Ψ, it can be shown that

$$\Psi \approx \frac{3\pi}{8} + \frac{\pi}{8}\left(\frac{a^2}{c^2}\right). \tag{A.12c}$$

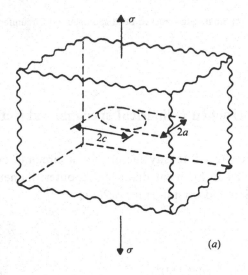

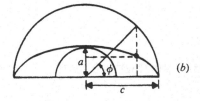

Fig. A.3. (*a*) A large plate containing an embedded elliptical crack. (*b*) Details of the crackfront.

When $a = c$, we obtain the solution for a circular (penny-shaped) crack. In this case, Eqs. A.12 reduce to

$$K_I = \frac{2}{\pi} \sigma\sqrt{\pi a}. \tag{A.13}$$

The solutions given by Eqs. A.12 can also be applied to the case of semi-elliptical surface cracks. For the semi-elliptical surface flaw (thumb-nail crack) in a finite size plate, the stress intensity factor at the mid-point (i.e. end of the minor axis, $\phi = \pi/2$) is

$$K_I = \frac{1.12\sigma\sqrt{\pi a}}{\sqrt{Q}}, \tag{A.14}$$

where the pre-multiplier 1.12 is the free-surface correction factor, Q is the flaw shape parameter extracted from Ψ in Eq. A.12b, $2c$ is the surface length of the crack, and a is the maximum depth (at $\phi = \pi/2$) of the crack into the material. $Q = \Psi^2$ in the elastic limit, $\sigma/\sigma_y \to 0$, where σ is the applied stress and σ_y is the yield strength of the material. Additional corrections may have to be made to Eq. A.14 to account for the proximity of the free surface to the crackfront (depending on the relative magnitudes of a and the specimen thickness B) and for crack-tip plasticity. The modified value of Q incorporating the plasticity correction is usually taken to be $Q \approx \Psi^2 - 0.212(\sigma^2/\sigma_y^2)$.

References

Abdel-Raouf, H. & Plumtree, A. (1971). The influence of deformation rate on the cyclic behavior of pure iron. *Metallurgical Transactions* **2A**, 1863–7.

Abel, A. & Ham, R.K. (1966). The cyclic strain behaviour of crystals of aluminium–4 wt.% copper–I. The Bauschinger effect. *Acta Metallurgica* **14**, 1489–94.

Ackermann, F., Kubin, L.P., Lepinoux, J. & Mughrabi, H. (1984). The dependence of dislocation microstructure on plastic strain amplitude in cyclically strained copper single crystals. *Acta Metallurgica* **32**, 715–25.

Aifantis, E.C. (1987). The physics of plastic deformation. *International Journal of Plasticity* **3**, 211–47.

Akhurst, K.N., Lindley, T.C. & Nix, K.J. (1983). The effect of mode III loading on fatigue crack growth in a rotating shaft. *Fatigue of Engineering Materials and Structures* **6**, 345–8.

Albert, W.A.J. (1838). Über Treibseile am Harz. *Archive für Mineralogie, Geognosie, Bergbau und Hüttenkunde* **10**, 215–34.

Allen, R.J. & Sinclair, J.C. (1982). The behaviour of short cracks. *Fatigue of Engineering Materials and Structures* **5**, 343–7.

Allison, J.E. (1979). Measurement of crack-tip stress distributions by X-ray diffraction. In *Fracture Mechanics*, Special Technical Publication 677, pp. 550–62. Philadelphia: American Society for Testing and Materials.

Anderson, T.L. (1995). *Fracture mechanics: fundamentals and applications*, second edition. Boca Raton, FL: CRC Press.

Andrews, E.H. (1974). A generalized theory of fracture mechanics. *Journal of Materials Science* **9**, 887–94.

Antolovich, S.D., Saxena, A. & Chanani, G.R. (1975). A model for fatigue crack propagation. *Engineering Fracture Mechanics* **7**, 649–52.

Antonopoulos, J.G., Brown, L.M. & Winter, A.T. (1976). Vacancy dipoles in fatigued copper. *Philosophical Magazine* **34**, 549–63.

Antonopoulos, J.G. & Winter, A.T. (1976). Weak-beam study of dislocation structures in fatigued copper. *Philosophical Magazine* **33**, 87–95.

Anzai, H. & Endo, T. (1979). On-site indication of fatigue damage under complex loading. *International Journal of Fatigue* **1**, 49–57.

Argon, A.S. (1980). Inelastic deformation and fracture in oxide, metallic, and polymeric glasses. *Glass Science and Technology* **5**, 79–132.

Argon, A.S. (1989). Sources of toughness in polymers. In *Advances in Fracture Research* (eds. K. Salama, K. Ravi-Chandar, D.M.R. Taplin & P. Rama Rao), vol. 4, pp. 2661–81. Oxford: Pergamon Press.

Argon, A.S. & Godrick, J.A. (1969). Fatigue of lithium fluoride crystals at elevated temperatures. In *Fracture 1969* (ed. P.L. Pratt), pp. 576–86. London: Chapman-Hall.

Argon, A.S. & Hannoosh, J.G. (1977). Initiation of crazes in polystyrene. *Philosophical Magazine* **36**, 1195–216.

Asaro, R.J. (1975). Elastic–plastic memory and kinematic-type hardening. *Acta Metallurgica* **23**, 1255–65.

Asaro, R.J., Hermann, L. & Baik, J.M. (1981). Transitions in fatigue crack closure in 2048 aluminum. *Metallurgical Transactions* **12A**, 1133–5.

Aswath, P.B., Suresh, S., Holm, D.K. & Blom, A.F. (1988). Load interaction effects on compression fatigue crack growth in ductile solids. *Journal of Engineering Materials and Technology* **110**, 278–85.

Atkinson, J.D., Brown, L.M. & Stobbs, W.M. (1974). The work-hardening of copper–silica. IV. The Bauschinger effect and plastic relaxation. *Philosophical Magazine* **30**, 1247–80.

Austen, I.M. & McIntyre, P. (1979). Corrosion-fatigue of high-strength steel in low-pressure hydrogen gas. *Metal Science* **13**, 420–8.

Aveston, J. & Kelly, A. (1973). Theory of multiple fracture of fibrous composites. *Journal of Materials Science* **8**, 352–62.

Azimi, H.R. (1994). Toughened epoxy polymers: fatigue crack propagation mechanisms. Ph.D. thesis. Bethlehem, PA: Lehigh University.

Azimi, H.R., Pearson, R.A. & Hertzberg, R.W. (1995). Role of crack tip shielding mechanisms in fatigue of hybrid composites containing rubber and solid glass spheres. *Journal of Applied Polymer Science* **58**, 449–63.

Bach, C. (1913). Die maschinen-elemente, 1. Band, 11. Auflage, Kröner Verlag Leipzig.

Baik, S. & Raj, R. (1982). Mechanisms of creep-fatigue interaction. *Metallurgical Transactions* **13A**, 1215–21.

Bairstow, L. (1910). The elastic limits of iron and steel under cyclic variations of stress. *Philosophical Transactions of the Royal Society, London* **210**, 35–55.

Banasiak, D.H., Grandt, A.F. & Montulli, L.T. (1977). Fatigue crack retardation in polycarbonate. *Journal of Applied Polymer Science* **21**, 1297–309.

Bannantine, J.A., Comer, J.J. & Handrock, J.L. (1990). *Fundamentals of Metal Fatigue Analysis*. Englewood Cliffs: Prentice-Hall.

Barenblatt, G.I. (1962). The mathematical theory of equilibrium cracks in brittle fracture. *Advances in Applied Mechanics* **7**, 55–129.

Barsom, J.M. (1972). Effect of cyclic-stress form on corrosion-fatigue crack propagation below K_{Iscc} in a high-yield-strength steel. In *Corrosion Fatigue: Chemistry, Mechanics and Microstructure* (eds. O.F. Devereux, A.J. McEvily & R.W. Staehle), vol. NACE-2, pp. 426–36. Houston: National Association of Corrosion Engineers.

Barsom, J.M. (1976). Fatigue crack growth under variable-amplitude loading in various bridge steels. In *Fatigue Crack Growth Under Spectrum Loads*, Special Technical Publication 595, pp. 217–35. Philadelphia: American Society for Testing and Materials.

Barsom, J.M. & McNicol, R.C. (1974). Effect of stress concentration on fatigue crack initiation in HY-130 steel. In *Fracture Toughness and Slow-Stable Cracking*, Special Technical Publication 559, pp. 183–204. Philadelphia: American Society for Testing and Materials.

Barsom, J.M. & Rolfe, S.T. (1987). *Fracture and Fatigue Control in Structures*, 2nd edn. Englewood Cliffs: Prentice-Hall.

Bartel, D.L., Bicknell, V.L. & Wright, T.M. (1986). The effect of conformity, thickness and material on stresses in ultra-high molecular weight components for total joint replacements. *Journal of Bone Joint Surgery* **68**, 1041–51.

Basinski, S.J., Basinski, Z.S. & Howie, A. (1969). Early stages of fatigue in copper single crystal. *Philosophical Magazine* **19**, 899–924.

Basinski, Z.S. & Basinski, S.J. (1984). Formation and growth of subcritical fatigue cracks. *Scripta Metallurgica* **18**, 851–6.

Basinski, Z.S., Korbel, A.S. & Basinski, S.J. (1980). The temperature dependence of the saturation stress and dislocation structure in fatigued copper single crystals. *Acta Metallurgica* **28**, 191–207.

Basinski, Z.S., Pascual, R. & Basinski, S.J. (1983). Low amplitude fatigue of copper single crystals – I. The role of the surface in fatigue failure. *Acta Metallurgica* **31**, 591–602.

Basquin, O.H. (1910). The exponential law of endurance tests. *Proceedings of the American Society for Testing and Materials* **10**, 625–30.

Bassani, J.L., Hawk, D.E. & Saxena, A. (1986). Evaluation of the C_t parameter for characterizing creep crack growth rate in the transient regime. In *Nonlinear Fracture Mechanics*, Special Technical Publication 995, pp. 7–26. Philadelphia: American Society for Testing and Materials.

Bassani, J.L. & McClintock, F.A. (1981). Creep relaxation of stress around a crack tip. *International Journal of Solids and Structures* **17**, 479–92.

Batte, A.D. (1983). Creep-fatigue life predictions. In *Fatigue at High Temperature* (ed. R.P. Skelton), pp. 365–401. London: Elsevier Applied Science.

Bauschinger, J. (1886). Ueber die Veränderungen der Elastizitätsgrenze und der festigkeit des Eisens und Stahls durch Strecken, Quetschen, Erwärmen Abkühlen und durch oftmals wiederholte Belastung. *Mitt: Mech-Tech Lab.*, XIII München.

Beahan, P., Bevis, M. & Hull, D. (1971). The morphology of crazes in polystyrene. *Philosophical Magazine* **24**, 1267–79.

Beardmore, P. & Rabinowitz, S. (1975). Fatigue deformation of polymers. *Treatise on Materials Science and Technology* **6**, 267–331.

Benoit, D., Namdar-Irani, R. & Tixier, R. (1980). Oxidation of fatigue fracture surfaces at low growth rates. *Materials Science and Engineering* **45**, 1–7.

Betagón, C. & Hancock, J.W. (1991). Two-parameter characterization of elastic–plastic crack-tip fields. *Journal of Applied Mechanics* **58**, 104–10.

Bilby, B.A., Cardew, G.E. & Howard, I.C. (1977). Stress intensity factors at the tips of kinked and forked cracks. In *Fracture 1977* (ed. D.M.R. Taplin), vol. 3, pp. 197–200. New York: Pergamon Press.

Bilby, B.A., Cottrell, A.H. & Swinden, K.H. (1963). The spread of plastic yield from a notch. *Proceedings of the Royal Society, London* **A272**, 304–14.

Bilby, B.A., Cardew, G.E., Goldthorpe, M.R. & Howard, I.C. (1986). A finite element investigation of the effect of specimen geometry on the fields of stress and strain at the tips of stationary cracks. In *Size Effects in Fracture*, pp. 37–46. London: The Institution of Mechanical Engineers.

Bingham, E.C. (1922). *Fluidity and plasticity*. New York, NY: McGraw–Hill Book Company.

Blankenship, C.P. & Starke, E.A. (1991). The fatigue crack growth behavior of the Al-Cu-Li alloy Weldalite 049. *Fatigue and Fracture of Engineering Materials and Structures* **14**, 103–14.

Blom, A.F. & Holm, D.K. (1985). An experimental and numerical study of fatigue crack closure. *Engineering Fracture Mechanics* **22**, 997–1011.

Bodner, S.R. & Partom, Y. (1975). Constitutive equations for elastic–viscoplastic strain-hardening materials. *Journal of Applied Mechanics* **42**, 385–95.

Bodner, S.R., Partom, I. & Partom, Y. (1979). Uniaxial cyclic loading of elastic–viscoplastic materials. *Journal of Applied Mechanics* **46**, 805–10.

Boettner, R.C., McEvily, A.J. & Liu, Y.C. (1964). On the formation of fatigue cracks at twin boundaries. *Philosophical Magazine* **10**, 95–106.

Bonnen, J.J., Allison, J.E. & Jones, J.W. (1991). Fatigue behavior of a 2XXX series aluminum alloy reinforced with 15 v/o SiC$_p$. *Metallurgical Transactions* **22A**, 1007–19.

Boudouard, O. (1911). Essai des métaux par amortissement de leurs vibrations. *Mémoires de la Société d'Encouragement à l'industrie Nationale*, pp. 545.

Boussinesq, J. (1885). Application des potentials à l'étude de l'équilibre et du mouvement des solides élastiques. Paris: Gauthier-Villars.

Bower, A.F. & Johnson, K.L. (1989). The influence of strain hardening on cumulative plastic deformation in rolling and sliding contact. *Journal of the Mechanics and Physics of Solids* **37**, 471–93.

Bower, A.F. & Ortiz, M. (1990). Solution of three dimensional crack problems by a finite perturbation method. *Journal of the Mechanics and Physics of Solids* **38**, 443–80.

Bowles, C.Q. & Schijve, J. (1973). The role of inclusions in fatigue crack initiation in an aluminum alloy. *International Journal of Fracture* **9**, 171–9.

Braithwaite, F. (1854). On the fatigue and consequent fracture of metals. *Proceedings of the Institution of Civil Engineers, London* **13**, 463–75.

Bretschneider, J., Holste, C. & Tippelt, B. (1997). Cyclic plasticity of nickel single crystals at elevated temperatures. *Acta Materialia* **45**, 1775–83.

Brett, S.J. & Doherty, R.D. (1978). Loss of solute at the fracture surface in fatigued aluminium precipitation-hardened alloys. *Materials Science and Engineering* **32**, 255–65.

Brockenbrough, J.R., Shih, C.F. & Suresh, S. (1991). Transient crack tip fields for mixed-mode power law creep. *International Journal of Fracture* **49**, 177–202.

Brockenbrough, J.R. & Suresh, S. (1987). Constitutive behavior of a microcracking brittle solid in cyclic compression. *Journal of the Mechanics and Physics of Solids* **35**, 721–42.

Broek, D. (1986). *Elementary Engineering Fracture Mechanics*, 4th edn. The Hague: Martinus Nijhoff.

Broom, T. & Ham, R.K. (1959). The hardening of copper single crystals by fatigue. *Proceedings of the Royal Society, London* **A251**, 186–99.

Broom, T., Mazza, J.A. & Whittaker, V.N. (1957). Structural changes caused by plastic strain and by fatigue in aluminium–zinc–magnesium copper alloys corresponding to DTD683. *Journal of the Institute of Metals* **86**, 17–23.

Brown, C.W. & Hicks, M.A. (1983). A study of short fatigue crack growth behaviour in titanium alloy IMI 685. *Fatigue of Engineering Materials and Structures* **6**, 67–76.

Brown, L.M. (1977). Dislocation substructures and the initiation of cracks by fatigue. *Metal Science* **11**, 315–20.

Brown, L.M. & Clarke, D.R. (1977). The work hardening of fibrous composites with particular reference to the copper–tungsten system. *Acta Metallurgica* **25**, 563–70.

Brown, L.M. & Ogin, S.L. (1985). Role of internal stresses in the nucleation of fatigue cracks. In *Fundamentals of Deformation and Fracture* (eds. B.A. Bilby, K.J. Miller & J.R. Willis), pp. 501–28. Cambridge University Press.

Brown, L.M. & Stobbs, W.M. (1971a). The work hardening of copper–silica I. A model based on internal stresses, with no plastic relaxation. *Philosophical Magazine* **23**, 1185–99.

Brown, L.M. & Stobbs, W.M. (1971b). The work hardening of copper–silica II. The role of plastic relaxation. *Philosophical Magazine* **23**, 1201–33.

Brown, M.W. & Miller, K.J. (1973). A theory for fatigue failure under multiaxial stress–strain conditions. *Proceedings of the Institute of Mechanical Engineers*, **187**, 745–55.

Bucci, R.J. (1970). Environment enhanced fatigue and stress corrosion cracking of a titanium alloy plus a simple model for the assessment of environmental influence on fatigue behavior. Ph.D. Thesis, Bethlehem: Lehigh University.

Bucci, R.J., Thakker, A.B., Sanders, T.H., Sawtell, R.R. & Staley, J.T. (1980). Ranking 7XXX aluminum alloy fatigue crack growth resistance under constant amplitude and spectrum loading. In *Effect of Load Spectrum Variables on Fatigue Crack Initiation and Propagation*, Special Technical Publication 714, pp. 41–78. Philadelphia: American Society for Testing and Materials.

Budiansky, B. & Hutchinson, J.W. (1978). Analysis of closure in fatigue crack growth. *Journal of Applied Mechanics* **45**, 267–76.

Budiansky, B., Hutchinson, J.W. & Lambropoulos, J.C. (1983). Continuum theory of dilatant transformation toughening in ceramics. *International Journal of Solids and Structures* **19**, 337–55.

Budiansky, B. & O'Connell, J. (1976). Elastic moduli of a cracked solid. *International Journal of Solids and Structures* **12**, 81–97.

Budiman, H. (1996). Mechanisms of damage tolerance and arrest in pressurized composite cylinders. Ph.D. Thesis. Cambridge, MA: Massachusetts Institute of Technology.

Calabrese, C. & Laird, C. (1974). Cyclic stress-strain response of two-phase alloys. Part I. Microstructures containing particles penetrable by dislocations. *Materials Science and Engineering* **13**, 141–57.

Carter, R.D., Lee, E.-W., Starke, E.A. & Beevers, C.J. (1984). Effect of microstructure and environment on fatigue crack closure in 7475 aluminum alloy. *Metallurgical Transactions* **15A**, 555–63.

Cattaneo, C. (1938). Sul contatto di due corpi elastici: distribuzion locale degli sforzi. *Reconditi dell Accademia nazionale dei Lincei* 27, 342–48, 434–6, 474–8.

Cerruti, V. (1882). Roma Accademia Lincei, Mem. fis. mat.

Chaboche, J.L. (1986). Time-dependent constitutive theories for cyclic plasticity. *International Journal of Plasticity* **2**, 149–88.

Chaki, T.K. & Li, J.C.M. (1984). Overload effect and fatigue crack propagation in amorphous metallic alloys. *Scripta Metallurgica* **18**, 703–8.

Chakrabortty, S.B. (1979). A model relating low cycle fatigue properties and microstructure to fatigue crack propagation rates. *Fatigue of Engineering Materials and Structures* **2**, 205–30.

Challenger, K.D. & Moteff, J. (1972). Characterization of the deformation substructure of AISI 316 stainless steel after high strain fatigue at elevated temperatures. *Metallurgical Transactions* **3A**, 1675–8.

Chanani, G.R., Antolovich, S.D. & Gerberich, W.W. (1972). Fatigue crack propagation in TRIP steels. *Metallurgical Transactions* **3A**, 2661–72.

Chang, R., Morris, W.L. & Buck, O. (1979). Fatigue crack nucleation at intermetallic particles in alloys – A dislocation pile-up model. *Scripta Metallurgica* **13**, 191–4.

Charalambides, P.G. & McMeeking, R.M. (1987). Finite element method simulation of crack propagation in a brittle microcracking solid. *Mechanics of Materials* **6**, 71–87.

Charsley, P. & Kuhlmann-Wilsdorf, D. (1981). Configurations of {100} dislocation walls formed during fatigue. *Philosophical Magazine A* **44**, 1351–61.

Charsley, P. & Thompson, N. (1963). The behaviour of slip lines on aluminium crystals under reversed stresses in tension and compression. *Philosophical Magazine* **8**, 77–86.

Chen. I.-W. & Reyes Morel, P.E. (1986). Implications of transformation plasticity in ZrO_2 containing ceramics: I. Shear and dilatation effects. *Journal of the American Ceramic Society* **69**, 181–9.

Cheng, A.S. & Laird, C. (1981). Fatigue life behavior of copper single crystals. Part I: Observations of crack nucleation. *Fatigue of Engineering Materials and Structures* **4**, 331–41.

Cherepanov, G.P. (1969). On crack propagation in solids. *International Journal of Solids and Structures* **5**, 863–71.

Chermahini, R.G., Shivakumar, K.N., Newman, J.C. & Blom, A.F. (1989). Three-dimensional aspects of plasticity-induced fatigue crack closure. *Engineering Fracture Mechanics* **34**, 393–402.

Christensen, R.H. (1959). *Metal Fatigue*. New York: McGraw-Hill.

Christman, T. & Suresh, S. (1986). Crack initiation under far-field cyclic compression and the study of short fatigue cracks. *Engineering Fracture Mechanics* **23**, 953–64.

Christman, T. & Suresh, S. (1989). Effects of SiC reinforcement and aging treatment on fatigue crack growth in an Al–SiC composite. *Materials Science and Engineering* **A102**, 211–16.

Chu, C.-C., Conley, F.A. & Bonnen, J.J.F. (1993). Multiaxial stress–strain modeling and life prediction of SAE axle shafts. In *Advances in Multiaxial Fatigue* (eds: D.L. McDowell and R. Ellis), pp. 37–54. Philadelphia, PA: American Society for Testing and Materials.

Clarke, D.R. (1987). On the equilibrium thickness of intergranular glass phases in ceramic materials. *Journal of the American Ceramic Society* **70**, 15–22.

Coffin, L.F. (1954). A study of the effects of cyclic thermal stresses on a ductile metal. *Transactions of the American Society of Mechanical Engineers* **76**, 931–50.

Coffin, L.F. (1973). Fatigue at high temperatures. In *Fatigue at Elevated Temperatures*, Special Technical Publication 520, pp. 5–36. Philadelphia: American Society for Testing and Materials.

Coffin, L.F. (1974). Fatigue at high temperature – prediction and interpretation. *Proceedings of the Institution of Mechanical Engineers* **188**, 109–27.

Coffin, L.F. (1979). Fatigue in machines and structures–Power generation. In *Fatigue and Microstructure* (ed. M. Meshii), pp. 1–27. Metals Park: American Society for Metals.

Coles, A. (1980). Material considerations for gas turbine engines. In *Mechanical Behaviour of Materials* (eds. K.J. Miller & R.F. Smith), vol. I, pp. 3–11. Oxford: Pergamon Press.

Constable, I., Williams, J.G. & Burns, D.J. (1970). Fatigue and cyclic thermal softening of thermoplastics. *Journal of Mechanical Engineering Science* **12**, 20–9.

Corlby, D.M. & Packman, P.F. (1973). On the influence of single and multiple peak overloads on fatigue crack propagation in 7075-T6511 aluminum. *Engineering Fracture Mechanics* **5**, 479–97.

Cotterell, B. & Rice, J.R. (1980). Slightly curved or kinked cracks. *International Journal of Fracture* **16**, 155–69.

Cottrell, A.H. (1958). Theory of brittle fracture in steel and similar metals. *Transactions of the Metallurgical Society*, 192–203.

Cottrell, A.H. & Hull, D. (1957). Extrusion and intrusion by cyclic slip in copper. *Proceedings of the Royal Society, London* **A242**, 211–13.

Cowles, B.A. (1988). Life Prediction in High Temperature Environments: Overview of a Current Gas Turbine Engine Approach. *Materials Science and Engineering* **A103**, 63–69.

Cowles, B. (1996). High cycle fatigue in aircraft gas turbines–an industry perspective. *International Journal of Fracture* **80**, 147–63.

Cox, B.N. (1996). Life prediction for bridged fatigue cracks. In *Life Prediction Methodology for Titanium Matrix Composites* (eds. W.S. Johnson, J.M. Larsen and B.N. Cox). Special Technical Publication 1253, pp. 552–72. Philadelphia: American Society for Testing and Materials.

Creager, M. & Paris, P.C. (1967). Elastic field equations for blunt cracks with reference to stress corrosion cracking. *International Journal of Fracture Mechanics* **3**, 247–52.

Crooker, T.W. (1971). Effects of tension–compression cycling on fatigue crack growth in high strength alloys. *Journal of Engineering for Industry* **93**, 893–6.

Crossland, B. (1956). Effect of large hydrostatic pressures on the torsional fatigue strength of an alloy steel. In *Proceedings of the International Conference on Fatigue of Metals*, pp. 138–49. London: The Institution of Mechanical Engineers.

Cuitiño, A.M. & Ortiz, M. (1992). Computational modeling of single crystals. *Modeling and Simulation in Materials Science and Engineering* **1**, 255–63.

Cummings, H.N., Stulen, F.B. & Schulte, W.C. (1958). Tentative fatigue strength reduction factors for silicate-type inclusions in high strength steels. *Proceedings of the American Society for Testing and Materials* **58**, 505–12.

Dafalias, Y.F. (1984). Modelling cyclic plasticity: simplicity versus sophistication. In *Mechanics of Engineering Materials* (eds. C.S. Desai & R.H. Gallagher), pp. 153–78. New York: Wiley.

Dafalias, Y.F. & Popov, E.P. (1975). A model of nonlinearly hardening materials for complex loading. *Acta Mechanica* **21**, 173–92.

Dang Van, K. (1973). Sur la résistance à la fatigue des métaux. Thèse de Doctorat ès Sciences. Paris: Scientifique et Technologique l'Armenent.

Dang Van, K. & Maitournam, M.H. (1994). Elasto-plastic calculations on the mechanical state in reciprocating moving contacts: application to fretting fatigue. In *Fretting Fatigue, European Structural Integrity Society Volume No. 18* (eds: R.B. Waterhouse and T.C. Lindley), pp. 161–8. London: Mechanical Engineering Publications.

Dauskardt, R.H. (1993). A frictional wear mechanism for fatigue-crack growth in crack bridging. *Acta Metallurgica et Materialia* **41**, 2165–81.

Dauskardt, R.H., Marshall, D.B. & Ritchie, R.O. (1990). Cyclic fatigue-crack propagation in magnesia-partially-stabilized zirconia. *Journal of the American Ceramic Society* **73**, 893–903.

Davidson, D.L. & Lankford, J. (1977). Determination of the energy of fatigue crack propagation and its alteration by wet air. In *Environment-Sensitive Fracture of Engineering Materials* (ed. Z.A. Foroulis), pp. 581–94. Warrendale: The Metallurgical Society of the American Institute of Mining, Metallurgical and Petroleum Engineers.

Davidson, D.L. & Lankford, J. (1981). The effect of water vapor on fatigue crack tip stress and strain range distribution and the energy required for crack propagation in low-carbon steel. *International Journal of Fracture* **17**, 257–75.

Davidson, D.L. & Lankford, J. (1986). High resolution techniques for the study of small cracks. In *Small Fatigue Cracks* (eds. R.O. Ritchie & J. Lankford), pp. 455–70. Warrendale: The Metallurgical Society of the American Institute of Mining, Metallurgical and Petroleum Engineers.

Davidson, D.L. & Suresh, S. (1984). Editors, *Fatigue Crack Growth Threshold Concepts*. Warrendale: The Metallurgical Society of the American Institute of Mining, Metallurgical and Petroleum Engineers.

Davis, L.A. (1975). Fracture of Ni–Fe base metallic glasses. *Journal of Materials Science* **10**, 1557–64.

Dempster, D.D. (1959). *The Tale of the Comet*. New York, NY: David McKay.

Deresiewicz, H. (1954). Contact of elastic spheres under an oscillating torsional couple. *Journal of Applied Mechanics* **21**, 52–6.

Desvaux, M.P.E. (1970). Strain bursts in Cu–12 at.% Al single crystals. *Zeitschrift für Metallkunde* **61**, 206–13.

Devincre, B. & Kubin, L.P. (1997). Mesoscopic simulations of dislocations and plasticity. *Materials Science and Engineering A* **A234–236**, 8–14.

Devincre, B. & Pontikis, V. (1993). Computer modelling of dynamically-induced dislocation patterning. In *Proceedings of the Materials Research Society Symposium*, vol. 291, pp. 555–60. Warrendale, PA: The Materials Research Society.

Dey, N., Socie, D.F. & Hsia, K.J. (1995). Modeling static and cyclic fatigue in ceramics containing a viscous grain boundary phase. *Acta Metallurgica et Materialia* **43**, 2163–75.

Dharan, C.H. (1975). Fatigue failure in graphite fibre and glass fibre–polymer composites. *Journal of Materials Science* **10**, 1665–70.

Dickson, J.I., Boutin, J. & L'Espérance, G. (1986). An explanation of labyrinth walls in fatigued f.c.c. metals. *Acta Metallurgica* **34**, 1505–14.

Differt, K. & Essmann, U. (1993). Dynamic model of the wall structure in persistent slip bands of fatigued metals. *Materials Science and Engineering A* **A164**, 295–9.

Differt, K., Essmann, U. & Mughrabi, H. (1986). A model of extrusions and intrusions in fatigued metals II. Surface roughening by random irreversible slip. *Philosophical Magazine A* **54**, 237–58.

Döker, H. & Marci, G. (1983). Threshold range and opening stress intensity factor in fatigue. *International Journal of Fatigue* **5**, 187–91.

Donahue, R.J., Clark, H.M., Atanmo, P., Kumble, R. & McEvily, A.J. (1972). Crack opening displacement and the rate of fatigue crack growth. *International Journal of Fracture* **8**, 209–19.

Dowling, N.E. (1972). Fatigue failure predictions for complicated stress–strain histories. *Journal of Materials* **7**, 71–87.

Dowling, N.E. (1977). Crack growth during low-cycle fatigue of smooth axial specimens. In *Cyclic Stress–Strain and Plastic Deformation Aspects of Fatigue Crack Growth*, Special Technical Publication 637, pp. 97–121. Philadelphia: American Society for Testing and Materials.

Dowling, N.E. (1979). Notched member fatigue life predictions combining crack initiation and propagation. *Fatigue of Engineering Materials and Structures* **2**, 129–38.

Dowling, N.E. (1993). *Mechanical Behavior of Materials: Engineering Methods for Deformation, Fracture and Fatigue*. Englewood Cliffs, NJ: Prentice–Hall.

Dowling, N.E. & Begley, J.A. (1976). Fatigue crack growth during gross plasticity and the *J*-integral. In *Mechanics of Crack Growth*, Special Technical Publication 590, pp. 82–103. Philadelphia: American Society for Testing and Materials.

Dowling, N.E., Brose, W.R. & Wilson, W.K. (1977). Notched member fatigue life predictions by local strain approach. In *Fatigue under Complex Loading – Analysis and Experiments*, pp. 55–84. Warrendale: Society of Automotive Engineers.

Dowling, Jr., W.E., Allison, A.E., Swank, L.R. & Sherman, A.M. (1993). TiAl-based alloys for exhaust valve applications. In *New Engine Design and Engine Component Technology*, Special Publication-972, pp. 31–38. Warrendale, PA: Society of Automotive Engineers.

Downing, S.D. & Socie, D.F. (1982). Simplified rainflow counting algorithms. *International Journal of Fatigue* **4**, 31–40.

Driver, J.H. (1971). The effect of boundary precipitates on the high temperature fatigue strength of austenitic stainless steels. *Metal Science* **5**, 47–50.

Drucker, D.C. & Palgen, L. (1981). On the stress-strain relations suitable for cyclic and other loading. *Journal of Applied Mechanics* **48**, 479–85.

Dugdale, D.S. (1960). Yielding of steel sheets containing slits. *Journal of the Mechanics and Physics of Solids* **8**, 100–8.

Duquette, D.J. (1979). Environmental effects I: General fatigue resistance and crack nucleation in metals and alloys. In *Fatigue and Microstructure* (ed. M. Meshii), pp. 335–63. Metals Park: American Society for Metals.

Duquette, D.J. & Uhlig, H.H. (1968). Effect of dissolved oxygen and NaCl on corrosion fatigue of 0.18% carbon steel. *Transactions of the American Society for Metals* **61**, 449–56.

Dutta, V.B., Suresh, S. & Ritchie, R.O. (1984). Fatigue crack propagation in dual phase steels: Effects of ferritic–martensitic microstructures on crack path morphology. *Metallurgical Transactions* **15A**, 1193–207.

Eden, E.M., Rose, W.N. & Cunningham, F.L. (1911). The endurance of metals. *Proceedings of the Institution of Mechanical Engineers* **4**, 839–974.

El Haddad, M.H., Dowling, N.E., Topper, T.H. & Smith, K.N. (1980). *J*-integral applications for short fatigue cracks at notches. *International Journal of Fracture* **16**, 15–30.

El Haddad, M.H., Topper, T.H. & Smith, K.N. (1979). Prediction of nonpropagating cracks. *Engineering Fracture Mechanics* **11**, 573–84.

Elber, W. (1970). Fatigue crack closure under cyclic tension. *Engineering Fracture Mechanics* **2**, 37–45.

Elber, W. (1971). The significance of fatigue crack closure. In *Damage Tolerance in Aircraft Structures*, Special Technical Publication 486, pp. 230–42. Philadelphia: American Society for Testing and Materials.

Elinck, J.P., Bauwens, J.C. & Homés, G. (1971). Fatigue crack growth in poly(vinyl chloride). *International Journal of Fracture* **7**, 277–87.

Endo, K., Okada, T., Hariya, T. (1972). Fatigue crack propagation in bearing metals lining of steel plates in lubricating oil. *Bulletin of the Japan Society of Mechanical Engineers* **15**, 439–45.

Erdogan, F. & Sih, G.C. (1963). On the crack extension in plates under plane loading and transverse shear. *Journal of Basic Engineering* **85**, 519–27.

Eshelby, J.D. (1956). The continuum theory of lattice defects. In *Progress in Solid State Physics* (eds. F. Seitz & D. Turnbull), vol. 3, pp. 79–144. New York: Academic Press.

Eshelby, J.D. (1957). The determination of the elastic field of an ellipsoidal inclusion and related problems. *Proceedings of the Royal Society, London* **A241**, 376–96.

Essmann, U., Gösele, U. & Mughrabi, H. (1981). A model of extrusions and intrusions in fatigued metals. I. Point-defect production and the growth of extrusions. *Philosophical Magazine A* **44**, 405–26.

Essmann, U. & Mughrabi, H. (1979). Annihilation of dislocations during tensile and cyclic deformation and limits of dislocation densities. *Philosophical Magazine A* **40**, 731–56.

von Euw, E.F.J., Hertzberg, R.W. & Roberts, R. (1972). Delay effects in fatigue crack propagation. In *Stress Analysis and Growth of Cracks*, Special Technical Publication 513, part I, pp. 230–59. Philadelphia: American Society for Testing and Materials.

Evans, A.G. (1978). Microfracture from thermal expansion anisotropy–I. Single phase systems. *Acta Metallurgica* **26**, 1845–53.

Evans, A.G. (1980). Fatigue in ceramics. *International Journal of Fracture* **16**, 485–98.

Evans, A.G. & Fuller, E.R. (1974). Crack propagation in ceramic materials under cyclic loading conditions. *Metallurgical Transactions* **5A**, 27–33.

Evans, A.G., Zok, F.W. & McMeeking, R.M. (1995). Fatigue of ceramic-matrix composites. *Acta Metallurgica et Materialia* **43**, 859–75.

Ewart, L. & Suresh, S. (1986). Dynamic fatigue crack growth in polycrystalline alumina under cyclic compressive loads. *Journal of Materials Science Letters* **5**, 774–8.

Ewart, L. & Suresh, S. (1987). Crack propagation in ceramics under cyclic loads. *Journal of Materials Science* **22**, 1173–92.

Ewart, L. & Suresh, S. (1992). Elevated-temperature crack growth in polycrystalline alumina under static and cyclic loads. *Journal of Materials Science* **27**, 5181–91.

Ewing, J.A. & Humfrey, J.C. (1903). The fracture of metals under rapid alterations of stress. *Philosophical Transactions of the Royal Society, London* **A200**, 241–50.

Ewing, J.A. & Rosenhain, W. (1900) Experiments in micro-metallurgy:–Effects of strain. Preliminary notice. *Philosophical Transactions of the Royal Society, London* **A199**, 85–90.

Faber, K.T. & Evans, A.G. (1983). Crack deflection processes: I. Theory. *Acta Metallurgica* **31**, 565–74.

Fairbairn, W. (1864). Experiments to determine the effect of impact, vibratory action, and long continued changes of load on wrought iron girders. *Philosophical Transactions of the Royal Society, London* **154**, 311.

Fatemi, A. & Socie, D.F. (1988). A critical plane approach to multiaxial fatigue damage including out-of-phase loading. *Fatigue and Fracture of Engineering Materials and Structures* **11**, 149–65.

Feltner, C.E. (1965). A debris mechanism of cyclic strain hardening for F.C.C. metals. *Philosophical Magazine A* **12**, 1229–48.

Feltner, C.E. & Laird, C. (1967a). Cyclic stress–strain response of F.C.C. metals and alloys–I. Phenomenological experiments. *Acta Metallurgica* **15**, 1621–32.

Feltner, C.E. & Laird, C. (1967b). Cyclic stress–strain response of F.C.C. metals and alloys–II. Dislocation structures and mechanisms. *Acta Metallurgica* **15**, 1633–53.

Figueroa, J.C. & Laird, C. (1983). Crack initiation mechanisms in copper polycrystals cycled under constant strain amplitudes and in step tests. *Materials Science and Engineering* **60**, 45–58.

Findley, W.N. (1959). A theory for the effect of mean stress on fatigue of metals under combined torsion and axial load or bending. *Journal of Engineering for Industry* **81**, 301–6.

Finney, J.M. & Laird, C. (1975). Strain localization in cyclic deformation of copper single crystals. *Philosophical Magazine A* **31**, 339–66.

Finot, M. & Suresh, S. (1994). Multitherm, Personal computer software for the thermomechanical modeling of multilayered and graded materials. Copyright Massachusetts Institute of Technology, Cambridge, MA. Further information on this program can be found in Finot, M. (1996). Deformation of multi-layered and graded materials: theory and experiments. Doctor of Science Thesis. Massachusetts Institute of Technology.

Fleck, N.A. (1988). Influence of stress state on crack growth retardation. In *Basic Questions in Fatigue*, Special Technical Publication 924, vol. I, pp. 157–83. Philadelphia: American Society for Testing and Materials.

Fleck, N.A. & Newman, J.C. (1988). Analysis of crack closure under plane strain conditions. In *Mechanics of Fatigue Crack Closure*, Special Technical Publication 982, pp. 319–41. Philadelphia: American Society for Testing and Materials.

Fleck, N.A. & Smith, R.A. (1982). Crack closure – is it just a surface phenomenon? *International Journal of Fatigue* **4**, 157–60.

Ford, F.P. & Silverman, M. (1979). Mechanistic aspects of environment-controlled crack propagation in steel/aqueous environment systems. *Report No. HTGE–451–8–12*. Schenectady, NY: General Electric Company.

Forman, R.G., Kearney, V.E. & Engle, R.M. (1967). Numerical analysis of crack propagation in cyclic-loaded structures. *Journal of Basic Engineering* **89**, 459–64.

Forsyth, P.J.E. (1953). Exudation of material from slip bands at the surface of fatigued crystals of an aluminium–copper alloy. *Nature* **171**, 172–3.

Forsyth, P.J.E. (1957). Slip band damage and extrusion. *Proceedings of the Royal Society, London* **A242**, 198–202.

Forsyth, P.J.E. (1962). A two stage process of fatigue crack growth. In *Crack Propagation: Proceedings of Cranfield Symposium*, pp. 76–94. London: Her Majesty's Stationery Office.

Forsyth, P.J.E. & Ryder, D.A. (1960). Fatigue fracture. *Aircraft Engineering* **32**, 96–9.

Forsyth, P.J.E. & Stubbington, A. (1955). The influence of substructure on the slip observed in pure aluminium and some aluminium alloys when subjected to fatigue stresses. *Journal of the Institute of Metals* **84**, 173–5.

Fouvry, S., Kapsa, Ph., Vincent, L. & Dang Van, K. (1996). Theoretical analysis of fatigue under dry friction for fretting loading conditions. *Wear* **195**, 21–34.

Franciosi, P. & Zaoui, A. (1982). Multislip in F.C.C. crystals: a theoretical approach compared with experimental data. *Acta Metallurgica* **30**, 1627–37.

Frost, N.E. (1960). Notch effects and the critical alternating stress required to propagate a crack in an aluminium alloy subject to fatigue loading. *Journal of Mechanical Engineering Science* **2**, 109–19.

Frost, N.E. (1966). The growth of fatigue cracks. In *Proceedings of the First International Conference on Fracture* (ed. T. Yokobori), pp. 1433–59. Sendai: The Japan Society for Strength and Fracture of Materials.

Frost, N.E. & Dugdale, D.S. (1957). Fatigue tests on notched mild steel plates with measurements of fatigue cracks. *Journal of the Mechanics and Physics of Solids* **5**, 182–92.

Frost, N.E., Marsh, K.I. & Pook, L.P. (1974). *Metal Fatigue*. Oxford: Clarendon Press.

Froustey, C. & Lasserre, S. (1989). Multiaxial fatigue endurance of 30Ncd16 steel. *International Journal of Fatigue* **11**, 169–75.

Fuchs, H.O. & Stephens, R.I. (1980). *Metal Fatigue in Engineering*. New York: John Wiley

Fujino, S. & Taira, S. (1980). Effect of thermal cycle on low cycle fatigue life of steels and grain boundary sliding characteristics. In *Mechanical Behaviour of Materials* (eds. K.J. Miller & R.F. Smith), vol. II, pp. 49–58. Oxford: Pergamon Press.

Fujita, F.E. (1963). Oxidation and dislocation mechanisms in fatigue crack formation. In *Fracture of Solids* (eds. D.C. Drucker & J.J. Gilman), pp. 657–70. New York: Interscience.

Gangloff, R.P. (1981). The criticality of crack size in aqueous corrosion fatigue. *Res Mechanica Letters* **1**, 299–306.

Gangloff, R.P. & Wei, R.P. (1986). Small crack–environment interactions: the hydrogen embrittlement perspective. In *Small Fatigue Cracks* (eds. R.O. Ritchie & J. Lankford), pp. 239–64. Warrendale: The Metallurgical Society of the American Institute of Mining, Metallurgical and Petroleum Engineers.

Gao, Hua, Alagok, N., Brown, M.W. & Miller, K.J. (1985). Growth of fatigue cracks under combined mode I and mode II loads. In *Multiaxial Fatigue*, Special Technical Publication 853, pp. 184–202. Philadelphia: American Society for Testing and Materials.

Garud, Y.S. (1981). Multiaxial fatigue: A survey of the state of the art. *Journal of Testing and Evaluation* **9**, 165–78.

Garvie, R.C., Hannick, R.H. & Pascoe, R.T. (1975). Ceramic steel? *Nature* **258**, 703–4.

Gassner, E. (1941). Auswirkung betriebsähnlicher Belastungsfolgen auf die Festigkeit von Flugzeugbauteilen. *Jahrbuch 1941 der deutschen Luftfahrtforschung*, 472–83.

Gent, A.N. (1970). Hypothetical mechanism of crazing in glassy plastics. *Journal of Materials Science* **5**, 925–32.

Gerber, H. (1874). Bestimmung der zulässigen Spannungen in Eisen-konstructionen. *Zeitschrift des Bayerischen Architekten und Ingenieur-Vereins* **6**, 101–10.

Gerber, T.L. & Fuchs, H.O. (1968). Analysis of non-propagating cracks in notched parts with compressive mean stress. *Journal of Materials* **3**, 359–74.

Gerberich, W.W. & Chen, Y.T. (1975). Hydrogen-controlled cracking–An approach to threshold stress intensity. *Metallurgical Transactions* **6A**, 271–8.

Gerold, V. & Karnthaler, H.P. (1989). On the origin of planar slip in F.C.C. alloys. *Acta Metallurgica* **37**, 2177–83.

Giannakopoulos, A.E., Lindley, T.C. & Suresh, S. (1998). Aspects of equivalence between contact mechanics and fracture mechanics: theoretical connections and a life prediction methodology for fretting-fatigue. *Acta Materialia* **46**, 2955–67.

Giannakopoulos, A.E. & Suresh, S. (1998). A three-dimensional analysis of fretting fatigue. *Acta Materialia* **46**, 177–92.

Gieseke, B. & Saxena, A. (1989). Correlation of creep-fatigue crack growth rates using crack-tip parameters. In *Advances in Fracture Research* (eds. K. Salama, K. Ravi-Chandar, D.M.R. Taplin & P. Rama Rao), vol. 1, pp. 189–96. Oxford: Pergamon Press.

Gladwell, G.M.L. (1980). *Contact problems in the classical theory of elasticity*. Amsterdam, The Netherlands: Sijthoff and Noordhoff.

Glazov, M., Llanes L.M. & Laird, C. (1995). Self-organized dislocation structures (SODS) in fatigued metals. *Physica Status Solidi (a)* **149**, 297-321.

Gong, B., Wang, Z., Zhang, Y.W. & Wang, Z.G. (1995). The cyclic deformation of copper single crystal oriented for double slip. *Materials Science and Engineering A* **A194**, 171–8.

Gong, B., Wang, Z.G. & Wang, Z. (1997). Cyclic deformation behavior and dislocation structures of [001] copper single crystals–I. Cyclic stress–strain response and surface feature. *Acta Materialia* **45**, 1365–77.

Goodman, J. (1899). *Mechanics Applied to Engineering*. London: Longmans Green.

Goranson, U.G. & Miller, M. (1989). Aging jet transport structural evaluation programs. In *Aeronautical Fatigue in the Electronic Era* (ed. A. Berkovits), pp. 319–53. Warley: Engineering Materials Advisory Services.

Gough, H.J. (1926). *The Fatigue of Metals*. London: Benn.

Gough, H.J. (1933). Crystalline structure in relation to failure of metals – especially by fatigue. Edgar Marburg Lecture. *Proceedings of the American Society for Testing and Materials* **33**, Part II, 3–114.

Gough, H.J. & Hanson, D. (1923). The behaviour of metals subjected to repeated stresses. *Proceedings of the Royal Society, London* **A104**, 535–65.

Gough, H.J., Pollard, H.V. & Clenshaw, W.J. (1951). Some experiments on the resistance of metals under combined stress. *Aeronautical Research Council Reports and Memoranda No. 2522*. London: Ministry of Supply, His Majesty's Stationery Office.

Gough, H.J. & Sopwith, D.G. (1932). Atmospheric action as a factor in fatigue of metals. *Journal of the Institute of Metals* **49**, 93–122.

Gould, D., Hirsch, P.B. & Humphreys, F.J. (1974). The Bauschinger effect, work-hardening and recovery in dispersion-hardened copper crystals. *Philosophical Magazine* **30**, 1353–77.

Grathwohl, G. (1988). Fatigue of ceramics under cyclic loading. *Materialwissenschaften Werkstofftechnik* **19**, 113–24.

Gray, G. T., Williams, J.C. & Thompson, A.W. (1983). Roughness-induced crack closure: an explanation for microstructurally sensitive fatigue crack growth. *Metallurgical Transactions* **14A**, 421–33.

Green, A.E. (1949). On Boussinesq's problem and penny-shaped cracks. *Proceedings, Cambridge Philosophical Society* **45**, 251–66.

Griffith, A.A. (1921). The phenomenon of rupture and flow in solids. *Philosophical Transactions of the Royal Society, London* **A221**, 163–97.

Gross, T.S. (1985). Frictional effects in mode III fatigue crack propagation. *Scripta Metallurgica* **19**, 1185–8.

Grosskreutz, J.C. & Mughrabi, H. (1975). Description of the work-hardened structure at low temperature in cyclic deformation. In *Constitutive Equations in Plasticity* (ed. A.S. Argon), pp. 251–326. Cambrige, Mass.: MIT Press.

Grosskreutz, J.C. & Shaw, G.C. (1969). Critical mechanisms in the development of fatigue cracks in 2024-T4 aluminum. In *Fracture 1969* (ed. P.L. Pratt), pp. 620–9, Oxford: Pergamon Press.

Gu, H., Guo, H., Chang, S. and Laird, C. (1994). Orientation dependence of cyclic deformation in high purity titanium single crystals. *Materials Science and Engineering A* **A188**, 23-36.

Guillct, A. (1910). Intervention de l'amortissement dans l'essai des fers. *Revue des Metall. Memoires*, pp. 885.

Guiu, F. & Anglada, M. (1980). Irreversible glide and shape changes in the cyclic deformation of b.c.c. crystals. *Philosophical Magazine A* **42**, 271–6.

Guiu, F., Dubniak, R. & Edward, R.C. (1982). On the nucleation of fatigue cracks in pure polycrystalline α–iron. *Fatigue of Engineering Materials and Structures* **5**, 311–21.

Guiu, F., Reece, M.J. & Vaughan, D.A.J. (1991). Cyclic fatigue in some structural ceramics. In *Fatigue of Advanced Materials* (eds. R.O. Ritchie, R.H. Dauskardt and B.N. Cox), pp. 193–210. Birmingham, England: Mechanical and Component Engineering Publications Ltd.

Gurney, T.R. (1968). *Fatigue of Welded Structures*. Cambridge University Press.

Haigh, B.P. (1915). Report on alternating stress tests of a sample of mild steel received from the British Association Stress Committee. Report of the British Association. **85**, 163–170.

Haigh, B.P. (1917). Experiments on the fatigue of brasses. *Journal of the Institute of Metals* **18**, 55–77.

Halford, G.R., Hirschberg, M.H. & Manson, S.S. (1973). Temperature effects on the strain range partitioning approach for creep-fatigue analysis. In *Fatigue at Elevated Temperatures*, Special Technical Publication 520, pp. 658–67. Philadelphia: American Society for Testing and Materials.

Halliday, M.D. & Beevers, C.J. (1981). Some aspects of crack closure in two contrasting titanium alloys. *Journal of Testing and Evaluation* **9**, 195–201.

Hamilton, G.M. (1983). Explicit equations for the stresses beneath a sliding spherical contact. *Proceedings of Institution of Mechanical Engineers* **197C**, 53–9.

Hamilton, G.M. & Goodman, L.E. (1966). The stress field created by a circular sliding contact. *Journal of Applied Mechanics* **33**, 371–6.

Han, L.X. & Suresh, S. (1989). High temperature crack growth in an alumina–silicon carbide whisker composite: Mechanisms of fatigue crack tip damage. *Journal of the American Ceramic Society* **72**, 1233–8.

Hancock, J.R. & Grosskreutz, J.C. (1969). Mechanisms of fatigue hardening in copper single crystals. *Acta Metallurgica* **17**, 77–97.

Harlin, G. & Willis, J.R. (1988). The influence of crack size on the ductile–brittle transition. *Proceedings of the Royal Society, London* **A415**, 197–226.

Harris, J.A. (1987). Engine component retirement for cause. Volume I–Executive summary. *Technical Report-87-4069*. Dayton: Air Force Wright Aeronautical Laboratories.

Hart, E.W. (1980). A theory for stable crack extension rates in ductile materials. *International Journal of Solids and Structures* **16**, 807–23.

Hartman, A. (1965). On the effects of oxygen and water vapor on the propagation of fatigue cracks in 2024-T3 Alclad sheet. *International Journal of Fracture* **1**, 167–88.

Hartman, A., Jacobs, F.J., Nederveen, A. & Derijk, R. (1967). National Laboratory for Research, Technical Note No. M. 2182, Amsterdam; cited in Wei, R.P. & Simmons, G.W. (1981).

Hartman, R.J. & Macherauch, E. (1963). Die Veränderung von röntgen-interferenzen Hysterese und obserflächenbild bei ein- und wechselseitig Beanspruchung von Messing, Nickel und Stahl. *Zeitschrift für Metallkunde* **54**, 161–72.

Hasegawa, T., Yakou, T. & Kocks, U.F. (1986). Forward and reverse arrangements of dislocations in tangled walls. *Materials Science and Engineering* **81**, 189–99.

Hashin, Z. & Rotem, A. (1973). A fatigue failure criterion for fiber reinforced composites. *Journal of Composite Materials* **7**, 448–64.

He, M.-Y. & Hutchinson, J.W. (1989). Kinking of a crack out of an interface. *Journal of Applied Mechanics* **56**, 270–8.

Helgeland, O. (1965). Cyclic hardening and fatigue of copper single crystals. *Journal of the Institute of Metals* **93**, 570–5.

Hellan, K. (1984). *Introduction to Fracture Mechanics*. New York: McGraw-Hill.

Henaff, G., Marchal, K. & Petit, J. (1995). On fatigue crack propagation enhancement by a gaseous atmosphere: experimental and theoretical aspects. *Acta Metallurgica et Materialia* **43**, 2931–42.

Herman, W.A., Hertzberg, R.W. & Jaccard, R. (1988). A simplified laboratory approach for the prediction of short crack behavior in engineering structures. *Fatigue and Fracture of Engineering Materials and Structures* **4**, 303–20.

Hertz, H. (1882). Über die Berührung fester elastischer Körper. (On the contact of elastic solids.) *J. reine und angewandte Mathematick* **92**, 156–71.

Hertzberg, R.W. (1995). *Deformation and Fracture Mechanics of Engineering Materials*, 4th edn. New York: Wiley.

Hertzberg, R.W. & Manson, J.A. (1973). Micromechanisms of fatigue-crack advance in PVC. *Journal of Materials Science* **8**, 1554–8.

Hertzberg, R.W. & Manson, J.A. (1980). *Fatigue of Engineering Plastics*. New York: Academic Press.

Hertzberg, R.W. & Manson, J.A. (1986). Fatigue. *Encyclopedia of Polymer Science and Engineering*, vol. 7, pp. 378–453. New York: Wiley.

Hertzberg, R.W. & Mills, W.J. (1976). Character of fatigue fracture surface micromorphology in ultra-low growth rate regime. In *Fractography–Microscopic Cracking Processes*, Special Technical Publication 600, pp. 220–34. Philadelphia: American Society for Testing and Materials.

Hertzberg, R.W., Nordberg, H. & Manson, J.A. (1970). Fatigue crack propagation in polymeric materials. *Journal of Materials Science* **5**, 521–6.

Hertzberg, R.W., Skibo, M.D. & Manson, J.A. (1979). Fatigue fracture mechanisms in engineering plastics. In *Fatigue Mechanisms*, Special Technical Publication 675, pp. 471–500. Philadelphia: American Society for Testing and Materials.

Hesselbarth, H.W. & Steck, E. (1992). A simulation of dislocation patterning by cellular automata. *Solid State Phenomena* **23 & 24**, 445–54.

Hill, R. (1950). *The Mathematical Theory of Plasticity.* Oxford: Clarendon Press.

Hills, D.A. & Ashelby, D.W. (1982). The influence of residual stresses on contact load bearing capacity. *Wear* **75**, 221–40.

Hills, D.A. & Nowell, D. (1994). *Mechanics of fretting fatigue.* Dordrecht, The Netherlands: Kluwer Academic Publishers.

Hills, D.A. & Sackfield, A. (1984). Yield and shakedown states in the contact of generally curved bodies. *Journal of Strain Analysis for Engineering Design* **19**, 9–14.

Himmelein, M.K. & Hillberry, B.M. (1976). Effects of stress ratio and overload ratio on fatigue crack delay and arrest behavior due to single peak overloads. In *Mechanics of Crack Growth*, Special Technical Publication 590, pp. 321–30. Philadelphia: American Society for Testing and Materials.

Hodgkinson, E.A. (1849). *Report of the Commissioners Appointed to Enquire into the Application of Iron to Railway Structures*, Command Paper No. 1123. London: His Majesty's Stationery Office.

Hohenemser, K. & Prager, W. (1932). Über die Ansätze der Mechanik isotroper Kontinua. *Zeits. angew.Math. u. Mech.* **12**, 216–26.

Hojo, M., Tanaka, K., Gustafson, C.G. & Hayashi, R. (1987). Effect of stress ratio on near-threshold propagation of delamination cracks in unidirectional CFRP. *Composites Science and Technology* **29**, 273–92.

Holm, D.K., Blom, A.F. & Suresh, S. (1986). Growth of cracks under far-field cyclic compressive loads: Numerical and experimental results. *Engineering Fracture Mechanics* **23**, 1097–106.

Holmes, J.W. (1991). Influence of stress ratio on the elevated temperature fatigue of a SiC fiber-reinforced Si_3N_4 composite. *Journal of the American Ceramic Society* **74**, 1539–45.

Holzwarth, U. & Essmann, U. (1993). The evolution of persistent slip bands in copper single crystals. *Applied Physics A* **A57**, 131–41.

Hopkins, S.W., Rau, C.A., Leverant, G.R. & Yuen, A. (1976). Effects of various program overloads on the threshold for high frequency fatigue crack growth. In *Fatigue Crack Growth under Spectrum Loads*, Special Technical Publication 595, pp. 125–41. Philadelphia: American Society for Testing and Materials.

Horibe, S. & Hirahara, R. (1991). Fatigue crack propagation of sintered silicon nitride in vacuum and air. *Fatigue and Fracture of Engineering Materials and Structures* **14**, 863–70.

Hornbogen, E. (1978a). Martensitic transformation at a propagating crack. *Acta Metallurgica* **26**, 147–52.

Hornbogen, E. (1978b). Shape change during the 19 °C-phase transformation of PTFE. *Progress in Colloid and Polymer Science* **64**, 125–31.

Hornbogen, E. & Zum Gahr, K.-H. (1976). Microstructure and fatigue crack growth in a γ–Fe–Ni–Al alloy. *Acta Metallurgica* **24**, 581–92.

Hourlier, F., d'Hondt, H., Truchon, M. & Pineau, A. (1985). Fatigue crack path behavior under polymodal fatigue. In *Multiaxial Fatigue*, Special Technical Publication 852, pp. 228–48. Philadelphia: American Society for Testing and Materials.

Hourlier, F., McLean, D. & Pineau, A. (1978). Fatigue crack growth behaviour of Ti–5Al–2.5Sn alloy under complex stress (mode I + steady mode III). *Metals Technology* **5**, 154–8.

Hu, Y.M. & Wang, Z.G. (1997). Cyclic stress–strain response and dislocation structures of [$\bar{3}45$]/[$\bar{1}17$] copper bicrystals. *Acta Materialia* **45**, 2655–70.

Hubbard, R.P. (1969). Crack growth under cyclic compression. *Journal of Basic Engineering* **91**, 625–31.

Hudak, S.J., Davidson, D.L., Chan, K.S., Howland, A.C. & Walsch, M.J. (1988). *Growth of Small Cracks in Aeroengine Disc Materials*, Report No. AFWAL–TR–88–4090. Dayton: Air Force Wright Aeronautical Laboratories.

Hudson, C.M. & Scardina, J.T. (1969). Effects of stress ratio on fatigue-crack growth in 7075-T6 aluminum-alloy sheet. *Engineering Fracture Mechanics* **1**, 429–46.

Hui, C.Y. & Riedel, H. (1981). The asymptotic stress and strain field near the tip of a growing crack under creep conditions. *International Journal of Fracture* **17**, 409–25.

Hunsche, A. & Neumann, P. (1986). Quantitative measurement of persisent slip band profiles and crack initiation. *Acta Metallurgica* **34**, 207–17.

Hurd, N.J. & Irving, P.E. (1982). Factors influencing propagation of mode III fatigue cracks under torsional loading. In *Design of Fatigue and Fracture Resistant Structures*, Special Technical Publication 761, pp. 212–23. Philadelphia: American Society for Testing and Materials.

Hutchinson, J.W. (1968). Singular behavior at the end of a tensile crack in a hardening material. *Journal of the Mechanics and Physics of Solids* **16**, 13–31.

Hutchinson, J.W. (1983). Fundamentals of the phenomenological theory of nonlinear fracture mechanics. *Journal of Applied Mechanics* **50**, 1042–51.

Hutchinson, J.W. (1987). Crack tip shielding by micro-cracking in brittle solids. *Acta Metallurgica* **35**, 1605–19.

Hutchinson, J.W. & Paris, P.C. (1979). Stability analysis of *J*-controlled crack growth. In *Elastic–Plastic Fracture*, Special Technical Publication 668, pp. 37–64. Philadelphia: American Society for Testing and Materials.

Hyzak, J.M. & Bernstein, I.M. (1982). The effect of defects on the fatigue crack initiation process in two P/M superalloys: Part I. Fatigue origins. *Metallurgical Transactions* **13A**, 33–43.

Ikeda, S., Izumi, Y. & Fine, M.E. (1977). Plastic work during fatigue crack propagation in a high strength alloy steel and in 7050 Al-Alloy. *Engineering Fracture Mechanics* **9**, 123–36.

Inglis, C.E. (1913). Stresses in a plate due to the presence of cracks and sharp corners. *Transactions of the Institute of Naval Architects* **55**, 219–41.

Irving, P.E. & Beevers, C.J. (1974). Microstructural influences on fatigue crack growth in Ti-6Al-4V. *Materials Science and Engineering* **14**, 229–38.

Irwin, G.R. (1956). Onset of fast crack propagation in high strength steel and aluminum alloys. In *Proceedings of the Second Sagamore Conference* vol. II, pp. 289–305. New York: Syracuse University.

Irwin, G.R. (1957). Analysis of stresses and strains near the end of a crack traversing a plate. *Journal of Applied Mechanics* **24**, 361–4.

Irwin, G.R. (1960). Plastic zone near a crack and fracture toughness. In *Proceedings of the Seventh Sagamore Ordnance Materials Conference*, vol. IV, pp. 63–78. New York: Syracuse University.

Irwin, G.R. (1964). Structural aspects of brittle fracture. *Applied Mechanics Research* **3**, 65–81.

Iwan, W.D. (1967). On a class of models for the yielding behavior of continuous and composite systems. *Journal of Applied Mechanics* **34**, 612–7.

Jacobs, D.S. & Chen, I.-W. (1995). Cyclic fatigue in ceramics: a balance between crack shielding accumulation and degradation. *Journal of the American Ceramic Society* **78**, 513–20.

James, L.A. (1972). The effect of frequency upon fatigue crack growth of Type 304 stainless steel at 1000°F. In *Stress Analysis and Growth of Cracks*, Special Technical Publication 513, pp. 218–29. Philadelphia: American Society for Testing and Materials.

James, M.N., Tait, R.B. & Mech, D.G. (1991). Compression fatigue in β–alumina. *Fatigue and Fracture of Engineering Materials and Structures* **14**, 227–35.

James, M.R. & Morris, W.L. (1983). Effect of fracture surface roughness on growth of short fatigue cracks. *Metallurgical Transactions* **14A**, 153–5.

Jenkin, C.F. (1923). The fatigue failure of metals. *Proceedings of the Royal Society, London* **A103**, 121–38.

Jin, N.Y. (1983). Dislocation structures in fatigued copper single crystals oriented for double-slip. *Philosophical Magazine A* **A48**, L33–L38.

Jin, N.Y. (1989). Formation of dislocation structures during cyclic deformation of F.C.C. crystals.–I. Formation of PSBs in crystals oriented for single-slip. *Acta Metallurgica* **37**, 2055–66.

Jin, N.Y. & Winter, A.T. (1984). Cyclic deformation of copper single crystals oriented for double slip. *Acta Metallurgica* **32**, 989–95.

Johnson, E.W. & Johnson, H.H. (1965). Imperfection density of fatigued and annealed copper via electrical resistivity measurements. *Transactions of the Metallurgical Society* **233**, 1333–9.

Johnson, K.L. (1962). A shakedown limit in rolling contact. In *Proceedings of the 4th US National Congress of Applied Mechanics*, pp. 971–80. New York, NY: American Society of Mechanical Engineers.

Johnson, K.L. (1983). Inelastic contact: plastic flow and shakedown. In *Contact Mechanics and Wear of Rail/Wheel Systems*, pp. 79–101. Waterloo, Ontario: University of Waterloo Press.

Johnson, K.L. (1985). *Contact Mechanics*. Cambridge: Cambridge University Press.

Johnson, K.L. & Jefferis, J.A. (1963). Plastic flow and residual stresses in rolling and sliding contact. In *Proceedings of the Institution of Mechanical Engineers*, pp. 50–9. London: Institution of Mechanical Engineers.

Johnston, T.L., Stokes, R.J. & Li, C.H. (1962). Crack nucleation in magnesium oxide bicrystals under compression. *Philosophical Magazine* **7**, 23–34.

de Jonge, J.B. & Nederveen, A. (1980). Effect of gust load alleviation on fatigue and crack growth in ALCLAD 2024-T3. In *Effect of Load Spectrum Variables on Fatigue Crack Initiation and Propagation*, Special Technical Publication 714, pp. 170–84. Philadelphia: American Society for Testing and Materials.

Kanazawa, K., Miller, K.J. & Brown, M.W. (1979). Cyclic deformation of 1%Cr–Mo–V steel under out-of-phase loads. *Fatigue of Engineering Materials and Structures* **2**, 217–28.

Kaneko, Y., Morita, Y. & Hashimoto, S. (1997). Change in shape of hysteresis loops during cyclic deformation on austenitic stainless steel single crystals oriented for single slip. *Scripta Materialia* **37**, 963–8.

Kanninen, M.L. & Popelar, C.H. (1985). *Advanced Fracture Mechanics*. New York, NY: Oxford University Press.

Katagiri, K., Omura, A., Koyanagi, K., Awatani, J., Shiraishi, T., & Kaneshiro, H. (1977). Early stage crack tip morphology in fatigued copper. *Metallurgical Transactions* **8A**, 1769–73.

Kayali, E.S. & Plumtree, A. (1982). Stress–substructure relationships in cyclically and monotonically deformed wavy slip mode metals. *Metallurgical Transactions* **13A**, 1033–41.

Kemsley, D.S. & Paterson, M.S. (1960). The influence of strain amplitude on the work hardening of copper crystals in alternating tension and compression. *Acta Metallurgica* **8**, 453–67.

Kennedy, A.J. (1963). *Processes of Creep and Fatigue in Metals*. New York: Wiley.

Kikukawa, M., Jono, M. & Mikami, S. (1982). Fatigue crack propagation and crack closure behavior under stationary varying loadings–test results of aluminum alloy (in Japanese). *Journal of the Society of Materials Science Japan* **31**, 438–87.

Kim, A.S. (1996). Fracture normal to layered and graded interfaces. Ph.D. thesis. Providence, RI: Brown University.

Kim, A.S., Besson, J. & Pineau, A. (1998). Global and local approaches to fracture normal to interfaces. *International Journal of Solids and Structures*, submitted.

Kim, A.S., Suresh, S. & Shih, C.F. (1997). Fracture normal to interfaces with homogeneous and graded compositions. *International Journal of Solids and Structures* **34**, 3415–32.

Kim, S.L., Skibo, M.D., Manson, J.A. & Hertzberg, R.W. (1977). Fatigue crack propagation in poly(methyl methacrylate): effect of molecular weight and internal plasticization. *Polymer Engineering and Science* **17**, 194–203.

Kim, W.H. & Laird, C. (1978). Crack nucleation and stage I propagation in high strain fatigue – II. Mechanism. *Acta Metallurgica* **26**, 789–99.

King, J. E. (1982). Surface damage and near-threshold fatigue crack growth in a Ni-base superalloy in vacuum. *Fatigue of Engineering Materials and Structures* **5**, 177–88.

Kishimoto, H., Ueno, A. & Kawamoto, H. (1987). Fatigue crack propagation in ceramics. *Society of Materials Science Japan*, **36**, 1122–7.

Kishimoto, H., Ueno, A., Okawara, S. & Kawamoto, H. (1994). Crack propagation behavior of polycrystalline alumina under static and cyclic load. *Journal of the American Ceramic Society* **77**, 1324–8.

Kitagawa, H. & Takahashi, S. (1976). Applicability of fracture mechanics to very small cracks or the cracks in the early stage. In *Proceedings of Second International Conference on Mechanical Behavior of Materials*, pp. 627–31. Metals Park: American Society for Metals.

Kitagawa, H. & Tanaka, T. (1990). Editors, *Fatigue 90*. Birmingham: Materials and Components Engineering Publications.

Kitagawa, H., Yuuki, R. & Ohira, T. (1975). Crack-morphological aspects in fracture mechanics. *Engineering Fracture Mechanics* **7**, 515–29.

Kitagawa, H., Yuuki, R., Tohgo, K. & Tanabe, M. (1985). ΔK dependency of fatigue crack growth of single and mixed-mode cracks under biaxial stresses. In *Multiaxial Fatigue*, Special Technical Publication 853, pp. 164–83. Philadelphia: American Society for Testing and Materials.

Knott, J.F. & Pickard, A.C. (1977). Effects of overloads on fatigue crack propagation: Aluminum alloys. *Metal Science* **11**, 399–404.

Kobayashi, H. (1987). On the examination report of the crashed Japan Airline Boeing 747 plane – Failure analysis of the rear pressure bulkhead (in Japanese). *Journal of the Japan Society for Safety Engineering* **26**, 363–72.

Kobayashi, H., Arai, Y. & Nakamura, H. (1987). Fatigue failure analysis of the rear pressure bulkhead of the crashed Boeing 747 jumbo jet plane (in Japanese). *Journal of the Society of Materials Science Japan* **36**, 1084–9.

Kobayashi, H., Ogawa, T., Nakamura, H. & Nakazawa, H. (1984). Measurement of crack closure and evaluation of near threshold characteristics in fatigue crack growth for A508-3 steel (in Japanese). *Transactions of the Japan Society of Mechanical Engineers* **50**, 309–19.

Koiter, W.T. (1956). *Koninkl. Ned. Ak. Wetenschap* **B59**, 24–50.

Könczöl, L., Schincker, M.G. & Döll, W. (1984). Interference optical measurements during continuous and discontinuous crack propagation in fatigue loaded polymers. *Journal of Materials Science* **19**, 1605–15.

Koo, G.P., Riddell, M.N. & O'Toole, J.L. (1967). Fatigue properties of polytetrafluoroethylene and related fluoropolymers. *Polymer Engineering and Science* **7**, 182–8.

Kramer, E.J. (1983). Microscopic and molecular fundamentals of crazing. In *Advances in Polymer Science* (ed. H.H. Kausch), vol. 52/53, pp. 1–56. Berlin: Springer-Verlag.

Kramer, E.J. & Berger, L.L. (1990). Fundamental processes of craze growth and fracture. In *Advances in Polymer Science* (ed. H.H. Kausch), vol. 91/92, pp. 1–68. Berlin: Springer-Verlag.

Kramer, E.J. & Hart, E.W. (1984). Theory of slow, steady state crack growth in polymeric glasses. *Polymer* **25**, 1667–78.

Krieg, R.D. (1975). A practical two-surface plasticity theory. *Journal of Applied Mechanics* **47**, 641–6.

Ku, R.C. & Johnston, T.L. (1964). Fracture strength of MgO bicrystals. *Philosophical Magazine* **9**, 231–47.

Kuhlmann-Wilsdorf, D. (1979). Dislocation behavior in fatigue. IV. Quantitative interpretation of friction stress and back stress derived from hysteresis loops. *Materials Science and Engineering* **39**, 231–45.

Kumar, V., German, M.D. & Shih, C.F. (1981). *An engineering approach to elastic-plastic analysis*, Report No. NP 1931. Palo Alto: Electric Power Research Institute.

Kung, C.Y. & Fine, M.E. (1979). Fatigue crack initiation and microcrack growth in 2024-T4 and 2124-T4 aluminum alloys. *Metallurgical Transactions* **10A**, 603–10.

Kuno, M., Waterhouse, R.B., Nowell, D. & Hills, D.A. (1989). Initiation and growth of fretting fatigue cracks in the partial slip regime. *Fatigue and Fracture of Engineering Materials and Structures* **12**, 387–98.

Kwadjo, R. & Brown, L.M. (1978). Cyclic hardening of magnesium single crystals. *Acta Metallurgica* **26**, 1117–32.

Laird, C. (1967). The influence of metallurgical structure on the mechanisms of fatigue crack propagation. In *Fatigue Crack Propagation*, Special Technical Publication 415, pp. 131–68. Philadelphia: The American Society for Testing and Materials.

Laird, C. (1976). The fatigue limits of metals. *Materials Science and Engineering* **22**, 231–6.

Laird, C., Finney, J.M. & Kuhlmann-Wilsdorf, D. (1981). Dislocation behavior in fatigue. VI. Variation in the localization of strain in persistent slip bands. *Materials Science and Engineering* **50**, 127–36.

Laird, C. & Smith, G.C. (1962). Crack propagation in high stress fatigue. *Philosophical Magazine* **8**, 847–57.

Laird, C. & Thomas, G. (1967). On fatigue-induced reversion and overaging in dispersion strengthened alloy systems. *International Journal of Fracture Mechanics* **3**, 81–97.

Lamba, H.S. & Sidebottom, O.M. (1978). Cyclic plasticity for nonproportional paths, Part 1. Cyclic hardening, erasure of memory, and subsequent strain hardening experiments. *Journal of Engineering Materials and Technology* **100**, 96–103.

Landes, J.D. & Begley, J.A. (1976). A fracture mechanics approach to creep crack growth. In *Mechanics of Crack Growth*, Special Technical Publication 590, pp. 128–48. Philadelphia: American Society for Testing and Materials.

Landgraf, R.W. (1978). Control of fatigue resistance through microstructure – Ferrous alloys. *Fatigue and Microstructure* (ed. M. Meshii), pp. 439–66. Metals Park: American Society for Metals.

Landgraf, R.W. & LaPointe, N.R. (1974). Cyclic stress–strain concepts applied to component fatigue life prediction. *Transactions of the Society of Automotive Engineers* **83**, 1198–207.

Langer, B.F. (1937). Fatigue failure from stress cycles of varying amplitude. *Journal of Applied Mechanics* **59**, 160–2.

Lankford, J. (1982). The growth of small fatigue cracks in 7075-T6 aluminum. *Fatigue of Engineering Materials and Structures* **5**, 233–48.

Lankford, J. (1983). The effect of environment on the growth of small fatigue cracks. *Fatigue of Engineering Materials and Structures* **6**, 15–31.

Lankford, J. & Davidson, D.L. (1981). The effect of overloads upon fatigue crack tip opening displacement and crack tip opening/closing loads in aluminum alloys. In *Advances in Fracture Research* (ed. D. Franccois), vol. 2, pp. 899–906. Oxford: Pergamon Press.

Lankford, J. & Davidson, D.L. (1983). Fatigue crack micromechanisms in ingot and powder metallurgy 7xxx aluminum alloys in air and vacuum. *Acta Metallurgica* **31**, 1273–84.

Lankford, J. & Kusenberger, F.N. (1973). Initiation of fatigue cracks in 4340 steel. *Metallurgical Transactions* **4A**, 553–9.

Lardner, R.W. (1967). A dislocation model for fatigue crack growth in metals. *Philosophical Magazine* **17**, 71–82.

Larsen, J.M., Nicholas, T., Thompson, A.W. & Williams, J.C. (1986). Small crack growth in titanium–aluminum alloys. In *Small Fatigue Cracks*, (eds. R.O. Ritchie & J. Lankford). pp. 499–512. Warrendale: The Metallurgical Society.

Larsson, S.G. & Carlsson, A.J. (1973). Influence of non-singular stress terms and specimen geometry on small-scale yielding at crack tip in elastic–plastic solids. *Journal of the Mechanics and Physics of Solids* **21**, 263–77.

Lathabai, S., Rödel, J. & Lawn, B.R. (1991). Cyclic fatigue from degradation at bridging grains in alumina. *Journal of the American Ceramic Society* **74**, 1340–8.

Laufer, E.E. & Roberts, W.N. (1966). Dislocations and persistent slip bands in fatigued copper. *Philosophical Magazine* **14**, 67–78.

Lawn, B.R., Padture, N.P., Guiberteau, F. & Cai, H. (1994). A model of microcrack initiation and propagation beneath Hertzian contacts in polycrystalline ceramics. *Acta Metallurgica et Materialia* **42**, 1683–93.

Lawn, B. & Wilshaw, T.R. (1975). *Fracture of Brittle Solids*. Cambridge University Press.

Lawn, B. (1993). *Fracture of Brittle Solids*, 2nd edition. Cambridge University Press.

Laws, N. & Brockenbrough, J.R. (1987). The effect of microcrack systems on the loss of stiffness of brittle solids. *International Journal of Solids and Structures* **23**, 1247–68.

Lee, E.H. (1969). Elastic–plastic deformation at finite strains. *Journal of Applied Mechanics* **36**, 1–6.

Lee, J.J. & Sharpe, W.N. (1986). Short fatigue cracks in notched aluminum specimens. In *Small Fatigue Cracks*, (eds. R.O. Ritchie & J. Lankford). pp. 323–39. Warrendale: The Metallurgical Society.

Lee, J.-K. & Laird, C. (1983). Strain localization during fatigue of precipitation hardened aluminum alloys. *Philosophical Magazine A* **47**, 579–97.

Lepisto, T.K. & Kettunen, P.O. (1986). Comparison of the cyclic stress–strain behavior of single- and {111} ⟨111⟩ multiple-slip oriented copper single crystals. *Materials Science and Engineering* **83**, 1–15.

Li, F.Z., Needleman, A. & Shih, C.F. (1988). Characterization of near-tip stress and deformation fields in creeping solids. *International Journal of Fracture* **36**, 163–86.

Li, X.W., Wang, Z.G., Li, G.Y., Wu, S.D. & Li, S.X. (1998). Cyclic stress–strain response and surface deformation features of [011] multiple-slip oriented copper single crystals. *Acta Materialia* **46**, in press.

Li, Y. & Laird, C. (1994). Cyclic response and dislocation structures of AISI 316L stainless steel. 2. Polycrystals fatigued at intermediate strain amplitude. *Materials Science and Engineering A* **A186**, 87–103.

Li, Y.C. & Wang, T.C. (1986). High-order asymptotic field of tensile plane strain nonlinear crack problems. *Scientia Sinica (Series A)* **29**, 941–55.

Liaw, P.K., Leax, T.R. & Logsdon, W.A. (1983). Near-threshold fatigue crack growth behavior in metals. *Acta Metallurgica* **31**, 1581–7.

Liaw, P.K., Leax, T.R., Williams, R.S. & Peck, M.G. (1982). Near-threshold fatigue crack growth behavior in copper. *Metallurgical Transactions* **13A**, 1607–18.

Lilholt, H. (1977). Hardening in two-phase materials–I. Strength contributions in fiber-reinforced copper-tungsten. *Acta Metallurgica* **25**, 571–85.

Lin, C.-K. J., Socie, D.F., Xu, Y. & Zangvil, A. (1992). Static and cyclic fatigue of alumina at high-temperatures. 2. Failure analysis. *Journal of the American Ceramic Society* **75**, 637–48.

Lin, T.H. & Ito, Y.M. (1969). Mechanics of fatigue crack nucleation mechanism. *Journal of the Mechanics and Physics of Solids* **17**, 511–23.

Lin, T.H. & Lin, S.R. (1979). Micromechanics theory of fatigue crack initiation applied to time-dependent fatigue. In *Fatigue Mechanisms*, Special Technical Publication 675, pp. 707–28. Philadelphia: American Society for Testing and Materials.

Lindigkeit, J., Terlinde, G., Gysler, A. & Lütjering, G. (1979). The effect of grain size on the fatigue crack propagation behavior of age-hardened alloys in inert and corrosive environments. *Acta Metallurgica* **27**, 1717–26.

Lindley, T.C. (1982). Near threshold fatigue crack growth: experimental methods, mechanisms and applications. In *Subcritical Crack Growth due to Fatigue, Stress*

Corrosion and Creep (ed. L.H. Larsson), pp. 167–213. London: Elsevier Applied Science.

Lindley, T.C. (1997). Fretting fatigue in engineering alloys. *International Journal of Fatigue*, **19**, Supplement Number 1, S39–S49.

Lindley, T.C. & Nix, K.J. (1991). Case studies in power engineering plant. In Fatigue Crack Initiation and Damage Counters. Paris: Societé de Metallurgie.

Lindley, T.C. & Nix, K.J. (1992). Fretting fatigue in the power generation industry. In *Standardization of Fretting Fatigue Test Methods and Equipment* (eds: H.A. Attia and R.B. Waterhouse) Special Technical Publication 1159, pp. 153–69. Philadelphia: American Society for Testing and Materials.

Lindley, T.C. & Richards, C.E. (1974). The relevance of crack closure to fatigue crack propagation. *Materials Science and Engineering* **14**, 281–93.

Lisiecki, L.L. & Weertman, J.R. (1990). Orientation effects on the elevated temperature fatigue of copper single crystals. *Acta Metallurgica et Materialia* **38**, 509–19.

Liu, S. -Y., Chen, I.-W. & Tien, T.-Y. (1994). Fatigue-crack growth of silicon-nitride at 1400 degrees C - A novel fatigue-induced crack-tip bridging phenomenon. *Journal of the American Ceramic Society* **77**, 137–42.

Llanes, L., Rollet, A.D., Bassani, J.D. & Laird, C. (1993). Effects of grain size and annealing texture on the cyclic response and its substructure evolution of polycrystalline copper. *Acta Metallurgica et Materialia*, **41**, pp. 2667–79.

Llorca, J., Needleman, A. & Suresh, S. (1990). The Bauschinger effect in whisker-reinforced metal-matrix composites. *Scripta Metallurgica et Materialia* **24**, 1203–8.

Llorca, J. & Sánchez Gálvez, V. (1990). Modelling plasticity-induced fatigue crack closure. *Engineering Fracture Mechanics* **37**, 185–96.

Lo, K.K. (1978). Analysis of branched cracks. *Journal of Applied Mechanics* **45**, 797–802.

Löhberg, K. & Mustofon, H. (1976). Einfluß einer mechanischen Wechselbeanspruchung auf die Zinnumwandlung. *Zeitschrift für Metallkunde* **67**, 333–7.

Ludwik, P. (1919). Über die Änderung der Festigkeitseigenschaften der Metalle bei wechselnder Beanspruchung. *Zeitschrift für Metallkunde* **11**, 157–68.

Lukáš, P., Klesnil, M. & Krejčí, J. (1968). Dislocations and P.S.B.s in copper single crystals fatigued at low strain amplitude. *Physica Status Solidi* **27**, 545–58.

Lukáš, P. & Kunz, L. (1985). Is there a plateau in the cyclic stress–strain curves of polycrystalline copper? *Materials Science and Engineering* **74**, L1–5.

Lynch, S.P. (1981). A comparative study of stress-corrosion cracking, hydrogen-assisted cracking and liquid-metal embrittlement in Al, Ni, Ti, and Fe-based alloys. In *Hydrogen Effects in Metals* (eds. I.M. Bernstein & A.W. Thompson), pp. 863–70. Warrendale: The Metallurgical Society of the American Institute of Mining, Metallurgical and Petroleum Engineers.

Lynch, S.P. (1988). Environmentally assisted cracking: Overview of evidence for an adsorption-induced localised-slip process. *Acta Metallurgica* **36**, 2639–61.

Ma, B.-T. & Laird, C. (1989a). Overview of fatigue behavior in copper single crystals – I. Surface morphology and stage I crack initiation sites for tests at constant strain amplitude. *Acta Metallurgica* **37**, 325–36.

Ma, B.-T. & Laird, C. (1989b). Overview of fatigue behavior in copper single crystals – II. Population, size, distribution and growth kinetics of stage I cracks for tests at constant strain amplitude. *Acta Metallurgica* **37**, 337–48.

McAdam, D.J. (1926). Stress–strain relationships and the corrosion fatigue of metals. In *Corrosion Fatigue*, Special Technical Publication 26, pp. 224–54. Philadelphia: American Society for Testing and Materials.

McCammon, R.D. & Rosenberg, H.M. (1957). The fatigue and ultimate tensile strength of metals between 4.2 and 293 K. *Proceedings of the Royal Society, London* **A242**, 203–11.

McCartney, L.N. (1976). The effect of periodic–random loading on fatigue crack growth. *International Journal of Fracture* **12**, 273–88.

McCartney, L.N. (1987). Mechanics of matrix cracking in brittle-matrix fibre-reinforced composites. *Proceedings of the Royal Society, London* **A409**, 329–50.

McCarver, J.F. & Ritchie, R.O. (1982). Fatigue crack propagation thresholds for long and short cracks in René 95 nickel-base superalloy. *Materials Science and Engineering* **55**, 63–7.

McClintock, F.A. (1963). On the plasticity of the growth of fatigue cracks. In *Fracture of Solids* (eds. D.C. Drucker & J.J. Gilman), vol. 20, pp. 65–102. New York: Wiley.

McClintock, F.A. (1971). Plasticity aspects of fracture. In *Fracture: An Advanced Treatise* (ed: H. Liebowitz), vol. III, pp. 47–225. New York: Academic Press.

McClintock, F.A. & Walsh, J.B. (1962). Fracture in compression of brittle solids. In *Proceedings of the Fourth U.S. Congress on Applied Mechanics*, vol. 2, pp. 1015–23. New York: American Society of Mechanical Engineers.

McClung, R.C. & Sehitoglu, H. (1989). On the finite element analysis of fatigue crack closure. *Engineering Fracture Mechanics* **33**, 237–72.

McDiarmid, D.L. (1994). A shear stress based critical-plane criterion of multiaxial fatigue failure for design and life prediction. *Fatigue and Fracture of Engineering Materials and Structures* **17**, 1475–84.

McDowell, J.R. (1953). Fretting corrosion tendencies of several combinations of materials. In *Fretting Corrosion*, Special Technical Publication 144, pp. 24–39. Philadelphia: American Society for Testing and Materials.

McEvily, A.J. (1977). Current aspects of fatigue. *Metal Science* **11**, 274–84.

McEvily, A.J. & Machlin, E.S. (1961). The effect of cyclic loading on MgO single crystals. *Transactions of the Metallurgical Society* **221**, 1086–8.

McEvily, A.J. & Wei, R.P. (1972). Fracture mechanics and corrosion fatigue. In *Corrosion-fatigue: Chemistry, Mechanics and Microstructure* (eds. O.F. Devereux, A.J. McEvily & R.W. Staehle), vol. NACE-2, pp. 381–95. Houston: National Association of Corrosion Engineers.

Mackay, M.E., Teng, T.-G. & Schultz, J.M. (1979). Craze roles in the fatigue of polycarbonate. *Journal of Materials Science* **14**, 221–7.

McMeeking, R.M. (1977). Finite deformation analysis of crack tip opening in elastic-plastic materials and implications for fracture initiation. *Journal of the Mechanics and Physics of Solids* **25**, 357–81.

McMeeking, R.M. & Evans, A.G. (1982). Mechanics of transformation toughening in brittle materials. *Journal of the American Ceramic Society* **65**, 242–5.

McMeeking, R.M. & Evans, A.G. (1990). Matrix fatigue cracking in fiber composites. *Mechanics of Materials* **9**, 217–32.

McMeeking, R.M. & Parks, D.M. (1979). On the criteria for *J* dominance of crack tip fields in large-scale yielding. In *Elastic–Plastic Fracture*, Special Technical Publication 668, pp. 175–94. Philadelphia: American Society for Testing and Materials.

McMillan, J.C. & Pelloux, R.M.N. (1967). Fatigue crack propagation under program and random loads. In *Fatigue Crack Propagation*, Special Technical Publication 415, pp. 505–35. Philadelphia: American Society for Testing and Materials.

Maier, H.J. & Christ, H.-J. (1996). Modelling of cyclic stress–strain behavior under thermomechanical fatigue conditions – a new approach based upon a multi-component model. *Scripta Materialia* **34**, 609–15.

Majumdar, B.S. & Burns, S.J. (1981). Crack tip shielding – An elastic theory of dislocations and dislocation arrays near a sharp crack. *Acta Metallurgica* **29**, 579–88.

Majumdar, B.S. & Burns, S.J. (1982). Push–pull fatigue of LiF at elevated temperatures – II. Microstructures. *Acta Metallurgica* **30**, 1751–60.

Majumdar, B.S. & Burns, S.J. (1987). TEM study of the recovery process of cyclically deformed MgO single crystals. *Journal of Materials Science* **22**, 1157–62.

Majumdar, S. & Maiya, P.S. (1980). A mechanistic model for time-dependent fatigue. *Journal of Engineering Materials and Technology* **102**, 159–67.

Majumdar, S. & Morrow, J.D. (1974). Correlation between fatigue crack propagation and low cycle fatigue properties. In *Fracture Toughness and Slow Stable Cracking*, Special Technical Publication 559, pp. 159–82. Philadelphia: American Society for Testing and Materials.

Malvern, L.E. (1969). *Introduction to the Mechanics of a Continuous Medium*. Englewood Cliffs, NJ: Prentice-Hall.

Mandell, J.F., McGarry, F.J., Huang, D.D. & Li, C.G. (1983). Some effects of matrix and interface properties on the fatigue of short fiber-reinforced thermoplastics. *Polymer Composites* **4**, 32–9.

Mann, J.Y. (1958). The historical development of research on the fatigue of materials and structures. *Journal of Australian Institute of Metals* **3**, 222–41.

Mann, J.Y. (1967). *Fatigue of Materials: an introductory text*. Melbourne University Press.

Manson, J.A., Hertzberg, R.W. & Bretz, P.E. (1981). Normalization of fatigue crack propagation behavior in polymers. In *Advances in Fracture Research* (ed. D. Francois), vol. I, pp. 443–8. Oxford: Pergamon Press.

Manson, J.A. & Sperling, L.H. (1976). *Polymer Blends and Composites*. New York: Plenum Press.

Manson, S.S. (1954). Behavior of materials under conditions of thermal stress. *National Advisory Commission on Aeronautics: Report 1170*. Cleveland: Lewis Flight Propulsion Laboratory.

Marissen, R., Trautmann, K.H. & Nowack, H. (1984). The influence of compression loads and of *dK/da* on the crack propagation under variable amplitude loading. *Engineering Fracture Mechanics* **19**, 863–79.

Marshall, D.B. (1986). Strength characteristics of transformation toughened zirconia. *Journal of the American Ceramic Society* **69**, 173–80.

Marshall, D.B. & Cox, B.N. (1987). Tensile fracture of brittle matrix composites: influence of fiber strength. *Acta Metallurgica* **35**, 2607–19.

Marshall, D.B., Cox, B.N. & Evans, A.G. (1985). The mechanics of matrix cracking in brittle-matrix fiber composites. *Acta Metallurgica* **33**, 2013–21.

Martin, J.W. (1980). *Micromechanisms in Particle-Hardened Alloys.* Cambridge University Press.

Masing, G. (1926). Eigenspannungen und verfestigung beim Messing. In *Proceedings of the Second International Conference of Applied Mechanics*, pp. 332–5. Zurich.

Masounave, J. & Baïlon, J.-P. (1976). Effect of grain size on threshold stress intensity factor in fatigue of a ferritic steel. *Scripta Metallurgica* **10**, 165–70.

Mathews, J.R. (1980). Indentation hardness and hot pressing. *Acta Metallurgica* **28**, 311–18.

May, A.N. (1960a). A model of metal fatigue. *Nature* **185**, 303–4.

May, A.N. (1960b). Random slip model of fatigue and Coffin's law. *Nature* **188**, 573–4.

Mecke, K. & Blochwitz, C. (1982). Saturation dislocation structures in cyclically deformed nickel single crystals of different orientations. *Crystal Research Technology* **17**, 743–58.

Mecke, K., Blochwitz, C. & Kremling, U. (1982). The development of the dislocation structures during the fatigue process of F.C.C. single crystals. *Crystal Research and Technology* **17**, 1557–70.

Melan, E. (1938). *Sitzungberichte der Ak. Wissenschaft*, Wien, Series 2A, **147**, 73–80.

Merkle, J.G. & Corten, H.T. (1974). A *J* integral analysis of the compact specimen, considering axial force as well as bending effects. *Journal of Pressure Vessel Technology* **96**, 286–92.

Meyn, D.A. (1968). Observations of micromechanisms of fatigue crack propagation in 2024 aluminum. *Transactions of the American Society for Metals* **61**, 42–51.

Miller, K.J. & de los Rios, E.R. (1986). Editors, *The Behaviour of Short Fatigue Cracks.* London: Mechanical Engineering Publications.

Milligan, W.W. & Antolovich, S.D. (1987). Deformation modeling and constitutive modeling for superalloys. NASA Report No. 4215 (unrestricted release). Cleveland, OH: National Aeronautics and Space Administration, Lewis Research Center.

Milligan, W.W. & Antolovich, S.D. (1991). The mechanisms and temperature dependence of superlattice stacking fault formation in the single-crystal superalloy PWA 1480. *Metallurgical Transactions A* **22A**, 2309–18.

Mills, N.J. & Walker, N. (1980). Fatigue crack initiation in glassy plastics in high strain fatigue tests. *Journal of Materials Science* **15**, 1832–40.

Min, B.K. & Raj, R. (1979). A mechanism for intergranular fracture during high temperature fatigue. In *Fatigue Mechanisms*, Special Technical Publication 675, pp. 569–91. Philadelphia: American Society for Testing and Materials.

Minakawa, K. & McEvily, A.J. (1981). On crack closure in the near-threshold region. *Scripta Metallurgica* **15**, 633–6.

Minakawa, K., Newman, J.C. & McEvily, A.J. (1983). A critical study of the crack closure effect on near-threshold fatigue crack growth. *Fatigue of Engineering Materials and Structures* **6**, 359–65.

Mindlin, R.D. (1949). Compliance of elastic bodies in contact. *Journal of Applied Mechanics* **16**, 259–68.

Mindlin, R.D. & Deresiewicz, H. (1953). Elastic spheres in contact under varying oblique forces. *Journal of Applied Mechanics* **75**, 327–44.

Mindlin, R.D., Mason, W.P., Osmer, J.F. & Deresiewicz, H. (1952). Effects of an oscillating tangential force on the contact surfaces of elastic spheres. In *Proceedings of the First*

US National Congress of Applied Mechanics, pp. 13–9. New York, NY: American Society of Mechanical Engineers.

Miner, M.A. (1945). Cumulative damage in fatigue. *Journal of Applied Mechanics* **12**, 159–64.

Mirra, J.M., Marder, R.A. & Amstuz, H.A. (1982). The pathology of failed total joint arthoplasty. *Clinical Orthopaedics* **170**, 175–83.

Mitchell, M. (1978). Fundamentals of modern fatigue analysis for design. In *Fatigue and Microstructure* (ed. M. Meshii), pp. 385–437. Metals Park: American Society for Metals.

Mitsche, R., Stanzl, S. & Burkert, D.G. (1973). Hoch-frequenzkinemato-graphie in der Metallforschung. *Wissenschaftlicher Film* **14**, 3–10.

Moliné, E., Piques, R. & Pineau, A. (1991). Behaviour of a 1Cr–1Mo–0.25V steel after long-term exposure–I. Charpy impact toughness and creep properties, II. Creep crack initiation and creep crack growth. *Engineering Fracture Mechanics* **14**, 531–63.

Moore, H.F. & Kommers, J.B. (1927). *The Fatigue of Metals*. New York: McGraw-Hill.

Moore, H.F. & Seeley, F.B. (1915). The failure of metals under repeated stress. *Proceedings of the American Society for Testing and Materials* **15**, part 2, 437–66.

Moosbrugger, J.C. & McDowell, D.L. (1989). On a class of kinematic hardening rules for nonproportional cyclic plasticity. *Journal of Engineering Materials and Technology* **111**, 87–98.

Mori, H., Tokuwame, M. & Miyazaki, T. (1979). Cyclic deformation of silicon–iron single crystals oriented for single glide. *Philosophical Magazine A* **40**, 409–22.

Morlet, J.G., Johnson, H.H. & Troiano, A.R. (1958). A new concept of hydrogen embrittlement in steel. *Journal of Iron and Steel Institute* **189**, 37–44.

Morris, W.L. (1979). Microcrack closure phenomena for Al 2219-T851. *Metallurgical Transactions* **10A**, 5–11.

Morris, W.L. (1980). The noncontinuum crack tip deformation behavior of surface microcracks. *Metallurgical Transactions* **11A**, 1117–23.

Morris, W.L. & James, M.N. (1980). Statistical aspects of fatigue crack nucleation from particles. *Metallurgical Transactions* **11A**, 850–1.

Morrow, J.D. (1968). *Fatigue Design Handbook–Advances in Engineering*, vol. 4, Sec. 3.2, pp. 21–29. Warrendale, PA: Society of Automotive Engineers.

Mott, N.F. (1958). A theory of the origin of fatigue cracks. *Acta Metallurgica* **6**, 195–7.

Mróz, Z. (1967). On the description of anisotropic work hardening. *Journal of the Mechanics and Physics of Solids* **15**, 163–75.

Mróz, Z. (1969). An attempt to describe the behavior of metals under cyclic loads using a more general workhardening model. *Acta Mechanica* **7**, 199–212.

Mudry, F. & Pineau, A. (1983). Rapport d'expertise corps MPT centrale de Violaines. Evry Cedex, France: Ecole Nationale Superieure des Mines de Paris.

Mughrabi, H. (1978). The cyclic hardening and saturation behaviour of copper single crystals. *Materials Science and Engineering* **33**, 207–23.

Mughrabi, H. (1980). Microscopic mechanisms of metal fatigue. In *The Strength of Metals and Alloys* (eds. P. Haasen, V. Gerold & G. Kostorz), vol. 3, pp. 1615–39. Oxford: Pergamon Press.

Mughrabi, H. (1981). Cyclic plasticity of matrix and persistent slip bands in fatigued metals. In *Continuum Models of Discrete Systems 4* (eds. O. Brulin & R.K.T. Hsieh), pp. 241–57. London: North-Holland.

Mughrabi, H. (1983). Dislocation wall and cell structures and long-range internal stresses in deformed metal crystals. *Acta Metallurgica* **31**, 1367–79.

Mughrabi, H., Ackermann, F. & Herz, K. (1979). Persistent slip bands in fatigued face-centered and body-centered cubic metals. In *Fatigue Mechanisms*, Special Technical Publication 675, pp. 69–105. Philadelphia: American Society for Testing and Materials.

Mughrabi, H., Herz, K. & Stark. X. (1976). The effect of strain-rate on the cyclic deformation properties of α-iron single crystals. *Acta Metallurgica* **24**, 659–68.

Mughrabi, H., Herz, K. & Stark, X. (1981). Cyclic deformation and fatigue behaviour of α-iron mono- and polycrystals. *International Journal of Fracture* **17**, 193–220.

Mughrabi, H. & Wang, R. (1981). Cyclic deformation of face-centered cubic polycrystals: a comparison with observations of single crystals. In *Deformation of Polycrystals: Mechanisms and Microstructures* (eds: N. Hansen, A. Horsewell, T. Leffers and H. Lilholt), Proceedings of the Second Risoe International Symposium, pp. 87–98. Roskilde, Denmark: Risoe National Laboratory.

Mughrabi, H., Wang, R., Differt, K. & Essmann, U. (1983). Fatigue crack initiation by cyclic slip irreversibilities in high-cycle fatigue. In *Quantitative Measurement of Physical Damage* (eds. J. Lankford, D.L. Davidson, W.L. Morris and R.P. Wei), Special Technical Publication 811, pp. 5–45. Philadelphia, PA: American Society for Testing and Materials.

Mughrabi, H. & Wüthrich, Ch. (1976). Asymmetry of slip and shape changes during cyclic deformation of α-iron single crystals. *Philosophical Magazine A* **33**, 963–84.

Murakami, Y, & Endo, M. (1986). Effects of hardness and crack geometries on ΔK_{th} of small cracks emanating from small defects. In *The Behaviour of Short Fatigue Cracks*, Publication 1 of the European Group on Fracture, pp. 275–93. London: Mechanical Engineering Publications.

Murakami, Y., Mura, T. & Kobayashi, M. (1988). Change of dislocation structures and macroscopic conditions from initial state to fatigue crack nucleation. In *Basic Questions in Fatigue*, Special Technical Publication 924, vol. I, pp. 39–63. Philadelphia: American Society for Testing and Materials.

Murakami, Y., Sakae, C. & Ichimaru, K. (1994). Three-dimensional fracture mechanics analysis of pit formation mechanism under lubricated rolling–sliding contact loading. *Tribology Transactions* **37**, 445–54.

Nabarro, F.R.N. (1952). Mathematical theory of stationary dislocations. *Advances in Physics* **1**, 269–395.

Nayeb-Hashemi, H., McClintock, F.A. & Ritchie, R.O. (1983a). Micromechanical modelling of mode III fatigue crack growth in rotor steels. *International Journal of Fracture* **23**, 163–85.

Nayeb-Hashemi, H., McClintock, F.A. & Ritchie, R.O. (1983b). Influence of overloads and block loading on mode III fatigue crack propagation in A469 rotor steel. *Engineering Fracture Mechanics* **18**, 736–83.

Nemat-Nasser, S. & Horii, H. (1982). Compression-induced nonplanar crack extension with application to splitting, exfoliation and bursting. *Journal of Geophysical Research* **87**, 6805–21.

Neuber, H. (1946). *Theory of Notch Stresses: Principle for Exact Stress Calculations*. Ann Arbor, MI: Edwards.

Neuber, H. (1961). Theory of stress concentration for shear-strained prismatical bodies with arbitrary nonlinear stress-strain law. *Journal of Applied Mechanics* **28**, 544–50.

Neumann, P. (1968). Strain bursts and coarse slip during cyclic deformation. *Zeitschrift für Metallkunde* **59**, 927–34.

Neumann, P. (1969). Coarse slip model of fatigue. *Acta Metallurgica* **17**, 1219–25.

Neumann, P. (1974). Modelling of changes in dislocation structure in cyclically deformed crystals. In *Constitutive Equations in Plasticity* (ed. A.S. Argon), pp. 251–326. Cambridge, Mass.: MIT Press.

Neumann, P. (1983). Fatigue. In *Physical Metallurgy* (eds. R.W. Cahn & P. Haasen), pp. 1554–93. Amsterdam: Elsevier Science.

Neumann, P. (1986). Low energy dislocation configurations: a possible key to the understanding of fatigue. *Materials Science and Engineering* **81**, 465–75.

Neumann, P. & Tönnessen, A. (1988). Crack initiation at grain boundaries in FCC materials. In *Strength of Metals and Alloys* (eds. P.O. Kettunen, T.K. Lepistö & M.E. Lehtonen), vol. 1, pp. 743–8. Oxford: Pergamon Press.

Neumann, R. (1975). Probenformänderungen von Niob-einkristallen bei symmetrischer Wechselbeanspruchung. *Zeitschrift für Metallkunde* **66**, 26–32.

Newman, J.C. (1971). An improved method of collocation for the stress analysis of cracked plates with various shaped boundaries. National Aeronautics and Space Administration Technical Note D-6376, Langley, Virginia.

Newman, J.C. (1976). A finite element analysis of fatigue crack closure. *Mechanics of Fatigue Crack Growth*, Special Technical Publication 590, pp. 281–301. Philadelphia: American Society for Testing and Materials.

Newman, J.C. (1983). A nonlinear fracture mechanics approach to the growth of small cracks. In *Behavior of Short Cracks in Airframe Components*, Proceedings 328, pp. 6.1–20. Neuilly sur Seine (France): Advisory Group for Aerospace Research and Development.

Newman, J.C. & Elber, W. (1988). Editors, *Mechanics of Fatigue Crack Closure*, Special Technical Publication 982. Philadelphia: American Society for Testing and Materials.

Nicholas, T., Heil, M.L. & Haritos, G.K. (1989). Predicting crack growth under thermo-mechanical loading. *International Journal of Fracture* **41**, 157–76.

Nicholas, T., Laflen, J.H., and VanStone, R.H. (1986). A damage tolerant design approach to turbine engine life prediction. *Proceedings: Conference on Life Predidction for High-Temperature Gas Turbine Materials* (eds.V. Weiss and W. T. Bakker), pp. 4-1–4-61. EPRI AP-4477, Syracuse, NY: Syracuse University,

Nine, H.D. (1973). Asymmetric deformation in fatigue of body-centered cubic single crystals. *Journal of Applied Physics* **44**, 4875–81.

Nisitani, H. & Takao, K. (1978). Fatigue crack acceleration and closure in rotating bending tests of 0.54% carbon steel. *Engineering Fracture Mechanics* **10**, 855–66.

Nowack, H., Trautmann, K.H., Schulte, K. & Lütjering, G. (1979). Sequence effects on fatigue crack propagation; mechanical and microstructural contributions. In *Fracture Mechanics*, Special Technical Publication 677, pp. 36–53. Philadelphia: American Society for Testing and Materials.

O'Dowd, N.P. & Shih, C.F. (1991). Family of crack-tip fields characterized by a triaxiality parameter: Part I – Structure of fields. *Journal of the Mechanics and Physics of Solids* **39**, 989–1015.

O'Dowd, N.P. & Shih, C.F. (1992). Family of crack-tip fields characterized by a triaxiality parameter: Part II – Fracture applications. *Journal of the Mechanics and Physics of Solids* **40**, 939–63.

Ogawa, T. (1992). Tensile fatigue crack growth of polycrystalline magnesia. In *Fracture Mechanics of Ceramics* (eds. R.C. Bradt, D.P.H. Hasselman, D. Munz, M. Sakai and V. Ya. Shevchenko), **9**, pp. 455–64. New York: Plenum Press.

Ogura, T., Fukushima, K. & Masumoto, T. (1975). Propagation of fatigue cracks in amorphous metals. *Scripta Metallurgica* **9**, 979–83.

Ohji, K., Ogura, K. & Kubo, S. (1980). Stress field and modified *J* integral near a crack tip under condition of confined creep deformation. *Journal of the Society of Materials Science Japan* **29**, 465–71.

Ohji, K., Ogura, K. & Yoshiji, O. (1975). Cyclic analysis of a propagating crack and its correlation with fatigue crack growth. *Engineering Fracture Mechanics* **7**, 457–64.

Ohji, T., Yamauchi, Y., Kanematsu, W. & Ito, S. (1990). Dependence of high-temperature tensile strength on displacement rate for hot-pressed silicon nitride. *Journal of Materials Science* **25**, 2990–6.

Ohuchida, H., Usami, S. & Nishioka, A. (1975). Fatigue limit of steel with cracks. *Bulletin of the Japan Society of Mechanical Engineers* **18**, 1185–93.

Okazaki, M. & Koizumi, T. (1983). Effect of strain wave shape on thermal-mechanical fatigue crack propagation in a cast low-alloy steel. *Journal of Engineering Materials and Technology* **105**, 81–7.

Olsson, M. (1998). Contact fatigue and tensile stresses. In *Engineering Against Fatigue* (eds: R.A. Smith and T.C. Lindley), Proceedings of an International Conference, Sheffield, March, 1997. In press.

Olsson, M., Giannakopoulos, A.E. & Suresh, S. (1995). Elastoplastic analysis of thermal cycling: ceramic particles in a metallic matrix. *Journal of the Mechanics and Physics of Solids* **43**, 1639–71.

Oriani, R.A. & Josephic, P.H. (1974). Equilibrium aspects of hydrogen-induced cracking in steels. *Acta Metallurgica* **22**, 1065–77.

Orowan, E. (1952). Fundamentals of brittle behavior of metals. In *Fatigue and Fracture of Metals* (ed. W. M. Murray), pp. 139–67. New York: Wiley.

Orowan, E. (1959). Causes and effects of internal stresses. In *Internal Stresses and Fatigue in Metals* (eds. G.M. Rassweiler and W.L. Grube), pp. 59–80. New York: Elsevier.

Ortiz, M. & Giannakopoulos, A.E. (1990). Mixed-mode crack tip fields in monolithic ceramics. *International Journal of Solids and Structures* **26**, 705–23.

Ortiz, M. & Popov, E.P. (1982). A statistical theory of polycrystalline plasticity. *Proceedings of the Royal Society, London* **A379**, 439–58.

Ortiz, M. & Suresh, S. (1993). Statistical properties of residual stresses and microfracture in ceramic materials. *Journal of Applied Mechanics* **60**, 244–9.

Osgood, C.C. (1982). *Fatigue Design*, 2nd edn. New York: Pergamon Press.

Palaniswamy, K. & Knauss, W.G. (1978). On the problem of crack extension in brittle solids under general loading. In *Mechanics Today* (ed. S. Nemat-Nasser), pp. 87–148. New York: Pergamon Press.

Palmgren, A. (1924). Die Lebensdauer von Kugellagern. *Zeitschrift des Vereins Deutscher Ingenieure* **68**, 339–41.

Papadopoulos, I.V., Davoli, P., Gorla, C., Filippini, M. & Bernasconi, A. (1997). A comparative study of multiaxial high-cycle fatigue criteria for metals. *International Journal of Fatigue* **19**, 219–35.

Paris, P.C. (1960). *The Growth of Cracks Due to Variations in Loads*. Ph.D. Thesis. Bethlehem: Lehigh University.

Paris, P.C., Bucci, R.J., Wessel, E.T., Clark, W.G. & Mager, T.R. (1972). Extensive study of low fatigue crack growth rates in A533 and A508 steels. In *Stress Analysis and Growth of Cracks*, Special Technical Publication 513, pp. 141–76. Philadelphia: American Society for Testing and Materials.

Paris, P.C. & Erdogan, F. (1963). A critical analysis of crack propagation laws. *Journal of Basic Engineering* **85**, 528–34.

Paris, P.C., Gomez, M.P. & Anderson, W.P. (1961). A rational analytic theory of fatigue. *The Trend in Engineering* **13**, 9–14.

Parks, D.M. (1992). Advances in characterization of elastic–plastic crack-tip fields. In *Topics in Fracture and Fatigue* (ed. A.S. Argon), pp. 59–98. Berlin: Springer–Verlag.

Partridge, P.G. (1969). Effect of cyclic stresses on the microstructures of hexagonal close packed metals. *Czech Journal of Physics* **B19**, 323–32.

Pearson, S. (1975). Initiation of fatigue cracks in commercial aluminum alloys and the subsequent propagation of very short cracks. *Engineering Fracture Mechanics* **7**, 235–47.

Pelloux, R.M.N. (1969). Mechanisms of formation of ductile fatigue striations. *Transactions of the American Society for Metals* **62**, 281–5.

Pelloux, R.M.N. (1970). Crack extension by alternating shear. *Engineering Fracture Mechanics* **1**, 697–704.

Peralta, P., Llanes, L., Czapka, A. & Laird, C. (1995). The effect of texture and grain size as independent factors in the cyclic behavior of polycrystalline copper. *Scripta Metallurgica et Materialia* **32**, 1877–81.

Petch, N.J. (1953). The cleavage strength of polycrystals. *Journal of the Iron and Steel Institute* **174**, 25–6.

Petch, N.J. (1956). Lowering of fracture stress due to surface adsorption. *Philosophical Magazine* **1**, 331–7.

Peterson, R.E. (1959). Notch sensitivity. In *Metal Fatigue* (eds. G. Sines and J.L. Waisman), pp. 293–306. New York: McGraw-Hill.

Petit, J., Davidson, D.L., Suresh, S. & Rabbe, P. (1988). Editors, *Fatigue Crack Growth under Variable Amplitude Loading*. London: Elsevier Applied Science.

Petroski, H. (1996). *Invention by Design: How Engineers Get from Thought to Thing*. Cambridge, MA: Harvard University Press.

Pineau, A. (1983). High temperature fatigue behaviour of engineering materials in relation to microstructure. In *Fatigue at High Temperature* (ed. R.P. Skelton), pp. 305–64. London: Elsevier Applied Science.

Pineau, A. (1986). Short fatigue crack behavior in relation to three-dimensional aspects and crack closure effect. In *Small Fatigue Cracks* (eds. R.O. Ritchie & J. Lankford), pp. 191–211. Warrendale: The Metallurgical Society of the American Institute of Mining, Metallurgical and Petroleum Engineers.

Pineau, A.G. & Pelloux, R.M.N. (1974). Influence of strain-induced martensitic transformations on fatigue crack growth rates in stainless steels. *Metallurgical Transactions* **5A**, 1103–12.

Pippan, R. (1987). The growth of short crack under cyclic compression. *Fatigue and Fracture of Engineering Materials and Structures* **9**, 319–28.

Plumtree, A. & Pawlus, L.D. (1988). Substructural developments during strain cycling of wavy slip mode metals. In *Basic Questions in Fatigue*, Special Technical Publication 924, vol. I, pp. 81–97. Philadelphia: American Society for Testing and Materials.

Pohl, K., Mayr, P. & Macherauch, E. (1980). Persistent slip bands in the interior of a fatigued low carbon steel. *Scripta Metallurgica* **14**, 1167–9.

Polák, J. (1970). The effect of intermediate annealing on the electrical resistivity and shear stress of fatigued copper. *Scripta Metallurgica* **4**, 761–4.

Polák, J., Klesnil, M. & Lukáš, P. (1974). High cycle plastic stress strain response of metals. *Materials Science and Engineering* **15**, 231–7.

Poncelet, J.V. (1839). *Introduction à la Mécanique, Industrielle, Physique ou Expérimentale.*, pp. 317–8. Deuxième édition, Imprimerie de Gauthier-Villars, Paris.

Ponter, A.R.S., Hearle, A.D. & Johnson, K.L. (1985). Application of the kinematical shakedown theorem to rolling and sliding point contacts. *Journal of the Mechanics and Physics of Solids* **33**, 339–62.

Pook, L.P. (1985). The fatigue crack direction and threshold behaviour of mild steel under mixed mode I and III loading. *International Journal of Fatigue* **7**, 21–30.

Pops, H. (1970). Stress-induced psuedo-elasticity in ternary Cu-Zn based beta prime phase alloys. *Metallurgical Transactions* **1A**, 251–8.

Porter, J. & Levy, J.C. (1960). The fatigue curves of copper. *Journal of the Institute of Metals* **89**, 86–9.

Prager, W. (1956). A new method of analyzing stresses and strains in work-hardening of plastic solids. *Journal of Applied Mechanics* **23**, 493–6.

Pratt, J.E. (1967). Dislocation sub-structure in strain-cycled copper as influenced by temperature. *Acta Metallurgica* **15**, 319–27.

Pruitt, L., Koo, J., Rimnac, C.M., Suresh, S. & Wright, T.M. (1995). Cyclic compressive loading results in fatigue cracks in ultra-high molecular weight polyethylene. *Journal of Bone Joint Surgery* **13**, 143–6.

Pruitt, L. & Suresh, S. (1993). Cyclic stress fields for fatigue cracks in amorphous solids–Experimental measurements and their implications. *Philosophical Magazine A* **67**, 1219–45.

Qian, J. & Fatemi, A. (1996). Mixed-mode fatigue crack growth: a literature survey. *Engineering Fracture Mechanics* **55**, 969–90.

Rabinowitz, S. & Beardmore, P. (1974). Cyclic deformation and fracture of polymers. *Journal of Materials Science* **9**, 81–99.

Ramamurty, U., Hansson, T. & Suresh, S. (1994). mechanisms of high-temperature crack growth in monolithic and SiC-reinforced Si3N4 under static and cyclic loads. *Journal of the American Ceramic Society* **77**, 2985–99.

Ramamurty, U., Suresh, S., & Petrovic, J.J. (1994). The effect of carbon addition on elevated temperature crack growth resistance in (Mo,W)Si2-SiCp composite. *Journal of the American Ceramic Society* **77**, 2681–88.

Rankine, W.J.M. (1843). On the causes of unexpected breakage of the journals of railway axles and the means of preventing such accidents by observing the law of continuity in their construction. *Proceedings of the Institute of Civil Engineers, London* **2**, 105–8.

Reece, M.J., Guiu, F. & Sammur, M.F.R. (1989). Cyclic fatigue crack propagation in alumina under direct tension–compression loading. *Journal of the American Ceramic Society* **72**, 348–52.

Reid, C.N., Williams, K. & Hermann, R. (1979). Fatigue in compression. *Fatigue of Engineering Materials and Structures* **1**, 267–70.

Rémy, L., Pineau, A. & Thomas, B. (1978). Temperature dependence of stacking fault energy in close-packed metals and alloys. *Materials Science and Engineering* **36**, 47–63.

Repetto, E.A. & Ortiz, M. (1997). A micromechanical model of cyclic deformation and fatigue crack nucleation in f.c.c. single crystals. *Acta Materialia* **45**, 2577–95.

Rice, J.R. (1967). Mechanics of crack tip deformation and extension by fatigue. In *Fatigue Crack Propagation*, Special Technical Publication 415, pp. 247–309. Philadelphia: American Society for Testing and Materials.

Rice, J.R. (1968). A path independent integral and the approximate analysis of strain concentrations by notches and cracks. *Journal of Applied Mechanics* **35**, 379–86.

Rice, J.R. (1971). Inelastic constitutive relations for solids: an internal-variable theory and its applications to metal plasticity. *Journal of the Mechanics and Physics of Solids* **19**, 433–57.

Rice, J.R. (1974). Limitations to the small scale yielding approximation for crack tip plasticity. *Journal of the Mechanics and Physics of Solids* **22**, 17–26.

Rice, J.R. (1985). First order variations in elastic fields due to variation in location of a planar crack front. *Journal of Applied Mechanics* **52**, 571–9.

Rice, J.R., Drugan, W.J. & Sham, T.-L. (1980). Elastic–plastic analysis of growing cracks. In *Fracture Mechanics: Twelfth Conference*, Special Technical Publication 700, pp. 189–221. Philadelphia: American Society for Testing and Materials.

Rice, J.R. & Rosengren, G.F. (1968). Plane strain deformation near a crack tip in a power law hardening material. *Journal of the Mechanics and Physics of Solids* **16**, 1–12.

Rich, D.L., Pinckert, R.E. & Christian, T.F. (1986). Fatigue and fracture mechanics analysis of compression loaded aircraft structure. In *Case Histories Involving Fatigue and Fracture Mechanics*, Special Technical Publication 918, pp. 243–58. Philadelphia: American Society for Testing and Materials.

Richards, C.E. & Lindley, T.C. (1972). The influence of stress intensity and microstructure on fatigue crack propagation in ferritic materials. *Engineering Fracture Mechanics* **4**, 951–78.

Ricker, R.E. & Duquette, D.J. (1988). The role of hydrogen in corrosion fatigue of high purity Al–Zn–Mg exposed to water vapor. *Metallurgical Transactions* **19A**, 1775–83.

Riddell, M.N., Koo, G.P. & O'Toole, J.L. (1967). Fatigue mechanisms in thermoplastics. *Polymer Engineering and Science* **6**, 363–8.

Riedel, H. (1983). Crack tip stress fields and crack growth under creep-fatigue conditions. In *Elastic Plastic Fracture*, Special Technical Publication 803, vol. I, pp. 505–20. Philadelphia: American Society for Testing and Materials.

Riedel, H. (1987). *Fracture at High Temperatures*. Berlin: Springer-Verlag.

Riedel, H. & Rice, J.R. (1980). Tensile cracks in creeping solids. In *Fracture Mechanics: Twelfth Conference*, Special Technical Publication 700, pp. 112–30. Philadelphia: American Society for Testing and Materials.

Rimnac, C.M., Hertzberg, R.W. & Manson, J.A. (1983). In *Fatigue in Polymers*, pp. 9.1–9.10. London: Plastics and Rubber Institute.

Rimnac, C.M., Wright, T.M., Bartel, D.L. & Burstein, A.H. (1986). Failure analysis of a total hip femoral component: A fracture mechanics approach. In *Case Histories Involving Fatigue and Fracture Mechanics*, Special Technical Publication 918, pp. 377–88. Philadelphia: American Society for Testing and Materials.

Ritchie, R.O. (1979). Near-threshold fatigue-crack propagation in steels. *International Metals Reviews* **20**, 205–30.

Ritchie, R.O. (1996). Fatigue and fracture of pyrolytic carbon: a damage-tolerant approach to structural integrity and life prediction in "ceramic*Æ* heart valve prosthesis. *Journal of Heart Valve Disease* **5**, Supplement 1, S-9–S-31.

Ritchie, R.O. & Knott, J.F. (1973). Mechanisms of fatigue crack growth in a low alloy steel. *Acta Metallurgica* **21**, 639–50.

Ritchie, R.O. & Lankford, J. (1986). Editors, *Small Fatigue Cracks*. Warrendale: The Metallurgical Society of the American Institute of Mining, Metallurgical and Petroleum Engineers.

Ritchie, R.O. & Suresh, S. (1981). Some considerations on fatigue crack closure at near-threshold stress intensities due to fracture surface morphology. *Metallurgical Transactions* **13A**, 937–40.

Ritchie, R.O., Suresh, S. & Moss, C.M. (1980). Near-threshold fatigue crack growth in $2\frac{1}{4}$ Cr–1 Mo pressure vessel steel in air and hydrogen. *Journal of Engineering Materials and Technology* **102**, 293–9.

Rivlin, R.S. & Thomas, A.G. (1953). Rupture of rubber. I. Characteristic energy for tearing. *Journal of Polymer Science* **10**, 291–318.

Roebben, G., Steen, M., Bressers, J. & Van der Biest, O. (1996). Mechanical fatigue in monolithic non-transforming ceramics. *Progress in Materials Science* **40**, 265–331.

Rooke, D.R. & Cartwright, D.J. (1976). *Compendium of Stress Intensity Factors*. Uxbridge: Hillingdon Press.

Roven, H.J., Langoy, M.A. & Nes, E. (1987). Striations and the fatigue crack growth mechanism in a micro-alloyed steel. In *Fatigue '87* (eds. R.O. Ritchie & E.A. Starke), vol. I, pp. 175–84. Warley: Engineering Materials Advisory Services.

Rühle, M., Clausen, N. & Heuer, A. (1986). Transformation and microcrack toughening as complementary processes in ZrO_2-toughened Al_2O_3. *Journal of the American Ceramic Society* **69**, 195–7.

Ruiz, C., Boddington, P.H.B. & Chen, K.C. (1984). An investigation of fatigue and fretting in a dovetail joint. *Experimental mechanics* **24**, 208–17.

Ruppen, J., Bhowal, P., Eylon, D. & McEvily, A.J. (1979). On the process of subsurface fatigue crack initiation in Ti–6Al–4V. In *Fatigue Mechanisms*, Special Technical Publication 675, pp. 47–68. Philadelphia: American Society for Testing and Materials.

Ryder, D.A. & Lynch, S.P. (1977). The effect of environment and frequency on crack nucleation, stage I and stage II crack growth in two aluminium zinc magnesium alloys. In *The Influence of Environments on Fatigue*, pp. 21–6. London: The Institute of Mechanical Engineers.

Ryder, D.A., Martin, M. & Abdullah, M. (1977). Some factors influencing stage I fatigue crack growth. *Journal of Metal Science* **11**, 340–4.

Saal, H. (1971). Fatigue crack growth in notched parts with a compressive mean load. *Journal of Basic Engineering* **93**, 1–6.

Sackfield, A. & Hills, D.A. (1983). A note on the Hertz contact problem: a correlation of standard formulae. *Journal of Strain Analysis* **18**, 195–7.

Sadananda, K. & Shahinian, P. (1977). Prediction of threshold stress intensity for fatigue crack growth using a dislocation model. *International Journal of Fracture* **13**, 585–94.

Sadananda, K. & Shahinian, P. (1979). A fracture mechanics approach to high temperature fatigue crack growth in Udimet 700. *Engineering Fracture Mechanics* **11**, 73–86.

Sadananda, K. & Shahinian, P. (1981). Creep-fatigue crack growth. In *Cavities and Cracks in Creep and Fatigue* (ed. J. Gittus), pp. 109–95. London: Elsevier Applied Science.

Saga, J., Hayashi, M. & Nishio, Y. (1977). Effect of grain size on fatigue damage in pure aluminum. *Journal of the Society of Materials Science Japan* **26**, 289–95.

Sanders, J.L. (1960). On the Griffith–Irwin fracture theory. *Journal of Applied Mechanics* **27**, 352–3.

Sandor, B.I. (1972). *Fundamentals of Cyclic Stress and Strain*. Madison: University of Wisconsin Press.

Sargent, C.M. & Purdy, G.R. (1974). Dissolution of small precipitates. *Scripta Metallurgica* **8**, 569–72.

Sauer, J.A. (1978). Static and dynamic properties of monodisperse polystyrenes: Influence of molecular weight. *Polymer* **19**, 859–60.

Sauer, J.A., Foden, E. & Morrow, D.R. (1977). The influence of molecular weight on fatigue behavior of polyethylene and polystyrene. *Polymer Engineering and Science* **17**, 246–50.

Saxena, A. (1988). A model for predicting the effect of frequency on fatigue crack growth at elevated temperature. *Fatigue of Engineering Materials and Structures* **3**, 247–55.

Saxena, A., Hudak, S.J., Donald, J.K. & Schmidt, D.W. (1978). Computer-controlled decreasing stress intensity technique for low rate fatigue crack growth testing. *Journal of Testing and Evaluation* **6**, 167–74.

Saxena, A., Wilson, W.K., Roth, L.D. & Liaw, P.K. (1985). The behavior of small fatigue cracks at notches in corrosive environments. *International Journal of Fracture* **28**, 69–82.

Schijve, J. (1973). Effect of load sequences on crack propagation under random and program loading. *Engineering Fracture Mechanics* **5**, 269–80.

Schijve, J. (1974). Fatigue damage accumulation and incompatible crack front orientation. *Engineering Fracture Mechanics* **6**, 245–52.

Schijve, J. (1980). Prediction methods for fatigue crack growth in aircraft material. In *Fracture Mechanics: Twelfth Conference*, Special Technical Publication 700, pp. 3–34. Philadelphia: American Society for Testing and Materials.

Schijve, J. (1982). Differences between the growth of small and large fatigue cracks in relation to threshold *K* values. In *Fatigue Thresholds* (eds. J. Backlund, A.F. Blom & C.J. Beevers), vol. 2, pp. 881–908. Warley: Engineering Materials Advisory Services.

Schmidt, R.A. & Paris, P.C. (1973). Threshold for fatigue crack propagation and the effects of load ratio and frequency. In *Progress in Flaw Growth and Fracture Testing*, Special Technical Publication 536, pp. 79–94. Philadelphia: American Society for Testing and Materials.

Schütz, W. (1996). A history of fatigue. *Engineering Fracture Mechanics* **54**, 263–300.

Sensmeier, M.D. & Wright, P.K. (1990). The effect of fiber bridging on fatigue crack growth in titanium matrix composites. In *Fundamental Relationships between Microstructures and Mechanical Properties of Metal Matrix Composites* (eds. P.K. Liaw & M.N. Gungor), pp. 441–57. Warrendale: The Minerals, Metals and Materials Society.

Shang, J.K. & Ritchie, R.O. (1989). On the particle-size dependence of fatigue-crack propagation thresholds in SiC-particulate-reinforced aluminum–alloy composites: role of crack closure and crack trapping. *Acta Metallurgica* **37**, 2267–78.

Sharma, S.M. & Aravas, N. (1991). Determination of higher-order terms in asymptotic elastoplastic crack tip solutions. *Journal of the Mechanics and Physics of Solids* **39**, 1043–72.

Sheldon, G.P., Cook, T.S., Jones, T.W. & Lankford, J. (1981). Some observations on small fatigue cracks in a superalloy. *Fatigue of Engineering Materials and Structures* **3**, 219–28.

Shen, H., Podlaseck, S.E. & Kramer, I.R. (1966). Effect of vacuum on the fatigue life of aluminum. *Acta Metallurgica* **14**, 341–6.

Shield, R.T. (1955). On plastic flow of metals under conditions of axial symmetry. *Proceedings of the Royal Society, London* **A223**, 267–87.

Shih, C.F. (1973). Elastic–plastic analysis of combined mode crack problems, Ph.D. Thesis. Cambridge, Mass.: Harvard University.

Shih, C.F. (1974). Small-scale yielding analysis of mixed-mode plane strain crack problems. In *Fracture Analysis*, Special Technical Publication 560, pp. 187–210. Philadelphia: American Society for Testing and Materials.

Shih, C.F. (1981). Relationships between the *J*-integral and the crack opening displacement for stationary and extending cracks. *Journal of the Mechanics and Physics of Solids* **29**, 305–26.

Shute, N. (1948). *No Highway*. London: Heinemann.

Sih, G.C. (1973). *Handbook of Stress Intensity Factors*. Bethlehem: Lehigh University.

Sih, G.C. (1974). Strain-energy-density factor applied to mixed-mode crack problems. *International Journal of Fracture* **10**, 305–20.

Sines, G. (1959). Behavior of metals under complex static and alternating stresses. In *Metal Fatigue* (eds. G. Sines and J.L. Waisman), pp. 145–69. New York: McGraw-Hill.

Skelton, R.P. (1983). Editor, *Fatigue at High Temperature*. London: Elsevier Applied Science.

Skelton, R.P. & Haigh, J.R. (1978). Fatigue crack growth rates and thresholds in steels under oxidising conditions. *Materials Science and Engineering* **36**, 17–25.

Skibo, M.D., Hertzberg, R.W. & Manson, J.A. (1976). Fatigue fracture processes in polystyrene. *Journal of Materials Science* **11**, 479–90.

Skibo, M.D., Hertzberg, R.W., Manson, J.A. & Kim, S. (1977). On the generality of discontinuous fatigue crack growth in glassy polymers. *Journal of Materials Science* **12**, 531–42.

Smith, J.H. (1910). Some experiments on the fatigue of metals. *Journal of Iron and Steel Institute* **91**, 365–97.

Smith, J.H. & Wedgwood, G.A. (1915). Stress–strain loops for steel in the cyclic state. *Journal of Iron and Steel Institute* **82**, 246–318.

Smith, M.C. & Smith, R.A. (1988). Toward an understanding of mode II fatigue crack growth. In *Basic Questions in Fatigue: Volume I*, Special Technical Publication 924, pp. 260–80. Philadelphia: American Society for Testing and Materials.

Smith, R.A. (1990). The Versailles railway accident of 1842 and the first research into metal fatigue. In *Fatigue 90* (eds. H. Kitagawa & T. Tanaka), vol. IV, pp. 2033–41. Birmingham: Materials and Component Engineering Publications.

Smith, R.A. & Miller, K.J. (1978). Prediction of fatigue regimes in notched components. *Journal of Mechanical Engineering Science* **20**, 201–6.

Smith, R.N., Watson, P. & Topper, T.H. (1970). A sress–strain parameter for the fatigue of metals. *Journal of Materials* **5**, 767–78.

Sobczyk, K. & Spencer, B.F. (1992). *Random fatigue: from data to theory*. San Diego, CA: Academic Press.

Sobeyejo, W.O., Lederich, R.J. & Sastry, S.M.L. (1995). Mechanical behavior of damage tolerant TiB whisker-reinforced *in-situ* titanium matrix composites. *Acta Metallurgica et Materialia* **42**, 2579–91.

Socie, D.F. (1977). Prediction of fatigue crack growth in notched members under variable amplitude loading histories. *Engineering Fracture Mechanics* **9**, 849–65.

Socie, D. (1993). Critical plane approaches for multiaxial fatigue damage assessment. In *Advances in Multiaxial Fatigue* (eds: D.L. McDowell and R. Ellis), Special Technical Publication 1191, pp. 7–36. Philadelphia, PA: American Society for Testing and Materials.

Socie, D. & Shield, T.W. (1984). Mean stress effects in biaxial fatigue of Inconel 718. *Journal of Engineering Materials and Technology*, **106**, 227–32.

Soderberg, C.R. (1939). Factor of safety and working stress. *Transactions of the American Society of Mechanical Engineers* **52**, 13–28.

Sproles, F.S. & Duquette, D.J. (1978). The mechanism of material removal in fretting. *Wear* **49**, 339–52.

Starke, E.A. & Williams, J.C. (1989). Microstructure and the fracture mechanics of fatigue crack propagation. In *Fracture Mechanics: Perspectives and Directions*, Special Technical Publication 1020, pp. 184–205. Philadelphia: American Society for Testing and Materials.

Stevenson, R. & Vander Sande, J.B. (1974). The cyclic deformation of magnesium single crystals. *Acta Metallurgica* **22**, 1079–86.

Stewart, A.T. (1980). The influence of environment and stress ratio on fatigue crack growth at near-threshold stress intensities in low alloy steels. *Engineering Fracture Mechanics* **13**, 463–78.

Stoloff, N.S. & Johnston, T.L. (1963). Crack propagation in liquid metal environment. *Acta Metallurgica* **11**, 251–6.

Stoltz, R.E. & Pelloux, R.M.N. (1976). The Bauschinger effect in preci-pitation strengthened aluminum alloys. *Metallurgical Transactions* **7A**, 1295–306.

Stoney, G.G. (1909). The tension of metallic films deposited by electrolysis. *Proceedings of the Royal Society, London* **A82**, 172–5.

Stulen, F.B. & Cummings, H.N. (1954). A failure criterion for multiaxial fatigue stresses. Special Technical Publication 54, pp. 822–35. Philadelphia: American Society for Testing and Materials.

Subramanium, K.N. & Washburn, J. (1963). Fatigue deformation of magnesium oxide. *Journal of Applied Physics* **34**, 3394–7.

Sugimura, Y., Grondin, L.& Suresh, S. (1995). Fatigue crack growth at arbitrary angles to bimaterial interfaces. *Scripta Metallurgica et Materialia* **33**, 2007–12.

Sugimura, Y., Lim, P.G., Shih, C.F. & Suresh, S. (1995). Fracture normal to a bimaterial interface: effects of plasticity on crack-tip shielding and amplification. *Acta Metallurgica et Materialia* **43**, 1157–69.

Sugimura, Y. & Suresh, S. (1992). Effects of SiC content on fatigue crack growth in aluminum alloys reinforced with SiC particles. *Metallurgical Transactions A* **23A**, 2231–42.

Sunder, R., Seetharam, S.A. & Bhaskaran, T.A. (1984). Cycle counting for fatigue crack growth analysis. *International Journal of Fatigue* **6**, 147–56.

Suresh, S. (1983a). Crack deflection: implications for the growth of long and short fatigue cracks. *Metallurgical Transactions* **14A**, 2375–85.

Suresh, S. (1983b). Micromechanisms of fatigue crack growth retardation following overloads. *Engineering Fracture Mechanics* **18**, 577–93.

Suresh, S. (1985a). Fatigue crack deflection and fracture surface contact: micromechanical models. *Metallurgical Transactions* **16A**, 249–60.

Suresh, S. (1985b). Crack initiation in cyclic compression and its applications. *Engineering Fracture Mechanics* **21**, 453–63.

Suresh, S. (1990a). Mechanics and micromechanisms of fatigue crack growth in brittle solids. *International Journal of Fracture* **42**, 41–56.

Suresh, S. (1990b). Fatigue crack growth in ceramic materials at ambient and elevated temperatures. In *Fatigue 90* (eds. H. Kitagawa & T. Tanaka), vol. II, pp. 759–68. Birmingham: Materials and Components Engineering Publications.

Suresh, S. & Brockenbrough, J.R. (1988). Theory and experiments of fracture in cyclic compression: single phase ceramics, transforming ceramics and ceramic composites. *Acta Metallurgica* **36**, 1455–70.

Suresh, S. & Brockenbrough, J.R. (1990). Cyclic damage zones ahead of tensile fatigue cracks in ceramic materials. In *Fatigue 90* (eds. H. Kitagawa & T. Tanaka), vol. II, pp. 739–44. Birmingham: Materials and Components Engineering Publications.

Suresh, S., Christman, T. & Bull, C. (1986). Crack initiation and growth under far-field cyclic compression: Theory, experiments and applications. In *Small Fatigue Cracks* (eds. R.O. Ritchie & J. Lankford), pp. 513–40. Warrendale: The Metallurgical Society of the American Institute of Mining, Mineral and Petroleum Engineers.

Suresh, S., Giannakopoulos, A.E. & Olsson, M. (1994). Elastoplastic analysis of thermal cycling: layered materials with sharp interfaces. *Journal of the Mechanics and Physics of Solids* **42**, 979–1018.

Suresh, S., Parks, D.M. & Ritchie, R.O. (1982). Crack tip oxide formation and its influence on fatigue thresholds. In *Fatigue Thresholds* (eds. A. F. Blom, J. Backlund & C.J. Beevers), vol. 1, pp. 391–408. Warley: Engineering Materials Advisory Services.

Suresh, S. & Ritchie, R.O. (1981). On the influence of fatigue underloads on cyclic crack growth at low stress intensities. *Materials Science and Engineering* **51**, 61–9.

Suresh, S. & Ritchie, R.O. (1982a). A geometric model for fatigue crack closure induced by fracture surface morphology. *Metallurgical Transactions* **13A**, 1627–31.

Suresh, S. & Ritchie, R.O. (1982b). Mechanistic dissimilarities between environmentally-influenced fatigue-crack propagation at near-threshold and higher growth rates in lower strength steels. *Metal Science* **16**, 529–38.

Suresh, S. & Ritchie, R.O. (1984a). Near-threshold fatigue crack propagation: a perspective on the role of crack closure. In *Fatigue Crack Growth Threshold Concepts* (eds. D.L. Davidson & S. Suresh), pp. 227–61. Warrendale: The Metallurgical Society of the American Institute of Mining, Mineral and Petroleum Engineers.

Suresh, S. & Ritchie, R.O. (1984b). Propagation of short fatigue cracks. *International Metals Reviews* **29**, 445–76.

Suresh, S. & Shih, C.F. (1986). Plastic near-tip fields for branched cracks. *International Journal of Fracture* **30**, 237–59.

Suresh, S., Shih, C.F., Morrone, A. & O'Dowd, N.P. (1990). Mixed-mode fracture toughness of ceramic materials. *Journal of the American Ceramic Society* **73**, 1257–67.

Suresh, S., Sugimura, Y. & Tschegg, E.K. (1992). Growth of a fatigue crack approaching a perpendicularly-oriented bimaterial interface. *Scripta Metallurgica et Materialia* **27**, 1189–94.

Suresh, S., Sugimura, Y. & Ogawa, T. (1993). Fatigue cracking in materials with brittle surface coatings. *Scripta Metallurgica et Materialia* **29**, 237–42.

Suresh, S., Tschegg, E.K. & Brockenbrough, J.R. (1989). Crack growth in cementitious materials under cyclic compressive loads. *Journal of Cement and Concrete Research* **19**, 827–33.

Suresh, S., Vasudevan, A.K. & Bretz, P.E. (1984). Mechanisms of slow fatigue crack growth in high strength aluminum alloys: Role of microstructure and environment. *Metallurgical Transactions* **15A**, 369–79.

Suresh, S., Vasudevan, A.K., Tosten, M.H. & Howell, P.R. (1987). Microscopic and macroscopic aspects of fracture in lithium-containing aluminum alloys. *Acta Metallurgica* **35**, 25–46.

Suresh, S., Zamiski, G.F. & Ritchie, R.O. (1981). Oxide-induced crack closure: an explanation for near-threshold corrosion fatigue crack growth behavior. *Metallurgical Transactions* **12A**, 1435–43.

Suzuki, H. & McEvily, A.J. (1979). Microstructural effects on fatigue crack growth in low carbon steels. *Metallurgical Transactions* **10A**, 475–81.

Swain, M.V. & Zelizko, V. (1988). Comparison of static and cyclic fatigue on Mg-PSZ alloys. In *Advances in Ceramics–Science and Technology of Zirconia III* (eds. S. Somiya, N. Yamamoto & H. Yanagida), vol. 24, pp. 595–606. Westerville: American Ceramic Society.

Sylva, L.A. & Suresh, S. (1989). Crack growth in transforming ceramics under cyclic tensile loads. *Journal of Materials Science* **24**, 1729–38.

Szolwinski, M.P. & Farris, T.N. (1996). Mechanics of fretting fatigue crack formation. *Wear* **198**, 93–107.

Tabor, D. (1951). *Hardness of Metals*. Oxford: Oxford University Press.

Tabor, D. (1955). The mechanism of rolling friction: the elastic range. *Proceedings of the Royal Society, London* **A251**, 378–90.

Tada, H., Paris, P.C. & Irwin, G.R. (1973). *Stress Analysis of Cracks Handbook*. Hellertown, PA: Del Research Corporation.

Taira, S., Inoue, T. & Yoshida, T. (1968). Low cycle fatigue under multiaxial stresses. In *Proceedings of the Eleventh Japan Congress on Materials Research*, pp. 60–65. Kyoto: Japan Society of Materials Science.

Taira, S. & Tanaka, K. (1979). Local residual stress near fatigue crack tip. *Transactions of the Iron and Steel Institute of Japan* **19**, 411–18.

Taira, S., Tanaka, K. & Hoshina, M. (1979). Grain size effect on crack nucleation and growth in long-life fatigue of low carbon steel, In *Fatigue Mechanisms*, Special Technical Publication 675, pp. 135–73. Philadelphia: American Society for Testing and Materials.

Takemori, M.T. (1982). Fatigue fracture of polycarbonate. *Polymer Engineering and Science* **22**, 937–45.

Takemori, M.T. (1984). Polymer fatigue. *Annual Review of Materials Science* **14**, 171–204.

Takemori, M.T. (1990). Competition between crazing and shear flow during fatigue. In *Advances in Polymer Science* (ed. H.H. Kausch), vol. 91/92, pp. 263–300. Berlin: Springer-Verlag.

Takemori, M.T. & Kambour, R.P. (1981). Discontinuous fatigue crack growth in polycarbonate. *Journal of Materials Science* **16**, 1108–10.

Talreja, R. (1987). *Fatigue of Composite Materials*. Lancaster, PA: Technomic.

Tanaka, K. (1974). Fatigue crack propagation from a crack inclined to the cyclic tensile axis. *Engineering Fracture Mechanics* **6**, 493–507.

Tanaka, K., Hojo, M. & Nakai, Y. (1983). Crack initiation and early propagation in 3% silicon iron. In *Fatigue Mechanisms: Advances in Quantitative Measurement of Fatigue Damage*, Special Technical Publication 811, pp. 207–32. Philadelphia: American Society for Testing and Materials.

Tanaka, K. & Mura, T. (1981). A dislocation model for fatigue crack initiation. *Journal of Applied Mechanics* **48**, 97–102.

Tanaka, K. & Mura, T. (1982). A theory of fatigue crack initiation at inclusions. *Metallurgical Transactions* **13A**, 117–23.

Tanaka, K. & Nakai, Y. (1983). Propagation and non-propagation of short fatigue cracks at a sharp notch. *Fatigue of Engineering Materials and Structures* **6**, 315–27.

Tanaka, K., Nakai, Y. & Yamashita, M. (1981). Fatigue growth threshold of small cracks. *International Journal of Fracture* **17**, 519–33.

Tangjitham, S. & Landgraf, R.W. (1993). Probability based methods for fatigue analysis. In *Fatigue Research and Applications*, Special Publication 1009, pp. 225–35. Warrendale, PA: Society of Automotive Engineers.

Taylor, D. (1989). *Fatigue Thresholds*. London: Butterworths.

Taylor, D. & Knott, J.F. (1981). Fatigue crack propagation behavior of short cracks: the effects of microstructure. *Fatigue of Engineering Materials and Structures* **4**, 147–55.

Taylor, G.I. (1934). The mechanism of plastic deformation of crystals. Part I. Theoretical. *Proceedings of the Royal Society, London* **A145**, 362–87.

Tennekes, H. (1996). *The Simple Science of Flight: From Insects to Jumbo Jets*. Cambridge, MA: The Massachusetts Institute of Technology Press.

Teodosiu, C. (1982). *Elastic Models of Crystal Defects*. New York, NY: Springer-Verlag.

Thompson, N., Wadsworth, N.J. & Louat, N. (1956). The origin of fatigue fracture in copper. *Philosophical Magazine* 1, 113–26.

Thouless, M.D. (1988). Bridging and damage zones in crack growth. *Journal of the American Ceramic Society* 71, 408–13.

Thum, A. (1939). Festigkeitsprüfung bei schwingender Beanspruchung. Siebel, Handbuch der Werkstoffprüfung, 2. Bd. S. 175/231. Berlin: Verlag Springer.

Timoshenko, S. & Goodier, J.N. (1951). *Theory of Elasticity*, 3rd edition. New York: McGraw-Hill.

Tomkins, B. & Wareing, J. (1977). Elevated temperature fatigue interactions in engineering materials. *Metal Science* 11, 414–24.

Tomlinson, G.A. (1927). The rusting of steel surfaces in contact. *Proceedings of the Royal Society, London* A115, 472–83.

Topper, T.H., Sandor, B.I. & Morrow, J. (1969). Cumulative damage under cyclic strain control. *Journal of Materials* 4, 189–99.

Topper, T.H. & Yu, M.T. (1985) The effect of overloads on threshold and crack growth. *International Journal of Fatigue* 7, 159–64.

Tschegg, E.K. (1983a). Mode III and Mode I fatigue crack propagation behaviour under torsional loading. *Journal of Materials Science* 18, 1604–14.

Tschegg, E.K. (1983b). Sliding mode crack closure and mode III fatigue crack growth in mild steel. *Acta Metallurgica* 31, 1323–30.

Tschegg, E.K. (1983c). The influence of the static load mode I and R ratio on mode III fatigue crack growth behaviour in mild steel. *Materials Science and Engineering* 59, 127–37.

Tschegg, E.K. & Stanzl, S. (1981). Fatigue crack propagation and threshold in b.c.c. and f.c.c. metals at 77 and 293 K. *Acta Metallurgica* 29, 33–40.

Tschegg, E.K. & Stanzl, S.E. (1988). The significance of sliding mode crack closure on mode III fatigue crack growth. In *Basic Questions in Fatigue: Volume I*, Special Technical Publication 924, pp. 214–32. Philadelphia: American Society for Testing and Materials.

Tschegg, E.K. & Suresh, S. (1988). Mode III fracture of 4340 steel: effects of tempering temperature and fracture surface interference. *Metallurgical Transactions* 19A, 3035–44.

Tu, L.K.L. & Seth, B.B. (1978). Threshold corrosion fatigue crack growth in steels. *Journal of Testing and Evaluation* 6, 66–74.

Tzou, J.-L., Suresh, S. & Ritchie, R.O. (1985). Fatigue crack propagation in oil environments: I. Crack growth behavior in silicone and paraffin oils. *Acta Metallurgica* 33, 105–16.

Usami, S. (1982). Application of threshold cyclic-plastic-zone-size criterion to some fatigue limit problems. In *Fatigue Thresholds* (eds. J. Backlund, A.F. Blom & C.J. Beevers), pp. 205–38. Warley: Engineering Materials Advisory Services.

Usami, S. & Shida, S. (1979). Elastic–plastic analysis of the fatigue limit for a material with small flaws. *Fatigue of Engineering Materials and Structures* 1, 471–81.

Valanis, K.C. (1980). Fundamental consequence of a new intrinsic time measure– Plasticity as a limit of the endocronic theory. *Archives of Mechanics* 32, 171–91.

Vasudevan, A.K. & Doherty, R. (1987). Grain boundary ductile fracture in precipitation hardened aluminum alloys. *Acta Metallurgica* 35, 1193–219.

Vasudevan, A.K., Liu, J. & Ricker, R.E. (1987). Mechanism of stress corrosion crack growth resistance of Al–Li–Cu alloys: Role of grain boundary precipitates. In *Environmental*

Degradation of Engineering Materials (eds. M.R. Louthan, R.P. McNitt & R.D. Sisson), pp. 321–7. University Park: Pennsylvania State University.

Vasudevan, A.K. & Sadananda, K. (1995). Classification of fatigue crack growth behavior. *Metallurgical Transaction* **26A**, 1221–34.

Vasudevan, A.K. & Suresh, S. (1982). Influence of corrosion deposits on near-threshold fatigue crack growth behavior in 2XXX and 7XXX series aluminum alloys. *Metallurgical Transactions* **13A**, 2271–80.

Veers, P.S. (1996). Statistical considerations in fatigue. In *Fatigue and Fracture*, ASM Handbook, Vol. 19, pp. 295–302. Materials Park, OH: The American Society of Materials, International.

Venkataraman, G., Chung, Y.-W., Nakasone, Y. & Mura, T. (1990). Free energy formulation of fatigue crack initiation along persistent slip bands: calculation of S-N curves and crack depths. *Acta Metallurgica et Materialia* **38**, 31–40.

Vingsbo, O. & Soderberg, D. (1988). On fretting maps. *Wear* **126**, 131–47.

Vosikovsky, O. (1976). Fatigue crack growth in an X65 pipe-line steel in sour crude oil. *Corrosion* **32**, 472–5.

Voss, H. & Walter, R. (1985). Fracture and fatigue of short glass-fibre reinforced polyethersulfone composites. *Journal of Materials Science Letters* **4**, 1174–7.

Wagner, L., Gregory, J.K., Gysler, A. & Lütjering, G. (1986). Propagation behavior of short cracks in Ti–8.6 Al alloy. In *Small Fatigue Cracks* (eds. R.O. Ritchie & J. Lankford), pp. 117–28. Warrendale: The Metallurgical Society of the American Institute of Mining, Mineral and Petroleum Engineers.

Walgraef, D. & Aifantis, E.C. (1985). Dislocation patterning in fatigued metals as a result of dynamical instabilities. *Journal of Applied Physics* **58**, 688–91.

Walker, K. (1970). The effect of stress ratio during crack propagation and fatigue for 2024-T3 and 7075-T6 aluminum. In *Effects of Environment and Complex Load History for Fatigue Life*, Special Technical Publication 462, pp. 1–14. Philadelphia: American Society for Testing and Materials.

Wang, Z., Gong, B. & Wang, Z. (1997). Cyclic deformation behavior and dislocation structures of [001] copper single crystals–II. Characteristics of dislocation structures. *Acta Materialia* **45**, 1379–91.

Wanhill, R.J.H. (1975). Formation of brittle fatigue striations. *Metallurgia* **31**, 66–71.

Warlow-Davies, E.J. (1941). Fretting corrosion and fatigue strength: brief results of preliminary experiments. *Proceedings of the Institution of Mechanical Engineers* **146**, 32–8.

Watanabe, T. (1985). Structural effects on grain boundary segregation, hardening and fracture. *Journal de Physique* **C4**, 555–66.

Waterhouse, R.B. (1992). Fretting fatigue. *International Materials Reviews* **37**, 77–97.

Waterhouse, R.B. & Lindley, T.C. (1994). *Fretting fatigue*, European Structural Integrity Society Publication Number 18. London: Mechanical Engineering Publications Ltd.

Watt, D.F., Embury, J.D. & Ham, R.K. (1968). The relation between surface and interior structures in low-amplitude fatigue. *Philosophical Magazine* **17**, 199–203.

Way, S. (1935). Pitting due to rolling contact. *Journal of Applied Mechanics* **2**, A49–58.

Weertman, J. (1966). Rate of growth of fatigue cracks calculated from the theory of infinitesimal dislocations distributed on a crack plane. *International Journal of Fracture* **2**, 460–7.

Weertman, J. (1973). Theory of fatigue crack growth based on a BCS crack theory with work hardening. *International Journal of Fracture* **9**, 125–31.

Wei, R.P. (1970). Some aspects of environment-enhanced fatigue-crack growth. *Engineering Fracture Mechanics* **1**, 633–51.

Wei, R.P. & Landes, J. (1970). Correlation between sustained-load and fatigue crack growth in high strength steels. *Materials Research and Standards* **9**, 25–7, 44–6.

Wei, R.P. & Simmons, G.W. (1981). Recent progress in understanding environment assisted fatigue crack growth. *International Journal of Fracture* **17**, 235–47.

Weibull, W. (1939). A statistical theory of the strength of materials. *Proceedings 151.* Stockholm: Royal Swedish Academy of Engineering Sciences.

Weir, T.W., Simmons, G.W., Hart, R.G. & Wei, R.P. (1980). A model for surface reaction and transport controlled fatigue crack growth. *Scripta Metallurgica* **14**, 357–64.

Weiss, B. & Stickler, R. (1972). Phase instabilities during high temperature exposure of 316 austenitic stainless steel. *Metallurgical Transactions* **3A**, 851–66.

Weiss, B., Stickler, R., Fembӧck, J.F. & Pfaffinger, K. (1979). High cycle fatigue and threshold behavior of powder metallurgical Mo and Mo alloys. *Fatigue of Engineering Materials and Structures* **2**, 73–84.

Weiss, V. & Lal, D.L. (1974). A note on the threshold condition for fatigue crack propagation. *Metallurgical Transactions* **5A**, 1946–9.

Wells, C.H. (1979). High-temperature fatigue. In *Fatigue and Microstructure* (ed. M. Meshii), pp. 307–31. Metals Park: American Society for Metals.

Westergaard, H.M. (1939). Bearing pressures and cracks. *Journal of Applied Mechanics* **61**, A49–53.

Westwood, A.R.C. & Kamdar, M.H. (1963). Concerning liquid metal embrittlement, particularly of zinc monocrystals in mercury. *Philosophical Magazine* **8**, 787–804.

Wheeler, O.E. (1972). Spectrum loading and crack growth. *Journal of Basic Engineering* **94**, 181–6.

White, C.S., Bronkhorst, C. & Anand, L. (1990). An improved isotropic–kinematic hardening model for moderate deformation metal plasticity. *Mechanics of Materials* **10**, 97–117.

Wilhelm, M. & Everwin, P. (1980). Cyclic deformation behavior of α-copper and underaged copper–cobalt alloy single crystal. In *The Strength of Metals and Alloys* (eds. P. Haasen, V. Gerold & G. Kostorz), vol. 2, pp. 1089–99. Oxford: Pergamon Press.

Wilkens, M., Herz, K. & Mughrabi, H. (1980). An X-ray diffraction study of cyclically and of unidirectionally deformed copper single crystals. *Zeitschrift für Metallkunde* **71**, 376–84.

Willenborg, J., Engle, R.M. & Wood, H. (1971). A crack growth retardation model using an effective stress intensity concept. *Technical Report TFR 71–701.* Los Angeles: North American Rockwell.

Williams, M.L. (1957). On the stress distribution at the base of a stationary crack. *Journal of Applied Mechanics* **24**, 109–14.

Wilson, D.V. (1965). Reversible work hardening in alloys of cubic metals. *Acta Metallurgica* **13**, 807–14.

Wilson, D.V. & Bate, P.S. (1986). Reversibility in the work hardening of spherodized steels. *Acta Metallurgica* **36**, 1107–20.

Winter, A.T. (1974). A model for the fatigue of copper at low plastic strain amplitudes. *Philosophical Magazine* **29**, 719–38.

Winter, A.T., Pederson, O.B. & Rasmussen, K.V. (1981). Dislocation microstructures in fatigued copper polycrystals. *Acta Metallurgica* **29**, 735–48.

Wirsching, P.H. (1995). Probabilistic fatigue analysis. In *Probabilistic Structural Mechanics Handbook* (ed: C. Sundarrajan), pp. 146–65. New York, NY: Chapman & Hall.

Wöhler, A. (1860). Versuche über die Festigkeit der Eisenbahnwagenachsen. *Zeitschrift für Bauwesen* **10**; English summary (1867). *Engineering* **4**, 160–1.

Wood, W.A. (1958). Formation of fatigue cracks. *Philosophical Magazine* **3**, 692–9.

Woods, P.J. (1973). Low-amplitude fatigue of copper and copper–5 at.% aluminium single crystals. *Philosophical Magazine* **28**, 155–91.

Wright, T.M., Burstein, A.H. & Bartel, D.L. (1985). Retrieval analysis of total joint replacements: a six year experience. In *Proceedings of the Second Symposium on Corrosion and Degradation of Implant Materials*, pp. 415–28. Philadelphia, PA: American Society for Testing and Materials.

Wüthrich, C. (1982). The extension of the *J*-integral concept to fatigue cracks. *International Journal of Fracture* **20**, R35–7.

Xia, L., Wang, T.C. & Shih, C.F. (1993). Higher-order analysis of crack-tip fields in elastic power-law hardening materials. *Journal of the Mechanics and Physics of Solids* **41**, 665–87.

Yan, B.D., Cheng, A.S., Buchinger, L., Stanzl, S. & Laird, C. (1986). The cyclic stress–strain response of single-crystals of Cu–16 at.% Al alloy. 1. Cyclic hardening and strain localization. *Materials Science and Engineering* **80**, 129–42.

Yoder, G.R., Cooley, Y.A. & Crooker, T.W. (1979). Quantitative analysis of microstructural effects on fatigue crack growth in Widmanstätten Ti–6Al–4V and Ti–8Al–1Mo–1V. *Engineering Fracture Mechanics* **11**, 805–16.

Yoder, G.R., Cooley, Y.A. & Crooker, T.W. (1983). A critical analysis of grain size and yield strength dependence of near-threshold fatigue crack growth in steels. In *Fracture Mechanics*, Special Technical Publication 791, pp. vol. I, 348–65. Philadelphia: American Society for Testing and Materials.

Yuen, J.L., Roy, P. & Nix, W. (1984). Effect of oxidation kinetics on the near threshold fatigue crack growth behavior of a nickel base superalloy. *Metallurgical Transactions* **15A**, 1769–75.

Zackay, V.F., Parker, E.R., Fahr, D. & Busch, R. (1967). The enhancement of ductility in high strength steels. *Transactions of the American Society for Metals* **60**, 252–9.

Zappfe, C. & Sims, C.E. (1941). Hydrogen embrittlement, internal stress and defects in steel. *Transactions of the American Institute of Mining, Metallurgical and Petroleum Engineers* **145**, 225–71.

Zappfe, C.A. & Worden, C.O. (1951). Fractographic registrations of fatigue. *Transactions of the American Society for Metals* **43**, 958–69.

Zener, C. (1948). Micromechanism of fracture. In *Fracturing of Metals*, pp. 3–12, Cleveland: American Society for Metals.

Zenner, H., Heidenreich, R. & Richter, I. (1985). Mehrachsige ermüdung. *Zeitschrift Werkstofftechnik* **16**, 101–12.

Zhang, Y.H., Edwards, L. & Plumbridge, W.J. (1998). Effect of crystallization on fatigue crack growth in a SiC-reinforced silicon nitride composite at 1200 °C. *Acta Materialia*, in press.

Ziegler, H. (1959). A modification of Prager's hardening rule. *Quarterly of Applied Mathematics* **17**, 55–65.

Author index

Subject index

Printed in the United States
By Bookmasters